Passivity of Metals and Semiconductors

THIN FILMS SCIENCE AND TECHNOLOGY

Advisory Editor: G. Siddall

Vol. 1   Langmuir-Blodgett Films (Barlow, Editor)
Vol. 2   Size Effects in Thin Films (Tellier and Tosser)
Vol. 3   Langmuir-Blodgett Films, 1982 (Roberts and Pitt, Editors)
Vol. 4   Passivity of Metals and Semiconductors (Froment, Editor)

THIN FILMS SCIENCE AND TECHNOLOGY, 4

Passivity of Metals and Semiconductors

Proceedings of the Fifth International Symposium on
Passivity, Bombannes, France, May 30-June 3, 1983,
Organized by the Société de Chimie Physique

Edited by

MICHEL FROMENT
Laboratoire de Physique des Liquides et Electrochemie (CNRS) — Université Pierre et Marie Curie, Paris, France

ORGANIZING COMMITTEE
M. Froment, Chairman
B. Agius
A. Hugot-Le Goff
M. Keddam
M.C. Petit

ELSEVIER, Amsterdam — Oxford — New York — Tokyo 1983

CHEMISTRY

7299-357

ELSEVIER SCIENCE PUBLISHERS B.V.
Molenwerf 1,
P.O. Box 211, 1000 AE Amsterdam, The Netherlands

Distributors for the United States and Canada:

ELSEVIER SCIENCE PUBLISHING COMPANY INC.
52, Vanderbilt Avenue
New York, NY 10017

Library of Congress Cataloging in Publication Data

International Symposium on Passivity (5th : 1983 :
    Bombannes)
    Passivity of metals and semiconductors.

    (Thin films science and technology ; 4)
    Bibliography: p.
    Includes index.
    1. Passivity (Chemistry)--Congresses. 2. Electro-
chemistry--Congresses. 3. Metals--Congresses.
4. Semiconductors--Congresses. 5. Corrosion and anti-
corrosives--Congresses. I. Froment, Michel.
II. Société de chimie physique. III. Title.
IV. Series.

QD501.I6345 1983   546'.3   83-16580
ISBN 0-444-42252-8 (v. 4)

ISBN 0-444-42252-8 (Vol. 4)
ISBN 0-444-41903-9 (Series)

Printed in The Netherlands

This volume is dedicated
to the memory of Doctor I. EPELBOIN.

Ce volume est dédié
à la mémoire du Docteur I. EPELBOIN.

ACKNOWLEDGMENTS

The Organizing Board wishes to express its thanks for the financial support of the following :

- Direction des Recherches Etudes et Techniques
- Conseil Général de la Gironde
- Conseil Régional d'Aquitaine
- Société Creusot Loire

SPONSORSHIPS

Deutsche Bunsen-Gesellschaft für Physikalische Chemie
International Society of Electrochemistry
Japan Society of Corrosion Engineering
The Electrochemical Society

SCIENTIFIC COMMITTEE

A.G. AKIMOV, Moscow, USSR
A.J. ARVIA, La Plata, Argentina
D.E. ASPNES, Murray Hill, USA
M. BIENFAIT, Marseille, France
L. BONORA, Genova, Italy
Mme M. JANIK-CZACHOR, Warsaw, Poland
R.P. FRANKENTHAL, Murray Hill, USA
S. GOTTESFELD, Tel Aviv, Israel
D.M. KOLB, Berlin, FRG
Y.M. KOLOTYRKIN, Moscow, USSR
J. KRUGER, Washington, USA
D. LANDOLT, Lausanne, Switzerland
Sir Nevill MOTT, United Kingdom
J. OUDAR, Paris, France
M. PRAZAK, Praha, Czechoslovakia
N. SATO, Sapporo, Japan
J.C. SCULLY, Leeds, United Kingdom
J.W. SCHULTZE, Düsseldorf, FRG
R.W. STAEHLE, Minneapolis, USA

F O R E W O R D

Passivity is the situation in which a metal is prevented from corroding by a thin film, despite thermodynamic conditions which foresee an instability of this metal towards the aggressive environment. Such films are very important because they frequently provide an efficient protection for highly reactive metals.

Nevertheless, the loss of passivity by metals or alloys gives rise to corrosion damage of either a general or localized form. Localized breakdown processes involved in crevice corrosion, pitting, stress corrosion, cracking ... are of considerable theoretical and practical interest.

This volume is composed of invited lectures and communications presented during the Fifth International Symposium on Passivity, held in Bombannes (France), from 30 May to 3 June 1983.

The previous symposia were essentially devoted to the passivity of metals such as iron, nickel, aluminium, titanium and their alloys, with special attention to stainless steels. Localized corrosion and film breakdown too were dealt with in a large number of conferences and communications. Undoubtedly, both topics, passivity of metallic materials and localized corrosion, remain of basic importance and have been treated in several sessions. Experimental techniques were also given full attention. In addition to classical approaches such as optical techniques and electron spectroscopy, EXAFS and Raman Spectroscopy now appear as new tools promising for in-situ characterization of passivity phenomenon ; these aspects have been developed in a full day session.

Moreover, the Organizing Board has estimated that basic and practical problems connected with the passivation of semiconductors are of growing importance. For instance, new advances in microelectronics devices and very large scale integrated circuits are to a large extent dependent on our knowledge of thin films on semiconductors such as silicon or gallium arsenide. It seems detrimental to both fields of investigation that the development of techniques and concepts is carried out separately. That is the reason why, by way of experiment, we have introduced semiconductor passivity as a new theme in the form of invited lectures. Recent advances in photoelectrochemical solar energy conversion are highly promising, but the application on an industrial scale of the more efficient junctions is restrained by their poor stability under illumination. Therefore we decided to have a session devoted to photoelectrochemistry. From another point of view, studies of the passive film in metals in terms of semiconducting material, although controversial, are certainly fruitful. For instance, theoretical descriptions of charge transfer at passive layers are likely to be improved by recent advances in photoelectrochemistry. Finally, spectrophotoelectrochemistry of passive layers can be regarded as an in-situ technique, easy to use and providing relevant data.

In view of the very large number of contributions, it was not possible to arrange an oral presentation for all contributors. Therefore the number of oral lectures was limited by the Organizing Committee and four Posters Sessions were organized. However, in order to allow thorough discussions and to enhance informal exchanges, each Poster Session was followed by a Round Table Discussion. These discussions were stimulated by the Chairman and the invited lecturers of the same session. Short transactions of these discussions have been included in this Proceedings Volume.

These transactions are in no way a precise recording of the poster discussions but only an attempt to report as far as possible the content of contributions, questions, answers and comments. The Organizing Committee apologizes for distortions with respect to an exact transcription.

The Organizing Committee would like to thank all those who contributed to the preparation of this Symposium. Special acknowledgments are due to the members of the Scientific Committee and to our colleagues approached during the elaboration of the programme and the referees of so many manuscripts. We are extremely grateful to invited lecturers and chairmen who were in charge of the animation of their session.

Undoubtedly, the many invited lectures and communications presented in Bombannes have established the advances achieved since the last Symposium held in Airlie (USA) in 1977, particularly with regard to the composition and some of the properties of the passive layers. However, as substantiated by the very active discussions which took place after the oral presentations and during the Round Tables, we are still far from understanding clearly the relationships between the physicochemical properties of the passive layers and their dynamic response to chemical attacks and mechanical stresses. The connection between mechanisms of formation and structure of the passive layers on metals such as iron or semiconductors such as silicon was scarcely dealt with in Bombannes and deserves further attention. I hope that further progress in these fields will be reported at the Sixth Symposium, to be held in Japan.

M. FROMENT

CONTENTS

Acknowledgements   vii
Sponsorship   viii
Scientific Committee   viii
Foreword   ix

SECTION I: CONCEPTS — COMPOSITION — TRANSPORT PHENOMENA
Transport of oxygen and water in oxide layers
   N.F. Mott   1
An overview of the kinetics of oxidation of silicon:
The very thin SiO_2 film growth regime
   E.A. Irene   11
Transfer reactions in oxides and passive layers
   W. Schmickler   23
Electronic properties of modified passive films
   C. Bartels, B. Danzfuss and J.W. Schultze   35
On the oxidation rate laws of metals that form nonstoichiometric oxides
   R.J. Good   43
An impedance approach of the passive film on iron in acidic media
   M. Keddam, J.F. Lizee, C. Pallotta and H. Takenouti   51
Electrochemical kinetics of iron and iron hydroxide in aqueous
electrolytes
   M.E. Vela, J.R. Vilche and A.J. Arvia   59
Iron passivation in borate solutions
   S.P. Tyfield   67
The role of inhibitor PAB at NiOOH-layers
   M. Goledzinowski and J.W. Schultze   73
Optical and impedance studies of nickel passivation film in
neutral solution
   T. Ohtsuka, K. Azumi and N. Sato   79
Electroreflectance for the study of the passivity of iron
   M. Froelicher, A. Hugot-Le Goff and V. Jovancicevic   85
Electrochemical and ellipsometric investigation of passive films
formed on iron in borate solution
   Z. Szklarska-Smialowska and W. Kozlowski   89
An ellipsometric study of the passivity of iron in neutral
borate electrolyte
   J.L. Ord   95

XII

Raman studies of passive films on iron

    M. Froelicher, A. Hugot-Le Goff, C. Pallotta, R. Dupeyrat

    and M. Masson   101

Raman spectroscopy of oxide layers on pure iron in electrochemical
environment

    J. Dünnwald, R. Lossy and A. Otto   107

Application of angle resolved XPS and AES depth profiling to the
study of transpassive films on nickel

    M. Datta, H.J. Mathieu and D. Landolt   113

Passivation of nickel—iron alloys. Influence of sulphur

    P. Marcus and J. Oudar   119

Anodic oxidation of UO_2. Part III. Electrochemical studies in
carbonate solutions

    D.W. Shoesmith, S. Sunder, M.G. Bailey and D.G. Owen   125

EXAFS as a probe of the passive film structure

    L. Bosio, R. Cortes, A. Defrain, M. Froment and A.M. Lebrun   131

Ex-situ and in-situ sample-and-detector chambers for the study
of passive films using surface EXAFS

    G.G. Long, J. Kruger and M. Kuriyama   139

Round Table Discussion (Concepts — Composition — Transport Phenomena)   145

SECTION II: TECHNIQUES

Study of passivity of iron by in-situ methods: Mössbauer and EXAFS

    R.W. Hoffman   147

Structural studies of passive films using surface EXAFS

    J. Kruger, G.G. Long, M. Kuriyama and A.I. Goldman   163

Structure simulation of surface films on iron and iron-based alloy
by radial distribution function (R.D.F.) method

    M. Kobayashi   169

The structure of passive films on single-phase nickel—molybdenum
alloys

    M.B. Ives, V. Mitrovic-Scepanovic and M. Moriya   175

Study of passive layers on stainless steels using low energy
electron induced X-ray spectroscopy (LEEIXS)

    E. Gaillard, M. Romand, A. Roche, M. Charbonnier, R. Bador

    and A. Desestret   181

Optical methods in the study of passive films

    B.D. Cahan   187

Reflectometry of iron passivation films in neutral borate and acidic
phosphate solutions
 T. Ohtsuka, K. Azumi and N. Sato   199
Combined ellipsometric and AC impedance measurements of oxide
films on ruthenium
 J. Rishpon, I. Reshef and S. Gottesfeld   205
Analysis of multiple layer surface films by modulated reflection
spectroscopy
 N. Hara and K. Sugimoto   211
A statistical approach to the study of localized corrosion
 D.E. Williams, C. Westcott and M. Fleischmann   217
Electrochemical noise measurements for the study of localized
corrosion and passivity breakdown
 U. Bertocci, J.L. Mullen and Y-X. Ye   229

SECTION III: AMORPHOUS METALS — BREAKDOWN AND REPASSIVATION
Passivation of amorphous metals
 K. Hashimoto   235
Supplement to Passivation of amorphous metals"
 K. Hashimoto   247
Depassivation and repassivation in localized corrosion
 J.C. Scully   253
Dissolution and passivation kinetics of Fe—Cr—Ni alloys during
localized corrosion
 R.C. Newman and H.S. Isaacs   269
Influence of electrode pretreatment on the pitting susceptibility
of Ni
 B. MacDougall   275
A kinetic theory of pit initiation and repassivation and its
relevance to the pitting corrosion of passive metals
 T. Okada   281
A comparison of models for localized breakdown of passivity
 P. Zaya and M.B. Ives   287
Stochastic aspects of mechanical and chemical breakdown of passivity
 C. Gabrielli, F. Huet, M. Keddam, R. Oltra and C. Pallotta   293
Pitting of Zircaloy 4: Statistical analysis of induction times
 G. Mankowski, P. Eygazier, Y. Roques, G. Chatainier and
 F. Dabosi   299

An electrochemical study of amorphous ion implanted stainless steels
C.R. Clayton, Y-F. Wang and G.K. Hubler 305

Electrochemical properties and passivation of amorphous $Fe_{80}P_{20}$ alloy
P. Cadet, M. Keddam and H. Takenouti 311

Effect of chromium on the passivation of amorphous alloy Fe—Ni—B—P—xCr.
Effect of a heat treatment
J. Crousier, K. Belmokre, Y. Massiani and J-P. Crousier 317

The effect of phosphorus on the corrosion resistance of amorphous
copper—zirconium alloys
T.D. Burleigh and R.M. Latanision 321

Passivation of iron, nickel and cobalt in concentrated nitric acid
solutions
E. Stupnisek-Lisac, M. Karsulin and H. Takenouti 327

Nickel passivation—depassivation in sulphuric acid: Influence on
the grain size
M. Cid and M.C. Petit 335

The passivation—depassivation behaviour of gold in a HCl—glycerol
solution and its application on electropolishing
J. Verlinden, J.P. Celis and J.R. Roos 341

Crevice corrosion test for stainless steels in chloride solutions
R.O. Müller 347

Initiation and inhibition of pitting corrosion on nickel
S.M. Abd El Haleem, M.G.A. Khedr, A.A. Abdel Fattah and H. Mabrok 353

Initial stages of film breakdown on passive austenitic stainless
steel in 3% sodium chloride
M.G.S. Ferreira and J.L. Dawson 359

Repassivation potential of corrosion pits in stainless steel
T. Hakkarainen 367

Study of the passive behaviour of a 304L stainless steel in 1 N
H_2SO_4 by a method of corrosion under friction
D. Boutard and J. Galland 373

Breakdown of passivity of iron and nickel by fluoride
H.-H. Strehblow and B.P. Löchel 379

Electrochemical behaviour of mild steel in sulphide and chloride
containing solutions
C.A. Acosta, R.C. Salvarezza, H.A. Videla and A.J. Arvia 387

The role of molybdenum in the inhibition of passivity breakdown in
neutral chloride solutions up to $250^{\circ}C$
P. Vanslembrouck, W. Bogaerts and A. van Haute 393

On the influence of both passive film composition and of non-
metallic inclusions on the initiation of localized corrosion
of stainless steel
    G. Hultquist, S. Zakipour and C. Leygraf   399
The effects of alloying on the resistance of ferritic stainless
steels to localized corrosion on Cl^- solutions
    W.R. Cieslak and D.J. Duquette   405
Effect of Al on the electrochemical behaviour of ternary brasses
    F. Terwinghe, J.P. Celis and J.R. Roos   413
Round Table Discussion (Amorphous metals — Breakdown and repassivation)   419

SECTION IV: SEMICONDUCTORS — PHOTOELECTROCHEMISTRY
Mechanisms of passivation and protection of semiconductors in
solar cells
    S.R. Morrison   425
Plasma oxidation of semiconductor and metal surfaces
    R.P.H. Chang   437
Thermal oxidation of niobium nitride films
    R.P. Frankenthal, D.J. Siconolfi, W.R. Sinclair and D.D. Bacon   445
Comparative study by XPS and LEED of nitridation processes on
silicon (111)
    C. Maillot, H. Roulet and G. Dufour   451
Structure and growth kinetics of SiO_2 ultra thin films on
Si(111) surface
    J. Derrien, F. Ringeisen and M. Commandre   457
Oxygen Transport studied by ^{18}O labelling in thin thermal silicon
oxide films in connection with their structural characteristics
    B. Agius, M. Froment, S. Rigo and F. Rochet   463
Growth of thin oxide layers on silicide compounds
    A. Cros   473
Properties of the passive film on iron electrodes by capacity and
photocurrent measurements
    U. Stimming   477
Coupled ellipsometric and capacitance measurements during growth
of anodic oxide onto GaAs
    P. Cléchet, J. Joseph, A. Gagnaire, D. Lamouche, J.R. Martin
    and E. Verney   483
Study by spectrophotoelectrochemistry (SPEC) of the passivity of
iron in neutral solutions
    M. Froelicher, A. Hugot-Le Goff and V. Jovancicevic   491

Influence of the anodization conditions on the electronic properties
and crystallographic structures of the corrosion layers on tungsten.
A photoelectrochemical approach
    F. Di Quarto, S. Piazza and C. Sunseri   497
Influence of ion implantation on the electrochemical behaviour of
passive titanium and hafnium electrodes
    B. Danzfuss, J.W. Schultze, U. Stimming and O. Meyer   503
Photoelectrochemical approach to passivity
    U. Stimming   509
Passivation and selective etching of III–V n-type semiconductors
under potential and pH control in H_2O_2 media
    P. Cléchet, E. Haroutiounian, D. Lamouche, J.R. Martin and
    J.P. Sandino   521
Round Table Discussion (semiconductors – photoelectrochemistry)   527

SECTION V: STAINLESS STEELS – VALVE METALS
Passivation and localized corrosion of stainless steels
    B. Baroux   531
Effect of chemical heterogeneity within the metal phase on the
stability of the passivating film on iron alloys
    M. Janik-Czachor and A. Szummer   547
The role of water in the kinetics of stainless steel dissolution
and passivation in organic media
    B. Elsener and H. Boehni   555
Quantitative ESCA analysis of the passive state of an Fe–Cr alloy
and an Fe–Cr–Mo alloy
    I. Olefjord and B. Brox   561
Mechanisms controlling passive film composition of Fe–Cr–Ni steels
in elevated temperature solutions
    W. Bogaerts, P. Vanslembrouck, A. Van Haute and M. Brabers   571
Characterizations of passive films formed on stainless steel in
high temperature water
    J.B. Lumsden and P.J. Stocker   579
Structure and stability of the anodically formed films on 304
stainless steel in sulfuric acid
    C.R. Clayton, K. Doss and J.B. Warren   585
Passivity and breakdown of passivity of directed energy sources
modified stainless steel
    P.L. de Anna, M. Bassoli, G. Cerisola, P.L. Bonora and P. Mazzoldi   591

On the mechanism for improved passiviation by additions of tungsten
to austenitic stainless steel
    N. Bui, A. Irhzo, F. Dabosi and Y. Limouzin-Maire   599
Passivation of steel in a borate buffer containing organic compounds
as a means for improving corrosion resistance
    H. Konno and H. Leidheiser, Jr.   607
Critical conditions in the passivation of Cu–Mo austenitic stainless
alloys in hot 20% H_2SO_4 acid
    J.C. Bavay, P. Damie, M. Traisnel and K. Vu Quang   613
A study of the film on austenitic stainless steel at the passive–
transpassive transition
    G. Rondelli, J. Kruger, J.J. Ritter and U. Bertocci   619
Etude du mecanisme de passivité d'aciers 17–13 a teneur variable en
molybdène dans des solutions d'acide phosphorique à 30% en P_2O_5
    A. Guenbour and A. Ben Bachir   625
Corrosion potential oscillations of stainless steels in aqueous
chloride solutions
    A. Atrens   631
Atmospheric corrosion of stainless steels
    S. Ito, M. Yabumoto, H. Omata and T. Murata   637
Electrochemical prevention of localized corrosion of stainless steels
    N. Azzerri, F. Mancia and A. Tamba   643
Anodic behaviour of Fe–31% Ni alloy in the multi steady states region
corresponding to the transition between active and passive states,
in normal sulphuric acid
    F. Wenger and J. Galland   649
Pit growth measurements on stainless steels
    F. Hunkeler and H. Boehni   655
Rotating ring–disc electrode studies of the passivation of low alloy
steels
    A.M. Riley and J.M. Sykes   661
Depassivation and repassivation of austenitic stainless steels
consequences on stress corrosion
    M. Helie, D. Desjardins, M. Puiggali and M.C. Petit   667
Hydrogen permeation through passive layers on austenitic high
temperature alloys
    H.P. Buchkremer, R. Hecker and D. Stöver   675
Impedance technique to study passivity of steel
    G.N. Mehta, T. Tsuru and S. Haruyama   681

The effect of chromium, nickel and molybdenum on passivity of iron
alloys in organic solutions of sulphuric acid
  J. Banaś   687
Electrochemical measurements on low-index planes of a Fe—14.5Cr—14.5Ni—2.5Mo
single crystal
  S. Tähtinen, H. Hänninen and T. Hakkarainen   695
Corrosion atmosphérique d'un acier pattinable soudé
  B.M. Rosales, E.S. Ayllon and C. Bonazzola   701
Composition, structure and properties of passivation layers on tinplate
  N. Azzerri and L. Splendorini   707
Passivation of tin using chromates, molybdates and tungstates
  D.R. Gabe and D. Bijimi   713
Galvanostatic formation and reduction of tin/tin oxides in Na_2SO_4
solutions
  C.V. D'Alkaine and J.M. da Silva   719
Comportement potentiodynamique d'une électrode d'iridium activée
dans l'acrylamide et H_2SO_4
T.F. Otero and M.S.L. Redondo   725
Repassivation of titanium and titanium alloys dependent on potential
and pH
  H.-J. Rätzer-Scheibe   731
Crystalline aluminum oxide films
  R.S. Alwitt and H. Takei   741
Auger electron spectrometry of porous layers of anodized aluminium
  H. Terryn and J. Vereecken   747
Etching and passivation of aluminum under alternating current.
Influence of temperature
  F. Brindel, R. Grynszpan, F. Bourelier,   J.C. Bavay and K. Vu Quang   753
Round Table Discussion (stainless steels — valve metals)   759

Subject index   763

Passivity of Metals and Semiconductors, edited by M. Froment
Elsevier Science Publishers B.V., Amsterdam — Printed in The Netherlands

TRANSPORT OF OXYGEN AND WATER IN OXIDE LAYERS

N.F. MOTT

Cavendish Laboratory, Madingley Road, Cambridge CB3 0HE, England.

ABSTRACT

In this paper a discussion is given of the mechanisms by which the oxidising agent passes through the layer of oxide already formed, with special reference to silicon, and to other materials where the oxide is vitreous. In silicon in dry oxygen O_2 appears to diffuse through the layer of SiO_2 with little exchange of oxygen with the network already formed. For oxidation in H_2O, though the mechanism is similar, there is considerable exchange for a reason which will be explained; thus some small charge transport by network oxygen can occur. For thin films transparent to electrons and in anodic oxidation, on the other hand, ionic motion can predominate; various mechanisms are discussed.

If an oxide film is formed on a metal or other material such as silicon, we have to understand how the oxidising agent passes through the film; if the film is protective, growth must virtually stop at a certain thickness. In this lecture, I discuss films of network-forming oxides, in which the oxide is non-crystalline. Silicon and aluminium are examples. Silicon has been so extensively studied that I shall pay particular attention to this material, in the hope that the discussion will yield insights into other materials.

The high temperature oxidation of silicon is described by the Deal-Grove[1] equation

$$X^2 + AX = B(t + t_o) \tag{1}$$

where X is the thickness after time t. The parabolic form $X^2 = Bt$ to which this tends for large thicknesses is governed by diffusion through the film; for the linear law at small thicknesses the reaction at the Si/SiO_2 interface determines the rate. It is known that in the parabolic regime the molecule O_2 is the diffusing species, the constant B being proportional to the pressure of O_2 (Doremus[2]; for discussions by the author see refs. [3,4]). Also for <u>dry</u> oxygen by the use of the isotope ^{18}O in O_2 Professor Amsel's group has shown that O_2 diffuses through the oxide without exchange with the oxygen already in the network, ^{18}O being deposited at the Si/SiO_2 interface[5].

The present author[3, 4] has pointed out that, to account for the solubility of O_2 and of rare gases in vitreous Si_2, it is necessary to assume the presence of a small number of sites to which the O_2 can be transferred from the gas with a small change on free energy. The potential energy should thus appear as in fig. 1

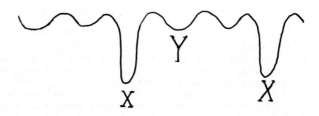

Fig. 1. Potential energy of O_2 in SiO_2

The nature and distance apart (26Å) of these deep sites is also discussed by Revesz and Schaeffer[6]. These deep sites, marked X in fig. 1, have no effect on the rate of oxidation; the constant B in equation (1) should contain the exponential $\exp\{-(W+U)/kT\}$, where W is the free energy required to bring an O_2 molecule from the gas to a site Y, and U is the activation energy to move it to a neighbouring site.

Schaeffer in earlier papers supposed that charge transport occurs also partly by diffusion of network oxygen, on the basis of results showing ^{18}O deposited throughout the film. In view however of these results from Professor Amsel's laboratory for dry oxygen, Revesz and Schaeffer[6] now believe that the earlier observations were due to traces of water vapour and that in its absence network diffusion now contributes to the oxygen transport one part in 300 or less.

We turn now to the effect of water vapour. As Deal and Grove[1] point out, the rate of oxidation is, as for O_2, proportional to the partial pressure of H_2O. As is well known, the presence of water vapour greatly increases the rate of oxidation, the main reason being that H_2O diffuses more rapidly than O_2; it is thought that hydrogen is released at the Si/SiO_2 interface and diffuses out of the specimen. On the other hand, the solubility of H_2O in vitreous SiO_2 is proportional to the square root of the pressure. Various authors (e.g. Roberts and Roberts[7]) have pointed out that this is because H_2O goes into solution as two non-bridging Si-OH bonds. These must have much lower mobility than interstitial H_2O, but this does not mean that they are not mobile at all. If they move, there can therefore be some movement of network oxygen and some plastic flow which may greatly increase the ductility.

An extremely interesting result from Prof. Amsel's group[8] is that, if $H_2^{18}O$ is used, the ^{18}O is now deposited at the gas/SiO_2 interface. Rapid exchange of O with the network already formed is thus indicated. The present author has suggested that this is because H_2O can react with the SiO_2 network to form the two non-bridging Si-OH bonds; there is no corresponding reaction for O_2. The formation of non-bridging Si-OH can give a mechanism of rapid exchange of ^{18}O with the network. At the same time it does not look as if these defects have a high mobility, or contribute much to the passage of oxygen, since the rate of oxidation is proportional to p, the pressure of H_2O. Irene and Ghez[9] give evidence that network diffusion does give some contribution, since the pressure dependence of the rate of oxidation is rather less than linear.

Many measurements confirm the approximate validity of the linear-parabolic model, for instance Irene and Van der Meulen[10] and Hopper[11], though the latter work shows different constants for thick and thin films.

We next examine the linear regime of the Deal-Grove equation (1), in which diffusion has produced a nearly constant concentration of O_2 or of H_2O in the oxide layer and the rate of the reaction is determined by the rate of reaction at the interface, with an activation energy comparable with that for diffusion (1.5 eV compared with 2.3 eV)[10]. O_2 attacks a clean silicon surface without activation energy - though the initial sticking probability is low[12] ($\sim 6 \times 10^{-3}$). The difference when oxidation is by O_2 in interstitial positions in SiO_2 is, of course, that all silicon atoms at the surface (or nearly all) form strong Si-O bonds, so the O_2 must diffuse through the surface layer and react at a point where the much weaker Si-Si bond is broken, presumably by thermal activation. This must be a complicated process in which there is a rearrangement of the atoms locally, and for which it is hardly possible to make a mathematical model.

SiO_2 occupies considerably more volume than Si in crystalline silicon; the effect of this volume change, in setting up strains in the oxide, has been considered by several authors (e.g. [13, 14]). The present author has in several

4

papers suggested that the volume change is of little importance. The process is illustrated in fig. 2 ;the Si atom, <u>below</u> the Si layer bonded to oxygen, is thought to be at a kink site. Reaction with oxygen is impossible unless

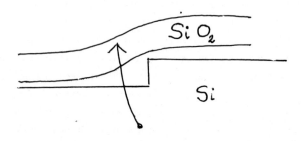

Fig. 2 Mechanism of film growth

considerable place-exchange occurs, allowing the SiO_2 to be incorporated into the network. If the latter is viscous, no strains will be set up, the extra volume being incorporated by a thickening of the oxide layer. But viscous flow may occur only under the very high stresses produced by the chemical reaction. If so, a small lateral strain may remain. We would expect this stress to decrease with increasing temperature. This seems in agreement with the observations of EerNisse[15, 16] who observes a (small) intrinsic stress as a result of oxidation at temperatures below $1000^{\circ}C$; above $1000^{\circ}C$. he supposes that viscous flow can occur (even for small stresses), as the oxide is formed. Irene et al find higher stress values with consequent higher density at lower temperatures (see also Irene et al[17], Taft[18]).

Irene and Dong[19] find a significant curvature in the Arrhenius plot of the linear rate constant, and ascribe this to the presence in the network at high temperatures of 0, the O_2 molecules being partly dissociated. It is supposed that the atomic 0 leads to a faster rate of reaction than O_2.

There is evidence[20] that the interface between Si and SiO_2 is sharp; there is no layer of SiO_x (x < 2) greater than $6\mathring{A}$ in width.

We now turn to mechanisms of low temperature oxidation, both for silicon

and metals, mechanisms which characteristically lead to a logarithmic growth law
and the formation of a protective film. Fehlner and Mott[21], in a review of
mechanisms of low-temperature oxidation, proposed that the first few layers are
produced by a mechanism of "place exchange", which can occur as a consequence of
the very strong fields which must exist. The process is considered further by
Fromhold[22] who calls the elementary excitations "hopons". Present models however are hard to compare with experiment.

Fromhold also suggests that place exchange may take place in anodic
oxidation of metals. There is extensive experimental evidence that in the
oxidation of aluminium the cation transport number is about 0.4 (see Skeldon et
al[23]). Pringle's[24] work on Tantalum gives 0.24 for the cations. Fromhold
points out that it is improbable that very different ions would have almost the
same activation energy for movement, and a correlated mechanism is indicated.
An alternative, however, if the activation energy for forming an interstitial
ion at a surface is what matters, is to assume that - say - the cation, arriving
at the oxide-electrolyte interface, excites on the average a given number of
anions into the oxide through the heat released in the oxidation process.

For further growth, the logarithmic growth law and the formation of protective films, a theory was given in 1938 by Cabrera and Mott[25], subsequently
developed by Fehlner and Mott[21]. This theory envisages oxidation by <u>charged</u>
species, normally anions, the film being transparent to electrons. This could
occur by electron tunnelling for films up to 20 - 30 Å thick or at high
temperatures by excitation of electrons from the metal (or silicon) into the
conduction band of the oxide. It is supposed that the energy W to form the
charged species at the oxide/gas interface is such that $\exp(-W/kT)$ is negligable,
so that normally the reaction will not take place; but if the film is transparent
to electrons a potential V_o will be set up between surface states at the oxide/
gas interface and the metal (or silicon). The rate at which the charged species
are formed will then be of the form

$$C \exp \left\{ - (W - qaV_o/x)/ kT \right\} \tag{2}$$

where q is the charge on the ion, <u>a</u> the hopping distance and x the thickness of
the film. Integration leads to a logarithmic or inverse logarithmic growth law.

The same model was applied to anodic oxidation by Verwey[26] in 1932, V_o
being now the applied voltage.

If W is greater than the activation energy for parabolic growth according to the Deal-Grove equation, then we would expect logarithmic growth always to be followed by a very slow parabolic growth. It is remarkable that this is found by Ponpon and Bourdon[27] in their work on the oxidation of hydrogenated amorphous silicon; logarithmic behaviour is shown for 10^5 seconds, followed by approximately parabolic growth. In crystalline silicon according to Lukes[28] logarithmic behaviour persists up to 10^7s. The smaller contribution from the logarithmic term in the amorphous material is to be expected if the crystal is n-type, because in the amorphous material, unless doped, the Fermi energy lies in mid-gap; therefore the potential V_o will be smaller.

Equation (2) would suggest an even smaller logarithmic term for crystalline p-type material; experiments in this regime would be of interest.

In the oxidation of silicon the work of Kamigaki and Itoh[29] is of interest; this was at $1000\,^{\circ}C$. While at high pressures parabolic growth was observed, at low pressures they claim to have observed logarithmic growth. Since according to the Cabrera-Mott model this will depend little on pressure (p), it should always take over at low enough values of p.

As regards the nature of the charged species, OH^-, O^- and O^{2-} are candidates. In anodic oxidation, the work of Schmidt and Ashner[30] and Croset and Dieumegard[31] show that water plays an overwhelming role, so OH^- is perhaps the most likely species. Defects, such as charged non-bridging oxygens, are also candidates; Raman spectra of steam-grown SiO_2 shows $\frac{1}{3}\%$ of OH (Galeener[32]). Here the work of Jorgensen[33], Mills and Kroger[34], and Barton[35] is relevant. Jorgensen showed that an applied electric field, using a permeable platinum electrode, can accelerate or retard the process. So if electrons are available from an open circuit, a process involving the movement of an anion is preferred. Also (at $800\,^{\circ}C$) a voltage of 1.78V blocks the process, while four times this is the energy, in electron volts per O_2 molecule, released in the reaction. Thus four electrons must flow around the circuit for each oxygen molecule, and this suggest that O^{2-} is formed through the reaction

$$O_2 + 4e \rightleftarrows 2O^{2-} \tag{3}$$

If O^{2-} is the mobile species it is not clear why the reaction rate is, according to Jorgensen, still proportional to the oxygen pressure; one would expect $p^{1/2}$ (Barton[35]). The discrepancy can perhaps be resolved by supposing that, when electrons are freely available at the SiO_2/O_2 interface, charged silicon vacancies are formed there and are responsible for the diffusion. Since one such vacancy is formed per O_2 molecule, the pressure-dependence is explained and if the vacancy carries four charges (one for each oxygen dangling bond) the blocking voltage is also as observed.

It must be emphasized that this model is applicable <u>only</u> if electrons are available, either from a platinum electrode or by tunnelling. To form the defect two Si-O bonds must be broken, and the large energy involved could come only from electrons falling from the Fermi level in the silicon into the oxygen orbitals.

In this model we need not suppose that neutral and charged species are in thermodynamical equilibrium, as postulated by Collins and Nakayama[36]; if electrons can get through the film, the charged defect is formed; if not, O_2 is the predominant mobile species.

Our hypothesis that a species carrying four charges is formed is certainly open to criticism; it should easily form a complex, for instance with hydrogen. We put it forward, however, as a very tentative explanation of Jorgensen's results.

Some new results do, however, shed some light on this process. Rochet[37] has found that, at $1000°C$ and in dry oxygen, a small amount of ^{18}O is deposited at the gas/oxide interface <u>for thin films</u>. The results are given in the table. Column I denotes the thickness of the oxide film before ^{18}O was substituted for ^{16}O; column II is, in equivalent thicknesses, the amount of deposition of ^{18}O at the gas/oxide interface, and column III is the deposition at the oxide/silicon interface (all thicknesses in Ångstroms). The residual water concentration was 1 to 3 p.p.m., and was not thought to have any effect.

I	II	III
2600	9	72
1800	11.5	92.5
270	13	150
500	15.4	188

If, as Rochet suggests (private communication), this is associated with some small passage of electrons through the film, and if this allows the formation of a negatively charged silicon vacancy, then clearly the oxygen will be built into the SiO_2 network immediately at the interface. This would explain these results.

Here again, we would have to assume that the large energy needed to break two silicon-oxygen bonds was provided by the electrons, with energies in defect states high in this energy gap of SiO_2, dropping to the defect. Since the width of the gap is perhaps 9 eV, much energy would be available.

Another model of interest is the proposal of Derrien and Commandre[38]; these authors also postulate for thin films the motion of an oxygen ion, which by "place exchange" produces a dipole layer increasing the work function (as observed for Al, Cr and Si), and which would hinder the motion of the ions and eventually bring oxidation to an end. These authors give calculations based on this model which give good agreement with experiment for oxidation at low pressures.

Finally we give a brief discussion of the formation of micropores (flaws) in oxidation. Gibson and Dong[39] give direct evidence for 1 nm micropores in dry thermal SiO_2 from high resolution transmission electron-microscopy, confirming earlier work by Irene (see this volume). Irene's hypothesis is that these form through misfit of "islands" formed initially. They are not observed in "wet" oxidation, and this could be due to the smaller viscosity of the oxide, for which   the hypothesis discussed earlier could account. In anodisation of metals similar "flaws" are observed[40]. They are thought to form on impurity segregates.

ACKNOWLEDGEMENTS

The author is grateful for discussions with F.P. Fehlner, E.A. Irene, M. Pepper, A.G. Revesz, S. Rigo, F. Rochet, G.E. Thompson and G.C. Wood.

REFERENCES

1. Deal, B.E. and Grove, A.S., 1965 J. Appl. Phys. 36, 3770.
2. Doremus, R.H., 1977 J. Phys. Chem. 8, 773.
3. Mott, N.F., 1981 Proc. R. Soc. Lond. A376, 207.
4. Mott, N.F., 1982 Phil. Mag. A45, 323.
5. Rosenscher, E., Straboni, A., Rigo, S. and Amsel, G., 1979 Appl. Phys. Lett. 34, 259.
6. Revesz, A.G. and Schaeffer, H.A., 1982 J. Electro-chem. Soc. 129, 357.
7. Roberts, G.J. and Roberts, J.P., 1966 Phys. Chem. Glasses 7, 82.
8. Rigo, S., Rochet, F., Straboni, A., Agius, B., 1980 Proc. Conference on the Physics of MOS Insulators, ed. G. Lucovsky, S.T. Pantelides and F.L. Galeener (N.Y., Pergamon Press, p. 167).
9. Irene, E.A. and Ghez, R., 1977 J. Electrochem. Soc. 124, 1757.
10. Irene, E.A. and van der Meulen, Y.J., 1976 J. Electrochem. Soc. 123, 1380.

11.   Hopper, M.A., Clark, R.A. and Young, L., 1975 J. Electrochem. Soc. <u>122</u>, 1216.
12.   Tougaarol, S., 1981 Surface Science <u>111</u>, 545.
13.   Tiller, W.A., 1980 J. Electrochem. Soc. <u>127</u>, 619.
14.   Irene, E.A., Tierney, E. and Angillelo, J. 1982 J. Electrochem. Soc. (in press).
15.   EerNisse, E.P. 1977 J. Applied Phys. Lett. <u>30</u>, 290.
16.   EerNisse, E.P. 1979 J. Applied Phys. Lett. <u>35</u>, 8.
17.   Irene, E.A., Dong, D.W. and Zeto, R.J. 1980, J. Electrochem. Soc. <u>127</u>, 396.
18.   Taft, E.P., 1978 J. Electrochem. Soc. <u>125</u>, 968.
19.   Irene, E.A. and Dong, D.W. 1978 J. Electrochem. Soc. <u>125</u>, 1146.
20.   Frenzel, H. and Balk, P. 1980 in "The Physics of MOS Insulators, ed. G. Lukovsky et al, Pergamon, P. 246.
21.   Fehlner, F.P. and Mott, N.F. 1970, Oxidation in metals, <u>2</u>, 59.
22.   Fromhold, A.T. 1980 J. Electrochem. Soc. <u>127</u>, 411.
23.   Skeldon, P., Shimizu, K., Thompson, G.E. and Wood, G.C. 1973, Nature (in press).
24.   Pringle, J.P.S. 1974, J. Electrochem. Soc. 121, 45.
25.   Cabrera, N. and Mott, N.F. 1948-49, Rep. Prog. Phys. <u>12</u>, 163.
26.   Verwey, E.J.W. 1932, Physica <u>2</u>, 1059.
27.   Ponpon, J.P. and Bourdon, B., 1981, Solid State Electronics <u>25</u>, 875.
28.   Lukes, F., 1972 Surface Science <u>30</u>, 91.
29.   Kamigaki, Y., and Itoh, Y., 1977 J. Applied Phys. <u>48</u>, 2.
30.   Schmidt, P.F. and Ashner, J.D., 1971 J. Electrochem. Soc. <u>118</u>, 325.
31.   Croset, M., and Dieumegard, D. 1971 J. Electrochem. Soc. <u>118</u>, 771.
32.   Galeener, F.L. and Mikkelson, J.C. 1981, Appl. Phys. Lett. <u>38</u>, (5) 330.
33.   Jorgensen, P.J. 1962 J. Chem. Phys. <u>37</u>, 874.
34.   Mills, T.G. and Kroger, F.A. 1973 J. Electrochem. Soc. <u>12</u>, 1582.
35.   Barton, R., 1980 Priv. Comm.
36.   Collins, F.C. and Nakayama, T. 1967 J. Electrochem. Soc. <u>118</u>, 771.
37.   Rochet, F., 1981 Thesis, Paris.
38.   Derrien, J. and Commandre, M. 1982 Surface Science, <u>118</u>, 690.
39.   Gibson, J.M. and Dong, D.W. 1980, J. Electrochem. Soc. <u>127</u>, 2728.
40.   Thompson, G.E. and Wood, G.C. 1983 "Anodic Films on Aluminium", ed. J.C. Scully, p. 205, Academic Press, London.

Note added in proof :

In view of Professor Good's model for logarithmic oxidation discussed at this meeting, I suggest that the Cabrera-Mott model may apply principally to amorphous oxides, which are probably much less susceptible to doping than crystals.

Dr. Rochet suggest that trivalent Si is a charged defect, already known from its ESR spectrum, which can be responsible for oxygen transport in SiO_2.

Passivity of Metals and Semiconductors, edited by M. Froment
Elsevier Science Publishers B.V., Amsterdam — Printed in The Netherlands

AN OVERVIEW OF THE KINETICS OF OXIDATION OF SILICON: THE VERY
THIN SIO$_2$ FILM GROWTH REGIME

E.A. Irene, Dept. of Chemistry
Univ. of North Carolina, Chapel Hill, N.C.   27514 (U.S.A.)

INTRODUCTION

Modern microelectronic devices such as bipolar and field
effect transitors are fabricated on silicon single crystal
surfaces using a planar technology. This kind of processing
utilizes primarily the surface of the semiconductor. The
individual devices operate based on the electrical properties of
the silicon surface. For reliable device operation, the surface
properties must be reproducible over large dimensions as compared
to the size of the devices, in order that large arrays of densely
packed devices can be produced.

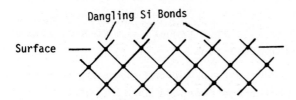

Fig. 1. Si (100) Surface with Dangling Bonds

It is intuitive that atoms on the surface of a solid are
environmentally different from atoms in the interior of the
material. Indeed, the earliest literature on semiconductor
physics reports this idea (1-3). Fig. 1 shows a cross-sectional
view of a Si crystal. It is seen that the surface contains atoms
with unsatisfied chemical bonds or so-called "dangling bonds". It
was predicted (1-3) that such dangling bonds would produce
electrically active states, ie. states in the Si band gap. This
prediction was confirmed. Such surface electronic states were
measured and the number of states was found to be in agreement
with the prediction. The dangling bond model for the surface

states has decisively important implications in semiconductor
device physics and technology. Firstly, the dangling bonds
represent active sites that can form chemical bonds to foreign
atoms in the vicinity. Secondly, the states can trap electrons.
Both of these consequences will alter the semiconductor surface
potential in an irreproducible manner. The result will be a
semiconductor with unpredictable and variable properties. This
situation is intolerable for the dense arrays of devices required
by modern computers.

Fortunately, nature has provided us with a material that
enables the stabilization of the silicon surface. The material is
silicon dioxide, SiO_2. SiO_2 grows quite naturally on a freshly
cleaved surface of Si. Even this natural growth in air serves to
reduce the number of Si surface states by several orders of
magnitude. However, to reduce the number of states to tolerable
levels from a device point of view, the oxidation of Si must be
done in a purposeful and deliberate manner. The kinetics of the
process by which the Si surface is passivated via oxidation is the
subject of this paper. The process step of the oxidation of
silicon to produce a film of SiO_2 is the most important step in
semiconductor processing, and the formation of the passivating
SiO_2 film with the added properties that the film is stable,
protective and an excellent insulator renders the $Si-SiO_2$ system
unequaled for electronics applications.

The first question to address is how can the SiO_2 film be
prepared? There are several common methods used in the
microelectronics industry that produce SiO_2 films with different
properties. First, there is the simple thermal oxidation of Si.
A thoroughly cleaned slice of polished single crystal Si is placed
in a high temperature furnace in which pure oxidant gas is
flowing. Usually oxygen or water is used. Under the assumption
that the environment is kept scrupulously clean, this method
produces the best electrically passivating film of SiO_2 and is
therefore chosen when the highest quality film is required, viz.
at active device regions. The next best method is chemical vapor
deposition, CVD. In this method, a reactive mixture of gases is
heated in the presence of the substrates to be coated. The
reaction takes place to produce the desired product which
condenses in thin film form onto the substrates. For the CVD
preparation of SiO_2, gaseous SiH_4 and N_2O are commonly used. It

is obvious that since there is relatively little reaction of the condensing film with the substrate, the passivating properties are not as good as for thermal oxidation. However, when CVD is performed at high temperatures or when the resulting film is annealed at high temperatures, a reasonably good SiO_2 film can be produced. The so-called physical vapor deposition, PVD, methods such as evaporation and sputtering are generally used to produce SiO_2 films for packaging applications, rather than for active device regions. Recently, there have been reports of the use of special techniques such as high oxidant pressures, and plasmas. Some of these techniques have produced reasonably good quality SiO_2 films. Thus far the best and consequently the most prevalent technique is the thermal oxidation of Si; this technique is the focus of this paper.

When considering the thermal oxidation kinetics of Si, it is useful to consider three oxidation regimes that are defined according to the SiO_2 film thickness. The thickest SiO_2 films from about several hundreds of angstroms upwards, grow according to linear-parabolic kinetics (4,5). We shall use the symbol L_o to represent the smallest SiO_2 film thickness in the linear-parabolic, L-P, regime. Below this oxidation regime the so-called "very thin" oxide regime exists which is central to this paper. This "very thin" SiO_2 regime can be viewed as made up of two regimes: one from 0 to about $10\overset{\circ}{A}$ SiO_2, 0 to L_n, and the other regime from L_n to L_o. L_n is used to denote the thickness of the native oxide that is formed after cleaving a Si crystal in the presence of oxidant.

Presently, the linear-parabolic regime is the most prominant regime from a technological point of view. Usually films in this regime are prepared by high temperature oxidation (>900$\overset{\circ}{C}$) in either O_2 or H_2O. The kinetics are pretty well understood for the higher temperatures and will be treated below. However, with the trend in microelectronics towards smaller, faster, and more densely packed device arrays, it is imperative that the thinner oxides be employed. With the use of thinner SiO_2 films, the speed of FET's increases and thinner SiO_2 films enable the use of lower processing temperatures. Lower oxidation temperatures preclude extensive solid state diffusion, thereby maintaining the sharpness of diffused or implanted junctions. This latter consideration is of prime importance for small devices in dense arrays. However, oxide films thinner than 50Å are of electron tunneling dimensions, and therefore only useful in some special memory device applications, viz. MIOS devices. SiO_2 films less than about 15Å have not been reliably prepared and hence have no present technological application. The regime from 0 to L_n is of great relevance in understanding the mechanism of oxidation and will be discussed below.

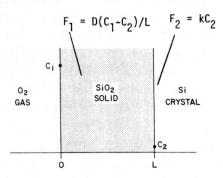

$$F_1 = D(C_1-C_2)/L \qquad F_2 = kC_2$$

Fig. 2. The Fluxes for the Linear-Parabolic Model.

THE LITERATURE

As mentioned above, the linear-parabolic, L-P, model quite reasonably describes the oxidation kinetics of thicker SiO_2 films grown at high temperatures. It is therefore useful to introduce this model so that the less well understood kinetics that apply to "very thin" SiO_2 films have some framework for comparison. Firstly, it has been established from radio tracer studies by Ligenza and Spitzer (6) and from chemical etch studies by Pliskin (7) that the reaction between Si and oxidant takes place at the Si-SiO_2 interface and therefore oxidant species are the predominant transported species. More recent studies by Rosencher et al (8) have confirmed the earlier work and demonstrated that transport occurs via the long range migration of molecular oxygen. This simple L-P model is best understood using Fig. 2. The model considers the transport of oxidant through the growing film to occur by Fickian diffusion and represented as F_1. The reaction of oxidant at the Si-SiO_2 interface produces a flux of SiO_2, F_2. As shown in Fig. 2, F_2 is represented by first order kinetics between Si and oxidant. Now it is clear that these series processes must be in a steady state so that:

$$F = F_1 = F_2 \tag{1}$$

The rate equation has the form:

$$\overset{\circ}{L} = F/\Omega \tag{2}$$

where Ω is the conversion factor for the number of oxidant molecules in a mole of SiO_2. The boundary conditions for the integration are from $t = t_o$ to t and from $L = L_o$ to L. The resultant equation has the following form:

$$t-t_o = (L-L_o)/k_1 + (L^2-L_o^2)/k_p \tag{3}$$

where k_1 and k_p represent the linear and parabolic rate constants, respectively and t_o, L_o represent an offset to the L-P model, viz. the very thin SiO_2 regime, that does not conform to L-P kinetics. Of primary importance in this review is the oxidation regime below t_o, L_o.

Firstly, we consider the ultra-thin regime of from 0 to L_n, which is usually referred to as the native oxide. This regime should yield considerable insight into the nature of the interface reaction that occurs between the oxidant and Si. However, the study of this regime is replete with experimental difficulties. Among the problems is the cleanliness of the Si surface, poor vacuum and unspecified surface conditions. For these reasons many of the studies in the literature have dubious merit and therefore much needs to be done in this regime.

Williams and Goodman (9) have measured the surface energy using a contact angle technique as the oxide thickness changed from 0 to 30Å. The change was found to be quite steep up to about 30Å, and too large to be explained by van der Waals forces. These authors suggested that perhaps a gradual compositional change of Si to SiO_2 occurs in a transition region. Sigmon et al (10) using Rutherford backscattering have found evidence for a transition layer of about three monolayers of SiO_2. They have calculated the excess Si in this region as compared to stoichiometric SiO_2 and found an excess of about $6 \times 10^{15} Si/cm^2$ in the transition region. This number is considerably larger that the interface charge and surface states measured near the Si-SiO_2 interface, and therefore they conclude that only a fraction are electrically active.

Furthermore, Chiaradia and Nannarone (11) have found from electron energy loss spectroscopy, EELS, and Auger spectroscopy, AES, that as the oxide coverage increased the number of surface states decreased. Recent work by Derrien and Commandre' (12) has shown that the oxide formed at low temperatures, ie. by exposing a cleaned Si surface to oxygen, is different in nature from oxide formed via high temperature oxidation. Also, a transition region extending to about 5$\overset{o}{A}$ exists. Similarly, Ibach et al (13) report that a different oxide is produced at temperatures below about 600$\overset{o}{C}$. They observe atomic O species in the high temperature produced oxide and molecular O_2 species in the low temperature material.

All of these careful studies point towards a first order understanding of this very early stage of oxidation. Firstly, it is clear that some sort of transition layer exists between pure Si and SiO_2. This is intuitive, since it would be difficult to reconcile an absolutely sharp boundary in a reaction zone. The fact that the surface state density decreases with coverage is also quite sensible in view of the success of the dangling bond model of the Si surface. The fact that the number of excess Si atoms does not correlate with the number of surface states is not troublesome at all and indeed may be actually more sensible. In order for the states to be electrically active, they must have energies in the band gap. It is certainly likely that different environments for the excess Si exist and therefore an associated spectrum of energies. Only a fraction of the states are in the gap. Finally, the oxide prepared in a purposeful way, ie. via high temperature oxidation, is a relaxed compound of Si and O atoms while lower temperature forms which comprise the native oxide, may not be purely oxide at all. This explains the irreproducible nature of the native oxide. This observation also suggests that for such thin oxides to be technologically useful, high vacuum techniques must be used to first clean the surfaces and then oxidize at sufficiently high temperatures.

The next regime of from L_n to L_o of the "very thin" regime is technologicaly relevant. The oxidation kinetics in this regime are also not well understood, but much more experimental data exists for this regime and more is understood with certainty than for the ultra-thin SiO_2 films.

In terms of the oxidation data itself, ie. SiO_2 film thickness, L, versus time, t, the results of Hopper et al (14) show that the shape of the data in this regime is also linear-parabolic but needs to be described with different rate constants than the thicker L-P regime. Irene (15) confirms this result and shows that L-P behavior is dominant for dry O_2, but more parabolic behavior is observed when H_2O is used as the oxidant. Besides some very strange initial curvature in their data which has not been reproduced, Smith and Carlan (16) have reported that the rate in this very thin regime scales with the square root of the oxidant pressure as opposed to the first power, as predicted if the conventional L-P model were to apply. van der Meulen (17) also shows that the linear rate constant scales with the square root of oxygen pressure. From this pressure dependence Ghez and van der Meulen (18) have derived a plausible mechanism for the surface reaction between O_2 and Si. They show that if both O_2 and O were present, then not only is the pressure dependence fulfilled but the correct form of the L vs. t data is obtained. From the observed non-Arrhenius temperature dependence of the linear rate constant, Irene (19) has shown that the Ghez and van der Meulen model is also consistent with the observed curvature. Blanc (20) has shown that the reaction with just atomic O gives the correct shaped data. However, Blanc's model is not in accord the pressure dependence. Hopper et al (14) did not fully agree with the Ghez and van der Meulen model because of the observed L dependence of the rate in this regime. Hopper et al argued that an additional transport term in parallel with the normal diffusional transport is necessary to fully explain the data. From data in this regime and dielectric breakdown histograms of the number of breakdowns vs. breakdown field, Irene (15) suggested that if micropores existed in these films both the kinetics and reliability results could be explained. In this study, Irene also provided transmission electron micrographs on chemically etched very thin SiO_2 films. These micrographs showed non-uniformities which should not be present in uniform amorphous SiO_2, and therefore is suggestive of the existence of micropores. It was pointed out, that pores are a rather common occurrence in ceramic type materials. Later, Gibson and Dong (21) have directly demonstrated the existence of micropores in dry O_2 grown SiO_2 films by the use of a sophisticated TEM technique. The pores were observed to be about 10Å in diameter and 100Å apart. In a private communication,

Mott (22) suggested that the origin of these pores may be related to the oxidation of non-uniform crystal surfaces. He cited a recent publication by Ponpon and Bourdon (23) as possible evidence. This publication reports parabolic oxidation at the outset of oxidation of amorphous Si as compared to rather linear initial kinetics for single crystal oxidation. The amorphous Si is more uniform due to the random arrangement of atoms while the crystalline material has periodic atomic differences across the surface as well as the possibility of having dislocations, steps, stacking faults and point defect clusters intersecting the surface.

The effect of impurities on Si oxidation has received considerable attention. Impurities have been shown to alter both the interfacial reaction between oxidant and Si, and the transport of oxidant across the film. An understanding of the role of the common impurities such as H_2O and Na is crucial towards obtaining reliable kinetics for interpretation. For example, Irene (24) has demonstrated that only 25ppm H_2O in O_2 can increase the overall rate of oxidation by 20%. Furthermore, Irene and Ghez (25) have shown that most of the kinetic acceleration occurs, percentage wise, with H_2O additions of only several hundred ppm to O_2, after which the effect saturates. In practical oxidation systems the occurrence of H_2O in amounts sufficient to alter the kinetics has been reported by Revesz and Evans (26). These authors have shown that both H_2O and Na can easily diffuse through the fused silica oxidation tubes typically used. Furthermore, they have shown that the effects on kinetics can be large. The above mentioned studies (24,25) have also shown that H_2O affects both the interfacial reaction and the transport of O_2. The recent studies of Rigo et al (27) and Pfeffer and Ohring (28) demonstrate that any H_2O in O_2 reacts with the SiO_2 network and causes the rapid exchange of O. On the other hand, there is no O exchange with the SiO_2 network when the O_2 is dry (see also Rosencher et al(8)). Doremus (29) also shows that from available oxidation and diffusion data it is likely that transport of oxidant species for both O_2 and H_2O oxidation is via molecular species.

There has been some controversy relative to the charged nature of the oxidant species that is transported during oxidation. If the transported species are charged than perhaps a mechanism as proposed by Mott (30) can be applied for very thin

SiO_2 films to be followed by a diffusion regime in which the diffusion of charged species occurs. Jorgenson (31) attempted experiments to test this possibility. Essentially, he applied an external electric field to a growing oxide film during oxidation. His results point to an alteration of the kinetics over the pure thermal case. However. Raleigh (32) has pointed out that the result is ambiguous, since the imposition of the external field provides an external circuit path, and hence superimposes an electrolysis upon normal thermal oxidation. Therefore, the question remains unresolved, as no evidence exists which compels the consideration of charges moities in the thermal oxidation of Si.

The recent report of viscoelastic effects during the thermal oxidation at low temperatures by Irene et al (33), as an explanation of the densification of SiO_2 films (34,35) and intrinsic stress (36), sheds some light on the behavior of the oxide during the film formation process. Essentially, Irene et al have shown that the large molar volume change that takes place as a result of the conversion of Si to SiO_2 should cause a large intrinsic stress to occur because of the confinement in the plane of the Si surface. For high temperature oxidation no intrinsic stress has been observed. However, at temperatures of $800\overset{o}{C}$ and below both an intrinsic stress and a higher density for the SiO_2 films are observed. These observations can be explained by considering that the SiO_2 viscosity at the higher temperatures is sufficiently low to allow viscoelastic flow of the oxide as it is formed, thereby precluding densification and stress resulting from the confinement of the oxide. At lower temperatures, however, the viscosity is too high to allow flow, hence the oxide densifies to reduce the volume requirement and simultaneously develops an intrinsic compressive stress. These low temperature phenomena may be related to the observation of different low temperature kinetics as evidenced by the non-Arrhenius behavior of the linear and parabolic rate constants mentioned previously.

Recently, Irene (37) has proposed a revised linear-parabolic model that considers transport of oxidant in micropores as well by diffusion and also includes the viscoelastic properties of the oxide. This model is obtained as represented in Fig. 3 from a steady state of three fluxes:

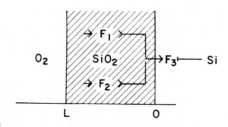

STEADY STATE :

$$F = F_1 + F_2 = F_3$$

Fig. 3 The Fluxes for a Revised Linear-Parabolic Model (37).

$$F = F_1 + F_2 = F_3 \tag{4}$$

F_1 is the normal diffusional flux and F_2 is for Knudsen-Poiseuille flow in micropores (38). F_3 is obtained by considering the rate of flow of oxide away from the Si-SiO$_2$ interface during oxidation as well as the reaction between Si and oxidant.

The form for the fluxes is given as:

$$F_1 = D(C_1-C_2)/L \tag{5}$$

$$F_2 = H(C_1-C_2)(L + L_o)/L^2 \tag{6}$$

$$F_3 = k'' C_2 C_{Si} \sigma_{xy}/\eta \tag{7}$$

where D is the diffusion constant, H is a composite of the constants in the Knudsen-Poiseuille equation plus the steady state number of micropores, k'' is a composite rate constant which includes the chemical reaction rate constant and geometrical effects, σ_{xy} is the intrinsic film stress in the plane of the Si surface and η is the SiO$_2$ viscosity. The number of micropores, N_p, is a function of L. N_p is a rapidly decreasing function of L and at L_o levels off to a steady state value. The final integrated equation has the L-P shape which has been shown to be valid:

$$t = \frac{\Omega}{2C_1(D+H)} (L^2-L_o^2) + \frac{\Omega}{C_1} \left[\frac{\eta}{k'' C_{Si} \sigma_{xy}} - \frac{HL_i}{(D+H)^2} \right] (L-L_o)$$

The change in the number of pores with L is then responsible for the rapid kinetics of from L_n to L_o which are not included in the unrevised L-P model. Hence there is no need for the t_o, L_o offset in the revised model. Additionally, one sees some coupling in the rate constants and the complex temperature dependence can now be more easily rationalized. It is clear that the revised model does not yet account for all the information available and the revised model needs testing but it does represent a step towards including some of the confirmed observations in a way consistent with the shape of the growth data.

SUMMARY

A variety of experimental and theoretical studies have been presented which have increased our understanding of very thin SiO_2 films. We now feel comfortable with the transport of O_2 by long range migration of molecular species. H_2O effects have been shown to cause rapid reaction with the SiO_2 network. These ideas in combination with the effect of micropores on transport, enable an explanation of the shape of the very thin oxidation regime. The attempts at understanding the ultra thin regime of oxidation have not yet provided us with startling results for incorporation into a model, but nonetheless the best studies confirm the hitherto intuitive notions concerning the interface, and therefore, provide a factual boundary to any ideas which are put forth. The controversy about temperature effects make us realize the oversimplicity of the conventional L-P model. The inclusion of mechanical effects into the interface kinetics provides another dimension for creative experiments and reflection. Indeed, the use of very thin SiO_2 films in the microelectronics industry is apparent. I therefore expect this area to be one of great relevance.

22

References

1. I. Tamm, Physik., Z. Sowjetunion, 1 (1932) 733.
2. W. Shockley, Phy. Rev., 56 (1939) 317.
3. M.M. Atalla, E. Tannenbaum and E.J. Scheibner, Bell System Tech. J., 38 (1959) 749.
4. B.E. Deal and A.S. Grove, J. Appl. Phys., 36 (1965) 3770.
5. W.A. Pliskin, IBM J. Res. and Develop. 10, (1966) 198.
6. J.R. Ligenza and W.G. Spitzer, J. Phys. Chem. Solids, 44 (1960) 131.
7. W.A. Pliskin and R.P. Gnall, J. Electrochem. Soc., 111 (1964), 872.
8. E. Rosencher, A. Straboni, S. Rigo and G. Amsel, Appl. Phys. Lett., 34 (1979) 254.
9. R. Williams and A.M. Goodman, Appl. Phys. Lett., 25 (1974) 531.
10. T.W. Sigmon, W.K. Chu, E. Lugujjo, and J.W. Mayer, Appl. Phys. Lett., 24 (1974) 105.
11. P. Chiaradia and S. Nannarone, Surface Science, 54 (1976) 547.
12. J. Derrien and M. Commandre, Surface Science, 118 (1982) 32.
13. H. Ibach, H.D. Bruchmann, and H. Wagner, Appl. Phys. A, 29 (1982) 113.
14. M.A. Hopper, R.A. Clarke, and L. Young, J. Electrochem. Soc., 122 (1975) 1216.
15. E.A. Irene, J. Electrochem. Soc., 125 (1978) 1708.
16. T. Smith and A.J. Carlan, J. Appl. Phys., 43 (1972) 2455.
17. Y. J. van der Meulen, J. Electrochem. Soc. 119 (1972) 530.
18. R. Ghez and Y.J. van der Meulen, J. Electrochem. Soc., 119 (1972) 1100.
19. E.A. Irene, Appl. Phys. Lett., 40 (1982) 74.
20. J. Blanc, Appl. Phys. Lett., 33 (1978) 424.
21. J.M. Gibson and D.W. Dong, J. Electrochem. Soc., 127 (1980) 2722.
22. N.F. Mott private communication, Dec. 1982.
23. J.P. Ponpon and B. Bourdon, Solid-State Electronics, 25 (1982) 875.
24. E.A. Irene, J. Electrochem. Soc., 121 (1974) 1613.
25. E.A. Irene and R. Ghez, J. Electrochem. Soc., 124 (1977) 1757.
26. A.G. Revesz and R.J. Evans, J. Phys. Chem. Solids, 30 (1969) 551.
27. S. Rigo, F. Rochet, B. Agius, and A. Straboni, J. Electrochem. Soc., 129 (1982) 867.
28. R. Pfeffer and M. Ohring, J. Appl. Phys., 52 (1981) 777.
29. R.H. Doremus, J. Phys. Chem., 80 (1976) 1773.
30. N.F. Mott, Phil. Mag. A., 45 (1982) 323.
31. P.J. Jorgenson, J. Chem. Phys., 37 (1962) 874.
32. D.O. Raleigh, J. Electrochem. Soc., 113 (1966) 782.
33. E.A. Irene, E. Tierney and J. Angillelo, J. Electrochem. Soc., 129 (1982) 2594.
34. E.A. Taft, J. Electrochem. Soc., 125 (1978) 968.
35. E.A. Irene, D.W. Dong and R.J. Zeto, J. Electrochem. Soc., 127 (1980) 396.
36. E.P. EerNisse, Appl. Phys. Lett., 35 (1979) 8.
37. E.A. Irene, Electrochemical Soc. Meeting, Oct. 1981, Abs. #371; submitted for publication 1983.
38. R.M. Barrer, "Diffusion In And Through Solids," Cambridge Press (1951) p. 54.

Passivity of Metals and Semiconductors, edited by M. Froment
Elsevier Science Publishers B.V., Amsterdam — Printed in The Netherlands

TRANSFER REACTIONS ON OXIDES AND PASSIVE LAYERS

W. SCHMICKLER
Institut für Physikalische Chemie der Universität Düsseldorf, D-4000 Düsseldorf, FRG

ABSTRACT
   A model for electron transfer reactions on film-covered electrodes is presented. When the film is badly conducting and sufficiently thin, the electron is
exchanged with the underlying metal; electron transfer through the film may be
effected by direct or by resonance tunnelling. When the film is thick, it acts
itself as an electron donor or acceptor; in the presence of a space charge
barrier tunnelling processes may again be important. Surface states may cata-
lyze the reaction by serving as electronic intermediate states.

INTRODUCTION
   Electron and ion transfer reactions can occur also on oxide electrodes and
passivated metal electrodes. While the mechanism of ion transfer reactions at
such electrodes, which are an inherent process in the formation of passive films,
has been well investigated, and is extensively reviewed in the literature
(ref. 1), systematic studies of electron transfer (ET) reactions have only been
undertaken during the last decade. On the few oxides that are good electronic
conductors ET reactions proceed in much the same way as on metal electrodes.In
the presence of a semiconducting or insulating film, however, the reaction rate
is generally reduced, and several mechanisms are possible; the reaction is very
sensitive to the electronic properties of the film, and even a small amount of
doping may enhance the rate by orders of magnitude.

   ET reactions on oxides and passive films are governed by the same quantum
mechanical principles as ET on metal electrodes and homogeneous exchange reactions. They are, however, more complicated in that tunnelling processes and
electronic intermediate states play an important role. In this work we shall
review our theory for ET on film-covered electrodes and present some new results for ET via surface states, thus presenting a framework for the interpretation of the current-potential characteristics of experimental systems.

   Two types of ET mechanisms operate on film covered metal electrodes. When
the film is badly conducting and sufficiently thin, the electron is exchanged
with the underlying metal, and the film acts as a barrier for the transferring
electron; we shall refer to this as the 'thin film' situation. At sufficiently
thick layers, the electron is exchanged with the film itself, and the metal

has no effect on the reaction rate ('thick film' situation); bulk oxide electrodes, of course, fall under the 'thick film' category. We shall first present the general principles of ET theory, and then treat both cases individually.

PRINCIPLES OF ET-THEORY

We consider ET from a solid electrode to a redox couple in solution. The spectrum of electronic states of the electrode is characterized by the electronic density of states $\rho(\varepsilon)$, where ε is the electronic energy, which we measure with respect to the Fermi-level. For metals, $\rho(\varepsilon)$ is broad, while for semiconductors it has the familiar band structure. The transfer of an electron to a redox couple is accompanied by a reorganization of the acceptor complex and the surrounding solvation sphere. This reorganization requires thermal activation, i. e. the system must be on the reaction hypersurface ('activated complex') for the reaction to occur. The rate $W(\varepsilon)$ of transfer of an electron of energy ε to the solution can then be written in the form (ref. 2, 3):

$$W(\varepsilon) = \rho(\varepsilon)\ n(\varepsilon)\ D_{ox}(\varepsilon)\ \kappa(\varepsilon) \tag{1}$$

$n(\varepsilon)$ is the Fermi-distribution; the product of the first two terms is thus the probability of finding an electron of energy ε on the electrode. $D_{ox}(\varepsilon)$ is the probability of finding an oxidized species on the reaction hypersurface for accepting an electron of energy ε. To a first approximation, the potential energy surfaces for the oxidized and reduced states are parabolic, and $D_{ox}(\varepsilon)$ has the familiar form of the Marcus theory (ref. 4):

$$D_{ox}(\varepsilon) = c_{ox}(\pi/\lambda kT)^{1/2}\ \exp\ -\ \frac{(\lambda - \varepsilon + e\eta)^2}{4\lambda kT} \tag{2}$$

where c_{ox} is the concentration of the oxidized species, λ the energy of reorganization of the redox couple, and η the overpotential. D_{ox} is sometimes referred to as the 'density of states of unoccupied levels in the solution' (ref. 3).

Finally, $\kappa(\varepsilon)$ is the square of the electronic matrix element of the interaction between the electrode and the redox couple, which determines the electronic transition probability. When there is an electronic potential energy barrier between the electrode and the redox couple, as will usually be the case in our systems, κ is proportional to the tunnelling probability P through the barrier: $\kappa(\varepsilon) = C \cdot P(\varepsilon)$, where C is a generally unknown constant.

The cathodic current density is obtained by integrating eq. (1) over the electronic energy spectrum ε of the electrode:

$$j^- = C \cdot F \int d\varepsilon\ \rho(\varepsilon)\ n(\varepsilon)\ D_{ox}(\varepsilon)\ P(\varepsilon) \tag{3}$$

A similar equation holds for the anodic current. Since the constant C is not

known, it is not possible to calculate absolute values of the current from
eq. (3); one can, however, calculate the dependence of the current on the various system parameters.

THIN FILMS

We idealize the situation by assuming that the thin, badly conducting film
covering the metal is homogeneous and non-porous. Such a film forms a potential
energy barrier for the transferring electron and thus primarily influences the
tunnelling probability $P(\varepsilon)$, which is strongly dependent on the thickness and
height of the barrier. When the film is a crystalline semiconductor, the barrier
height is determined by the bottom of the conduction band E_c (see fig. 1); in
the case of an amorpheous semiconductor, one has to speak of an effective barrier, whose upper edge is probably near the mobility edge. From solid state

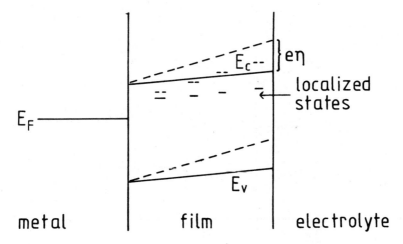

Fig. 1. Schematic representation of the thin film situation at the equilibrium
potential (solid lines) and at an anodic overpotential η (broken lines).
E_F: metal Fermi-level; E_c: lower edge of conduction band, E_v: upper edge of
valence band.

physics several elastic and inelastic tunnelling mechanisms through thin films
are known (ref. 5, 6). In electrochemical systems only two mechanisms: direct
elastic and elastic resonance tunnelling, seem to have been observed.

Direct elastic tunnelling

The conceptually simplest transfer mechanism through the barrier is direct
elastic tunnelling, in which the electron transverses the barrier in one step
and without loss of energy. The corresponding transition probability can be
calculated in the WKB-approximation:

$$P(\varepsilon) = \exp \left\{ - \frac{2\sqrt{m^*}}{\hbar} \int \left[E_c(x) - \varepsilon^{1/2} \right] dx \right\} \tag{4}$$

where m^* is the effective mass of the electron in the barrier, and x is the co-ordinate perpendicular to the electrode surface. The integral is to be performed over the classically forbidden region. We can now substitute eqs. (2) and (4) into eq. (3) and note, that the product $n(\varepsilon) \, D_{ox}(\varepsilon) \, P(\varepsilon)$ is centred near the Fermi-level with a width of several tenths of an eV; the metal density of states can be considered as constant over this range, and be taken in front of the integration sign. The integral over ε can then be performed numerically, and current-potential curves can be calculated.

Typical Tafel plots for the partial current densities are shown in fig. 2.

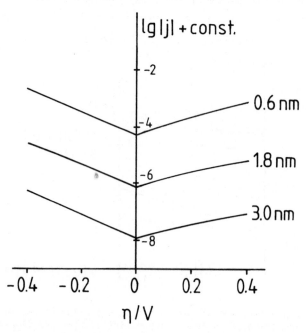

Fig. 2. Tafel plots of the partial current densities for direct tunnelling through thin films of different thicknesses. E_C was assumed to be flat at $\eta = 0$ and 0.6 eV above E_F; $\lambda = 1$ eV.

The current decreases strongly with the film thickness, and with the average barrier height. Since the application of an anodic overpotential increases the barrier height, the anodic transfer coefficients are somewhat smaller than 0.5; conversely, the cathodic coefficients are somewhat greater. For a more detailed discussion we refer to ref. 7.

Resonance tunnelling

In many cases the film contains localized electronic levels within the barrier. Such states can be due to impurities, to oxygen or metal ion vacancies, or in the case of an amorpheous film due to local disorder. They can serve as short-lived intermediate states in the tunnelling process, a mechanism known as 'resonance tunnelling' in solid state physics, where it was first observed (ref. 8). The tunnelling probability for this process is:

$$P_{res}(\varepsilon) = \delta(\varepsilon-\varepsilon_r) \frac{P_1(\varepsilon)\ P_2(\varepsilon)}{P_1(\varepsilon)+P_2(\varepsilon)} \tag{5}$$

where the δ-function accounts for the fact that the energy ε of the transferring electron must equal the energy ε_r of the localized level for resonance to occur. P_1 and P_2 are the tunnelling probabilities from the metal to the localized state, and from the metal to the redox couple; they can be calculated in the WKB-approximation. Inserting eq. (5) into eq. (3) gives the contribution of a particular localized state to the current density; to obtain the total current, we have to sum over all localized states that are present.

The optimum energy for a resonance state is close to the Fermi-level, where the product $n(\varepsilon)\ D_{ox}(\varepsilon)$ has its maximum; its optimum position is determined by the condition $P_1 = P_2$ (see eq. (5)), which is generally fulfilled near the centre of the barrier (ref. 9).

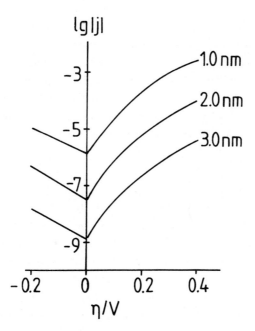

Fig. 3. Tafel plots for resonance tunnelling through thin films of various thicknesses; $\lambda = 1$ eV

Fig. 3 shows Tafel plots for a resonance tunnelling mechanism. These calculations were performed for a semiconducting film, whose flat band potential coincides with the equilibrium potential for the reaction, and which is uniformly
doped with localized states of an energy 0.3 eV below the conduction band, which
was taken as 0.2 eV above the Fermi-level at the flat band potential. The Tafel
plots show a marked assymmetry between the anodic and cathodic branches, and a
change of transfer coefficient on the anodic side. The application of a cathodic overpotential shifts the resonance states, which at $\eta = 0$ are 0.1 eV, below
Fermi-level, even further below the Fermi-level and thus away from the most favourable energy; consequently, the cathodic transfer coefficients are small.
Conversely, application of an anodic overpotential shifts the resonance states
towards the Fermi-level, so that the anodic transfer coefficients are high at
low overpotentials. However, at high anodic overpotentials most resonance states
have passed through the Fermi-level and are shifted further away from it with
increasing overpotential; then, the anodic coefficients are also small. The
change in transfer coefficient occurs roughly at $\eta = 0.2$ eV, where the states
near the centre of the barrier pass through the Fermi-level. It is obvious from
this discussion, that the shape of the current-potential characteristics depends
critically on the electronic characteristics of the film, particularly on the
energy of the intermediate states. For further examples and discussions we refer
to the literature (ref. 9, 10).

Experiments

Space limitations do not permit a discussion of experimental work. Instead,
we cite several systems and indicate, in terms of which mechanism they can be
understood: reduction of Ce^{4+} on oxide covered platinum (ref. 11) (direct tunnelling); oxygen evolution on oxide-covered platinum (ref. 12, 13) (direct and
resonance tunnelling); $Fe(CN)_6^{3-/4-}$ on passive iron (ref. 14, 15) (resonance
and direct tunnelling), Fe^{2+}/Fe^{3+} on platinum doped passive titanium (ref. 10)
(resonance tunnelling).

THICK FILMS

With increasing film thickness tunnelling through the film becomes less probable, and electron exchange with the film itself becomes the dominant process.
In a semiconducting film, the density $\rho(\varepsilon)$ of delocalized electronic states vanishes outside the allowed energy bands, and the current splits into contributions from the conduction and from the valence bands. Most metal oxides that
are formed by passivation are n-type semiconductors; the donors are provided by
oxygen vacancies. Then the Fermi-level is close to the conduction band, and the
latter will generally give the dominant contribution to the current. Here, we
shall therefore only consider the typical case of an n-type semiconductor

exchanging electrons via a conduction band mechanism.

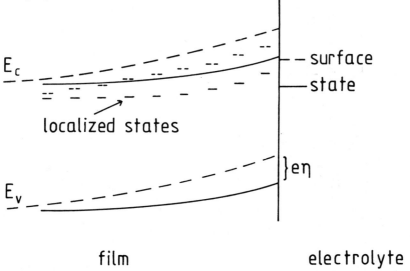

E_c

— surface

— state

localized states

$}e\eta$

E_v

film   electrolyte

Fig. 4. Space charge barrier at the interface between a thick, semiconducting
film and a redox electrolyte: solid lines: equilibrium potential; broken lines:
at an anodic overpotential

At potentials above the flat band potential, a space charge barrier exists at
the film/electrolyte interface; its height and thickness are determined by the
applied potential and by the Debye-length L_W of the film (see fig. 4). The transferring electron either tunnels through this barrier or passes over it. The most
important tunnelling mechanisms are again direct elastic tunnelling and resonance tunnelling; the corresponding tunnelling probabilities are calculated in
the same way as for the thin film mechanism.

Electron exchange between the conduction band and a redox couple can thus be
calculated from Eq. (3) by performing the integral over the energy range of the
conduction band. Typical Tafel plots for both tunnelling mechanisms are shown
in fig. 5. In both cases, the anodic transfer coefficients are small, since an
anodic overpotential makes the space charge barrier higher and thicker; conversely, the cathodic coefficients are large. The current due to resonance tunnelling has a maximum on the cathodic branch, beyond which it drops to zero:
with increasing cathodic overpotential more and more localized states are shifted to energies below the bottom of the conduction band, and can thus no longer
serve as intermediate states. In practice, this can lead to a transition from
resonance to direct tunnelling, which can show up as shoulders in the Tafel
plots (ref. 16).

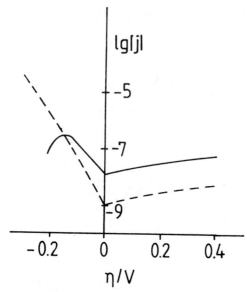

Fig. 5. Typical Tafel plots for direct tunnelling (broken lines) and resonance tunnelling (solid lines) through a space charge barrier. The energy of the conduction band was taken as E_c^b = 0.2 eV in the bulk and E_c^s = 0.6 eV at the surface

Experimental studies have been performed, for instance, on passive niobium (ref. 17), passive titanium (ref. 17, 18) and on SnO_2 (ref. 19). Both direct and resonance tunnelling seem to occur in practice (ref. 16).

ET VIA SURFACE STATES

Finally, we want to discuss ET via surface states, which have, for instance, been investigated experimentally by Peter et al. (ref. 20), and which are often discussed in photoelectrochemical reactions. This process is similar to resonance tunnelling in that it involves electronic intermediate states, but it is complicated by the fact that the transfer of an electron from the electrode to a surface state generally involves reorganization of a complex and/or the solvent. While the general framework of Eqs. (1) and (3) still applies, D_{ox} is now given by (ref. 21):

$$D_{ox} = \frac{1}{2kT} D^{-1/2} \exp \left\{ -\left[\lambda_{13}(\epsilon_s - \epsilon)^2 + \lambda_{13}(e\eta - \epsilon)^2 - 2\bar{\lambda}(\epsilon_s - e\eta)(e\eta - \epsilon) \right. \right.$$

$$\left. \left. + 2\lambda_{13}(\lambda_{12} - \bar{\lambda})(\epsilon_s - \epsilon) + 2\lambda_{12}(\lambda_{13} - \bar{\lambda})(e\eta - \epsilon) + \lambda_{12}\lambda_{13}\lambda_{23} \right] / DkT \right\} \qquad (6)$$

where $D = 2\lambda_{12}\lambda_{23} + 2\lambda_{12}\lambda_{13} + 2\lambda_{23}\lambda_{13} - \lambda_{12}^2 - \lambda_{13}^2 - \lambda_{23}^2$

and $\bar{\lambda} = (\lambda_{12} + \lambda_{13} - \lambda_{23})/2$

λ_{12}, λ_{13}, λ_{23} are the energies of reorganization for ET from the electrode to the surface state, from the latter to the redox couple, and from the electrode to the redox couple, resp.; ε_s is the energy of the surface state with respect to the Fermi level. D_{ox} is again the probability of finding the system on the reaction hypersurface, which is here determined by the common intersection of the potential energy surfaces of all three electronic states involved. In the particular case, where $\lambda_{12} = 0$, we recover the expression for resonance tunnelling (ref. 21).

As an example we consider electron exchange between the conduction band of a bulk oxide, or a thick film, and a redox couple via surface states in the presence of a space charge barrier (see fig. 4). Assuming that the coupling of the surface state with the redox couple is stronger than that with the bulk electronic states, from which it is separated by the barrier, the electronic factor $\kappa(\varepsilon)$ is proprotional to the tunnelling probability through the barrier.

A typical set of Tafel plots for different energies ε_s of the surface state is shown in Fig. 6. Here, λ_{12} was taken as 0.2 eV, since the interaction of a

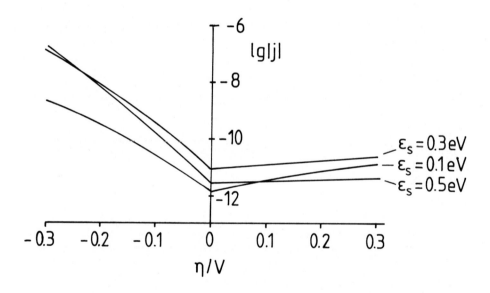

Fig. 6. Tafel plots for ET via surface states of various energies

surface state with a solvent is relatively small, and $\lambda_{23} = \lambda_{13} = 1$ eV, a value typical for moderately fast redox reactions; the bulk value E_c^b of the conduction band was taken as 0.2 eV, its value E_c^S at the surface as 0.6 eV. With such a small value of λ_{12} we should expect that the most favourable energy ε_s of the surface state is close to E_c^b. Over most of the potential region the surface state with $\varepsilon_s^0 = 0.3$ eV (at $\eta = 0$) gives the highest current, since it is situated close to E_c^b through the investigated potential range; only at high cathodic overpotentials is ε_s^o = 0.5 eV more favourable. The current for ε_s^o = 0.1 eV is smaller than that for the other two values on the cathodic branch, since there its energy is below E_c^b; but it has the highest anodic transfer coefficient, since with increasing anodic overpotential it is shifted towards E_c^b.

In a real system more than one type of surface states may be present. For instance, when small metal clusters are deposited on the surface, as was done by Peter et al., a whole energy spectrum of intermediate states is available, so that at every overpotential the transferring electrons have intermediate states of a suitable energy available to them. Also, we should expect that the energy of reorganization λ_{12} for small metal clusters is particularly small. So, our present model offers a ready explanation, why small metal clusters on the surface should be particularly good catalysts.

CONCLUSION

Electron transfer reactions on oxides and passive films can proceed via several different mechanisms. Particularly noteworth are the occurance of tunnelling processes and the participation of electronic intermediate states. The present state of the theory allows us to calculate the dependence of the reaction rate on the electronic characteristics of the film, but not its absolute value. Conversely, the theory may be used to estimate the electronic properties of a film from the experimentally observed current-potential curves of a redox reaction.

ACKNOWLEDGEMENT

Financial support by the Deutsche Forschungsgemeinschaft is gratefully acknowledged.

REFERENCES

1   L. Young, Anodic Oxide Films, Academic Press, New York, 1961
    B. E. Conway, Electrode Processes, Ronald Press, New York, 1965
2   V. G. Levich, Kinetics of reactions with charge transfer, in: Eyring, Henderson, Jost, Physical Chemistry, Vol. IXb, Academic Press, New York 1970
3   H. Gerischer, Z. phys. Chem. NF 26 (1960) 233, 325
4   R. A. Marcus, J. Chem. Phys. 43 (1965) 679
5   C. B. Duke, Tunnelling in Solids, Academic Press, New York 1969
6   T. Wolfram, Inelastic Electron Tunnelling Spectroscopy, Springer Verlag, Berlin 1978
7   W. Schmickler and J. Ulstrup, Chem. Phys. 19 (1977) 217
8   J. W. Gadzuk, J. Appl. Phys. 41 (1970) 286
    C. B. Duke and M. E. Alferieff, J. Chem. Phys. 46 (1967) 923
9   W. Schmickler, J. Electroanal. Chem. 83 (1977) 387
10 W. Schmickler and U. Stimming, Thin Solid Films, 75 (1981) 331
11 P. Kohl and J. W. Schultze, Ber. Bunsenges. Phys. Chem. 77 (1973) 953
12 J. W. Schultze and M. Haga, Z.Phys. Chem. NF 104 (1977) 73
13 W. Schmickler and J. W. Schultze, Z. Phys. Chem. NF 110 (1978) 277
14 J. W. Schultze and U. Stimming, Z. Phys. Chem. NF 98 (1978) 277
15 R. V. Moshtev, Electrochimica Acta 16 /1962) 2039
16 W. Schmickler, Ber. Bunsenges. Phys. Chem. 82 (1978) 477

17 K. E. Heusler and R. S. Yun, Electrochim. Acta 22 (1977) 977
18 J. W. Schultze, U. Stimming, and J. Weise, Ber. Bunsenges. Phys. Chem.,
    in press
19 S. Kapusta, N. Hackermann, J. Electrochem. Soc. 128 (1981) 327
20 L. M. Peter, W. Dürr, P. Bindra, and H. Gerischer, J. Electroanal. Chem.
    71 (1976) 31
21 W. Schmickler, J. Electroanal. Chem. 137 (1982) 189

Passivity of Metals and Semiconductors, edited by M. Froment
Elsevier Science Publishers B.V., Amsterdam — Printed in The Netherlands

ELECTRONIC PROPERTIES OF MODIFIED PASSIVE FILMS

C. BARTELS, B. DANZFUSS, and J. W. SCHULTZE
Institut für Physikalische Chemie, Universität Düsseldorf, Universitätsstraße 1
D-4000 Düsseldorf, FRG

ABSTRACT

Passive films of titanium (n-type semiconductor) and hafnium (insulator) were modified by ion implantation (impox) and cathodic deposition of gold (diodes). The modified electrodes were characterized by their repassivation behaviour, the electrode capacity and the rate of electron transfer reactions. Ion implantation causes an increase of electrode capacity (increase of donor concentration and dielectric constant), but diodes show an increase of capacity only if there are short-circuits within the film. The rate of electron transfer reactions is enhanced for impox as well as diodes. Especially the anodic current density, which is negligible on all valve metals, increases by 3-4 orders of magnitude. This catalysis is due to the increase of electron states at the oxide surface in case of diodes and due to a decrease of the thickness of the space charge layer and an increased tunnel probability in the case of impox. Diffusion limited currents are observed in the case of short circuits of the passive film.

INTRODUCTION

Electronic properties of passive films determine the corrosion behaviour of metals as well as their applicability as electrocatalysts (ref. 1, 2). Therefore, the modification of these properties by protective films of metals, oxides or organic compounds is of great interest. A change of electronic properties is important for the passivity of alloys, too. But with alloys, it is often not clear, which changes can be caused by an enrichment within or at the surface of the film. Hence, a defined film modification is of interest. To distinguish the influence of metallic states at the surface, and that of donors, metall clusters, or even shortcircuits within the film, we measured the electrode capacity and the rate of electron transfer reactions of passive films after cathodic deposition of metals (diodes (ref. 3) and after ion implantation into the oxide (ref. 4). As model system, we chose the semiconducting film of TiO_2 and the insulating film of HfO_2.

PREPARATION

Oxide films were prepared by anodic polarization of titanium or hafnium in 1 N H_2SO_4 or $HClO_4$ up to the formation potential ε_f. For ion implantation, the electrodes (ε_f = 20 V, d = 30-50 nm) were bombarded with Pd^+ ions of 20-220 keV to get an almost constant concentration profile throughout the passive film (ref. 4). Thickness and concentration profiles were checked by Rutherford backscattering. The concentration of implanted ions was varied from 0.1-10 %. Diodes were prepared by cathodic galvanostatic deposition of Au on passive films (ε_f = 1 to 7V, d = 3 to 10 nm) from $HAuCl_4$ solution. The coverage Θ and size of nuclei (10^5-10^7 cm, $\emptyset \approx$ 1-10 μm) was varied by the polarization time. Θ was determined by XPS measurements, SEM pictures and by calculation from the cathodic deposition charge (ref. 3).

REPASSIVATION

Both types of modified passive electrodes are more active, i. e. the capacity increases and the electron transfer reactions are enhanced. This influence decreases slightly with increasing time due to slow repassivation of the film. Quantitative repassivation (reformation) was achieved with Ti-diodes by anodic polarization up to ε_f. It takes place by short breakdown pulses and causes a decrease of capacity and an increase of the overvoltage of electron transfer reactions. The repassivation of Hf-diodes was more difficult and took place during longtime-polarization only. In general, repassivation seems to repair local defects and yields a more homogeneous film.

Repassivation of impox is possible for small concentrations of palladium and was carried out by potentiodynamic anodic polarization which caused a small, but continuous anodic current. The repassivation charge increased with implantation concentration and was less than that of the oxide formation. At higher concentrations of Pd, the oxygen evolution is enhanced so much, that the electrode cannot be polarized beyond 2 V. Hence, a complete repassivation of these electrodes was impossible.

CAPACITY

The electrode capacity C was measured under potentiostatic or potentiodynamic conditions at 1 kHz (ref. 3). Results for Hf-electrodes are shown in Fig. 1 in dependence on the potential. The dotted lines show the results for the pure passive film, for which the capacity is independent of ε and proportional to the reciprocal thickness:

$$C = D \cdot D_0 / d \tag{1}$$

For HfO_2, D = 14, ion implantation causes an increase of C by a factor of 3 to 5, but C remains independent of ε. Donor terms of Pd cannot be detected by C-measurements. Hence, the increase of C can be explained by an increase of the dielectric constant only, D = 40 to 80, which may be correlated to an increase of crystallinity (ref. 5). The Hf-diodes show a different behaviour. The capacity is slightly increased and shows a small hysteresis due to gold oxide formation and

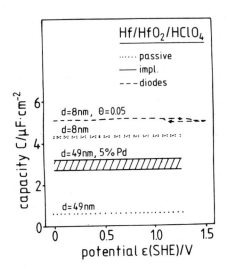

Fig. 1: Electrode capacity of
pure and modified passive films
on Hf-electrodes in dependence
on the electrode potential ε
at 1 kHz in 1 N HClO₄.

reduction. This can be explained by a shortcircuit of the passive film.

The capacity of TiO_2-films shows the typical Schottky-Mott behaviour of an
n-type semiconductor film (ref. 6). Freshly prepared diodes have a larger capacity due to shortcircuits, but after repassivation the curve agrees with the
original one inspite of the presence of Au-layer. Impox have a much larger capacity, too, but repassivation causes a smaller decrease (fig. 2). Schottky-Mott-

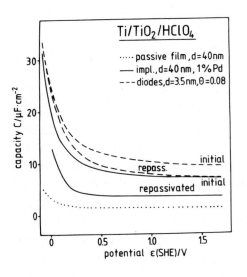

Fig. 2: Electrode capacity of
pure and modified passive films
of TiO_2 in dependence on the
electrode potential at 1 kHz in
1 N HClO₄.

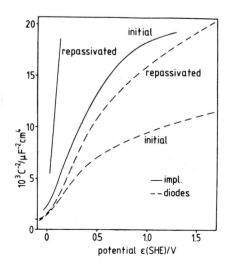

Fig. 3: Schottky-Mott diagram
of the data given in Fig. 2

plots of Fig. 3 show, that the flat band potential is very similar for all TiO_2
films. Repassivated diodes have the same slope as the pure films, but impox have
a smaller slope. Since the slope yields the product $D \cdot N$, the increase of D may
contribute as well as that of N. The almost constant capacity of impox at potentials exceeding 1 V indicates an exhaustion of the total film and consequently
an increase from $D \approx 20$ up to $D \approx 100$. The increase of donor concentrations from
10^{19} to 10^{20} per cm^3, which are presumably TiO_2 dislocations and Pd atoms
(ref. 4), contributes, too.

ELECTRON TRANSFER REACTIONS (ETR)

The rate of cathodic and anodic ETR was measured at pure and modified films
by slow potentiodynamic cycles in Fe^{2+}/Fe^{3+} solutions. Results are shown in
Fig. 4 for passive Ti and Hf, resp. For the pure passive film of Ti (fig. 4a),
only cathodic currents could be measured (dotted lines). The anodic current
was negligible ($i < 1$ $\mu A/cm^2$). Impox (solid lines) as well as diodes (broken
line) show a strong acceleration of ETR by 1 to 2 orders of magnitude on the
anodic branch, but the properties of a gold electrode are not reached. The Tafel
lines of impox are steeper than those of diodes, but the rate of ETR is similar
for both modifications, in spite of the much larger thickness of oxide films of
impox. Fig. 4b shows the Tafel diagram for Hf-electrodes. Without modification,
ETR cannot be detected, but the exchange current densities exceed $\mu A/cm^2$ for
diodes and impox. There is a strong difference between both modifications: the
current potential curves of diodes resemble those of small metallic electrodes
with diffusion overvoltage, which means that the electron transfer at defects in
the diode is fast and not rate determining. The current potential curves of

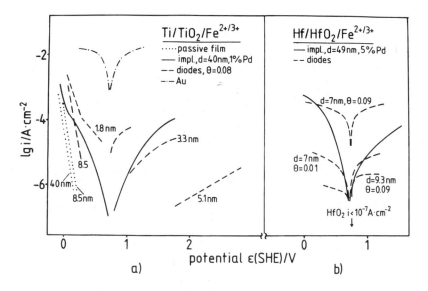

Fig. 4: Tafel diagram of the redox system Fe^{2+}/Fe^{3+} in 1 N $HClO_4$ at pure and modified passive films on a) titanium and b) hafnium.

implanted Hf-films resemble those of an n-type passive film: the exchange current density is smaller than on metal electrodes, and the cathodic lines are much steeper than the anodic ones, that means that the cathodic transfer coefficient exceeds the anodic one.

DISCUSSION

Ion implantation and metal deposition cause similar modifications of the electrode behaviour, but the physical reason differs completely. Ion implantation causes a damage of the film (probably with some changes in stoichiometry and crystallinity), and incorporation of foreign ions. The non-stoichiometry of the oxide can cause a drastic increase of D as well as introduction of donor levels, which are important for the increase of ETR. The oxide-electrolyte interface and the potential drop in the Helmholtz-layer are not changed.

Metal deposits, on the other hand, cause an increase of electron states at the oxide surface, enhancing the probability of ETR, but the passive film itself has constant electronic properties. Besides this expected change, there is another influence of defects with a local increase of conductivity. The important difference between impox and diodes for practical applications is the influence of thickness. Since tunnel distances must be less than 2 nm, metal depositions are important only for thickness values less than 6 nm. Thick insulating films cannot be made conductive by metal deposition. Ion implantation, on the other

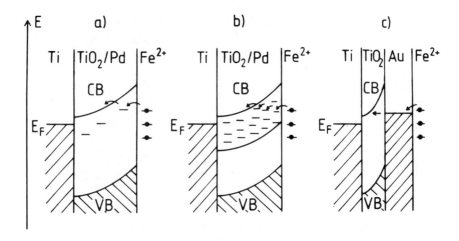

Fig. 5: Schematic diagram of the electron terms at the metal/oxide/electrolyte interface for a) and b) impox with small and high concentration of donors, c) diodes during anodic electron transfer.

hand, is possible up to d = 100 nm and may be useful to increase the electronic conductivity. The catalysis by modification can be explained by the schematic energy diagram of Fig. 5, using the Gurney-Gerischer theory of ETR (ref. 7). For the anodic ETR from the electrolyte to the solution, the current density $i_+(E)$ at a special electron energy E is given by the distribution function of occupied electron terms at the oxide surface $D_{occ}(E)$, the density of vacant terms in the conduction band of the oxide $D_{vac}^{ox}(E)$ and the tunnel probability

$$W = \exp(-2d_t \cdot \sqrt{2m\ E}/\hbar) \qquad (2)$$

and an electron frequency ν

$$i_+(E) = e_0 \cdot D_{occ}(E) \cdot D_{vac}^{ox}(E) \cdot W \cdot \nu \qquad (3)$$

Here we will confine the discussion to an arbitrarily chosen energy E in the height of the conduction band of the oxide. An extended discussion will be given later (ref. 8). For this discussion, we will use the current i_p of the reaction $Fe^{2+} \rightarrow Fe^{3+}$ at the pure passive film as reference. Then, $D_{occ} = D_{Fe^{2+}}(E)$, and the tunnel distance $d_t < d$. Compared with a i_m at a metal, we obtain a small ratio from eq. (3)

$$\frac{i_p}{i_m}(E) = \frac{D_{vac}^{ox}}{D_{vac}^m}(E) \cdot \frac{W_p}{W_m} \approx \frac{D_{vac}^{ox}}{D_{vac}^m}(E) \cdot \exp(-\frac{d_t}{d_0}) \ll 1 \qquad (4)$$

with $W_m = 1$ and $d_0 \approx$ const. In the case of impox, W will increase exponentially, since resonance tunnel processes can proceed via smaller distances W_r (ref. 9).

$$W_r = W_1 \cdot W_2/(W_1 + W_2) \approx \exp(-\frac{d_i}{d_o}) \qquad (5)$$

with the rate determining largest tunnel distance $d_i < d_t$. For impox, $d_i \approx d/n$ with n = number of implanted atomic layers of foreign atoms, and the current ratio will be increased by various orders of magnitude:

$$\frac{i_{impox}}{i_p}(E) = \frac{W_{impox}}{W_p} \approx \exp(-\frac{d_i}{d_{o,imp}} + \frac{d_t}{d_{o,p}}) \gg 1 \qquad or \qquad (6)$$

$$lg(i_{impox}/i_p) \approx -\frac{d/n - d_t}{d_o} \leq \frac{d_t}{d_o} > 0 \qquad (7)$$

Even in the case of large implantation, the rate i_{impox} will be less than i_m, since the term D_{vac}^{ox}/D_{vac}^{m} of eq. (4) cannot be compensated.

For diodes, the current i_{diode} will be larger than on pure passive films

$$\frac{i_{diode}}{i_p}(E) = \frac{D_{Au}}{D_{Fe^{2+}}} \gg 1 \qquad (8)$$

since $D_{occ}^{diode}(E)=D_{Au}=D_{Au}^{1} \cdot \Theta_{Au} \gg D_{Fe^{2+}}$ (ref. 3), but it will be smaller than on metals due to $W_{diode} \ll W_m \approx 1$.

The role of short circuits, which are most important for the passive films of Hf is trivial. In this case, the electron transfer through the diode is very fast, and the reaction becomes diffusion limited around such short circuited nuclei. If we compare the observed diffusion limited current densities with that of a metal electrode, we can estimate the coverage of these nuclei. A schematic explanation of the changes of electronic properties is given in Fig. 5.

ACKNOWLEDGEMENT

The support of this work by the Bundesministerium for Forschung und Technologie is gratefully acknowledged.

REFERENCES

1 J. W. Schultze, Proceedings of the 8th intern. Congr. on Metallic Corr., Dechema, Frankfurt 1981, 1
2 J. W. Schultze, M. A. Habib, J. Appl. Electrochem. 9 (1979) 255
3 J. W. Schultze, C. Bartels, J. Electroanal. Chem., in press
4 B. Danzfuß, O. Meyer, J. W. Schultze, U. Stimming, Thin solid films, in press
5 H. M. Naguib, R. Kelly, Radiation Effects 25 (1975) 1
6 W. Schottky, Z. Phys. 113 (1939) 367; 118 (1942) 539
7 H. Gerischer, Z. phys. Chem. NF 26 (1960) 223, 26 (1960) 325, 27 (1961) 48
8 J. W. Schultze et al. ECS Symposium on Modified Films, Washington 1983
9 W. Schmickler in Frankenthal, Kruger: "Passivity of Metals", The Electrochem. Soc., Princeton

Passivity of Metals and Semiconductors, edited by M. Froment
Elsevier Science Publishers B.V., Amsterdam — Printed in The Netherlands

ON THE OXIDATION RATE LAWS OF METALS THAT FORM NONSTOICHIOMETRIC OXIDES

Robert J. Good[1]

[1]Department of Chemical Engineering, State University of New York at Buffalo

ABSTRACT

It is shown that the electronic conductivity of a nonstoichiometric oxide that is an n-type semiconductor (i.e., a reduction semiconductor) can control the cation transport through the oxide, and hence, the rate of oxidation. This argument leads to a mathematical derivation of the logarithmic oxidation rate law.

If the oxide is an oxidation semiconductor, it is p-type, and the hole conductivity will be sufficiently great that the rate-limiting property will be ion transport, and not electronic conduction. In this case, the Wagner ion-electron theory of the parabolic oxidation rate law will apply.

The microscopic nature of the crystalline solids is discussed, including the gradients of composition and of conductivity across the oxide layer, and how these effects influence the oxidation rate.

I. INTRODUCTION

The majority of studies of the mechanism of oxidation of metals have followed Wagner [1-3] in concentrating directly on ion transport mechanisms in the oxide layer. We will, here, investigate the fact that the electronic transport in a nonstoichiometric oxide semiconductor can exercise a direct limitation on the mass transport, and can thereby over-ride the transport mechanisms of ion diffusion in an electric field.

It has been generally assumed—see ref.[2], p. 14 and ref. [4], p. 80-- that the electronic conductivity of the oxide is orders of magnitude greater than the ionic conductivity, and hence, that variations in electronic conductivity need not be examined explicitly, for a particular oxide, in regard to the oxidation rate laws. Tomashov, however has remarked, (ref. [4], pp. 78-80), that "not only a parabolic relationship but also linear and logarithmic laws of metal oxidation kinetics can be obtained from the ion-electron theory." To obtain a logarithmic law, "some supplementary retardation" must be assumed.

We will show that, while direct ionic-transport control of oxidation processes is important when the oxide is a p-type semiconductor, this mechanism is of limited importance in the case of n-type nonstoichiometric oxides.

Cabrera and Mott [5] have developed a theory of the logarithmic oxidation rate law, that explains (as an inverse logarithmic law) many cases, but does not adequately predict the cases in which a logarithmic time dependence will be absent. The Cabrera-Mott theory assumes the rate-limiting step to be charge transfer at one of the phase boundaries of the oxide; this is postulated to be very strongly influenced by the space charge just beneath the oxide surface. There is not much doubt that this mechanism is important in the "rectifying" action of certain metal/oxide interfaces. However, it is not so clear that it is the complete explanation for the logarithmic oxidation law. There is a correlation between the two phenomena, rectification and logarithmic oxidation rate; but there are important exceptions to the correlation. The Cabrera-Mott theory has been discussed extensively by Hauffe [2], and reviewed by Fehlner and Mott [6].

Over four decades ago, Mott and Gurney [7] described nonstoichiometric oxides on metals as being either oxidation semiconductors or reduction semiconductors. This classification has been recognized by Tomashov [4], Hauffe [2], and others. Yet, as a direct causal factor in passivity and in the logarithmic oxidation rate law, this characterization of metal oxides has, in effect, lain dormant. The facts that metal oxides may be n- or p-type semiconductors on account of impurity ions (or atoms, as in the case of ZnO) and that this may influence oxidation behavior, have been recognized for a number of years. Effects due to exogenous impurities, i.e. metal ions derived from elements other than those of the host lattice, such as Li^+ or Al^{+3} in ZnO, have been studied extensively; see ref. 2, pp. 14-25. The role of anion impurities has also been investigated [8]. Endogenous impurities, i.e. impurities whose atomic composition is the same as the ions of the host lattice, differ from exogenous impurities in that their supply is, effectively, unlimited. For example, endogenous impurity cations can be formed by oxidation or reduction of cations of the host lattice, and also by the introduction of cations transferred from the underlying metal. It is even more important, that endogenous impurity cations may cease to exist as such, on account of oxidation or reduction converting them to ions identical with those of the host lattice. These facts lead to a different dependence of the local conductivity due to endogenous impurity ions (or atoms) upon location and on time, from the conductivity due to exogenous ions or atoms.

Cahan and Chen [9] have proposed a model for the oxide film on iron in the passive condition, that is related to the model described in the previous paragraph. They proposed the term, "chemi-conductor." They did not apply their model to dry oxidation and the logarithmic rate law. They did, however, note that the characterization of an extrinsic semiconductor, with fixed dopant concentration, does not lead to an adequate electronic description of the conduction behavior of the oxide film on iron.

II. THEORY

        The operational characteristic of an oxidation semiconductor is, that
as the degree of oxidizing character of the environment increases (e.g., by an
increase in oxygen partial pressure, or by the electrode potential becoming
more positive) the conductivity increases. The opposite is true if the reducing character of the environment increases. The operational behavior of a
reduction semiconductor is that its conductivity increases as the degree of
reducing character of the environment increases, but is suppressed if the
environment becomes more strongly oxidizing.

        The changes in conductivity of a crystal, with alteration of the oxidizing/
reducing character of the environment, cannot occur instantaneously. If the
oxide is in contact with the metal from which it derives its cations, then
there is, in effect, a steady reducing environment at the metal side of the
oxide. As oxidation proceeds, the oxide layer (which we will assume to be continuous and adherent) thickens. Hence, for an element of volume at a specified
distance from the oxide/gas interface, the degree of access (by electronic and
ionic transport) to the reducing environment decreases. It is well established
that, for an oxide such as FeO on iron, the approach to stoichiometry is
closest in the layer nearest to the metal [4].

        So, for an element of volume in the oxide that is stationary with respect
to the external oxide surface, the local net oxidizing character at that place
in space will increase, as the oxide thickens. The local conductivity there
will increase with time, if the oxide is an oxidation semiconductor. If it is
a reduction semiconductor, the conductivity will decrease.

        We need to enlarge upon these concepts by examining the microscopic nature
of extrinsic semiconductors with endogenous impurities. Consider a metal with
two ionic states, M^{+2} and M^{+3}, whose oxides have large band gaps. The lower
oxide, assumed to have the empirical formula $M_{1-\alpha}O$, may have a deficit of metal
ions (i.e., cation lattice vacancies) so that it may be written, $(M^{+2}_{1-3\alpha}, M^{+3}_{2\alpha})O$.

        Alternatively, the oxide might have a metal ion at essentially every lat-
tice cation side, and a number of interstitial, excess oxide ions. A corresponding number of metal ions would be M^{+3} ions, at lattice sites. Because the
O^{-2} ion is considerably larger than cations such as Fe^{++}, Al^{+3}, etc., this condition is probably not important for metals in the first two or three rows of
the periodic table. But for oxides of heavy metals, it is a very real possibility, which would have to be considered explicitly in a detailed analysis. For
the purpose of developing the typology of the semiconductors, it is not
necessary to consider, separately, oxides of the type, $MO_{1+\beta} = (M^{+2}_{1-\beta} M^{+3}_{\beta})O_{1+\beta}$,
since the presence of an additional kind of defect (interstitial oxide ions)
will not change the argument as to whether the semiconductor is n- or p-type.

        For the assumed lower oxide, the M^{+3} ions can act as acceptor sites.

Valence band electrons can be promoted into acceptor levels; and the oxide is a p-type semiconductor. The concentration of M^{+3} sites will increase with increasingly oxidizing character of the external environment; see above. The electronic (in this case, hole) conductivity will increase. Thus, we have given an atomic-level description of an oxidation semiconductor.

Next, we consider the higher oxide, $M_{2+\alpha}O_3$ or $M_2O_{3-\beta}$. The first empirical formula corresponds to the presence of an essentially perfect oxide lattice with interstitial excess cations: $(M^{+3}_{2-2\alpha},M^{+2}_{3\alpha})O_3$. It is immaterial, for present purposes, whether the M^{+3} ions are all at lattice sites and the M^{+2} ions are interstitials, or if the lattice sites are occupied by either M^{+3} or M^{+2} ions and some or all of the occupied interstitial sites contain M^{+3} ions. The second empirical formula corresponds to the existence of anion lattice vacancies, with electroneutrality being maintained by reduction of the corresponding number of M^{+3} ions to M^{+2}: $M^{+3}_{2-2\beta}M^{+2}_{2\beta}O_{3-\beta}$. In either case (i.e., whether the solid has an excess of cations or a deficit of anions, and also in cases where the oxide may have an even more complex structure) the M^{+2} ions can function as donor sites. Electrons in donor levels can, of course, be promoted into the conduction band; and the oxide will be an n-type semiconductor. The electronic conductivity, σ_e, will decrease with decreasing concentration of defect sites. Hence, an increase in the oxidizing character of the environment will lead to a decrease in σ_e. And so, we have given an atomic-level description of a reduction semiconductor.

We can now show, qualitatively, under what conditions the magnitude of the electronic conductivity, relative to the cation and anion transport properties, limits the oxidation rate. First, in the limit of vanishingly small electronic conductivity, the rate of ionic mass transport across the oxide layer is controlled by the mobility of the less-mobile species. If transport of the more mobile species gets ahead of that of the less mobile species, a space charge is set up that retards the transport of the more mobile species and accelerates the less mobile ions. Usually, the cations are considerably more mobile than the oxide ions. Second, if the electronic conductivity is large enough, then the transport of the more mobile ion species (e.g. cations) is not retarded by the negative charge transport. The mathematics of the ion-electron theory, then, leads directly to the parabolic rate law [1,2,4]. And third, if the electronic conductivity is appreciably larger than the anion conductivity, but smaller than the cation conductivity, then the electronic conduction (rather than the anion transport) will be the rate-limiting property of the solid. It is this third condition, in regard to the magnitudes of the three transport properties, that we will show leads to a logarithmic rate law.

It should be noted that, if a duplex oxide film exists, e.g., MO next to the metal, M_2O_3 next to the gas [10,11], the region of lowest conductivity will lie in the M_2O_3 layer. If the charge transport mechanisms meet the third condition, in a lamina of that layer, then the oxidation rate will follow the predictions derived below, regardless of the conduction behavior of the MO sublayer.

In order to derive the logarithmic law, we need to establish the variation, within the oxide film, of the local concentration of reduction defects in a higher oxide on a metal. For simplicity, we may consider the reduction defects, in an M_2O_3 crystal, to be M^{+2} ions in interstitial sites of a cation-excess solid. These defect ions may be formed by ionizing metal atoms and moving them to interstitial sites at various distances below the oxide/gas interface. Compare ref. 7, pp. 259-61. We let $\Delta\phi$ be the potential drop between the metal and the oxide/gas interface. If the oxide thickness is ℓ, then the potential within the oxide, at a small distance $\delta\ell$ from the surface, will be approximately $(\delta\ell/\ell)\Delta\phi$, or $\lambda\Delta\phi$. (This will be true if there is not an appreciable space charge.) The free energy required to form an interstitial M^{+2} ion close to the oxide/gas interface will be, approximately,

$$nF\lambda\Delta\phi = -RT\ln(N'/N_{is}) \tag{1}$$

where N' is the number of reduction defects, N_{is} is the number of interstitial sites where M^{+2} atoms can be placed, F is the Faraday constant, and n is the ion charge—in this case, 2. (An expression of this sort should be applicable whatever the nature of the reduction defect.) N_{ce}, the number of conduction electrons, will be proportional to N'. The electronic conductivity, σ, at distance $\delta\ell$ from the surface, will be related to N_{ce} at that location, and to the electron mobility, n_{ce}, by

$$\sigma = \sigma_{ce} = N_{ce}n_{ce}q_e \tag{2}$$

where q_e is the electron charge. σ will be related to the potential drop by

$$\sigma \propto \exp(-F\lambda\Delta\phi/RT) \tag{3}$$

The electronic current i, through the oxide, which is equal to the cation current in the same direction, will be

$$i \propto \Delta\phi\exp(-nF\Delta\phi/RT). \tag{4}$$

$$\propto \Delta\phi N'/N_{is} \tag{5}$$

When an electron dissociates from an interstitial M^{+2} ion, to become a conduction electron, the M^{+2} ion becomes an interstitial M^{+3} ion. The local chemical potential of M^{+3} ions will be raised, and these ions will tend to diffuse to the external surface, leaving the interior of the oxide depleted of sites that have the potentiality of providing donor levels. Hence N' will decrease, as will N_{ce}, and i. Thus, the mechanism which produces electronic conductivity also functions to destroy the current-producing sites. This decrease in conductivity will lead to a decrease in the rate of ionic mass transport. Hence, there will be a decrease in the rate of oxidation that is more rapid than the decrease that would come about by simple diffusion control, with increasing thickness.

Mathematically, we may assume a reasonable form of time dependence of electronic conductivity:

$$\sigma_e = \sigma_{eo} - Cit \tag{6}$$

where C is a constant and σ_e is the conductivity in the region where conductivity is lowest. The rate of metal oxidation, dm/dt, is proportional to the current:

$$\frac{dm}{dt} \propto i = \sigma_e \Delta\phi \tag{7}$$

So we obtain, for the time dependence of σ_e,

$$\sigma_e = \sigma_{eo} - C't(dm/dt) \tag{8}$$

where C' is another constant. Combining Eqs. (7) and (8),

$$\frac{dm}{dt} = C''\Delta\phi(\sigma_{eo} - C't\frac{dm}{dt}) \tag{9}$$

Rearranging (lumping the constants) and integrating,

$$dm = \frac{Bdt}{1 + At} \tag{10}$$

$$m(t) - m_o = (1/B)\ln(1 + At). \tag{11}$$

Thus, we have obtained a logarithmic oxidation rate law.

III. DISCUSSION

The rate-limiting property of the reduction semiconductor is the low electronic conductivity of the solid in the interior region just below the oxide/oxidizing gas interface. It is there that σ_e will be the lowest; and, provided that region is not so thin that quantum mechanical tunnelling through it could be important, it will provide a "barrier." We have neglected space charge, up to this point; but the effect of space charge should be (logarithmically) additive with that due to the semiconducting behavior described above. The space charge will not develop without a low electronic conductivity; and so we can consider the reduction semiconductor pattern of behavior, and the presence of an oxidizing environment, to be jointly causal in bringing about the space charge.

The theory developed, above, of the role of nonstoichiometry of the oxide, M_2O_3, in the logarithmic rate law, appears to depend upon the existence of a phase with nominal composition MO, so that M^{+2} defect sites can be formed in the M_2O_3. But it is not necessary to postulate that an MO phase is thermo-dynamically stable. All that is needed is that M^{+2} ions can be formed, in the M_2O_3 lattice; and it is quite possible for this to occur (at least, by the localization of an electron in the lattice) in spite of the absence of an MO phase from the phase diagram at the temperature under consideration. The Born-Madelung lattice energy of an MO crystal is considerably smaller than that of an M_2O_3 crystal. So bulk MO could disproportionate spontaneously to $M + M_2O_3$, in spite of the M^{+2} ions being stable enough to exist as defect sites in an M_2O_3 lattice. Certainly, the entropy of mixing of M^{+2} ions with lattice M^{+3} ions would favor the formation of M^{+2} sites, particularly in small concentration. And a small concentration is all that is needed, for reduction semi-conductor behavior in which the conductivity would be suppressed by oxidizing conditions.

A further complication, in regard to microscopic structure, is the exist-ence of surface structures that differ appreciably from the bulk oxide struc-ture, and the presence of occupied electronic "surface levels" in the band gap. Since we have indicated, above, that the region just below the external surface is where the electronic conductivity will be lowest, it seems likely that there will be a considerable interaction of the surface electronic levels with the subsurface region of the oxide.

We may now suggest a reason (in addition to causes that involve oxide phase stability) for the switch-over from a logarithmic to a parabolic oxida-tion law with increasing temperature. It is that, as temperature rises, the electronic conductivity of the semiconductor increases until it surpasses the cation conductivity, even in the part of the oxide layer that is most accessi-ble to the oxidizing environment. Then the limited electronic conductivity will cease to over-ride the direct rate control by ionic transport.

It is clear that our theory is relevant to passivity. See Ref. [9], in which Cahan and Chen have reported upon investigations of the passive film on iron.

The aqueous medium that is necessarily present in systems where passivity is exhibited, makes the oxide structure far more complex than is the case with an oxide in the absence of water. The presence of protons, or water of hy-dration, in γ-Fe_2O_3 in a passive film [12] leads to a distinct structural difference from the film that is present in dry conditions. Cahan and Chen [9] have found evidence for Fe^{+4}, or an equivalent depletion of protons, in the outermost region of the oxide. They state that this leads to a considerable decrease in resistivity of the outermost portion of the passive film, which is conducive to oxygen evolution at high potentials. A further complication,

50

with passive metals, is the potential jump between the oxide and the aqueous phase. These structural, compositional and electrical considerations, however, do not vitiate our model, in regard to the qualitative explanation of passivity as being due to the reduction-semiconductor behavior of the metal oxide.

A difference that is, perhaps, more important, as between logarithmic dry oxidation and passivity, is that in the latter case, the electrolyte generally dissolves the oxide at a rate such that a steady-state thickness of oxide is quickly achieved. However, Kruger and Calvert [13] have examined the film thickness as a function of time, for iron in near-neutral solution with the electrode potential held in the passive region. Under these conditions, the oxide did not dissolve; and a logarithmic growth of film thickness was found.

Acknowledgement

The author thanks Dr. Jerome Kruger for advice and criticism of the manuscript.

REFERENCES

1   C. Wagner, Z. Phys. Chem., 213, 25 (1933).
2   K. Hauffe, "Oxidation of Metals," Plenum Press, New York 1965.
3   O. Kubachewski and B. E. Hopkins, "Oxidation of Metals and Alloys" Butterworth & Co., London, 2nd ed., 1962.
4   N. D. Tomashov, "Theory of Corrosion and Protection of Metals," The Macmillan Company, N.Y. 1966.
5   N. Cabrera and N. F. Mott, Rep. Prog. Phys., 12, 163 (1948-9).
6   P. F. Fehlner and N. F. Mott, Oxidation of Metals, 2, 59 (1970).
7   N. F. Mott and R. W. Gurney, "Electronic Processes in Ionic Crystals," 1st ed., Oxford, 1940.
8   A. K. Vijh, Corrosion Science, 12, 105 (1972).
9   B. D. Cahan and C.-T. Chen, in "Corrosion and Corrosion Protection," R. P. Frankenthal and F. Mansfield, eds., Proc. Electrochem. Soc., 81-8, p. 37.
10 R. E. Pawel, J. Electrochem. Soc., 126, 1111 (1979).
11 A. T. Fromhold, Jr., Physics Letters, 84A, 219 (1981).
12 H. T. Yolken, J. Kruger and J. P. Calvert, Corrosion Science, 8, 103 (1968).
13 J. Kruger and J. P. Calvert, J. Electrochem. Soc., 114, 43, 1266 (1967).

Passivity of Metals and Semiconductors, edited by M. Froment
Elsevier Science Publishers B.V., Amsterdam — Printed in The Netherlands

AN IMPEDANCE APPROACH OF THE PASSIVE FILM ON IRON IN ACIDIC MEDIA

M. KEDDAM, J.F. LIZEE, C. PALLOTTA[*], H. TAKENOUTI
Groupe de Recherche n° 4 du CNRS "Physique des Liquides et Electrochimie",
associé à l'Université Pierre et Marie Curie, 4 place Jussieu,
75230 Paris Cedex 05 (France)

INTRODUCTION

Several models tried to account for the steady state passive current and the kinetics of film growth an iron (ref.1,2,3). An assessment of these models requires a knowledge on the film properties as a function of potential, solution composition, temperature, etc... It is possible to separate the steady state properties of the film from the growth process and to obtain relevant data on the potential distribution across the interface by impedance methods performed over a wide frequency range.

In the case of iron electrodes in strongly acid media this kind of experiments is extremely difficult since a undermining attack occurs rapidly between metal and embedding material (ref.4). Available data are restricted to the acoustic frequency range. In this work the electrode was very carefully prepared with a cured epoxy-phenolic resin avoiding any undermining dissolution. This allowed us to define a constant steady state passivation current density i_p (c.a. $7\mu A.cm^{-2}$ on Fe/1M H_2SO_4) during several days. In these conditions it is possible to access to very low frequencies (V.L.F.).

EXPERIMENTAL

Iron was used as provided by Johnson-Matthey. Solutions were made with ion exchanged water and "p.a." reagents. Measurements were performed after 24 hours at a given passivation potential with a Solartron Schlumberger equipment (Interface 1186, TFA 1250). It is verified that 5 mV r.m.s. fullfilled linearity conditions. VLF data (<0.1Hz) were improved by F.F.T. analysis of a step response (ref.5).

RESULTS AND DISCUSSION

The passive current depends on the electrolyte composition and increases as the charge to size ratio of anions increases ($PO_4^{3-} > SO_4^{2-} > NO_3^- > ClO_4^-$) as

[*] Fellowship of the Consejo Nacional de Investigaciones Cientificas de la Republica Argentina

shown in table 1. However, the same shape of complex impedance diagram (Fig. 1) is found in the fully passive potentials region between 0.2 and 1.0 V (SSE). It consists of 3 frequency domains clearly separated. By decreasing frequency : capacitive (HF), inductive (MF) and capacitive (VLF).

The HF domain looks like a parallel R-C circuit but from the data analysis one can see that there is a clear distribution of time constants. The capacitive value is close to the electrochemical double layer capacitance for noble metals (50 μF.cm^{-2}). The R_{HF} resistance increases linearly with the potential as can be seen in Fig. 2 according to :

$$R_{HF} = R_{HF}^{FLADE} + 7.0 \text{ K}\Omega \text{ cm}^2 \text{ V}^{-1} (E - E^{FLADE}) \text{ for } H_2SO_4 \text{ 1M at } 25°C$$

The product ip. R_{HF} (c.a. 100 mv) remains approximately independent of the experimental conditions as shown in table 1. In the Flade potential region the product ip. R_{HF} attains only 70 mV and the low frequency domain of the complex impedance diagram is modified (ref.8).

The HF domain is related to the ferrous ion injection into the film and to its oxidation to ferric ions during their transport through the bulk at constant film thickness. Simultaneously, protons adsorbed to the film/solution interface are desorbed in an equivalent amount.

The presence of an inductive domain is a characteristic feature, not previously reported, of the iron/acid media system in the fully passive state. For potentials lower than 0.2 V the passive current is potential dependent and the inductive loop no longer exists. The relaxation time constant τ relative to the inductive loop is inversely proportional to ip determined for different acids : τ.ip = 72 μC.cm^{-2}. This value explainswhy the inductive behaviour is no longer observed when ip < 1 μA.cm^{-2} (Stain less steel, Fe/HClO$_4$, Fe/Borate Buffer pH : 8.3).

TABLE 1

Solution	T(°C)	$i_p(\mu A/cm^2)$	$R_{HF}(K\Omega)$	$R_{HF} \cdot i_p(mV)$
H$_3$PO$_4$ 1M	25	58.0	10.7	125
	37	130.0	4.9	127
H$_2$SO$_4$ 1M	25	8.2	73.5	121
	50	55.0	10.5	116
HNO$_3$ 1M	25	3.9	117.0	91.3
	37	8.65	55.0	95.2
HClO$_4$ 1M	25	0.22	2.257.0	.97
	45	2.0	251.4	100

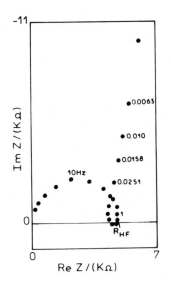

Fig. 1. Complex impedance diagram
Fe/1M H_3PO_4, 37°C.
Electrode surface 0.2 cm^2

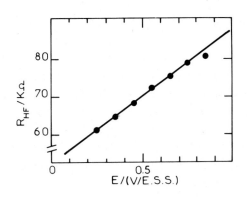

Fig. 2. Potential dependence of R_{HF},
Fe/1M H_2SO_4, 25°C.

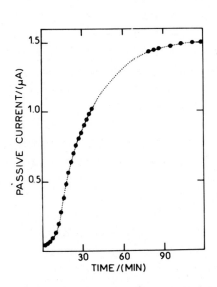

Fig. 3. Current response to a step-
wise change of electrolyte composition
from 1M $HClO_4$ to 0.5M $HClO_4$, 0.5M H_2SO_4.
E = 0.35 V(SSE)

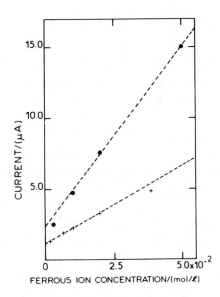

Fig. 4. Relation between ferrous ion
concentration and ip, Fe/0.5M H_2SO_4.
E = 0.5 V..., E = 0.25 V +++

EFFECTS OF IONS ON THE PASSIVE STATE

Once the electrode is passivated in a given electrolyte it is possible to change ip by adding increasing amounts of differents acids to the solution. Care was taken in order to avoid pH changes. For instance if H_2SO_4 acid is added to $Fe/HClO_4$ one molar acid system (ip = 0.1 µA), ip increases and finally reaches that of the Fe/H_2SO_4 one molar acid system (ip = 1.5 µA) at a sulfate concentration as low as 0.3 M. The surface adsorption equilibrium (ref.9) is achieved while the film thickness remains constant and consequently the film dissolution current increases. Then the film thickness decreases in order to equilibrate the new dissolution current. As can be seen in fig. 3 the time constant of the current change is related to a frequency of c.a. 1 mHz. It is not possible to accomplish the opposite change if $HClO_4$ is added to the Fe/H_2SO_4 one molar system.

The passive current increases almost instantaneously if ferrous ions are added to the electrolyte in the potential region above 0.15 V (ref.10). Fig. 4 shows that ip increases with ferrous ion concentration with a potential dependent slope. Contrarily to the observation in alkaline solution (pH : 8.3) (ref.10) ip increases linearly with potential up to 0.7 V (Fig.5). At more anodic potentials a Tafel relationship is verified.

The impedance diagram measured at E < 0.7 V is characterized by a 45° straight line at the low frequency domain (Fig. 6) suggesting a diffusion limited process. Simultaneously, the apparent activation energy calculated from the ip/temperature dependence changes for $14K.cal.mol^{-1}$ to $7.7 Kcal.mol^{-1}$ in presence of ferrous ions. The passive current does not depend on the rotation speed of the electrode. It is clear that the ferrous/ferric oxidation reaction is controlled by the diffusion of ferrous ions from the bulk to the oxide/solution interface were ferrous ion concentration is considered to be very small. However, it is not easy to accept the formation of an hydrated oxide like film (ref.10) as a result of the ferrous oxidation reaction at pH near zero. Probably the diffusion control is related to a partially blocked interface (ref.11). At potentials E > 0.7V the reaction rate is activation controlled with a transfer coefficient of 0.4.

LOW FREQUENCY DOMAIN

The low frequency domain is achieved most easily in H_3PO_4 1M solutions at temperatures higher than 25°C (Fig. 1). The application of FFT techniques to the current transient response to potential steps in anodic and cathodic directions allows to obtain reliable information at frequencies lower than 1 mHz (ref.5). The low frequency capacitance is obviously related to the change of the film thickness under the effect of potential changes. The current/time curves obtained by applying potentials steps of 0.1 V in both directions are asymmetric,

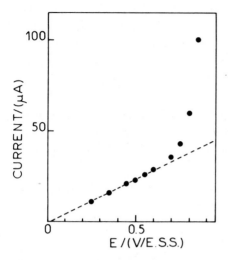

Fig. 5. Relation between potential and current for Fe/0.5M H_2SO_4 + 0.138M Fe(II), T = 25°C.

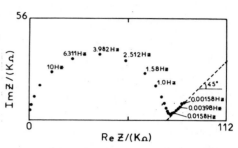

Fig. 6. Complex impedance diagram Fe/0.5M H_2SO_4 + 0.05M Fe(II), E = 0.5V, T = 25°C.

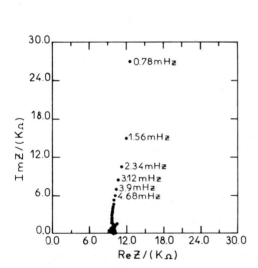

Fig. 7. VLF impedance calculated by FFT processing of potentiostatic step experiments. The impedance is an average of values from a cathodic (0.7V - 0.6V) and an anodic (0.6V - 0.7V) response. Fe/1M H_3PO_4. T = 25°C.

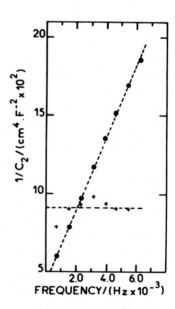

Fig. 8. Low frequency capacitance/frequency relationship calculated for anodic... and cathodic +++ 0.1 potentials steps (Fig. 7).

dealing with two different impedance diagrams. For $\Delta E < 15$ mV the current/time profiles are symetric. Fig. 7 shows the average FFT low frequency impedance diagram whose shape and real axis position is similar to the equivalent domain in the impedance diagram shown in fig. 1. Owing to the accuracy of the FFT technique in the low frequency range it is concluded that the passive current is not diffusion controlled.

The capacitance calculated from the cathodic direction step is independent of frequency and its value is of c.a. 25 mF.cm^{-2} for the Fe/H$_3$PO$_4$ 1M at 25°C. On the contrary, as can be seen in Fig. 8, the capacitance calculated from the anodic direction step decreases as frequency increases. As described by Schuhmann (ref.12), the $1/c^2$ frequency relationship and the low frequency impedance diagram shows a migration control process.

CONCLUSION

Above the Flade potential ferric ions are present as film constituents and their activity is constant and related to an oxide like stoichiometry. The ferrous ion activity is almost zero at the film solution interface and increases in the metal/film interface were the ferric activity decreases. The stoichiometry, conductivity and dissolution rates are not changed by changing the potential. Only thickness changes with potential in order to maintain the ferric ions flux constant at the film/solution interface by restoring the electric field across the film. The ferric ions flux is fixed by the film dissolution rate at the film/solution interface whose value depends on the electrolyte composition by an anion adsorption/desorption process. The adsorption energy of the Fe/anion complex (c.a. 14 Kcal.mol^{-1}) determines the passive current value.

Although the passive current is attained in several minutes the film stabilization requires 24 hours as indicated by the spacing of frequencies on the complex impedance diagram. This aging process may be related to bulk properties of the film which does not affect the passive current.

The passive film has a good electronic conductivity as a consequence the ion transport is the rate determining step for film growth. Among the transport mechanisms the ion migration is supposed to control as indicated by the frequency dependence of the low frequency capacitance.

For large potential steps the film growth and dissolution processes are asymetric and they are related to controlling mechanisms of totally different nature.

The inductive domain originates probably in a competition between the ferric/ferrous ions migration across the film and their interconversion by electrochemical reaction. It may also be ascribed to the build-up of a surface charge in the film/solution interface related to the film formation (ref.7).

Our conclusions support the most significant aspects of the model proposed by Vetter et al. (ref.9) for the steady and non steady behaviours of the passive film on iron in acid media.

Accurate data provided by impedance techniques offer the possibility of designing a more quantitative description in the near future.

REFERENCES
1. K.J. VETTER
    Electrochimica Acta, 16, 1923 (1971).
2. M.D. REIGEVERTS, ANDREEVA, SUKHOTIN
    Elektrokhimiya, 15, 972 (1979).
3. CHAO, LIN, MAC DONALD
    J. Electrochem. Soc., 129, 1874 (1982).
4. K.J. VETTER, F. GORN
    Werkst. und Korr., 21, 703 (1970).
5. C. GABRIELLI, F. HUET, M. KEDDAM, J.F. LIZEE
    J. Electroanal. Chem., 138, 201 (1982).
6. M. DIGNAM
    "Oxides and oxide films", vol. 1, M. DEKKER, N.Y. (1972).
7. H.J. de WITT, C. WIJENBERG, C. CREVECOEUR
    J. Electrochem. Soc., 126, 779 (1979).
8. M. BADDI
    Thesis, Paris 1977.
9. K.J. VETTER and F. GORN
    Electrochimica Acta, 18, 321 (1973).
10. M. NAGAYAMA and S. KAWAMURA
    Electrochim. Acta, 12, 1109 (1967).
11. A. CAPRANI and Ph. MOREL
    J. Applied Electrochem., 7, 65 (1977).
12. D. SCHUHMANN
    Thesis, Paris 1964.
    Publ. Scient. et Techn. du Ministère de l'Air (1965) n° UT 154.

Passivity of Metals and Semiconductors, edited by M. Froment
Elsevier Science Publishers B.V., Amsterdam — Printed in The Netherlands

ELECTROCHEMICAL KINETICS OF IRON AND IRON HYDROXIDE IN AQUEOUS ELECTROLYTES

M.E. VELA, J.R. VILCHE and A.J. ARVIA

Instituto de Investigaciones Fisicoquímicas Teóricas y Aplicadas (INIFTA). Casilla de Correo 16, Sucursal 4, 1900 La Plata, ARGENTINA.

ABSTRACT

The electrochemical behaviour of both polycrystalline iron and precipitated iron hydroxide electrodes is studied in different aqueous media. The dynamic response of the interfaces can be interpreted through a complex reaction pattern involving different surface species.

INTRODUCTION

The kinetics of both the electrodissolution and the passivation of iron in aqueous media under quasi-stationary and transient conditions, including triangularly modulated triangular potential sweeps (TMTPS) and impedance measurements have been interpreted through complex reaction mechanisms[1-5]. The electrodissolution process implies at least the participation of two reaction intermediates and the passivation process involves a complicated reaction pathway where the electrochemical interface exhibits dynamic characteristics. Recent works on this subject are coincident in admitting the formation of various hydroxides and oxides at the Fe/solution interphase, their hydration degree strongly depending both on the applied anodic potential and on the composition of the electrolyte solution, and the participation of ageing processes of the surface compounds[1,2,6,7].

The present work contributes to the knowledge of the intermediate stages of the iron electrodissolution and passivation through the investigation of colloidal precipitated iron hydroxides in diluted alkaline solutions, the kinetics of the active to passive transition of iron in neutral media, particularly in those containing borate, sulphate and carbonate ions, and the study of the anodic dissolution in acid electrolytes using as perturbation technique pseudorandom signals in a parametric identification method by spectral analysis.

RESULTS AND DISCUSSION

1. The potentiodynamic behaviour of iron hydroxide on different conducting substrates

Under proper perturbation (either repetitive, RTPS, or single, STPS, triangular potential sweeps combined with potential steps) the electrochemical response

60

of thin colloidal iron hydroxide films supported on conducting substrates (Pt, Au, vitreous carbon)[8] shows a reasonable separation of a part of the stages involved in the passivation of iron. Thus, in alkaline solutions within the potential range where Fe(II) and Fe(III) species are stable, for a constant anodic charge, the distribution of the cathodic charge along the electroreduction E/I profile depends considerably on the potential scan rate (Fig. 1). When the electroreduction at $E_{s,c}$ is completed ($\tau > 1$ min) the charge involved in the successive E/I profiles remains practically constant (Fig.2) and the potentiodynamic

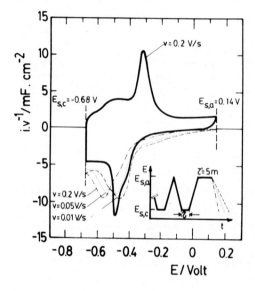

Fig. 1. Influence of the cathodic scan rate on the E/I profile. Pt/ Fe(OH)$_2$/0.01 M KOH + 0.33 M K$_2$SO$_4$. 25°C. The electrode was prepared from 20 alternative immersions (N) in 0.002 M FeSO$_4$ and in 0.01 M KOH + 0.33 M K$_2$SO$_4$ (solution II). The perturbation programme is also shown in the figure.

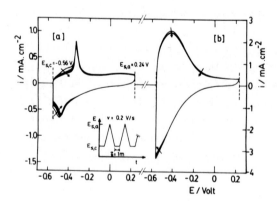

Fig. 2. Influence of N on the E/I profile of an electrode prepared from 0.02 M FeSO$_4$ and solution II. (a) N = 5; (b) N = 50,

response which is characteristic of Fe in alkaline electrolyte is approached with the number of cycles (Fig. 3).

The potentiostatic ageing also reveals that rearrangement processes within the film are coupled to the electron transfer reactions. The electroreduction reactions of Fe(III) species appear as relatively slow processes, as compared for instance to Ni(III)-oxyhydroxide species[1]. This behaviour confirms the response of the Fe/alkaline solution interphase obtained through the TMTPS technique, where three redox couples related to Fe(II)/Fe(III) were found[7].

2. Fe electrodes in neutral electrolyte solutions

The voltammograms run with sodium borate-boric acid buffer (pH 9.3) under RTPS (3 mV/s < v < 100 mV/s) in the -1.0 V to 1 V (vs. NHE) range (Fig. 4), show

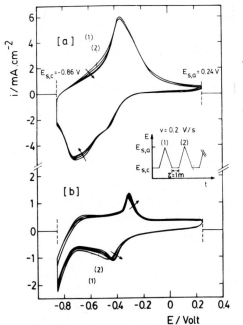

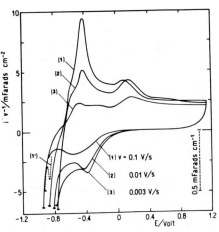

Fig. 4. Stabilized RTPS E/I profiles at different v obtained with the electrode preanodized at 1.1 V during 1 min. Fe/0.15 M H_3BO_3 + 0.075 M $Na_2B_4O_7$ 25°C.

Fig. 3. Influence of N on the E/I profile of an electrode prepared from 0.2 M $FeSO_4$ and solution II. (a) N = 50; (b) N = 5.

three anodic and two cathodic current peaks related to the Fe(0)/Fe(II) and Fe(II)/Fe(III) conjugated couples as seen also through TMTPS experiments. Both the anodic and cathodic peaks overlap to a great extent and their locations are remarkably dependent on the switching potentials (Fig. 5).

The anodic and cathodic current peaks associated with the Fe(0)/Fe(II) couple

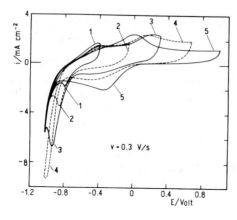

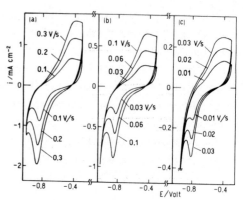

Fig. 5. Influence of the progressively increased $E_{s,a}$ on the E/I displays. Fe/0.15 M H_3BO_3 + 0.075 M $Na_2B_4O_7$. 25°C.

Fig. 6. Influence of v on the stabilized RTPS E/I displays. Fe/0.15 M H_3BO_3 + 0.075 M $Na_2B_4O_7$. 25°C.

exhibit linear i_p vs. v and E_p vs. ln v relationships. The slope of the latter is RT/F, and the charge involved in both peaks decreases as v increases approaching 1.5-1.8 mC.cm^{-2} (Fig. 6). Ageing effects in this case are not detected by the usual perturbation techniques. Nevertheless, the overall stabilized E/I profiles approach those recorded in alkaline electrolytes.

The charge associated with the reactions taking place in the potential range of the Fe(II)/Fe(III) conjugated couple is much smaller than that earlier obtained in alkaline solutions (pH > 11.5) probably due to the greater passivation causing a decrease in the base metal electrodissolution.

The results obtained with borate-boric acid buffers within the pH range 7.6 < pH < 9.3 exhibit similar potentiodynamic behaviours, although the current peaks are shifted -0.06 V per pH unit.

The voltammograms obtained in potassium carbonate/bicarbonate buffer (pH 8.9) and in neutral solutions containing sodium sulphate fit, in general, the above mentioned description. However, in the presence of bicarbonate ions, the precipitation of iron carbonate contributes as a first stage in the prepassivation of the metal surface. Measurements performed with a Fe rotating disk electrode reveal an acceleration of the prepassive surface film dissolution as the rotation speed increases (Fig. 7). This effect can be interpreted through equilibrium reactions between surface species and ions in solution. Carbonate ions influences the process at the outer surface layer, and the formation and stability of the inner Fe(OH)$_2$ layer appears independent of the anion nature.

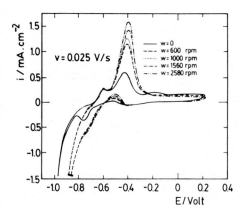

Fig. 7. Influence of w on
the E/I profile. Fe/0.05 M
K_2CO_3 + 0.75 M $KHCO_3$. 25°C.

3. Fe electrodes in acid electrolytes

The response of Fe in x H_2SO_4 + y K_2SO_4 (x + y = 0.5 M) (0.3 < pH < 5) to
small amplitude pseudo-random binary signals (generated by a PDP 11/20 minicomputer) superimposed to a base signal corresponding to a steady-state condition can be compared to both the impedance spectrum resulting from frequency
response analysers using sine wave perturbations[3,4], and to the TMTPS measurements[2,10]. The digital analysis of the input and output signals were carried
out by the use of a fast Fourier transform algorithm employing power spectral
techniques. Sampling of excitation and response signals was simultaneously performed and the data alternatively acquired were corrected by software. The complex impedance within the 0.04-3000 Hz range was analyzed in three selected frequency regions varying accordingly to the sampling rate. The squared coherency
spectrum was evaluated for each measurement to test the signal to noise ratio.

Figs. 8 and 9 show the impedance diagrams for Fe in 0.5 M H_2SO_4 and Fe in 10^{-2}
M H_2SO_4 + 0.49 M K_2SO_4 after ohmic drop correction, using the identification
technique by spectral analysis in the active dissolution potential region. According to earlier results the stationary E/log i curves are characterized at
low overpotentials by Tafel slopes of ca. 0.04 V/decade and electrochemical reaction order with respect to OH^- concentration close to 1. The complex impedance
diagrams are similar, in principle, to those described in the literature[3,4]
employing frequency response analyser of the Solartron type for transfer locus
measurements. The kinetic parameters, however, are different from those earlier
reported. In the conditions of the experiments a single capacitive loop and two
inductive loops are only observed. Results in slight acid solutions can be com-

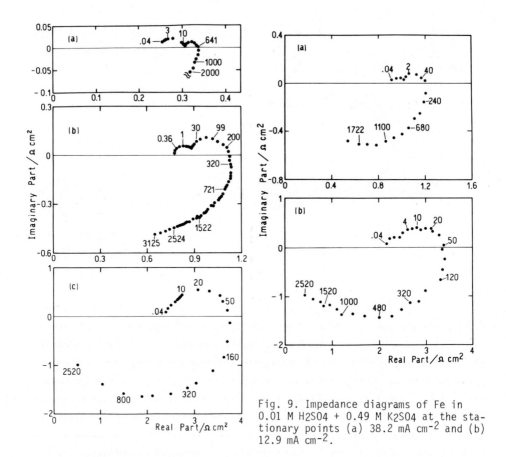

Fig. 8. Impedance diagrams of Fe in 0.5 M H2SO4 at different stationary current densities. (a) 136 mA cm^{-2}; (b) 48.3 mA cm^{-2}; (c) 10.4 mA cm^{-2}. Frequencies in Hz.

Fig. 9. Impedance diagrams of Fe in 0.01 M H2SO4 + 0.49 M K2SO4 at the stationary points (a) 38.2 mA cm^{-2} and (b) 12.9 mA cm^{-2}.

pared with the electrochemical response under complex fast potentiodynamic perturbations using the TMTPS technique[2]. The first charge transfer step appears as a reversible step. No third inductive contributions could be observed within the frequency domain limits considered.

CONCLUSIONS

The present results independently of the electrolyte composition (including pH) the corrosion and passivation of iron in aqueous solutions in the absence of oxygen and surface active substances fit a generalized reaction scheme involving the following stages[9,10] occurring through a series of consecutive and alternative steps: i) the electroformation of the first adsorbed O-containing layer; ii) O-containing multilayer growth; iii) ageing of film forming species; iv) base metal electrodissolution through the iron-protective surface film; v) oxide phase formation; and vi) hydrolysis reactions including iron salt solid phase formation.

The influence of the electrolyte components manifests in the various stages in different ways, such as a modification of the double layer structure either at the inner or at the outer plane, ionic adsorption at the oxide layer, change in chemical composition and structure of the passive layer including degree of hydration[11], crystallographic and electrical characteristics.

The described procedure for the dynamic study of electrochemical interfaces applying pseudo-random binary sequence perturbations appears to a powerful technique to acquire impedance data from short time measurements covering a wide frequency range, even for the case of a rather complex electrodissolution process involving relaxing effects of the electrogenerated surface species as the anodic iron dissolution. The electrochemical behaviour can be interpreted on basis of the complex reaction pattern earlier postulated[2,10].

REFERENCES

1. R.S. Schrebler Guzmán, J.R. Vilche and A.J. Arvía, J.Appl.Electrochem., 11, 551 (1981).
2. J.O. Zerbino, J.R. Vilche and A.J. Arvía, J.Appl.Electrochem., 11, 703 (1981).
3. H. Schweickert, W.J. Lorenz and H. Friedburg, J.Electrochem.Soc., 127, 1693 (1980).
4. M. Keddam, O.R. Mattos and H. Takenouti, J.Electrochem.Soc., 128, 257, 266 (1981).
5. B.D. Cahan and C.T. Chen, J.Electrochem.Soc., 129, 17, 474, 921 (1982).
6. R.S. Schrebler Guzmán, J.R. Vilche and A.J. Arvía, Electrochim.Acta, 24, 395 (1979).
7. J.R. Vilche and A.J. Arvía, Acta Cient.Venez., 31, 408 (1980).
8. V.A. Macagno, J.R. Vilche and A.J. Arvía, J.Appl.Electrochem., 11, 417 (1981).
9. J.R. Vilche and A.J. Arvía, Proc. 6th ICMC, Rio de Janeiro, pp. 245-256 (1978).
10. J.R. Vilche and A.J. Arvía, Anal.Acad.Cs.Ex.Fis.Nat., Buenos Aires, 33, 33 (1981).
11. O.J. Murphy, J.O'M. Bockris, T.E. Pou, D.L. Cocke and G. Sparrow, J.Electrochem.Soc. 129, 2149 (1982).

IRON PASSIVATION IN BORATE SOLUTIONS

S.P. TYFIELD
CEGB, Technology Planning and Research Division, Berkeley Nuclear Laboratories,
Berkeley, Gloucestershire, GL13 9PB, UK

ABSTRACT

The role of borate as an inhibitor for iron is considered on the basis of
an open circuit and potentiodynamic study of an abraded iron (99.57%) in borate
solutions as a function of pH (with 1250 ppm total boron), boron content (at pH
6.3) and the presence and absence of oxygen. Borate is considered to be a
complexant with iron. Raising the solution borate content is interpreted to
enhance borate adsorption, whilst the hydroxide concentration at the iron
surface is maintained due to the consequent pH buffer capacity increase. The
inhibition of iron corrosion in borate solutions with and without oxygen is
resolved in terms of the primary passivation model of Ogura and Sato (ref. 1).

INTRODUCTION

CEGB spent AGR fuel ponds are dosed with the neutron absorber boron

(1250 ppm) in the form of boric acid neutralised by sodium hydroxide to pH 7.

Borate solutions are recognised inhibitors of iron and steel (ref. 2), but

despite a variety of electrochemical and surface analytical studies their mode

of action remains unresolved. The controversy is whether borates function

simply because of their pH buffering capability or whether they are inhibitors

in their own right. The behaviour of borate solutions is complex because of the

presence of polyborates (ref. 3). In this study the borate content is described

in terms of the total boron content and solution pH.

METHODS

Abraded iron discs (wt % C 0.028, S 0.026, Mn 0.12, P 0.012, Si 0.01, B

0.002, Ni 0.036, Cr 0.015, Mo 0.01, Ti 0.01) were mounted in a Tefzel specimen

holder fitted with a Kalrez sealing washer that exposed 1 cm^2. The holder was

assembled with two high density graphite electrodes in a litre flask that could

be purged by high purity nitrogen or air. A saturated calomel reference

electrode was connected via a Luggin probe. Borate solutions were prepared from

Analar reagents and triply distilled water. All experiments were conducted at

20 °C and the solutions were stirred.

Open circuit potentials were monitored for 24 hours. Polarisation curves

were subsequently determined with a Parc 173 potentiostat and 175 programmer using a scan rate of 1 mV sec^{-1}.

RESULTS

Open circuit corrosion potential of iron

Air saturated solutions. The time variation of the open circuit corrosion potential of iron in borate solutions (1250 ppm boron) obtained from boric acid and sodium hydroxide (pH 5 to 9.2) is presented in Fig. 1a. The steady state values are given in Table 1.

TABLE 1

Corrosion potential, E_{corr}, primary passivation potential, E_{pp}, and primary passivation current density, i_{crit}, of borate solutions (1250 ppm boron) at 20 °C. Potentials in mV vs SCE, current density in A.m^{-2}

Solution pH	Steady state open circuit E_{corr}		Potentiodynamic Results			
			Aerated Solution	De-aerated Solution		
	Aerated Solution	De-aerated Solution	E_{corr}	E_{corr}	E_{pp}	i_{crit}
5.0	−595	−580	−547	−610	−	−
5.3	+50	−	+19	−	−	−
5.6	+20	−	−39	−	−	−
	0	−	−45	−	−	−
6.0[+]	+35	−	−	−	−	−
6.3	−110	−638	−100	−645	−	−
6.3[*]	−140	−	−127	−	−	−
6.3[**]	−590	−702	−620, −294, −268	−710	−300	7.3
7.0	−	−660	−	−690	−321	3.7
7.2	−100	−	−100	−	−	−
	−142	−	−167	−	−	−
7.3	−	−718	−	−135	−418	1.8
7.8	−	−754	−	−766	−487	1.3
8.0	−148	−	−202	−	−	−
8.4	−184	−	−225	−	−	−
9.2	−220	−786	−300	−787	−672	0.059

[+]Distilled water, [*]125 ppm boron, [**]6250 ppm boron.

Surface films were not visible on iron that spontaneously ennobled whereas a yellow-orange corrosion film formed in the boric acid and low pH borate solutions (pH ≤ 5.6). In distilled water a yellow film around the sealing ring was indicative of crevice corrosion.

The effect of varying borate content was studied a pH 6.3 (Fig. 1a, Table 1). In the 6250 ppm boron solution the iron was coated with a yellow film, whereas in the 125 ppm boron solution crevice corrosion was apparent.

Nitrogen saturated solutions. The corrosion potential rapidly declined to a steady active value that decreased with increase in pH (Table 1).

The iron did not lose its metallic lustre in the de-aerated borate solutions.

Potentiodynamic scans

The results are summarised in Table 1.

Nitrogen saturated solutions. The corrosion potentials determined by the forward anodic scan were active ($E_{corr} < E_{pp}$, Table 1). The onset of anodic passivation was only noted above a pH of 7 for borate solutions containing 1250 ppm boron. The effect of boron content at pH 6.3 (Fig. 2a) indicates that anodic passivation may be established at a lower pH by increasing the borate content. The corrosion potentials determined potentiodynamically were in fair accord with the steady state open circuit results (Fig. 1b).

Air saturated solutions. The anodic polarisation scans confirm that the corrosion potential dependence on pH, above pH 5.6, is $- 59$ mV pH^{-1} (Fig. 1b). The effect of increasing the borate level at pH 6.3 was to lower the corrosion potential and promote active corrosion of iron (Fig. 2b).

DISCUSSION

Critical pH for passivation

This study confirms that the spontaneous passivation of iron occurs in borate solutions provided oxygen is present and the pH exceeds a critical value. In this study pH_{crit} is noted to be sensitive to the borate content. With 1250 ppm boron present it is about pH 6 but exceeds pH 6.3 with 6250 ppm boron present. pH_{crit} of 7 and 8 have been noted for air filmed and cathodically cleaned iron respectively in aerated borate solutions (ref. 4). Iron is reported to be uninhibited at pH 9 (ref. 5) but passivated at pH 10 (ref. 6) in 0.1 N air saturated borate solutions. Evidently the concept of a critical pH of iron passivation in borate media is not straightforward. Surface preparation and impurity content are undoubtedly important but so is the borate content.

pH dependence of E_{corr}

The effect of pH on corrosion potential depends on the relative polarisation changes with pH of the component anodic and cathodic processes that determine the corrosion potential.

In the de-aerated solutions the corrosion potential of the air filmed iron is considered to be primarily controlled by the reduction of iron ($Fe^{3+} \rightarrow Fe^{2+}$) which is coupled to iron oxidation ($Fe \rightarrow Fe^{2+}$). The Fe^{2+}/Fe^{3+} redox potential change with pH in anaerobic non-oxidising inhibitor anion (benzoate, acetate) solutions (ref. 7) closely resembles the corrosion potential variation of iron in de-aerated borate solution noted in this study and by Fischer et al (ref. 4). It is proposed that the noted E_{corr}-pH variations ($\frac{dE_{corr}}{dpH} \simeq - 29$ mV.pH^{-1} at pH < 6, $\frac{dE_{corr}}{dpH} \simeq - 59$ mV. pH^{-1} at pH > 7.3) are determined by the change with pH

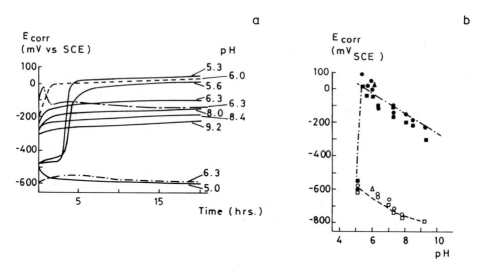

Fig. 1. Corrosion potential of iron in borate solutions at 20 °C. (a) Open circuit E_{corr} versus time showing pH effect ——(1250 ppm boron) and boron effect (-- 0 ppm, —— 125 ppm, —·—6250 ppm) at pH 6.3. (b) Steady state open circuit E_{corr} (● with air, ○ without air, 1250 ppm boron; ▲ with air, △ without air, distilled water) and potentiodynamic E_{corr} (■ with air, □ without air, 1250 ppm boron) versus solution pH.

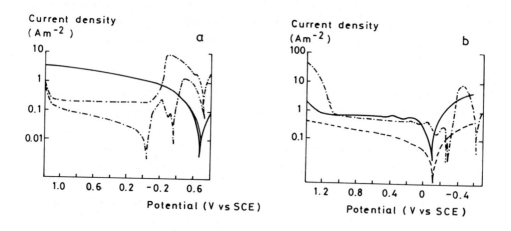

Fig. 2. Voltammagrams of iron in pH 6.3 borate solutions at 20 °C showing effect of borate level defined by boron content (—·—6250 ppm, —1250 ppm, ———125 ppm) (a) nitrogen saturated and (b) air saturated solutions. Scan rate 1 mV sec^{-1}.

of the Fe^{2+}/Fe^{3+} redox potential. The redox potential would change with pH, on the basis of the Nernst equation by -29 mV.pH^{-1} when half a hydroxide ion is lost per electron gained per iron and by -59 mV.pH^{-1} when one hydroxide is lost per electron gained per iron. The Fe^{2+} and Fe^{3+} species involved in the redox reaction are likely to involve hydroxide and borate ions. It is speculated that the Fe^{2+}/Fe^{3+} redox reaction in borate solutions varies with pH with the predominant reaction at pH < 6:

$$\left[Fe^{III} (OH)_x L_y\right]^z + e^- = \left[Fe^{II} (OH)_{x-1} L_y\right]^z + OH^-$$

and at pH > 7.3,

$$2\left[Fe^{III} (OH)_x L_y\right]^z + 2e^- = 2\left[Fe^{II} (OH)_{x-1/2} L_y\right]^{z-1/2} + OH^-$$

where L is borate (or non-oxidising anion) inhibitor. The explanation in terms of the equilibrium potential (ref. 4) only accounts for the higher pH results and implies that the corrosion potential is controlled by the Fe/Fe^{3+} redox couple rather than the Fe^{2+}/Fe^{3+} couple.

In the presence of oxygen, the corrosion potential of the iron is considered to depend on the oxidation of iron ($Fe^{2+} \rightarrow Fe^{3+}$) and the reduction of oxygen ($O_2 + 2H_2O + 4e^- = 4 OH^-$). Since the anodic Tafel slope of passivated iron is infinite, the change with pH of the corrosion potential of passivated iron (pH $\gtrsim$ 6) is interpreted to be determined by the pH dependence of oxygen reduction. This is supported by the similar (-59 mV pH^{-1}) dependence of oxygen reduction at a platinum electrode (ref. 9), where the Tafel slope of the anode process is considered to be also infinite.

Anodic polarisation of iron

In de-aerated borate solutions the anodic polarisation curve for iron features one major peak at a scan rate of 1 mV sec^{-1}, however two extra more noble peaks are discernible at higher scan rates (ref. 1). These peaks are accounted for by the passivation mechanism of Ogura and Sato (ref. 1). The primary passivation reaction, which is associated with the major anodic peak, is proposed to be:

$$2 Fe + H_2O + L^- \rightarrow \left[Fe (H_2O)\right]^{2+} + \left[Fe L_{ads}\right] + 3e^- \qquad (1)$$

where L represents hydroxide or borate anions.

The effect of varying the boron content and consequently the borate content as studied at pH 6.3 may be resolved in terms of this mechanism.

The anodic polarisation of iron in the absence of oxygen resulted in passivation in the high borate solution (6250 ppm B) but not in the lower borate solutions (< 1250 ppm B) (Fig. 2b). Evidently a high borate level favours anodic passivation without oxygen. Several factors are relevant. First, increasing borate content would promote adsorption according to (1) and so promote primary passivation. This implies that there is a critical borate content for a particular pH value to ensure passivation. Secondly, borate solutions function as pH buffers whose pH buffer capacity at any pH is raised with the borate content in the pH range 5 to 9.2 (ref. 9). Increasing the borate content consequently assists in the maintenance of the solution pH adjacent to iron and the complexation of hydroxide with iron (as in (1)) may occur.

The corrosion behaviour of iron in aerated borate solutions is considered to depend on the relative magnitude of the diffusion restricted oxygen current density and the iron primary passivation critical current density, which is affected by both borate and oxygen content. The effect of borate on the cathodic limited current density is considered to be similar to that proposed for benzoate (ref. 11), although this is not generally accepted (ref. 12). The anodic polarisation of iron is simultaneously modified by the combined effect of borate and oxygen. The oxygen contribution may depend on the cathodic production of hydroxide (ref. 13), which would promote reaction (1). Alternatively oxygen may be adsorbed in preference to hydroxide and borate anions in reaction (1). The control of iron passivation in terms of the cited primary passivation scheme does not support the proposal that the reduction of oxygen by reaction with ferrous surface complexes, so stimulating the oxidation of iron (ref. 4), is the primary effect of oxygen. The action of borate is concluded to be akin to that of other non-oxidising anion inhibitors except that its chemistry is complicated by polyborate formation.

REFERENCES

1 K. Ogura and K. Sato, Passivity of Metals, Proc. 4th Internatl. Sym. on Passivity, Electrochem. Soc., Princeton, 1978, pp. 463-478.
2 Z. Szklarska-Smialowska, ibid, pp. 443-462.
3 C.F. Baer, jr. and R.E. Mesmer, Hydrolysis of Cations, Wiley, New York, 1976, pp. 104-111
4 M. Fischer, W. Gruner and G. Reinhard, Corros. Sci., 15, 1975, pp. 279-293.
5 D. Gilroy and J.E.O. Mayne, Brit. Corros. J., 1, 1966, pp. 161-165.
6 J.G.N. Thomas and T.J. Nurse, Brit. Corros. J., 2, 1967, pp. 13-20.
7 J.E.O. Mayne and S. Turgoose, Brit. Corros. J., 10, 1975, pp. 44-46.
8 D.B. Sepa, M.V. Vojnovic and Damjanovic, Electrochim. Acta, 26, 1981, pp. 781-793.
9 S.P. Tyfield, M. Dearneley and P.W.G. Simpson, CEGB Report to be published.
10 H. Kaesche, Korrosion des Metalle, Springer, Berlin, 1966.
11 D.E. Davies and Q.J.M. Slaiman, Corros. Sci., 13, 1973, pp. 891-905.
12 W. Forker, G. Reinhard and D. Rahner, Corros. Sci., 19, 1979, pp. 745-751.
13 Yu. N. Mikhailovskii and V.M. Popova, Dokl. Akad. Nauk (Phys. Chem.) Eng. Trans., 255, 1980, pp. 1034-1037.

Passivity of Metals and Semiconductors, edited by M. Froment
Elsevier Science Publishers B.V., Amsterdam — Printed in The Netherlands

THE ROLE OF INHIBITOR PAB AT NiOOH-LAYERS

M. GOLEDZINOWSKI* and J. W. SCHULTZE

Institut für Physikalische Chemie, Universität Düsseldorf, Universitätsstraße 1,
D-4000 Düsseldorf, FRG

ABSTRACT

The formation and properties of NiOOH-layers were investigated by electrochemical measurements in presence and absence of 2-pentylamino-benzimidazole (PAB).
Various positions of the inhibitor inside (in), outside (out) and in the total
film (tot) can be distinguished, but concentration profiles are not stable,
since the inhibitor can diffuse within the film. Therefore, the inhibition of
various ion and electron transfer reactions within the film increases with polarization time and number of redox cycles. The NiOOH/PAB-layers are presumably
conducting for electrons, and H^+ or OH^-. The inhibition of redox reactions of
the NiOOH is explained by a chemical transformation of the NiOOH, the inhibition
of the oxide growth by an inhibition of the Ni^{2+}-transfer reaction.

INTRODUCTION

For many corrosion systems, the stabilization of passive films by inhibitors
is of great interest (ref. 1, 2). The system passive film/inhibitor, however, is
very complicated, since the inhibitor can be located inside (in), outside (out),
and in the total film (tot). Recently, we have shown by XPS-measurements that
all three types of inhibitor layers can be prepared in the system NiOOH/PAB
(2-pentylamino-benzimidazole) on inert gold electrodes (ref. 3). In the present
paper, the electrochemical behaviour of the layers and the influence of the
inhibitor will be discussed.

The anodic layer formation
$$Ni^{2+} + 2 H_2O \rightarrow NiOOH + 3 H^+ + e^- \tag{1}$$
consists of three steps involving ion transfer reactions of 2- or 3-valent nickel
ions (Ni^{z+}) and OH^- between the aqueous phase (aq) and the NiOOH (ox) and an
electron transfer reaction at the interphase:
$$Ni^{z+}(aq) \rightarrow Ni^{z+}(ox) \tag{1a}$$
$$Ni^{2+}(aq,ox) \rightarrow Ni^{3+}(aq,ox) + e^- \tag{1b}$$
$$OH^-(aq) \rightarrow OH^-(ox) \tag{1c}$$
Simultaneously, the anodic oxygen evolution takes place
$$2 H_2O \rightarrow O_2 + 4 H^+ + e^- \tag{2}$$
which involves reaction (1c). Finally, the layer can be reduced and reoxidized:
$$Ni(OH)_2 \leftrightarrow NiOOH + H^+ + e^- \tag{3}$$

* on leave of absence from Warsaw University

EXPERIMENTAL

The electrolyte was a 1 M solution of NaClO$_4$ with a 0.1 M borate buffer of pH 7.2. The concentration of Ni^{2+} was c=0 and 10^{-4}M, that of PAB was 0 and 10^{-3}M respectively. NiOOH layers with an average thickness of 5 to 20 nm were deposited on a gold wire. The electrochemical cell and the equipment was described elsewhere (ref. 3). Electrode potentials are referred to the hydrogen electrode in the same solution (HESS). Analytical and XPS-measurements were carried out on gold foils after complete removing of the electrolyte by rinsing with water. They showed that the layer 3out consists of NiOOH covered by a PAB-layer of a thickness d ≈ 3 nm. The layer 3tot is covered by PAB too, but PAB is also incorporated and forms NiOOH/PAB complexes (ref. 3).

RESULTS

Oxide growth

To study the influence of PAB on ion transfer reactions, the anodic oxide growth according (1) was investigated at various potentials ε. At ε ≥ 1.65 V, nuclei of NiOOH are formed which grow continuously with time. Therefore, the current density which is mainly due to reaction (2) increases with time, i ∼ t^n (ref. 4).

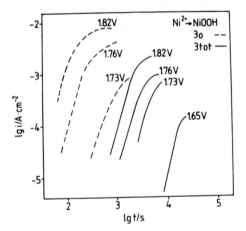

Fig. 1: Double logarithmic plot of the current density i in dependence on the polarization time t of NiOOH deposition at different potentials at $c_{Ni^{2+}} = 10^{-4}$ M

--- 3o-layers
___ 3tot-layers

Fig. 1 shows the double logarithmic plot lgi/lgt which yields at all potentials straight lines with a slope n = 2.8 to 3.5, bending to n = 1 or even zero, if the nuclei form continuous films (ref. 4). This type of oxide growth is observed as well in absence, which yields the pure 3-valent hydroxide 3o, as in presence of PAB, which yields the film 3tot. The only difference is the shift to longer

times which is due to the inhibition of reaction (1) and (2). This means that
the growth of 3tot takes place also by the mechanism of progressive nucleation
and growth of thin lenses (ref. 4). Unfortunately, the evaluation of i_{ox} (rate
of (1)) is not possible as in the case of 3o (ref. 4), but from the similar
shape and distance of the lines, we can assume that a similar charge transfer
step is rate determining for 3tot as well as 3o.

Reduction and Reoxidation

Pure NiOOH-films (3o) and with inner PAB (3in) can be reduced and reoxidized
according to reaction (3) many times. In presence of PAB in the solution, however, the charge of reduction Q decreases with increasing number of cycles.

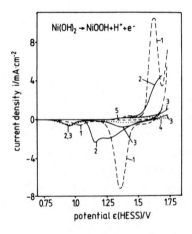
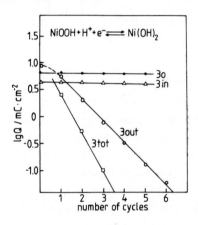

Fig. 2: Potentiodynamic reduction and
reoxidation of NiOOH-layers with $d\varepsilon/dt$
= 100 mV/s, d=20nm. $c_{Ni^{2+}}$ = 0. Curve 1:
3o-layer (c_{PAB} = 0), 2: 3out-layer
($c_{PAB} \approx 10^{-3}$ M), 3,4,5: following cycles

Fig. 3: Logarithm of reduction charge Q
in dependence on the number of cycles
for different NiOOH/PAB layers: 3o: d=
17 nm, 3in: $\approx$ 5+5 nm, 3out=21 nm,
3tot = 8 nm.

Fig. 2 shows some curves for 3o and 3out as an example. After 4 cycles, Q becomes almost negligible. The decrease is even faster for films of 3tot, i. e.
NiOOH/PAB layers with incorporated PAB. Fig. 3 shows that lgQ decreases for 3out
and 3tot linearly with the number of redox cycles, i. e. according to a reaction
of first order. Simultaneously, the electrode capacity decreases (see Fig. 4).
In spite of the negligible Q, Ni is present at the gold surface as was shown by
analytical measurements. The inhibition of (3) increasing with cycles (and time)
suggests a diffusion of PAB into the layer which was justified by XPS-measurements (ref. 3).

The reduction charge Q is almost constant for 3in (in the absence of PAB in
solution). This means that the inner NiOOH/PAB-layer does not inhibit reaction
(3).

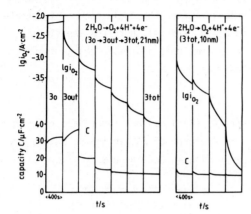

Fig. 4: Oxygen evolution and capacity in dependence on time for 3out and 3tot layers. Vertical lines indicate a potentiodynamic reduction/reoxidation cycle, $C_{Ni^{2+}} = 0$, $C_{PAB} = 10^{-3}$ M for 3out and 3tot.

Oxygen evolution

The anodic oxygen evolution (2) was used as an example of an electron transfer reaction. It can be investigated best in absence of Ni^{2+} at a constant film thickness. Steady state measurements in presence of PAB, however, show a slow decrease with time (Fig. 4). After short redox-cycles, i_{O_2} decreases again faster indicating a strong film change during the redox cycle.

The potential dependence of i_{O_2} is shown in the Tafel plot in Fig. 5 for various films after short polarization times. The catalysis of reaction (2) by

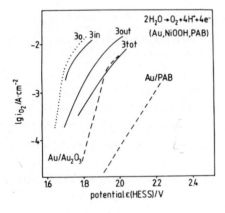

Fig. 5: Tafel plots of oxygen evolution at different NiOOH/PAB layers. d = 20 nm for 3o and 3out, d ≈ 5,5 + 10 nm for 3in, d ≈ 8 nm for 3tot

pure NiOOH (3o) is not changed by an inner film (3in), but it decreases for the film 3out and even more for 3tot. In any case, however, the rate of (2) exceeds always the rate on pure Au in presence of PAB. The inhibition is combined not only with a decrease of exchange current densities but also with a decrease of

slope, i. e. an increase of the Tafel factor b from 50 mV (3o) to about
180 mV (3tot) or a corresponding decrease of the apparent transfer coefficient
from 1.2 to 0.3.

DISCUSSION
Electron transfer reaction

Figs. 1 to 5 show that all reactions (1) to (3) are inhibited by PAB, but various electron and ion transfer reactions must be discussed in detail. As can be
obtained from Fig. 1, reactions(1b) as well as (2) take place with a reasonable
rate at thick films ($i_{1b} > 1$ µA/cm², and $i_2 > 300$ µA/cm² at 1.73 V). Hence, the
bulk film seems to be conducting for electrons. This assumption is supported by
the large rate of (2) on films of 3in, but the inner film is thin (d = 5.5 nm)
and the homogeneity of the inner and outer films could not be proved independently. Another argument in favour of the electronic conductivity is the large electrode capacity which exceeds always 10 µF/cm². On the other hand, the increase
of the b-factors of reaction (2) indicate the presence of a space charge layer
at the oxide surface or within the outer PAB-layer. The inhibition of reaction
(3) does not indicate insulating properties of 3tot-layers, since these films
are not pure NiOOH but contain an appreciable amount of PAB complexes of Ni.

Ion Transfer Reactions of OH^- and Ni^{2+}

Even in presence of PAB, the rate of (2) is always faster on 3out and 3tot
than on Au (Fig. 5). This means that reaction (1c) must be fast enough. Then,
the inhibition of the overall reaction (2) must be explained by a decreasing
activity of OH^- at the surface or by the inhibition of other reaction steps.
Considering reaction (1) we can conclude that the reaction (1a) must be rate determining, since (1b) and (1c) are fast enough. Hence, the inhibition of the
overall reaction (1) by PAB must be explained mainly by the inhibition of reaction (1a) by the outer layer of PAB which has a thickness of about 3 nm (ref.3).

ACKNOWLEDGEMENT

The support of this work by the Alexander-von-Humboldt-Stiftung and the FE-
KKS-program of the Bundesministerium für Forschung und Technologie is gratefully
acknowledged.

REFERENCES
1   M. M. Lohrengel, K. Schubert, J. W. Schultze, Werkst. Korr. 32 (1981) 13
2   D. Kuron, H.-J. Rother, H. Gräfen, Werkst. Korr. 32 (1981) 409
3   M. Goledzinowski, S. Haupt, J. W. Schultze, Electrochim. Acta, submitted for
    publication
4   J. W. Schultze, M. M. Lohrengel, D. Roß, Electrochim. Acta, in press

Passivity of Metals and Semiconductors, edited by M. Froment
Elsevier Science Publishers B.V., Amsterdam — Printed in The Netherlands

OPTICAL AND IMPEDANCE STUDIES OF NICKEL
PASSIVATION FILM IN NEUTRAL SOLUTION

TOSHIAKI OHTSUKA, KAZUHISA AZUMI, and NORIO SATO
ELECTROCHEMISTRY LABORATORY, FACULTY OF ENGINEERING,
HOKKAIDO UNIVERSITY, 060 SAPPORO, JAPAN

ABSTRACT

The passive film on nickel was studied by 3-parameter reflectometry and impedance measurements. It was found that the complex refractive index and the dielectric constant of the film increase with the anodic potential of film formation, indicating that the nonstoichiometric composition of the film is potential-dependent. The electrode impedance of passivated nickel consists of a film capacitor and an electric double layer capacitor at the film surface.

INTRODUCTION

The in-situ light-reflection methods such as ellipsometry and reflectometry have been proved powerful for the study of thin surface layers on metals.
There have been a few ellipsometric study of the passive oxide film on nickel
(ref. 1-3), which showed the presence of a very thin oxide film about 1 nm
thick on the passivated nickel in neutral solution. In these previous studies,
the optical constant of the film was assumed to be independent of the potential
of film formation. A recent paper (ref. 4), however, revealed that the optical
constant and hence the film composition varied depending on the potential.
In this study, the 3 parameter reflectometry developed by one of the authors
(ref. 5) has been employed to simultaneously determine the three unknowns of
thickness, refractive index and extinction index of the passive film on nickel.
Furthermore, the impedance of passivated nickel electrode has been measured to
reveal the dielectric property of the film.

EXPERIMENTAL

The fundamentals and calculations of the 3P reflectometry are found
elsewhere (ref. 5). The film-free reference surface was obtained by potentiostatic reduction at E = -0.10 V (vs. Hydrogen Electrode in the Same Solution;
HESS) in 0.15 M sulphuric acid solution (pH 0.8). The reflectivity measurements started with this surface. After exchanging the solution to a neutral
borate solution at pH 8.4, the electrode was kept for 1 h at a certain
potential in the passivation range, which was then followed by the cathodic

tial. This is in agreement with a nonstoichimetric oxide NiO_{1+x} model for the passive film on nickel proposed by Vetter (ref. 6) who assumed that the nonstoichiometry x increases with potential.

Recently, Chao et al. (ref. 7) measured the impedance spectrum for a passivated nickel electrode in a frequency range from 0.1 to 1 KHz in weakly basic borate and phosphate solutions and interpreted a Warburg-type impedance spectrum obtained at low frequencies in terms of a point defect model for the passive film. In Fig. 2 we assume an equivalent circuit consisting of a paralell R_h - C_h combination at high frequencies, which represents the film/ solution interface, and a paralell R_1 - C_1 combination at low fraquencies, which represents the passive film itself.

The paralell R_h - C_h circuit may correspond to a combination of a faradaic impedance with a double layer capacity at the film/solution interface. The most probable reaction which determines the faradaic impedance at the film/ solution interface would be the adsorption-desorption of OH^- ion;

$$OH_{aq}^- \rightleftarrows OH_{ad(oxide)}^- ,$$

The time constant estimated from R_h and C_h in the passive potential region is about 5×10^{-5}s, which is likely to correspond to the relaxation time of the OH^- ion adsorption-desorption reaction at the film/solution interface.

The paralell R_1 - C_1 circuit that manifests itself at low frequencies is related to the passive film on nickel, and the inverse of capacitance C_1 would be proportional to the film thickness. Fig. 4 shows the film thickness, d, and the inverse capacitance, C_1^{-1}, as a function of potential. Both d and C_1^{-1} first increase and then decrease with increasing anodic potential, each showing a maximum at around +0.9 V, which is close to the potential at which R_1 reaches a peak value (Fig. 3). It is noted that this potential at +0.9 V agrees with the potential where the potential-dependent, transpassive dissolution commences in acid solution (ref. 5 and 6).

If the capacitance C_1 represents the paralell capacitor of the passive oxide film, we could estimate from d and C_1 the dielectric constant, ϵ, of the film. The dielectric constant thus estimated as a function of potential is shown in Fig. 5. It is seen that the dielectric constant is about 19 in the vicinity of the passivation potential and increases with anodic potential up to 40 at 1.4 V. A steep rise of ϵ observed at potentials more noble than 1.5 V is due to an electronic current flowing through the film for the oxygen evolution reaction. It appears from the potential-dependent dielectric cons- tant of the film that the film composition changes with the potential at which the film has been formed.

reduction again in 0.15 M sulphuric acid solution at E = -0.10 V . The reflectivity change was then measured between the film-free reference surface and the passivated surface. When the reflectivity of the surface reduced after passivation did not recover into the original reflectivity, the data were discarded.

Impedance measurments of the passivated nickel electrode were carried out by using a two-phase lock-in amplifier under potentiostatic conditions at an AC voltage of 4.5 mV (peak to peak) in a frequency range from 10 to 10^5 Hz.

RESULT

Fig. 1 shows the results obtained from the 3P reflectometry, in which the refractive and extinction indices (n_2 and k_2), film thickness (d) and anodic c.d. (i) after 1 h oxidation are plotted as a function of potential. The data show some scatter probably because of the film thickness which is too thin to quantitatively analyse from the 3P reflectometry. The optical constants, n_2 and k_2, increase with increasing anodic potential, particularly k_2 markedly increasing at potentials more positive than 1.2 V . The film thickness appears around 1 nm in the passive potential region. At potentials more positive than 1.6 V where the second passivation occurs, the film thickness reaches more than 10 nm and the complex refractive index of the film is $\bar{n}_2 = 1.6 - 0.5$ i .

Fig. 2 shows an example of the impedance measurements of the nickel electrode passivated at E = 0.94 V in neutral borate solution, where the amplitude and phase angle of impedance are plotted against the frequency (Bode plot). The simulated curve is drawn from a calculation of the least-square method by assuming the equivalent circuit shown in Fig. 2. Such impedance spectra were measured at various constant potentials, and the results are given in Fig. 3, where the values of C_1, R_1, C_h, R_h and i are plotted as a function of potential. The dependence of these values on the potential is so complex that its quantitative analysis can not be easily made.

DISCUSSION

From the refractive index of the film which increases with the anodic potential (Fig. 1), it is evident that the composition of the passive oxide film on nickel varies with the potential at which the film has been formed. According to the Lorenz-Lorenz law for the refractive index, an increased refractive index will result from an increase in the density of solids. It follows therefore that the film becomes denser at more anodic potentials in the passive region from 0.3 V to 1.1 V .

The increase of the extinction index k_2 with anodic potential (Fig. 1) implies that the electronic conductivity of the film increases with the poten-

REFERENCES

1 J.O'M.Bockris, A.K.Reddy and B.Rao, J. Electrochem. Soc., 113, 1133(1966)
2 N.Sato and K.Kudo, Electrochem. Acta, 19, 461(1974)
3 J.L.Ord, J.C.Clayton, and D.J.DeSmet, J. Electrochem. Soc., 124, 1714(1977)
4 T.Ohtsuka and K.F.Heusler, J. Electroanal. Chem., 102, 175(1979)
5 T.Ohtsuka and K.E.Heusler, J. Electroanal. Chem., 100, 319(1979):
   K.E.Heusler and T.Ohtsuka, Surface Sci., 101, 194(1980)
6 K.J.Vetter, J. Electrochem. Soc., 110, 597(1963)
7 C.Y.Chao, L.F.Lin and D.D.Macdonald, J. Electrochem. Soc., 129, 1874(1982)

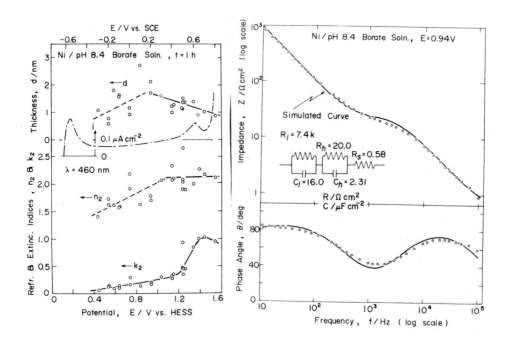

Fig. 1. Complex refractive index ($\bar{n}_2 = n_2 - ik_2$) and thickness (d) of the passive film on nickel.

Fig. 2. Amplitude and phase angle of impedance of a passivated nickel electrode as a function of frequency.

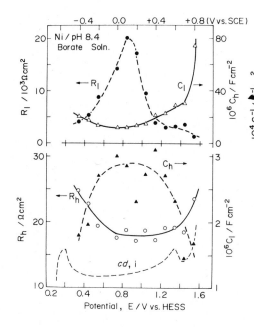

Fig. 3. Values of C_l, R_l, C_h, and R_h simulated from the equivalent circuit as a function of potential.

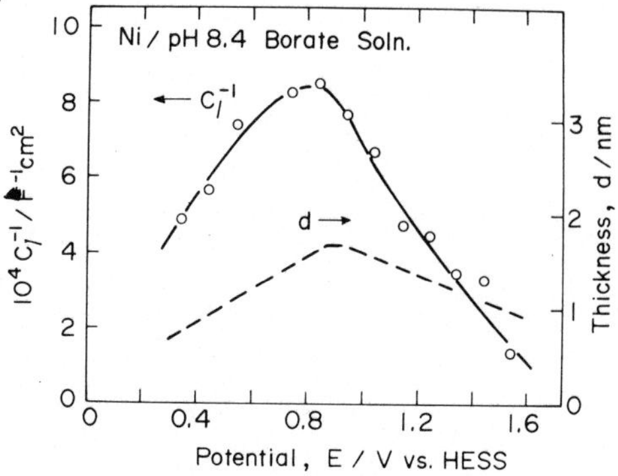

Fig. 4. Comparison between inverse capacitance, C_l^{-1}, and film thickness, d, as a function of potential.

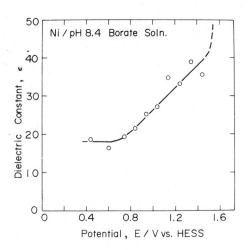

Fig. 5. Dielectric constant of passive film on nickel as a function of potential. It was estimated from values of C_l and d.

Passivity of Metals and Semiconductors, edited by M. Froment
Elsevier Science Publishers B.V., Amsterdam — Printed in The Netherlands

ELECTROREFLECTANCE FOR THE STUDY OF THE PASSIVITY OF IRON

M. FROELICHER, A. HUGOT-LE GOFF, V. JOVANCICEVIC
Groupe de Recherche n° 4 du C.N.R.S. "Physique des Liquides et Electrochimie",
associé à l'Université Pierre et Marie Curie,
4 place Jussieu, 75230 Paris Cedex 05, France.

ABSTRACT
Electroreflectance has been applied to the study of the passivity of iron in
buffered borate or sulfuric acid solutions, compared with the behaviour of an
α-Fe_2O_3 single-crystal. One can see that two different domains appear in the
passive range, in which the modulation of the potential plays very different
roles.

Electroreflectance (ER) is an optical technique with a very high sensitivity,
able to give informations about adsorbates on a metal electrode (1), as well as
about solid films when they have the crystallographic properties of a bulk modification (2). But, the interpretation of the results is not easy because it
requires a theoretical knowledge of the relation between the field modulation
and the optical properties modulation which remains a rather hard problem even
in the simplest case of a bulk semiconductor. In the case of a thin oxide film
for which the mechanism of conduction is not so evident, particularly in the
case of iron oxide since the density of carriers as estimated from Mott-Schottky
plots is too high for a semiconductor model, one has also to take in account
the problem of the penetration of the electric field in the system metal/oxide/
electrolyte, and one is in front of a very complicated system. It is not surprising that different mechanisms of modulation are able to occur.

Here, we will try to explain the differences in the ER spectra obtained in
the range of passivity of iron, with references to the results obtained on a
doped hematite crystal.

EXPERIMENTAL
The experimental equipment has been described elsewhere (1,2). The modulation
potential added to the dc polarisation potential of the sample has a frequency of 30 Hz, and an amplitude equal to 100 mV (hematite) or 50 mV (passive
films). Here, the in-phase part of the reflectance modulated component is plotted. Generally the out-of-phase part is very little.

RESULTS

Hematite single-crystal

The single-crystal has been polarised at 0 V/SSE in a 0.1N H_2SO_4. Unfortunately, we do not dispose of a well doped crystal, large enough to allow the optical measurements. To ensure the electrical conductivity, we have doped the crystal used here with carbon put on the external side, but the optical results prove that this preparation is not convenient and does not create an homogeneous doping but a surface reduced layer. Figure 1 shows the ER spectra obtained at various incidence angles. At an incidence angle of 75 degrees, the spectrum can be described by a broad peak centered at 3.1 eV , and shoulders at 2.6 and 3.9 eV. At 45 degrees, a change of the sign of the spectrum at 3.2 eV is observed. At intermediate angles, the optical responses are intermediate between the two shapes. One can explain these results by the presence on the outer side of the crystal of a perturbated layer having about 200 Å of thickness. At 45 degrees, 75 % of the incident light is able to reach the unperturbated underlying Fe_2O_3, and at 75 degrees, only 30 %. The spectrum obtained at 45 degrees represents then the actual ER spectrum of hematite, and the passage by zero at 3.2 eV agrees well with the maximum of optical absorption, related to the first charge transfer transition (O 2p → 3 t_{2g}) given in the literature at 3.31 eV (3) or 3-3.1 (4) . The different shoulders in the spectrum agrees satisfactorily with the theoretical transitions computed in 3 at 2.53 and 2.95 eV (crystal-field transitions), and at 3.94 and 4.82 eV (charge transfer transitions).

The other type of spectrum, obtained at high incidence angles, passes by a maximum at the energy of this first intense charge transfer transition and presents also a smaller maximum at the energy of the second O 2p → 3 t_{2g} transition. It is clear that in this case one is in presence of an other mode of modulation, not leading to a first derivative effect, and that we have to relate to the existence of disordered Fe^{2+} ions in the lattice.

Passive films

On figures 2 and 3 are seen the ER spectra obtained at an incidence angle of 45 degrees on iron passivated in 0.1 N sulfuric acid (figure 2) and buffered borate (figure 3), at different potential in the passive range ; the same spectra have been found in 1M Na_2SO_4 solution (5). On figure 2, the dotted curve has been obtained in the active region . In this case, we have found an out-of-phase part of the optical response, with a amplitude equal to the inphase part (i.e., a phase shift of 45 degrees), which means that one has to do with an adsorbate and not a solid layer. In the passive region, one finds again the two types of spectra that we have described in the case of the single crystal. The spectra obtained at the end of the range shows a first structure between 3 and

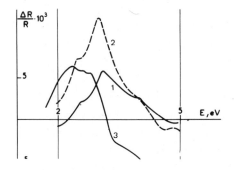

Fig. 1. ER spectra of α-Fe$_2$O$_3$
studied at different
incidence angles :
1 : 75 degrees
2 : 60 degrees
3 : 45 degrees

0.1N H$_2$SO$_4$; V = 0.0 V/SSE ;
ΔV = 100 mV, 30 Hz ; p. Pol.

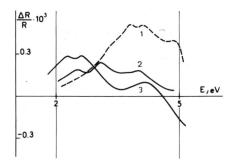

Fig. 2. ER spectra of passivated
iron.
1 : V = - 1.0 V/SSE
2 : V = - 0.5 V/SSE
3 : V = + 0.5 V/SSE

Buffered borate solution ;
ΔV = 50 mV, 30 Hz ;
45 degrees, p. Pol.

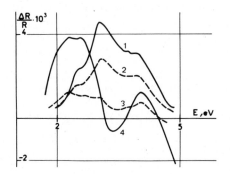

Fig. 3. ER spectra of passive film
on iron.
1 : V = - 0.0 V/SSE
2 : V = + 0.2 V/SSE
3 : V = + 0.7 V/SSE
4 : V = + 1.0 V/SSE

0.1N H$_2$SO$_4$
ΔV = 50 mV, 30 Hz ;
45 degrees, p. Pol.

4 eV, sharper in the case of sulfuric medium with a first passage by zero at 3.2 eV, and a final change of the sign of spectrum at 4.65 eV, value which corresponds to the maximum of absorption of the wustite, i.e., to a transition between Fe^{2+} ions and oxygen. Between 2 and 3 eV, the shoulders due to the crystal field transitions indicate the presence of Fe^{3+} ions. One is then in presence of a crystallographically defined species, containing Fe^{2+} and Fe^{3+} ions, and we assign to Fe_3O_4 such spectra. Elsewhere, this spectrum corresponds satisfactorily to the optical properties of magnetite, as described in (6). Then, we assume that Fe_3O_4 is formed in the second part of the passive range, which corroborates the results obtained by SPEC (7). The spectrum obtained at the end of the passive range in the sulfuric solution could correspond to a mixture of Fe_2O_3 and Fe_3O_4 since this mixture has been found in this case by Raman spectroscopy (8).

In the first part of the passive region, one finds again the shapes that we have assigned to a reduced, disordered layer. In this region, SPEC (7) indicates the presence of only a ferrous species. In this case, the field modulation acts differently than in a "bulk" form. A mechanism similar to the mechanism proposed in (9), that is, a modulation in the concentration of one of the ionic carriers, could take place. It seems that the peak at 3.2 eV be proportional to the concentration of Fe^{2+}, since it increases when the potential decreases, but the reason why remains unclear, and requires a theoretical support more elaborated than in (5) or in (9).

CONCLUSION

ER is an highly sensitive method, to the study of thin films on electrodes, since it allows to see easily a monolayer but a quantitative "chemical" interpretation of the results remains questionable. We have clearly established the difference of nature of the films at the beginning and at the end of the passive region. In the second part, a solid phase which can be Fe_3O_4 exists, but in the first part where the film can have "polymeric" structure, ER is not sufficient to advance in the identification of the film.

REFERENCES

1 M. Froelicher, A. Hugot-Le Goff, V. Jovancicevic, Thin Solid Films, 82 (1981) 81.
2 G. Blondeau, M. Froelicher, A. Hugot-Le Goff, V. Jovancicevic, Surf. Sci., 80 (1979) 151.
3 L. Marusak, R. Messier, W. White, J. Phys. Chem. Solids, 41 (1980) 981.
4 P. Bailey, J. Appl. Phys., 31 (1960) Suppl. 39s - 40s.
5 N. Hara, K. Sugimoto, J. Electrochem. Soc., 126 (1979) 1328.
6 A. Schlegel, S. Alvarado, P. Wachter, J. Phys. C, 12 (1979) 1157.
7 M. Froelicher, A. Hugot-Le Goff, V. Jovancicevic, This Journal.
8 R. Dupeyrat, M. Froelicher, A. Hugot-Le Goff, M. Masson, This Journal.
9 D. Wheeler, B. Cahan, C. Chen, E. Yeager, "Passivity of metals", R. Frankenthal & J.Kruger eds, The Electrochem. Society, Princeton,NJ (1978).

Passivity of Metals and Semiconductors, edited by M. Froment   89
Elsevier Science Publishers B.V., Amsterdam — Printed in The Netherlands

ELECTROCHEMICAL AND ELLIPSOMETRIC INVESTIGATION OF PASSIVE FILMS FORMED ON
IRON IN BORATE SOLUTION

Z. SZKLARSKA-SMIALOWSKA[*] and W. KOZLOWSKI
Institute of Physical Chemistry, Polish Academy of Sciences Warsaw, Poland

ABSTRACT
   The film growth on iron has been found to occur in four distinct stages.
The electric field strength across the film is not constant but decreases with
progressing film growth. At certain characteristic thicknesses of the passive
film associated with successive changes in the rate of film formation,
inflections on the reduction curves potential vs time and potential vs film
thickness are observed.

INTRODUCTION
   In the majority of previous ellipsometric studies of passive film formation
in iron, the measurements began approximately one minute after a constant
anodic potential was applied, i.e. at a time subsequent to the development of
a major part of the film. As far as we are aware, only in two studies
(refs.1,2) have attempts been made to determine the rate of film growth during
the intial minute folowing the application of the potential. Kruger and
Calvert (ref.1) have found that film growth occurs in three stages. In the
first stage, a linear relationship exists between the film thickness and the
square root of time. During the second stage, a number of complex processes
occur which cannot be characterized by any simple rate law. The third stage,
i.e. that of the steady-state growth, can be described equally well by either
a logarithimic or inverse logarithimic law.
   Anodic films grown on iron in borate solutions have been studied by a
number of investigators using the cathodic reduction method. If a low
constant current density is applied, two plateaux appear in the graph
of consumed charge as a functon of potential. The first plateau is commonly
attributed to the reduction of the external film layer, and the second plateau
to the reduction of the internal layer of Fe_3O_4 (refs.3,4). According to
others (refs.5,6) the anodic film contains a single layer of γ-Fe_2O_3 which is
reduced to Fe^{2+} and Fe^0. It has also been assumed that the reductive

---
[*] Z. Szklarska-Smialowska is at present Professor of Metallurgical Engineering,
The Ohio State University, Columbus, Ohio, U.S.A.

dissolution of the film is accompanied by other reduction processes occuring
in the solid phase. For example, such mechanisms have been proposed by Sato
et al. (ref.5) and Chen and Cahan (ref.7) for galvanostatic and potentiostatic
reduction respectively.

For calculations of film thickness based on the cathodic reduction current,
arbitrary values are taken for the molecular weight of the compound undergoing
reduction, density of this compound, roughness coefficient, and number of
electrons used. Therefore, the results of these calculations are uncertain
and must be treated with caution.

The objective of the present work was to characterize the kinetics of film
growth on Armco iron in borate buffer solution. Particular emphasis has been
placed on correlating the results obtained using the electrochemical and
ellipsometric techniques, so that a consistent picture can be developed of the
evolution of the passive film and its reduction as a function of time. The
results obtained by anodic oxidation of iron were compared with those of the
cathodic reduction of surface film.

EXPERIMENTAL PROCEDURE

Armco iron specimens utilized in this study were carefully polished,
finally with 1 μm diamond paste, and then electropolished in a mixture of
glacial acetic and perchloric acids.

Borate buffer solution of pH 8.45 was used: 0.02 M H_3BO_3 + 0.005 M $Na_2B_4O_7$.
The solution was deaerated by bubbling argon (99.99%) for at least 24 hr
before the experiment. All measurements were made at 22°C.

The ellipsometer used was a Rudolf Research Model 2000 equipped with a
revolving analyzer. The light source was a tungsten iodine lamp from which
monochromatic light of 546.1 nm wavelength was obtained using an interference
filter.

Optical parameters were used for films formed at constant potential at 100
mV intervals over the range -0.5 V_{SCE} to 0.8 V_{SCE}. The kinetics of film
growth were studied for periods extending from about 0.2 s to 60 min after
application of the potential during which time changes in the optical
parameters and current densities were recorded.

The cathodic reduction was commenced in the same electrolyte solution
immediately after the passivation experiment was completed. A constant
current of 17 μA was used. During reduction, the potential, cathodic charge,
and changes in optical constants were recorded. This procedure was chosed
because during the time required to change the electrolyte, taking into
consideration that the cell had a volume of 0.5 liter, the exposure of the

electrode to the argon atmosphere could result in modification of the surface film. However the procedure employed had also some potential disadvantages, namely that if ferric ions passed into solution during oxidation, they could be reduced to ferrous, which would lead to an excessive consumption of cathodic charge. Calculations showed that this factor would rise to a negligible error in the cathodic charge consumed during film dissolution.

RESULTS and DISCUSSION

The optical parameters of films obtained after passivation of iron for 60 min are given in Fig. 1. As shown, n and K (refractive and adsorption coefficients) change with the passivation potential. In the first region, which occurs at more negative potentials, the composition and properties of the film as, characterized by its optical parameters, change more significantly with passivation potential. Also, the refractive indeces of the film are relatively low suggesting the formation of hydroxides or oxyhydroxides. In the second region, at more anodic potentials, the effect of potential is less pronounced, and the optical parameters suggest that the film is composed of oxide and hydroxides.

An example of the relationship between film thickness d as determined ellipsometrically and log time t is plotted for different potentials in Figure 2, where the four consecutive steps of film formation (marked I-IV at 0.8 V) can be distinguished. From these data it is apparent that at a given anodic potential, 80-90% of the film thickness attained in one hour's passivation is formed within the initial 10 to 30 s. The film formed in stage I is very thin, probably a monolayer or less. The film formed in stage II thickens with increasing anodic potential, but the effect of potential is insignificant in both stages III and IV.

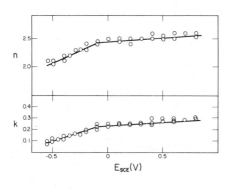

Figure 1.

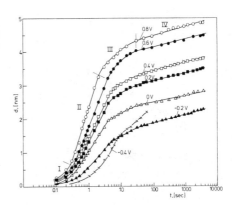

Figure 2.

In certain parts of stages II, III and IV, the kinetics of film growth have been found to obey inverse logarithmic law, but the calculated jump distances are 10^{-2} Å, 0.1 Å, and ∿9.4 Å respectively, instead of the predicted value of ∿3 Å.

Figure 3 shows the effect of film thickness at the end of each consecutive passivation stage. Based on the film thickness as a function of potential data shown in Figure 3, the field strength of the passive film in Stages I, I+II, I+II+III, and I+II+III+IV was $1.8 \times 10^7 \text{Vcm}^{-1}$, $4.8 \times 10^6 \text{Vcm}^{-1}$, and $3.9 \times 10^6 \text{Vcm}^{-1}$ respectively.

The totality of results obtained indicates that the process of passive film growth on iron cannot be described satisfactorily by any of the existing models.

Assuming that the film growth kinetics is controlled by a diffusion process with no ion migration, the apparent diffusion coefficients that are effictive during the various oxidation stages observed up to 60 minutes were calculated, from the equation $d = \sqrt{6Dt}$, where D is the apparent diffusion coefficient, d = the film thickness, and t = oxidation time. Calculations show that the apparent diffusivities are dependent on the film thickness and in Stage I+II are four orders of magnitude larger, and in Stage I+II+III are three orders of magnitude larger than d values computed for the whole passivation time (Fig. 4). It can be concluded, therefore, that the outermost film layer formed in Stage IV is primarily responsible for the sharp decline of the film growth rate and for the increased passivity of iron.

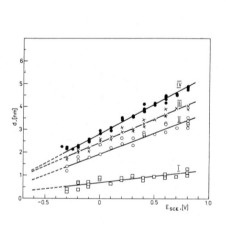

Figure 3.

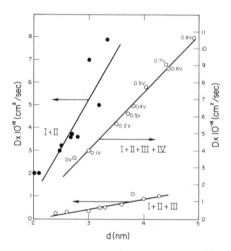

Figure 4.

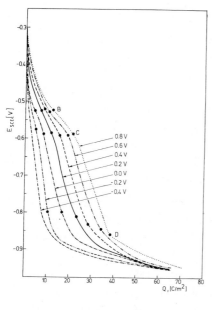

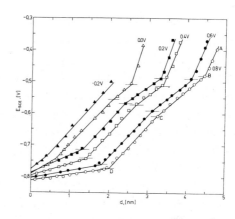

Figure 5.   Figure 6.

During the cathodic reduction of the film also four distinct stages are
observed (Figs. 5 and 6). The A-B step showing reduction of the outermost
film layer corresponds to the oxidation Stage IV, B-C corresponds to StageIII,
C-D to Stage II. At film thicknesses corresponding to the characteristic
points B and C (Fig. 6), the reduction potentials are independent of both
oxidation potential and time. Because at the characteristic points B and C
the film thicknesses obtained from measurements of the changes of d vs t
during oxidation and those of d vs E during cathodic reduction are virtually
the same, it can be deduced that during the oxidation Stages II, III, and IV,
at a given anodic potential, no substantial changes in film composition occur
irrespective of the oxidation and reduction times.

Using ellipsometric and electrochemical data, anodic charges for film
reduction, have been calculated. By comparing the theoretical values of $\Delta Q/\Delta d$
during passivation with experimental data, it can be assumed that within the
first potential region, between -0.5 and 0.0V, the film being formed contains
iron in the oxidation stages +2 and +3. The charges per unit volume
correspond approximately to those for $Fe(OH)_3$ and $Fe(OH)_2$. The increase of
$\Delta Q/\Delta d$ with increasing potential in Stages I+II can be attributed to a larger
Fe^{3+}/Fe^{2+} ratio and to a reduced content of water in the film. For both
Stages III and IV, the $\Delta Q/\Delta d$ values greatly exceed those theoretically
predicted for any oxide and hydroxide of iron. One possible reason for this
discrepancy might be the omission of the surface roughness coeficient.

94

Another reason may be associated with passive film dissolution in addition to film formation. (It was found by chemical analysis that growth of anodic films on iron is accompanied by the transmission of iron ions to the electrolyte). The values of $\Delta Q/\Delta d$ obtained during cathodic reduction are also higher than those theoretically predicted. The longer the passivation time of iron at a given potential, and the higher the anodic potential for the given passivation time, the higher is the charge consumed in film reduction. (Since it is possible that, at potentials more negative than those corresponding to point D, Fe^{3+} could be reduced to Fe^0, and also that a fraction of the cathodic charge could be consumed in reduction of hydrogen ions, only steps A-B and B-D, during which the above two reactions could not occur, were taken into consideration.) Among different possible explanations of these results, the most reasonable seems to the assumption of the occurence in films of iron at a valency higher than 3+.

The occurrence of highly oxidized iron compound in passive films was presumed by several authors (refs.2,3,8). According to Nagayama and Cohen (ref. 3) the internal (i.e., iron-adherent) layer of the film is composed of Fe_3O_4, while the external layer is composed of Fe_2O_3, but containing iron oxidized to Fe^{6+} on the surface. Konno and Nagayama (ref.8) claim that in the outermost film layer the average oxidation degree significantly exceeds that of Fe^{3+}. Using XPS method, they have found an average valence of iron in the film equal to 3.59.

Taking into consideration the occurrence of higher oxidized iron (above Fe^{3+}) in the external part of film, average oxidation degrees (v) were calculated assuming different values of the roughness coefficient (r). For r=1, high unrealistic values were obtained. If r=1.5 was assumed, the computed v value was slightly higher than 3+ for the reduction step and slightly lower than 3+ for the B-D step. At longer oxidation times, and higher anodic potentials, higher v values were obtained. The v values obtained for r=1.5 seem probable, and are lower than those found by Konno and Nagayama (ref. 8).

REFERENCES

1 J. Kruger and J.P. Calvert, J. Electrochem. Soc., 114, 43 (1967).
2 C.T. Chen, B.D. Cahan and E. Yeager, Technical Report No. 48, Case Western Reserve University, Cleveland, Ohio (1979).
3 M. Nagayama and M. Cohen, J. Electrochem. Soc., 109, 781 (1962).
4 T. Tsuru and S. Haruyama, Corr. Sci., 16, 623 (1976).
5 N. Sato, K. Kudo and T. Noda, Corr. Sci., 10, 785 (1970).
6 K. Ogura and T. Ohama, Corrosion, 37, 569 (1981).
7 C.T. Chen and B.D. Cahan, J. Electrochem. Soc., 129, 17 (1982).
8 H. Konno and M. Nagayama, Passivity of Metals, Ed. R.P. Frankenthal and J. Kruger, The Electrochem. Soc., Princeton, N.Y., p. 585, 1978.

Passivity of Metals and Semiconductors, edited by M. Froment
Elsevier Science Publishers B.V., Amsterdam — Printed in The Netherlands

AN ELLIPSOMETRIC STUDY OF THE PASSIVITY OF IRON IN NEUTRAL BORATE ELECTROLYTE

JACK L. ORD

Department of Physics, University of Waterloo, Waterloo, Ontario, Canada N2L 3G1

ABSTRACT

Measurements in deaerated neutral borate electrolyte which is free of ferrous ions support the duplex film model in which the outer layer gives the electrode its passive properties, and the current is limited by the field in the inner layer across which the overpotential appears. Different results are obtained in neutral electroytes which contain ferrous ions or are saturated with air, but there is no sign in neutral electrolyte of the layer-conversion reactions which take place in strong basic electrolytes.

INTRODUCTION

At the Fourth Passivity Symposium (ref.1) we looked at why the ellipsometric literature on the passivity of iron presents such a contradictory picture of the phenomenon. We concluded that the choice of experimental conditions can determine the nature and thickness of the layer which forms on the passive electrode, particularly in the popular neutral borate electrolyte where active-state reaction products can play an important role.

In this paper we look first at the buildup and removal of the passive layer which forms on iron in a deaerated borate electrolyte free of ferrous ions, and use an optical technique to test the passivity of the electrode as the layer is removed. We compare the results of galvanostatic experiments with results of experiments which include extended potentiostatic intervals. We then show how the experimental results are altered if the experiment is performed without removing the ferrous ions produced on passivation prior to raising the potential above the FeOOH deposition potential. We show also the results which are obtained when an iron electrode is cycled in air-saturated borate electrolyte in which ferrous ions are not stable.

Although we interpret our results in terms of a duplex film whose component layers undergo simultaneous growth during passive-state oxidation and sequential removal during cathodic reduction, there is a substantial body of opinion which argues that the film is a single layer whose oxidation state rises progressively as it grows, and lowers during the first stage of cathodic reduction. We address this question here by showing results obtained in sodium hydroxide electrolyte where such processes do occur, and comparing them with the results obtained in dearated borate electrolyte free of ferrous ions.

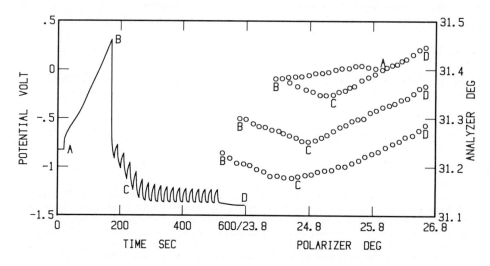

Fig.1. Galvanostatic oxidation-reduction cycle for iron in deaerated borate electrolyte free of ferrous ions. The optical data are at the upper right, and optical reduction data from cycles with potentiostatic periods of 1 hour and 24 hours at point B are at the middle and lower right respectively. The cycling during the reduction tests the passivity of the electrode.

RESULTS AND DISCUSSION

The results of experiments performed in deaerated sodium borate/boric acid electrolyte are shown in the first two figures. The first figure shows the results of experiments in which the iron electrode is passivated at 1 milliamp/cm^2 to a potential of -0.825 volts (relative to a mercury/mercurous sulfate reference electrode) which is below the FeOOH deposition potential. The electrode is held at this potential (with brief open-circuit periods during two changes of electrolyte) until the current density falls below 2 microamp/cm^2 at which point (point A in the figure) a 20 microamp/cm^2 oxidation current is applied. The current is reversed when the potential reaches .3 volts at point B, and at 10-second intervals during the reduction a 2 microamp/cm^2 anodic current is applied for 10-second intervals to test the passivity of the electrode. The locus traced out by the null settings of the ellipsometer's polarizer and analyzer prisms are plotted at the upper right in Fig.1 with the analyzer scale magnified 7.5 times over the polarizer scale. On a plot of this type, buildup or removal of a single-phase film produces a straight line through the bare surface point (point D), and sequential removal of the component layers of a two-phase film produces the B-C-D two-line plot traced out by the data in Fig.1. Passive-state oxidation causes proportionate growth of both component layers, but the growth curve does not extrapolate through D because of the inner layer growth which occured in the active state before

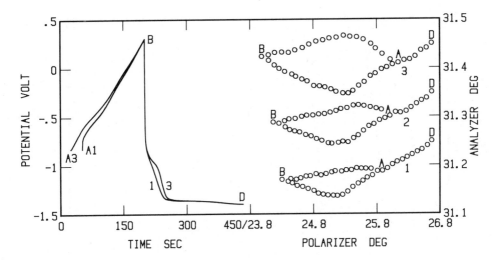

Fig.2. Galvanostatic oxidation-reduction cycles for iron in deaerated borate electrolyte with (1) and without (2,3) electrolyte change after passivation. 1 and 2 (potential-time curve not shown) have a potentiostatic interval after passivation, 3 does not.

formation of outer layer. The optical oxidation data exhibit some structure in the region of point A which may mark the completion of a monolayer of the outer film. (The reduction curve from point A shows that the electrode is passive with only a fraction of a monolayer of the outer film present on the surface). The optical data do not identify the components of the passive film, but they do show that at point B neither the passive film nor its outer layer are composed of FeOOH, a material whose refractive index has been determined by deposition experiments (ref.2). The duplex film has been identified by other techniques as consisting of an outer layer of Fe_2O_3 on top of an inner layer of Fe_3O_4 (ref.3).

The curves offset vertically at the middle and lower right of Fig.1 show how the optical reduction data are affected if the electrode is held at .3 volts for periods of 1 hour (middle curve) and 24 hours (lower curve) before reduction. The two-line structure is still clearly defined after 1 hour, and the layers have thickened by the expected amount, but after 24 hours the reduction curve has an intermediate linear region which washes out the structure and makes any decomposition into linear regions of doubtful significance. It is not at all surprising that electrochemical measurements on electrodes which have been allowed to 'stablize' for 24 hours after passivation bear little resemblance to measurements made during galvanostatic oxidation across the passive region.

The anodic segments along the reduction curve allow the passivity of the

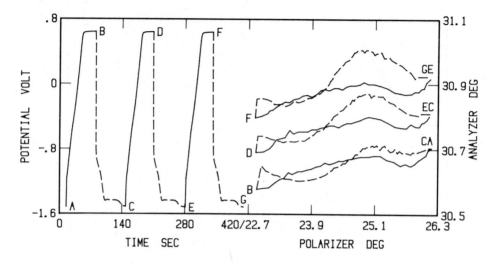

Fig.3. Galvanostatic oxidation-reduction cycles for iron in air-saturated borate electrolyte.

electrode to be tested optically. We define the initially passive film-covered electrode to be active once the application of a small anodic current begins to cause film dissolution, and use dP/dt, the time rate of change of the polarizer null setting, to detect film dissolution. Least squares analysis of the first 6 anodic segments shows no film dissolution for the first three segments, and an approximately constant rate of film dissolution (dP/dt = 3 millidegrees/sec) for the next four segments beginning at C, the point at which the removal of outer layer is complete. This result confirms our assertion that it is the outer layer which renders the electrode passive.

Fig.2 shows how ferrous ions affect film growth in the neutral borate electrolyte. Experiment 1 is the same experiment as in Fig. 1 without the anodic segments on the reduction curve. Experiment 2 uses the same potentiostatic period after passivation as 1, but the electrolyte is not changed, and experiment 3 ends the potentiostatic period after passivation at 20 rather than 2 microamp/cm^2, again with no change of electrolyte. For clarity, the V-t curve for experiment 2 which lies between the curves 1 and 3 is not plotted, and the optical data are offset vertically to avoid overlap. FeOOH deposition causes a rapid upward shift in the optical oxidation curve once the deposition potential is crossed, and results in a thicker outer layer which has a more sharply defined reduction plateau. As the results of experiment 2 indicate, effects due to ferrous ions can be reduced by allowing time for the ions to diffuse into the bulk of the electrolyte before raising the potential above the deposition potential.

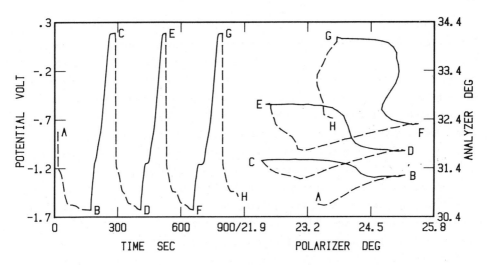

Fig.4. Galvanostatic oxidation-reduction cycles for iron in air-saturated sodium hydroxide electrolyte.

Fig.3 shows the results of cycling an iron electrode in air-saturated borate electrolyte at 180 microamp/cm^2. The reduction cycle removes the film completely from the surface if the current density is high enough to dominate spontaneous oxidation, and although there is some change in the structure of the optical data (which is again offset to avoid overlap), the cycling does not produce the progressive increase in film thickness which occurs in a basic electrolyte.

If we choose to regard effects due to extended passivation periods, FeOOH deposition, and oxidation by dissolved oxygen as phenomena which may be interesting in their own right, but play no essential role in the passivity of iron in neutral borate electrolyte, we are left with the A-B-C-D optical data pattern in Fig.1 as the cornerstone for models of the passive film. Although we have interpreted this result as supporting a duplex film model, and economy of parameters arguments support this interpretation, other interpretations are possible. For example, a model based on a single homogeneous surface film can account for the optical results if the index of the film changes appropriately during the experiment. If one postulates a value of 1.7, the refractive index of FeOOH, for the real component of the film index, analysis of the optical data will yield graphs showing how the thickness and absorption coefficient of the film vary during the experiment. We have used such models ourselves when studying the conversion processes associated with electrochromism in iridium oxide films (ref.4), but it is difficult to defend the validity of the results unless they are shown to be physically reasonable on the basis of independent measurements of some kind. It is not difficult, however, to see what sort of optical results are obtained when iron oxide does undergo conversion reactions.

Fig.4 shows the results obtained when an iron electrode, oxidized to point B in Fig.1 in deaerated borate electrolyte, is placed in a cell containing air-saturated 1 molar sodium hydroxide. A 75 microamp/cm^2 current is applied, and the electrode is first reduced, then alternately oxidized and reduced through three cycles, with the third cycle terminating at the end of the first reduction plateau. Note that Fig.4 does not use the highly expanded analyzer scale used in the first three figures. The endpoints of the reduction cycles, points B, D, and F do not superimpose at the bare surface point. Instead, they trace out a locus which shows that cycling causes the progressive growth of a film on the surface of the reduced electrode. Points C, E, and G trace out the growth curve for the oxidized form of the film. The individual cycles each involve both film growth and conversion processes. On the first cycle the conversion plateaus are not clearly defined, but with the progressive growth of the surface film they become more distinct on both the potential-time and optical data plots. Before optical data as complicated as the data in Fig.4 can be analyzed, it is necessary to perform a series of experiments with cycles designed to study the individual regions of the curves in Fig.4. These experiments form part of our on-going research on electrochromism and battery processes, but are not within the scope of the present paper. For our purposes here it is sufficient to note that the optical 'signature' of film-conversion processes, both as it appears in Fig.4 and in figures showing the time rate of change of the optical data, is quite different from anything that is seen in neutral borate electrolyte. Although this does not rule out absolutely the possibiltiy that the film is first reduced to a lower oxidation state in region B-C of Fig. 1, then removed from the surface in region C-D, we regard the accumulated evidence in support of a duplex film model as conclusive, and regard a single-layer model as only a remote possibility.

ACKNOWLEDGEMENT

This work received partial support from the Natural Sciences and Engineering Research Council Canada under Grant No. A-1151.

REFERENCES
1. J.L. Ord, in R.P. Frankenthal and J. Kruger (Eds.), Passivity of Metals, The Electrochemical Society, Princeton, 1978, pp. 273-284.
2. J.L. Ord and D.J. DeSmet, J. Electrochem. Soc., 118 (1971) 206-209.
3. M. Cohen and K. Hashimoto, ibid., 121 (1974) 42-45.
4. J.L. Ord, ibid., 129 (1982) 335-339.

Passivity of Metals and Semiconductors, edited by M. Froment
Elsevier Science Publishers B.V., Amsterdam — Printed in The Netherlands

RAMAN STUDIES OF PASSIVE FILMS ON IRON

M. FROELICHER, A. HUGOT-LE GOFF, C. PALLOTTA
Groupe de Recherche n° 4 du C.N.R.S. "Physique des Liquides et Electrochimie"

R. DUPEYRAT, M. MASSON
Département de Recherches Physiques, L.A. 71 du C.N.R.S.

Université Pierre et Marie Curie
4, place Jussieu, 75230 PARIS CEDEX 05, FRANCE.

ABSTRACT

Raman spectroscopy has been applied to the in-situ study of thin oxide films grown on iron in different electrolytic solutions . The resulting films were artificially thicker than the usual passive films, from oxidation-reduction cycles, or from a polarization maintained for a long time (2 days). Their thickness was about 100-200 Å and their spectra compared to those of thermally grown oxide films having about the same thickness.

All the classic optical techniques have failed to give precise chemical iden-tification of the thin layers grown during the passivity of iron because the knowledge of the various optical constants of the oxides is not sufficient to discriminate between the numerous possible compounds which could be formed. In spite of the very little amount of material present (the thickness of the passive film being 30-50 Å), Raman spectroscopy seems promising, since one is now able to sum repeated scans of the spectrum.

The results given here, show that it is possible to discriminate between different oxides and oxihydroxides by a comparison with spectra obtained from thermally grown oxide films, but the feasibility was obtained on in-situ films thicker than the passive films (about in the ratio of 5 to 1).

EXPERIMENTAL

Following the method outlined by one of us (C.P.) at the University of La Plata, we produced films of increased oxide thickness by subjecting an iron electrode to very fast oxidation-reduction (O.R.) cycles in Na and K phosphates solution (pH 7), in buffer borate solution (pH 8.4), in basic solution (KOH 1M), and in acidic solution (H_2SO_4 0.5M). The potential varied between that where a noticeable reduction current appeared and 0.4 or 0.6 V/SSE, the exact point being a function of the electrolyte. The sweeping rate was 36 V/sec. In another experiment we maintained the iron electrode potentiostatically at a potential of 0.6 V/SSE for 60 hours. Under both these conditions we obtained oxide layers having thicknesses of about 100-200 Å. These thicknesses are estimated from the interferences appearing in the reflectance of oxidized iron.

Spectra were then recorded with a Dilor RT 30 spectrometer, with an interface specially designed for an APPLE 2 computer, which allows among other things for the accumulation of spectra, and the substraction of background (to substract for instance the spectrum of the electrolytic solution, in spite of the fact that, as we will see later, it is generally not disturbing). The excitation wavelength was 4880 A, and the available power at this laser wavelength was 400 mW. The sample was placed flat in an electrolytic cell, below the lens of the microscope and at the entry of the Raman spectrometer, so that the Raman light had to pass through about 1 mm of electrolyte.

We have used some thermally grown iron oxide films as reference samples, where we have shown (1) that at an oxidation temperature of above 300°C, one obtains hematite $\alpha\text{-Fe}_2\text{O}_3$, and when below 200°C, one obtains magnetite Fe_3O_4. A range exists between these two temperatures, where a mixture of the two modifications is obtained.

Table I gives the assignments of the various Raman peaks.

RESULTS

Thermal oxides

First we will for reference show the Raman spectrum of a mixture of magnetite and hematite, under conditions where an oxihydroxide cannot appear. The results are plotted in figure 1a and b. Now in a pure single crystal of hematite, the peak at 412 cm^{-1} is known to be greater that the peak at 299 cm^{-1}, but this intensity ratio is not found in the thermally grown films : in certain cases, the magnitude of peaks can even inverse. Further the 550 cm^{-1} peak of magnetite tends to disappear, in spite of the stability of the peak at 676 cm^{-1}. We have not been able to obtain single crystals of $\alpha\text{-FeOOH}$, but, if one trusts the results of literature (2), the Raman spectrum is not dramatically different from the spectra of the mixture of hematite and magnetite, apart the position of the main peak now at 385 cm^{-1} instead of 413 cm^{-1}. A careful assignment of this peak in the passive films will therefore be necessary.

Anodic films

Figure 2 shows the spectrum obtained on a sample after O.R. cycle in KOH 1M solution. Here the background due to the presence of the electrolytic solution has not been substracted. This presence does not create any noticeable disturbances above 300 cm^{-1} (excepted in the case of sulfuric acid solution, which displays, in the useful range, several very intense peaks which prevent absolutely any in-situ observation). One finds again the spectrum of water : a very large bump with a faint broad maximum at 450 cm^{-1}, on which is superimposed the spectrum of oxide. Further we know that there is a peak at about 400 cm^{-1} which allows one to discriminate between oxide and oxihydroxide. On the contrary, below 300 cm^{-1}, the very high background of the water obscures the oxide spectrum.

A problem would exist in the case of γ-FeOOH, which is characterized essentially by a peak at 252 cm^{-1} (3). Finally, in all the conditions where we apply O.R. cycles, one observes, more or less clearly, the peak at 676 cm^{-1} characteristical of Fe_3O_4.

Passive films

In figure 3 is shown the spectrum obtained ex-situ after oxidation (for the reasons given above) on an iron sample polarized for 60 hours in a H_2SO_4 0.5M solution. It indicates clearly a mixture of Fe_2O_3 and Fe_3O_4. In any case, it cannot be confused with α-FeOOH. The sample has a faint interference color, which proves that a main effect of ageing is to increase the thickness of the passive film. This is possible in a sulfuric acid solution because the passivating current (i.e., the residual transport of matter) is noticeable higher than for instance in a buffered borate solution.

DISCUSSION

We wish to present two conclusions from this study : 1) our experimental results, and 2) the future of the Raman spectroscopy for the study of passive films.

From our experimental results, we can say that we have observed neither oxihydroxides nor purely trivalent species. This is opposite to a large majority of the theories on iron passivation, but is in agreement with our results from photoelectrochemistry (3).

What is the actual meaning of these mixtures of magnetite and hematite ? Juxtaposition of patches of one and another modification ? A new spinel not found up to now in bulk form ? Or, which appears as the more reasonable assumption, growth of a "sandwich layer", Fe_2O_3 being formed by the reoxidation during ageing of the inner layer of Fe_3O_4 ?

Regarding the future of Raman spectroscopy in the study of the passivation, one can do the following remarks :

From the interference colors, one can estimate film thicknesses to be up to a maximum of 200 Å, the thickness of the film being artificially increased by the sulfuric acid solution. Further a single scan was sufficient to have a good identification of the film. In figure 3 where 4 scans were summed, the sharpness of the peaks is as good as that on a bulk material. If one has a "true" passive layer, with a lower thickness, by about a ratio of 5, it would still be

104

detectable by summing, for instance, 10 scans. Since in fact, our experiment was unsuccessful, and one could well ask if the passive layer is not too poorly crystallized to give a good Raman response. In any case, the role of ageing (and, probably, of O.R. cycle) could be to bring about a slow recrystallization of the film. This could account for the difference. Since sensitivity is not a problem further work will be directed towards the application of Raman spectroscopy to the study and measurement of thin passivating films.

REFERENCES

1 R. Dupeyrat, M. Froelicher, A. Hugot-Le Goff, V. Jovancicevic, M. Masson, in "Raman Spectroscopy", J. Lascombe, D.V. Huong (eds), J. Wiley (1982) 679. M. Froelicher, A. Hugot-Le Goff, Le Vide, Les Couches Minces, 38 (1983) 159.
2 J. Keiser, C. Brown, R. Heidersbach, J. Electrochem. Soc. 129 (1982) 2686.
3 M. Froelicher, A. Hugot-Le Goff, V. Jovancicevic, this volume.

TABLE 1

Fe_2O_3	Fe_3O_4	α-FeOOH	γ-FeOOH
(1)	(1)	(2)	(2)
293			
299	550	300	252
412	676	385	
613			

Only the large peaks are given. The largest is underlined.

(1) Theoretical values, Hart T., Temkinsl H., Adams S. "Light scattering in solids", Balkansky ed. (1978), p. 254 - 259.

(2) Experimental values, Ref.(2).

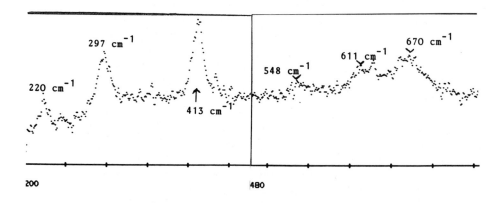

<u>FIGURE 1 a and b</u>

Thermal oxide grown at 300°C on iron. 4 scans. Mixture of
hematite and magnetite

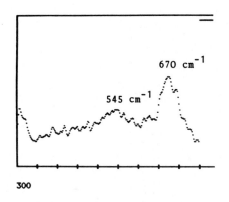

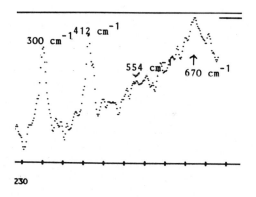

<u>FIGURE 2</u>

Iron studied in-situ, submitted to
ORCycles in KOH 1M, 4 scans, film
thickness ∿ 200 Å. Pure magnetite.
The background due to the solution
(maximum at 450 cm^{-1}) is not obvious
here.

<u>FIGURE 3</u>

Iron polarised in H_2SO_4 1M at
V = 0.6 V/SSE during 60 hours,
studied ex-situ, 4 scans, film
thickness ∿ 200 Å.
Mixture of hematite and magnetite.

Passivity of Metals and Semiconductors, edited by M. Froment
Elsevier Science Publishers B.V., Amsterdam — Printed in The Netherlands

RAMAN SPECTROSCOPY OF OXIDE LAYERS ON PURE IRON IN ELECTROCHEMICAL ENVIRONMENT

J. DÜNNWALD, R. LOSSY, and A. OTTO

Physikalisches Institut III, Universität Düsseldorf, D-4000 Düsseldorf 1,
Fed. Rep. of Germany

ABSTRACT

In situ Raman spectroscopy of the oxide films formed on pure iron in 1 M KOH electrolyte after several special oxidation reduction cycles shows a Raman spectrum similar to that of Fe_3O_4 when the sample appears metallic, and a spectrum similar to that of α-FeOOH when the sample appears "rusty". We tentatively assign the spectra of the first kind to a passivation layer.

INTRODUCTION

In the last years, Raman spectroscopy has become a tool to investigate the oxide films on iron and steel. Thibeau et al. (ref. 1) investigated the Raman spectra of Armco iron oxidized in air at 250 ^{O}C and identified the surface oxide film as Fe_3O_4. This was corroborated by Dupeyrat et al. (ref. 2). Farrow et al. (ref. 3) reported the Raman spectra from stainless steel with and without titanium additions, exposed to air at 980 ^{O}C for 1000 h. The vibrations of the surface layers were assigned to α-Cr_2O_3 and TiO_2. Corresponding in situ measurements (ref. 4 and 5) during oxidation between 300 and 850 ^{O}C showed mainly α-Fe_2O_3 and Fe_3O_4 below 300 ^{O}C, above 600 ^{O}C the predominant oxide was Cr_2O_3. Raman spectroscopy combined with ion bombardment allows to measure the depth profile of the oxide films (ref. 6).

Keiser et al. (ref. 7) studied by Raman spectroscopy the oxidation of a Fe_3O_4 paste on carbon steel and weathering steel to α- or γ-FeOOH under a variety of conditions similar to atmospheric (wet) exposures.

Here we report Raman spectroscopic investigations of the controlled electrochemical oxidation of pure iron in situ, based in part on ref. 8. Our aim was the detection of a Raman signal from the passivation layer. In all the work cited above, oxide layers were thicker than 150 Å.

EXPERIMENT

Polycrystalline iron of 99.99% purity was mechanically abraded with 16 μ carbide grains, polished with 9 and 3 μ Al_2O_3 lapping film and etched for more than 10 minutes in HNO_3. The sample was transferred into a specially designed

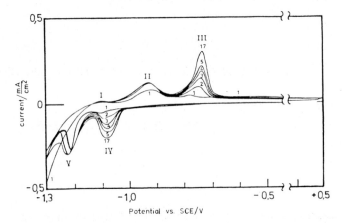

Fig. 1. Voltammograms of an iron electrode in 1 M KOH electrolyte, sweep rate 1 mV/sec. Potential is given versus SCE. Numbers refer to number of cycle.

electrochemical cell, which was purged by a constant flow of N_2 gas, and sealed against air. After insertion into the 1 M KOH electrolyte, the sample was reduced for about 10 minutes at -1.3 V versus a saturated calomel electrode (SCE). After this, the sample was repeatedly cycled from -1.3 to 0.5 V_{SCE} with 1 mV/sec. Examples of the corresponding voltammograms are shown in Fig. 1.

Raman spectra were recorded with ca. 150 mW laser light of 514.5 nm wavelength incident at about 45° degrees - the polarization could be choosen parallel or perpendicular to the plane of incidence. Raman scattered light was collected around the normal of the sample, polarized parallel to the plane of incidence.

Raman spectra in situ after one and five cycles did not show any sign of vibrational bands attributable to an oxidic film. In order to avoid the background of the translational and vibrational modes of water (ref. 9) which might mask the searched for bands, the electrolyte was removed from the sample under continuing electrostatic control at 0.3 V_{SCE} (by having an iron bar connected to the sample always submerged in the electrolyte) within the sealed cell. The emerged electrode was rinsed with destilled water and blown dry by a nitrogen stream, all within the sealed cell. In this way, the sample was always kept under potential control in oxygen free environment. This procedure yielded no results after one and five cycles, but good results after twelve additional cycles (alltogether 17 cycles).

Fig. 2 shows the in situ Raman spectra of the iron electrode, the electrolyte touching the sample surface or being removed, for the 2 different states of incident polarization.

The new bands at ca. 543 and 674 cm^{-1} are better defined for incident s-polarization. In this case, the Rayleigh wing extending out for 800 cm^{-1} is weaker

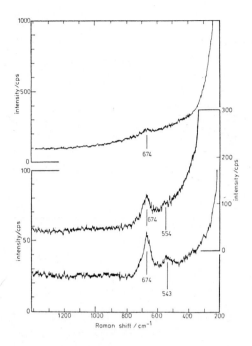

Fig. 2. In situ Raman spectra of an iron electrode after 17 cycles. a) Iron electrode in contact with the electrolyte at 0.3 V_{SCE}, p-polarized incident light; b) after removal of the electrolyte at 0.3 V_{SCE} (see text), p-polarized incident light; c) like b, but s-polarized incident light.

due to the perpendicular polarization of incident and elastically scattered light. Visually, no change of the metallic appearance of the sample was observed after 17 cycles. Preliminary experiments revealed a spectrum similar to Fig. 2 after 20 sweeps from -1.2 to -0.85 V_{SCE} with 1 mm/sec. (In this case, the reduction peaks at about -1.1 V_{SCE} was absent.) In a third series of experiments, the sample was cycled at 50 mV/sec. Examples of the corresponding voltammograms are displayed in Fig. 3. Raman spectra of the "dry sample" at 0.3 V_{SCE} in situ already showed the 680 cm^{-1} band after one completed cycle. After ten completed cycles the 674 cm^{-1} band was comparable in strength to that shown in Fig. 2. Visually, no change of the metallic appearance of the sample was observed up to about 20 cycles, when the maximum current peak at about -0.6 V_{SCE} was observed. After 20 cycles, this current peak decreases again and the sample develops a reddish brown "rusty" appearance and a different Raman spectrum at 0.3 V_{SCE} (normal incidence, no analysis of polarization) was observed, see Fig. 4. No OH stretch vibrations near 3400 cm^{-1} could be detected, as observed by Neugebauer et al. (ref. 10) by infrared spectroscopy. Estimates of charge recovery during a cycle in Fig. 3 show that not more than one monolayer of iron may remain unreduced after one cycle. The reddish brown "rusty" layer could be removed from

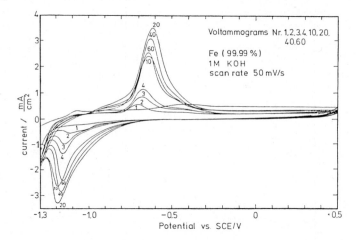

Fig. 3. Like Fig. 1, but sweep rate 50 mV/sec.

the surface by a strong water jet inside the cell under N_2 atmosphere, after which the sample appeared metallic again. A Raman spectra taken from this sample in situ was similar to the ones displayed in Fig. 2. When the sample was brought to the ambient air and a spectrum was recorded after one day, no changes were observed. This indicates, that exposure to gaseous oxygen is uncritical in our experiments.

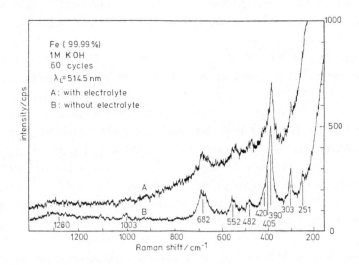

Fig. 4. In situ Raman spectra of an iron electrode after 60 cycles as shown in Fig. 3.

DISCUSSION

The spectra in Fig. 2, in particular spectrum c) resembles the spectrum of Fe_3O_4 which has Raman bands at 667 and 543 cm^{-1} (refs. 8 and 11). δ-FeOOH in a KBr pellet yields only a band at 663 cm^{-1} (ref. 11), FeO (wüstite) a band at 663 cm^{-1} (ref. 1). α-FeOOH and γ-FeOOH (refs. 1 and 8) and α-Fe_2O_3 (ref. 1) have very different Raman spectra from the ones in Fig. 2. The only interpretation of the spectra in Fig. 2 compatible with full oxidation to Fe^{+++} would be an assignment to amorphous δ-FeOOH. The most likely assignment of the spectra in Fig. 2 is to magnetite Fe_3O_4.

The effect of repeated fast cycles (see Fig. 3) is the formation of a secondary oxide layer, only weakly attached to the surface. This reddish brown "rusty" layer is probalby mainly α-FeOOH, as the spectrum in Fig. 4 resembles closely those of α-FeOOH (refs. 1 and 8).

The fact, that a spectrum like in Fig. 2 is observed when the oxide films corresponding to the oxidation peak at -0.9 V_{SCE} (peak II in ref. 13) is present only (either by oxidizing only to -0.85 V_{SCE} or by floating off the rusty film formed by the oxidation peak III (ref. 13)) leads to assign the spectrum of Fig. 2 to the film formed by oxidation peak II (probably magnetite). This film which according to the voltammogram in Fig. 1 corresponds to a charge of about 13 mC/ cm^2 passivates the iron sample - the current at a constant potential of 0.85 V_{SCE} is smaller than 0.5 $\mu A/cm^2$. Therefore, it is fair to say, that the spectrum of Fig. 2 originates from a passivating layer.

The spectrum in Fig. 4 is assigned to a film (most probably α-FeOOH) on top of the passivating layer whose oxidation is reflected by peak III in the voltammograms. We tentatively explain its apparent growing thickness with the number of cycles (see Fig. 1) by the transformation of iron in each cycle in the active region from -1.3 to about -1.0 into a porous, spongy and hydrated $Fe(OH)_2$ film, which is oxidized further to α-FeOOH at higher potential (peak III) but never again reduced to iron.

With this second film on top of the passivating layer, the current density at +0.4 V is 2 - 3 $\mu A/cm^2$.

In summary, we have presented evidence, that oxidation peak II corresponds to a passivating layer of magnetite, and peak III to a secondary layer, which is oxidized to nearly nonadhesive α-FeOOH. This should be compared to previous assignments based on purely electrochemical evidence (refs. 12 and 13). Peak I (see Fig. 1) was assigned to the formation of $Fe(OH)_2$ (refs. 12 and 13), peak II to Fe_3O_4 (ref. 12) or $Fe(OH)_2$ (ref. 13), peak III to $Fe(OH)_3$ or Fe_2O_3 (ref. 12) or FeOOH (ref. 13). Our work shows, how Raman spectroscopy may help to decide between these different assignments.

ACKNOWLEDGEMENT

We thank Dr. G. Nauer for samples of α-, δ-, and γ-FeOOH, $Fe(OH)_3$ and Fe_3O_4 and Prof. Dr. J.W. Schultze for a discussion.

REFERENCES

1 R.J. Thibeau, C.W. Brown, R.H. Heidersbach, Appl. Spectr. 32, 532 (1978)
2 R. Dupeyrat, M. Froelicher, A. Hugot-Le Goff, I. Jovancicevic, M. Masson in
   Raman Spectroscopy, J. Lascombe, D.V. Huong (eds.), J. Wiley 1982, p. 679
3 R.L. Farrow, P.L. Mattern, A.S. Nagelberg, Appl. Phys. Lett. 36, 212 (1980)
4 R.L. Farrow, A.S. Nagelberg, Appl. Phys. Lett. 36, 945 (1980)
5 R.L. Farrow, R.E. Benner, A.S. Nagelberg, P.L. Mattern, This Solid Films
   73, 353 (1980)
6 J.C. Hamilton, R.E. Benner, B.E. Mills, J. Vac. Sci. Technol. 20, 946 (1982)
7 J.T. Keiser, C.W. Brown, R.H. Heidersbach, Corrosion 38, 357 (1982)
8 R. Lossy, Diplomarbeit "Aufbau eines ramanspektroskopischen Meßplatzes,
   erste Messungen zur Passivierung von Eisen", Düsseldorf, Oct. 1982
9 M. Moskovits, K.H. Michaelian, J. Chem. Phys. 69, 2306 (1978)
10 H. Neugebauer, G. Nauer, N. Brinda-Konopik, G. Gidaly, J. Electroanal. Chem.
   122, 381 (1981)
11 Thibeau et al. (ref. 1) reported bands at 663 cm^{-1} and 616 cm^{-1} for Fe_3O_4
   and FeO (wüstite). However, the bands at 616 cm^{-1} may be due to a grating
   ghost (ref. 1).
12 D.D. Macdonald, D. Owen, J. Electrochem. Soc. 120, 317 (1973)
13 R.S. Schrebler Guzman, J.R. Vilche, A.J. Arvia, Electrochim. Acta 24, 395
   (1979)

Passivity of Metals and Semiconductors, edited by M. Froment
Elsevier Science Publishers B.V., Amsterdam — Printed in The Netherlands

APPLICATION OF ANGLE RESOLVED XPS AND AES DEPTH PROFILING TO THE STUDY OF TRANSPASSIVE FILMS ON NICKEL

M. Datta, H.J. Mathieu, D. Landolt
Materials Department, Swiss Federal Institute of Technology,
CH-1007 Lausanne, Switzerland

ABSTRACT

Variation of the take-off angle in XPS permits one to obtain information on chemical composition of very thin films as a function of depth without the use of ion sputtering. The method together with Auger depth profiling was applied to the study of nickel surfaces subjected to transpassive dissolution in nitrate solution. Obtained results indicate the presence of thin oxide films containing nitrogen which is present in a reduced form.

INTRODUCTION

It has been shown previously by Auger Electron Spectroscopy (AES) and by coulometry that upon increasing anodic polarisation of a nickel electrode in nitrate solution into the transpassive potential region the passive film thickness passes through a maximum (ref. 1). High rate transpassive metal dissolution takes place in the potential region beyond the maximum where the apparent film thickness is independent of potential. The purpose of the present study is to use angle resolved X-ray Photoelectron Spectroscopy (XPS) to further characterise electrode surfaces produced by transpassive dissolution of nickel in nitrate solution and to compare results to those obtained by AES depth profiling using ion sputtering.

Angle resolved XPS is a non-destructive technique suitable for the investigation of the in-depth composition of very thin surface layers, i.e. layers of thickness smaller or comparable to the mean free path of the X-ray induced photoelectrons (ref. 2). The magnitude of the mean free path of electrons in solids depends on their kinetic energy and on the matrix (ref. 3). For X-ray induced photoelectrons it is typically 1 - 2.5 nm. The principle of angle resolved XPS is shown in figure 1. The analysis information depth is equal to the electron escape depth, Λ, which is related to the electron mean free path, λ, by equation 1 in which ϕ is the take-off angle with respect to the surface normal.

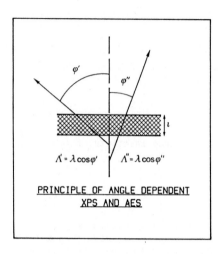

Figure 1

$$\Lambda = \lambda \cos \phi \qquad\qquad\qquad (1)$$

It follows from this that by varying ϕ the depth of information can be varied which allows one in principle to study composition as a function of depth of very thin films without the need for ion sputtering.

EXPERIMENTAL

Transpassive dissolution experiments were carried out in a flow channel cell (ref. 4) at a constant electrolyte velocity of 2.5 m/s using a 6M $NaNO_3$+0.1M NaOH solution. A constant anodic current density of 3 A/cm^2 was applied for 400ms corresponding to a maximum depth of dissolution of 0.4 μm. Single crystal nickel anodes (99.95%, orientation [111]) were employed. Before being subjected to dissolution they were mechanically polished with diamond spray upto 6 μm and finally using alumina to 1/4 μm finish. They were then cleaned and cathodically pre-polarised in the working electrolyte at 15 mA/cm^2 for 5 s. After dissolution the anodes were removed from the flow channel cell, dried and stored in a dessicater before introducing into the UHV chamber for analysis. Typical transfer time between flow channel and UHV was 30-60 min.

Surface analysis was performed in a combined AES/XPS system (Physical Electronics Model 590/550 A) including a double pass cylindrical mirror analyser (CMA), a differentially pumped raster ion gun and an Al X-ray source. AES depth profile analysis was performed by rastering a Kr$^+$ beam over an area of 1.5x1.5mm.

XPS spectra were recorded at 50 V pass energy with an energy resolution of 1 eV. Take-off angle was varied using an angle resolving aperture of 12° which could be rotated with respect to the CMA axis (ref. 5). The samples were oriented at $\alpha = 42^o$ between the surface normal and the CMA axis. The take-off angle ϕ is given by

$$\cos\phi = \cos \phi_A \cos\alpha - \sin \phi_A \sin\alpha \cos\delta \tag{2}$$

were ϕ_A is the CMA acceptance angle (42.3^o) and δ is the externally adjustable azimuth angle.

RESULTS

Figure 2 shows AES depth profiles of nickel electrodes before and after being subjected to transpassive dissolution. The sample (a) was mechanically polished then ultrasonically cleaned in water. Sample (b) was subjected to transpassive dissolution under the conditions described above. The sputter rate is the same for the two samples, namely 1.2 nm Ta_2O_5/min[*]). The data of figure 2 show that

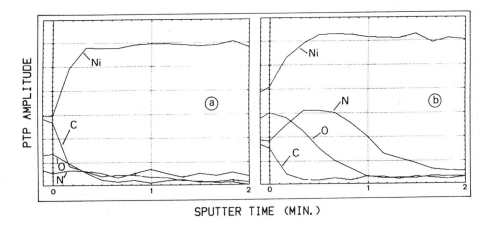

SPUTTER TIME (MIN.)

Fig. 2   AES depth profiles of nickel anodes   (a) after pretreatment
(b) after transpassive dissolution. N amplitude is 8x.

---

[*]) The sputter rate of a single crystal NiO is approx. 1.5 times that of Ta_2O_5. The sputter rate of the surface oxide film is not known but may be assumed to be similar.

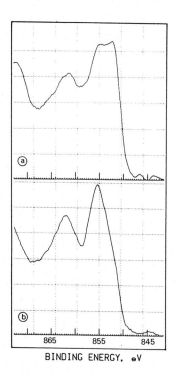

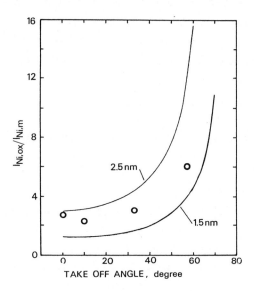

Fig. 4 Peak intensity ratio of nickel
as a function of take-off angle
o : experimental, — calculated

Fig. 3 Ni2p$_{3/2}$ spectra at (a) 0°,

(b) 57° take-off angle

the surface composition after transpassive dissolution differs from that of the
not dissolved sample in that more oxygen and nitrogen are present. The nitrogen
exhibits a maximum at the oxide-metal interface consistent with previous observations (ref. 1).

XPS was performed at different take-off angles varying from $\phi = 0°$ to $\phi = 57°$.
The Ni2p$_{3/2}$, O1s, N1s and C1s peaks were monitored. All binding energies were
determined with respect to C1s = 284.6 eV. Figure 3 shows that the shape of the
Ni2p$_{3/2}$ peak depends on take-off angle. At $\phi = 0°$ a double peak with maxima at
852.1 eV and 854.7 eV binding energy is observed attributed to metallic nickel
and oxidized nickel (NiO) respectively (ref. 6,7). At $\phi = 57°$ the metallic Ni
peak is present only as a weak shoulder since at this angle most of the information comes from the surface film. This confirms the surface sensibility of angle
resolved XPS.

Peak deconvolution and background subtraction were carried out and the peak area was calculated using manufacturer supplied software (Perkin Elmer program no. 005 A). Figure 4 shows the variation of the intensity ratio $R_{Ni}=I_{Ni,ox}/I_{Ni,m}$ as a function of take-off angle. Here $I_{Ni,ox}$ is the peak intensity (area) of oxidized nickel and $I_{Ni,m}$ that of metallic nickel.

Assuming a homogeneous film of thickness ℓ covering a flat surface the intensity ratio R_{Ni} is given by equation (3).

$$R_{Ni} = \frac{n_f}{n_m} \frac{\lambda_f}{\lambda_m} \frac{1 - \exp(-\frac{\ell}{\lambda_f \cos\phi})}{\exp(-\frac{\ell}{\lambda_f \cos\phi})} \tag{3}$$

Here n_f and n_m is the nickel atomic concentration in the film and the substrate, respectively, λ_f and λ_m is the electron mean free path in the film and the substrate. Theoretical curves calculated for different values of ℓ are given in Fig. 4. They were obtained by using the numerical values $n_f = 0.089$ mol/cm^3, $n_m = 0.152$ mol/cm^3, $\lambda_f = 1.74$ nm, $\lambda_m = 1.07$ nm. The values of λ_f and λ_m were calculated using the formulas for non metallic and metallic solids given in ref. 3. The data in figure 4 suggest a film thickness of $\sim$ 2 nm.

A further question concerns the oxidation state of the nitrogen present. The measured XPS N1s peak is shown in figure 5. Its maximum is at 397.5 eV. This may be compared with the binding energy values listed in Table I which were either

TABLE I N1s binding energies of different compounds containing nitrogen referred to: (*) Pt4f$_{7/2}$ = 71.1 eV and (**) C1s = 284.6 eV.

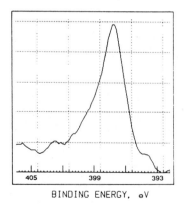

BINDING ENERGY, eV

Fig. 5 N1s spectra at 0° take-off angle

Compound	N1s Binding Energy, eV		
	Ref. 8 (*)	Ref. 9 (**)	Present work (**)
NaNO$_3$	407.3	407.0	407.1
NaNO$_2$	403.3	403.3	403.6
NH$_4$NO$_3$	406.0 (400.9)	-	-
NH$_4$Cl	-	-	401.8
Iron nitride	-	-	398.0
NaN$_3$	398.6	-	-
Transpassive film			397.5

taken from the literature (ref. 8,9) or measured in the present work. It follows
that the N1s binding energy increases with increasing oxidation state. The observed binding energy value of figure 5 corresponds to a reduced form of nitrogen
and is closest to the value for nitride.

DISCUSSION AND CONCLUSIONS

The present angle resolved XPS data and AES depth profiles suggest the presence on transpassively dissolved nickel anodes of a surface film the thickness
of which is of the order of the mean free path of the photoelectrons. In part
the observed film may have been formed after completion of the dissolution experiments but the difference observed in film thickness and composition between
anodically dissolved and non dissolved samples suggests that transpassive dissolution indeed affects surface composition. The nitrogen found by AES analysis
has been identified by XPS as a reduced form, possibly nitride. This observation
can be explained in two ways, (i) by specific surface interactions during dissolution, similarly as has been observed during transpassive dissolution of
nickel and titanium in ClO_3^- and ClO_4^- solutions (ref. 10,11) or (ii) by a
reduction reaction taking place after switching off the anodic current. A more
detailed account of the results presented here and a more quantitative evaluation of angle resolved XPS and AES depth profiling for the characterisation of
very thin films will be given elsewhere (ref. 12).

ACKNOWLEDGEMENT : This work was financially supported by Fonds national suisse
pour la recherche scientifique, Bern, Switzerland.

REFERENCES

1 M. Datta, H.J. Mathieu, D. Landolt, Electrochim. Acta 24, 843 (1979)
2 M.F. Ebel, Surface and Interface Analysis, 3, 173 (1981)
3 M.P. Seah, A.W. Dench, Surface and Interface Analysis, 1, 2 (1979)
4 M. Datta, D. Landolt, Electrochim. Acta, 25, 1255 (1980)
5 H.J. Mathieu, D. Landolt, Appl. Surf. Sci. 10, 455 (1982)
6 Handbook of X-Ray photoelectron Spectroscopy, C.D. Wagner, W.M. Riggs,
   L.E. Davis, J.F. Moulder, G.E. Muilenberg, Eds., Perkin Elmer Corp, 1979
7 P. Marcus, J. Oudar, I. Olefjord, J. Microsc. Spectrosc. Electron.,4, 63 (1979)
8 B. Folkesson, Acta Chemica Scandinavica, 27, 287 (1973)
9 H. Schultheiss, E. Fluck, J. Inorg. Nucl. Chem., 37, 2109 (1975)
10 M. Datta, D. Landolt, Electrochim. Acta, 25, 1265 (1980)
11 J.B. Mathieu, H.J. Mathieu, D. Landolt, J. Electrochem. Soc., 125, 1039 (1978)
12 M. Datta, H.J. Mathieu, D. Landolt, to be published.

Passivity of Metals and Semiconductors, edited by M. Froment
Elsevier Science Publishers B.V., Amsterdam — Printed in The Netherlands

PASSIVATION OF NICKEL - IRON ALLOYS. INFLUENCE OF SULPHUR.

P. MARCUS and J. OUDAR

Laboratoire de Physico-Chimie des Surfaces associé au C.N.R.S.

Ecole Nationale Supérieure de Chimie de Paris

11, rue Pierre et Marie Curie

75005 PARIS (FRANCE)

ABSTRACT

   The influence of sulphur on the passivation of nickel-iron alloys was
investigated. Particular attention was given to the relation between the surface
composition and the electrochemical behaviour. A model of the passive film on
a Ni/Fe alloy is suggested. The main result regarding the role of sulphur is the
impossibility for the passive film to be formed when the surface is covered by a
monolayer of adsorbed sulphur. The involved mechanisms, i.e. blocking and dilution effects on the adsorption of hydroxyl groups, are discussed. On sulphur-
containing nickel-iron alloys, sulphur segregates on the surface during the dissolution of the metallic elements. Above a critical sulphur content a non-protective thin sulphide film is formed on the surface instead of the oxide passive
film.

INTRODUCTION

   The general frame for the work presented here is the attempt to relate the
structure and composition of alloy surfaces to their reactivity in aqueous
medium. Indeed the reactivity is strongly dependent on the surface composition,
which may strikingly differ from the composition of the matrix. The problem is
complex because the initial surface composition may deviate from the bulk composition, due to thermodynamic factors or lattice strains effects, and may be
further modified by exposure to the electrolyte. Therefore, it is of primary
importance to analyse the surface composition at each step, i.e. after surface
preparation, exposure to the electrolyte at the rest potential, polarization in
the active region, and after passivation. Surface sensitive techniques AES and
ESCA have proved to be able to fulfill this requirement. In the same way, if the
influence of impurities, such as sulphur, is to be investigated, the coverage of
the surface by the adsorbed impurity must be known at the various stages of the
reaction. For sulphur, this may be achieved by means of the radiotracer technique.
The aim of this paper is to emphasize some aspects of the passivation of Ni/Fe
alloys and the influence of sulphur. Experimental results on the passivation of
Ni [1,2], a Ni-25% Fe alloy [3] and the influence of sulphur on Ni [1,4] and the
alloy [5] have been published in the recent past by MARCUS, OUDAR and OLEFJORD.
These results, as well as additional results, will provide the basis of the
discussions in this paper.

INITIAL SURFACE COMPOSITION OF THE Ni/Fe ALLOYS

   The surface composition of Ni/Fe alloys (Ni-25at% Fe and Ni-50at% Fe single
crystals) were analysed by AES[6] and ESCA [7] after electro-polishing and
subsequent argon ion etching or annealing in purified hydrogen. Both techniques
show that the surface composition of the sulphur-free alloys is identical to the

matrix composition. This was expected considering the similarity of Ni and Fe with regard to the sublimation enthalpy and the atomic radii [6]. In contrast, the presence of adsorbed sulphur may induce the segregation of a metal element to the surface. In the case of Ni-25% Fe only the sulphur-covered (111) face exhibits an iron segregation at elevated temperature, but the normal composition is obtained at room temperature [6]. On the sulphur-covered Ni-50at.% Fe (100) iron is segregated at high temperature and the segregation is retained at room temperature. On the sulphur-free alloys, a reproducible surface composition, identical to the matrix composition, was obtained prior to the exposure to the electrolyte. A transfer system in which the sample was under protective gas was used to avoid oxidation and contamination of the surface.

DISSOLUTION AND PASSIVATION

Rest potential and dissolution

Exposure of the alloy surfaces to 0.1N H_2SO_4 causes a marked nickel enrichment on the surface of all investigated alloys. On Ni-25at.% Fe alloy, the nickel content of the first plane, i.e. the plane in contact with the solution, was found to be 90 - 100%, i.e. about a complete nickel plane [3]. On Ni-50at.% Fe alloy, the nickel concentration of the surface plane is 75 - 85% [8]. On Ni - 75at.%Fe [8], the nickel surface content obtained by exposure to the acid is 50 - 60%. It is most interesting to connect these results to the measurements of the rest potential for a series of Ni/Fe alloys [9]. Figure 1 shows the relation between the bulk nickel content of the alloy, the surface nickel content, and the rest potential normalized to the rest potential of iron in 0.1N H_2SO_4 ($E_r - E_r(Fe)$). It appears clearly that the rest potential does not vary linearly

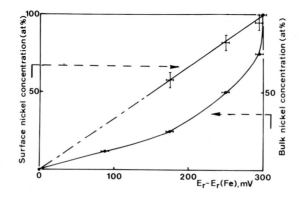

Figure 1. Relations between the surface nickel concentration, the bulk nickel concentration and the rest potential in 0.1N H_2SO_4 (normalized to the rest potential of iron) for Ni/Fe alloys.

with the nickel concentration in the matrix. Two interpretations may be considered : (i) the existence of an alloying effect causing a behaviour that is not the simple superposition of the behaviour of the individual metal elements, (ii) changes in the surface composition upon exposure to the electrolyte. Straighforward evidence for the second factor is given by figure 1. Indeed, considering the surface composition, the relation between composition and the rest potential is linear for the studied range of compositions. Work is in progress to check the validity of this result for the Fe-rich alloys.
During polarization of the alloy in the active region, the nickel enrichment in the outermost metallic plane is retained. The dissolution process takes place at

a nickel-enriched surface. The extent of the enrichment remains limited to the first atomic plane, i.e. there is no further accumulation of one of the metal elements, and the ratio of dissolved metals is identical to the ratio of the metals in the alloy.

Passivation

In a paper published some years ago [2], MARCUS, OUDAR and OLEFJORD presented a model for the passive film on pure nickel, which was supported by ESCA-analyses, a technique that is particularly suitable to distinguish between metallic, oxide and hydroxide states. ESCA experiments performed recently by MARCUS and OLEJFORD [3] allow us to suggest a similar model for the passive film on Ni-25at.% Fe. Indeed it was shown that the passive film on the alloy consists of an inner layer of nickel oxide NiO and an outer layer of Ni and Fe hydroxides. The thickness of the passive film was estimated to ~ 10 Å. It was also shown that the alloy/oxide interface is enriched with nickel in the metallic state. A model of the passive film is given in Figure 2.

OH   OH   OH

OH Ni OH Fe OH Ni

Ni O Ni O Ni O

O Ni O Ni O Ni

nickel-enriched plane

Substrate:

Ni : Fe = 3 : 1

Figure 2. A model of the passive film on $Ni_{75}Fe_{25}$ alloy.

This model is consistent with a growth mechanism, for the passive film, involving the recombination of hydroxyl groups, a disproportionation reaction of the

type 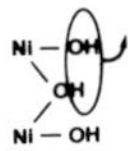 which was previously suggested for pure nickel.

The surface hydroxylation already takes place in the active region, but a higher potential is required to break the O-H bond of an adsorbed OH group and recombine with an OH group adsorbed on an adjacent site. Studies of oxide catalyst, such as MgO, have revealed the presence of surface hydroxyl groups [10], and the properties of oxide catalysts are known to be strongly dependent on the presence of OH on the surface. More recently, studies of the adsorption of water from a gas phase onto Ag(110) [11] and Ni(110) [12] clearly show that the surface hydroxylation takes place by exposure to water, and that the adsorbed hydroxyls may recombine via a disproportionation reaction and yield the oxide.

INFLUENCE OF SULPHUR

The search for the mechanisms that govern the influence of sulphur on the passivation behaviour of Ni/Fe alloys is obviously motivated by the high detrimental effect of this element on the corrosion resistance of metallic alloys. Careful control of the chemical state and the coverage of the surface by sulphur allowed us to elucidate the role of sulphur in the dissolution and passivation of nickel [1,4]. Similar effects were observed on a Ni-25at.%Fe alloy [5]. When sulphur is pre-adsorbed in an $H_2S - H_2$ mixture under appropriate conditions

(see Fig. 3) leading to the saturation of the surface by a complete monolayer * (6), the electrochemical behaviour of the alloy is markedly changed.
Figure 3a shows the i-E curves recorded with the sulphur-free and the sulphur-
covered Ni$_3$Fe (110)alloy.

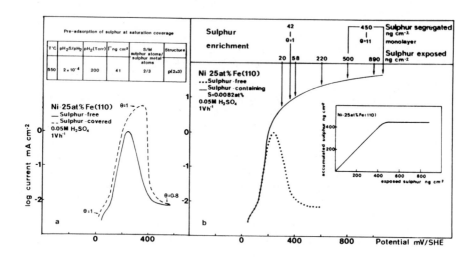

Figure 3. Anodic polarization curves of Ni$_{75}$Fe$_{25}$ (110). a) sulphur-free and
sulphur-covered surfaces ; b) sulphur-free (same as in (a)) and sulphur-doped
alloy. Plot in insert in (b) shows the surface content of segregated sulphur
(in ng cm^{-2}) vs the amount of sulphur exposed by the dissolution of the
sulphur-containing matrix.

The presence of sulphur promotes the dissolution of the alloy. The adsorbed
sulphur remains on the surface while the alloy is dissolved, hence the sulphur
acts as a catalyst for the anodic dissolution reaction. Also the passivation is
delayed until the sulphur coverage is lowered, by desorption into the electrolyte, to a value of 0.8 (i.e. 80% of the complete monolayer).
When sulphur is in solid solution in the alloy, it segregates on the surface
during the dissolution of the metal elements. Due to the segregated sulphur,
the electrochemical behaviour of the surface is completely changed :
the active/passive transition is precluded and the alloy dissolves in the whole
potential range where a sulphur-free alloy would be passivated. This effect is
clearly shown in Fig. 3b, where the parameters for the anodic sulphur segregation are given. Above the saturation coverage for the adsorbed sulphur monolayer,
a sulphide film, the composition of which is Ni$_{2.5}$Fe$_{0.5}$S$_2$, grows on the surface
up to a stationary thickness of ∿ 25A.This thin sulphide film has no protective
effect, in contrast with the oxide film formed on the sulphur-free surface. It
is important to notice that the three-dimensional sulphide observed after polarization was not pre-existent, but was formed by a surface reaction during the

---

*The complete monolayer is defined as the saturation of available sites for
adsorption of sulphur. The coverage θ is then equal to 1. On the (110) face of
Ni-25at.%Fe alloy, it corresponds to a ratio of sulphur atoms to metal atoms
in the first atomic layer (S/M) of 2/3.

anodic dissolution process. Figure 4 illustrates the relation between the surface composition of the alloy and its electrochemical behaviour.

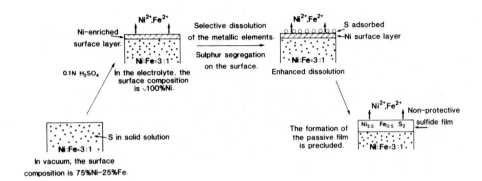

Figure 4. The effect of sulphur on the passivation behaviour of Ni/Fe alloy.

On an atomistic point of view, the effects of sulphur on the dissolution/
passivation behaviour of nickel and nickel-iron alloys may be attributed to the
following factors :

1. The adsorption of sulphur weakens the metallic bonds of the nearest metal
   atoms. The adsorbed sulphur atoms draw charge out of the metal-metal bonds
   and weaken them. Recent quantum mechanical cluster calculations on Ni_4S
   and Fe_4S [13] support this statement. Weakening the metal-metal bonds
   will favor the dissolution of the metal. However, this implies that the
   metal-sulphur bond is heteropolar, and this is not confirmed by work
   function measurements on S/Ni (100) [14], which indicate a covalent character of the sulphur-nickel bonding.Nevertheless, even the formation of
   a covalent S-M bond will cause a disruption of the neighboured M-M metallic bonds (4s for Ni [15]) and the metal bond strength will be reduced.

2. The adsorbed sulphur atoms block adsorption sites for hydroxyls. OH groups
   still adsorb to some extent on the sulphur-covered surface, thus allowing
   the dissolution to take place, but the disproportionation reaction involved in the passivation process is limited, due to an increased mean distance between the adsorbed OH.

Work is now in progress to characterize the effect of other alloying elements,
such as molybdenum, which are likely to counteract the detrimental effect of
sulphur on the passivation of nickel-based alloys.

ACKNOWLEDGMENTS

The authors are grateful to Pr. I. Olefjord for the continuous and fruitful
collaboration on the ESCA-studies and the related discussions.
Financial support from the C.N.R.S. is gratefully acknowledged.

REFERENCES

1. J. OUDAR and P. MARCUS — Appl. Surface Science 3(1979) 48
2. P. MARCUS, J. OUDAR and I. OLEFJORD — J. Microsc. Spectrosc. Electron. 4(1979) 63
3. P. MARCUS and I. OLEFJORD — Surface and Interface Anal. 4(1982) 29
4. P. MARCUS, J. OUDAR and I. OLEFJORD — Mater. Sci. Eng. 42(1980) 191
5. a) P. MARCUS, A. TEISSIER and J. OUDAR to be published in Corrosion Science
   b) P. MARCUS, I. OLEFJORD and J. OUDAR to be published in Corrosion Science
6. P. MARCUS, A. TEISSIER and J. OUDAR — Surface Science (in press)
7. I. OLEFJORD and P. MARCUS — Surface and Interface Anal. 4(1982) 25
8. P. MARCUS and I. OLEFJORD to be published
9. P. MARCUS, A. BOURUET and J. OUDAR — Mem. Sci. Rev. Met. 9(1981) 509
10. H. KNOZINGER — Advances in Catalysis, Academic Press (1976) vol. 25 P. 214
11. E.M. STUVE, R.J. MADIX and B.A. SEXTON — Surface Science 111(1981) 11
12. C. BENNDORF, C. NOBL, M. RUSENBERG and F. THIEME — Surface Science 111(1981) 87
13. R.P. MESSMER and C.L. BRIANT — Acta Met. 30(1982) 457
14. a) G.E. BECKER and H.D. HAGSTRUM — Surface Sci. 30(1972) 505
    b) J.E. DEMUTH and T.N. RHODIN — Surface Sci. 45(1974) 249
15. S.P. WALCH and W.A. GODDARD — Surface Sci. 72(1978) 645

Passivity of Metals and Semiconductors, edited by M. Froment
Elsevier Science Publishers B.V., Amsterdam — Printed in The Netherlands

ANODIC OXIDATION OF UO_2-PART III. ELECTROCHEMICAL STUDIES IN CARBONATE SOLUTIONS

D.W. SHOESMITH, S. SUNDER, M.G. BAILEY and D.G. OWEN
Research Chemistry Branch, Whiteshell Nuclear Research Establishment, Atomic
Energy of Canada Limited, Pinawa, Manitoba ROE 1L0 (Canada)*

ABSTRACT
    The oxidative dissolution of UO_2 has been studied in aqueous carbonate solu-
tions using steady-state potentiostatic and cyclic voltammetric techniques. The
stoichiometry of the UO_2 electrode was determined by X-ray photoelectron spectroscopy. Dissolution occurs from a $UO_{2.33}$ surface layer at low anodic potentials and from a $UO_{2.5}$ layer at higher potentials. Steady-state dissolution
currents, as a function of carbonate concentration and applied potential, show a
variable reaction order with respect to carbonate concentrations, and a change
from a two-electron to a one-electron rate-determining step as the potential
increases.

INTRODUCTION
    Earlier experiments in neutral (ref.1-3) and alkaline sulphate solutions
(ref.2-4) show that the anodic oxidation of UO_2 progresses through a series of
surface oxide phases, accompanied by dissolution (as UO_2^{2+}) at higher potentials
($E \gtrsim$ +0.200 V vs. SCE), see below. The dissolution of UO_2 in carbonate solutions under oxidizing conditions has been studied earlier (ref.3,5) but the
details of the surface oxidation and dissolution mechanism are not complete. In
this paper we present an account of our steady-state and cyclic voltammetric
experiments in aqueous carbonate solutions.

$$UO_2 \xrightarrow{(a)} UO_{2+x}(\text{monolayer}) \xrightarrow{(b)} UO_{2+x} \xrightarrow{(c)} UO_{2.33}$$

$$\xrightarrow{(d)} UO_{2.5} \xrightarrow{(e)} UO_{2.67} \xrightarrow{(f)} UO_3$$

$$UO_{2.33} + (UO_2^{2+})_{ads}$$

$$\xrightarrow{(g)} (UO_2^{2+})_{surf} \xrightarrow{(j)} UO_3 \cdot zH_2O$$

$$\xrightarrow{(h)} (UO_2^{2+})_{bulk}$$

Reaction scheme for the oxidative dissolution of UO_2 in the absence of strongly
complexing anions; from references (1) and (4).

* AECL #7956

126

EXPERIMENTAL

Details on equipment and procedure have been published previously (ref.1,6). Sodium perchlorate and sodium sulphate were used as base electrolytes with no observed difference in behaviour. All potentials are measured against the saturated calomel electrode (SCE). Steady-state currents were recorded at a disk electrode rotating at w = 16.7 Hz, an angular velocity sufficient to avoid any stirring dependence, by increasing the potential at constant carbonate concentration, and increasing the concentration at constant potential.

RESULTS

Figure 1 shows a series of voltammograms depicting the effects of carbonate on the various film-formation and reduction processes. Two early stages of oxidation, (a) and (b), corresponding to the formation of UO_{2+x} (monolayer) and UO_{2+x}, respectively, are observed prior to the major rise in anodic current and the onset of UO_2^{2+} dissolution. The major reduction peak is a doublet and can be attributed to the reduction of the $UO_{2.5}/UO_{2.33}/UO_{2+x}$ layers. This peak was incorrectly assigned to the reduction of UO_2CO_3 by Nicol and Needes (ref.5). The film reduced at peak (e) is formed by precipitation of dissolved UO_2^{2+} as $UO_3 \cdot zH_2O$ (ref.1).

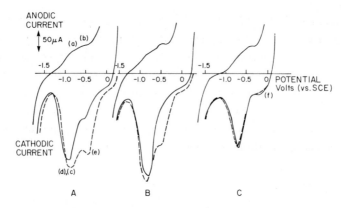

Fig. 1. Effect of carbonate on voltammograms recorded in 0.5 mol·dm^{-3} Na$_2$SO$_4$; pH = 10.5, v = 20 mV·s^{-1}. (A) $[CO_3]_t^*$ = 0 mol·dm^{-3}; (B) 10^{-3} mol·dm^{-3}; (C) 5 x 10^{-1} mol·dm^{-3}; —— Anodic limit +0.500 V; ---- Anodic limit = +0.600 V.

Peaks (a) and (b) are unaffected in the presence of carbonate. In 10^{-3} mol·dm^{-3} carbonate the reduction peak doublet, (c) and (d), is larger, and peak (e) is vestigial unless the anodic sweep is allowed to reach a potential (E_A) equal to +0.600 V before reversal. In 5 x 10^{-1} mol·dm^{-3} carbonate the cathodic

---

* $[CO_3]_t$ denotes the total concentration of dissolved carbonate plus bicarbonate.

127

profile is simpler, only peak (c) of the doublet and a new peak (f) being observed.

Figures 2A and 2B show the anodic (Q_A) and cathodic (Q_C) charges, obtained by integration of the voltammograms, as a function of sweep rate (v) and $[CO_3]_t$. These data show that carbonate has little impact on the extent of dissolution ($Q_D = Q_A - Q_C$) for $[CO_3]_t \lesssim 10^{-3}$ mol·dm^{-3} especially at higher sweep rates. As $[CO_3]_t$ increases, Q_A increases and Q_C decreases indicating the increased importance of dissolution over film formation at higher carbonate concentrations. The peak in the Q_C-log v plot at lower $[CO_3]_t$ (Fig. 2B) is a consequence of the competing rates of reactions (g), (h) and (j) in the reaction scheme. At intermediate sweep rates, the rate of dissolution plus precipitation is high enough to maximize the $UO_3 \cdot zH_2O$ precipitation. Fig. 2B shows that $UO_3 \cdot zH_2O$ formation is negligible for $[CO_3]_t \gtrsim 10^{-3}$ mol·dm^{-3}.

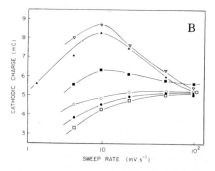 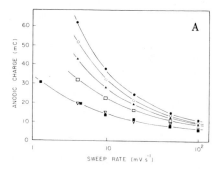

Fig. 2. (A) Anodic charge, Q_A (from voltammograms) as a function of sweep rate (v) for various $[CO_3]_t$: ■ 0 mol·dm^{-3}; ▽ 10^{-3}; □ 3 x 10^{-3}; ▲ 10^{-1}; o 2 x 10^{-1}; ● 5 x 10^{-1}. (B) Cathodic charge, Q_C, (from voltammograms) as a function of sweep rate (v) for various $[CO_3]_t$: ▲ 0 mol·dm^{-3}; ▽ 10^{-3}; ■ 3 x 10^{-3}; o 10^{-1}; ● 2 x 10^{-1}; □ 5 x 10^{-1}.

Figure 3 shows Tafel plots for various $[CO_3]_t$. At low potentials a slope of between 75 and 62 mV/decade is obtained compared to a slope of between 95 and 155 mV/decade at higher potentials. At E $\gtrsim$ +0.300 V, and low $[CO_3]_t$, the current tends to become independent of potential. Potentiostatic transients, at stationary and rotating electrodes, in this potential region suggest that carbonate transport to the electrode surface is partially rate-controlling. Voltammetry to various anodic limits (E_A) shows that peak (f) (Fig. 1) appears for $E_A \gtrsim$ +0.200 V. Nicol and Needes observed (ref.5) a similar current plateau at higher potentials for the higher $[CO_3]_t$.

Figure 4 shows the steady-state dissolution current as a function of $[CO_3]_t$ for various potentials. Three regions are apparent: (i) For $[CO_3]_t \lesssim 10^{-4}$ mol· dm^{-3}, the current appears independent of $[CO_3]_t$, the concentration being controlled at this level by dissolved atmospheric CO_2. (ii) For intermediate $[CO_3]_t$, the dissolution current is first order in carbonate for E $\geq$ +0.250 V but

128

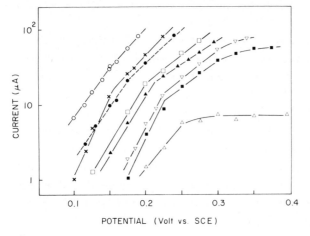

Fig. 3. Tafel plots recorded in 0.1 mol·dm^{-3} NaClO$_4$ (pH = 9.5) plus various amounts of carbonate: o 5 x 10^{-1} mol·dm^{-3}; x 10^{-1}; • 5 x 10^{-2}; □ 10^{-2}; ▲ 5 x 10^{-3}; ▽ 3 x 10^{-3}; ■ 10^{-3}; △ 3 x 10^{-4}.

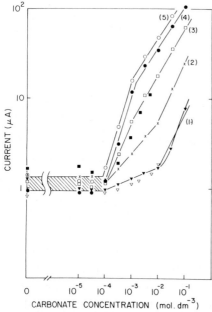

Fig. 4. Steady-state dissolution currents recorded in 0.1 mol·dm^{-3} NaClO$_4$ (pH = 9.5) as a function of carbonate concentration at various potentials. ▼ +0.100 V; x +0.150; ■ +0.200; • +0.250; o +0.300; ▽ +0.100 V in 0.5 mol·dm^{-3} Na$_2$SO$_4$; □ +0.200 V in 0.5 mol·dm^{-3} Na$_2$SO$_4$.

is less than first order at lower potentials. (iii) At higher $[CO_3]_t$ the reaction order decreases at high potentials and increases at lower potentials.

A number of XPS experiments were performed on UO$_2$ electrodes oxidized in solutions containing between 10^{-4} and 10^{-2} mol·dm^{-3} carbonate (ref.7). After 20 min of potentiostatic oxidation at +0.100 V (i.e. region (i) of steady-state plots), a surface composition close to UO$_{2.33}$ is obtained over the whole $[CO_3]_t$ range. After 20 min at +0.300 V (i.e. region (iii) of steady-state plots), a composition close to UO$_{2.5}$ is obtained for $[CO_3]_t$ < 10^{-3} mol·dm^{-3}, whereas $[CO_3]_t$ ≥ 10^{-2} mol·dm^{-3} the composition is close to UO$_{2.33}$. For intermediate $[CO_3]_t$ the composition varies between these two limits.

DISCUSSION

The first three reaction stages shown in the reaction scheme are not affected by the addition of carbonate, the XPS results indicating that the $UO_{2.33}$ surface layer is present during dissolution even in concentrated carbonate solutions. The subsequent stages of film formation (steps (a) to (f)) are dependent on $[CO_3]_t$. For $[CO_3]_t < 10^{-3}$ mol·dm^{-3} the progression from $UO_{2.33}$ to $UO_{2.5}$ can occur at sufficient large anodic potentials (i.e. $\geq$ +0.300 V), as demonstrated by XPS. At higher $[CO_3]_t$, the extent of dissolution is increased and $UO_{2.5}$ formation does not occur.

The Tafel plots of Figure 3 and the reaction order plots of Figure 4 suggest a complex dissolution reaction. At low potentials, a Tafel slope of between 62 and 75 mV/decade and a potential-dependent reaction order of << 1.0 with respect to carbonate can be explained by the reaction sequence

$$UO_2 \overset{k_1}{\rightleftharpoons} UO_2^{2+} + 2e \tag{1}$$

$$UO_2^{2+} + 3HCO_3^- \rightarrow UO_2(CO_3)_3^{4-} + 3H^+ \tag{2}$$

If dissolution were controlled by reaction (1) then the dissolution current would be independent of $[CO_3]_t$. This is almost the case at +0.100 V, where XPS shows dissolution is occurring from a base layer of $UO_{2.33}$. At higher $[CO_3]_t$ at this potential, the reaction tends to first order suggesting the reaction scheme proposed by Nicol and Needes (ref.5),

$$UO_2 + HCO_3^- \overset{k_3}{\rightleftharpoons} UO_2CO_3 + H^+ + 2e \tag{3}$$

$$UO_2CO_3 + 2HCO_3^- \overset{k_4}{\rightarrow} UO_2(CO_3)_3^{4-} + 2H^+ \tag{4}$$

If reaction (3) is rate-determining, then the dissolution current at steady-state would be given by

$$I_d = 2Fk_3[HCO_3^-]\exp(1-\beta_3)2FE/RT \tag{5}$$

At higher potentials (+0.150 V and +0.200 V), reactions (3) and (4) predominate over (1) and (2), accounting for the potential-dependent reaction order (Fig. 4).

At higher potentials there appears to be a change in mechanism from a single two-electron transfer reaction to one involving two consecutive one-electron transfer reactions. The reaction scheme

$$UO_2 + OH^- \overset{k_6}{\rightleftharpoons} UO_{2.5} + 0.5H_2O + e \tag{6}$$

$$UO_{2.5} + HCO_3^- \rightarrow UO_2CO_3 + 0.5H_2O + e \tag{7}$$

followed by reaction (4) could explain the behaviour observed at +0.250 V and +0.300 V. Thus, at low $[CO_3]_t$, and with reaction (6) rate determining,

$$I_d = 2Fk_6[HCO_3^-]\exp(1-\beta_6)FE/RT \tag{8}$$

This predicts a first-order dependence on $[HCO_3^-]$, and a Tafel slope of 120 mV^{-1} ($\beta_6 = 0.5$), as observed at +0.250 and +0.300 V in the concentration range 10^{-4} to 10^{-3} $mol \cdot dm^{-3}$. It may be implied from reaction (6) that a U^V intermediate is formed on the electrode surface prior to dissolution as U^{VI}. Our XPS results show that, under these conditions, a layer of $UO_{2.5}$ is present. At higher $[CO_3]_t$, the $UO_{2.5}$ film is not observed by XPS, and the Tafel slope tends to decrease (i.e. to 95 mV/decade at $[CO_3]_t = 5 \times 10^{-1}$ $mol \cdot dm^{-3}$).

At higher $[CO_3]_t$, a thin layer of UO_2CO_3 is present and reaction (4) might be expected to be rate-determining and the dissolution current to be given by

$$I_d = 2Fk_5[HCO_3^-]^m \qquad (9)$$

This could account for the change of reaction order with respect to carbonate for $[CO_3]_t > 10^{-3}$ $mol \cdot dm^{-3}$ and E > +0.200 V. The data of Figure 4 suggest a value of m = 0.5. Nicol and Needes (ref.5) obtained a value of 0.7. Peak (f) (see Fig. 1), is observed for $[CO_3]_t > 10^{-3}$ $mol \cdot dm^{-3}$ and E $\gtrsim$ +0.200 V, and could be attributed to the electrochemical reduction of UO_2CO_3, adding credence to our conclusion that the dissolution of a thin layer of UO_2CO_3 could be rate-controlling under these circumstances. Nicol and Needes (ref.5) claim a much thicker layer of UO_2CO_3 on the basis of open-circuit dissolution experiments at a rotating ring-disk electrode. However, we suspect they are observing the slow dissolution of occluded UO_2CO_3 in the pores of a severely roughened electrode. Reactions (1), (3) and (6) can only be considered as partially representing the first electron transfer step. As demonstrated by our XPS results, dissolution is occurring from a layer of $UO_{2.33}$ and not from the UO_2 substrate.

REFERENCES

1 S. Sunder, D.W. Shoesmith, M.G. Bailey, F.W. Stanchell and N.S. McIntyre, J. Electroanal. Chem. 130 (1981) 163.
2 M.J. Nicol and C.R.S. Needes, Electrochim Acta, 20 (1975) 585.
3 C.R.S. Needes, M.J. Nicol and N.P. Finkelstein, in Leaching Reduct. Hydrometall. ed't by A.R. Burkin, IMM (London), 12 (1975) (and references therein).
4 S. Sunder, D.W. Shoesmith, M.G. Bailey and G.J. Wallace, J. Electroanal. Chem. (in press.)
5 M.J. Nicol and C.R.S. Needes, Electrochim Acta, 22 (1977) 1381.
6 N.S. McIntyre, S. Sunder, D.W. Shoesmith and F.W. Stanchell, J. Vac. Sci. Technol. 18 (1981) 714.
7 S. Sunder, D.W. Shoesmith, G.J. Wallace and M.G. Bailey, to be published.

EXAFS AS A PROBE OF THE PASSIVE FILM STRUCTURE

L. BOSIO, R. CORTES, A. DEFRAIN, M. FROMENT, A.M. LEBRUN

Groupe de Recherche n° 4 du C.N.R.S. "Physique des Liquides et Electrochimie",
associé à l'Université Pierre et Marie Curie,
4 place Jussieu, 75230 Paris Cedex 05, France.

ABSTRACT

Various techniques for acquiring EXAFS data are examined and their relative
advantages and disadvantages delineated ; particular attention is focused on
in-situ methods. Finally a survey is presented of potentiality and limitation
in the determination of the local environment in passive layers by EXAFS
spectroscopy.

INTRODUCTION

Despite the intensive research about the processes involved during the passivation phenomena, using modern and powerful techniques to determine the
passivating species, there is as yet a lack of agreement concerning the nature
of the passive films. A probable explanation for the discrepancies between the
reported results may be related to the passive film preparation but also to the
modification occuring during the layer studies. For instance, nitrite-formed
film on iron can differ from an anodic oxidation film obtained in a borate
solution owing to the possible incorporation of some atom or ion species as
glass forming elements ; on the other hand, there is no guarantee that the
passive film remains unaltered during, say, ionic bombardment before an AES
measurement or simply by removal the sample from electrolyte to the vacuum environment of a spectrometer.

The present-day tendency is to develop new techniques allowing in-situ
studies, free from destructive effects. Since there is no commonly accepted
view concerning either the chemical composition or the structural order of the
passive film, EXAFS spectroscopy seems to obtain many workers' interest. In
this paper we attempt to review some EXAFS devices suitable for studying structural parameters of thin films and thus applicable to passive layers.

DESCRIPTION OF THE METHODS

In addition to the conclusions one can draw about the chemical bonding from
the shape and the energy shift of the absorption edge the modulating part
-called the Extended X-ray Absorption Fine Structure (EXAFS)- observed up to
1000 eV above the threshold of the absorption coefficient may provide a new

method for local order determination, in situation where diffraction fails.

The most direct and commonly used method to acquire EXAFS data is a simple transmission experiment on samples of convenient thickness. Indeed, for our purposes, such experimental conditions reduce the usefulness of the technique since the contribution of the surface is negligible with respect to the signal from the bulk. Fortunately some specially designed systems can overcome the limitation of the method and allow surface studies.

At usual glancing angles, for X-ray energies higher than about 7 keV and for moderately heavy atoms, the penetration depth reaches several µm so that the application of EXAFS to surface studies requires the use of i) either a detection mode which is sensitive to a small number of atomic layers within the surface (this is the case in photoelectron detection) ii) or a very thin film deposited onto a substrate free from atom species identical to the absorbing atoms under investigation ; these experiments thus employ fluorescence detection. An alternative method is based on an extended analysis of reflectance data (ReflEXAFS) ; in this case the penetration depth for grazing angles lower than 0.5° is low, say <50 Å, and bulk samples can be used. This method has been developed in our laboratory and preliminary results will be given further.

1) Ionization chamber employing photocathode emissions

When X-rays strike a sample, electrons are ejected from the sample. The observation of photoelectrons emitted by the surface in the absorption process is routinely performed in SEXAFS spectroscopy (1). However these techniques using Auger electron signal or the total electron yield are not suitable for the passive film studies since they require a vacuum environment. But, an interesting variant of these methods has been developed (2) : in an ionization detector filled of a rare gas at atmospheric pressure, the flat sample takes place of the cathode (Fig. 1).

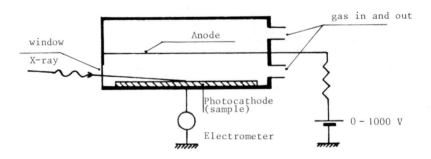

Fig. 1 : Schematic diagram of photocathode ionization chamber

When the X-ray beam impinges surface at grazing incidence, electrons are emitted from the photocathode and amplified by impact ionization as they pass through gas to the collecting anode. In addition to the amplification part, the use of helium gas helps to keep small the background current due to absorption length of the incident X-rays (note that great care has to be taken to keep constant the pressure inside because the chamber works as a sensitive manometer).

One of the advantages of the device is to allow experiments on bulk samples since the electrons emerging from the cathode are only generated near the surface. Incidentally, it is worthy of note that the detector current is a function of the illuminated area and the roughness of the flat sample is of little effect as far as the grazing angles are not too small.

Another advantage of the photocathode chamber is its good efficiency (at least an order of magnitude higher than that of the fluorescence detection described below) so that house built EXAFS design can be used. Actually, extended fine structure spectra from iron passivated in either a potassium chromate or a sodium nitrite passivating solution were already obtained by such a technique (3).

Of course, in-situ measurements on passive layers are ruled out owing to the short electron mean free path in liquids, but loss of chemically bound water is avoided since the surface analysis does not require exposure to vacuum.

2) Fluorescence detection upon ultra thin film

Fluorescence and in some few cases also luminescence, in principle, permits EXAFS spectra measurements. In these experiments, the incident radiation impinges on the sample at an angle θ_i relative to the front plane of the sample ; the resulting fluorescent radiation is monitored by a detector that subtends a solid angle as great as possible (Fig. 2). The simplest device employs scintillation counters in conjunction, if possible, with a filter assembly (4).

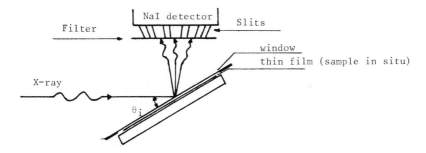

Fig. 2 : Experimental arrangement for obtaining EXAFS data on thin films using fluorescence detection.

The first limitation of the method is related to the rather low signal emitted from such surfaces since samples consist in thin films. However, surface sensitivity can be increased by making θ_i as small as possible and it has been shown (5) that EXAFS spectra could be obtained from films 30 Å in thickness. The small absorption coefficient of electrolyte for X-ray wavelength of about 1.5 Å makes possible to obtain EXAFS data from an in-situ passive film under a thin layer of electrolyte, with of course a loss of sensitivity (18,19).

The second limitation is related to the sample preparation. In addition to the fact that very thin films may exhibit properties different from those in the bulk (on account of texture, impurities ...) which depend on formation mode, great care has to be taken to passivate the entire film thus preventing the EXAFS signal from the surface layer to be swamped by that from the remaining unaltered underlayers.

By combining this method to the specular reflectivity of X-rays, some of the above limitations can be overcome.

3) <u>EXAFS study on superficial regions by means of specular</u>
   <u>reflection (ReflEXAFS)</u>

At X-ray frequencies the refractive index of condensed matter is less than unity ; up to the critical angle θ_c, the reflectivity of a flat surface is high and the penetration depth low (6). In such experimental conditions, the reflectivity depends on the absorption coefficient and therefore contains the extended-X ray absorption fine structure (7,8,9).

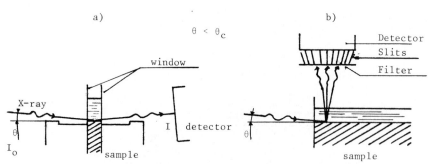

Fig. 3 : Experimental arrangement for obtaining ReflEXAFS data
a) in the reflected beam   b) using fluorescence detection

By measuring either the reflected beam intensity (Fig. 3a) or the fluorescence intensity (Fig. 3b), this technique overcomes the difficulties noted in the previous method since bulk samples can be used, on reserve that their surface

would be smooth enough. On the other hand we have performed measurement of the
reflected intensity versus angles of incidence (at constant energy) in order to
provide further informations concerning some physical properties of the superficial region : as an example, the passivative film, formed after cycling polished
iron sample in phosphoric acid solution 1N (10),   exhibits a rather low
apparent density, 2.8 g.cm^{-3}, as deduced from the reflectivity curve shown in
Fig. 4. In addition to such a result, the lack of Kiessig interferences and the
shape of the curve give evidence of a porous passive film with inequal thickness.

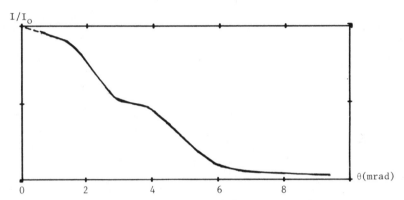

Fig. 4 : Angular-dependent reflectivity of a passivated
iron sample (X-ray energy : 8395 eV).

The ReflEXAFS spectrum carried out from this sample, at grazing angle slightly
higher than the critical angle θ_c relative to the passive film, resembles, as
expected, the EXAFS from pure iron whereas at small angle, say 2.8 mrad, the
fine structure modulations are related to the local order in the passive layer.
The Fourier transform of the EXAFS signal is shown in Fig. 5 : the first peak
is related to the Fe-O distances and the second peak to the Fe-Fe distances.The
positions of these maxima are similar to those found in experiments performed on
thermally oxided iron or, using this time transmission mode, from Fe_2O_3 and
Fe_3O_4 powders.
    We have obtained same results from Ni samples passivated in concentrated
sulfuric acid solutions (12 N). In this case the EXAFS spectra exhibited only
one distinct oscillation thus indicating a rather hith structural disorder in
the oxided layer.
    Cells for in-situ studies using both fluorescence and reflected radiation has
been already described (11).
    One of the major advantage of the ReflEXAFS method is its high efficiency
allowing measurements on a laboratory built design equipped with a conventional

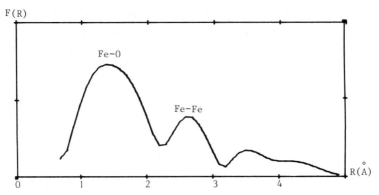

Fig. 5 : Fourier transform of the EXAFS spectrum obtained on passivated iron sample.

X-ray tube in conjunction with a bent mirror (12). Such a sensitivity largely compensates the difficulties in the alignment process ; indeed, the specular re-flection at low angles (< 10 mrad) requires high precision in the sample settle-ment.

CAPABILITY AND LIMITATION OF EXAFS IN PASSIVATING FILM STUDIES

The range applications of EXAFS spectroscopy has been discussed at length in the literature (see for instance (13) and ref. therein). It is reasonable to base hopes on this method which probes directly the position and identity of the nearest neighbors of the excited atom. The techniques described above showed that extended fine structure can be used in the passive layer studies. Thus the limitations arise essentially from the EXAFS method itself.

Structural informations require measurements over a large wave vector range (say, 2.5 to 15 $\AA^{-1}$) and knowledge of several factors which express the compli-cated interactions of the photoelectron with the backscattering atoms. Usually, these parameters ab initio calculated (14) have to be readjusted from standard materials data assuming the transferability concepts. Nevertheless, some diffi-culties arise in spectra analysis ; for instance in iron the first and the second shells in the real space are not resolved and peaks at larger distances differ appreciably both in amplitude and in position in regard to the expected values (15,16). Thus the phase shifts determination from α-Fe study has to be carefully analysed before transfering them to unknown systems.

Another crucial problem arises from the non-Gaussian peak shapes in the pair correlation function : actually, the simplest model often used to analyse EXAFS data is based on the assumption that the Ni neighbors in a shell i are confined in a Gaussian distribution about $\bar{r}_i$, the mean radial position of the atoms ; with such a distribution the analysis of data are readily tractable. However, in most cases including glasses and disordered materials (a probable occurence for passive layers) the nearest peaks are strongly asymmetric in form ; it is obvious that the usual analysis is unable to lead either to the real mean nearest-neighbor distance or to the coordination number (13).

Furthermore the possible coexistence of various oxides and other compounds in the passivative film may complicate the data reduction and the identification process. For instance, EXAFS spectra on γ-Fe_2O_3 and Fe_3O_4 have been obtained but the differences between the two curves are not very clear and, in practice, identification of these compounds will not be easy. Fortunately, as said above, the position and the shape of the absorption edge can provide some further informations (17).

These limitations notwithstanding, significant advantages will make EXAFS a valuable and complementary structural probe of passive layers : particularly small amount of structural variations as function of any electrochemical para-meter will be detected with good accuracy. Encouraging results were already obtained from ex-situ experiments and, certainly, other papers will be published in this topic (18,19).

REFERENCES

1 J. Stöhr, R. Jaeger and S. Brennam, Surface Science, 117 (1982) 503.
2 N.J. Shevchik and D.A. Fischer, Rev. Sci. Instrum., 50 (1979) 577.
3 G.G. Long, J. Kruger, D.R. Black and M. Kuriyama, J. Electrochem. Soc., 130 (1983) 240.
4 E.A. Stern and S.M. Heald, Rev. Sci. Instrum., 50 (1979) 1579.
5 J.A. Del Cueto and N.J. Shevchik, J. Phys. C : Solid State Phys., 11 (1978) L833.
6 L.G. Parratt, Phys. Rev., 95 (1954) 359.
7 R. Barchewitz, M. Cremonese-Visicato and G. Onori, J. Phys. C : Solid State Phys., 11 (1978) 4439.
8 R. Fox and S.J. Gurman, J. Phys. C : Solid State Phys., 13 (1980) L249.
9 G. Martens and P. Rabe, J. Phys. C : Solid State Phys., 14 (1981) 1523.
10 C. Pallotta, Thesis, Buenos Aires 1981.
11 L. Bosio, R. Cortes, A. Defrain and P. Gomes da Costa, 33rd Meeting Intern. Society of Electrochemistry, Lyon (sept. 1982).
12 L. Bosio, R. Cortes and G. Folcher, To be published.
13 T.M. Hayes and J.B. Boyce, Solid State Physics, 37 (1982) 173.
14 B.K. Teo and P.A. Lee, J. Am. Chem. Soc., 101 (1979) 2815.
15 G.S. Knapp and P. Georgopoulos, "Crystal, Growth, properties and applica-tions, Springer-Verlag, Berlin Heidelberg, 7 (1982) 82.
16 N. Motta, M. de Crescenzi and A. Balzarotti, Phys. Rev., B27 (1983) 4712.
17 M. Belli, A. Scafati, A. Bianconi, S. Mobilio, L. Palladino, A. Reale and E. Burattini, Solid State Communications, 35 (1980) 355.
18 J. Kruger, G. Long and M. Kuriyama, This volume,
19 R.W. Hoffman, This volume,

Passivity of Metals and Semiconductors, edited by M. Froment
Elsevier Science Publishers B.V., Amsterdam — Printed in The Netherlands

EX-SITU AND IN-SITU SAMPLE-AND-DETECTOR CHAMBERS FOR THE STUDY OF
PASSIVE FILMS USING SURFACE EXAFS

G. G. LONG, J. KRUGER and M. KURIYAMA
Center for Materials Science, National Bureau of Standards, Washington, DC 20234

ABSTRACT

Two sample-and-detector chambers for the study of surface films on metals
using x-ray absorption spectroscopy are described. Results have been obtained
using both a high intensity rotating anode x-ray generator and using the Cornell
High Energy Synchrotron Source (CHESS).

INTRODUCTION

The structural study of surface passive films on metals requires a high
sensitivity experiment which examines the film in an environment as near as
possible to that in which the real films are used. An extended x-ray absorption
fine structure (EXAFS) experiment can be used to probe the near neighbor
distribution in the film, and the near edge spectral features are sensitive to
the electronic structure in the film. The high sensitivity to the surface
signal is achieved here by having an ultrathin substrate layer and by using an
innovative detection system. For the in solution (in situ) experiment, the x-
ray fluorescence product of the K-absorption event is detected, while for the
ex situ (but non-vacuum) experiment the total yield electron product is detected.

THE EXAFS EXPERIMENT

The x-ray absorption experiment requires that a monochromatic (i.e. ΔE of
the order of 1-10 eV) x-ray beam be continuously tuned through energies from
several eV below a K- or L-absorption edge to about 1000 eV on the high energy
side of the edge. The basic x-ray spectrometers - to be used in EXAFS optics
as "monochromators" - are divided into two classes: flat-crystal instruments
and bent-crystal instruments. The flat-crystal instruments with two or more
crystals generally deliver the higher resolution. The bent-crystal instrument
collects and diffracts photons within an energy band pass of the incident beam
and focuses them to a position on the Rowland circle. This method of
monochromatization reconcentrates the beam at the exit slit of the instrument
and is very efficient (ref. 1).

EX SITU X-RAY ABSORPTION DETECTOR

A high-pass photocathode x-ray ionization chamber, especially designed for surface EXAFS (refs. 2 and 3) was adapted for structural studies of passive films on metals. The sample, of which less than 10% may be the substrate metal or alloy, serves as the conducting cathode plate inside the helium filled detector. A filter assembly in front of the detector limits the energy band pass of the detector to the region between the absorption edge of the photo-cathode and that of the filter. This means, for example, that an iron K-edge absorption experiment would make use of a cobalt filter, so that the high bandpass of the detector would be between 7.1 and 7.7 keV.

A schematic of the sample-and-detector chamber is shown in Fig. 1. The x-rays enter through an aluminized mylar window and impinge on the sample (which is also the photocathode) at grazing incidence. The photocathode plate absorbs the x-rays and emits electrons, which are amplified through field intensified ionization in the helium gas. The current passing through the detector is measured by an electrometer connected to the plate. The result is an experiment similar to total yield spectroscopy in which the electrons are collected as a function of the energy of the incident x-ray beam.

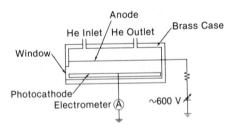

Fig. 1   The ex situ EXAFS detector and sample chamber.

Four kinds of electrons emerge from the photocathode; photoelectrons, secondaries, inelastically scattered electrons, and Auger electrons. The photoelectrons are excited simultaneously with the absorption of the x-rays, but the kinetic energy of these electrons near the absorption edge is very low, and thus they will not be able to produce much detector current. The secondary electrons that have been excited by collisions, or the inelastic electrons that are photo- or Auger electrons that have suffered inelastic collisions, have a wide spectrum of energies. The Auger electrons are emitted by the decay of the core hole states created by the photoabsorption. These electrons have large kinetic energies. The subsequent decay of the L shell holes is usually dominated by Auger emission so that it is possible that nearly one electron may be excited for each photon absorbed.

The electrons generated near the surface of the cathode undergo field intensified ionization, increasing the current by a multiplying factor. Further sensitivity is achieved in that this detector collects over nearly the entire 2π steradian hemisphere. Using a 12 kW laboratory x-ray generator, focusing optics (ref. 1) and the present detector, the thin film EXAFS spectra of passive films shown in Fig. 2 were each obtained in the approximately 4 hours of scanning. The magnitudes of the Fourier transforms of these data are given in Fig. 3.

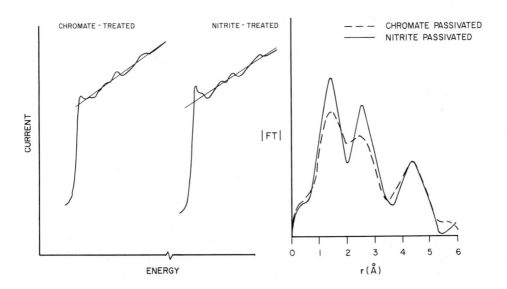

Fig. 2   EXAFS spectra of the films   Fig. 3   Magnitude of the Fourier
          formed in each of two   transforms for the data
          passivating solutions.   shown in Fig. 2.

IN SITU X-RAY ABSORPTION CELL

The in situ EXAFS measurement requires that the x-ray absorption spectra be taken while the thin film is immersed in an aqueous solution where electrochemical studies are performed. Fortunately, water becomes transparent again in the hard x-ray region of the electromagnetic spectrum. In this case, the x-ray absorption event is followed by measuring the photon fluorescence product of the decay of the core hole. Data collection is somewhat hampered by a large scattering component from the solution. This is brought under control through the use of an x-ray filter assembly in front of each x-ray detector, so that the signal is dominated by the fluorescence photons.

A schematic of the cell is shown in Fig. 4. All of the elements of the cell are embedded in an inert epoxy so that the surface of the film under investigation can be brought nearly into contact with the x-ray window. The reference electrode is connected to a Luggin capillary embedded in the epoxy with its opening in the plane of the specimen. The Pt counter electrode also lies in the same plane of the specimen and is connected through the epoxy to the rear of the cell. The metal sample is deposited on the surface of a glass slide and connection is made through a plug of bulk metal through the glass.

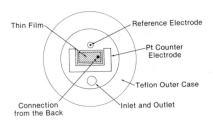

Fig. 4   The in situ EXAFS cell.

It is desirable to do the electrochemistry in a rather larger volume of solution, so the entire cell is mounted in a piston arrangement in a teflon outer case. This permits us to reduce the "as received" sample and grow the film on a bare metal surface in a large volume of solution. Once the film has formed, the current is very small, and the sample can safely be brought to within a small fraction of a mm of the x-ray window for measurement while under potentiostatic control.

The cell was tested to demonstrate that a passive film could be formed and maintained under anodic polarization. The cell was also tested to demonstrate that spectra could be obtained through the x-ray window and the aqueous electrolyte. In situ and ex situ data were taken using the C-1 monochromator at CHESS. Fig. 5 shows the magnitude of the Fourier transform for the (a) ex situ and (b) in situ data, both taken in high resolution geometry. Fig. 5(a) is to be compared with Fig. 3 which was derived from laboratory data using the ex situ detector. Data collection times were similar.

143

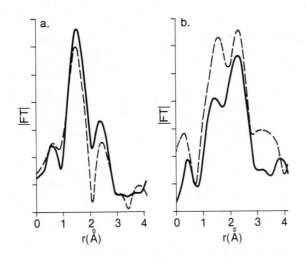

Fig. 5   Magnitudes of the Fourier transforms for the (a) ex situ and (b) in
         situ passive films. The nitrite-formed film results are given by the
         dashed lines and the chromate-formed film results by the solid lines.

REFERENCES

1 G. G. Cohen, D. A. Fischer, J. Colbert and N. J. Shevchik, Rev. Sci. Instrum.
   51 (1980) 273 pp.
2 N. J. Shevchik and D. A. Fischer, Rev. Sci. Instrum. 50 (1979) 577 pp.
3 D. A. Fischer, G. G. Cohen and N. J. Shevchik, J. Phys. F 10 (1980) L139 pp.

Passivity of Metals and Semiconductors, edited by M. Froment
Elsevier Science Publishers B.V., Amsterdam — Printed in The Netherlands

ROUND TABLE DISCUSSION

CONCEPTS - COMPOSITION - TRANSPORT PHENOMENA

N. MOTT asked whether Schultze's results indicate the presence of deep donors, i.e. does the addition of Pd to the passive film make it become more metallic.

W. SCHULTZE agreed that over 10 % Pd produced a film that was nearly metallic. If 1 % is added the film's electronic properties are improved ; 3 % produce a film with optimal electronic properties especially those that affect transfer reactions.

U. STIMMING suggested that Mott's idea may not be entirely correct. He indicated that the dielectric constant should decrease. Pointed out that Xe implantation also increased apparent dielectric constant. An idea that Mott agreed may be valid.

N. MOTT also suggested that Good's model is an improvement to the Mott and Cabrera for crystalline solids but not as good for amorphous solids because shallow donors are not treated.

B. CAHAN suggested that the implantation of Xe ions in Schultze's studies introduces damage into a film's lattice. He felt that this damage was more important than the concentration of new species in the lattice. The Xe ions make the outside of the film a conductor and the inside an insulator. Schultze answered that Rutherford backscattering experiments (examining the thickness of the film) showed that Cahan's ideas are not valid. Two reasons : (a) In electron transfer reactions electrons were formed to go through the film rapidly - an impossibility if part of the film is an insulator ; (b) A study of the repassivation behavior found that the film containing 10 % of the implanted ions could not be repassivated but films containing less than 10 % could be repassivated. In other words the damage layer contributed only slightly.

R.P. FRANKENTHAL suggested that XPS may be a good way to look at this problem and to examine O/He ratio at the surface after implantation.

M. FROMENT initiated the discussion on the poster session.
"What is new in the case of in-situ measurements ? What kind of new data will be available ? "

W. SCHULTZE : Mainly information about structure.

J. KRUGER : The crucial goal and, indeed, dream of passivation studies it to be able to make in-situ determinations of the chemical composition and structure of the passive film. Thus far, with a few exceptions, we have been able to do this ex-situ in a vacuum where the film changes.

A short discussion takes place about the surface selectivity of in-situ and ex-situ techniques.

P. MARCUS : True surface sensitive techniques (Auger, XPS) can only be used in vacuum. In-situ techniques are not surface selective in that sense.

R. CORTES : EXAFS is very sensitive to surface effects.

R.W. HOFFMAN : The in-situ and ex-situ experiments to bridge the electrochemical and vacuum environments.

R.P. FRANKENTHAL : In relation with the interesting poster by Mrs. Hugot and coworkers on Raman techniques, can the technique be applied to thinner layers of the order of say 20 Å ?

A. HUGOT : By accumulating data over many spectra it is possible to identify very thin films except if they are not crystalline.

B. MAC DOUGALL : XPS, Auger, SEMS are very sensitive but we must be extremely careful with possible impurities effects. Calibrating the instrument is the problem.

B. CAHAN : We are using sophisticated techniques and we try to push them to their limit. None of these techniques can solve the problem completely but a negative result is as useful as a positive one.

H.Jr. LEIDHEISER : We are talking about positive or anodic passivity. Has anybody an example of negative passivity ?

J. YAHALOM : The deposition of Cr.

J. KRUGER : Active-passive behaviour of Al.

Passivity of Metals and Semiconductors, edited by M. Froment
Elsevier Science Publishers B.V., Amsterdam — Printed in The Netherlands

STUDY OF PASSIVITY OF IRON BY IN SITU METHODS: MOSSBAUER AND EXAFS

R. W. HOFFMAN

Dept. of Physics and Case Center for Electrochemical Studies, Case Western
Reserve University, Cleveland, Ohio 44106 U.S.A.

ABSTRACT

Considerable controversy exists as to the nature of the passive film on
iron both in chemical composition and degree of structural order. Since the
structure of the passive layer is believed to change on removal, we chose to
study the passive layer in-situ, using MES and EXAFS spectroscopies.

As neither technique is inherently surface sensitive on the scale of the nm
passive layer at the surface of the iron sample, signal-to-background enhancement was necessary. For MES high counting rates and long times were necessary
in transmission. X-ray backscatter provides background suppression and was used
for low temperature data. A rotating electrode within a conversion electron detector has been developed to optimize the surface sensitivity for electrochemical studies.

For the EXAFS data it was necessary to observe the iron x-ray fluorescence
signal that is scattered at 90° from the incident synchrotron radiation beam
using a detector geometry with fan collimators to eliminate background radiation
and a thick MnO_2 filter to absorb much of the radiation at energies lower than
FeK_α. In addition, because the spectra represent the superposition of the passive layer and the remaining iron in the specimen, careful thickness control of
the 2 nm vacuum deposited iron was required.

The passivations were carried out in a buffered borate solution pH 8.4
which was deaereated with nitrogen. Any air formed oxide was reduced at a
cathodic potential of -350 mV RHE for 30 minutes followed by a step to -200 mV
to allow any hydrogen to escape from the surface. At this point the borate buffer is again replaced to remove any ferrous ions which may be in the solution
and the working electrode is stepped to an anodic potential in the passive
region, commonly +1350 mV RHE. Results for the in-situ transmission MES from
several laboratories indicate an isomer shift of Fe^{3+} and the large quadrupole
splitting and linewidth characteristic of disordered systems.

Experiments were performed to look at spectra of iron species which might
precipitate from solution at anodic potentials. Iron gels anodically deposited
gave doublets with smaller quadrupole splittings then for the passive film. A
natural Fe film passivated in a ^{57}Fe-containing buffered borate produced no
Mossbauer spectrum. Data for other passivating electrolytes is reported.

Reference EXAFS spectra were obtained for crystalline αFe_2O_3, Fe_3O_4, FeO,
$Fe(NO_3)_3 \cdot 9(H_2O)$, the oxy-hydroxides, α, γ, and $\delta FeOOH$ and several Fe containing
gels.

Fits to the EXAFS spectra of a 6-nm cathodically protected film showed
8.7 ± 0.5 iron atoms 0.248 ± 0.001 nm from the central atom and virtually no
oxygen, in agreement with bulk iron metal which has 8 iron atoms at 0.248 nm.

Iron atoms in the passive film appear to have a local environment of only
oxygen nearest neighbors at an average distance of 0.203 nm. Furthermore, the
EXAFS spectra of the passive films also indicate the absence of bulk crystalline oxides.

INTRODUCTION

Because of the nature in which thin passive layers are grown in electro-chemical cells it has long been a concern of whether the structure and hence properties of the film are the same under electrochemical environment as they are when subsequent measurements are made. Much of our information comes from the electrochemical measurements themselves, especially the interpretation of valence state and thickness information. In this paper we review in-situ tech-niques of Mossbauer effect spectroscopy (MES) and extended x-ray absorption fine structure (EXAFS) as practiced for passive films grown on iron. Both of these techniques use photons in the energy range from 6-15 KeV as the incident probe and photons or electrons for detection. As the structural information is carried by these photons or electrons, experiments using thin layer cells make it possible to use a number of different geometries and detection schemes to make the techniques near surface sensitive. In addition to the in-situ measure-ments samples may be removed from the electrolyte and measured in an environ-ment comparable to that used for other structural measurements.

Neither MES nor EXAFS has been widely applied to the study of passive films on iron. If we strictly restrict ourselves to in-situ measurements the litera-ture contains less than five papers for each technique. We shall give the necessary background to design and carry out an in-situ experiment but also call upon the complete literature from these techniques when we discuss the results.

Because MES and EXAFS in-situ electrochemical experiments are not common we shall first give a brief description of the physics of each spectroscopy follow-ed by the experimental design requirement. Neither technique is inherently sur-face sensitive. Thus the geometry and detection technique must enhance the signal-to-noise in order to probe a thin layer. MES and EXAFS have a common background in that both use a comparable range of photon and electron energies. Hence, similar electrochemical cell design concepts may be used although in practice we have found the EXAFS case to be more difficult.

We shall then present the data that is available both from our laboratory as well as from the literature. Finally, we examine the consequences of the new information available from the in-situ measurements with respect to existing models for the structure of the passive film.

Mossbauer Effect Spectroscopy

The Mossbauer effect, now 25 years old, provides a sensitive and highly localized probe in the form of the Mossbauer isotope nucleus, in the case of iron, ^{57}Fe. As is reviewed in many places (ref.1,2), the physics of hyperfine interactions is revealed by the form of the absorption or emission spectra as a function of small shifts in the energy of the nuclear energy levels resulting from its local environment.

Three terms are commonly identified. The isomer shift is evidenced by a shift of the center of gravity of the spectrum and physically may be related to the valence. The quadrupole splitting arises from the interaction of the nuclear electric quadrupole moment and the electric field gradient at the nuclear site. For a given Mossbauer isotope it is determined by the magnitude and sign of the electric field gradient and hence, is a measure of the deviation from cubic site symmetries. The nuclear Zeeman effect gives rise to a sextet which in turn measures the magnitude and direction of the effective magnetic field at the nuclear site. Paramagnetic or small particle relaxation effects give rise to additional spectral features which may be a useful complication in the present problem. A detailed interpretation of the spectral line shapes and intensities is required for the present problem and puts severe requirements on the statistical quality of the data and fitting procedures. However, the valence and site geometry information obtained is invaluable.

Extended X-Ray Absorption Fine Structure

The term EXAFS describes the oscillatory portion of the x-ray absorption coefficient for the energy region from roughly 50 eV to 1000 eV higher than the absorption edge. The basic physical explanation resides in modifications of the final state of the photoelectron by the solid. The interest in EXAFS as a structural tool resulted from the availability of tuneable sources of high intensity and the recognition that a single scattering short range order theory is usually sufficient.

EXAFS is a rapidly changing field but we reference here only the first modern paper (ref.3) and a recent review (ref.4). EXAFS has the capability of determining the bond distances and the number and species of the atoms for several shells surrounding the element examined.

As the EXAFS oscillations result from the interference of the outgoing photoelectron wave with those backscattered by nearby atoms, an independent knowledge of phase and amplitude functions with wave vector is required in the analysis. Theory is helpful (ref.5), but these parameters are usually determined from model compounds whose structure is known from x-ray diffraction.

EXPERIMENTAL DESIGN FOR IN-SITU MES AND EXAFS
General Considerations

Passive layers are generally only a few nanometers thick and data of sufficient quality for quantitative analyses are not commonly obtainable from MES and EXAFS in short periods of time. Hence, at every stage one must use all techniques available to maximize the signal and minimize the background noise. We shall discuss passivation cell detector and source optimization and some of the compromises that must enter the design. The same general concepts apply to both MES and EXAFS, but we have used different cells in practice.

It would be desirable for the technique to have near atomic spatial resolution. However, we must settle for samples having areas of $\sim$ 1 cm^2 exposed to the incident radiation. For some geometries monolayer depth resolution is possible, but not yet practiced. For a true surface technique, UHV and electron escape depth limitations are usually required. A UHV MES system has been designed with submonolayer sensitivity in the emission or 'source' geometry (ref. 6) for surface magnetization experiments. SEXAFS or surface EXAFS is accomlished using Auger or secondary electron detection for surface sensitivity, (ref. 7,8). Neither qualify as electrochemical in-situ techniques.

Electrochemical Cell Design Transmission

Both techniques may be designed as in-situ adsorption or emission (scattering) spectroscopies. In the usual case of MES transmission, the isotopic resonant adsorption cross-section and concentration must be sufficient to be statistically significant below the off-resonance background. The recent trends are to strong sources (50-100 mCi) and fast counting rates (10^5-10^6c/s) (ref.9) in order to obtain the $\sim$ 2 x 10^7 counts per velocity channel required to quantify spectral details. The cell design is a choice between well collimated geometry to reduce linewidths and preserve Lorentzian shapes and increasing counting time beyond the $\sim$ 20 hours needed to obtain a spectra. In spite of this "brute-force" approach most of the in-situ MES data has been obtained this way because the cell may be used to electroplate the metallic film prior to passivation. This type of cell was designed by O'Grady (ref.10) and provides an adjustable electrolyte thickness in addition to iron (sample) working, gold counter and PdH reference electrodes necessary for the potientiostatic electrochemistry.

If synchrotron radiation sources are used for transmission EXAFS, the ion chamber detectors are exposed to sufficient radiation that the quality of the data is not normally limited by nuclear counting statistics. The sample thickness is adjusted such that $\mu x \sim 2.5$ for best signal-to-noise (ref.4). The many subleties that exist in the design of monochromaters to minimize multiple Bragg reflections and mirrors which act as low pass filters are beyond the scope of this paper, but must be examined for our problem. Transmission EXAFS does not have the sensitivity necessary for passive layer detection because μx is so small that the transmitted intensity is indistinguishable from the incident intensity (ref.11).

Scattering. The sensitivity may be greatly increased by using scattering geometry. For both MES and EXAFS, it is important to realize the abcissa of the spectra is measured by the velocity of the source drive for MES and the monochrometer position for EXAFS. Thus the energy axis is not determined by the detector resolution. As a result, it is only necessary to detect a signal proportional to the absorption event to produce the desired spectra. Crudely, the scattering geometry allows the "background" to go through the cell undetected and a proportional signal to be detected. For MES, either the FeK_α x-rays on the $\sim$ 5 KeV conversion electrons may be used (ref.12,13). In addition to the improved signal-to-noise resulting from the background suppression and even more efficient electron counters the scattering geometry may take advantage of large area (in principle approaching 4π) detectors without the loss of resolution. In the case of ^{57}Fe, an approximate tenfold increase in counting rate over transmission results from the increased probability of internal conversion following the nuclear absorption. Some depth resolution may be obtained from energy analyzing the conversion electrons. The source-detector geometry is commonly such that the radiation scattered through angles greater than 90°; hence the terminology "backscatter detector". Equations for the signal-to-noise ratios are given in reference 13.

In the case of EXAFS a favorable geometry has the incident and fluorescent beams leaving the front surface of the sample symmetrically at 45°. This 90° geometry takes advantage of the polarized synchrotron radiation. Additional background suppression can be obtained by the use of filters and focussing collinators (ref.14).

An important consideration in the construction of an in-situ cell is the window material. For the 14.4 KeV radiation used in transmission MES, the signal loss in traversing two windows and perhaps a millimeter of electrolyte is acceptable. However, the softer $K\alpha$ radiation requires cells with thinner windows and electrolyte layers. The present status of fluorescent EXAFS allows the observation of the Cl edge (ref.15). Practical sensitivities, expressed as monolayers of Fe, are given in Table I.

TABLE I
Practical sensitivity limits for in situ MES and EXAFS geometries expressed as monolayers of Fe

Cell	Detection		Sensitivity for	
	MES	EXAFS	MES	EXAFS
Transmission	14.4 KeV p	7-8 KeV p	5	> 1000
Scattering	6.3 KeV p	6.3 KeV p	1	5
CEMS	5 KeV e		0.5	?
Emission	14.4 KeV p		0.01	

In addition to minimizing the window absorption to the incoming and fluorescent radiation, conducting but chemically inert electrodes are required for cell construction. 20 nm thick evaporated Au films on Melinex®-505, a polyester film treated to promote adhesion for printing applications, satisfies these multiple requirements and does not lose adhesion under hydrogen or oxygen evolution (ref. 16). The Au coated Melinex® may serve both as a counter electrode or as a substrate upon which an Fe film is electrodeposited or evaporated.

Examples of in-situ Cells. We have found that using cells with flexible walls gives a freedom of design that serves many applications. This "bag" concept has been used for a number of in- and ex-situ MES adsorption studies (ref. 17) and has been incorporated into the present fluorescence EXAFS geometry.

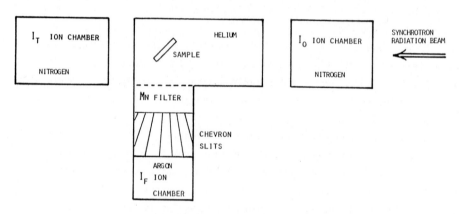

Fig. 1 a. In situ fluorescence EXAFS scattering chamber

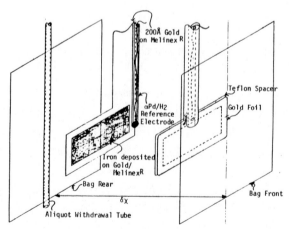

Fig. 1 b. Detail of bag passivation cell

A schematic view of the experimental geometry is presented in Figure 1a (ref.18). The incident x-rays were generated by bending the trajectory of the electrons in the storage ring (SPEAR) of SSRL. The incident x-ray intensity was detected in the I_0 ionization chamber; the beam then impinged on the sample, located at 45° to the x-ray source. The absorption spectra was obtained by comparing the transmitted x-ray intensity (measured at I_T) with the incident intensity I_0. The fluorescence spectra was obtained simultaneously by comparing the fluorescent x-ray intensity (I_F) with I_0. Three channels of data were collected: I_0, I_T, and I_F.

Figure 1b illustrates an exploded detail view of the electrochemical passivation cell. The iron sample was vapor deposited on a Gold/Melinex® substrate. This assembly became the working electrode where the passivation occured. The counter electrode was gold foil (ca. 0.05mm), separated from the working electrode by a teflon spacer. The exposed area of the gold working electrode was 40% larger than the iron sample area to ensure even current flow. The reference electrode was comprised of hydrogen adsorbed on a palladium tip, charged prior to passivation. This type of reference electrode is advantageous since it does not contaminate the electrolyte. After passivation, the gold counterelectrode assembly was raised, the sample cell thickness was reduced, and the EXAFS spectrum was taken. If necessary multiple 20 min scans were taken for later signal averaging.

The horizontal translation device to adjust the electrolyte conductance when large currents are needed during reduction or initial stages of passivation has been omitted from the figure. The cell must be flushed with new buffered borate solution following reduction in order to remove the iron in solution, in part because it adds to the EXAFS signal. MES is insensitive to any iron in solution as the recoilless fraction is zero; this difference may be exploited.

We have also developed a new device for CEMS of solution grown passive films that avoids the difficulties of thick electrolyte layers and detector windows under almost in-situ conditions (ref.19).

The new method involves the electrochemical formation of the passive film on a substrate that can be continually emersed from the passivation cell, withdrawn into the conversion electron detector and subsequently returned to the electrolyte. The electrode to be passivated is a disk with its lower half under potential control (or open circuit) in the electrolyte. The emersed half is in the conversion electron detector, surrounded by the (mostly helium) counting gas. The disk is mounted on a motor that continually rotates freshly passivated surface into the beam of the γ-ray source and returns the previously measured area to the solution. The geometry is shown in Figure 2.

We have determined that the rotation of the substrate and the presence of moisture in the detector as well as liquid remaining on the electrode surface do

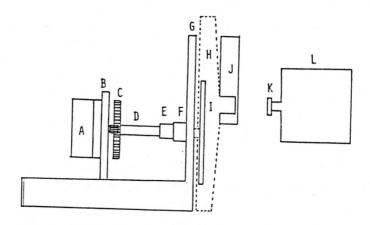

Fig. 2. Schematic diagram of rotating passivation cell for conversion electron detection. A. motor B. shaft bearing C. reduction gear D. electrically insulating shaft E. electrical contact F. teflon bushing G. support H. plastic bag cell I counter disk electrode J. conversion electron detector K. [57]Co MES source L. Mossbauer doppler velocity transducer.

not interfere with CEMS measurements. Recent work by Gordon (ref.20) indicates the electrolyte residue to be below 50 Å on silver films and to show some variation with potential. Whether potential control is maintained on the upper half of the electrode has not been determined, although Hansen and Kolb (ref.21) report numerous examples of successfully emersed electrodes.

The effect of transverse motion on the Mossbauer parameters has been calculated by Kirichenko and Chekin (ref.22). Preliminary data indicate that our device may be used at tangential velocities an order of magnitude larger than the current 0.72 mm/sec. and possibly greater velocities with improved collimation. This geometry appears to be especially promising as it increases the surface sensitivity as well as reducing the data acquisition time to less than 10 min. for usable MES spectra. We will attempt to extend the rotation/detection system to EXAFS and other electron detection spectroscopies.

EXPERIMENTAL RESULTS

We summarize the in-situ literature and present some new data.

Passivation in Buffered Borate

In-situ Mossbauer transmission spectra on ∿ 50 nm Å electroplated [57]Fe enriched films passivated in pH 8.4 buffered borate have been reported (ref.23) and confirm the major findings of the first in-situ data on Fe by O'Grady (ref. 10). Only one other in-situ study is known to us (ref.24) which concentrated on

the chemical breakdown after introduction of chloride ions. This MES study
identified superparamagnetic γFe_2O_3 as the corrosion product and demonstrated
a substantial difference from the oxide film formed in the absence of chloride
ions.

Mossbauer parameters for in-situ passive films are summarized in Table II.
The essential results are:

1. The isomer shift (I.S.) of 0.64 $\pm$ 0.01 nm/sec vs. SNP clearly indicates
   iron in the passive film is in the high spin Fe^{3+} state within a sensitivity
   limit of about 0.2 nm of oxide.

2. The passive film is reduced after applying a potential of -350 mV RHE.

3. An unexpected increase in the oxide resonant area occurs with cyclic repassivation.

4. No MES evidence is found for deposition of Fe from Fe^{2+} in solution after
   stepped passivation.

5. The large quadrupole splitting rules out bulk crystalline oxides, but not
   necessarily small particle oxides (ref.25) or disordered structures (ref.
   26).

6. The large linewidths observed are consistent with non-crystalline structure.

7. A small (15%) increase in resonant area with increasing potential is found.

8. Significant decreases in QS are found after removal of the passive film
   from the electrolyte and subsequent drying.

We have made further investigations using conversion electron, x-ray backscatter, and emission geometries sensitivity. These data are also given in
Table II.

Thinner iron films were prepared so that the fraction of the total resonant
area due to the passive film would be greater. One approach was just to electroplate thinner ^{57}Fe films. About 10 nm was the smallest film thickness for
which the electroplated film appeared optically uniform. The I.S. and Q.S. for
such a film passivated at +1300 mV SCE in pH 8.4 buffered borate are listed in
Table I. These values for the passive film are somewhat lower than those previously obtained for thicker samples. There was concern, however, that the non-
flat morphology of the electroplated films as revealed by SEM could significantly affect the passive film. SEM photographs showed the electroplated films to
consist of densely packed spheres about 50 nm in diameter and that the gold substrate was exposed in several places. To solve this problem, ^{57}Fe was deposited
onto gold on Melinex®. SEM photographs of these films showed no structure; uniform metallic films under 100 Å can be produced. The I.S. and Q.S. from a transmission spectrum of a 20 nm evaporated film passivated in borate are also listed
in the table.

Backscatter spectra of 7.5 nm (evaporated) ^{57}Fe film passivated in borate
show improved statistics for the passive film and   give an idea of its

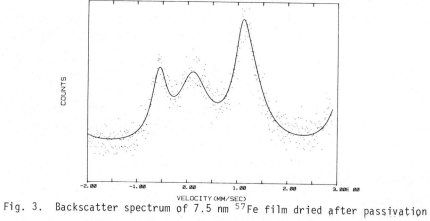

Fig. 3. Backscatter spectrum of 7.5 nm ^{57}Fe film dried after passivation

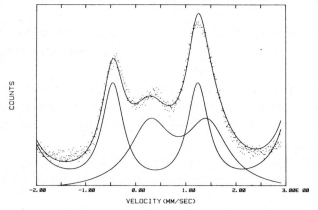

Fig. 4. In situ spectrum at +1300 RHE taken with rotation CEMS system

TABLE II. Mossbauer spectral parameters for various thin oxide films

Sample	Isomer shift mm/sec vs SNP	Quadrupole splitting mm/sec	Linewidth mm/sec	Ref.
Electroplated 50 nm				
O'Grady	0.70 ± 0.01	1.02 ± 0.07	0.86 ± 0.08	10
Furtak	0.62 ± 0.01	0.94 ± 0.06	0.55 ± 0.08	24
Our study	0.64 ± 0.01	1.14 ± 0.09	0.85 ± 0.15	23
Electroplated 10 nm				
Air oxidized	0.57	0.81	0.34	
Passivated +1350 mV	0.68	1.07	0.94	
Evap. 7.5nm pass.				
vacuum dried BS	0.64	1.11	0.88	
Phosphate passivation				
+1550 mV RHE	0.64	1.14	0.63	
+1750 mV RHE	0.64	1.14	0.67	
Dried	0.64	1.00	0.69	
Open circuit in H_2O_2	0.63	0.68	0.44	
Electrodeposited gel				
Room Temp.	0.62	0.80	0.49	
80 K	0.64	0.80	0.74	

relative thickness. The Q.S. and I.S. are about the same as for previous runs
with thicker films, as shown in Fig. 3. The resonant area from the passive film
is approximately one-fifth that from the entire film. If recoiless fractions
were equal, this would indicate about one-tenth of the iron is in the passive
film. This corresponds to about 1.5 nm of metallic iron or roughly 3.0 nm
oxide thickness.

The most distinguishing feature of the spectrum of the passive film is its
large Q.S. compared to other oxide films studies. For example, an air-formed
oxide formed on the 10 nm electroplated ^{57}Fe film before it was reduced gave a
Q.S. of 0.81 mm/s compared to the 0.92 mm/s for the subsequently passivated film.

Figure 4 shows the central two lines of the metallic iron CEMS spectrum and
the two line contribution from the passive film obtained with the rotating emer-
sion system. The sample was a 50 mm diameter, 1 mm thick Armco low carbon disk
mechanically polished and then electroplated with about 100 nm of ^{57}Fe cathodically reduced and passivated. These data set an upper limit for Fe^{2+} ions incorporated in the passive film of 1 monolayer of ^{57}Fe, assuming a doublet.

Additional Passivating Electrolytes and Iron Oxide Gels

Passivation in phosphate solutions was carried out; the role of the phosphate ion is expected to be very similar to that of the borate ion. An approximately 50 nm electroplated ^{57}Fe film was passivated in pH 7.0 sodium phosphate
at +900 mV SCE and then later stepped to +1100 mV SCE. The I.S. and Q.S. for
the passive film are identical to the passive film in the borate system (see
Table II). The film passivated in phosphate underwent a clear change upon air-
drying; there was a sharp decrease in the Q.S. Another film was passivated in
pH 10 Na_3PO_4; similar parameters were obtained for the passive film.

Passivation in electrolytes which should behave differently than borate was
examined. Open circuit passivation in 7.5% H_2O_2 gave a very prominent doublet
quite different from those seen in films passivated in borate or phosphate. The
much greater resonant area indicates a much thicker oxide film produced by the
peroxide than borate or phosphate.

Electrodeposited iron oxide gels were prepared following the method of
Leibenguth and Cohen (ref.27). These films should be similar to an overlayer
which might form on top of a passive film if ferrous ions were present in the
solution. The Mossbauer spectrum of the electrodeposited gel gave a symmetric
doublet with a Q.S. of 0.80 mm/s which is significantly lower than that obtained
for the passive film, but higher than the values reported for bulk γFeOOH.

In-Situ EXAFS Data

The first evidence that in-situ EXAFS data could be obtained for passive
layers (ref.11) demonstrated that the fluorescence mode had adequate sensitivity

and that any hydrated electrons did not adversely affect the passivation process. To our knowledge no other in-situ EXAFS data exists in the literature, but we call attention to the ex-situ papers for chromate and nitrate passivations by the NBS group (refs.28,29) and the in-situ paper at this conference (ref.31). These data support the disordered nature of the passive film and interpret near edge data and edge shifts.

Our efforts have been confined to the EXAFS region, in part because of the $\sim$ 5 eV x-ray energy resolution. Fluorescence EXAFS spectra have been reported (ref.30) for 2 and 6 nm passivated films, both in-situ and emersed, 1 and 10 nm air-oxidized metallic films, a 6 nm cathodically protected (-0.35 V RHE) metallic film, and iron in solution. Reference EXAFS spectra were obtained for αFe, α and γFe_2O_3, αFe_3O_4, FeO, $Fe(NO_3)_3 \cdot 9(H_2O)$ and the oxy-hydroxides, α, γ, and δ FeOOH. A transmission MES spectrum was obtained for an air-oxidized 1- nm iron film, afterwards, the film was reduced and passivated and another spectrum recorded with the film under potential control.

In order to analyze the EXAFS data, a good phase and amplitude reference is required with well separated peaks in a stable material whose structure is known. Initially none of the oxide or oxyhydroxide references gave satisfactory numbers for the coordination numbers, but the distances were within $\pm$ 0.05 Å for the standards. By using phase and amplitude information from FeO plus empirical scaling to γFe_2O_3 we obtained consistently correct values for the first shell in the crystalline standards (ref.18) but at a sacrifice of poorer values for the distance. EXAFS χ data uncorrected for phase shifts for α, δ, and γ precursor gels of iron are shown in Fig. 5. Upon dessication, these gels form the corresponding FeOOH crystalline species used for references. The gels have a local structure even though they are dispersed in an aqueous media. The gels were chemically produced following the technique of Misawa (ref.32). Comparison of the χ data and the Fourier transforms of the gels with their crystalline counterparts shows a shorter range character in the gels. The δ gel had only one major peak in the transform while the γ gel was found to have > 2% iron in the gel liquor by iron colorimetry, possibly accentuating the single shell coordination found in the EXAFS data for the γ gel.

There is little attenuation of the incident and fluorescent beams in the specimen; hence the data averages over the iron remaining in the substratum, as well as the iron in the passive layer, and of course any iron present in the electrolyte or as impurities in any materials seeing the beam. Spectral subtraction is then required to isolate the contribution of the passive layer. We have modelled EXAFS χ functions by mixing experimental dried passive film (on its $\sim$ 2 nm substrate) spectrum with a cathodically reduced (essentially metallic

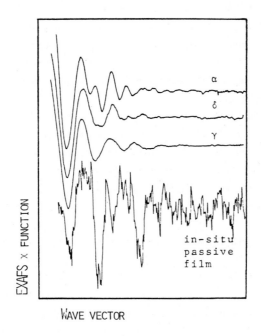

Fig. 5. EXAFS χ functions for α, δ, and γ precursor gels and 2.0 nm Fe film passivated at +1350 mV RHE.

iron) 6 nm spectrum. Because these spectra are quite different in periodicities, it is possible to recognize less than a 10% contribution from the iron spectrum.

However, the observed composite spectra from the iron film, passive film, and iron in solution (ref.31) could not be deconvoluted, but the redesigned cell described earlier in this paper has the features needed. Careful iron colorimetry with quartz crystal microbalance measurements confirm that a 2.0 nm vapor deposited iron film on a gold/Melinex(R) substratum is the optimal thickness. Of the original iron thickness about 0.7 nm air-oxidizes before the experiment and is removed during cathodic reduction, a thickness dependent 0.4 - 0.8 nm is converted to the passive film, and the remainder is left as the iron base. The equivalent passive layer thickness appears to be smaller than that determined in other experiments (ref.33). The edge height in the EXAFS spectra should also serve as a fluorescence measurement of the iron if the spectra can be successfully subtracted.

The lowest curve of Fig. 5 is a noisy in-situ spectra of a 2.0 nm film under potential control at 1300 MV RHE in buffered borate. Most of the noise resulted from unstable orbit conditions and represent an experimental variable not found in MES or laboratory EXAFS. Nevertheless, the overall features are discernable, and the high frequency noise may be filtered. We have obtained similar data for 2.0 nm iron layers that have an air oxidized or a dried passive layer. While we are reluctant to quote final values, we have treated these data in a consistent manner and find transforms similar to the crystalline γFe_2O_3 for the dried passive layer and $\alpha Fe_2O_3/Fe_3O_4$ with αFe for the air oxidized specimen. In all cases the transforms are indicative of a somewhat smeared atomic distribution. The passive film, treated as a single homogeneous layer, shows most similarity to the δ FeOOH gel.

While these in-situ EXAFS experiments are in their infancy, there is no doubt their being a viable technique. Careful sample preparation is required, and improved sensitivity would be desirable. Nevertheless, we can state:

1. In-situ structural information about the electrochemically produced passive layer can be obtained using EXAFS spectroscopy, if: (a) the electrolyte thickness is minimized, and (b) the iron in electrolyte concentration is minimized.

2. The passive film exhibits structural differences from a similar film analyzed after drying.

3. The overall form of the spectra are indicative of Fe-O coordination with $\sim$ 6.0 neighbors at 2.0 ± 0.1 Å separation. Little contribution from high shells is noted, and the spectra are typical of disordered or solution structures.

4. Spectral substraction or modelling is required for a detailed analysis.

5. Chemically produced gels can be analyzed and have similar spectral features to the thin layer data.

SUMMARY

The important question is whether significant new information has come from in-situ experiments. Various models for the structure of the passive film have been prepared (ref.34). The structural considerations from in-situ data confirm that significant changes in the direction of crystallization take place upon removal from solution and subsequent long term drying. To date, the quality of the information have not allowed sensitive spectral subtraction, and this puts limitations on the distinction of multilayer structural models. Within the caviats given in the paper, we can rule out even a modest quantity of Fe^{2+} in the film, and certainly any ferromagnetic oxide at room temperature. Helium temperature experiments are required to examine the spin systems possible in amorphous or superparamagnetism.

On the basis of the literature and some 80 K measurements, we suspect no major contributions from superparamagnetic blocking but the issue is not yet settled.

Refinements, especially in electron detection to improve the sensivity and in modelling of the structures to simulate the observed spectra should yield new structural information within the next few years.

ACKNOWLEDGEMENTS

I am pleased to acknowledge the many persons who have shared in this research effort: M. E. Kordesch, J. Eldridge, J. M. Fine, J. J. Rusek, J. Wainright, P. Abel, E. Yeager, B. D. Cahan, and R. Hehemann at Case Western Reserve University, and D. R. Sandstrom and C. Marcus at Washington State University. The research was supported by the Office on Naval Research Grant N00014-79C-0795. SSRL is supported by the National Science Foundation and the U. S. Department of Energy.

REFERENCES

1. Mossbauer Isomer Shifts, edited by G. K. Shenoy and F. E. Wagner, North Holland (1978).
2. Applications of Mossbauer Spectroscopy, R. L. Cohen, ed., Vol. 1 and 2, Academic Press (1980).
3. D. E. Sayers, E. A. Stern, and F. W. Lytle, Phys. Rev. Lett, 27, (1971) 1204.
4. P. A. Lee, P. H. Citrin, P. Eisenberger, and B. M. Kincaid, Rev. Mod. Phys. 54 (1981) 769-806.
5. B. K. Teo and P. A. Lee, J. Am. Chem. Soc. 101 (1979) 2815.
6. C. R. Anderson, B. G. Richards, and R. W. Hoffman, J. Vac. Sci. Technol. 16 (1979) 466.
7. P. H. Citrin, P. Eisenberger, and R. E. Hewitt, Phys. Rev. Lett. 41 (1978) 309.
8. J. Stohr, L. Johansson, I. Lindau, and P. Pianetta, Phys. Rev. B20 (1979) 664.
9. R. J. Semper, C. R. Guarnieri, and J. C. Walker, Nuc. Instr. and Methods 129 (1975) 447.
10. W. E. O'Grady, J. Electrochem. Soc. 127 (1980) 555.
11. R. W. Hoffman, J. M. Fine, J. A. Mann, J. J. Rusek, and D. R. Sandstrom, SSRL Report 82/01 (1982) VIII-129.
12. See, for example, the chapters by G. W. Simmons and H. Leidheiser, Jr. or G. P. Huffman in ref. 2.
13. R. Oswald and M. Ohring, J. Vac. Sci. Technol. 13 (1976).
14. E. A. Stern and S. M. Heald, Rev. Sci. Instrum. 50 (1979) 1579.
15. D. R. Sandstrom, private communication.
16. M. E. Kordesch and R. W. Hoffman, International Conf. on Metallurgical Coatings, San Diego, April, 1983.
17. D. A. Sherson, S. B. Yao, E. B. Yeager, J. Eldridge, M. E. Kordesch, and R. W. Hoffman, J. Phys. Chem. 87 (1983) 932.
18. J. J. Rusek, Ph.D. thesis, Case Western Reserve University (1983).
19. M. E. Kordesch, J. Eldridge, D. A. Sherson, and R. W. Hoffman, Abstract No. 55, Extended Abstracts Vol. 83-1. The Electrochemical Society (1982).
20. J. G. Gordon, II, Abstract No. 658, Extended Abstracts, Vol. 82-1, The Electrochemical Society (1982).
21. W. N. Hansen, Abstract No. 657, Extended Abstracts, Vol. 82-1, The Electrochemical Society (1982) and references therein.
22. V. G. Kirichenko, V. V. Chekin, Zavodskaya Laboratoriya, Vol. 46, No. 7, (1980) 608.
23. J. Eldridge, M. E. Kordesch, and R. W. Hoffman, J. Vac. Sci. Technol. (1982) 934.
24. M. C. Lin, R. G. Barnes, and J. E. Furtak in AIP Conference Proceedings 84 - Physics in the Steel Industry (1982).
25. S. Morup, J. A. Dumesic, and H. Topsoe in ref. 2.
26. J.M.D. Coey, Phys. Rev. B6 (1972) 3240.
27. J. L. Leibenguth and M. Cohen, J. Electrochem. Soc. 119 (1972) 987.
28. G. G. Long, J. Kruger, D. R. Block, and M. Kuriyama, J. Electrochem. Soc. (In press).

162

29. G. G. Long, J. Kruger, D. R. Block, and M. Kuriyama, J. Electroanal. Chem. and Interfacial Electrochem. (In press).
30. J. Kruger, G. G. Long, and M. Kuriyama, This Conference.
31. J. M. Fine, J. J. Rusek, J. Eldridge, M. E. Kordesch, J. A. Mann, R. W. Hoffman, and D. R. Sandstrom, J. Vac. Sci. Technol. A1 (1983) (In press).
32. T. Misawa, K. Hashimoto, and S. Shinodaira, Corrosion Science 14 (1974) 131.
33. M. Nagayama and M. Cohen, J. Electrochem. Soc., 109 (1962) 781.
34. M. Cohen, in Passivity of Metals, R. P. Frakenthal and J. Kruger, eds. The Electrochemical Society (1978).

Passivity of Metals and Semiconductors, edited by M. Froment
Elsevier Science Publishers B.V., Amsterdam — Printed in The Netherlands

STRUCTURAL STUDIES OF PASSIVE FILMS USING SURFACE EXAFS

Jerome Kruger,[1] Gabrielle G. Long,[1] Masao Kuriyama,[1] and Alan I. Goldman[2]
[1]Center for Materials Science, National Bureau of Standards,
   Washington, DC 20234, (U.S.A.)
[2]Dept. of Physics, State Univ. of N.Y., Stony Brook, N.Y. 11794 (U.S.A.)

ABSTRACT

Iron K-absorption edge spectra were obtained from the passive films on iron for the dried films in air (ex situ) and for the films in the passivating solutions (in situ). The ex situ results demonstrate that, while the structures of the films are more disordered than the spinel-like iron oxides (e.g. γ-Fe_2O_3), they are nevertheless closely related to these crystalline oxides. The in situ data shows evidence of a quite different structure, which may be due to the accommodation of hydrogen containing species into the structure.

INTRODUCTION

At the Fourth International Symposium on Passivity, Revesz and Kruger (ref. 1) pointed out that a vitreous structure provides a more effective passive film, and that the incorporation of "glass forming" elements, e.g., chromium, promotes the formation of such a desirable structure. This paper describes studies whose aim was to determine the structure of passive films on iron, and to explore the effect of glass formers such as chromium and additives such as hydrogen on the structure of these films using both near-edge and extended x-ray absorption fine structure (EXAFS).

Our earlier EXAFS studies (refs. 2 and 3), involving a specially designed system (ref. 4) to maximize surface sensitivity, were able to produce ex situ (but non vacuum) EXAFS spectra of the passive films formed on iron in nitrite and chromate passivating solutions. Those studies found that the EXAFS signatures of the passive films resemble those of the spinel-like iron oxides, but that the chromate-formed film, which was found to contain chromium, may be more vitreous than the spinel-type oxides.

Our approach to the structural study of passive layers, as described in the present paper, aims to apply x-ray absorption spectroscopy to the in situ as well as ex situ investigation of the surface films on iron formed in two passivating solutions, one containing a glass forming element (a chromate solution), and the other not containing a glass forming element (a nitrite solution). These studies extend our previous work in the following ways: 1. a synchrotron source with high resolution optics was used, enabling us to examine the absorption edge profiles; 2. thinner iron substrates were used, resulting in film spectra much less contaminated by the iron substrate signal; and 3. in situ spectra were obtained.

Two kinds of valuable information were extracted from the x-ray absorption spectra. First, the spectral features within ~ 30 eV of the iron K-absorption edge are sensitive to effects such as excitons and unoccupied final states of p-character. The positions in energy space of the K-edge of of the thin films, relative to that for bulk iron, are directly related to the oxidation state, the coordination, and the degree of covalency in the bonding. These are important clues concerning the electronic structure of the films. The second kind of information derives from the fact that the extended x-ray absorption fine structure (from ~ 30 to 1000 eV above the absorption edge) is sensitive to the geometric structure in the vicinity of the absorbing atom. EXAFS spectra of model iron oxides have been taken to serve as empirical measures of the iron-oxygen and iron-iron phase shifts in these materials. Then, thin films on iron, formed in the inorganic passivating solutions, were measured, yielding structural parameters different from the model structures.

EXPERIMENT

The substrates used in these experiments were ~ 4 nm iron films deposited on glass slides in an ultra high vacuum system (base pressure ~ 1×10^{-8} torr). The iron films were immersed immediately after removal from the vacuum system into either 0.005 M potassium chromate or 0.1 M sodium nitrite. Passive films formed upon exposures of over one week to the passivating solutions have been shown (refs. 5 and 6) to have structure, composition and formation kinetics similar to those passive layers that form in Fe^{++} - free solutions by anodic oxidation.

After immersion in the passivating solutions for more than a week, the slides were removed, rinsed in distilled water and spectroscopic grade ethanol, and dried in a stream of pure nitrogen. Near-edge and EXAFS measurements were carried out on the passivated specimens either in air (ex situ) or in a specially constructed cell (ref.4) for in situ measurements. The spectra were taken at the C-1 station at the Cornell High Energy Synchrotron Source (CHESS). This station has a double flat crystal Si(111) monochromator with fixed exit position, suitable for this experiment. Data were taken in the fluorescence mode using MnO_2 filters in front of the detectors to reduce the signal due to scattering. An array of three detectors was used to accumulate data simultaneously during each scan. Data near the edges were taken in 0.5 eV steps, while EXAFS data was taken in 3 eV steps. Data collection time for each set required more than one CHESS fill (2-4 hours of running time), and duplicate sets were acquired for each in situ and ex situ case.

RESULTS

The edge structures we measured are shown in Figs. 1 (ex situ and in situ nitrite-formed films) and 2 (ex situ and in situ chromate-formed films). The shapes of the in-situ and ex-situ film profiles are very different from each other. One can examine how different they are by taking the slope and plotting the derivative spectra. Fig. 3 shows the derivative spectra of the profiles in Fig. 1, and Fig. 4 shows those for the profiles in Fig. 2. It is rather striking that the edges from the two ex situ samples resemble one another much more closely than they resemble their respective in situ spectra. Furthermore, the ex situ edge results are also very similar to our edge data for γ-Fe_2O_3, indicating again a close relationship between these structures. For the in situ derivative results, we see two peaks for the chromate-formed film and one peak and a diminished second peak for the nitrite-formed film. While these two results resemble one another more closely than they do the respective ex situ data, there remain differences between them. Further analyses of this data will be forthcoming in a future publication (ref. 7).

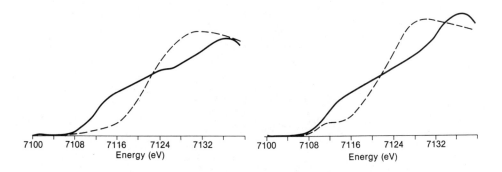

Fig. 1. K-edge profiles versus energy for the nitrite-passivated ex situ (--) and in situ (—) films.

Fig. 2. K-edge profiles versus energy for the chromate-passivated ex situ (--) and in situ (—) films.

The magnitudes of the Fourier transforms of the in situ and ex situ
nitrite data and chromate data are shown in Figs. 5(a) and (b). As was
shown earlier (refs. 2 and 3) the ex situ results for the passive films
formed in the two different passivating solutions are quite similar.
Since there is less bulk iron signal here than in refs. 2 and 3, the
resemblance between these passive film results and the data for the
crystalline γ-Fe$_2$O$_3$ is much stronger. Both films appear to be more
disordered than the model γ-Fe$_2$O$_3$, with the chromate-formed film
showing the less order of the two. The in situ results, shown in Fig.
5(b), suggest that these structures, again while related to each other,
are different from the ex situ structures. Also, it appears that the in
situ structures differ from one another more than the ex situ structures
do. All of this is consistent with the observed changes in the
Fe K-absorption edges and the derivative spectra shown in Figs. 1-4.
The details of the analysis will be presented elsewhere (ref. 7) since
the present purpose is to give an overview of these results.

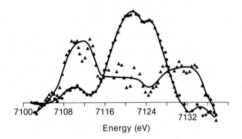

Fig. 3. Derivative spectra of the in situ (triangles)
and of the ex situ (dots) nitrite edge spectra.

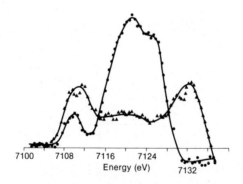

Fig. 4. Derivative spectra of the in situ (triangles)
and of the ex situ (dots) chromate edge spectra.

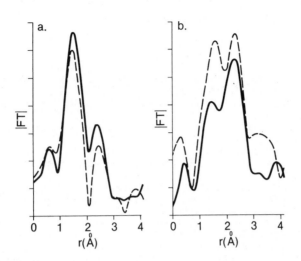

Fig. 5. Magnitudes of the Fourier transforms for the (a) ex situ and
(b) in situ passive films. The nitrite film results are given by the dashed
lines and the chromate film results by the solid lines.

DISCUSSION AND SUMMARY

The use of thinner iron substrates in this work yielded a clearer picture
than before of the close relationship between the ex situ structures of
the passive films on iron. Nevertheless, the chromate-formed structures
were still found to be less well ordered than the nitrite-formed structures,
and this result is interpreted as being due to the incorporation of
chromium into the chromate-formed film, promoting a more glassy structure.

When the passive films were immersed in their respective passivating
solutions, large differences in both the near-edge and the EXAFS spectra were
observed. Both the bonding distances and the coordination are seen to change.
The in situ structures, while again they are related to each other, show
greater differences between their respective structures than do the pair
of ex situ structures. This suggests that hydrogen containing species are
part of the in situ structures, and that the influence of chromium incorporation may be greater for the in situ case. The hydrogen containing species
have been suggested to appear in the structure as protons (ref. 8), water
(ref. 9) or hydroxyl ions (ref. 9), through experiments involving Mossbauer
spectroscopy (ref. 10), SIMS (ref. 11) and radiotracer methods (ref. 8).

The structures that produce these x-ray absorption spectra reflect the
profound influence that glass formers such as chromium and additives such
as hydrogen may have on the passive film on iron.

168

ACKNOWLEDGMENT

The contribution of J. Kruger was supported by the Office of Naval Research under contract NAONR 18-69 NRO 36-082.

REFERENCES

1   A. G. Revesz and J. Kruger, in Passivity of Metals, R. P. Frankenthal and J. Kruger, Eds., Electrochem Soc., Princeton, NJ (1978) 137 pp.
2   G. G. Long, J. Kruger, D. R. Black, and M. Kuriyama, J. Electrochem. Soc. 130 (1983) 240 pp.
3   G. G. Long, J. Kruger, D. R. Black, and M. Kuriyama, J. Interfacial and Electroanal. Chem., in press.
4   G. G. Long and M. Kuriyama, Proc. V Int'l Symposium on Passivity.
5   C. L. Foley, J. Kruger and C. J. Bechtoldt, J. Electrochem. Soc. 114 (1967) 994 pp.
6   J. Kruger, J. Electrochem. Soc. 110 (1963) 654 pp.
7   G. G. Long, J. Kruger, M. Kuriyama, A. I. Goldman, D. R. Black, E. Farabaugh and D. Sanders, to be published.
8   H. T. Yolken, J. Kruger, and J. P. Calvert, Corrosion Science 8(1968) 103 pp
9   T. Noda, K. Kudo and N. Sato, Z. Phys. Chem. N. F., 98 (1975) 271 pp.
10 W. F. O'Grady, J. Electrochem Soc., 127 (1980) 555 pp.
11 D. J. Murphy, J. O'M. Bockris, T. E. Pou, D. L. Cocke, and G. Sparrow, J. Electrochem. Soc. 129 (1982) 2149 pp.

Passivity of Metals and Semiconductors, edited by M. Froment
Elsevier Science Publishers B.V., Amsterdam — Printed in The Netherlands

STRUCTURE SIMULATION OF SURFACE FILMS ON IRON AND IRON BASED ALLOY BY
RADIAL DISTRIBUTION FUNCTION (R.D.F.) METHOD

M. KOBAYASHI

Solid State Chemistry Laboratory, The Institute of Physical and Chemical
Research, 2-1 Hirosawa, Wako-shi, Saitama, 351 (Japan)

ABSTRACT

Anodic passive film and chemical corrosion film on iron, and air-oxidized
film on iron based alloy were arranged as surface films. They are amorphous, as
characterized by two diffuse halos of "d-spacings" of ~2.5 and ~1.5 A in their
electron diffraction patterns. The short range order structures were derived
through a Radial Distribution Function (R.D.F.) method applied to the haloed patterns and simulated well with the aid of a "gel structure model" (G.S.M.). The
G.S.M. has recently been proposed for an amorphous ferric oxyhydroxide gel which
gives an X-ray haloed pattern with similar "d-spacings" as above. The G.S.M. has
a modified γ-Fe_2O_3 structure in which 32e lattice sites are equally occupied by
16 O^{2-} and 16 OH^- ions, and 16 Fe^{3+} ions in all are statistically situated on 8a
tetragonal (AS) and 16d octahedral (BS) sites with variable occupation fractions,
where $BS = 1 - AS/2$ or composition is FeOOH.

INTRODUCTION

Surface films on iron and iron based alloy, which are formed by primary oxidation at room and at low temperatures, by chemical and electrochemical passivation, and by chemical corrosion etc., are apt to be amorphous in their initial
growth steps. Their electron diffraction patterns are characterized by two
diffuse halos with "d-spacings" of ~2.5 and ~1.5 Å which are generally known as
"two diffuse halos" (ref.1,2) and "two ring substance" (ref.3). These halos have
been interpreted as "randomly oriented micro-crystalline due to the broadening
of the (113) and (440) reflections in a spinel type structure or amorphous"
(ref.1). It is the latter, however, that comes to the point since the haloed
patterns are similarly observed (ref.4,5) irrespective of a reflection electron
diffraction method and of a transmission. It is shown in Table 1 the "d-spacings"
of 18-8 stainless steels. The short range order (S.R.O.) structure for these
films will be made clear by means of a Radial Distribution Function (R.D.F.)
method applied to the haloed patterns without using crystal diffractometry. In
this case an S.R.O. structure of properly selected or deduced structure model
for these films is needed for comparison. Recently a "gel structure model"
(G.S.M.) (ref.6) has been proposed for an amorphous ferric oxyhydroxide gel
which gives an X-ray haloed pattern with similar "d-spacings" as above. The

G.S.M. reproduces observed X-ray scattering intensities and the reduced $R.D.F.$: $G(r)$, and further explains well the room temperature Mössbauer parameters and density of the gel. The S.R.O. structures for the surface films are also expected to be explained by the R.D.F. method with the aid of the G.S.M., and simulations of $G(r)$'s derived from electron diffraction patterns of surface films on iron and iron based alloy are made here.

EXPERIMENTAL

Sample preparation

   EP: an anodic oxidation film on iron[*]. The specimen (10×10 mm^2 with small handle) was electropolished followed by cathodic reduction (~ 7 μA/cm^2) and polarized at $+ 0.4$ V (vs. S.C.E.) for 1 h in a de-aerated boric acid-sodium borate buffer solution (pH = 8.4). The film maintaining current density was ~ 60 nA/cm^2.

   PP: an anodized film on iron[*]. The specimen was anodized[**] directly at $+ 0.7$ V for 16 h in the same solution. The current density was ~ 3 μA/cm^2.

   CC: a chemical corrosion film on iron[*] (ref.7). The specimen (10×10 mm^2) was etched with concentrated HCl for a few seconds and washed with distilled water, dehydrated in acetone and in ethyl alcohol, wiped dry and immersed in an aqueous 1 % NaCl solution for 14 h.

   SS: a stripped oxide film (ref.7) from 18-8 stainless steel[***]. The specimen (10×10 mm^2) was emery paper polished 0/10, air-oxidized at 600 °C for 30 s and quenched to room temperature, and stripped off in a Br$_2$-methanol solution.

Structure model

   G.S.M. (ref.6). A modified γ-Fe$_2$O$_3$ structure in which 32e lattice sites are equally occupied by 16 O^{2-} and 16 OH$^-$ ions, and 16 Fe^{3+} ions in all are statistically situated on 8a tetragonal (AS) and 16d octahedral (BS) sites with variable occupation fractions where $BS = 1 - AS/2$ or the composition is FeOOH.

Electron diffraction

   An electron microscope was operated under an accelerating voltage of 50 kV. The electron diffraction patterns of specimens EP$\sim$CC and SS were taken by reflection and transmission methods, respectively. The observed patterns for the specimens were as follows, EP: halos + α-Fe, PP: halos + γ-FeOOH, and CC and SS:

---

[*]99.4 % purity iron sheet (C, 0.013; Si, 0.02; Mn, 0.29; P, 0.025; S, 0.004 %), 1.2 mm in thickness, emery paper polished 0/10.
[**]This was a preliminary experiment and a set of apparatus used was different from that employed for EP.
[***]1.0 mm thick sheet with chemical composition (C, 0.04; Si, 0.58; Mn, 1.51; P, 0.028; S, 0.005; Ni, 9.04; Cr, 18.49 %).

halos. The optical densities of the photographic plates as read through a micro-photometer were converted into the diffraction intensities by the Karle and Karle method (ref.8,9).

DATA REDUCTION

Reduced R.D.F. (observed): $G(r)_{obs}$ for observed intensities

The equations used here are the same as those employed in a previous paper (ref.7). The interference intensity function and $G(r)_{obs}$ are written as

$$i'(s) = [(I'_{obs} - I'_{back})/f^2_{av}] \times \exp(- As^2),\qquad(1)$$

$$G(r)_{obs} = 4\pi r\Sigma_{uc} K_i[g'(r) - g'_0] = (2/\pi)\int_0^{s_{max}} si'(s)\ \sin\ rs\ ds \qquad(2)$$

where notations are common to the usual R.D.F. method. The primed terms mean their arbitrariness with respect to unit. I'_{obs} and I'_{back} correspond to the observed electron diffraction intensity and the independent scattering intensity or background. f_{av} is the average atomic scattering factor per single nuclear charge and given as $f_{av} = \Sigma_{uc} f_i/\Sigma_{uc} Z_i$ wherein f_i and Z_i are the atomic scatter-ing amplitude for incident electrons and the atomic number of the ith atom. Σ_{uc} indicates the sum for the assumed unit of composition. $\exp(- As^2)$ is an "artifi-cial temperature factor" and introduced to minimize a termination-of-series errors at the experimental upper limit, s_{max}. The value of A in the artificial temperature factor was determined so as to the factor become 0.1 at s_{max}. s is the scattering function defined as $s = 4\pi \sin \theta/\lambda$ in which 2θ is the scattering angle and λ is the electron wavelength. K_i is an effective mean number of nuclear charge for the atom and is supposed to be the average value of f_i/f_{av} values over the experimentally observed ranges of s. $g'(r)$ and g'_0 are the nuclear charge density function and its average, respectively. g'_0 is finally normalized to g_0 of the assumed structure model.

Reduced R.D.F. (caluculated): $G(r)_{calc}$ for structure model

The Debye's scattering intensity equation is somewhat modified as

$$I''(s)_{calc} = \sum_i f^2_i + \sum_{ij}^{i\neq j} f_i f_j\ \sin\ sr_{ij}/sr_{ij} \times \exp(- Bs^2) \times D(r_{ij}),\qquad(3)$$

$$I(s)_{calc} = I''(s)_{calc}/N \qquad(4)$$

where r_{ij} = the interatomic distance between atoms i and j, B = a "temperature factor" and $D(r_{ij})$ is a size factor (ref.10) written as

$$D(r_{ij}) = 1.0 - 1.5 \ RT + 0.5 \ RT^3. \tag{5}$$

Here $RT = r_{ij}/D$, D = a diameter of an atomic cluster model and N is a number of all the ferric ions: Fe^{3+}. By introducing $I(s)_{calc}$ of eq. (4) and $\Sigma_{uc} \ f_i^2$ to I'_{obs} and I'_{back} in eq. (1), $G(r)_{calc}$ is obtained according to eq. (2).

RESULTS AND DISCUSSION

Figure 1 shows observed $G(r)_{obs}$'s (solid lines) and simulated $G(r)_{calc}$'s (discrete lines) for the specimens: EP, PP, CC, and SS. Here ordinates and abscissae are in the nuclear charge units expressed as + $ne^2/Å^2$ and for the radial distance: r (Å), respectively. GF is $G(r)_{calc}$ for γ-FeOOH model. The numbers insetted for each specimen and the model are parameters employed in the calculation of the $G(r)_{calc}$: AS (an AS site's occupation fraction by Fe^{3+} in the G.S.M.), B (a "temperature factor") and D (a diameter of the atomic cluster). For the samples EP, PP, and CC and the GF, the values of D's are estimated as 25~30 Å, while the value of D for the sample SS becomes infinity since the stripped film thickness (~1000 Å) is large enough compared to "the S.R.O. distances" (~10 Å). As is evident from the figure, undulations of the $G(r)_{obs}$'s for the samples accord well with those of the $G(r)_{calc}$'s obtained through the G.S.M., in which the sample PP seems as a composite of the G.S.M. and of the GF (γ-FeOOH). This is in accord with the electron diffraction pattern described above. It seems strange that the air-oxidized sample SS is explained well by the G.S.M. which has bound water like ferric oxyhydroxide (FeOOH). However the sample SS has nearly the same "d-spacings" as those of chemically passivated specimens as is shown in Table 1. As for the latter, the existence of water and gel-like prop-

TABLE 1

"d-spacings" of surface films on 18-8 stainless steels observed by an electron diffraction (Å).

Air-oxidation		Chemical passivation		
(present)	(ref.11)	(ref.5)	(ref.12)	(ref.5)
2.54(vs)	2.55(s)	2.53-2.48	2.50	2.53
2.08(m)	2.05(w)	2.08-2.05	2.07	2.12
1.48(s)	1.47(m)	1.47-1.44	1.46	1.48
1.20(vw)	1.20(vw)	1.18-1.16	1.17	
a	a	b	b	c

[a] 600 °C, 30 s, transmission electron diffraction (T.E.D.).
[b] 5 % HNO_3-0.5 % $K_2Cr_2O_7$, 60 °C 30 min (T.E.D).
[c] 5 % HNO_3-0.5 % $K_2Cr_2O_7$, 60 °C 1 h (reflection E.D.).

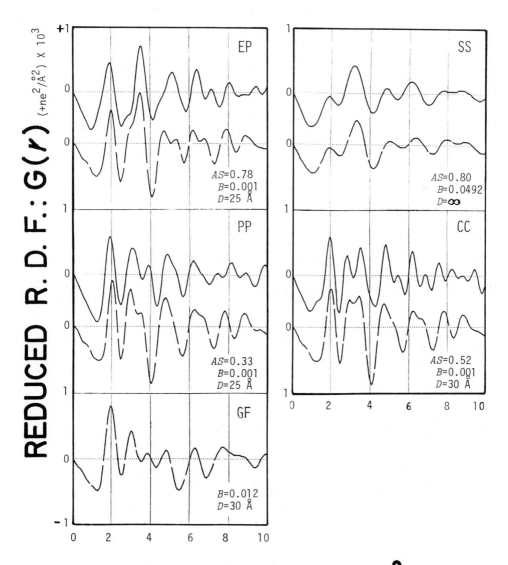

Fig. 1. Comparison of observed $G(r)$obs's (solid lines) and simulated $G(r)$calc's (discrete lines) for samples: EP, PP, CC, and SS. GF is $G(r)$calc for γ- FeOOH. Ordinates are in units of nuclear charge $+ne^2/Å^2$ and abscissae are of radial distance: r (Å). The insetted numerals are parameters employed in the calculation of $G(r)$calc's; AS: AS site's occupation fraction by Fe^{3+}, B: a "temperature factor", and D: a diameter of an atomic cluster model, respectively.

erties were indicated (ref.12). Besides this, many works reporting the existence
of water, and proposing the composition of the passive films on iron and iron
based alloys to be hydrated oxyhydroxide or oxyhydroxide have appeared as pertinent instrumental analyses have been carried out, such as XPS (ref.13-15), AES
(ref.16), SIMS (ref.17) and Mössbauer spectroscopy (ref.18). These support the
G.S.M. as to its composition. On the other hand, AS values of 0.8 and 0.33,
estimated for the samples EP, SS and PP are also found to be similar to those of
the amorphous ferric oxyhydroxide gel (ref.6). The former numeral is the same as
that of the slowly desiccated gel (AS = 0.8) and the latter one is almost
similar to that of the rapidly dried gel (AS = 0.3). In other words, it may be
concluded that the S.R.O. structure of the passive film is affected by its
growing condition.

ACKNOWLEDGEMENT

The author wishes to express his sincere thanks to Dr. M. Uda under whose
kind guidance and valuable advices this work has been conducted and also to
Prof. S. Suzuki of Science University of Tokyo for his encouragement and support
to this work.

REFERENCES

1 P.B. Sewell, C.D. Stockbridge and M. Cohen, J. Electrochem. Soc., 108 (1961)
   933-941.
2 H.R. Nelson, J. Chem. Phys., 5 (1937) 252-259.
3 S. Miyake, Sci. Pap. I.P.C.R., 36 (1939) 363-370.
4 M. Cohen and A.F. Beck, Z. f. Elektrochem., 62 (1958) 696-699.
5 E.M. Mahla and N.A. Nielsen, Trans. Electrochem. Soc., 93 (1948) 1-16.
6 M. Kobayashi and M. Uda, BOSHOKU GIJUTSU (Corrosion Engineering), 31 (1982)
   582-590.
7 M. Kobayashi and M. Uda, J. Non-Crystalline Solids, 41 (1980) 241-249.
8 I.L. Karle and J. Karle, J. Chem. Phys., 17 (1949) 1052-1058.
9 J. Karle and I.L. Karle, ibid., 18 (1950) 957-962.
10 G. Mason, Nature, 217 (1968) 733-735.
11 Y. Tsuji, private communication.
12 T.N. Rhodin, Corrosion, 12 (1956) 123t-135t.
13 M.W. Roberts and P.R. Wood, J. Electr. Spect. and Related Phenomena, 11
    (1977) 431-437.
14 K. Asami, K. Hashimoto and S. Shimodaira, Corros. Sci., 18 (1978) 151-160.
15 K. Hashimoto, M. Naka, K. Asami and T. Masumoto, ibid., 19 (1979) 165-170.
16 R.W. Revie, B.G. Baker and J.O'M. Bockris, J. Electrochem. Soc., 122 (1975)
    1460-1466.
17 O.J. Murphy, J.O'M. Bockris and T.E. Pou, ibid., 129 (1982) 2149-2151.
18 W.E. O'Grady, ibid., 127 (1980) 555-563.

Passivity of Metals and Semiconductors, edited by M. Froment
Elsevier Science Publishers B.V., Amsterdam — Printed in The Netherlands

THE STRUCTURE OF PASSIVE FILMS ON SINGLE-PHASE NICKEL-MOLYBDENUM ALLOYS

M.B. IVES[1], V. MITROVIC-SCEPANOVIC[2] and M. MORIYA[3]

[1]Institute for Materials Research, McMaster University, Hamilton, Ontario,
L8S 4M1, (Canada)

[2]Chemistry Division, National Research Council, Ottawa, Ontario, K1A OR6,
(Canada)

[3]Faculty of Engineering, Hokkaido University, Sapporo (Japan)

ABSTRACT

The effect of molybdenum alloying additions on the structure and properties
of passive films formed on nickel in acid sulphate solution has been studied
by a combination of electrochemical measurements and physical surface analytical
techniques. In particular, scanning transmission electron microscopy has been
possible on anodized thin films of the alloys. Molybdenum is detected on the
anodized surfaces after removal from solution, but both its distribution in the
films and its presence in a crystalline phase has not been resolved.

INTRODUCTION

While there have been many studies of the nature of passive films on pure
metals, the effect of alloying additions to metals on the subsequent anodic film
formation is not well characterized. In particular, passivation will depend on
whether an alloying element is incorporated into the anodic film (dissolved in
the base metal film or as a separate phase) or provides a new solution species
which modifies the corrosion processes by the adsorption of aquo-complexes etc.
The addition of molybdenum to improve the resistance of iron- and nickel- base
alloys to localized corrosion is a good example of alloying for corrosion prevention, but the mechanism of its action is not clear. Is Mo incorporated into
the anodic films (ref.1), or is it not (ref.2), and/or is it most effective
when present in the aggressive environment (ref.3)?

Detailed studies are reported here of the passivating films formed on Ni-Mo
alloys in 0.15N sodium sulphate solutions, at pH 2.8. Analysis has involved
potentiostatic anodic polarization, open-circuit potential decay (OCPD) measurements, and surface analysis using Auger electron spectroscopy (AES), high
energy reflection electron diffraction (RHEED), X-ray photoelectron spectroscopy (XPS), and scanning transmission electron microscopy (STEM).

ELECTROCHEMICAL MEASUREMENTS

The anodic polarization behaviour of a range of Ni-Mo alloys in acid

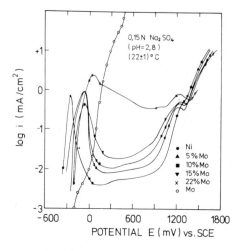

Fig. 1. Anodic polarization curves using stepwise potentiostatic technique.

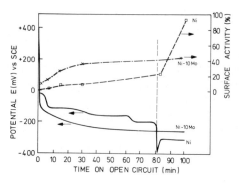

Fig. 2. Open circuit potential decay after anodic polarization at 500 mV(SCE) for 10 min, with the corresponding surface activites.

solution (ref.4) is summarized in Fig.1, with the curves for pure Ni and pure Mo. Mo additions to Ni shift the passivation potential by a constant 200mV in the noble direction, and monotonically increase the passive current densities. Such effects would be interpreted as Mo decreasing the state of passivity of the nickel. Details of this are reported in ref. 4.

OCPD measurements of samples anodized at passive potentials suggest, however, that the films formed on Ni-Mo alloys are more difficult to reduce completely than is the passive film on nickel. Fig.2 shows an example of OCPD, along with measurements of surface activity, defined and measured in the manner first described by MacDougall and Cohen (ref.5). In contrast to pure Ni, the alloy films are never removed completely, so that the potential does not return to the corrosion potential, even though the initial decay of potential is signif- icantly greater for the alloys than for pure nickel. Consonant with MacDougall's (ref.6) conclusion that the potential arrests for nickel correspond to various degrees of defectiveness, it could be that the alloy films are considerably more defective than those of nickel. It has been established (ref.4) that no oxides of Mo or Ni, mixed or separate, have equilibrium potentials in the OCPD range of potentials.

SURFACE ANALYSIS

The anodic films have also been subject to a number of surface analytical techniques. The elemental composition of as-anodized surfaces, as obtained by AES, is summarized in Table I. This shows surface enrichment of molybdenum in the anodic film formed at 500 mV(SCE) on a Ni-13w/o Mo alloy.

TABLE I

Anodization time(h)	Ni a/o	Mo a/o	O a/o	(Mo/Ni)$_{surface}$	$\dfrac{(Mo/Ni)_{surface}}{(Mo/Ni)_{alloy}}$
1	42.0	5.0	53.0	0.119	1.3
2	40.0	5.0	55.0	0.125	1.4
4	39.0	6.0	56.0	0.154	1.7
48	39.0	5.0	56.0	0.128	1.4

Unfortunately, AES sputter profile measurements are complicated by large differences in the sputter efficiencies of Ni and Mo, and these have not been precisely determined. Consequently the profile exemplified by Fig. 3 cannot be used to deduce the concentration profiles of the elements in real space, and no conclusion is possible on the distribution of Mo in the films. The AES profiling can provide a measure of film thickness, however, by measuring the sputter times necessary to reduce the oxygen level to an arbitrary low level (5 a/o), as shown in Fig. 4. The films grown at passive potentials on the Ni-Mo alloys are seen to thicken an order of magnitude faster than they do on pure nickel. AES analysis of alloy surfaces following OCPD (ref.7) confirm that substantial quantities of film remain even after long decay times, consistent with the surface activity measurements noted above in Fig. 2.

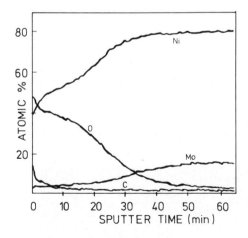

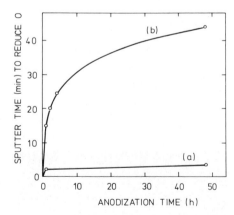

Fig. 3. AES sputter profile of Ni-13Mo after 48h at +500mV (SCE), pH 2.8.

Fig. 4. Anodic film growth kinetics from AES sputter profiles, plotted as time to reduce oxygen signal to 5 a/o. (a) pure Ni; (b) Ni - 13w/o Mo.

STRUCTURAL ANALYSIS

To obtain structural information on the passive films various techniques have
been adopted. RHEED patterns(ref.7) exhibit a single-phase ring structure, with
d-spacings appropriate to NiO. XPS analysis provides indirect chemical information. Fig. 5 shows the Ni(2p) spectrum from a Ni-13w/oMo alloy after anodizing
at 500mV(SCE) for 2h. Since Ni(2p) binding energies reported for pure NiO range
from 854.0 to 854.5 $\pm$ 0.2 eV, and the peak of Fig. 5 is at 855.6 $\pm$ 0.2 eV, there
is a likelihood that the film is not NiO. If Mo were dissolved in NiO a chemical shift could result (ref. 8) and thereby explain the discrepancy.

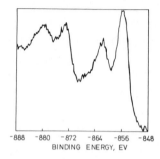

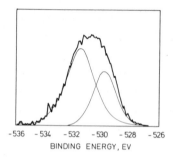

Fig. 5. Ni(2p) spectrum of Ni-13Mo after
2h at 500mV (SCE); Au(4f$_{7/2}$) as standard.

Fig. 6. O(1s) peak for same sample,
with a trial deconvolution

A clue to the possibility of a two-phase film, or of at least two kinds of
oxygen, is provided by the broad O(1s) peak in the anodic film spectra. Fig. 6
shows one deconvolution exercise providing peaks at 529.8 eV (100% Gaussian) and
531.5 eV (77.6% Gaussian). It is possible that one of the peaks corresponds to
bound water in the films (ref. 9).

A more detailed analysis of the alloy anodic films has been made by the
application of high-resolution STEM. In particular a "dedicated" STEM (Vacuum
Generators, Model HB-5) has been used to analyse in-situ anodic films formed on
electrochemically thinned alloy films (ref. 10). Fig. 7 shows a bright field.
(BF), and Fig. 8 the corresponding annular dark field (ADF) image of a Ni-13
w/oMo foil anodized at 500mV (SCE) for 24 h. Examinations at higher magnifi-

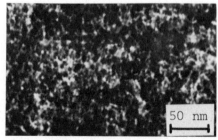

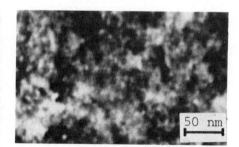

Fig.7. Bright-field STEM image.

Fig.8. Corresponding ADF image.

cations (ref. 11) show an irregularity of film thickness, including holes where the underlying alloy has been corroded away. These images suggest a mosaic structure with epitaxy between film and alloy, as shown in Fig. 9. The spots attributed to the film are just inside those due to the alloy, and are indexable as a cubic lattice expanded $3.5 \pm 0.7\%$ over that for NiO. Similar patterns obtained under identical conditions from films grown on pure nickel foil suggest a film lattice parameter expanded only 1% over that for NiO (ref. 10). The additional expansion for alloy anodic films could be due to Mo in solid solution in NiO, forming (NiMo)O. That the epitaxial relationship is not perfect is demonstrated in Fig. 10, by superpositioning SAD patterns from many different places on the same substrate grain. A range of film reflections is recorded for each diffracted beam.

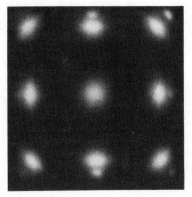

Fig. 9. 001 microdiffraction pattern after 24h at 500mV(SCE)

An estimate of mosaic size has been obtained from diffraction pattern halfwidths (ref. 12), and suggests a 3 nm grain size, consistent with Figs. 7 and 8.

Energy-dispersive X-ray spectrometry in the STEM confirms the AES conclusions (Table I) that Mo is en-

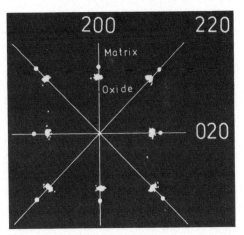

Fig. 10. Superimposed 002 SAD patterns from film of Fig. 9.

riched in the anodized surfaces to a concentration of almost 12 a/o for the 8.4 a/o Mo alloy (ref. 11). It is unlikely that all the excess Mo is incorporated into NiO, since the Mo would have to displace 1/8th of the Ni ions and the excess charge on the Mo^{6+} ions would cause large cation deficiency. The excess Mo more likely exists in some other form. However, no additional crystalline phase been detected in the diffraction patterns. Furthermore, point-to-point measurement of the Mo X-ray spectrum does not reveal any lateral segregation of Mo (ref. 11) to a resolution of 3 nm.

When anodized samples were vacuum annealed at 400°C for 40 min, distinct changes were observed in the BF (Fig. 11) and SAD (Fig. 12) images. A new crystalline phase appears, tentatively indexed as $NiMoO_4$ (ref. 11). This same phase

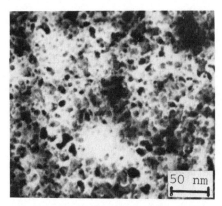

Fig. 11. BF image of Ni-13Mo anodized   Fig. 12. SAD pattern from same sample.
film after vacuum anneal, 400°C, 40 min.

appears when this alloy is oxidized by heating in air at 400°C (ref. 11).

CONCLUSIONS

The passivating anodic films formed on single-phase Ni-Mo alloys in acid
solution appear to comprise a fine mosaic of epitaxial (NiMo)O, with a grain
size of 3 nm, along with a molybdenum-rich phase which is either amorphous or
present as unresolvable finely-dispersed crystals. Such a structure would
explain the effect of Mo on increasing the growth kinetics of nickel passive
films (Fig. 4), and on reducing the passivating action of such films (Fig. 1)
by providing interphase boundaries for increased cation transport.

Higher resolution is still needed to characterize the Mo-rich constituent
of the films. In particular, its presence at either the electrolyte or alloy
interface will indicate if the Mo is incorporated directly into the film by
metal dissolution, or if it is redeposited from solution.

REFERENCES

1 K. Sugimoto and Y. Sawada, Corrosion Science 17 (1977) 425.
2 J.B.Lumsden and R.W. Staehle, Scripta Metallurgica 6 (1972) 1205.
3 T.Kodama and J.R.Ambrose, Corrosion 35 (1977) 155.
4 V.Mitrovic-Scepanovic and M. B.Ives, J. Electrochem. Soc. 127 (1980) 1903.
5 B.MacDougall and M.Cohen, ibid. 123 (1976) 191.
6 B.MacDougall, ibid. 127 (1980) 789.
7 V.Mitrovic-Scepanovic and M.B.Ives, Corrosion, in press.
8 K.Tachibana and M.B.Ives, in R.P. Frankenthal and J.Kruger (Eds.), Passivity
   of Metals, Electrochemical Society, Princeton, New Jersey, 1978, pp.878-897.
9 K.Asami, K.Hashimoto and S.Shinodaira, Corrosion Science 17 (1977) 713.
10 M.Moriya and M.B.Ives, J.Microscopy, in press.
11 M.Moriya and M.B.Ives, Corrosion, in press.
12 Z.G.Pinsker, "Electron Diffraction", Butterworths, London, 1953, p.202.

Passivity of Metals and Semiconductors, edited by M. Froment
Elsevier Science Publishers B.V., Amsterdam — Printed in The Netherlands

STUDY OF PASSIVE LAYERS ON STAINLESS STEELS USING LOW ENERGY ELECTRON INDUCED
X-RAY SPECTROSCOPY (LEEIXS)

F. GAILLARD, M. ROMAND, A. ROCHE, M. CHARBONNIER[1], R. BADOR[2], A. DESESTRET[3]

[1]Applied Chemistry Department (C.N.R.S., ERA 300), University Claude-Bernard,
LYON I, 69622 Villeurbanne Cedex (France)

[2]Biophysics Laboratory, Pharmaceutical Sciences Unit, University Claude-Bernard
LYON I, 69008 Lyon (France)

[3]Creusot-Loire, Centre de Recherches d'Unieux, 42701 Firminy (France)

ABSTRACT

This study deals with the use of low-energy electron-induced X-ray spectroscopy (LEEIXS) for investigating surfaces prior to and after a passivation process. After a brief introduction to LEEIXS fundamentals and apparatus, experimental results are given, regarding the characterization of films produced by
passivation treatments (5 % H_2SO_4 at 60°C solution, boiling $MgCl_2$ solution) on
stainless steels.

INTRODUCTION

The use of physical methods, in order to analyze and characterize solid
surfaces and interfaces, has made these recent years a great stride. In a general way, these methods use an excitation process by photons or charged particles (electrons, ions). The probed depth depends, among other things, upon the
choice of the excitation method and the nature of the analyzed radiations or
particles.

Both Auger electrons (AES) and soft X-rays (SXS) emission spectroscopies
are techniques which are based on the excitation by low-energy electrons. The
former is, today, a well known tool for surface analysis. Inversely, the possibilities of the later are widely under exploited.

The aim of the present work is to describe characteristics and performances
of low-energy electron induced X-ray spectrometry (LEEIXS) (réf. 1).

METHODS

The device used is a wavelength-dispersive X-ray emission spectrometer,
equipped with a cold cathode tube (CCT). The CCT is used to bombard the sample
surface with a quasi-monoenergetic electron beam, selectable over the range from
0.5 to 5 keV. The probed surface area is typically less than 1 cm^2.

The capabilities of the LEEIXS technique as a near-surface analysis tool

(ref. 2-4) are linked to the low penetration depth of primary electrons in
solids and therefore to the low depth in which characteristic soft and ultra-
soft X-rays are emitted.

RESULTS

Figure 1 shows spectra of an AISI-304 surface a) prior to and b) after a
15 minutes treatment in 5 % H_2SO_4 at 60°C. According to the primary electron
beam energy used (1.5 keV), the probed depth is in the range 13-23 nm, respectively for the metallic substrate and for a bulk oxide. The comparison of these
spectra shows, consecutively to the passivation process, a nickel surface enrichment. This one is revealed by the decrease of the FeLα/NiLα intensity ratio.
Figure 2 shows depth profiles obtained by glow discharge optical spectrometry
(GDOS). The study of intensity variations for the metallic elements between the
matrix (part a) and the oxide layer (part b, characterized by an higher oxygen
concentration) confirms the LEEIXS information indicating a surface nickel enrichment.

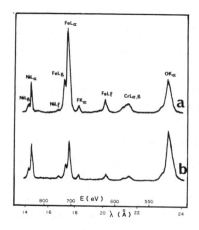

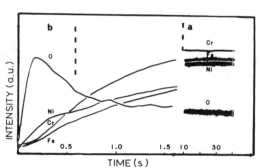

Fig. 1. LEEIXS spectra of AISI-304   Fig. 2. GDOS depth profiles of AISI-
a) prior to and b) after a 5 % H_2SO_4 304 after a 5 % H_2SO_4 at 60°C treat-
at 60°C treatment.   ment.

Figures 3 and 4 are relative to an AISI-304 stainless steel subjected to a
boiling (142°C) concentrated and aerated $MgCl_2$ solution, for various treatment

durations. Measurements of MgKα, ClKα , OKα intensities are performed using
electron energies high enough to probe the whole surface layers. These curves
show, for treatment durations higher than 6 hours, a modification of the oxidation process kinetic and of the surface layer nature. Obviously, this process
kinetic is highly dependent upon the surface conditioning of the alloy and
upon its purity.

The LEEIXS data also emphasize (see Fig. 4 and 5) a surface silicon enrichment of the sample for treatment durations longer than 6 hours. The SiKα spectra
in figure 5 are relative to a) an electropolished AISI-304 surface and b) the
surface after à 16 hours exposure to the MgCl$_2$ medium.

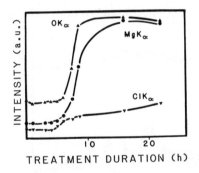

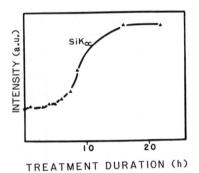

Fig.3. OKα, MgK α, ClKα intensities,
measured by LEEIXS, versus treatment
duration in the MgCl$_2$ medium.

Fig. 4. SiKα intensity, measured by
LEEIXS versus treatment duration in the
MgCl$_2$ medium.

The SiKβ spectra shown (see Fig. 6) are representative of silicon a) in a
glass (silicon bonded to oxygen), b) in a silicon substrate and c) at the surface of the sample after the MgCl$_2$ passivation process. The fine structure Kβ',
exhibited by spectrum c) shows that silicon is bonded to oxygen in the passive
film. An electron beam of 2.2 keV is used for these investigations by LEEIXS
(see Fig. 4, 5, 6c) ; the silicon probed depth is then lower than 10 nm.

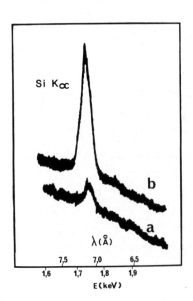

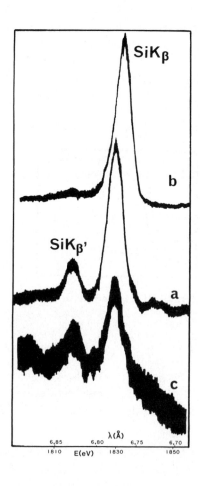

Fig. 5. SiKα spectra relative to AISI-304 a) before and b) after exposure (16 hours) to the $MgCl_2$ medium.

Fig. 6. SiKβ spectra relative to a) a glass, b) a silicon substrate, c) the surface of the sample after the exposure to the $MgCl_2$ medium.

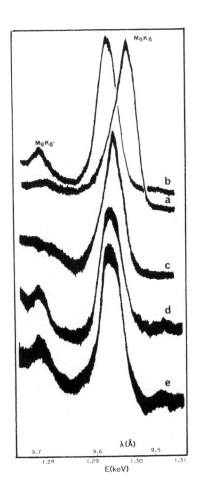

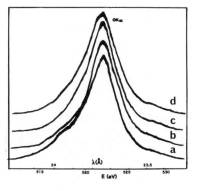

Fig. 7. MgKβ spectra relative to
a) a magnesium substrate, b) MgO,
c) $MgCl_2$, d) a magnesium silicate
and e) AISI-304 surface after $MgCl_2$
treatment.

Fig. 8. OKα spectra relative to a) SiO_2,
b) MgO, c) a magnesium silicate,
d) AISI-304 surface after $MgCl_2$ treatment.

The MgKβ spectra shown (see Fig. 7) are representative of magnesium a) in a magnesium substrate, b) in MgO, c) in $MgCl_2$, d) in a magnesium silicate (forsterite) and e) at the surface of the sample after the $MgCl_2$ passivation process. The presence of the Kβ' structure and the position of the peak maximum indicate clearly that magnesium is bonded to oxygen in the film formed at the surface of AISI-304. Moreover, the emission bands (d) and (e) have quite similar apparent full width at half maximum (FWHM) and asymmetry .

The OKα spectra shown (see Fig. 8) are representative of oxygen a) in SiO_2, b) in MgO, c) in a magnesium silicate and d) at the surface of AISI-304 after the $MgCl_2$ treatment. Considering the parameters (FWHM and asymmetry) of the OKα emission band, we can conclude that the spectrum d) is quite different from the a) and b) ones but is very similar to the c) one.

The LEEIXS data relative to the study of X-ray emission bands of magnesium, silicon, oxygen which are present at the surface of AISI-304 after a 16 hours $MgCl_2$ treatment are in good agreement. They all show that the main component of the surface film formed in such a medium is probably magnesium silicate-like. This would confirm an hypothesis which has been previously put forward (réf. 5).

CONCLUSION

The above results demonstrate that soft X-rays can be used for studying the near surface chemical composition of solids. More particularly, LEEIXS appears to be a suitable tool for investigating metallic surfaces which have undergone specific processes or treatments in order to determine the characteristics of the solid with respect to corrosion (réf. 4), adhesion (réf. 6), passivation...

REFERENCES

1 M. ROMAND, M. CHARBONNIER, A. ROCHE, F. GAILLARD, R. BADOR, Colloque Rayons X, Montpellier, Siemens Ed., 1981, pp. 160-176.
2 R. BADOR, M. ROMAND, M. CHARBONNIER, A. ROCHE, Adv. X-Ray Anal., 24, (1981), 351.
3 A. ROCHE, M. CHARBONNIER, F. GAILLARD, M. ROMAND, R. BADOR, Appl. Surf. Sci., 9 (1981), 227.
4 B. HAGE-CHAHINE, F. GAILLARD, M. ROMAND, M. CHARBONNIER, A. ROCHE, R. BADOR, 33rd Meeting of the International Society of Electrochemistry, Lyon, 1982, p. 911.
5 H. WIEGAND, F.W. HIRTH, T. GRESS, K.SCHWITZGEBEL, Metalloberflache , 22, (1968), 353.
6 F. GAILLARD, A.A. ROCHE, M.J. ROMAND, Int. Symposium on Adhesive Joints, Kansas City. Organic Coatings and Applied Polymer Science Proceedings, American Chemical Society Ed., 1982, vol. 47, p. 100

Passivity of Metals and Semiconductors, edited by M. Froment
Elsevier Science Publishers B.V., Amsterdam — Printed in The Netherlands

OPTICAL METHODS IN THE STUDY OF PASSIVE FILMS

B.D. CAHAN

Chemistry Department, Case Western Reserve University, Cleveland, Oh. 44106 USA

ABSTRACT

   A discussion of the strengths and weaknesses of several optical techniques useful in studying the passive film is presented. The paper deals primarily with the methodology rather than interpretation. Examples from the literature and the authors research are used for illustration.

INTRODUCTION

   It is unfortunate indeed that electrons have a very short mean free path in aqueous electrolytes, and that water and ultra-high vacuum systems are usually incompatible. If this were not the case, a wide range of very powerful suface sensitive analytical tools would be available for the study of the electrochemical interface. Nature being what it is however, we are forced to find other in-situ methods for looking at the surfaces. Fortunately, the photon does have a sufficient mean free path to be used as an in situ probe in most electrolytes although the energy window available is narrower than we would like. A number of methods have been developed which use the interaction between photons and the interface to produce an effect which is diagnostic of the surface. Unfortunately, even the massless photon does have energy and can change the surface during the measurement. Our task is to find methods which change the surface as little as possible or to recognize that changes will occur and use them to our advantage. The primary objective of this paper will be a discussion of the merits and drawbacks of optical techniques for looking at the passive interface with optical techniques, such as ellipsometry, electroreflectance, Raman, and photo-voltaic spectroscopy, focusing on the weaknesses as well as on the strong points of the several methods. Many excellent reviews can be found in the literature (ref.1,2) which deal with the history, development and applications of optical techniques and any rehashing of these would be of little value. No attempt will be made to discuss data or interpretation from the literature, except for purposes of illustration.

   One of the more useful properties of light lies in the wide wavelength range available to the experimenter. Aqueous solutions however restrict the range of utility to about two octaves unless special tricks are employed. Within this window lie enough molecular and atomic transitions to allow some clues to the details of the processes occuring at the surface. However, the results obtained with any one of these techniques will not usually be sufficient to identify them uniquely. The simultaneous or concurrent use of a second technique will permit reasonable guesses as to the results of the first.

ELLIPSOMETRY

A beam of polarized monochromatic light, the most general form of which is elliptical, can be described in terms of several parameters which can be defined in many different ways depending on the experimental setup and frame of reference. Looking down the light beam, the tip of the electric vector describes an ellipse (hence the name ellipsometry) defined with respect to the plane of the surface and the plane normal to the surface in which the beam lies. The most common definition is stated in terms of Ψ, the ratio of the amplitudes of the electric field vector, **E**, in the parallel and perpendicular directions, and Δ, the phase difference between the two. Other definitions may be more convenient for use with different ellipsometers. For example, the Rudolph RR2000 automatic ellipsometer (ref.3) uses geometrical parameters which describe the ellipse in terms of the azimuth α and ellipticity ε, while yet another automatic instrument (ref.4) uses the coefficients which result from a discrete Fourier transform.

For a film free _ideal_ interface, Δ and Ψ can be rigorously related (ref.1) to the complex optical constants ($n_c = n - ik$) of the reflecting surface and the optical constants of the ambient (usually real). Of interest to the worker in passivation is the fact that a thin film at the interface will cause changes in the measured Δ and Ψ. These changes are of the order of a degree per monolayer and the experimental technique is easily capable of resolving changes of the order of $0.01°$. If we know the optical constants of the bare surface and if the overlying film is transparent ($k = 0$), one can calculate the index of refraction, n (which is real), and the thickness, t, of the film.

Unfortunately, surfaces are never ideal and passive films are rarely perfectly transparent. For ellipsometry to be quantitatively useful, the surface should be flat, perfectly smooth and specularly reflecting. Flatness is required so that the angle of incidence is well defined. One of the best ways of aligning a surface in an ellipsometer is by means of an auto-collimator (ref.5) which requires a flat surface for proper focus. A rough surface appears to the ellipsometer like a film of some intermediate index. Several excellent treatments are to be found in the literature such as the ones by McCrackin (ref.6) or by Lukes (ref.7). Proper use of these treatments require a detailed foreknowledge of the topology of the surface which is almost never available. Muller (ref.8) has presented a number of fascinating papers with elaborate computer fitting of models of highly structured surface layers that correlate well with electron microscopy of these surfaces. The treatment requires a large number of fitting parameters and it appears uncertain whether these numbers are of value in a fundamental sense. 'Give me a digital computer as a fulcrum and a long enough lever of adjustable parameters and I can model the world'.

Classical ellipsometry measures only two parameters to characterize the reflected polarization state (2-P ellipsometry). With an unknown film of unknown thickness three quantities (namely n, k and t or their equivalents) must be computed and the system is underdetermined. The assumption that literature data exists and is <u>reliable</u> is fraught with danger. In the past authors have often resorted to one or more of the following devices.

1. The thickness was assumed to be known from some other independant measurement like coulometry. If the material of the film is not known precisely its density is likewise unknown and the thickness can be off by a factor of two or more, resulting in gross errors in the calculations.

2. The film was assumed to be non-absorbing (k small). In this case the problem reduces to a two unknown situation. Even materials which are transparent in the bulk usually contain enough inclusions, defects and imperfections to appear absorbing. For films transparent at one wavelength and absorbing at others one can calculate the thickness at the first wavelength and use that thickness for calculating the optical constants at another (ref.9). Most passive films are sufficiently absorbing to make this method invalid.

3. Measurements were made at several angles of incidence. While occasionally yielding satisfactory results the two sets of equations are quasi-homogeneous and at best give a wide range of possible solutions (ref.10). The precision required for this method to work is usually beyond the limit of experimental feasibility. This technique <u>can</u> be useful for determining if measurements made with other procedures are at least self-consistent.

4. Attempts have been made to replace the ambient medium (e.g. the electrolyte) with a medium of signifigantly different refractive index. The new liquid must be sufficiently miscible with the old that a film of the original liquid doesn't remain on the surface. A replacement of the liquid by air obviously doesn't fulfill this requirement, and suffers additionally from the problem of loss of potential control which can cause changes in the film.

5. A set of integro-differential equations known as the Kramers-Kronig relations relate the real part of a complex function to the imaginary part. One can use these relations to calculate a consistent set of thickness and values for the complex dielectric function if the e.g. Δ and ψ are known for <u>all</u> wavelengths (ref.11). Since 'all' requires about two decades above and below the wavelength of interest, and we only have about two octaves, we must make drastic assumptions about the values outside our range. This procedure may be questionable.

In addition to the two parameters of classical ellipsometry, the absolute intensity is also required to characterize the beam fully. This parameter is normally thrown away in conventional nulling instruments and most automatics. It was recognized (ref.12,13) that reflectivity could serve as an independent third parameter (3-P). The Rudolph RR2000 was designed (ref.3) using a rotating analyzer which determined the ellipsometric parameters by analysis of the waveform produced by rotating the analyzer in the beam reflected from the surface. This technique had several advantages. Since the system makes a measurement each revolution of the motor (which spins at 54 r.p.s.) time resolution improves from the five minutes or so required to make a manual measurement to less than 20 milliseconds. Nulling ellipsometry requires a quarter-wave plate (or other compensator) to operate. In practice these are neither perfect nor achromatic and must be calabrated for each wavelength used. A rotating analyzer system does not need a compensator.

The relative intensity of the reflected beam can be determined by an integration over precisely one revolution. With a sufficiently stable light source it is now possible to do 3-P measurements (ref.12). In the case of an ideal two-layer system, the reflectivity of the interface is, in principle, calculable from the n and k of the substrate which, in turn, are determined directly from ellipsometric measurments and the n of the medium. It is a common practice to calculate pseudo-constants for a surface which has a film, but is not in general true that the reflectivity calculated using the pseudo-constants will be that actually produced by the interface with a film. If a measurement can be made of a film free substrate and then of the substrate with a film, the change in reflectivity can be used to calculate semi-unambiguously the optical constants _and_ thickness of the film. In contrast to 2-P ellipsometry which will give continuous families of solutions, 3-P ellipsometry gives a limited set of unique solutions most of which can usually be rejected on other grounds. The problem is still not completely solved because even with 3-P ellipsometry a precise location of the roots may require a greater precision than is practically obtainable (ref.5).

A significant benefit of 3-P ellipsometry is a result of the uniqueness of the solutions. Not all sets of delta, psi and reflectivity yield solutions that are physically meaningful. In such cases, it can be shown that something _must_ be wrong with the assumed model for the system. In the case of iron (ref.14), for example, sets of measurements were made by voltammetric cycling of an iron electrode in the electrolyte between potentials where the electrode was alternately cathodically reduced and anodically passivated, while measurements were made by the ellipsometer under computer control. We were chagrined that when we tried to solve the system of equations using 3-P methods, for no consistant solutions at all were obtained. We were able to show that the lack of convergence was traceable to at least two phenomena. First, the cycling of the electrode caused the surface to roughen. This was obvious from the increased non-specular scatter from an originally mirror-smooth surface. Second, it became apparent that soluble iron species produced during the cycling were redepositing on the surface as an over-layer. Switching from potential sweeps to potential steps solved this problem. Using the experimental values obtained this way we were able to obtain unique, consistant solutions for the passive film _in situ_ over a wide range of wavelengths.

Of interest here is not the particular values obtained (although this data was vital in the development of the Chemi-conductor (ref. 15) model) but the fact that 3-P ellipsometry can often be used as a diagnostic tool for determining the validity of an experiment. With 2-P systems a set of roots are always found and assumed to be correct. The use of 3-P ellipsometry can be helpful in pinpointing problems that otherwise might have gone undetected.

The rapid rate of data acquisition of an automatic ellipsometer makes their use in a 'spectroscopic' mode easy. Aspnes (ref.16) has used this mode extensively for determining the complex optical dielectric function of several metals and semiconductors. With the additional help of 3-P ellipsometry, Cahan and Chen (ref.14) were able to determine an in-situ spectrum of the passive film on iron, and to interpret this spectrum in terms of fundamental physical properties (ref. 17,18). Ellipsometry has thus developed from a tool for simply detecting and measuring film thickness, but is now capable of yielding information on a molecular or atomic basis.

The use of more than one wavelength is a necessity rather than a convenience. During the reduction of iron two distinct species are generated that happen to have almost identical optical properties at 546.1 nm (the green Hg line used most often in ellipsometry). This transition appears to have been missed by researchers who worked at only one wavelength (ref.19).

REFLECTANCE SPECTROSCOPY

This topic embraces the set of techniques where primary attention is paid to the intensity rather than the polarization state after reflection from an interface. The reflectivity of a surface will be affected by its electronic structure. Any atomic or molecular processes which will absorb light will of necessity change the reflectivity. With suitable techniques the wavelength of a beam of light may be scanned and a spectrum recorded which can be related back to the molecular structure of the system if one has some idea about the causes of the absorption. Usually, there is not sufficient data in a spectrum obtained over a limited range to make any quantitative statement about the physical processes involved, but qualitative detection of absorption edges and maxima can be made. With a wide frequency range extending from the far UV to the far IR, use could be made of the Kramers-Kronig relations mentioned above to determine the complex optical parameters of a simple surface (ref.20), but an extension of this technique to cover films is very complex, and not applicable to surfaces in liquids because of the limited light pass band.

Most measurements in this class are the result of some type of modulation. The use of frequency selective detection can greatly improve the signal-to-noise ratio (S/N) of a measurement. Some method is chosen to perturb the optical behavior of the interface, and the resultant change in the reflection of the light is detected, usually with a synchronous detector like a phase-lock amplifier. Numerous techniques for this modulation can be found in the literature (ref. 21), including modulation of the electrode potential (or charge) in which case the technique is called Electro-reflectance.

The improved S/N permits several orders of magnitude greater sensitivity than is practical with integral techniques, allowing us to observe small changes which would otherwise be undetectable. The signal obtained is a derivative, and is not always clearly relateable back to a specific phenomenon. Any external perturbation which will alter the optical structure will affect the response function. For example, the concentration of an absorbing species at the interface can change or the band structure or free elctron concentration of either the substrate or of the film. It is by no means a trivial matter to determine if an optical absorption is caused by a change in the substrate, in the film itself or even by an adsorbed species.

Electroreflectance changes are readily detected on many bare metals, even in the absence of a film. Gold electrodes exhibit electroreflectance changes (ref.22) of the order of those to be expected from the formation of a monolayer of oxide (ref.23). The changes are attributed,for example, to the uncovering of the d-electrons as the s-electrons are pushed in and out by the electric field (ref.24), or to the shifting of the band edge in the metal surface (ref.25). Control experiments must be made on the bare metal, but even these are not proof that field perturbations in the metal are the same as when a film is present on the surface (ref.26). Similar effects can be expected in oxide films, but in addition chemical changes are possible (ref.27) which can affect light absorption. Since oxides are in general much less absorbing than metals the depth over which optical effects can be observed is much greater.

Modulated electroreflectance can be used in combination with other techniques to greatly enhance sensitivity and/or selectivity. Horkans et al. (ref.28,29) derived the relations between Δ, ψ, and $\delta R/R$ and the intensity measured at three settings of the analyzer of an ellipsometer. This technique was originally used with total intensities recorded on separate sequential sweeps of a voltammetric scan, but could easily be used with an AC modulated signal. Consider a conventional manual ellipsometer, with the analyzer mounted in a stepper motor, or some electro-optical means of changing the effective angle of the analyzer in known discrete steps. From the amplitude of the light modulation at three analyzer settings one can calculate $\delta\Delta$, $\delta\Psi$, and $\delta R/R$. The sensitivity of an ellipsometer could thus be increased from .01 - .001° to at least 10^{-5} degrees for small changes of the surface. Absolute information may be lost, however, as with any derivative technique. If a Transfer Function Analyzer is used as the phase sensitive detector, the relative intensity can be recorded simultaneously, and the original integral equations (ref.29) used. A phase lock amplifier, of the type which has several stages of analog integration for noise elimination would not be as satisfactory for this technique, because of the long settling and/or overload recovery time required to accomodate the abrupt change of intensity at the time of analyzer angle switching.

One of the more powerful techniques available to the passivation researcher today relates to the use of interfacial impedances (ref.30). This method is not as selective as one might like, since the results of measurement include the totality of the interphase, namely the metal sufaces, films, reactions, and double layer. With electroreflectance, specific portions of the interphase can be detected separately by the use of specific wavelengths 'tuned' to a particular species or process. In cases where it can be shown that a change of absorbance is caused by a change of concentration produced by an electrochemical reaction, it has been shown (ref.27) that a linear relation exists between the electroreflectivity coefficient, ρ, and the capacitance, C. By selecting suitable wavelengths corresponding to specific transitions, and simultaneously measuring the light modulation, voltage and current as a function of modulation frequency, it is possible to dissect an impedance curve into its several processes. Not only does this aid in resolving the 'true' equivalent circuit into its individual elements, but details about the process such as diffusion or reaction rates can be determined.

While analysis of electroreflectance data may not be as straightforward as the interpretation of ellipsometric data (which is itself not a trivial matter), reflectance equipment is certainly less expensive and easier to assemble. Even though the data may not be as quantitative, spectral features are often detectable that can be recognized in terms of fundamental phenomena.

RAMAN AND INFRA-RED

A description of the principles of Raman or infra-red is beyond the scope of this paper. However a brief comparison of the several techniques is relevant. In Raman studies, a specimen is irradiated with a source of <u>intense</u> monochromatic light. Most of the light is simply reflected, scattered or transmitted unchanged, but a few photons interact with some of the molecular or crystalline modes of the sample. The photon is the re-emitted with a new energy (wavelength) which is the difference (or sum) of the original photon and the energy of the e.g. vibration energy of the molecule. When this scattered light is passed through a high resolution monochromator, these <u>new</u> wavelengths show up as <u>small</u> peaks on a recorded spectrum. In most other infra-red techniques a direct absorption of the radiation by a species is involved. Although the principles behind these two techniques and the 'selection rules' involved are quite different, for many purposes the two types of spectra yield similar information. Experimentally, a significant difference is that most infrared measurement involve low energy photons to which few detectors are very sensitive, and in Raman, much higher energy photons are used and ordinary photomultipliers that are sensitiuive to individual photons can be used.

Unfortunately, the Raman process is extremely inefficient and great quantities of highly monochromatic photons must be used. It was a simple matter to calculate that a thin suface film just wouldn't scatter enough to make a measurement of the surface. Some foolhardy people tried it anyway (ref.31), and found that some surfaces enhanced the Raman scattering process by many orders of magnitude, and one could actually obtain Surface Enhanced Raman Spectra (SERS) (ref.32) from a monolayer or two on the suface.

With the development of SERS technology, great excitement was generated over the prospects for the examination of the passive film. Unfortunately, very little if any surface enhancement is produced by an iron surface in electrolyte. Attempts at several laboratories (including ours at CWRU) have yielded only a few highly suspect curves. Measurements have been made (ref.33) on a number of standard samples of those materials likely to be found in the passive film. Several studies have been succesful in producing spectra from thicker films produced by high temperature oxidation (ref.34). Interest is not dead, however, and studies are still underway to further improve the signal-to-noise ratio enough to make such measurements a reality.

Several problems still remain to be solved before this technique can be used succesfully with the passive film on iron. The most important is the problem of adequate signal. Once the experiment has been improved to the point that individual photons are detected, only a relatively small and asymptotic advantage can be gained by improving the geometric collection efficiency. This leaves the temporal efficiency which can be improved by increasing the intensity of the incident light, or the mode of photon collection. An optical multichannel analyzer (OMA) replaces the usual slit of a scanning monochromator by many parallel microdiode detectors and can record an entire spectrum simultaneously. Statistics limit the advantage to $n^{1/2}$ or a factor of about 30 for 1000 detectors. Physical (and financial) constraints of current semiconductor technology limit the number of detectors in a given sized array. This leaves integration time and light intensity as available degrees of freedom. These two are not as independent as might appear at first glance.

Two factors limit the amount of light that can be used. An absorbing sample is heated by the incoming beam,which must be focused to a spot size comparable to the entrance slit of the monochromator. Large local temperature rises can be encountered even in high thermal conductivity systems. Thibeau et al.. (ref.33) state that 'After a period of exposure, the surface of a pellet of [Fe_3O_4 or FeO] may be visibly changed'. An even greater potential problem is related to the fact that photoelectrochemical reactions can and do occur at many metal solution interfaces, including passive iron. Since exposures of several hours may required to acquire a useable signal, the film from which one obtains a spectrum may not be the same as the one we started with. Nazri (ref.34) reported a spectrum from Fe_2O_3 in air that is the same as those found in the literature, but the spectrum from the same material in buffered borate solution was completely different. It is still not clear which spectrum is correct! Did the material in electrolyte undergo a photochemical reaction, or did the presence of the electrolyte cool the sample enough to prevent thermal decomposition? Yeager et al. (ref.35) showed that a silver electrode on which Ag_2O was formed under potential control (where AgO should not be present) showed only a Raman signal of AgO, with a blackening of the spot illuminated by the beam.

The effect of temperature rise may be minimised by rotation of the sample,but even this may not suffice. If changes are produced photochemically, the effects may spread over a larger area than that hit by the beam. It may be possible to minimise these effects with the use of a pulsed laser and a gated photon detector, in an attempt to get the signal from the material before it has time to change. The sample could be moved between pulses so that each succesive pulse would be taken from a fresh spot.

While SERS does not appear to have been useful, other types of resonance phenomena can be utilized. The use of a laser 'tuned' to wavelengths at which the film absorbs may enhance Raman scattering, but it is not yet known whether the absorption by passive oxides is of the right type or sharpness of resonance. A tuneable laser may also help by using wavelengths below the threshold for photovoltaic effects, but unfortunately, the Raman cross-section also decreases with the light frequency.

In a desperate attempt to use SERS for the passive film, some measurements were made (ref.34) of a film produced by co-evaporating a film of 90% Ag and 10% Fe. These two metals are insoluble in each other, and it was hoped that a passive film could be grown on the iron domains and detected on the silver. It did not work too well, although a few tentative peaks did appear if one used a healthy dose of imagination.

Infra-red measurements covering the equivalent range are very difficult since the solution is so very absorbing at these wavelengths. At our laboratories (ref.34) an iron film 10 nm thick was sputtered onto a sheet of

KRS-5 (which is IR transparent) and the film was passivated chemically in 70% HNO3. A reasonable spectrum was obtained on these films in air in a Fourier Transform IR system from both the front surface by reflection and the back surface by transmission, but not to date from films passivated electrochemically in aqueous solutions.

PHOTO-VOLTAIC AND RELATED TECHNIQUES

One of the drawbacks of the use of photons is that they can cause photochemical reactions at the surface. We can turn this possibility to our advantage for the study of the nature of the passive film. A number of elegant experiments have been devised to try to use the photo-response to characterize the passive film. Many of these have been based on the tacit assumption that the film is a semi-conductor and the experiments have been oriented to the characterization and quantification of these materials and adapted from the field of semi-conductor technology. For example, and incident light beam can be chopped and the resultant photo-currents are sychronously detected and recorded as a function of wavelength, voltage, solution compositon or pH. At a more sophisticated level the entire transient wave form is recorded and analyzed. Thus for example, the wavelength at which photo-currents begin to appear is taken as the band gap of the semi-conductor. The transients produced are interpreted in terms of slow states, deep states, surface states, etc.

It has been recognized for a number of years that light can cause changes at the surface of a metal in electrolyte. A number of years ago, while trying to observe the _in situ_ formation of lead sulfate on a metallograph (ref.36) it was found that PbO_2 could be produced on a lead surface in H_2SO_4 when light was shined on the surface even at potentials 0.5 volts or more cathodic than that which would have been expected in the dark. Similiar effects have been found (ref.35) on silver. Hackerman (ref.37) found photo-currents on the passive film and on an air formed film on iron.

Many authors have taken the position that these photo-currents are _proof_ of the semi-conductor nature of the passive interface. This is a necessary but not a sufficient condition, however. One can conceive of other possible mechanisms for the production of such photo-currents. A photon with sufficient energy can always excite an electron out of the energy level, orbital or band in which it happens to be sitting. If there is a potential gradient of some kind in the region of that electron it can be removed from the physical vicinity of the residual electron deficiency and can be collected by an available external circuit. The 'hole' left behind is an ion which can be neutralized by another electron falling into it or by a chemical reaction with its environment.

Since the currents produced are typically fairly small, it can be difficult to differentiate photochemical changes from simple photo-currents. In the case of PbO$_2$ or AgO the new phases produced are sufficiently colored that their formation is readily detectable with the naked eye. Many other passive films are transparent enough in the thickness required for passivation that they are not as readily detected. Some other methods must then be utilized to identify the fundamental principles involved.

When photo-currents are produced in a classical semi-conductor, light excites an electron from the valence band into the conduction band. This electron can either fall back into the hole or swept away under the presence of some field or by a 'built-in' or induced junction. The photon must have a threshold energy greater than the band gap which should be independent of the applied potential.

If we recongize that the surface of a metal or a semi-conductor or a passive film can also be a reactant, then alternatives to the semi-conductor model for photo-current production are possible. A photo-electron which leaves behind an excited species may have been emitted near the very strong field across the Helmholtz layer or in an electrostatic potential gradient in a slightly conducting passive film or in a chemical potential gradient of a chemi-conducting film. If this electron is captured and removed from the vicinity the resultant excited state can react chemically with its environment. One of the ways this reaction can occur is with the formation of additional film. It can also react with any other mobile species in the film or with a neighboring atom or group of atoms releasing a mobile species. An atom at the surface can be given enough energy to 'kick it' the rest of the way over the barrier at the interface. This can occur especially easy in a disordered material which has no well-defined discrete band structure. In such material the band structure picture must be replaced by something closer to the density of states model used for materials like amorphous silicon. For such material one might expect a variation of the photo-threshold with applied potential. One also might expect a variation of the time response to a pulse of light with potential since the density of states in the material would be affected by the changes of defect density or stoichiometry produced by the photo-current itself. Froelicher et al. (ref.38) have reported such changes in the photo-threshold and the photo-response of iron passive films.

SUMMARY

Optical techniques while not perfect or universally applicable still remain the best in situ tools for looking at passive interfaces in aqueous electrolytes. Great care must be exercised in analysis of the data and in the experimental arrangement. There is still considerable room for improvement in these techniques and we look forward to many advances in methodology and understanding in the future.

ACKNOWLEDGEMENT

The author would like to thank Peter J. Pearson for significant help in the preparation of this manuscript.

REFERENCES

1 R.H. Muller, in Adv. Electrochem. Electrochem. Eng., P. Delahay and
   C.W. Tobias (Eds.), 9 (1973) 167.
2 J.D.E.McIntyre in Ref.1, pp. 61.
3 B.D.Cahan and R.F. Spanier, Surf. Sci. 16 (1969) 166.
4 D.E.Aspnes, A.A. Studna, Appl. Opt. 14 (1975) 220.
5 C.T. Chen, PhD. Thesis, Case Western Reserve University, 1979.
6 C.A. Fenstermaker and F.L. McCrakin, Surf. Sci. 16 (1969) 85.
7 I. Ohlidal and F. Lukes, Optical Acta 19 (1972) 817.
8 R.H.Muller, Surf. Sci. 96 (1980).
9 L.S. Bartell and D. Churchill, J. Phys. Chem. 65 (1961) 2242.
10 F.L. McCrackin and J.P. Colson, in Ellipsometry in the Measurement of
   Surfaces and thin films, E. Passaglia, R.R. Stromberg and J. Kruger, Eds.,
   Natl. Bur. Stand. Misc. Publ. 256, Washington, D.C. (1964).
11 W.Plieth and K. Naegele, Surf. Sci. 50 (1975) 53.
12 B.D. Cahan, Surf Sci. 56 (1976) 354.
13 W-K Paik and J. O'M. Bockris, Surf. Sci., 28 (1971) 61.
14 C.T.Chen and B.D. Cahan, J. Electrochem. Soc., 129 (1982) 17.
15 B.D. Cahan and C.T. Chen, J. Electrochem. Soc., 129 (1982) 921.
16 D.E. Aspnes, Phys. Rev. B12 (1975) 4008.
17 D.M. Radman, M. Sc. Thesis, Case Western Reserve University, 1982.
18 N.C. Debnath and A.B. Anderson, J. Electrochem. Soc (1982).
19 J.L. Ord and D.J. DeSmet, J. Electrochem Soc. 113 (1966) 1258.
20 F.Wooten, Optical Properties of Solids, Academic Press, N.Y., 1972.
21 M. Cardona, Modulation Spectroscopy, Academic Press, N.Y., 1969.
22 J. Feinleib, Phys. Rev. Lett., 16 (1966) 1200.
23 B.D. Cahan, J. Horkans and E. Yeager, Symp. Faraday Soc., 4 (1970) 36.
24 J.D.E. McIntyre and D.E. Aspnes, Bull. Amer. Phys. Soc., 15 (1970) 366.
25 A. Prostak and W.N. Hansen, Phys. Rev., 160 (1967) 600.
26 B.E. Conway, H. Angerstein-Kozlowska and L.H. Laliberte, J. Electrochem.
   Soc., 121 (1974) 1596.
27 D. Wheeler, B.D.Cahan, C.T. Chen and E. Yeager in Passivity of Metals,
   R.P. Frankenthal and J. Kruger, Eds., The Electrochemical Soc.,
   Princeton, N.J., 1978.
28 W.J. Horkans, Ph.D. Thesis, Case Western Reserve University, 1973.
29 W.J. Horkans, B.D. Cahan and E. Yeager, Surf. Sci., 46 (1974) 1.
30 B.D.Cahan and C.T.Chen, J. Electrochem. Soc., 129 (1982) 476.
31 M. Fleischmann, P. Hendra and A.J. McQuillan, Chem. Phys. Lett.,
   26 (1976) 163.
32 R.P. Van Duyne, D.L. Jeanmaire, D.L. Suchanski, W. Wallace and T. Cape,
   in Resonna Roman Spectroelectrochemistry, National Meeting, The Electorchem.
   Soc., Washnington D.C., 1976 Paper 357.
33 R.J. Tibeau, C.W. Brown and R.H. Heidersbach, Appl. Spect. 32 (1978) 532,
34 A. Nazri, Ph.D. Thesis, Case Western Reserve University, 198 .
35 R. Kotz and E. Yeager, J. Electroanal. Chem., 111 (1980) 105.
36 B.D.Cahan and J. McGivern, Unpublished results.
37 F.M. Delnick and N. Hackerman, J. Electrochem. Soc., 126 (1979) 732.
38 M. Froelicher, A. Hugot-Le Goff, V. Jovancicevic, M.Maja and P. Spinelli,
   ISE Meeting, Lyon, France, 1982, Abstract no. IV C 21.

Passivity of Metals and Semiconductors, edited by M. Froment
Elsevier Science Publishers B.V., Amsterdam — Printed in The Netherlands

REFLECTOMETRY OF IRON PASSIVATION FILMS IN NEUTRAL BORATE AND ACIDIC
PHOSPHATE SOLUTIONS

TOSHIAKI OHTSUKA, KAZUHISA AZUMI, and NORIO SATO
Electrochemistry Laboratory, Faculty of Engineering, Hokkaido University,
060 Sapporo, Japan

ABSTRACT
     3P reflectometry was applied to measure the passive oxide film on iron.
In neutral solution the refractive index of the film changes at around the
Flaae potential where the film composition appears to change. As the
solution pH decreases, the optical density of the film increases indicating
that the film is more dehydrated in the more acidic solution. The absorption
spectrum of the film shows two absorption edges at 2 eV and at 2.6 eV
corresponum to two electron energy gaps in the passive film.

INTRODUCTION
     A number of optical techniques have recently been employed in studying
metallic passivity. Among them ellipsometry has frequently been used for
measuring the passive film on metals. Ellipsometry, however, suffers one
disadvantage in that three unknown parameteters (refractive index n_2,
extinction index k_2, and thickness d) of the surface film have to be estimated
from two measurable parameters. This disadvantage is now overcome by a modern
ellipsometric technique in which the average reflectivity change can be
measured in addition to the two ellipsometric..parameters, Ψ and Δ (ref. 1).
On the same principle as the modern ellipsometric technique, it is possible to
estimate the three unknowns of the surface film from measurements of three
reflectivity changes at three different states of light polarization
independent of each other, which hereafter is referred to as 3P reflectometry
(ref. 2 and 3). In this paper, the 3P reflectometry is applied to the passive
films on iron formed in neutral borate and acidic phosphate solutions to reveal
the optical and spectroscopic properties of the films.

EXPERIMENTAL
     The apparatus used for 3P reflectometry consisted of a light source,
a spectrometer, a polarizer, a sample electrode, an analyser, and a photomultiplier. For stabilizing the apparatus, the light source was made of a
tungsten lamp connected to a light-feedback system, and the detection of light

reflectivity was made with AC amplification using a light chopper and lock-in amplifier. Three reflectivity changes during the film growth were measured at three azimuth settings of the analyser (A=0°, 45°. and 90°) at a fixed azimuth of the polarizer (P = 40°). The incidence angle of light was $\phi=55.06°$. Estimation of the three unknowns of n_2, k_2 and d from the three reflectivity changes was made by Newton-Raphson method of successive approximations (ref.2 and 3).

RESULT

An example of 3P reflectometry measurements of the passive film on iron is shown in Fig. 1, in which the film growth at constant anodic current followed by the film reduction at constant cathodic current in neutral borate solution at pH 8.4 is plotted. The film thickness increases with the anodic charge passed and then decreases with the cathodic charge. These changes in the film thickness are accompanied by small variation in the refractive index of the film, indicating that the film composition changes during the galvanostatic film growth.

Fig. 2 shows the complex refractive index ($\tilde{n}_2 = n_2 - ik_2$) and the thickness of the passive film formed in neutral borate solution at pH 8.4 by potentiostatic oxidation for one hour. The refractive index changes at around 0.6 V (vs Hydrogen Electrode at the Same Solutions; HESS), while the thickness increases almost linearly with the electrode potential. It is noted that the potential of 0.6 V coincides with the Flade potential (E_F = 0.58 V vs HESS) measured in acidic solution (ref. 4). It is also noted from comparison between the results of potentiostatic oxidation (Fig.2) and galvanostatic oxidation (Fig. 1) that the refractive index, which is related to the film composition or structure, is dependent on the film formation process.

Fig. 3 shows the complex refractive index and the thickness of the passive film formed in acidic phosphate solution at pH 3.1 by potentiostatic oxidation for one hour. Compared with the film in neutral solution, the film in the acidic solution is thinner than that in neutral solution, but its refractive index is larger. The refractive index of the film in the acidic solution (Fig. 3) appears close to that of the film formed during galvanostatic oxidation in the neutral borate solution (Fig. 1).

Fig. 4 shows the spectrum of the complex refractive index of the passive film on iron, the spectrum which was calculated from the three reflectivity changes at constant wave length. There is significant difference between the films formed above and below the Flade potential (0.6 V) in neutral solution and also between the films formed in the acidic and in the neutral solution.

DISCUSSION

Previous ellipsometric measurements(ref. 5) have suggested the presence of a hydrated layer in the outer part of the passive film on iron in neutral solution. Recent measurments by Chen and Cahan (ref. 1) who used 3P ellipsometry, however, have failed to find any hydrated outer layer of low optical density in the passive film on iron. In this discussion, therefore, we assume a single layer model for the film.

The thickness and optical constants of the passive films formed in neutral solution(Fig. 2) coincide with those measured in a previous study (ref. 6), except for the variation in the refractive index of the film that occurs at around the Flade potential, E_F. This variation of the refractive index implies that the film composition changes at the Flade potential. The film formed at potentials less noble than E_F does function as a barrier against iron dissolution in neutral solution but it does not cause the passivation of iron in acid solution.

Fig. 5 shows the effect of pH on the refractive index n_2 and extinction index k_2 at a wave length of light λ = 460 nm and on the thickness of the film formed at 1.54 V. As the solution pH increases, n_2 decreases, while k_2 remains constant. This indicates that the optical density of the film decreases with pH, probably because the extent of film hydration increases with pH.

From the spectra of k_2 (Fig. 4), we can estimate the light-absorption coefficient, $2\alpha = 4\pi k_2 \lambda^{-1}$, as a function of wave length of light. The absorption spectra thus estimated are shown in Fig. 5. There are two absorption edges, one at 2 eV for weak absorption and the other at 2.6 eV for strong absorption. The light absorption edge at 2 eV corresponds to the critical photon energy where the initial increase in the photocurrent was observed by Wilhelm et. al. (ref. 7). They also found a steep increase in the photocurrent at wave lengths less than 460 nm and estimated the band gap energy at 2.5 eV for the passive film on iron. The absorption spectra shown in Fig. 6 are in fairly good agreement with the photocurrent spectra obtained by Wilhelm et. al. It is evident from Fig. 6 that there are two critical energy gaps at 2.0 eV and at 2.6 eV for electron excitation in the passive oxide film on iron in acidic and neutral solution.

REFERENCES

1 C.T. Chen and B.D. Cahan, J. Electrochem. Soc., 129, 17(1982)

2 T. Ohtsuka and K.E. Heusler, J. Electroanal. Chem., 100, 319(1979)

3 K.E. Heusler and T. Ohtsuka, Surf. Sci., 101, 192(1980)

4 U.F. Frank, Z. Naturforsch., 4a, 398(1949)

5. N. Sato, K. Kudo and T. Noda, Z. Phys. Chem. N.F., 98, 271(1975)

6 N. Sato and K. Kudo, Electrochem. Acta, 16, 447(1971)

7 S.M. Wilhelm, K.S. Yun, L.W. Ballenger and N. Hackerman, J. Electrochem. Soc. 126, 419(1979): S.M. Wilhelm and N. Hackerman, ibid, 128, 1668(1981)

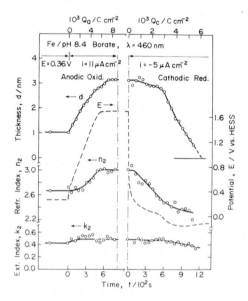

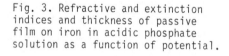

Fig. 1. Change in refractive and extinction indices and in thickness of passive film on iron during galvanostatic transient in neutral borate solution.

Fig. 2. Refractive and extinction indices and thickness of passive film on iron in neutral borate solution as a function of potential.

Fig. 3. Refractive and extinction indices and thickness of passive film on iron in acidic phosphate solution as a function of potential.

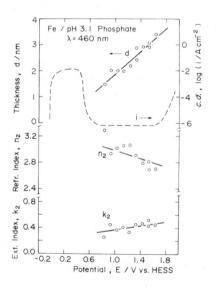

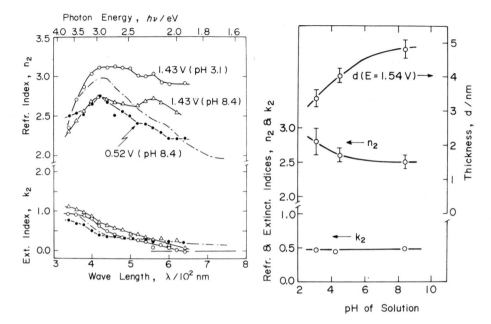

Fig. 4. Spectra of refractive and extinction indices of passive film on iron. The dotdash line shows for comparison the result measured in neutral borate solution by Cahan et.al[1]).

Fig. 5. Effect of pH for film formation on the refractive index n_2 and extinction index k_2 at wave length $\lambda = 460$ nm and on the film thickness formed at 1.54 V vs. HESS.

Fig. 6. Spectra of absorption coefficient, $2a$, of the passive films on iron. The absorption coefficient was calculated from k_2 value shown in Fig. 4.

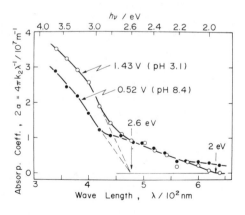

Passivity of Metals and Semiconductors, edited by M. Froment
Elsevier Science Publishers B.V., Amsterdam — Printed in The Netherlands

COMBINED ELLIPSOMETRIC AND AC IMPEDANCE MEASUREMENTS OF OXIDE FILMS ON RUTHENIUM

J. RISHPON, I. RESHEF and S. GOTTESFELD
Chemistry Dept., University of Tel Aviv, Ramat Aviv 69978, Israel

ABSTRACT

An instrumental setup for the in-situ characterization of surface films is described, in which a microprocessor is used for experiment control, data acquisition and data analysis in automated measurements of the ellipsometric parameters and of the ac impedance ("network analysis"). Results for oxide films grown on Ru metal are described. The mode of film growth is shown to have a clear effect on the optical properties of the oxide, reflecting the degree of porosity and hydration. The ac impedance measurements in electrolytes of different acid concentrations show that the larger part of the Ru oxide charge capacity depends on the supply of protons from solution. The extent of reversible charging is not negligible, however, in electrolytes with very low concentrations of free protons, testifying probably for a limited injection of ions other than H^+ into the oxide structure.

INTRODUCTION

Oxides of ruthenium formed by various techniques are well known for their high electrocatalytic activity in the anodic processes of O_2 and Cl_2 evolution. This activity is structure dependent, and is expected to be affected by the composition and oxidation state at the oxide-electrolyte interface, as well as by the electronic and ionic conductivities within the oxide phase [1]. The combined effects of these factors may be elucidated by examination of the process of oxygen or OH electrosorption on an oxide surface site, believed to be the first step in the overall reaction of O_2 evolution at the oxide surface[2]:

$$Ru^{+n}---- OH_{2,ads} \rightleftharpoons Ru^{+n+1}- OH_{ads} + H^+_{aq} + e \tag{1}$$

Equation (1) implies that the anodic formation of the adsorbed intermediate is aided by a Ru^{+n}/Ru^{+n+1} redox couple, located at the surface (including the "inner" surface) of the oxide. The rate of this redox process depends, in turn, according to eq.(1) on rates of exchange of both electrons and protons at the surface site, and may be thus limited by the electronic and protonic conduct-

ivities of the oxide phase, or by the rate of the interfacial process itself.
The degree of surface hydration can be seen from eq.(1) to facilitate the release of a proton while forming an adsorbed OH intermediate, and should be,
therefore, another important factor in electrocatalysis, at least for the O_2
evolution process. Similar oxidative processes of surface metal ions bonded to
solvent molecules and exchanging ions and electrons with the solution and metal
substrate, respectively, determine also the reactivity of such oxides in dissolution processes. Reactivity-structure relationships derived for such films bear,
therefore, relevance to oxide stability as well.

EXPERIMENTAL

Figure 1 shows a schematic diagram of the system for automated ellipsometry
and ac impedance measurements, employed in this work. A S-100 microcomputer is
used for experiment control and for data acquisition and analysis in both the
ellipsometric and the impedance measurements. In the ellipsometric measurement,
the microcomputer controls the rate of revolution of a rotating analyzer with an
accuracy of 1:5000 ($\omega=250\pm0.05$ Hz). This level of stability is achieved by on-
off control of a simple DC motor, applied according to the cycle time continuously clocked by the computer. This stability of revolution allows to obtain a
precision of 10^{-2} deg in phase angle reading after averaging for less than a
second. The sinusoidal output of the photomultiplier is digitized by an A/D converter, and analyzed by the microcomputer, using a fast algorithm which yields
the three fourier parameters for the waveform after each single cycle. The values of Δ and ψ for the reflecting surface are then calculated by the computer
from these parameters, and further numerical analysis gives film thickness and
refractive index.

In the impedance measurement mode the microcomputer scans the frequency of
an oscillator between $5\cdot10^{-3}-4\cdot10^{3}$Hz by a voltage ramp applied from a D/A unit.
(The higher frequency limit is set by the response of the existing A/D). Exactly
the same algorithm employed for the analysis of the optical sinusoidal waveform
is employed also to evaluate the amplitude and phase of the periodic current
signal resulting from the ac potential perturbation.

The Ru electrodes used were in the form of rods inserted in teflon or in
epoxy housings, so as to expose a single surface of a disk to the solution.
Potentials are reported in each case vs. the reversible hydrogen electrode in
the same solution (RHE).

DISCUSSION

The optical properties evaluated for oxide films grown on Ru, at constant
applied potential or by cyclic potential multipulsing, satisfactorily resembled

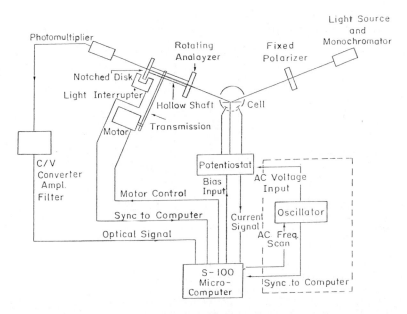

Fig. 1. Schematic diagram of the experimental setup for automated ellipsometry and ac impedance measurements.

the results obtained before with a commercial ellipsometer [2]. Fig.2 shows, for example, ellipsometric results for oxide growth by potential multipulsing, together with the best computer fitting to a uniform film growth curve [2]. Films grown at a constant applied anodic potential exhibited a different complex refractive index: $n_{film} = (3.5 \pm 0.2) - (0.9 \pm 0.2)i$. These results mean that the films grown by multipulsing are more porous and hydrous. The enhanced rate of Ru dissolution induced by cyclic multipulsing has been reported before [3]. It is very likely, therefore, that during multipulsing the film is formed by a dissolution-precipitation mechanism. The breakdown of passivity induced by an alternating applied voltage seems to be a general phenomenon, which can be used for the growth of thicker oxide films on relatively passive metal surfaces. (The thickness of the film formed on Ru at a constant anodic potential did not exceed 250-300 Å).

Fig.3 shows two complex capacitance plots for an oxide film grown by constant potential anodization of a Ru metal electrode in 0.5M H_2SO_4, recorded at two different applied potentials of 0.72 and 0.22V. The data in Fig.3 is <u>not</u> corrected for solution resistance. The plots contain a semicircle in the higher frequency region, which corresponds to a series RC branch, where the value of C was found to increase with film thickness. The values of ω_{max} for the semicircles always corresponded to the charging of the capacitors via a resistance practically equal to the solution resistance, as calculated for a disk geometry.

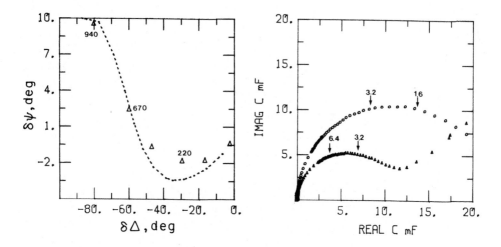

Fig. 2. (left). Experimental ellipsometric points and the computer fitted curve for uniform film growth, obtained for oxide growth on Ru in 0.5M H_2SO_4 by potential multipulsing between 0 and 1.47V at 0.5Hz. The film refractive index evaluated is 1.65-0.39i (546nm), measured at 0.72V. Thickness in Å is depicted.

Fig. 3. (right). Complex capacitance plots measured at 0.72V (upper) and at 0.22V (lower) in 0.5M H_2SO_4, following oxide film growth on Ru at 1.47V for 10 min. (solution resistance not corrected). Disk area - 0.52 cm^2.

This was true up to the highest film thicknesses generated. The charging of these large surface capacitors is thus a very fast process in 0.5M H_2SO_4, and it can be stated only that the interfacial resistance associated with it is significantly smaller than 0.1 Ωcm^2. Very similar capacitance plots were obtained for the films grown by multipulsing, and also when $HClO_4$ was used instead of H_2SO_4. The lowering of C at 0.22V is most probably simply due to the shifting of the potential away from the central region of the Ru$^{+3/+4}$ redox activity. Optical measurements revealed, however, some variation in ellipsometric readings due to cathodic applied potentials, which could be interpreted by some thickening of the oxide film and a simultaneous lowering of the absorption coefficient. Such effects are expected for enhanced hydration.

The origin of the additional capacitance dispersion obtained at the lowest frequencies (Fig.3) has remained unclear. The insensitivity of these ac currents to the anodic oxide thickness or to solution stirring show that they may reflect either a slow process in an underlying thin and compact oxide film, or, more probably, leakage currents due to minute solution penetration between the Ru disk and the teflon or epoxy housing. (A similar capacitance dispersion at very low frequencies has been reported for a Ni/Ni oxide electrode of a similar configuration [4]).

A plot of oxide capacitance (measured according to the main semicircular

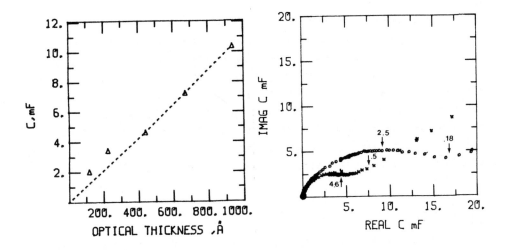

Fig. 4. (left). Oxide capacitance at 0.72V in 0.5M H_2SO_4 (measured as the diameter of the semicircle in the complex C plot) plotted vs. the optically evaluated film thickness, for film growth by multipulsing for 2,4,8,12 and 16 min. Disk area 0.22 cm^2.

Fig. 5. (right). Complex capacitance plots for a Ru/Ru oxide electrode. Oxide grown in 0.5M H_2SO_4 for 10 min at 1.47V, and the electrode then transferred to a 5mM H_2SO_4+0.5M Na_2SO_4 solution (upper plot), or to a 0.5M Na_2SO_4 solution (lower plot). Measurements taken at 0.72V RHE in each solution. Disk area - 0.52 cm^2. (Solution resistance has been subtracted).

feature) vs. the optically evaluated film thickness is given in Fig.4 for films grown by potential multipulsing. The deviation from linearity found for the lower film thicknesses is most likely due to a less satisfactory fit of the ellipsometric results to a single film model in the early stages of growth. This can actually be realized from Fig.2.

Fig.5 exhibits the complex C plot obtained in a 5mM H_2SO_4+0.5M Na_2SO_4 solution, at a potential of 0.72V. Ellipsometric monitoring of film properties did not reveal any significant variation due to the transfer from 0.5M H_2SO_4 to less acidic solutions. Following correction for the solution resistance, the plot has the shape of a circular arc, expected for a diffusion-controlled charging of a capacitor, where the total film capacitance remains the same as that measured in 0.5M H_2SO_4 for the same oxide film thickness (Fig.3). The value of ω_{max}, the frequency corresponding to the maximum of C_{imag}, leads in this case to a calculated diffusion coefficient of $1\times10^{-5} cm^2 sec^{-1}$, in agreement with control by diffusion of a proton donor in the solution phase. This result means that the reversibly charging sites in the Ru oxide film, or at least most of them (see below), require protons as the injected ion in the ion-electron double injection process.

There is, however, a significant fraction (ca. 20%) of the sites in the oxide which can be charged at a relatively high rate even in the absence of excess protons. This can be concluded from the other plot in Fig.5, obtained in a Na_2SO_4 solution with no added acid. This result resembles the one found for anodic Ir oxide under positive applied potentials [5]. We suggest that it reflects the extent of the basic double-layer charging process in which e.g. Na^+_{aq} may serve as a counter ion. This last form of charging may be limited, however, to sites which belong to the outer part of the oxide film, as well as sites inside the oxide structure in contact with larger solution filled pores.

REFERENCES

1. S. Trasatti and W.E. O'Grady, in Advances in Electrochemistry and Electro-chemical Engineering, Vol. 12, H. Gerischer and C.W. Tobias Ed., John Wiley 1981, and references therein.

2. S. Gottesfeld, J. Rishpon and S. Srinivasan, in Proc. Symp. on Electro-catalysis, W.E. O'Grady, P.N. Ross and F.G. Will Ed., The Electrochemical Society, 1982.

3. S. Hadzi-Jordanov, H. Angerstein-Kozlowska, M. Vukovic and B.E. Conway, J. Electrochem. Soc. 125 (1978) 1471.

4. S.H. Glarum and J.H. Marshall, J. Electrochem. Soc. 129 (1982) 535.

5. S.H. Glarum and J.H. Marshall, J. Electrochem. Soc., 127 (1980) 1467.

Passivity of Metals and Semiconductors, edited by M. Froment
Elsevier Science Publishers B.V., Amsterdam — Printed in The Netherlands

ANALYSIS OF MULTIPLE LAYER SURFACE FILMS BY MODULATED REFLECTION SPECTROSCOPY

N. HARA and K. SUGIMOTO

Department of Metallurgy, Faculty of Engineering, Tohoku University, Sendai 980
(Japan)

ABSTRACT

Methods for qualitative and quantitative analyses of thin multiple layer surface films by modulated reflection spectroscopy have been studied using single-
and two-layer films of γ-Fe_2O_3 and Cr_2O_3 and single-layer films of composite oxide
of Fe_2O_3-Cr_2O_3 prepared by CVD technique. The application of the methods to the
in situ analysis of passive films is demonstrated on Fe-Cr alloys in solutions.

INTRODUCTION

Modulated reflection spectroscopy (MRS) can provide data for the analysis of
very thin films on metals in aqueous environments (refs.1 and 2). Hitherto,
in situ studies of passive films on Co (ref.3), Mo (ref.4), Ni (refs.5 and 6),
Cu-Zn (ref.7), Fe-Cr (ref.8) and Fe-Cr-Ni alloys (ref.9) by MRS have been
reported. Results of these studies, however, seem to be still in qualitative
stages and there is some uncertainty about the location of the region in a film
from which spectral evidence can be obtained.

The purpose of the present investigation is to establish the methods for qualitative and quantitative analyses of multiple layer passive films on alloys by
MRS. To realize it, the change in MRS spectrum as a function of composition and
thickness of the layer has been measured on single- and two-layer oxide films
with known compositions and thicknesses. The availability of the methods
developed is demonstrated in the analysis of passive films on Fe-Cr alloys in
Na_2SO_4 solutions.

EXPERIMENTAL

Specimen preparation

Single- and two-layer films of γ-Fe_2O_3 and Cr_2O_3 and single-layer composite
films of Fe_2O_3-Cr_2O_3, which were all coated on Pt plates 25mm x 15mm x 0.5mm by
CVD technique (ref.10), were used as passive-film-simulated specimens.
$Fe(O_2C_5H_7)_3$ and $Cr(O_2C_5H_7)_3$ powder were employed as vapor sources for CVD. The
thickness of each layer was examined by ellipsometry. The composition of composite films was analyzed by atomic absorption spectroscopy after the films were
dissolved in 5 ml of 1 $kmol \cdot m^{-3}$ H_2SO_4. The cationic mass fraction of Cr^{3+} ions,

X_{Cr}, of the films was calculated from the equation, $X_{Cr} = W_{Cr} / (W_{Fe} + W_{Cr})$ where W_{Cr} and W_{Fe} were respectively the Cr and the Fe content of the films. Plates 25mm x 15mm x 2mm of Fe-15Cr (Cr:15.3 mass%) and Fe-19Cr (Cr:19.1 mass%) alloys were also used as specimens. These were used for the analysis of real passive films. All the specimens were covered with an epoxy resin except for the electrode area of 1×10^{-4} m^2.

Solutions

Deaerated 1 kmol·m^{-3} Na$_2$SO$_4$ adjusted to pH 2.0 and 6.0 were used as solutions. The solutions contained small amounts of H$_2$SO$_4$ and NaH$_2$PO$_4$ + Na$_2$HPO$_4$, respectively. The temperature of the solutions were kept at 293 K.

Measurement of modulated reflection spectra

The apparatus and experimental conditions used for MRS were the same as those described in a previous paper (ref.11). The spectrum measurements were performed on the specimens under potentiostatic or galvanostatic control. In order to modulate the electrode potential, a sinusoidal a-c voltage with a frequency of 65 Hz was superimposed on the d-c bias potential. The amplitude of the a-c voltage, ΔE, was kept less than 100 mV$_{p-p}$ under potentiostatic control and 50 mV$_{p-p}$ under galvanostatic control. The measured reflectivity change, $-\Delta R/R$, was normalized to the condition of $\Delta E = 100$ mV$_{p-p}$ and then plotted versus the photon energy of incident light, $\hbar\omega$. A saturated calomel electrode was used as the reference electrode.

RESULTS AND DISCUSSION
Qualitative analysis

Spectra for a single-layer γ-Fe$_2$O$_3$ film 2.9 nm thick and a single-layer Cr$_2$O$_3$ film 2.0 nm thick in 1 kmol·m^{-3} Na$_2$SO$_4$ of pH 6.0 are shown in Figs. 1 and 2. The structure and intensity of spectra for both films change with potential. This is presumably due to the change in the energy band structure of the oxides with the change in field strength across the films, since no changes in thickness and composition of the films occurred in the potential ranges examined. In Figs. 1 and 2, characteristic peak maxima appear at 3.2 eV for the γ-Fe$_2$O$_3$ film at potentials between -0.10 and 0.40 V and at 5.0 eV for the Cr$_2$O$_3$ film at potentials between -0.60 and 0.60 V. These peak maxima can be used for the qualitative analysis of Fe^{3+} and Cr^{3+} ions in films with unknown compositions.

Quantitative analysis

Spectra for single-layer composite films of Fe$_2$O$_3$-Cr$_2$O$_3$ with various X_{Cr} are shown in Fig. 3. They were measured at 0.00 V in 1 kmol·m^{-3} Na$_2$SO$_4$ of pH 6.0. It can be seen that the photon energy of peak maximum, $\hbar\omega_{pm}$, in the spectrum

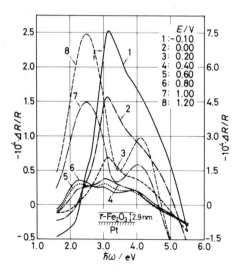

Fig. 1. Change in MRS spectrum for γ-Fe₂O₃(2.9nm)/Pt electrode with potential E.

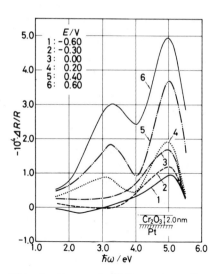

Fig. 2. Change in MRS spectrum for Cr₂O₃(2.0nm)/Pt electrode with potential E.

changes with X_{Cr}. The intensities of the spectrum, $-\Delta R/R$, at 3.2 and 5.0 eV also depend on X_{Cr}. Therefore, both the $\hbar\omega_{pm}$ and the ratio of $-\Delta R/R$ at 5.0 eV to $-\Delta R/R$ at 3.2 eV, $r_{5.0eV/3.2eV}$, can be used as measures of X_{Cr}. The $\hbar\omega_{pm}$ vs. X_{Cr} and the $r_{5.0eV/3.2eV}$ vs. X_{Cr} relation are given in Fig. 4. If such relations are known at adequate potentials, it is possible to determine values of X_{Cr} for passive films on Fe-Cr alloys at the corresponding potentials.

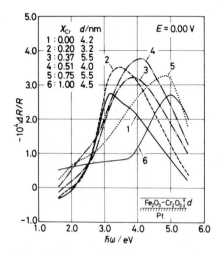

Fig. 3. Change in MRS spectrum for Fe₂O₃-Cr₂O₃/Pt electrode with cationic mass fraction of Cr^{3+}, X_{Cr}.

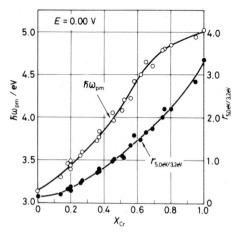

Fig. 4. Changes in $\hbar\omega_{pm}$ and $r_{5.0eV/3.2eV}$ as a function of X_{Cr} for composite film of Fe₂O₃-Cr₂O₃.

214

Spectral information depth

Spectra for two-layer films, which have various thicknesses of the outer Cr_2O_3 layer, d_{OL}, on the inner γ-Fe_2O_3 layer 2.2 nm thick, are shown in Fig. 5. They were measured at 0.00 V in 1 kmol·m^{-3} Na_2SO_4 of pH 6.0. In Fig. 5, only the peak maxima at 5.0 eV for the outer Cr_2O_3 layer, with no peak maxima at 3.2 eV for the inner γ-Fe_2O_3 layer, can be seen in the spectra until d_{OL} decreases to 0.9 nm. Spectra for two-layer films, which have various thicknesses of the outer γ-Fe_2O_3 layer, d_{OL}, on the inner Cr_2O_3 layer 2.5 nm thick, are given in Fig.6. They were measured in the same condition as that in Fig. 5. Again, only the peak maxima at 3.2 eV for the outer γ-Fe_2O_3 layer, with no peak maxima at 5.0 eV for the inner Cr_2O_3 layer, can be seen in the spectra until d_{OL} decreases to 0.9 nm. These results suggest that the spectral information depth for the oxides should be less than 0.9 nm. Such a value for the depth seems to be close to the thicknesses of space charge regions, i.e., ca. 0.5 nm, found in the single-layer films of γ-Fe_2O_3 and Cr_2O_3 (ref.11). Consequently, the change in reflectivity caused by the potential modulation is thought to be ascribed to the change in the optical constant of the space charge region varying with the field strength across the region.

Quantitative analysis of real passive films

The change in the spectrum for Fe-19Cr alloy with changing potential was measured in 1 kmol·m^{-3} Na_2SO_4 of pH 2.0 and 6.0. The potential was changed at intervals of 100 mV in the noble direction. After retention at each potential for 3.6 ks, the current and spectrum were measured. The result for the pH 6.0

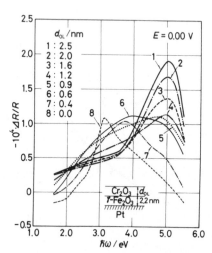 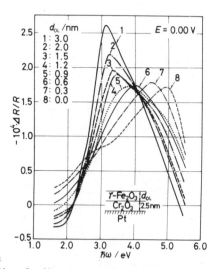

Fig. 5. Change in MRS spectrum for Cr_2O_3/γ-Fe_2O_3(2.2nm)/Pt electrode with thickness of the Cr_2O_3 layer, d_{OL}.

Fig. 6. Change in MRS spectrum for γ-Fe_2O_3/Cr_2O_3(2.5nm)/Pt electrode with thickness of the γ-Fe_2O_3 layer, d_{OL}.

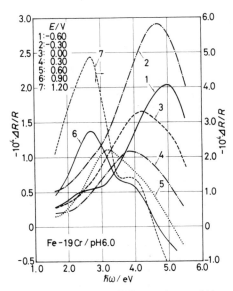

Fig. 7. Change in MRS spectrum with potential E for Fe-19Cr alloy in 1 kmol·m⁻³ Na₂SO₄ of pH 6.0.

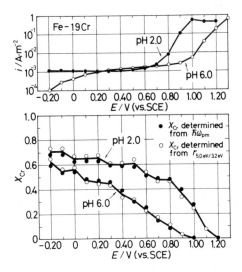

Fig. 8. Anodic polarization curves (upper) and X_{Cr} vs. E curves (lower) for Fe-19Cr alloy in 1 kmol·m⁻³ Na₂SO₄ of pH 2.0 and 6.0.

solution is shown in Fig. 7, as an example. Changes in X_{Cr} for passive films formed at pH 2.0 and 6.0 as a function of potential, E, are shown in Fig. 8. No difference can be seen between the value of X_{Cr} determined from $\hbar\omega_{pm}$ and that from $r_{5.0eV/3.2eV}$. The film formed at lower pH has higher X_{Cr} than that formed at higher pH. The X_{Cr} for each film gradually decreases with increasing potential. For example, the value of X_{Cr} for the film formed at pH 6.0 changes from 0.62 at -0.20 V to 0.00 at 1.00 V, indicating that no Cr exists in the film at 1.00 V. It should be noted, however, that the value thus determined is for a very thin region close to the film surface.

Figure 9 shows the change in spectrum with charge, Q, passed during the galvanostatic reduction of a passive film 3.7 nm thick, which was formed on Fe-15Cr alloy at 0.90 V in 1 kmol·m⁻³ Na₂SO₄ of pH 6.0. The reduction of the film proceeded while the potential stayed at -0.15 V. But a film 0.9 nm thick remained at final stages of the reduction at the potential (at No.4 in Fig. 9a). When the cathodic charge was applied further, the transformation of the remained film occurred without changing thickness (at No.5 in Fig. 9a). The value of X_{Cr} at each stage of reduction can be determined from the corresponding spectrum. The spectrum No.5, however, should be disregarded, because this came from the trans- formed film. The change in X_{Cr} as a function of film depth is shown in Fig. 10. The film depth was obtained from the film thickness determined by ellipsometry at each stage of the reduction. Results for films formed at 0.10 and 0.50 V are also given. An increase in the value of X_{Cr} with increasing depth can be seen

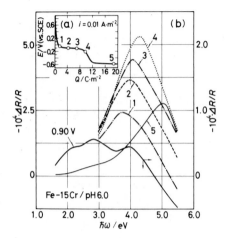

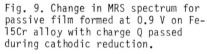

Fig. 9. Change in MRS spectrum for passive film formed at 0.9 V on Fe-15Cr alloy with charge Q passed during cathodic reduction.

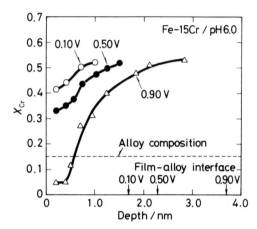

Fig. 10. Change in X_{Cr} as a function of film depth for passive films formed at various potentials on Fe-15Cr alloy.

for each film. It is interesting that any films formed at different potentials have approximately the same value of ca. 0.52 of X_{Cr} near the end of reduction. This means that there should be no potential-dependent change in X_{Cr} in the region near the film-alloy interface.

CONCLUSION

(1) Peak maxima at 3.2 and 5.0 eV in MRS spectra can be used for the qualitative analysis of Fe^{3+} and Cr^{3+} ions in passive films, respectively.

(2) Both $\hbar\omega_{pm}$ vs. X_{Cr} and $r_{5.0eV/3.2eV}$ vs. X_{Cr} relation can be used for the quantitative analysis of the Cr content of passive films.

(3) The spectral information depth in MRS of γ-Fe_2O_3 and Cr_2O_3 should be less than 0.9 nm.

(4) The X_{Cr} for the outermost of a passive film on an Fe-Cr alloy changes in dependence on potential, whereas that for the innermost should be virtually independent of potential.

REFERENCES

1   W. Paatsch, Metalloberfläche, 34 (1980) 24-28.
2   N. Hara and K. Sugimoto, Bull. Japan Inst. Metals, 20 (1981) 599-605.
3   N. Hara and K. Sugimoto, J. Japan Inst. Metals, 44 (1980) 915-924.
4   N. Hara and K. Sugimoto, ibid., 44 (1980) 1312-1321.
5   N. Hara and K. Sugimoto, ibid., 47 (1983) 31-39.
6   N. Hara and K. Sugimoto, Trans. Japan Inst. Metals, 24 (1983) 236-245.
7   A.G. Akimov, Elektrokhimiya, 16 (1980) 117-120.
8   N. Hara and K. Sugimoto, J. Electrochem. Soc., 126 (1979) 1328-1334.
9   A.G. Akimov, M.G. Astaf'ev and I.L. Rozenfel'd, Zash. Met., 15 (1979) 557-560.
10 K.L. Hardee and A.J. Bard, J. Electrochem. Soc., 122 (1965) 739-742.
11 N. Hara and K. Sugimoto, J. Japan Inst. Metals, 43(1979) 992-999.

Passivity of Metals and Semiconductors, edited by M. Froment
Elsevier Science Publishers B.V., Amsterdam — Printed in The Netherlands

A STATISTICAL APPROACH TO THE STUDY OF LOCALIZED CORROSION.

D.E. Williams[1], C. Westcott[1] & M. Fleischmann[1]

[1] Materials Development Division, AERE, Harwell, Oxon. OX11 ORA. UK.

[2] Chemistry Department, Southampton University, Southampton, SO9 5NH. UK.

ABSTRACT

The initiation of pitting corrosion of a stainless steel in dilute NaCl solution is approached via study of the statistics of ensembles of current-time transients.

The powerful effect on pit initiation of varying the conductivity and buffer capacity of the electrolyte is shown. The data can be interpreted using very simple ideas relating to the need to establish and maintain gradients of potential and pH on the scale of the surface roughness of the specimen in order to initiate pits. Current fluctuations are related to fluctuations in the thickness of a solution boundary layer at the metal surface. A stochastic model for pit initiation is outlined. The model permits the derivation of the probabilities of birth and death of pits. A consistent interpretation also involves a critical age beyond which pits do not die and a precursor state which must be reached before a current increase is observed.

1. INTRODUCTION

Despite the probabilistic nature of localised corrosion[1],

particularly pitting corrosion, recent literature has often attempted to

define critical potentials for pit initiation and repassivation. In fact,

pitting potentials and repassivation (protection) potentials in any given

medium are statistical variables, with repassivation potentials showing

less variability from one specimen to another than pitting potentials. One

attempt to define critical potentials has used the potential dependence of

induction time for pit initiation[2]. The induction time, however, is

also a statistical variable, a study of its variability directly yielding

information about the pitting probability [3]. It is also dependent on

the surface preparation of the specimen. Localised corrosion is in fact a

rare event; some experiments to determine parameters such as critical

pitting potentials can terminate when only one pit is formed on the

specimen. It is to be expected that the results from such experiments

would be extremely variable. Given such considerations, we have chosen to treat the variability of experimental results as a source of information about the phenomenon of localized corrosion in contrast with the more usual approach, which seeks to define conditions which reduce the variability to a minimum.

The probabilistic nature of localized corrosion shows up in the observation of current fluctuations at constant potential[4]. Differences in the mode of fluctuation for different types of localized corrosion have been remarked[5]. One rationalization to explain such fluctuations could obviously be framed in terms of random nucleation events.

Pitting has indeed been described previously as a nucleation process, characterized by a probability, or frequency, of nucleation[3]. It seems reasonable to propose that once nucleation has occurred, the subsequent time evolution of the current at the nucleation site follows some definite rule, determined perhaps by the shape of the evolving defect. The theory describing the statistics of current-time transients resulting from a random nucleation process followed by a deterministic time evolution of current to the resulting nuclei, has been developed to describe electrocrystallization [6,7]. The analogy between electrocrystallization, in which a nucleus grows out of the electrode surface, and pitting corrosion, in which the nucleus grows into the surface, seems clear. It is one object of this paper to indicate to what extent nucleation theory as developed to describe electrocrystallization, can be applied to the discussion of pitting corrosion.

Several different mechanisms for the initiation of corrosion have been proposed. These have been reviewed recently [8,9]. It is agreed that corrosion pits propagate as a result of the development and maintainance of an elevated local acidity. As far as the nucleation of pits is concerned, authors have variously emphasized, amongst other phenomena, inhomogeneity in the metal, cracking and slow healing of the passive film [8,9], development of critical acidity levels in microscopic flaws [10] and defect transport in passive films[11]. The experimental data presented in this paper shows that the development of gradients of pH and electric potential in local regions, defined by the surface roughness of the metal, is central to the nucleation of pits.

It is proposed that the critical material-dependent parameter for

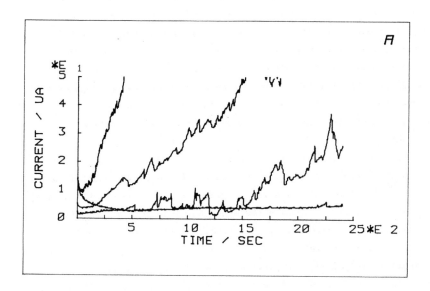

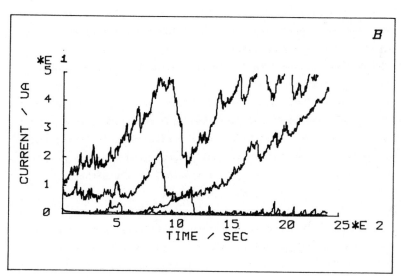

Figure 1   Representative current-time transients for 18Cr13Ni1Nb stainless
steel in 0.028M NaCl, + 50 mV sce. Specimen area, 5cm^2

(a) stepped from rest potential (-200mV sce)
(b) ramped at 10mV/s from -200 mV sce.

pitting is the dissolution current in the passive state, as has previously
been postulated for crevice corrosion[12]. We propose that mechanisms
used to describe the dissolution of oxides [13,14] can be adapted(as in
ref(11), for example) to relate the passive dissolution current to the
acidity and chloride concentration in the solution at the metal surface.
Simple diffusion calculations then show how an unstable situation can
develop. An outline of this theory is presented here. It is proposed that
the observed current fluctuations may be related to fluctuations in the
thickness of a boundary layer in the solution at the metal surface.

2.   EXPERIMENTAL

Disk specimens 2.5cm in diameter were stamped out from a single sheet
(2mx2mx3mm) of 18Cr13Ni1Nb stainless steel, set in an epoxy mounting resin,
polished to a 240 grit finish and rinsed with acetone. The electrochemical
cell was formed simply by clamping the specimen mount onto a 25mm diameter
glass flange joint set vertically. Gas feed, Luggin capillary, and Pt
secondary electrode were introduced from the top. The solution contained
in the assembly was purged with purified argon for 30 min before each
experiment, and a slow flow of Ar was maintained during each experiment.
Current was measured as a voltage across a 100Ω resistor in the secondary
electrode lead, and was logged from four cells simultaneously using a reed
relay multiplexer and autoranging digital voltmeter(Datron, 1μV
resolution) controlled via the IEEE 488 interface bus by a microcomputer
(CBM, PET 4032). The sampling frequency was 1Hz on each cell. The cells
were independently controlled (H.B. Thompson "Microstat"); application of
the control potential, and potential ramp, was also under the control of
the microcomputer. Data were transferred to a minicomputer for analysis.

Solutions were prepared with analytical reagent grade chemicals and
distilled water. The buffer was tris (hydroxymethyl) amino methane,
("tris"), adjusted to pH7.0 with 1M $HClO_4$ solution.

3.   RESULTS

Figure 1(a) shows representative current-time transients at constant
potential, independently determined on different stainless steel specimens,
obtained after a potential step from the rest potential (-200mV sce) into
the regime of pitting corrosion (+50mV sce), in a very dilute (0.028M) NaCl
solution. The transients comprise fluctuations on a rising background.

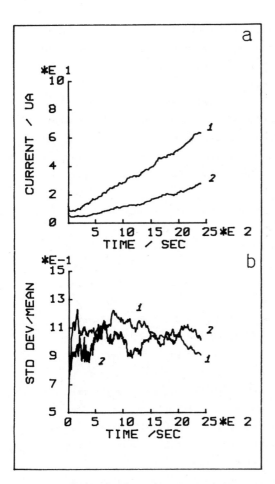

Figure 2

(a) Ensemble mean current
for ramp and step
start

(b) Ratio of ensemble mean
to standard deviation
for ramp and step start

1 - step start
2 - ramp start

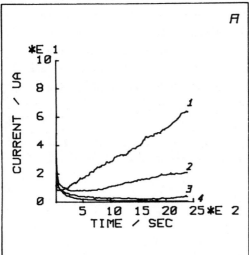

Figure 3

Effect of solution
conditions on ensemble
statistics (step start,
+50 mV sce)

1 : 0.028M NaCl
2 : 0.028M NaCl buffered
with 0.005M tris
3 : 0.028M NaCl buffered
with 0.05M tris
4 : 0.028M NaCl + 1M $NaClO_4$

(a) mean transients

The fluctuations had a characteristic form of slow rise and sharp fall.
The charge passed during a fluctuation was typically $10-500\mu C$, equivalent
to the dissolution of 10^{-10} to 10^{-8} cm^3 of metal – a cube of
typical linear dimension about 10μ m.

The transients themselves could be divided into two different classes.
The first class compised those transients in which the current rose steeply
immediately following the step; specimens which displayed this behaviour
were found to have developed crevice corrosion at the boundary between the
specimen and the mounting material. The second class comprised those in
which the current rose at some later time; only pitting corrosion was
observed on these specimens. A large proportion of the specimens showed no
corrosion and no current increase at all. It has been remarked in the
past[15] that the balance between pitting and crevice corrosion around
the specimen mount is affected by the surface conditon of the specimen,
rougher specimens promoting pitting and smoother ones crevice corrosion.

The potential step gives rise to a short current pulse, readily seen
in Figure 1(a), associated with passive film growth and double layer
charging, and it was this current pulse which was responsible for the
occasional immediate nucleation of crevice corrosion. This was
demonstrated by collecting an ensemble of over thirty current–time
transients at constant potential in which the transition from the rest
potential to the passive state (+50mV sce) was accomplished by a potential
ramp at 10mV/s. The initial current pulse was thereby eliminated and the
immediate current rise associated with crevice corrosion never occurred.
Some typical transients are shown in Figure 1(b).

In Figure 2(a) the ensemble mean transients for the step and ramp
initiation methods are compared. The mean current for the ramp start was
only half that for the step start over the time span followed. The
diagnostic parameter arising from the stochastic theory described later is
the ratio of the ensemble standard deviation to the ensemble mean: Figure
2(b) shows that this ratio, in contrast to the effect observed with the
mean transient, was more or less the same for the ramp and step start and
had a similar time variation.

Figure 3(a) shows the effect of solution conditions on the ensemble
statistics. Increasing the buffer capacity of the solution delayed the

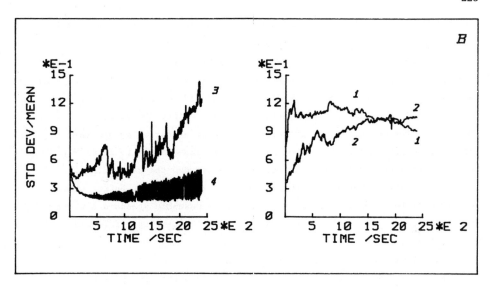

Figure 3

Effect of solution conditions on ensemble statistics (step start, +50 mV sce)

(b) Standard deviation/mean. 1-4 as Figure 3(a).

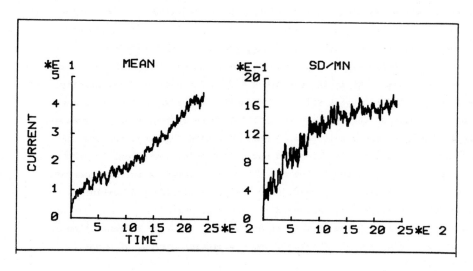

Figure 4   Simulation result.

Parameters: birth probability: $1/20$ cm $^{-2}$ s^{-1}
death probability: $1/20$ s^{-1}
absorbing state attained at age: 180 s
current after nucleation: $i = 0.1t$, for each pit

onset of, and, at sufficiently high buffer concentration, eliminated all but a small amount of pitting corrosion.

Addition of a backing electrolyte, 1M NaClO₄, had a dramatic effect, completely eliminating pitting corrosion. This effect persisted up to high potentials in chloride solutions of up to 0.3M and for times in excess of two days[16]. Addition of the backing electrolyte into the dilute chloride solution during the propagation of pitting corrosion had the effect of stopping the pits – the current fell immediately to zero[16]. Addition of a sufficient concentration of buffer had the same effect. Figure 3(b) demonstrates that adding backing electrolyte or buffer changed the form of the time variation of the ratio of ensemble standard deviation to mean from concave with time to convex, over the time span followed.

4.   DISCUSSION

   4.1 Parameters Controlling Pit Initiation

   A rational interpretation of the effects of buffer and backing electrolyte on pit initiation can be obtained by proposing that pits initiate as a consequence of the generation and maintainance of gradients of potential and acidity on the scale of the surface roughness of the specimen. The effect of the backing electrolyte is presumed to be simply to increase the conductivity of the solution (by factor of ten in this case) and hence increase the distance scale for a given potential change. If the potential gradient in the solution phase required to localize the dissolution process can be developed on the scale of the surface roughness of the metal, then pitting can initiate, whereas if the solution conductivity is too high to allow this, then pitting cannot initiate. The effect of a step start is to generate a sudden increase in local acidity, as a result of the current pulse. The local acidity would tend to greatest round the edges of the specimen because the pulse current would be greatest there. The gradient developed in a small occluded region at the edge of the specimen could be sufficient to nucleate crevice corrosion.

   Here, we focus attention on the effect of a variable surface boundary layer on the dissolution process, essentially developing and combining ideas due to Galvele[10], MacDonald[11] and Oldfield[12].
Thus, by analogy with the effects of acidity and chloride concentration in the rates of dissolution of oxides, the dissolution current of a passive metal will in

general be function of the surface concentrations of chloride and hydrogen
ions:

$$i = k_1, f([Cl^-]_s, [H^+]_s) \exp (\frac{\Delta \phi_{fs}}{RT})$$

(1)

where $\Delta \phi_{fs}$ represents the potential difference across the passive
film - solution interface. The rate constant k_1, can, in the most
general case, be considered to be variable over the metal surface, because
of inhomogeneities or inclusions - for example.

Concentrations at the film-solution interface are determined by
diffusional fluxes from the bulk solution and by the hydrolysis
equilibrium:

$$FeCl^+ + H_2O \longleftrightarrow Fe(OH)^+ + H^+ + Cl^-$$

(2)

$$K_2 = 5 \times 10^{-8} \ mol^2 l^{-2}$$

Iron in solution would be essentially entirely present as $FeCl^+$, and the
build-up of the species at the interface can be approximately written:

$$\frac{d[FeCl^+]s}{dt} = \frac{i_d}{2F} - \frac{D[FeCl^+]s}{\delta}$$

(3)

where δ denotes the diffusion layer thickness.

If i_d is large enough and the bulk concentrations are small enough,
then the surface concentrations of H^+ & Cl^- are fixed only by reaction
(2):

$$[H^+]_s \simeq [Cl^-]_s \simeq \sqrt[3]{K_2 [FeCl^+]_s}$$

(4)

Equations (1) - (4) yield a relationship between current and time,
and, depending only on the form of equation (1), the current could increase
very rapidly with time. For example, should equation (1) be of the common
form:

$$i_d = K_5 [Cl^-]_s (1 + \frac{[H^+]s}{K_5})$$

(5)

then the current will increase rapidly with time if $\left[H^+\right]_s$ becomes of the order of K_5. In equation 5, K_5 could be interpreted as an equilibrium constant for protonation of the surface.

The critical parameter in this formulation, as in Galvele's description[10] is the diffusion length, δ. We view the diffusion length as being made up of two parts. Firstly there is the thickness of the solution boundary layer, which is prone to local fluctuations and dependent on stirring conditions. Secondly there is the depth of the surface roughness, or any mark or defect in the metal surface. We suppose that if the two parts of δ are of similar magnitude, then any local increases of acidity are capable of being removed as a result of random decreases in the solution boundary layer thickness. Thus any newly nucleated pit can die as a result of fluctuations in the solution boundary layer thickness wiping out the local concentration gradients. However, if a pit grows far enough that its depth becomes significantly greater than the boundary layer thickness, then the gradients developed become insensitive to boundary layer variations and the pit becomes stable. In the terminology of stochastic processes, an absorbing state is attained. This model accounts for effects of surface preparation on pitting and predicts effects of stirring and specimen orientation on pit nucleation.

It identifies the passive current density and its variation with Cl^- and H^+ concentration as the most important material dependent parameter[12]. It is a parameter which can be measured independently, and whose variation with solution composition is susceptible to detailed mechanistic description.

4.2 Stochastic Models of Pit Growth

Using a simple numerical simulation procedure, stochastic models may be compared with the experimental results. A stochastic model involving a simple Poisson birth and death process for nuclei combined with a deterministic law for growth of the nuclei yields a form for the mean transient and for the standard deviation to mean ratio which is quite different from the experimental forms shown in figures 2 & 3. Such a stochastic model predicts a mean transient which rises to a final steady value, and a standard deviation to mean ratio which decreases steadily with time (as $t^{-\frac{1}{2}}$). However, introduction of an absorbing state, a

condition that pits of more than a certain age cannot die, yields results
which follow the form of the experimental transients (Figure 4). The form
of the time variation of the standard deviation to mean ratio depends on
the values of the parameters of the model – the nucleation frequency, the
death probability and the critical age beyond which a pit is stable.

Inspection of the experimental transients (Figure 1) indicates that
the nucleation rate, given by the frequency of occurrence of current peaks,
varies widely from one specimen to another. A way of formulating this,
which relates clearly to the concentration fluctuation model given above,
is to postulate that a precursor state, the attainment of a critical local
acidity, must be attained before the current starts to increase. The
concentration fluctuation model, outlined in the previous section, implies
that the probability of attaining the precursor state would depend on the
surface roughness of the specimen. A simple simulation shows clearly that
the apparent nucleation rate is very sensitive to variations in the
probability of attainment of the precursor state.

5. CONCLUSION

The development of gradients of pH and electric potential in local
regions, defined by the surface roughness of the metal, is central to the
nucleation of pits, the critical material-dependent parameter being the
dissolution current in the passive state. The observed current
fluctuations may be related to fluctuations in the thickness of a boundary
layer in the solution at the metal surface.

The stochastic description of the process requires the introduction of
an absorbing state, a condition that pits greater than a certain age do not
die. A full description also requires the introduction of a precursor
state, a condition which must be attained before a current increase occurs.
The concentration fluctuation model provides a rational, simple
interpretation of these states.

REFERENCES

1. U.R. Evans, "Corrosion & Oxidation of Metals", Edward Arnold, London,
1960

2. K. Pessal & C. Liu, Electrochim. Acta, 16 (1971) 1987

3. T. Shibata & T. Takeyama, Corrosion, 33 (1977) 243

228

4. H. Saito, T. Shibata & G. Okamoto, Corros. Sci, 19 (1979) 693

5. K. Hladky & J.L. Dawson, Corros. Sci, 21 (1981) 317

6. M. Fleischmann, M. Labram, C. Gabrielli& A. Sattar, Surface Sci, 101
(1980) 583

7. P. Bindra, M. Fleischmann, J.W. Oldfield & D. Singleton
Faraday Disc. Chem. Soc. 56 (1974) 180

8. M. Janik-Czachor, G.C. Wood & G.E. Thompson, Br. Corros. J., 15 (1980)
155

9. M. Janik-Czachor, J. Electrochem. Soc. 128 (1981) 513C

10. J.R. Galvele, J. Electrochem. Soc., 123 (1976) 464

11. L.F. Lin, C.Y. Chao & D.D. MacDonald, J. Electrochem. Soc. 128 (1981)
1194

12. J.W. Oldfield & W.H. Sutton, Br. Corros. J. 15 (1980) 31

13. N. Valverde, Ber. Bunsenges. Physik. Chem. 80 (1976) 333

14. J.W. Diggle in J.W. Diggle (Ed.), Oxides, & Oxide Films, Vol. 2,
Marcel Dekker, New York, 1973, p281

15. R.W. Staehle, B.F. Brown, J. Kruger & A. Agrawal (Eds),
            Localized Corrosion, NACE, Houston 1974

16. C. Westcott, M. Fleischmann & D.E. Williams, to be published.

Passivity of Metals and Semiconductors, edited by M. Froment
Elsevier Science Publishers B.V., Amsterdam — Printed in The Netherlands

ELECTROCHEMICAL NOISE MEASUREMENTS FOR THE STUDY OF LOCALIZED CORROSION
AND PASSIVITY BREAKDOWN

U. BERTOCCI, J.L. MULLEN, and Y-X. Ye*
Center for Materials Science, NBS, Washington, DC   20234

ABSTRACT

   Measurement of the random fluctuations of the current for Al, Fe-Cr and
amorphous Fe-Ni-Cr electrodes are reported, both in the form of time records and
frequency spectra. Comparison between the noise measured when no pitting could
occur and when pitting was possible, showed that detectable fluctuations were
present only in the second case. For the amorphous alloy, which is not susceptible to pitting, little noise could be measured even when the electrode was
undergoing transpassive dissolution. Examples of random noise used for measuring electrode impedance are also given.

INTRODUCTION

   The analysis of the noise, that is small amplitude incoherent fluctuations,
in electrical signals can give important information on the characteristics of
the system generating them, and has been employed extensively in the study of
semiconductors [1]. Applications to electrochemical systems are still rare [2],
but are more common for the study of transport in membranes [3]. Noise analysis
has been applied to corrosion [4], including studies on the stability of passive
films [5], beginning with the pioneering work of Okamoto and coworkers [6].

   Noise analysis, besides being a convenient way to obtain impedance data, can
also provide information concerning the breakdown of protective films on electrodes, which can cause small fluctuations in either the electrode potential or
the current. As already recognized by Okamoto [6], the examination of the
fluctuations of the current under potentiostatic conditions is to be preferred to
the study of the voltage noise in the galvanostatic mode, since there is a
better control of the conditions leading to localized attack, although the experimental difficulties are probably greater. In this paper we will present
some results obtained in our laboratory concerning the use of noise measurements
for the study of the electrochemical behavior of passive films and particularly
their breakdown leading to pit formation.

---

*Guest Scientist, People's Republic of China

EXPERIMENTAL TECHNIQUES

The fluctuations in the current of a potentiostatted cell can be the result
of noise in the control voltage, or can be caused by random changes in the electrode impedance. Their interpretation, therefore, is difficult, until it can
be established which of the two causes is at work in a particular instance.
Once the source of the current noise is known, in either case its analysis can
be useful: in the first instance the electrode impedance can be derived; in the
second, information concerning the stability of the electrode properties, and
possibly the time constants of the fluctuations can be obtained.

The traditional way to insure knowledge of the source of noise is that of
making sure that one of the two is much larger than the others: in the first
case this can be achieved by feeding random noise of the sufficient amplitude
as the input signal of the potentiostat.

To measure the stochastic part of the current noise, however, it is necessary
to reduce all instabilities of the instrumentation to a minimum, but it is often
difficult to be sure that the desired conditions have been met.

The separation of the two components of the current noise, however, is possible in the frequency domain, if both electrode potential and cell current are
monitored simultaneously with a two-channel spectrum analyzer capable of calculating the cross-power spectrum of the two signals. The details of the method
have already been published [7]. All spectra presented here have been obtained
by processing the data in this way. It is not only possible to know the origin
of noise, but if its two components have comparable amplitudes, both kinds of
noise data can be obtained at the same time. One of the consequences of this
data processing is also that it is easy to know when the random current fluctuations are too small compared to the instrumental noise, and what is the minimum
amplitude that can be reliably detected.

RESULTS

1. Aluminum

Spectra of the passive current in chloride-containing solutions above and
below the pitting potential have been reported [6]. In the latter case, no
fluctuations above the instrumental noise could be detected, suggesting that the
passive film did not have any tendency to spontaneous breakdown below the pitting potential. Similar measurements have been repeated, with the aim of
detecting and analyzing the fluctuations preceding pitting.

Fig. 1 shows two noise spectra. One was taken below the pitting potential;
the amplitude is below the instrumental noise level, indicating that no measurable fluctuations occurred. The second spectrum was taken above the pitting
potential, but before pit formation, at the same time as the series of current

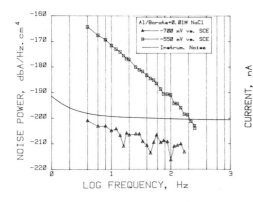

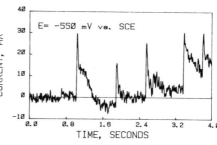

Fig. 1. Spectra of random current
fluctuations for Al in borate
buffer + 0.01 mol/L NaCl,
above and below pitting potential.

Fig. 2. Time record of current fluctua-
tions in same system as Fig. 1.
Sampling rate 2 ms/pt. H. P.
Filter 30 mHz.

transients shown in Fig. 2. These transients decay roughly exponentially, in
agreement with the slope (-20 db/decade) of the spectrum. The test was
interrupted before observing any change in the dc current. Examination of the
electrode by optical microscopy failed to find any visible pit.

2. Fe-Cr Alloys

A detailed study of the current fluctuations in the presence and in the absence of chlorides, before and after pit initiation for Fe-Cr alloys of varying
Cr content [8] has shown no indication of detectable passive film breakdown in
the absence of chlorides. Here we will present some results relative to the 12%
Cr alloy, which contains a sufficient amount of Cr to confer corrosion resistance. The random part of the current noise at the same potential was measured
in the presence and in the absence of chlorides in solution. Random fluctuations above the instrumental noise were found only in the chloride solution, and
they appear as sudden current spikes, even before the formation of the first
pit, which is indicated both by a linear increase of the average d.c. current
and by a substantial increase in the noise. The isolated current spikes as
seen through a 30 mHz high pass filter, have the form shown in Fig. 3A. Part B
of the same figure shows the continuation of the same record including the instant when pitting begins. In Fig. 4 the spectrum of the fluctuations during
the induction period is shown together with that taken after pit initiation.
The instrumental noise level is also shown, demonstrating that the noise in the
presence of chloride is well above the limit of detection. The flatness of the
spectrum taken before pitting initiation is a consequence of the suddenness of
the current spikes, within the frequency ranges investigated.

232

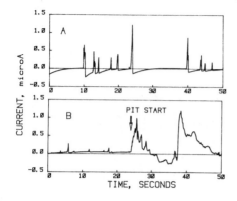

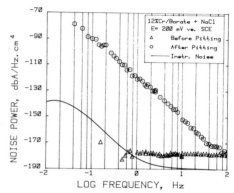

Fig. 3. Time records of current fluctuations for Fe-12%Cr alloy in
borate + 0.1 mol/L NaCl. E=200
mV vs. SCE. A) Transients during induction period. B) Start
of pitting.

Fig. 4. Spectra of random current
fluctuations in same system
as Fig. 3, before and after
pit initiation.

3. Amorphous metals

Surface imperfections are often the initiation points of localized corrosion,
and the enhanced resistance to pitting exhibited by amorphous alloys of stainless steel-like composition has been attributed to the greater uniformity of the
passive film on them.

The amplitude of the current fluctuations gives an indication of the degree
of localization of the anodic attack, as shown by the large noise current found
during pitting. Previous work [5] had shown that the noise during transpassive
dissolution of a Fe-Ni-Cr amorphous alloy was much smaller than the noise for
the same material after recrystallization. These results, however, were obtained in a solution which did not contain chloride, and therefore did not
promote pitting. Also, separation of the two noise components had not been
carried out.

The material employed was a rapidly cooled alloy with the composition
$Fe_{32}Ni_{36}Cr_{14}B_6F_{12}$ and the solution was 1 mol/L H_2SO_4 with 0.5 mol/L NaCl added.
The electrochemical behavior of such an alloy has been extensively studied [9],
but less is known concerning its transpassive dissolution. As a way to
characterize its behavior, impedance measurements were taken in the passive
and in the transpassive state using random noise as the input signal.

Bode plots of the electrode impedance at two potentials in the passive range
are shown in Fig. 5. These data were obtained by applying white noise of
amplitude ranging from 0.3mV (at the lowest frequencies) to 1μv. The passive
electrode behaves largely like a capacitor, with a capacitance of the order of
$20μF/cm^2$. At higher potentials, transpassive dissolution begins, and under

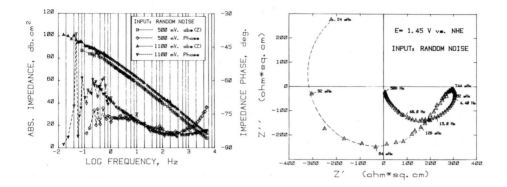

Fig. 5. Bode plots of the impedance of a Fe-Ni-Cr-P-B amorphous alloy in 1M H₂SO₄ + 0.1M NaCl in the passive range. Input: white noise <300 μV.

Fig. 6. Nyquist plot of the impedance of the same system as in Fig. 6 in the transpassive range. Input: white noise <300 μV.

galvanostatic conditions, periodic, very regular oscillations with a period inversely proportional to the applied current density (ranging from 2 to 30 s) are observed. In this transpassive range, Nyquist plots of the impedance show negative resistance loops, indicating the existence of instabilities. One of these plots is shown in Fig. 6.

Current noise in this potential range was measured, but its amplitude was found to be very small and to increase very slightly with the increase in transpassive current. Random noise spectra are shown in Fig. 7. Even at an anodic current of 2 mA/cm^2, the fluctuations in the current were at least 10^{-5} times smaller. These results confirm that transpassive dissolution on these materials occurs uniformly and there is no tendency to pitting in spite of the presence of chlorides in solution. The lack of pit formation was confirmed by optical microscopy after the electrochemical measurements.

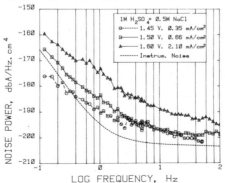

Fig. 7. Spectra of random current fluctuations of same system as Fig. 5, during transpassive dissolution.

234

DISCUSSION

The results presented here aim to show how noise analysis can be employed as an additional tool for electrochemical studies, particularly when examining passivity and its breakdown. On one side, random noise as an input signal, as pointed out by other authors before [10], can be used conveniently for a rapid and straightforward acquisition of a.c. impedance data. On the other, random fluctuations in the current are an excellent indication of pitting, and their study can contribute to our knowledge of the gradual breakdown during the induction period of the passive films which lead to localized attack.

The results obtained both with Al and Fe-Cr provide evidence, perhaps for the first time, that chlorides must be present for film breakdown to begin, and that their most important role might be that of promoting breakdown rather than that of affecting the kinetics of repassivation, since no spontaneous breakdown episodes were detected in the absence of chlorides. This does not imply that chlorides do not affect the repassivation process as well.

The low noise observed on the amorphous Fe-Ni-Cr alloy, even during dissolution at substantial rates, confirms that the amplitude of the fluctuations is well correlated with inhomogeneous attack, and can be used as a diagnostic criterion for susceptibility to localized corrosion.

REFERENCES

1 A. van der Ziel, "Fluctuation Phenomena in Semiconductors", London (1959)
2 V. A. Tyagai, Electrochim. Acta, 16, 1647 (1971); G. C. Barker, J. Electro-anal. Chem. 21, 127 (1969), 82, 145 (1977); G. Blanc, I. Epelboin. C. Gabrielli, M. Keddam, J. Electroanal. Chem. 62, 59 (1975)
3 L. J. DeFelice, "Introduction to Membrane Noise", New York (1976)
4 W. P. Iverson, J. Electrochem. Soc. 115, 617 (1968): G. Blanc, C. Gabrielli, M. Keddam, C. R. Acad. Sci. Paris, 283C, 107 (1979); U. Bertocci. Proc. 7th Internat. Congress on Metallic Corrosion, 2010, Rio de Janeiro (1978): K. Hladky, J. L. Dawson, Corrosion Sci. 22, 231 (1982)
5 U. Bertocci, J. Kruger, Surf. Sci. 101, 608 (1980)
6 G. Okamoto, T. Sugita, S. Nishiyama, K. Tachibana, Boshoku Gijutsu 23, 439. 445 (1974)
7 U. Bertocci, J. Electrochem. Soc. 123, 520 (1981)
8 U. Bertocci, Y-X. Ye, J. Electrochem. Soc. submitted for publication.
9 K. Hashimoto, K. Osada, T. Masumoto, S. Shimodaira, Corrosion Sci. 16, 71 (1976); K. Hashimoto, K. Asami, M. Naka, T. Masumoto, Corrosion Sci. 19, 857 (1979)
10 S. C. Creason, D. E. Smith, J. Electroanal. Chem. 36 (1972) App. 1; S. C. Creason, J. W. Hayes, D. E. Smith, J. Electroanal. Chem. 47, 9 (1973); G. Blanc, I. Epelboin, C. Gabrielli, M. Keddam, Electrochim. Acta, 20, 599 (1975)

PASSIVATION OF AMORPHOUS METALS

K. HASHIMOTO

The Research Institute for Iron, Steel and Other Metals, Tohoku University,
Sendai 980 (Japan)

ABSTRACT

The passivation behavior of metals with the amorphous structure is reviewed.
Some amorphous alloys show extraordinarily high corrosion resistance by spontaneous passivation in very aggressive solutions such as hot concentrated hydrochloric acids. The high corrosion resistance is interpreted in terms of chemically homogeneous nature and high reactivity of the alloys. The former is based
on the amorphous structure and the latter provides rapid formation of a passive
film in which beneficial elements are remarkably concentrated.

INTRODUCTION

Amorphous alloys are novel metallic materials with various unusual and
attractive properties. When they are prepared by rapid quenching from the
liquid state, they always possess remarkably high mechanical strength in combination with considerable toughness. Some amorphous alloys are ferromagnets with
the soft magnetic behavior. Their chemical properties are also unique: Some
amorphous iron-nickel-metalloid alloys are the effective catalysts for hydrogen-
ation of carbon monoxide (refs.1-3). Some amorphous palladium-based alloys are
appropriate to the anode for production of chlorine by electrolysis of sodium
chloride solutions (refs.4-6) and to the electrode for electrochemical oxidation
of methanol for methanol-air fuel cells (ref.7,8). The most interesting characteristic in their chemical properties is extremely high corrosion resistance,
which has never been obtained by crystalline metals and is quite helpful for
us to gain fresh insight in the nature of passivity and its breakdown.

CORROSION RESISTANCE

The corrosion rate of amorphous iron-metalloid alloys without a second metallic element is higher than that of crystalline iron metal. However, the corrosion rate is decreased by addition of almost all metallic elements such as titanium, zirconium, vanadium, niobium, tantalum, chromium, molybdenum, tungsten,
cobalt, nickel, copper, ruthenium, rhodium, palladium and platinum (ref.9).
In particular, the addition of chromium is quite effective. As shown in Fig. 1,
the Fe-8Cr-13P-7C alloy* passivates spontaneously in 2 N HCl at room temperature
(ref.10). The Fe-3Cr-2X-13P-7C alloys containing 2 atomic percent of third met-

allic element, X, are passivated by anodic polarization in 1 N HCl at room temperature (ref.11). They do not suffer pitting corrosion in these solutions even by anodic polarization up to the transpassive region of chromium. The addition of chromium to amorphous cobalt- and nickel-base alloys is also effective in improving the corrosion resistance as shown in Fig. 2 (ref.12).

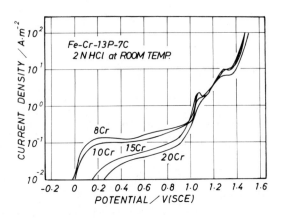

Fig. 1. Anodic polarization curves of amorphous Fe-Cr-13P-7C alloys measured in 2 N HCl at room temperature (ref.10).

Fig. 2. Changes in corrosion rates of amorphous Fe-, Co- and Ni-based alloys measured in 1 N HCl at room temperature as a function of alloy chromium content (ref.12).

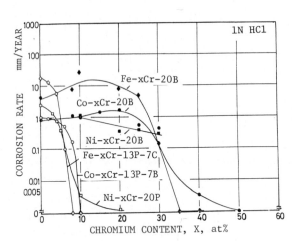

The combined addition of chromium and molybdenum is further effective in increasing the corrosion resistance (refs.10,13,14). As shown in Fig. 3 if sufficient amounts of chromium and molybdenum are added, the amorphous Fe-Cr-Mo-13P-7C and Fe-Cr-Mo-18C alloys passivate spontaneously in 12 N

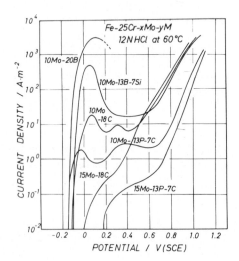

Fig. 3. Anodic polarization curves of amorphous Fe-25Cr-Mo-metalloid alloys in 12 N HCl at 60°C. The Fe-25Cr-15Mo-13P-7C and Fe-25Cr-15Mo-18C alloys passivate spontaneously, but the other alloys require anodic polarization for passivation (ref.13).

*The number attached to the respective element in an alloy formula denotes the content of the element in at.% unless otherwise stated.

HCl at 60°C and do not suffer pitting corrosion even by anodic polarization. In this manner, the amorphous Fe-Cr-Mo-metalloid alloys containing sufficient amounts of chromium and molybdenum passivate spontaneously in hydrochloric acids of various concentrations and temperatures.

Fig. 4 illustrates the critical concentrations of

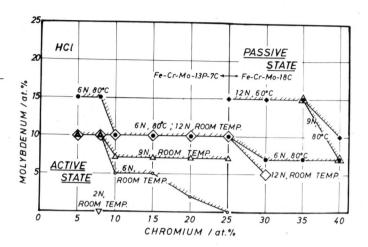

Fig. 4. Critical concentrations of chromium and molybdenum necessary for spontaneous passivation of amorphous Fe-Cr-Mo-13P-7C and Fe-Cr-Mo-18C alloys in hydrochloric acids of various concentrations and temperatures (ref.13)

chromium and molybdenum necessary for spontaneous passivation of amorphous Fe-Cr-Mo-13P-7C and Fe-Cr-Mo-18C alloys in hydrochloric acids of various concnetrations and temperatures (ref.13). Such a high corrosion resistance of the amorphous alloys is in marked contrast to the fact that any commercial, corrosion-resistant, crystalline metals except tantalum and niobium cannot passivate spontaneously even in diluted hydrochloric acids such as 1 N HCl at room temperature.

PASSIVITY

A decrease in the active dissolution rate by alloy additions is often responsible for an improvement of corrosion resistance of various amorphous alloys (ref.15). However, the extremely high corrosion resistance of various amorphous metal-metalloid alloys containing chromium results from spontaneous passivation even if solutions are very aggressive, such as hot concentrated hydrochloric acids.

According to our X-ray photoelectron spectroscopic studies, passivation of various amorphous alloys containing chromium takes place by the formation of a passive film consisting mainly of hydrated chromium oxyhydroxide, $CrO_x(OH)_{2-2x} \cdot nH_2O$ (ref.16). This is also the major constituent of the passive films on crystalline ordinary stainless steels. In this connection, addition of molybdenum to amorphous and crystalline alloys is effective for improvement of corrosion resistance and for prevention of pitting corrosion. Molybdenum assists passivation and repassivation by the formation of hydrated chromium or iron

TABLE 1

Cationic fraction of chromium in the passive films formed on amorphous alloys and stainless steels in 1 N HCl at room temperature.

Alloy	Fraction of Cr^{3+} in Film	Passivation Procedure	References
Amorphous alloy			
Fe-10Cr-13P-7C	0.97	spontaneous	22
Fe-3Cr-2Mo-13P-7C	0.57	anodic polarization	11
Co-10Cr-20P	0.95	spontaneous	23
Ni-9Cr-15P-5B	0.74	spontaneous	24
Ferritic stainless steel			
Fe-30Cr	0.75	anodic polarization	19
Fe-19Cr	0.58	anodic polarization	20

oxyhydroxide film but does not generally constitute the passive film (ref. 13-15,17-20). In other words, the molybdenum addition does not change the passive film composition but facilitates passivation. The protective quality of the passive film increases when the concentration of hydrated chromium oxyhydroxide in the passive film increases (ref.21). Table 1 shows the difference in the degree of chromium enrichment in the passive films formed on amorphous alloys and crystalline ferritic stainless steels. The amorphous metal-metalloid alloys are able to concentrate remarkably the beneficial chromic ions in the passive film, and accordingly have the high ability for spontaneous passivation with a consequent superior corrosion resistance.

PASSIVATION RATE

When passivation occurs the amor-

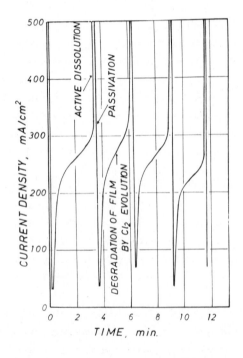

Fig. 5. Change in current density showing repetition of active dissolution, passivation and depassivation during potentiostatic polarization of the amorphous Pd-16Ti-19P alloy at 1.5 V(SCE) in 4 N NaCl solution of pH 1.5 and 80°C (ref.25).

phous metal-metalloid alloys are significantly stable. The passive film is a corrosion product in which chemically stable species are concentrated. In order to accumulate the beneficial metallic ions and hydroxyl ions at the alloy-solution interface prior to the passive film formation, it is necessary to dissolve alloy components useless for film formation. An example that active dissolution leads to passivation is shown in Fig. 5. Anodic polarization of amorphous Pd-16Ti-19P alloy at 1.5 V(SCE) in a 4 N NaCl solution of pH 1.5 and 80°C leads repeatedly to active dissolution and passivation because nascent chlorine formed on the passive film degradates the passive film (ref.25). In other words, even under the conditions where the passive film is chemically broken by nascent chlorine, a certain time of active dissolution results in accumulation of beneficial species such as titanium ions and hydroxyl ions at the alloy-solution interface with a consequent onset of passivity.

In general, the passivating ability of amorphous metal-metalloid alloys containing passivating elements such as chromium is higher than that of crystalline alloys containing the similar amount of the passivating elements. The high passivating ability is attributable to the high reactivity of their unfilmed surface which leads to rapid enrichment of beneficial species at the alloy-solution interface prior to the passive film formation and subsequently to a high concentration of beneficial species in the developping passive film. Accordingly, the high passivating ability is partly responsible for the high corrosion resistance of amorphous alloys. The advantage of the high reactivity of the alloy for passivation is clearly seen in Fig. 2: When chromium is not sufficiently added, the corrosion rate of the alloys, which are in the active state, decreases in decreasing order of the alloy activity, that is, in the order of Fe-Cr-20B, Co-Cr-20B and Ni-Cr-20B alloys. When the alloy chromium content is raised, the corrosion rate of the Fe-Cr-20B alloys which are most active among these three series of alloys stays almost unchanged up to about 25 atomic percent chromium due to active dissolution, but becomes undetectable by the addition of the least amount of chromium of 35 atomic percent, because the most active iron-based alloys are able to concentrate most rapidly the beneficial chromic ions at the alloy-solution interface with a consequent onset of passivity. The Co-Cr-20B alloys are less active than the Fe-Cr-20B alloys and hence require the addition of 50 atomic percent chromium for spontaneous passivation.

Metalloid elements, particularly phosphorus, contained in the amorphous metal-metalloid alloys serve the purpose to stimulating rapid initial dissolution of the unfilmed surface prior to the passive film formation (ref.26). The difference in the passivating ability with the difference in metalloid elements contained in the alloys can be seen in Fig. 3 and is partly attributable to the difference in the active dissolution rates of the unfilmed surfaces. Consequently, the amorphous metal-metalloid alloys are characterized by the ability for rapid formation of the passive film containing a high concentration of beneficial species which have a high stability and protective quality.

CHEMICAL HOMOGENEITY

The amorphous alloys are free of defects associated with the crystalline state. In addition, rapid quenching from the liquid state prevents solid state diffusion during quenching. Accordingly they are also free of defects such as precipitates and segregates which are apt to be formed during slow cooling or heat treatment. Therefore, the amorphous alloys prepred by rapid quenching of the molten alloys can be regarded as the ideally homogeneous solid solu-

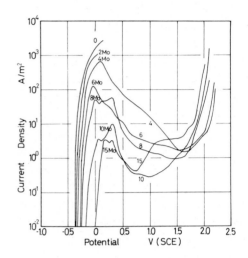

Fig. 6. Anodic polarization curves of amorphous Fe-Mo-13P-7C alloys measured in 1 N HCl at room temperature (ref.27).

tion alloys. In general, the corrosion resistance is dependent also upon the uniformity of the passive film because corrosion begins to occur at weak points in the film. In this regard, the homogeneous nature of amorphous alloys ensures the formation of the uniform passive film and hence the high corrosion resistance. As shown in Fig. 6, when the alloy molybdenum content is raised, amorphous iron-molybdenum-metalloid alloys are passivated by anodic polarization in 1 N HCl at room temperature (refs.18,27). For example, the average composition of the passive film formed on the amorphous Fe-15Mo-13P-7C alloy at 0.75 V(SCE) in 1 N HCl at room temperature can be expressed as:

$$(Fe^{II}_{0.186}Fe^{III}_{0.798}Mo^{IV-VI}_{0.016})(PO_4)_{0.030}Cl_{0.005}(OH)_{0.251-0.283}O_{0.177-0.145}H_2O.$$

Since the contents of molybdenum, phosphate and chloride ions are very low, this passive film consists essentially of hydrated iron oxyhydroxide (ref.17). This film composition is almost the same as that of the passive film formed on the crystalline iron metal in 2 N H_2SO_4 (ref.28). Accordingly, the passive hydrated iron oxyhydroxide film is not unstable in 1 N HCl. Nevertheless, crystalline iron metal cannot be passivated in 1 N HCl by anodic polarization and suffers pitting corrosion. The hydrated iron oxyhydroxide film may be formed on iron even in 1 N HCl similarly to the amorphous iron-molybdenum-metalloid alloys, but because the crystalline iron metal is not chemically homogeneous the hydrated iron oxyhydroxide film cannot cover the entire surface of the iron specimen in 1 N HCl, and hence localized corrosion attack on chemically heterogeneous sites appears to prevent passivation of iron in hydrochloric acids.

It has been known that the high corrosion resistance of the amorphous metal-metalloid alloys disappears by heat treatment for crystallization (ref.29-32). As shown in Fig. 7, the amorphous Fe-10Cr-13P-7C alloy passivates spontaneously in 1 N HCl at room temperature, but once a crystalline metastable phase is formed in the amorphous matrix by heat treatment at 430°C for about 100 min, the alloy is no longer passive. Even if detectable crystalline phase is not formed, heat treatment of amorphous Fe-20B and Fe-40Ni-20B alloys at 247 °C for 10 days results in embrittle-

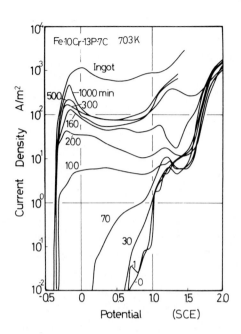

Fig. 7. Effect of heat treatment of amorphous Fe-10Cr-13P-7C alloy at 430 °C on the anodic polarization curve measured in 1 N HCl at room temperature. The number in the figure denotes time of heat treatment in min. (ref.32).

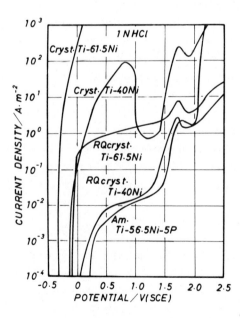

Fig. 8. Anodic polarization curves of crystalline Ti-Ni alloys measured in 1 N HCl at room temperature. RQ crysalline alloys were prepared by rapid quenching from the liquid state (ref.37).

ment of the alloys and significantly increases the anodic currente density in sulfate solutions due possibly to short range decomposition into two amorphous phases (ref.33).

It is interesting to clarify whether a dominant factor in decreasing the corrosion resistance is the chemical heterogeneity or the structural heterogeneity. According to Anthony and Cline (ref.34), rapidly quenched crystalline type 304 stainless steel after laser surface-melting possesses a higher corrosion resitance than solution-treated conventional type 304 stinless steel since various compositional fluctuations are removed by rapid quenching from the molten state. The high resistance against pitting and crevice corrosion in chloride-containing solutions has also been observed for microcrystalline austenitic stainless steel rods prepared by the procedure of rapid quenching from the liquid state, cold compaction, annealing at 593 ℃ for 48 to 72 hours and hot extrusion at 1065ºC (ref.35,36). The high corrosion resistance has been interpreted in terms of the high degree of homogeneity characteristic of the microcrystalline alloys, which results from rapid solidification. . As shown in Fig. 8, microcrystalline, single phase Ti-40Ni and Ti-61.5Ni alloys prepared by rapid quenching of the alloy melts have the high corrosion resistance and passivating ability in 1 N HCl at room temperature, whereas crystalline coase-grained Ti-40Ni and Ti-61.5Ni alloys are composed of two phase mixture and are hardly passivated by anodic polarization in 1 N HCl (ref.37). On the other hand, vacuum-annealing of amorphous Fe-10Cr-13P-7C alloy at 350ºC for 3.5 months gives rise to the formation of microcrystalline supersaturated solid solution phase, whose average composition has been thought to be the same as that of the surrounding amorphous matrix. The current density in the passive potential region of the alloy measured in 1 M H_2SO_4 solution increases about two orders of magnitude due to the formation of the supersaturated solid solution phase in the amorphous matrix (ref.30). In this manner, rapidly quenched single phase alloys show significantly high corrosion resistance even if they are crystalline, and the formation of the crystalline phase with the same composition as that of the surrounding amorphous matrix by heat treatment considerably decreases the corrosion resistance. Because rapid quenching from the liquid state prevents introduction of various compositional fluctuations into the alloys, crystal defects in the rapidly quenched crystalline alloys may be chemically clean due to no segregation and precipitation and may not act as preferential active surface sites with respect to corrosion. The chemically homogeneous single phase nature without compositional fluctuations is essential for the high corrosion resistance of the amorphous alloys.

LOCALIZED CORROSION RESISTANCE

It has widely been found that the amorphous alloys doe not suffer pitting corrosion even by anodic polarization in strongly acidic solutions containing chloride ions. The chemically homogeneous single phase nature of the amorphous alloys without compositional fluctuations is primarily responsible for the high resistance against localized corrosion. However, crevice corrosion does not require surface heterogeneity for initiation, and hence the susceptibility of amorphous alloys to crevice corrosion is dependent upon the passivating ability of the alloys.

Diegle (refs.38,39) has studied crevice corrosion of amorphous Fe-30Ni-Cr-

14P-6B alloys containing various concentrations of chromium: Cold rolling of the alloy with 16 atomic percent chromium to 38 percent reduction in thickness considerably increases the current densities in both the active and passive regions in 1 N NaCl of pH 7 with a consequent initiation of pitting corrosion at the intersections of surface microcracks formed by cold rolling (refs.38,40). However, during potentiostatic polarization for 1 hour at 0.6 V(SCE), which is significantly anodic to promote pitting and crevice corrosion of crystalline ordinary stainless steels of similar chromium contents, the current density of the amorphous alloy decreases rather than increases, the latter being expected if pitting and crevice corrosion grow deeply. At the end of the potentiostatic polarization at 0.6 V(SCE), a backward potential sweep leads to observation of a negative current in the active region of the alloy (ref.38). These results indicate that localized corrosion occurs as a result of formation of tight crevices due to severe cold-rolling, but pits thus formed are readily repassivated by potentiostatic polarization at the high potential, suggesting a high passivating ability of the alloy. Furthermore, when the amorphous Fe-30Ni-Cr-14P-6B alloys are placed inside an artificial crevice, crevice corrosion hardly occurs unless the specimens are anodically polarized at significantly high potentials (refs.38,39). As shown in Fig. 9, potentiostatic polarization of the amorphous Fe-30Ni-16Cr-14P-6B-2V alloy inside the crevice at 1.4 V(SCE) in 1 N NaCl of pH 7 results in initiation of crevice corrosion and pH decrease inside

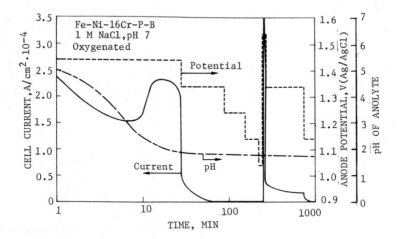

Fig. 9. Changes in crevice cell current and anolyte pH inside an artificial crevice with applied potential. The amorphous Fe-30Ni-16Cr-14P-6B-2V alloy is placed in the crevice immersed in an oxygenated 1 N NaCl solution of pH 7 and 22°C. (From R. B. Diegle, Corrosion 36 (1980) 362. Reprinted by permission of the author and the National Association of Corrosion Engineers, Houston, Texas.)

the crevice, but a lowering of applied potential of 0.1 V leads to a decrease in the crevice cell current to nearly zero. The cell current reappears by polarization at 1.5 V(SCE) but again decreases by potential decrease to 1.3 V(SCE). Accordingly, crevice corrosion begins to occur on the amorphous alloy at very high potentials, but the difference between the initiation and protection potentials of crevice corrosion is less than 0.1 V. Consequently, the very corrosive conditions are required for the onset of crevice corrosion for the amorphous alloys in comparison to crystalline stainless steels of similar chromium contents, and the amorphous alloys repassivate by slight potential decrease since they have considerably high passivating ability due primarily to the beneficial effect of metalloid elements as is mentioned previously.

When the corrosion-resistant chromium-bearing amorphous alloys are polarized anodically at high potentials in concentrated hydrochloric acids, the passive hydrated chromium oxyhydroxide film is sometimes dissolved in the form of general corrosion, showing a sharp increase in the anodic current density. As shown in Fig. 10, the critical potential for the sharp current increase is only dependent upon the alloy chromium content, regardless of the molybdenum content and passivating ability of the alloys. For instance, the amorphous Fe-5Cr-Mo-13P-7C alloys passivate spontaneously in 6 N HCl at 80°C when 15 atomic percent or more molybdenum is added (ref.41). The corrosion potential of the amorphous Fe-5Cr-15Mo-13P-7C alloys is about 0.09 V(SCE), but the anodic polarization of this alloy from the corrosion potential results in a sharp increase in the current density as if pitting corrosion begins to occur. The X-ray photoelectron spectroscopy has been applied to analyze the surface film formed on this alloy during potentiostatic polarization. Fig. 11 shows fractions of cations in the surface film formed on this alloy in 6 N HCl at 60°C. This alloy contains only 5 atomic percent chromium, but chromium ions comprize about 65 percent of cations in the passive film formed spontaneously. However, at 0.4 V(SCE) where the current density of this alloy is of the order of 10^2 A/m^2, the chromium enrichment in the surface film becomes insignificant. After potentiostatic polariza-

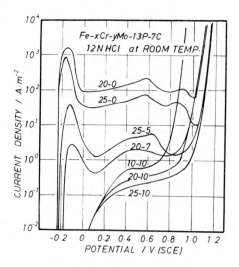

Fig. 10. Anodic polarization curves of amorphous Fe-Cr-Mo-13P-7C alloys measured in 12 N HCl at room temperature (ref.13).

tion at high anodic current densities
any traces of localized corrosion attack are not found. The critical potential of the sharp current increase
for such spontaneously passivating
low chromium alloys is lower than the
potential for transpassive dissolution of chromium. This is a characteristic of the amorphous alloys in
very aggressive concentrated hydrochloric acids where any crystalline
corrosion-resistant ferrous alloys
suffer readily pitting corrosion at
far low potentials. Consequently,
the sharp anodic current increase may
be attributed to passivity breakdown

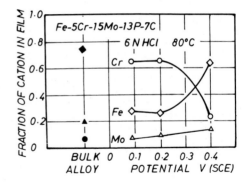

Fig. 11. XPS result of cationic
fractions in the surface film formed
on the amorphous Fe-5Cr-15Mo-13P-7C
alloy by polarization in 6 N HCl at
80°C.

in the form of general corrosion on the homogeneous alloys which do not possess
localized active surface sites. As is described previously, passivation of
chromium-bearing alloys occurs by the formation of a passive hydrated chromium
oxyhydroxide film, and molybdenum in the alloys assists the passive film formation without constituting the passive film. Consequently, the critical potential for passivity breakdown on the homogeneous alloys covered by the uniform
passive film should increase with increasing alloy chromium content but be
independent of the alloy molybdenum content since the passive hydrated chromium
oxyhydroxide film appears to be dissolved into the hydrochloric acids at
potentials higher than the critical potential.

REFERENCES

1 A. Yokoyama, M. Komiyama, H, Inoue, T. Masumoto and H. M. Kimura, J. Catalysis 68 (1981) 355.
2 A. Yokoyama, M. Komiyama, H. Inoue, T. Masumoto and H. M. Kimura, Scripta
   Met., 15 (1981) 365.
3 A. Yokoyama, H. Komiyama, H. Inoue, T. Masumoto and H. M. Kimura, "Rapidly
   Quenched Metals", Proc. 4th Int. Conf. Rapidly Quenched Metals, T. Masumoto
   and K. Suzuki, Eds., Vol. II, p.1419. The Japan Institute of Metals, Sendai
   (1982).
4 M. Hara, K. Hashimot and T. Masumoto, "Rapidly Quenched Metals", Proc. 4th
   Int. Conf. Rapidly Quenched Metals, T. Masumoto and K. Suzuki, Eds., Vol. II
   p.1423. The Japan Institute of Metals, Sendai (1982).
5 M. Hara, K. Hashimoto and T. Masumoto, J. Non-Cryst. Solids, 54 (1983) 1.
6 M. Hara, K. Hashimoto and T. Masumoto, J. Appl. Electrochem., 13 (1983).
7 A. Kawashima and K. Hashimoto, "Rapidly Quenched Metals", Proc. 4the Int.
   Conf. Rapidly Quenched Metals, T. Masumoto and K. Suzuki, Eds., Vol. II,
   p.1427. The Japan Institute of Metals, Sendai (1982).
8. A. Kawashima and K. Hashimoto, Sci. Rep. Res. Inst. Tohoku University, A-31

(1983) 174.

9   M. Naka, K. Hashimoto and T. Masumoto, J. Non-Cryst. Solids, 31 (1979) 355.
10   K. Kobayashi, K. Asami and K. Hashimoto, "Rapidly Quenched Metals", Proc.
     4th Int. Conf. Rapidly Quenched Metals, T. Masumoto and K. Suzuki, Eds.,
     Vol. II, p.1443. The Japan Institute of Metals, Sendai (1982).
11   K. Hashimoto, M. Naka, J. Noguchi, K. Asami and T. Masumoto, "Passivity of
     Metals", R. P. Frankenthal and J. Kruger, Eds., p.156. The Electrochemical
     Society Monograph Series, Princeton, N.J. (1978).
12   M. Naka, K. Hashimoto and T. Masumoto, J. Non-Cryst. Solids, 34 (1979) 357.
13   K. Hashimoto, K. Kobayashi, K. Asami and T. Masumoto, "Metallic Corrosion",
     Proc. 8th Int. Cong. Metallic Corrosion, Vol. I, p.70, DECHEMA, Frankfurt
     (1981).
14   K. Asami, M. Naka, K. Hashimoto and T. Masumoto, J. Electrochem. Soc., 127
     (1980) 2130.
15   K. Hashimoto, K. Asami, M. Naka and T. Masumoto, Corros. Sci., 19 (1979) 857.
16   K. Hashimoto, Suppl. to Sci. Rep. Res. Inst. Tohoku University, A-28 (1980)
     201.
17   K. Hashimoto, M. Naka, K. Asami and T. Masumoto, Corros. Sci., 19 (1979) 165.
18   K. Hashimoto, K. Asami, M. Naka and T. Masumoto, Sci. Rep. Res. Inst. Tohoku
     University, A-27 (1979) 237.
19   K. Hashimoto, K. Asami and K. Teramoto, Corros. Sci., 19 (1979) 3.
20   K. Hashimoto and K. Asami, Corros. Sci., 19 (1979) 251.
21   K. Asami and K. Hashimoto, Corros. Sci., 19 (1979) 1007.
22   K. Asami, K. Hashimoto T. Masumoto and S. Shimodaira, Corros. Sci., 16
     (1976) 909.
23   K. Hashimoto, K. Asami, M. Naka and T. Masumoto, Corros. Engng. (Boshoku
     Gijutsu), 28 (1979) 271.
24   K. Hashimoto, M. Kasaya, K. Asami and T. Masumoto, Corros. Engng. (Boshoku
     Gijutsu), 26 (1977) 445.
25   M. Hara, K. Hashimoto and T. Masumoto, Electrochim. Acta, 25 (1980) 1215.
26   K. Hashimoto, M. Naka, K. Asami and T. Masumoto, Corros. Engng. (Boshoku
     Gijutsu), 27 (1978) 279.
27   M. Naka, K. Hashimoto and T. Masumoto, J. Non-Cryst. Solids, 29 (1978) 61.
28   K. Asami, K. Hashimoto and S. Shimodaira, Corros. Sci., 18 (1978) 151.
29   R. B. Diegle and J. E. Slater, Corrosion 32 (1976) 155.
30   K. Hashimoto, K. Osada, T. Masumoto and S. Shimodaira, Corros. Sci., 16
     (1976) 71.
31   T. Kulik, J. Baszkiewicz, M. Kaminski, J. Latuszkiewicz and H. Matyja,
     Corros. Sci., 19 (1979) 1001.
32   M. Naka, K. Hashimoto and T. Masumoto, Corrosion, 36 (1980) 679.
33   S. Kapusta and K. E. Heusler, Z. Metallkde., 72 (1981) 785.
34   T. R. Anthony and H. E. Cline, J. Appl. Phys., 49 (1978) 1248.
35   T. Tsuru, S.-X. Zhang and R. M. Latanision, "Rapidly Quenched Metals, Proc.
     4th Int. Conf. Rapidly Quenched Metals, T. Masumoto and K. Suzuki, Eds.,
     Vol. II, p.1437. The Japan Institute of Metals, Sendai (1982).
36   T. Tsuru and R. M. Latanision, J. Electrochem. Soc., 129 (1982) 1402.
37   M. Naka, K. Asami, K. Hashimoto and T. Masumoto, "TITANIUM'80", Proc. 4th
     Int. Conf. Titanium, H. Kimura and O. Izumi, Eds., Vol. IV, p.2679.   The
     Metallurgical Society of AIME, Warrendale, Pa. (1980).
38   R. B. Diegle, Corrosion, 35 (1979) 250.
39   R. B. Diegle, Corrosion, 36 (1980) 362.
40   T. M. Devine, J. Electrochem. Soc., 124 (1977) 38.
41   K. Kobayashi, K. Hashimoto and T. Masumoto, Sci. Rep. Res. Inst. Tohoku
     University, A-29 (1981) 284.

Passivity of Metals and Semiconductors, edited by M. Froment
Elsevier Science Publishers B.V., Amsterdam — Printed in The Netherlands

SUPPLEMENT TO "PASSIVATION OF AMORPHOUS METALS"

K. HASHIMOTO

The Research Institute for Iron, Steel and Other Metals, Tohoku University,
Sendai, 980 (Japan)

Following the suggestion given by Dr. J. Kruger and many other participants at
the symposium, the author supplements the following illustrations.

UNUNIFORMITY OF PASSIVE FILM

Passivation of alloys containing sufficient amounts of chromium occurs by the
formation of a passive hydrated chromium oxyhydroxide film. As shown in Fig. 1*,
the passive film formed on the heterogeneous alloy is not uniform but contains
various weak points because the stable passive film cannot be formed on the chemically heterogeneous sites of the alloy surface. Passivity breakdown and repas-

Fig. 1. The Chromiums have to overstrain themselves to join their hands on unsafe
points.

* Oxygen contained in the passive film plays an important role in the protective
quality of the film but is not shown in the figures in this paper.

248

sivation occur repeatedly at various weak points and the current observed in the passive state is the leakage current passing through the weak points as shown in Fig. 2. In other words, the activity of the weak points is responsible for the current observed in the passive state. This can be seen from various experimental results (refs.1-5). When passivity breakdown at weak points cannot be followed by repassivation, localized corrosion takes place as shown in Fig. 3.

ROLE OF MOLYBDENUM

Molybdenum contained in the alloy assists passivation. When passivity breakdown occurs at weak points of a passive film, the localized active dissolution of all the alloy constituents including those necessary for passivation leads to formation of a corrosion product film in which molybdenum ions are significantly concentrated (refs.1,6,7). Molybdenum ions in the film appear to be in the tetra valent state (ref.7). The molybdenum-enriched corrosion product film is formed due probably to the fact that molybdenum is passive in the active region of chromium and iron, and acts as an effective diffusion barrier against further active dissolution of the alloy. Accordingly, as shown in Fig. 4, the molybdenum-enriched corrosion product film protects the alloy constituents necessary for passivation from active dissolution, and consequently the passive hydrated chromium or iron oxyhydroxide film grows under the corrosion product film.

Fig. 2. When a boy called Chromium defending one of unsafe points is killed by a devil Chloride, another boy of the Chromiums usually covers him, although losing the boy Chromium is so sad that the family shed current.

Fig. 3. When new Chromium fails to cover the boy Chromium killed by the devil Chloride during defending one of unsafe points, they are attacked catastrophically by the devils and bleed profuse current.

Once repassivation occurs due to passive film growth at the weak points, the molybdenum-enriched corrosion product film dissolves into the solution as shown in Fig. 5, because formation of the molybdenum-enriched corrosion product film is no more than the result of a slower dissolution rate of molybdenum than the other alloy constituents during rapid active dissolution of all the alloy constituents, and because the film is not stable at high potentials in the passive region of chromium and iron due mainly to transpassive dissolution of molybdenum.

Consequently, the molybdenum addition increases the passivating ability of the alloy and decreases the activity of the weak points of the passive film. In this manner, molybdenum assists passivation but does not constitute the passive film. In addition, when molybdenum ions are contained in the passive film, the film is unstable at relatively high potentials in the passive region of the alloy due to transpassive dissolution of molybdenum (ref.7).

ROLE OF METALLOIDS IN ALLOYS

In general, the composition of the substrate alloy immediately under the chromium- or iron-enriched passive film is almost the same as the composition of the bulk alloy. Passivation therefore occurs as a result of preferential dissolution of the alloy constituents unnecessary for passive film formation. The amorphous metal-metalloid alloys consisting at least of two kinds of metallic elements have

Fig. 4. When an elder brother Chromium defending one of unsafe points is killed by a devil Chloride, a mother Mrs. Molybdenum immediately comes to the unsafe point and protects baby Chromium from the devil Chlorides.

Fig. 5. When Chromium held in mother's arms becomes a big boy so as to join hands with other Chromiums, the aged mother Mrs. Molybdenum departs from this life.

a significantly higher passivating ability than the corresponding crystalline (ref.8) and amorphous (refs.9,10) metals without metalloid addition. This is due to the fact that metalloids contained in the amorphous alloys accelerate remarkably preferential dissolution of the alloy constituents unnecessary for passive film formation with a consequent enhancement of accumulation of the beneficial elements in the passive film (ref.11) as shown in Fig. 6.

When the environment is especially corrosive, or when the passivating ability of the alloy is not sufficiently high, e.g. because of the low chromium content of the alloy, phosphorus is often contained in the

Fig. 6. Laudable Phosphorus is going away taking others who will encumber Chromiums, since Chromiums have to construct the strongest defense line consisting only of them.

surface film as phosphate. In other words, when an alloy-environment combination is particularly corrosive, all the alloy constituents dissolve at a high rate, and consequently a corrosion product film containing various ions is formed since the concentrations of some ions of the alloy constituents exceed their solubility limits at the alloy solution interface. The surface film containing phosphate thus formed is less protective than the stable passive film as shown in Fig. 7.

On the other hand, when boron or silicon is contained in an alloy, they tend to be contained in the passive film as borate and silicate, and hence the protective quality of such a film is lower than that of the passive hydrated oxyhydroxide film. This is the minor detrimental effect of boron and silicon. However, these metalloids accelerate greatly an enrichment of beneficial ions ion the surface film, and hence they are effective in increasing the passivating ability.

Fig. 7. When attacked violently by the devil Chlorides, they have to construct, the defense line together with children who fail to escape, although the defense line is not strong.

ACKNOWLEDGEMENT

The author expresses his thanks to his wife Yasuko Hashimoto for her drawing the illustrations.

REFERENCES

1 K. Hashimoto, K. Asami and K. Teramoto, Corros. Sci. 19 (1979) 3.
2 M. Naka, K. Hashimoto and T. Masumoto, Corrosion 36 (1980) 679.
3 K. Hashimoto, K. Osada, T. Masumoto and S. Shimodaira, Corros. Sci. 16 (1976) 71.
4 R. B. Diegle and J. E. Slater, Corrosion 32 (1976) 155.
5 R. B. Diegle, "Rapidly Quenched Metals", Proc. 4th Int. Conf. Rapidly Quench-ed Metals, T. Masumoto and K. Suzuki, Eds. Vol. II, p.1475. The Japan Institute of Metals, Sendai (1982).
6 K. Hashimoto, K. Asami, M. Naka and T. Masumoto, Corros. Sci. 19 (1979) 165.
7 K. Asami, M. Naka, K. Hashimoto and T. Masumoto, J. Electrochem. Soc. 127 (1980) 2130.
8 K. Hashimoto, K. Asami, M. Naka and T. Masumoto, Corros. Sci. 19 (1979) 857.
9 M. Naka, K. Hashimoto and T. Masumoto, J. Non-Cryst. Solids 30 (1978) 29.
10 M. Naka, K. Asami, K. Hashimoto and T. Masumoto, "Titanium 80", Proc. 4th Int. Conf. Titanium, H. Kimura and O. Izumi, Eds. Vol.4, p.2677. The Metal-lurgical Society of AIME, Warrendale, Pa. (1981).
11 K. Hashimoto, M. Naka, K. Asami and T. Masumoto, Corros. Engng. (Boshoku Gi-jutsu) 27 (1978) 279.

Passivity of Metals and Semiconductors, edited by M. Froment
Elsevier Science Publishers B.V., Amsterdam — Printed in The Netherlands

DEPASSIVATION AND REPASSIVATION IN LOCALIZED CORROSION

J.C. SCULLY

Department of Metallurgy, The University of Leeds, Leeds, LS2 9JT, England.

ABSTRACT
  Loss of passivity (depassivation) may occur over the complete surface of a metal but, in practice, depassivation in localized regions constitutes major forms of corrosion: crevice and pitting corrosion and stress corrosion cracking. Depassivation occurs as a result of the migration of aggressive ions through a passive film, or down pores or by interfering with the repassivation process in films that exhibit a crack/heal mechanism. Repassivation rates exhibit a number of different time dependencies. In some cases of pitting the repassivation rate is the determinant factor. The pitting potential corresponds to that potential at which the measured repassivation rate has a specific value and will become more active or noble depending upon an increase in the concentration of aggressive or inhibitive ions. In stress corrosion cracking propagation rates and crack morphologies are controlled by repassivation rates.

INTRODUCTION

  Depassivation of metals occurs when the passive film becomes unstable for what may be chemical or electrochemical reasons. The processes by which the film is stabilised alter, perhaps only slightly, but sufficiently, and loss of film (depassivation) occurs. Such an event can occur over a whole surface or be confined to highly localised regions. When it occurs over a complete surface general corrosion will then follow. Characteristically, this occurs under open circuit conditions as the bulk pH value of a solution is lowered or raised and at values which will depend not only upon the pH value alone but also upon the specific metal being examined and upon environmental factors such as solution anion type and temperature. General depassivation conditions are clearly defined theoretically according to the thermodynamic diagrams of Pourbaix (ref.1) and they have been illustrated practically in numerous studies e.g. on a 304 stainless steel by Leckie (ref.2) who examined breakdown at low pH values at room temperature in sulphate and chloride solutions. When breakdown occurs at just a few points on a metal surface particular forms of corrosion are observed: crevice corrosion, pitting corrosion and stress corrosion cracking. This paper is mainly concerned with processes of localized breakdown. These are the most commonly encountered with passive metals and they are also in many respects the most difficult to understand. The depassivation process is first discussed in relation to pitting corrosion. Repassivation is then considered in some general detail and its importance in pitting corrosion is described. Finally depassivation and repassivation processes in relation to stress corrosion cracking are discussed.

PITTING CORROSION

This is a very complicated subject which has been reviewed extensively (ref.3). While it can occur with a number of different anion types the Cl⁻ ion is the most widely encountered and investigated. Discussion is limited to pitting caused by this species.

For reasons that are not well understood Cl⁻ ions under particular conditions cause localized depassivation. The specific function of Cl⁻ ions in this process has been the subject of prolonged discussion. Any proposed mechanism must include an assumption about the nature of the passive film about which there is still much to be determined. It is possible to attempt an approximate division of the depassivation processes proposed according to whether the film is (a) compact, stable and three-dimensional, (b) compact but porous, and (c) unstable and subject to continuous breakdown and repair. The possibility that passivity is the result of an adsorbed monolayer is not intentionally discounted but for the alloys considered below the passive film observed exhibits a thickness greater than a monolayer and in considering depassivation mechanisms whether the thick film is the cause or merely the consequence of passivity is not important.

Compact Films

Much discussion in the past has centred around the idea that Cl⁻ ions migrate through the passive film on to which they have adsorbed. Whether they have adsorbed relatively uniformly and penetrated at specifically vulnerable spots or have adsorbed only at such spots is not usually specified. Despite attempts to observe such penetration it has proved remarkably difficult to provide data about such processes. The migration of Cl⁻ ions into anodized aluminium films has been reported upon by Beck et al (ref.4) although this effect has been disputed (ref.5). Abd Rabbo et al (ref.6) studied Cl⁻ ion reaction with various films on aluminium. While a SIMS analysis revealed a substantial amount of Cl⁻ ion at the oxide/solution interface, no Cl⁻ ion was detected within the bulk of the oxide. Penetration by chromate and phosphate ions was detected. Kuodeltova et al (ref.7) and others (ref.8) reported Cl⁻ ion penetration of films on aluminium by investigation with ESCA techniques.

On ferrous alloys the situation is perhaps less unclear. McBee and Kruger (ref.9) reported that optical changes produced in oxide films on iron exposed to chloride solutions were reversible if the chloride ions were removed. Similar smaller changes observed on Fe-Cr alloys were not reversible. On some Fe-Cr alloys Cl⁻ ions exhibited only a small amount of penetration to a few atom layers according to an AES study (ref.10) while on iron the chloride ion appears to be only weakly bonded to the passive film.

Porous Films

Discussion of Cl^- ion penetration depends upon the configuration of the surface and this is usually not clearly discernible. Once it is proposed that films exhibit porosity any proposed breakdown will be modified according to the type and degree of porosity, that is envisaged. Often this is not explained since it is very difficult to obtain direct evidence for its existence. With any type of pore, on a purely geometric basis, the thickness of the film will be less at the bottom of a pore than elsewhere. Depending on the ratios of thickness, film conductivity and degree of porosity the passivating current density will be unevenly distributed across the surface, with the major part coming from within the pores which can be expected therefore to contain liquid rich in metal cations. In chloride solutions pits may develop at the deepest set of pores where the operative current density is highest because the film is thinnest (or non-existent) and therefore the cation flux and chloride ion migration rates are at maximum values. The associated hydrolysis and saturated acid chloride salt solutions that develop within pits would then proceed, provided that certain conditions were attained. Such conditions have been developed quantitatively by Galvele (refs.11 and 12) who has argued that one important parameter for pit (and crevice) development is the multiple of the pore base current density and the pore length. Only when that multiple exceeds a specific value for a given metal and solution will either form of corrosion ensue. He has argued that the pore length will be limited to a maximum value, corresponding to the thickness of the passive film, on the assumption of a simple one-dimensional pit, whereas a crevice can be any length, so that there is no maximum value. Thus the multiple of c.d.xl can be exceeded for any metal solution. On that simple basis he concludes that any metal can exhibit crevice corrosion, provided that the crevice is long enough, whereas only a few exhibit pitting corrosion.

Among the ideas of pores that have been published must be included the work of Richardson et al (ref.13). They concluded that pores and defects allow instantaneous penetration and localized corrosion to occur within the film. Breakdown occurs at the beginning of immersion by preferential metal dissolution at flaw bases in the surface film. Any measured induction time represents the time over which pits are too small to create any changes in the total measured currents.

With this type of model where developments within pores are of significance it is possible to suppose that other anions are likely to be inhibitive of pitting either because they cause passive film growth and/or because they form

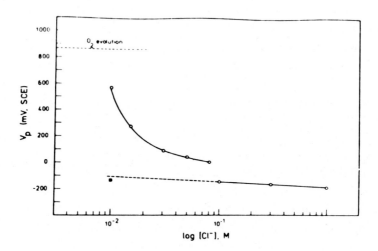

Fig. 1. Variation of the pitting potential, E_p, of AISI 304 stainless steel with $[Cl^-]$. Solution pH 2. $KCl + 6.5 \times 10^{-3} M H_2SO_4$. $10^{-2} M$ HCl. (Ref.14)

insoluble metal salts and thereby effectively block the pore. The effect of such anions will depend upon their concentration vis-a-vis the Cl^- ion concentration. An example is shown in Fig. 1 (ref.14).

This shows the dependence of the pitting potential, E_p, upon the logarithm of the Cl^- ion concentration. For $[Cl^-] \geq 10^{-1}$ M the change is linear, according to:-

$$E_p = a - b \log [Cl^-] \text{ mV (SCE)} \tag{1}$$

with $a = -195$ mV and $b = 45$ mV. This is the type of relationship usually observed e.g. for stainless steels (e.g. ref.15) and other metals (e.g. ref.16). The curve in Fig. 1 exhibits a discontinuity in the region of $0.1M[Cl^-]$ and at $10^{-3}M$ (which is not shown in Fig. 1) the steel was found to be immune to pitting. In the curved region of the relationship E_p and $[Cl^-]$ are related by the empirical expression:-

$$E_p = K_o + K_1 [Cl^-]^{-m} \text{ mV (SCE)} \tag{2}$$

with K_o = -351 mV, K_1 = 179 mV and m = 1.75. Assuming that this expression applies up to the potential for oxygen evolution, the minimum [Cl⁻] that causes pitting can be evaluated by making equation (2) E_p = E_{O_2} = 868.5 mV. By making suitable calculations employing activity coefficients this relationship is in good agreement with results reported by Leckie and Uhlig (ref.15).

$$\log c_{Cl^-} = 9.85 \log a_{SO_4^{2-}} -0.05 \tag{3}$$

The interpretation (ref.17) that competitive adsorption leads to an expression of the form of equation (3) is not necessarily the only interpretation. Similar equations have been derived assuming either ion exchange equilibria in the surface oxide (ref.18) or transport-controlled distribution of aggressive and inhibitor ions inside the pits (ref.19).

On Fig. 1 is also plotted E_p in a sulphate-free 10^{-2} M HCl solution. Within the limits of experimental error the value of E_p fits equation (1) which is valid for H_2SO_4 solutions with [Cl⁻] ⩾ 10^{-1} M. At [Cl⁻] 10^{-2} M there is a discontinuity consisting of a 150 mV increase in the value of E_p. These two points show that SO_4^{2-} ions are ineffective in affecting pitting with [Cl⁻] > 10^{-1} M. At lower concentrations sulphate ions become increasingly effective.

The mechanism of SO_4^{2-} inhibition is probably associated with metal salt formation. In Fig. 2 the polarization curve is drawn. A small increase in current density occurs at -140 mV (SCE). Since in the absence of SO_4^{2-} ions E_p would be -146 mV (SCE), the increase observed in the presence of SO_4^{2-} is most probably associated with activation at some points of the steel surface. Between -140 and 0 mV (SCE) partial repassivation occurs as indicated by a reduction in the rate of current density increase with rise in potential.

The effect of Cl⁻ ions has been examined in an indirect way by investigating how these ions accelerate film removal of the type that occurs at the Flade Potential (ref.20). Specimens of iron were polarized anodically at +400 mV (SCE) in 0.5N H_2SO_4. Additions of Cl⁻ ions were then made and surface reactivation brought about by switching off potentiostatic control before pitting initiated. At low pH values (1.6) reactivation was accelerated by Cl⁻ ion additions, as is shown in Fig. 3.

This effect may arise from the prior existence of iron chloride solution inside the pores while specimens were under potentiostatic control or from participation in the film reduction process at the reactivation potential. The prior existence of iron salts inside pores was the interpretation put upon results obtained at higher values of solution pH. Low levels of chloride addition increased the reaction time, a result readily attributed to the precipitation of $FeSO_4$ within the pores as a result of an increased flux of

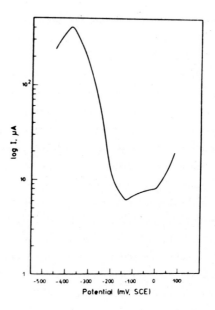

Fig. 2. Anodic polarization curve for AISI 304 stainless steel in 8.04×10^{-2}M
KCl + H_2SO_4. (Ref.14)

-Fe^{2+} created by the presence of the Cl$^-$ ions. High levels of chloride
additions promoted pitting. The transition is probably observed when the
Cl$^-$/SO$_4^{2-}$ ratio is exceeded, a point already discussed in relation to Fig. 1.
The existence of pores was thought to be shown by the observation that adding
additional SO$_4^{2-}$ ions to the solution in which a specimen had already been
passivated resulted in a fall in the passivating current density and an
increase in the time to reactivation.

The Continual Breakdown of Films

On some passive surfaces there is evidence that the film is not stable.
Potential and current oscillations, under galvanostatic and potentiostatic
conditions respectively, have been reported for aluminium (ref.21) and
stainless steels (ref.22). Aziz (ref.21), for example, showed that when an
aluminium surface is just exposed to a solution, corrosion is initiated as a
very high density of sites most or all of which become sealed and inactive,
depending upon the relationship between the pitting and corrosion potentials.

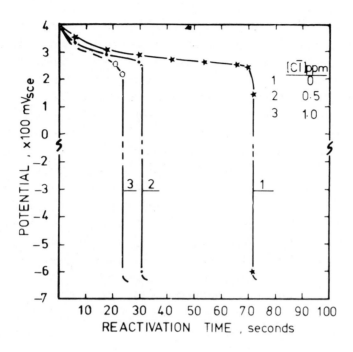

Fig. 3. The effect of chloride ion additions upon the reactivation of Fe in a de-aerated H_2SO_4 solution of pH 1.6. (Ref.20)

Most recently a crack/heal mechanism has been developed (ref.13) with film breakdown and repair occurring continually as already indicated. Film fracture and repair are competing processes in any such mechanism. Slowing down the repair process by, for example, raising the potential or by increasing the chloride ion concentration, eventually allows at least one fracture site to develop sufficient hydrolysis following sufficient metal dissolution to stabilize the unfilmed site and result in at least one pit. This is a simple description. Metal salt precipitation as well as oxide formation must be considered, for example. Where such a model applies, chloride ion penetration of bulk oxide, over which there is conflicting evidence as indicated briefly above, or pore penetration by chloride ions, both become unnecessary to a pitting mechanism.

The explanation for periodic breakdown of the passive film in chloride solutions is not clear. Sato (ref.23) has calculated that the electrostriction pressure due to the electric field across the film, 10^6-10^7 V/cm, produces a compressive stress which could exceed the fracture stress of the oxide film. He proposed that the surface tension stabilizes the anodic oxide film but that the effect decreases with increasing film thickness. Thus there is a critical thickness above which the mechanical deformation or breakdown of the oxide film will occur. He also advocated that anion adsorption on the oxide lowers the surface tension and thereby decreases the initial thickness for breakdown. Such an idea might account for the effect of specific anions. Other sources of breakdown have been proposed. A possible role for hydrogen (ref.24) has been advocated. There is microscopic evidence that hydrogen molecules form at the metal/oxide interface. Penetration of the oxide by atoms or protons results in molecule formation. Proton transport would not require electron transport through films and would be easier to propose on metals whose oxides have low conductivities. A detailed explanation for how chloride ions promote this reaction is not available although it may be that it is a common feature of the exposure of films to many different types of anions. The uniqueness of the chloride ion may lie in the effect upon the rate of repair rather than any unusual capacity to cause breakdown in the first place.

REPASSIVATION

The brief discussion above has drawn attention to film repair processes, which are summarized as a repassivation process. The kinetics of repassivation have been examined by a number of authors who have employed scratching and straining methods and potential switching techniques. The object of such work has included the development of new methods for examining pitting corrosion as well as an attempt to obtain fundamental information on the initiation of films on unfilmed metal surfaces.

The growth of passive films has been analyzed by many authors. A dissolution-precipitation may apply for some systems but a solid state transformation is the most widely invoked and has been treated extensively by Armstrong et al (ref.25). Their analysis of oxide platelet growth for instantaneous nucleation has been applied to studies of rapid fracture on titanium (ref.26). Such growth occurs laterally and surfaces will be covered with a monolayer within 10-20 ms. Most studies of repassivation therefore incorporate growth of films on completely covered surfaces rather than surfaces consisting of a mixture of an increasing proportion of filmed area associated with spreading plates and a decreasing proportion of dissolved unfilmed area.

Film growth kinetics are usually studied by either coulometric or ellipsometric techniques. In the former the assumption is made that all the charge is converted into film thickness, which will be valid only if the dissolution rate is negligible in comparison with the film forming current. Adjustments have to be made where this is not true. Similar problems do not arise in ellipsometry but problems in interpretation can arise if surface roughening or changes in the refractive index of the solution or film occur together with the formation of non-uniform films and other events.

The rate laws observed in repassivation take a number of forms which are derived theoretically from quite different growth mechanisms. Growth by ion migration under high yield conditions yields an inverse logarithmic relationship between i and t whereas a place exchange mechanism is thought to yield a direct logarithmic law. While the theoretical arguments are outside the scope of this paper such growth rates have been observed by a number of investigators. For iron in borate solutions an inverse law has been reported (ref.27) as well as a direct law (ref.28) as well as a transition from one to the other (ref.29) on aluminium. Logarithmic laws have been reported for a number of ferrous alloys (ref.30) including stainless steels (ref.31). On iron a direct logarithmic law has been observed in Na_2SO_4 solutions at potential values most noble than 500 mV (SHE). At lower potentials films grew linearly and were not protective (ref. 32).

The Time Dependence of Repassivation Processes

The kinetics of repassivation have been revealed in many investigations by determining the relationship between the current, measured under potentiostatic conditions, and time.

The current density associated with film growth, i_f, which is proportional to the change of thickness, δ, with time t, can be expressed (ref.28) as:-

$$i_f = K \exp \left(\beta V - \frac{\delta}{B} \right) \qquad (4)$$

where K, β and B are constants. Integration of this gives the direct logarithmic law:-

$$\delta = k_o \ln \left(\frac{t}{\tau} + 1 \right) \qquad (5)$$

where k_o and τ are constants.

For values of $t \gg \tau$

$$i_f = at^{-1} \tag{6}$$

Wagner (ref.33) pointed out that a continual transition should be expected from the limiting case of thin films to that of thick films in which the well established laws of diffusion would operate giving parabolic oxidation so that a general expression would be obtained

$$i_f = at^{-b} \tag{7}$$

with $0.5 \leqslant b \leqslant 1$. This type of current decay has been quite widely observed: in Ti-6Al-4V in halide solutions (ref.34), 316 steel in 5N H_2SO_4 (ref.31), Fe in borate solutions (ref.28), Co in borate solutions (ref.45) and Bi in various solutions (ref.36).

In the absence of film dissolution film thickness would proceed indefinitely and no stationary state would be reached, but in practice a balance is commonly reached between film formation and dissolution which may occur by either chemical or electrochemical processes. In addition electrochemical dissolution of the metal by transport of metal ions through the film may occur. Any interpretation of the measured current must take into account the corrosion of the metal that occurs during film initiation and growth. Reported results divide into two categories in which there is either no dissolution or a significant amount. Iron in borate solutions, for example, passivates with all the charge converted into film thickening (refs.28,37) although contradictory work on the same system indicates that only 10% of the recorded current density causes film thickening (ref.38), a result similar to the 15% observed for iron in sulphuric acid. Other systems generally indicate significant amounts of corrosion. Iron in nitrate results in a lot of corrosion, moreso than in nitrite solutions (ref.39). On titanium (ref.26) and on a Cr-Ni steel (ref.40) a large proportion of the current has been reported as causing dissolution. Because of the complexity of such processes it is often difficult to compare one set of results with another and it is important not to over-emphasize apparent contradictions since they may arise from emphasis being placed on different measuring procedures.

Some authors have reported an exponential decay of current of the form:-

$$i = i_o \exp(-\beta t) \tag{8}$$

This has been found on a stainless steel (ref.41) and on aluminium alloys (ref.42). No particular explanation has been put forward to account for this but a simple computer model of a scratching electrode has been devised (ref.53) to replicate events on the assumption that all the charge goes into making the film. The essence of the analysis was that a logarithmic decay of the form indicated in equation (6) would be expected to be obeyed after a time had elapsed at the end of the scratching procedure. During that time period, however, either equation (5) or (8) would be equally accurately applicable since the most accurate fit arose from a third equation. Such complexity arises from the combination of scratching and repair of scratched areas going on simultaneously.

REPASSIVATION AND PITTING CORROSION AND STRESS CORROSION CRACKING

One particular study on aluminium (ref.42) has demonstrated that repassivation rates are closely connected with the initiation of pitting corrosion. Scratching electrode experiments revealed an exponential decay of current according to equation (8). Figure 4 shows the values of β as a function of potential obtained in a number of inhibited solutions containing a fixed amount of chloride (3×10^3 ppm). In all cases the pitting potential for each solution corresponded to a specific value of β (indicated by a dashed vertical line). The constant varied according to potential as has also been shown for stainless steel (ref.41).

For stainless steels logarithmic repassivation has been observed (ref.44). The value of b varies as shown in Fig. 5 but it is not necessarily so useful in determining the pitting potential. Many such alloys contain inclusions, particularly MnS, and these serve as nucleation sites for pitting. Even when scratching tests reveal complete repair of films subsequent maintenance of the potential may eventually lead to pitting at such inclusions. The values of b does reduce at noble potentials, indicating a less protective film but a specific value has not been associated with the real value of the pitting potential. In Fig. 5 the quasi-stationary technique indicated a pitting potential of +630 mV (SCE) but pitting occurred at +350 mV (SCE) after 50h.

In stress corrosion crack propagation the role of repassivation processes has been analyzed in some detail (ref.45). In chloride solutions the crack tip solution has a composition similar to that reported in pits but cracking occurs at and below the pitting potential. A dynamic picture has been portrayed. At the crack tip there is a strain-rate created by the stress which continually creates fresh metal area. Also the fresh metal area will be repassivating.

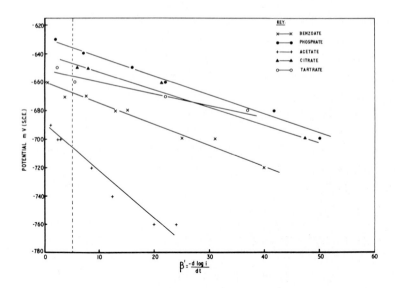

Fig. 4. The repassivation constant β as a function of potential in solutions containing 3×10^3 Cl^- + 0.08M of one of five different inhibitors for Al scratched while under potentiostatic control. The pitting potential for each solution corresponds to the vertical dashed line. (Ref.42)

For a given strain-rate too high a repassivation rate will result in crack arrest and too low a strain rate will result in too much corrosion, giving crack blunting and some form of pitting in place of crack propagation. Considering the two rate processes there will be sets of values that will lead to crack propagation. These are included within the area shown in Fig. 6 which attempts to summarize the effects of potential, inhibitor and chloride ion level and strain-rate.

A further subdivision within the area is possible since it has also been analyzed that minor differences in the crack tip solution composition affect crack morphology. Thus in low strength steels, austenitic stainless steels and alpha brasses cracking is intergranular in solutions in which repassivation is relatively rapid, whereas it is transgranular in solutions in which it is relatively slow. The effects of solution composition and potential on crack morphology can all be explained readily in this way. The inference is that grain boundaries are more difficult to repassivate than emergent slip steps and that is not unlikely.

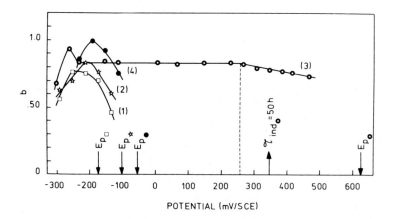

Fig. 5. The repassivation index, b, as a function of potential for AISI 304 stainless steel scratched while under potentiostatic control. Solutions 1-3 contain $5 \times 10^{-3}M$ SO_4^{2-} + (1) 1M KCl, (2) 0.1M KCl, and (3) 0.01M KCl. Solution 4 is 0.01M HCl. In solution (3) the pitting potential is +630mV with a quasi-stationary technique and +350mV under potentiostatic conditions with an induction time of 50h. (Ref.44)

CONCLUSION

Depassivation and repassivation together constitute a very large area of interest. Such processes are affected by both electrochemical and chemical processes in addition to metallurgical factors such as alloy composition and by hydrolysis kinetics which will depend upon the amount of metal cations forming during film initiation and, possibly, growth. In the summary above only a simple picture has been presented and it has been necessary to omit many possible references, both these restraints being necessary in the interests of length. Much more work is required in this area because it constitutes an integral part of many important processes.

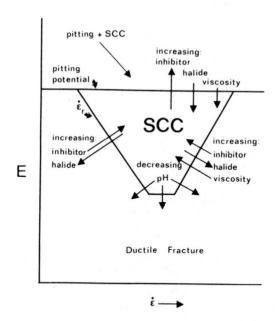

Fig. 6. Schematic diagram of the relationship between electrode potential and crack tip strain-rate, $\dot{\varepsilon}$, for a solution in which passivation is possible. The boundaries are determined by solution composition, particularly inhibitors, halide ion concentration, pH and solution viscosity. (Ref.45)

REFERENCES

1. M. Pourbaix, Atlas of Electrochemical Equilibria in Aqueous Solutions, Pergamon Press, Oxford (1966).
2. H.P. Leckie, Corrosion, 24 (1968) 70-4.
3. J.R. Galvele, in R.P. Frankenthal and J. Kruger (Eds.) Passivity of Metals, The Electrochem. Soc., Princeton, New Jersey (1978) pp.285-327.
4. A.F.Beck, M.A. Heine, D.S. Keir, D. van Rooyen and M.J. Pryor, Corros. Sci. 2(1962) 133-45.
5. G.C. Wood, W.H. Sutton, T.N.K. Riley and A.G. Malherbe, in R.W. Staehle, B.F. Brown, J. Kruger and A. Agrawal (Eds.) Localized Corrosion, N.A.C.E., Houston (1974) pp.526-39.
6. M.F. Abd Rabbo, J.A. Richardson and G.C. Wood, Corros. Sci. 16 (1976) 677-88.
7. M. Kuodeltova, J. Augustynski and H. Berthou, J. electrochem. Soc. 124(1977) 1165-8.
8. J. Augustynski, in ref.3 pp.989-1002.
9. C.L. McBee and J. Kruger, ref.5 pp.252-62.
10. J.B. Lumsden and R.W. Staehle, Paper No. 122, preprinted from Corrosion/73 N.A.C.E., Houston (1973).
11. J.R. Galvele, J. electrochem. Soc. 123 (1976) 464-74.

12. J.R. Galvele, Corros. Sci. 21 (1981) 551-80.
13. J.A. Richardson and G.C. Wood, Corros. Sci. 10 (1970) 313-24.
14. M. Barbosa and J.C. Scully, Corros. Sci. 22 (1982) 1025-36.
15. H.P. Leckie and H.H. Uhlig, J. electrochem. Soc. 113 (1966) 1262-7.
16. H.H. Strehblow and B.B. Titze, Corros. Sci. 17 (1977) 461-72.
17. H.H. Uhlig and J.R. Gilman, Corrosion 20 (1964) 289-92.
18. D.M. Brasher, D. Reichenberg and A.D. Mercer, Br. Corros. J. 3 (1968) 144-50.
19. B.G. Ateya and H.W. Pickering in ref. 3 pp.350-68.
20. A. Cakir and J.C. Scully, in ref.3 pp.385-402.
21. P.M. Aziz, J. electrochem. Soc. 101 (1954) 120-9.
22. H. Saito, T. Shibata and G. Okamoto, Corros. Sci. 19 (1979) 693-708.
23. N. Sato, Electrochim. Acta, 16 (1971) 1683-92.
24. C.B. Bargeron and R.B. Givens, Corrosion 36 (1980) 618-25.
25. R.D. Armstrong, J.A. Harrison and H.R. Thirsk, Corros. Sci. 10 (1970) 679-92.
26. T.R. Beck, Electrochim. Acta, 18 (1973) 815-23.
27. R.V. Moshtev, Ber. Bunsengesel. Phys. Chem. 71 (1967) 1079-86.
28. N. Sato and M. Cohen, J. electrochem. Soc. 111 (1964) 512-9.
29. C. Benndorf, H. Seidel and F. Thieme, Surf. Sci. 67 (1977) 469-79.
30. K.N. Gosmami and R.W. Staehle, Electrochim. Acta 16 (1971) 1895-1907.
31. G.M. Bulman and A.C.C. Tseung, Corros. Sci. 13 (1973) 531-44.
32. S. Smialowska and G. Mrowcznski, Br. Corros. J. 10 (1975) 187-92.
33. C. Wagner, Corros. Sci. 13 (1973) 23-52.
34. H. Buhl, Corros. Sci. 13 (1973) 639-46.
35. G.W. Simmons, E. Kellerman and H. Leidheiser, Jr., in R.W. Staehle and H. Okada (Eds.) Passivity and its Breakdown on Iron and Iron Base Alloys, N.A.C.E., Houston (1976) pp.65-73.
36. I.A. Ammar, S. Darwish and M.W. Khalil, Corrosion 32 (1976) 173-9.
37. J. Kruger and J.P. Calvert, J. electrochem. Soc. 114 (1967) 43-9.
38. V. Brusic, Ph.D Thesis, University of Pennsylvania, (1971).
39. J.R. Ambrose and J. Kruger, Corrosion 28 (1972) 30-5.
40. V.A. Makarov and T.A. Zagorel'eva, Prof. Met. (Zaschchita Met.) 12 (1976) 381-8.
41. P. Engseth and J.C. Scully, Corros. Sci. 15 (1975) 505-20.
42. W.J. Rudd and J.C. Scully, Corros. Sci. 20 (1980) 611-32.
43. M. Barbosa and J.C. Scully in Z.A. Foroulis (Ed.) Environment-Sensitive Fracture of Engineering Materials, A.I.M.E., Warrendale, Pa. (1979) pp.91-115.
44. M. Barbosa and J.C. Scully, ref.3 pp.403-6.
45. J.C. Scully, Corros. Sci. 20 (1980) 997-1016.

Passivity of Metals and Semiconductors, edited by M. Froment   269
Elsevier Science Publishers B.V., Amsterdam — Printed in The Netherlands

DISSOLUTION AND PASSIVATION KINETICS OF Fe-Cr-Ni ALLOYS
DURING LOCALIZED CORROSION

R.C. NEWMAN and H.S. ISAACS
Corrosion Science Group, Brookhaven National Laboratory, Upton, NY 11973, USA

ABSTRACT

An artificial pit technique has been used to study the anodic and cathodic
behavior of Fe-Cr-Ni alloys during localized corrosion in neutral chloride solutions. The active dissolution kinetics have been determined as a function of the
concentration of the localized environment, and the conditions for passivation
established. Platinum microelectrodes have been used to analyze the pit
solutions. The cathodic behavior of a pit on type 304L stainless steel is
dominated by copper ions dissolved in the pit. A high purity Fe-Cr-Ni alloy
behaves like stainless steel if copper is separately dissolved in the pit
environment. The variation of dissolution rate with concentration of corrosion
products is used to show that localized corrosion can involve as many as five
diffusion-coupled steady states. The stability of these states is discussed.

INTRODUCTION

Studies of localized corrosion in stainless alloys have rarely sought to establish the actual electrode kinetics of the metal in its localized environment.
These kinetics must, however, be known in order to predict the conditions under
which localized corrosion will occur. They also provide a basis for understanding the influence of alloying additions or inhibitors. We have used a
1-dimensional artificial pit technique to generate the appropriate localized
environments in situ. Although this technique involves the application of
relatively oxidizing potentials, it works well with alloys containing iron,
chromium and nickel because of the stability of the Fe^{2+}, Cr^{3+} and Ni^{2+} ions.

EXPERIMENTAL METHOD

The artificial pit technique has been described previously (refs. 1-3). 1 mm
wires of two alloys were prepared: type 304L stainless steel (18.50Cr, 8.99Ni,
0.013C, 1.61Mn, 0.33Cu, 0.62Si, 0.32Mo, 0.021S, 0.17Co, 0.136N) and a high
purity iron-19Cr-10Ni alloy (18.7Cr, 10.3Ni, others <0.01). They were mounted
in rods of epoxy resin and corroded from one end. The metal surface was recessed to the desired depth (usually 0.5-1.0 mm) by prolonged dissolution above
the pitting potential in 1M NaCl. The electrochemical cell was a 100 ml glass
container with a platinum counter electrode, saturated calomel electrode (SCE)
and a nitrogen bubbler. The specimen surface was flat and faced upwards.

The electrolyte was 1M NaCl at $25 \pm 1°C$. The localized environments were generated by dissolving the recessed metal at +400 mV (SCE); uniform dissolution was rapidly achieved through coalescence of pits. Several tests were carried out using lower pre-dissolution potentials (e.g. +100 mV) to ensure that high oxidation states such as Fe^{3+} were not being generated at +400 mV. Once a salt film is established, it absorbs a large part of the applied potential (ref. 1) and the dissolution is diffusion controlled. A syringe with a hypodermic needle was used to flush the pit between experiments. Cyclic voltammetry on a platinum wire (275 μm diameter, tip exposed) was used to analyze the pit solutions. Corrections were made for the IR error introduced by the current flowing from the alloy electrode. The electrochemical instrumentation consisted of two Stonehart BC 1200 potentio-stats, a PAR 175 function generator and a Nicolet digital oscilloscope (model 2090). A simulated pit solution was prepared (2.8M $FeCl_2$, 0.7M $CrCl_3$, 0.35M $NiCl_2$) and pre-electrolyzed between a rotating platinum cathode and a station-ary anode. Its pH was initially 1.1, and fell to 0.1 within a few days.

RESULTS AND DISCUSSION

pH and Cation Content of the Pit Environment

Cyclic voltammetry of the platinum microelectrode revealed differences be-tween the alloys when the electrode was positioned as near as possible to the bottom of a pit undergoing diffusion-controlled dissolution. For the high purity alloy, the result indicated a pH approximately equal to that of the simulated pit solution after aging (Figure 1a-b). The only other clear feature was the Fe^{2+}/Fe^{3+} reaction. For the stainless steel, however, a pair of features appeared at about -350 and -200 mV, indicating cathodic and anodic processes respectively (Figure 1c). These results can be interpreted as follows:

1. The pH of the near-saturated pit environment is about 0.5. The complex hydrogen evolution kinetics are probably due to underpotential deposition of nickel and/or iron (ref. 4).

2. In the pit grown on the commercial stainless steel, there is cathodic deposition of a constituent of the solution in the pit.

3. This constituent shows a rather reversible deposition-dissolution behavior with a reversible potential close to -270 mV (SCE).

Examination of the composition of the steel indicated that Cu ions could be re-sponsible for this behavior. When the Pt wire was held at -340 mV for 90 min in an active pit, subsequent energy dispersive X-ray analysis detected high concentrations of Cu, Cr, Si and Cl. The Cr, Si and Cl are presumably present as hydrolysis products; the Pt electrode reduces H^+ ions, and therefore locally raises the pH.

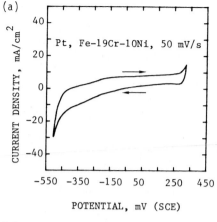

(a)

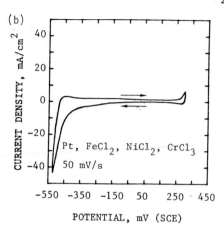

(b)

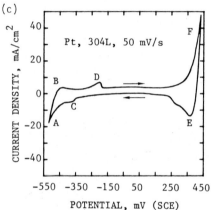

(c)

Fig. 1. Cyclic voltammograms for Pt in
near-saturated metal chloride solutions
(a) In artificial pit on high purity
    alloy.
(b) Simulated pit solution.
(c) In artificial pit on 304L stainless
    steel. A,B : $H^+ \rightleftharpoons \frac{1}{2}H_2$
            C,D : $Cu^+ \rightleftharpoons Cu$
            E,F : $Fe^{3+} \rightleftharpoons Fe^{2+}$

Copper Ions in the Pit Environment

Data for the Cu(I) complex $CuCl_2^-$ (ref. 5) indicate a stability constant
of ~10^6. Using other data from Pourbaix (ref. 6), we obtain:

$$Cu + 2Cl^- = CuCl_2^- + e^- \quad \ldots \quad E_{o,H} = 0.164 + 0.059 \log_{10}[a(CuCl_2^-)/a(Cl^-)^2] \text{ V} \quad (1)$$

Taking [Cl^-] as 8M, and the mean ionic activity coefficient as 1.7 (ref. 7), we
obtain E_o= -308 mV (SCE) for 0.013 M $CuCl_2^-$. This compares reasonably well with
-270 mV (estimated from Figure 1c), which contains an unknown diffusion
potential.

When a stainless steel wire was placed in the simulated pit solution, it
adopted a stable potential of -410 mV. This contrasts with the open circuit
potential decays for presaturated pits (Figure 2). The potential of the
stainless steel remained close to -270 mV for about 1000 s, while the concentration of dissolution products decreased by a factor of about 50 (ref. 8).
This is due to copper deposition. The high purity alloy quickly adopted a low
potential, but if a 250 µm copper wire was simultaneously dissolved near the

bottom of the pit (-100 mV, 10 μA, 60 s), the open circuit behavior was similar to that of the stainless steel. Figure 3 shows cathodic polarization curves for active pits; copper deposition precedes hydrogen ion reduction on the steel, but not on the high purity alloy.

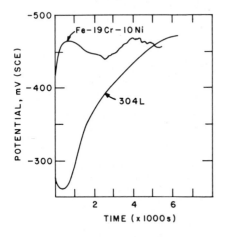

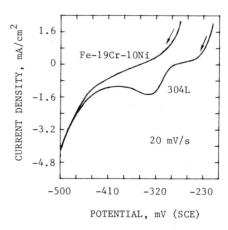

Fig. 2. Open circuit potential variation for artificial pits, 0.85 mm deep, after reaching diffusion steady state at +400 mV.

Fig. 3. Cathodic portions of polarization curves for 0.85 mm artificial pits after reaching diffusion steady state at +400 mV.

Active Dissolution Kinetics in Pit Environments

The artificial pit technique produces a roughly linear concentration gradient of metal ions. This is modified by electrical forces and by the decreased ionic diffusivity in concentrated solutions (ref. 2). If a lower potential is subsequently applied, and the new dissolution rate is less than the diffusion-controlled rate, the dissolution kinetics can be determined over a range of metal ion concentrations (and therefore pH, via Cr^{3+} hydrolysis). Figure 4 shows current density/potential curves for near-saturated pit solutions. By extrapolating the Tafel line for the stainless steel to below the corrosion potential (controlled by Cu^+ reduction), one can rationalize the occurrence of crevice corrosion between -400 and -300 mV (ref. 9). This crevice corrosion cannot involve Cu dissolution; Cu^+ is therefore not available as a cathodic reactant in active crevices, whereas it is present in pits and influences the "protection potential".

The concentration of dissolution products during localized corrosion is controlled by the dissolution kinetics (as modified by solution resistance) and the diffusion boundary conditions (which vary with time). When a 1-dimensional pit of depth δ is undergoing diffusion-controlled dissolution and the current density drops to a much lower value, the <u>initial</u> variation of the interfacial

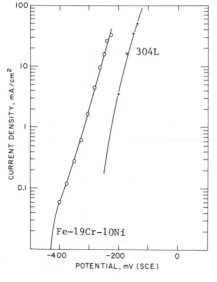

Fig. 4. IR-corrected polarization diagrams for artificial pit electrodes in approximately 95% saturated solutions of dissolution products, measured by pulsing potential after reaching diffusion steady state at +400 mV.

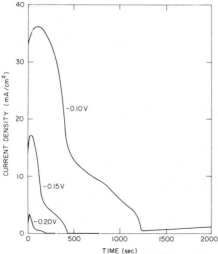

Fig. 5. Variation of current density for 0.8 mm artificial pit on stainless steel, at 3 different potentials after pulsing from steady state at +400 mV.

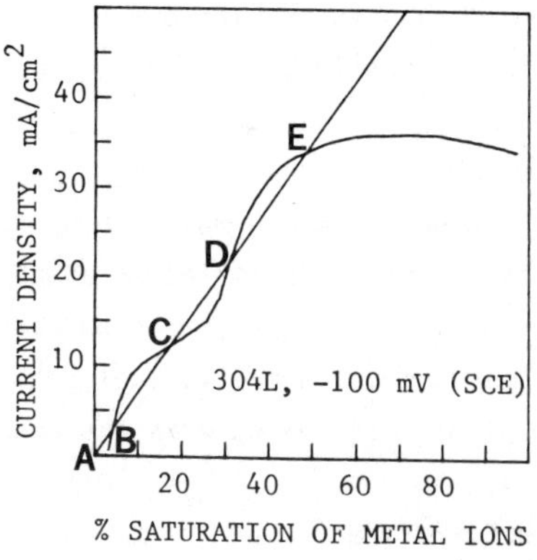

Fig. 6. Approximate concentration variation of current density at -100 mV (from Fig. 5 and equation 3), showing location of multiple steady states A to E.

concentration C_o of metal ions is not much affected by the new, lower, current density. We therefore have, from the diffusion equation:

$$C_o = 8C_s/\pi^2 \sum_1^\infty \frac{1}{(2n-1)^2} \exp\left\{-\left[\frac{(2n-1)\pi}{2\delta}\right]^2 Dt\right\} \tag{2}$$

where C_s is the initial (saturated) concentration. This expression uses a constant diffusivity D, which is not strictly accurate for concentrated solutions. However, the high order terms in the Fourier series (n>1) partially compensate for this at short times. For practical purposes, we obtain a useful rule:

$$\log_{10}(C_0/C_s) \simeq -t/\tau \cdots \cdots \text{ where } \tau = \delta^2/D \tag{3}$$

We have previously presented potentiostatic current decays for a pit on
the high purity alloy (ref. 8). The results for the stainless steel were more
complex (Figure 5), and imply an important multiplicity of steady states.
Figure 6 shows one of these current decays with the time axis converted to concentration using equation 3. The straight line superimposed on this curve
defines steady states for one-dimensional diffusion with a constant diffusivity
from a localized corrosion site of a particular geometry. The steady states of
the diffusion-coupled dissolution system are therefore the intersections of the
straight line and the curve. The line has been drawn so that there are no less
than five steady states. The states A, C and E are attractors--that is, they
are stable against small fluctuations in the system parameters. The states B
and D are at best metastable--that is, if placed at B or D the system would
eventually pass to A, C or E. With the incorporation of complicating factors
such as moving boundary conditions or time-varying values of current density at
constant concentration, such a system could rise to oscillations or pseudo-
random fluctuations. These may be a component of electrochemical "noise" in
localized corrosion.

CONCLUSIONS

1. Copper ions are mainly responsible for the anomalous cathodic behavior
of pits on type 304L stainless steel.

2. Multiple steady states and pseudo-random electrochemical noise during
localized corrosion are logical consequences of the way in which the
dissolution rate varies with the concentration of dissolution products.

ACKNOWLEDGEMENTS

This work was supported by the U. S. Department of Energy, Division
of Basic Energy Sciences, under contract DE-AC02-CH00016.

REFERENCES

1 H.S. Isaacs, J. Electrochem. Soc., 120 (1973), 1456.
2 J.W. Tester and H.S. Isaacs, J. Electrochem. Soc., 122 (1975), 1438.
3 H.S. Isaacs and R.C. Newman, in Corrosion and Corrosion Protection, The
   Electrochemical Society, 1981, p. 120.
4 D.M. Drazic and L.Z. Vorkapic, Corros. Sci., 18 (1978), 907.
5 L.G. Sillen and A.E. Martell, Stability Constants of Metal Ion Complexes,
   Supplement Number 1, The Chemical Society, London, 1971, p. 176.
6 M. Pourbaix, Atlas of Electrochemical Equilibria in Aqueous Solutions,
   Pergamon Press, London, 1966, p. 384.
7 E. McCafferty, J. Electrochem. Soc., 128 (1981), 39.
8 R.C. Newman and H.S. Isaacs, J. Electrochem. Soc., (1983, in press).
9 J.N. Wanklyn, Corros. Sci., 21 (1981), 211.

Passivity of Metals and Semiconductors, edited by M. Froment
Elsevier Science Publishers B.V., Amsterdam — Printed in The Netherlands

INFLUENCE OF ELECTRODE PRETREATMENT ON THE PITTING SUSCEPTIBILITY OF NI

B. MACDOUGALL

National Research Council of Canada, Ottawa, Ontario, Canada (K1A OR9)

ABSTRACT

The pitting susceptibility of nickel electrodes prepassivated in pH 4.0 Na_2SO_4 is determined by measuring the induction time, τ_{ind}, for pitting at a constant potential in 0.08 M Cl^-. For constant pretreatment conditions, τ_{ind} varies exponentially with potential so that $1/\tau_{ind} \propto \exp V$. Reproducibility of induction times is quite good and this permits a careful study of the influence of prior electrochemical treatment on the pitting susceptibility of nickel. The resistance to pitting increases with increasing time and potential of prepassivation in Cl^--free solution, i.e. with increasing perfection of the passive oxide film. Pretreatment in the transpassive region leads to an increase in pitting susceptibility, in agreement with the increased defectiveness of the prior oxide. The results suggest that a potential dependent modification of the oxide film occurs prior to pitting in the 0.08 M Cl^- solution and that this modification governs the induction time. The rate of this modification, which is a measure of the pitting susceptibility, is highly dependent on the prior state of the oxide film. A simple Cl^- incorporation into the oxide lattice does not appear to be the mechanism of oxide modification since prior incorporation of Cl^- actually increases the pitting resistance.

INTRODUCTION

The pitting susceptibility of a metal in an aqueous Cl^- solution depends on such parameters as $[Cl^-]$, solution pH and temperature, electrode pretreatment and the potential at which the pitting experiment is conducted. Frequently, pitting experiments are performed by keeping the $[Cl^-]$, solution pH and temperature constant and potentiostatically measuring the induction time for pitting at different potentials (ref. 1-4). The induction time changes very rapidly with anodic potential and this has been suggested as a possible reason for the observed scatter in the induction time data (ref. 1). Several workers have criticized the standard method for obtaining pitting induction times (ref. 5-7), arguing that the change in electrode potential changes both the electrochemical driving force for pitting and, simultaneously, the state of the surface oxide. Results obtained with stainless steel electrodes indicated that the susceptibility to pitting increased with increasing stability of the oxide towards open-circuit dissolution (ref. 5). While these results were explained in terms of the defect state of the oxide (and the presence or absence of "bound H_2O"), the observed changes could have been simply due to a potential

dependent variation of the oxide composition since an Fe-Cr-Ni alloy was being
studied.

In the present work, the influence of state of the surface oxide film on
pitting susceptibility is investigated by using a pure nickel electrode where
it is known that the composition of the passive oxide in neutral and acid solutions is NiO (ref. 8). The only parameter that appears to change with
potential and time of anodization is the defect state of the oxide (ref. 9),
making nickel an excellent system for studying the influence of oxide defect
density on pitting susceptibility. The significance of the results in terms of
the mechanism of pit initiation will be discussed.

EXPERIMENTAL

All experiments were performed in pH 4.0 Na_2SO_4 at 25°C either with or
without added Cl^- and all potentials are quoted against the saturated calomel
electrode (SCE). 5 cm^2 nickel samples were electropolished in 57 vol. % H_2SO_4
before stepping the potential into the passive region. After this pretreatment, the pitting susceptibility was determined in one of two different ways:
(i) the potential was adjusted to some lower value in the passive region (where
pitting does not occur), sufficient NaCl was added to give a 0.08M Cl^- solution
and the potential was then stepped to the value at which induction time, τ_{ind},
was to be measured; (ii) the electrode was removed from the solution and
reimmersed in a fresh 0.08M Cl^- solution (pH 4.0) at a fixed potential and τ_{ind}
measured. In this way, τ_{ind} at a constant pitting potential could be measured
as a function of extent of prior anodic treatment in a Cl^--free solution.

RESULTS AND DISCUSSION

(i) Method of determining pitting susceptibility of prepassivated surfaces

Fig. 1 shows the current-time response of a nickel electrode which was
prepassivated at 0.67V for 15 min. and then reimmersed in a 0.08M Cl^- solution
at 0.49 V. Pit inititation is indicated by the sharp rise in anodic current 27
min. after immersion. Induction times obtained in this way are shown as a
function of potential of reimmersion in Fig. 2, a constant pretreatment condition of 15 min. at 0.67V being used in each case. The results obtained on two
different 5 cm^2 Ni electrodes are shown and, while there is some scatter of the
induction time data, the reproducibility is surprisingly good. The induction
time for prepassivated electrodes is a sensitive function of the potential at
which the pitting experiment is performed, increasing exponentially with
decreasing anodic potential. It should be noted that τ_{ind} can change by
several orders of magnitude for a potential change as small as 0.03V, a result
that can easily give rise to an apparent critical pitting potential if the

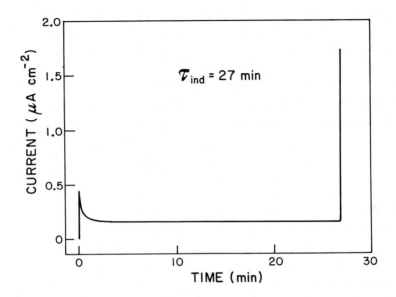

Fig. 1. Current-time response for a prepassivated Ni electrode in 0.08M Cl⁻ at 0.49V; prepassivation was in pH 4.0 Na_2SO_4 at 0.67V for 15 min.

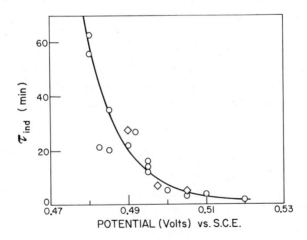

Fig. 2. Induction time for pitting, τ_{ind}, versus potential of anodization in 0.08M Cl⁻; samples pretreated at 0.67V for 15 min. in pH 4.0 Na_2SO_4. Results from two different 5 cm² Ni samples are shown.

pitting experiments are not done with sufficient sensitivity. The magnitude of τ_{ind} at a particular potential will be used as a measure of the pitting suscep- tibility of the electrode and in this way the influence of prior treatment on susceptibility will be studied.

(ii) Influence of electrode pretreatment on pitting susceptibility

A very important question in pitting concerns the nature of τ_{ind} and its exponential dependence on potential for a prepassivated surface, ie. why does it take 1 hr. to pit at 0.48V but only 2 min. at 0.52V (Fig. 2) and what processes are occurring during the 1 hr. time period at 0.48V? Is the pitting susceptibility of the system increasing during the potentiostatic holding at 0.48V so that after 1 hr. the susceptibility has reached the point where pit- ting can occur? If this is indeed the case, the most likely property of the system which is changing during the induction time is the nature of the passive oxide film, eg. by aging or Cl^- uptake during exposure to the 0.08M Cl^- solution. The rate of such changes should be potential dependent and this could explain the strong influence of potential on τ_{ind}. In terms of an aging of the film, it is interesting to note that the anodic passive current continues to decrease with time even in the 0.08M Cl^- solution indicating that the state of perfection of the oxide is increasing (10). This would seem to mean that the more perfect the film in terms of its defect density, the more susceptible it is to pitting and that τ_{ind} is simply the time required to reach a critical state of film perfection (cf ref. 5-7).

To check this interpretation, the pitting susceptibility under constant pit- ting conditions was determined as a function of prior electrode treatment in a non- Cl^- solution. The results are given in Table I for various prepassivation times at both 0.505V and 0.67V, pitting susceptibility being determined in all cases by measuring τ_{ind} in 0.08M Cl^- at 0.505V. It is evident that the pitting

TABLE I

Influence of electrode pretreatment in pH 4.0 Na_2SO_4 on pitting susceptibility of Ni; τ_{ind} measured in 0.08M Cl^- at 0.505V.

Potential and Time of Pretreatment	τ_{ind} (min)
0.505V for 15 min	1.5
0.505V for 11 hrs	20
0.67V for 15 min	5
0.67V for 3 hrs	15
0.67V for 11 hrs	47;56

susceptibility is influenced by electrode pretreatment, decreasing with increasing strength of anodic prepassivation treatment. Since these potentials are within the passive region, the susceptibility to pitting is decreasing with increasing state of perfection of the oxide film. This is just the opposite result to that observed by Okamoto et al. (ref. 5-7) for 304 stainless steel and it is not compatible with the first explanation for τ_{ind} given above, ie. in terms of an aging or perfecting of the film.

(iii) Influence of prior Cl⁻ treatment on pitting susceptibility

Another possible explanation for the existence of a potential dependent induction time is that exposure to the 0.08M Cl⁻ solution results in a time dependent modification of the film which increases the pitting susceptibility. The induction time would then be the time needed for Cl⁻ to modify the film to the point where pitting can occur, eg. by incorporation into the oxide lattice. It is known from previous work that when a passive film is exposed to a Cl⁻ solution, Cl⁻ does incorporate into the NiO lattice; the extent of this incorporation is potential dependent and its rate decreases with increasing perfection of the prior passive film (ref. 11). While these observations are consistent with the results reported here on the influence of electrode pretreatment on pitting susceptibility, and the mechanism itself is certainly a reasonable one, experiments where Cl⁻ was purposely incorporated in the film by preanodization in a 1.0M Cl⁻ solution indicated that such films were actually more resistant to pit initiation (ref. 12). In fact, for a constant τ_{ind}, the films pretreated in a Cl⁻ solution had pitting potentials almost 0.05V more anodic than those treated in Cl⁻-free solutions. This is a significant change in the pitting susceptibility and in the wrong direction from that suggested by the above mechanism. It does not appear that simple incorporation into the oxide film is the explanation for the observed induction times. Further experiments, eg. analysis of the film just prior to pitting by SIMS and XPS , are required to identify the exact nature of the potential dependent modification of the oxide film.

(iv) Influence of experimental technique on pitting susceptibility

In the section on experimental procedure, it was stated that the pitting susceptibility was determined in two different ways. All the results presented so far have been with reimmersion in the 0.08M Cl⁻ solution, after pretreatment, at that potential where τ_{ind} is being measured. The other approach is not to remove the electrode at all but to lower the potential after passivation to 0.42V and then add the Cl⁻ and, after it is well mixed, step the potential to that value at which τ_{ind} will be measured. This procedure also gives a

280

potential dependent τ_{ind}, however, the profile of Fig. 2 is shifted by ~0.035V in the anodic direction, ie. pitting susceptibility is somewhat lower with this procedure. The reason for this change in susceptibility is not clear but it is probably associated with a difference in the nature of the oxide film. Whatever the reason, it should be noted that the absolute value of pitting susceptibility is strongly dependent on the nature of the oxide film and subsequently on the experimental procedure adopted. Table II shows the results obtained using this procedure, for τ_{ind}'s measured as a function of prior electrode treatment in a non- Cl⁻ solution. The trends are exactly the same as those given in Table I (using the reimmersion procedure) and indicate that the resistance to pit initiation increases with increasing state of perfection of the oxide film. The increase in pitting susceptibility noted at 0.74V is understandable since this potential is close to that for oxygen evolution (0.745V vs. SCE) and it may therefore be in the early stages of transpassive dissolution. This would result in a decrease in oxide perfection and the corresponding increase in pitting susceptibility.

TABLE II

Influence of prior electrode treatment in pH 4.0 Na_2SO_4 on the τ_{ind} for pitting of nickel; pitting was initiated at 0.54V (vs. SCE) in 0.08 M Cl⁻.

Prior Electrode Treatment	τ_{ind}
5 min at 0.54V	30 sec
15 min at 0.54V	2.5 min
3½ hr at 0.54V	12 min; 12.5 min
3½ hr at 0.64V	70 min
10 3/4 hr at 0.74V	4 min

REFERENCES

1 K.E. Heusler and L. Fischer, Werkstoffe and Korrosion 27, 551 (1976).
2 M. Janik-Czachor, ibid. 30, 255 (1979).
3 T. Shibata and T. Takeyama, Corrosion 33, 243 (1977).
4 N. Sato, T. Nakagawa, K. Kudo and M. Sakashita, Trans. Japan Inst. Metals 13, 103 (1972).
5 Go Okamoto, Corrosion Sci. 13, 471 (1973).
6 Go Okamoto and T. Shibata, ibid. 10, 371 (1970).
7 H. Saito, T. Shibata and Go Okamoto, ibid. 19, 693 (1979).
8 B. MacDougall, D.F. Mitchell and M.J. Graham, Corrosion 38, 85 (1982).
9 B. MacDougall, J. Electrochem. Soc. 127, 789 (1980).
10 B. MacDougall, ibid. 130, 114 (1983).
11 B. MacDougall, D.F. Mitchell, G.I. Sproule and M.J. Graham, ibid. 130, 543 (1983).
12 B. MacDougall and M.J. Graham, ibid., in preparation.

Passivity of Metals and Semiconductors, edited by M. Froment
Elsevier Science Publishers B.V., Amsterdam — Printed in The Netherlands

A KINETIC THEORY OF PIT INITIATION AND REPASSIVATION AND ITS RELEVANCE TO THE PITTING CORROSION OF PASSIVE METALS

T. OKADA

Industrial Products Research Institute, M.I.T.I., Tsukuba, Ibaraki 305 (Japan)

ABSTRACT

A kinetic approach is used to test the theory of pit initiation and repassivation in the pitting corrosion of passive metals. It follows from consideration of irreversible thermodynamic stability that the pit initiation process is a probability event in which hemi-spherical halide nuclei grow to a critical radius r* in a passive film in order for pitting to take place. The repassivation process starts as soon as the monolayer formation of the metal oxide becomes stable at the bottom of the pit. It is shown that the two characteristic values of the pitting phenomenon, pit initiation and the repassivation potential are intrinsically different from each other, and coincide only when the passive film becomes very thin. The present study suggests the theoretical significance of the repassivation potential as a critical value of pitting.

INTRODUCTION

The anodic breakdown of passivity in a solution containing aggressive ions X^- ($X^- = Cl^-, Br^-, I^-$), leading to pitting has been the subject of much attention (ref. 1,2). Theoretical considerations include X^- adsorption (ref.1), X^- penetration through passive film flaws (ref.2-4), mechanical factors (ref.5), ion transport (ref.6) and concentration polarization (ref.7,8). However, very few of the theories so far proposed can satisfactorily explain the kinetic features of the breakdown phenomena of passivity. For example, these theories cannot explicitly predict induction time (ref.2), pit localization or the appearance of the breakdown and the repassivation potential at different values (ref.9,10).

The "transitional halide complex" proposed by Hoar and Jacob (ref.11), and "two-dimentional halide islands" proposed by Heusler and Fischer (ref.12) are attractive models because of their microscopic nature. However the mechanism of halide penetration into the passive film is still not yet clarified.

This study was intended to kinetically explain the nature of pit initiation and the repassivation processes and to determine the conditions under which pitting occurs, from the irreversible thermodynamic point of view.

Critical pitting potential

Suppose that an oxide is just about to form on a metal surface at the bottom of a pit, where the metal is strongly corroding. If the rate of formation i_1 of the monolayer of the oxide M → MO overcomes the rate i_2 where it changes to the halide MO → MX, this would bring about an inhibition of the pitting reaction. The potential E_{cp} at which $i_1 = i_2$ is uniquely determined as

$$E_{cp} = \frac{RT}{\alpha_{MO}zF} \, ln(\frac{k_{OX}}{k_{MO}}) + \frac{1}{\alpha_{MO}zF} [\, \mu^*_{MO} - \{ \frac{z(\alpha_{MO} + \alpha_{OX})}{2} + \nu_W \} \mu^*_W + \nu_X \mu^*_X + z(\alpha_{MO} + \alpha_{OX})\mu^*_H \,]$$
$$+ \frac{RT}{F} \, ln \frac{\gamma_X \, c^*_X}{\gamma^s_X \, c^s_X} \tag{1}$$

in which k_{MO}, k_{OX} and α_{MO}, α_{OX} = rate constants and the transfer coefficients for i_1 and i_2, respectively, μ^*_{MO} = chemical potential of $M_2O_{z(\alpha_{MO} + \alpha_{OX})}$ at the metal surface, μ^*_W, μ^*_X and μ^*_H = critical chemical potentials of H_2O for formation of MO, of X^- and H^+ for pitting, respectively, c^*_X and c^s_X = the critical X^- concentration and the X^- concentration in the bulk of the solution, ν = stoichiometric numbers and γ = activity coefficients. E_{cp} will be more noble than the "Flade potential", because of aggressive environment of the pit (ref.13). The critical situation at $E = E_{cp}$ resembles the case of anodic brightening, as Hoar $et.$ $al.$ (ref.3) pointed out, but the non-protective film which covers the dissolving metal is not the same for each case.

Pit initiation

At $E > E_{cp}$, pitting can occur. However, if it starts with metal passivated beforehand, the aggressive ion must break through a pre-existing passive film of thickness L. The hemi-spherical halide nuclei are supposed to form on the surface of the passive film and grow inwards. Whether the nuclei grow continuously or not can be determined by referring to the general evolution criterion of irreversible thermodynamics (ref.14).

$$\delta_X P = \int \Sigma_k J_k \, d_t X_k \, dV \leqslant 0 \tag{2}$$

where J_k = flux and $d_t X_k$ = the time derivative of the conjugated force at fixed boundary conditions.

Designating M^{z+} and O^{2-} ions as M and O respectively, the flux within the oxide film is expressed as (ref.15)

$$J_i = \frac{z_i}{|z_i|} f_i n_i a \, exp \, (- \frac{U_i - \alpha_i |z_i| \, F a \Delta\Phi_0/\delta}{RT}) \quad , \; i = M,O \tag{3}$$

where f_i = vibrational frequency, z_i = valence of i, n_i = concentration of i, a = "jump distance", α_i = "transfer coefficient", $\Delta\Phi_0$ = potential difference within the oxide layer and δ = thickness of the oxide layer. The force corresponding to each flux is, assuming a strong field,

$$X_i = \frac{1}{T} z_i F \frac{\Delta\Phi_O}{L - r'} \quad , \quad i = M, O \tag{4}$$

in which r' = the "thickness" of the metal halide in the simplified model: $r' = \gamma r$, and γ = the ratio of the area occupied by the halide nuclei on the surface of the passive metal.

The partial current densities for the lattice metal and oxygen dissolution at the interface between the passive oxide and the halide nuclei can be expressed as

$$i_M = k_M \exp\left(\frac{\mu_H}{RT}\right) \exp\left(\frac{z\mu_X}{RT}\right) \exp\left(\frac{\alpha_M zF\Delta\Phi_{MO/MX}}{RT}\right) \tag{5.1}$$

$$i_O = k_O \exp\left(\frac{\mu_O}{RT}\right) \exp\left(\frac{2\mu_H}{RT}\right) \exp\left(\frac{2\alpha_O F\Delta\Phi_{MO/MX}}{RT}\right) \tag{5.2}$$

where μ_M and μ_O = chemical potentials of M^{z+} and O^{2-}, respectively, in the passive film, μ_X and μ_H = chemical potentials of X^- and H^+, respectively, in the halide nuclei and $\Delta\Phi_{MO/MX}$ = the Galvani potential difference at the interface.

Transport of M^{z+}, X^-, H^+ and H_2O (designated as W) occurs in the halide nuclei. A linear relation is assumed between the fluxes and forces (ref.16) because of the high ionic conductance of the halide:

$$J'_i = L'_{iW} X_W + L'_{ik} X_{k \neq i} + L'_{ii} X_i \quad , \quad i = W, M, X, H \tag{6}$$

in which L' = phenomenological coefficients, and

$$X_i = \frac{1}{T} \frac{\Delta\tilde{\mu}_i(MX)}{r} = \frac{1}{T}\left\{ \frac{\Delta\mu_i(MX)}{r} + \frac{z_i F\Delta\Phi_X}{r} \right\} \quad , \quad i = W, M, X, H \tag{7}$$

and $\Delta\mu_i(MX)$ = the average chemical potential difference of i, and $\Delta\Phi_X$ = the potential difference, within the halide. For simplicity, it is assumed that

$$L'_{iW} \frac{X_W}{X_i} + L'_{ik} \frac{X_{k \neq i}}{X_i} + L'_{ii} = \text{const} = T\bar{n}'_i v'_i \quad , \quad i = M, X, H \tag{8}$$

where $\bar{n}'_i$ and v'_i = the average concentration and the mobility per unit force of i, respectively, within the halide.

The partial current densities, i'_M and i'_X, of M^{z+} and X^- dissolution at the interface between the halide and the solution is expressed by a relationship similar to that shown in eq.(5).

Suppose that, at the passive film-solution interface, the film does not change its thickness, so that $\delta\Delta\Phi_{MO/S} \fallingdotseq 0$.

The total current I° from the metal to the solution can be divided into two parts: I, which flows through the halide nuclei into the solution, and I', which flows directly through the oxide-solution interface. The former is much larger than the latter. Assuming no accumulation of charge, the following continuity equations should hold:

$$I^\circ = I_M + I_O = (zFJ_M - 2FJ_O) S \tag{9.1}$$

$$I = 2\pi r^2 NS(i_M + i_O) = \pi r^2 NS(zFJ_M' - FJ_X' + FJ_H') = \pi r^2 NS(i_M' + i_X') \tag{9.2}$$

where S = the area of the metal which contacts the solution and N = the number of halide nuclei per unit area. The chemical potentials of the species are assumed to be nearly in equilibrium in the solid phase:

$$2\delta\mu_M + z\delta\mu_O = 0 \;, \quad \delta\mu_H + \delta\mu_X = 0 \;, \quad \delta\mu_M + z\delta\mu_X = 0 \;. \tag{10}$$

Ignoring the flux of H_2O, and further assuming that

$$I = A\pi r^X NS \;, \quad A \doteqdot i = k_X(C_X)^n \, exp\left(-\frac{\xi FE}{RT}\right) \tag{11}$$

where i = the dissolution current density at the halide nuclei, χ = a constant ranging from 1 to 2 and E = potential of the metal in the solution, then eq.(2) can be expressed as follows, for the increase δr in the radius of the nuclei

$$\delta_X P = F(L,r,i)\,\delta r \leqslant 0 \tag{12.1}$$

with $F(L,r,i) \equiv \dfrac{I}{r}\dfrac{R}{F}\left[\lambda_L \chi\left(\dfrac{L}{a} - \dfrac{\gamma}{a}r\right) - (2-\chi)\lambda_I - \dfrac{\lambda_X}{\Lambda_X NS}\dfrac{F}{RT}i\pi r^{X-1}\right]$ (12.2)

where λ = positive constants of the order of magnitude of 1, and $\Lambda_X \equiv z^2 F^2 \bar{n}_M' v_M' + F^2 \bar{n}_X' v_X' + F^2 \bar{n}_H' v_H'$, the specific conductance of the halide. In the case of $\chi = 2$, where the current coming out of the metal substrate concentrates in the halide nuclei, the critical radius r* which the halide nuclei must surpass for growth and which leads to pitting, is obtained as

$$r* = \frac{2\lambda_L \dfrac{L}{a}}{2\lambda_L \dfrac{\gamma}{a} + \pi \dfrac{\lambda_X}{\Lambda_X}\dfrac{F}{RT}i} \;. \tag{13}$$

Pitting will begin when r* falls short of r_{max}, the maximum statistically attainable value, and occurs most easily when the passive film is thin and the specimen is more anodically polarized. The pit initiation potential E_{pi} is expressed as

$$E_{pi} = const. - \frac{nRT}{\xi F}\, ln\, C_X \;. \tag{14}$$

The induction time τ is the time required for the halide nuclei to become larger than the size r*:

$$ln \ \tau = const. - 2 n \ ln \ C_X - \frac{2\xi FE}{RT} \ .$$ (15)

Repassivation

Suppose that pitting is in progress with a non-protective film of thickness l at the pit bottom. H_2O in this film plays an important role (ref.17) for pit repassivation. Consider that the oxide film, with an area s_0 and a thickness d, is produced at the pit bottom in contact with the metal surface. In the case where d changes to d+δd, with s_0 constant, neglecting the change in $\Delta\Phi$ at the oxide film-hydrous halide and the hydrous halide-solution interfaces and in the diffusion layer in the solution,

$$T \ \delta_X^d P = F^d(l,d,i,J_W')s_0 \delta i + G^d(i,J_W') \ is_0 \ \delta(l-d)$$ (16.1)

$$\text{with } F^d(l,d,i,J_W') \equiv \frac{RT}{F} \{ \lambda_0 \frac{d}{a} + \frac{i(l-d)}{\Lambda_X} \} + (\lambda_{OW} \frac{i_0}{F} + J_W') \ c_I' \ (l-d)$$ (16.2)

$$G^d(i,J_W') \equiv (\lambda_{OW} \frac{i_0}{F} + J_W') \ c_I' - \lambda_{WI} J_W' c_I' - \frac{\lambda_X i}{\Lambda_X}$$ (16.3)

where λ = positive constants of the order of magnitude of 1, J_W' = the flux of H_2O through the hydrous halide and c_I' = the interaction coefficient of the chemical potential gradient of H_2O with the current. The electrical conductance of the passive film is assumed to be very low, so that

$$\delta i = - \frac{zF}{RT} \ \frac{\alpha a}{d} \ \Delta\Phi_0 i \ \frac{\delta d}{d} \ .$$ (17)

Eqs.(16) and (17) result in $\delta_X^d P < 0$ for $\delta d > 0$, so that the oxide film grows in thickness stably until it reaches a steady state and stops growing.

Consider next, that the oxide extends its area from s_0 to $s_0+\delta s_0$, its thickness d remaining constant. At the δs_0 surface, the reaction M → MX at rate i_3 vanishes and the reaction M → MO at rate i_1 and MO → MX at rate i_2 ($=i_1$, because d is constant) appears. In this case eq.(2) is expressed by the sum of the components due to the change in the surface reaction and material transport as

$$T \ \delta_X^s P = F^s(l,d,i,J_W')(1-\varepsilon)(i_1 - i_3)\delta s_0 - \frac{1}{2} G^s(\Delta\Phi)(i_1 + i_3)\delta s_0$$ (18.1)

$$\text{with } F^s(l,d,i,J_W') \equiv (J_W' - \frac{i}{2F}) \ c_I' \ (l-d) + \frac{i}{\Lambda_X}(l-d) + \lambda \frac{RT}{F}$$ (18.2)

$$G^s(\Delta\Phi) \equiv - \frac{1}{F} \delta\mu_H - \lambda_I (1-\varepsilon) \ \delta\Delta\Phi_{MX/S} - (1-\varepsilon) \ \delta\Delta\Phi_S - (\Delta\Phi_{M/MO} - \Delta\Phi_{M/MX}) \ .$$ (18.3)

where $i = (i_1 + i_3)/2$, ε = the correction due to the tributary current in the shadowed area by s_0, and $\Delta\Phi_s$ = the potential drop in the diffusion layer in the solution. If the oxide film is very thin, $\Delta\Phi_{M/MO} = \Delta\Phi_{M/MX}$ and $T\delta^s_X P < 0$ for $\delta s_0 > 0$, so that the thin oxide film which is formed at the pit bottom extends stably.

The above considerations suggest that repassivation proceeds *via* the extension of the very thin oxide, followed by its thickening until a steady state is attained, resulting in the thorough suppression of the pitting reaction.

When $E < E_{cp}$ and $i_1 > i_2$ at d = monolayer, the oxide film starts to form at the pit bottom, though it is not sufficient to stop pitting until it brings about the condition $\mu_X < \mu_X^*$ and $\mu_H < \mu_H^*$ at the pit bottom. Then the current falls off and the metal is protected from dissolution.

It can thus be seen that the repassivation potential E_{pr} is very close to E_{cp}, whereas the pit initiation potential E_{pi} shifts more noble than E_{cp} by some tenths of a volt. This was confirmed experimentally for the pitting of stainless steel (ref.9) and of titanium (ref.10). Pit initiation of titanium was suppressed to a great extent when the passive film thickness increased. It is evident from eq.(13) that E_{pi} coincides with E_{cp} only when the passive film is very thin, as experimentally observed by the "scratch method" (ref.18). These results indicate the significance of the repassivation potential as a characteristic value of pitting (ref.9,10).

REFERENCES

1 Ja.M. Kolotyrkin, Corrosion, 19 (1963) 261t-268t.
2 M. Janik-Czachor, G.C. Wood and G.E. Thompson, Br. Corros. J., 15 (1980) 154 -161.
3 T.P. Hoar, D.C. Mears and G.P. Pothwell, Corros. Sci., 5 (1965) 279-289.
4 J.A. Richardson and G.C. Wood, Corros. Sci., 10 (1970) 313-323.
5 N. Sato, Electrochim. Acta, 19 (1971) 1683-1692.
6 H. Kaesche, Z. Physik. Chem. N.F., 34 (1962) 87-108.
7 D.A. Vermilyea, J. Electrochem. Soc., 118 (1971) 529-531.
8 N. Sato, J. Electrochem. Soc., 129 (1982) 260-264.
9 M. Pourbaix, Corrosion, 26 (1970) 431-438.
10 T. Okada, DENKI KAGAKU, 49 (1981) 584-588.
11 T.P. Hoar and W.R. Jacob, Nature, 216 (1967) 1299-1301.
12 K.E. Heusler and L. Fischer, Werkst. Korros., 27 (1976) 551-556.
13 T. Okada and T. Hashino, Corros. Sci., 17 (1977) 671-689.
14 P. Glansdorff and I. Prigogine, Thermodynamic Theory of Structure, Stability and Fluctuations, Wiley-Interscience, 1974, Chap.9.
15 T.P. Hoar, in J.O'M. Bockris (Ed.), Modern Aspects of Electrochemistry, Butterworths, London, No.2, 1959, 295 pp.
16 D.D. Fitts, Nonequilibrium Thermodynamics, McGraw-Hill, 1962, Chap.4.
17 T.P. Hoar, J. Electrochem. Soc., 117 (1970) 17c-22c.
18 N. Pessal and C. Liu, Electrochim. Acta, 16 (1971) 1987-2003.

Passivity of Metals and Semiconductors, edited by M. Froment
Elsevier Science Publishers B.V., Amsterdam — Printed in The Netherlands

A COMPARISON OF MODELS FOR LOCALIZED BREAKDOWN OF PASSIVITY

P. ZAYA[1] and M.B. IVES[2]

[1]Alcan International Research Laboratory, Kingston, Ontario, K7L 4Z4 (Canada)

[2]Institute for Materials Research, McMaster University, Hamilton, Ontario,
L8S 4M1 (Canada)

ABSTRACT

Mass transport equations are solved to compute the potential and concentrations at the metal surface at the base of a flaw in a passive film. Three criteria, corresponding to different theories of pitting corrision, have been investigated by simulating experimental conditions known to be critical for the onset of pitting corrosion. The results indicate that the most probable mechanism controlling pitting is the adsorption of chloride ions at the metal surface.

INTRODUCTION

Various authors (refs.1-5) have suggested that the initiation of corrosion pits does not depend only on the rupturing of the passive layer, but also on the opportunity for the flaw to subsequently heal. Several models have been proposed to explain the irreversible changes that take place within a damaged film and result in either repassivation or pitting (refs. 5-12).

The purpose of this work was to compare three such models by simulating the conditions prevalent at the bottom of a pore in a passive film, employing a modification of the simulations of Pickering and Frankenthal (ref. 7) and Galvele (ref. 10).

MODELLING

The geometry of the simulated pore is shown in Fig. 1. The passive film is considered to be inert. The metal sample is polarized at constant potential and we are considering the situation just after the rupture of the passive film when the metal at the

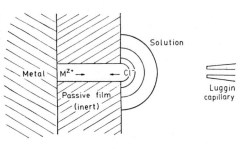

Fig. 1. Model for the simulations.

bottom has just contacted the solution. It is expected that this small area of bare metal (c. 1 nm^2) will be subjected to very high (c. 10^4 A/cm^2) current density[*] during a very short time (c. 1 μs) before being covered by an adsorbed layer or by a 3-dimensional film.

------

[*] Current densities of such magnitudes have been measured (ref. 18).

It is this initial period which has been studied here. It must be realized that the increase in current (10^{-9} A) will be too small and too brief to be recorded adequately. On the other hand, all the experimental data available (e.g. potentiostatic polarization) have been obtained after the initial period.

The dissolution of iron in NaCl - NaOH - $B(OH)_3$ has been simulated in this study. Chemical reactions occuring are metal hydrolysis, metal complexing with Cl^-, the decomposition of water and hydrolysis of boric acid.

The concentrations of all species present at all points in the pore and the adjacent electrolyte after initial contact are computed using mass transport equations involving diffusion, electro-migration and the products of chemical reactions, following Newman (ref. 13). Convection is neglected in the model. Boundary conditions were chosen such that the concentrations are constant in the unaffected region of the solution and that Faraday's Law applies to metal dissolution at the bottom of the pore.

The current is computed assuming that the only electrochemical reaction is the dissolution of the metal according to a Tafel-type relationship. For this purpose, the potential at the bottom of the pore is calculated by subtracting from the applied potential the ohmic drop, the potential drop due to the concentration gradient and the charging of the double layer (assumed to be in parallel with the dissolution).

Numerical values for the Faradaic current-potential relationship were obtained from the "bare surface polarization" experiments of Burstein and Davies (refs. 14,15). The Tafel slope was taken as $b_a = (2RT)/F$, and the function,

$$i = \{K_1 + K_2/(K_3 + [H^+]) + K_4 \cdot [Cl^-]\} \cdot \exp(E/b_a)$$

was fitted to the Burstein and Davies data to evaluate the constants.

The system of partial differential equations obtained from mass transport considerations is approximated by finite difference to another system of ordinary differential equations. This latter system is solved using the LSODE package (ref. 16) for the integration of stiff differential equations.

The original feature of this approach is that it allows the computation of the concentrations and potential at any time, unlike the previous models which assumed steady state (refs. 7,10). It also does not need to assume that the solution contains a supporting electrolyte, thereby giving a better estimate of the potential at the pore bottom. The models of Pickering and Frankenthal (ref. 7) and of Galvele (ref. 10) were used to determine the conditions of pit stability, while here we are studying the conditions of pit initiation.

PITTING THEORY CRITERIA

Three pitting theories were analyzed in the following manner. The solution compositions and potentials from reported experimental determinations of pitting potentials were used as initial conditions, and the situation at the bottom of the pore evaluated at pseudo steady-state. If a theory is correctly describing the mechanism controlling pitting corrosion in the range of variables considered, then the associated criterion should be constant, and independent of the particular set of experimental conditions (e.g. pH, Cl^- concentration in the bulk solution), since all the calculations were made for conditions corresponding to the pitting potential, i.e. conditions borderline between pitting and passivity.

In the following, $[Fe^{2+}]$, $[Cl^-]$, $[OH^-]$ and E are calculated concentrations and potential at the bottom of the pore.

The three theories under consideration are:

Localized acidification   Pitting will occur if the local pH is too low for hydroxide precipitation, that is

$$[Fe^{2+}] \cdot [OH^-]^2 / K_s(1) < K_{rp}$$

where $K_s(1)$ is the solubility product of $Fe(OH)_2$, K_{rp} is the supersaturation ratio.

Salt film formation   Pitting will occur if the solubility of the metal chloride is exceeded at the surface, precipitating a salt film,

$$[Fe^{2+}] \cdot [Cl^-]^2 / K_s(2) > K_{sf}$$

where $K_2(2)$ is the solubility product of $FeCl_2$, K_{sf} is the supersaturation ratio.

Aggressive ion adsorption   Pitting will occur if a critical coverage of Cl^- chemisorbed on iron is exceeded. The dissolution of iron is considered (ref.15) to take place according to the following reaction scheme,

$$Fe + Cl^- \rightarrow (FeCl)_{ads} + e^- \quad \text{(rate determining step)}$$

$$(FeCl)_{ads} \leftrightarrows Fe^{2+} + Cl^- + e^-$$

The criterion in this case would be

$$[Fe^{2+}] \cdot [Cl^-] \cdot \exp(EF/RT) > \theta \cdot \exp(E_o F/RT) = K_{ads}$$

where θ is the critical coverage, E_o the potential corresponding to the equilibrium reaction, and K_{ads} a parameter taken as the criterion for this mechanism

RESULTS

To demonstrate the type of information obtainable from the computer program simulations, Figs. 2 and 3 show the concentrations of each chemical species as a function of position in the pore and time. These results were obtained from an early test solution, comprising a simple 1M NaCl solution, and are introduced here for exemplary purposes only.

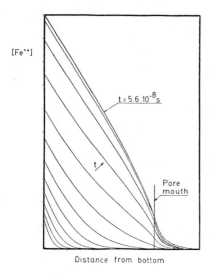

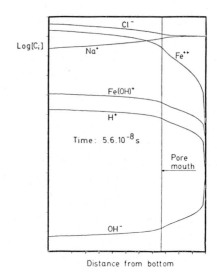

Fig. 2. Example of Fe^{2+} distribution as a function of position and time.

Fig. 3. Example of species distribution after reaching pseudo steady-state.

Twenty three simulations were performed in the Fe-NaCl-NaOH-borate system, using the experimental pitting potentials and solution conditions reported by Janik-Czachor (ref. 17) and Heusler and Fischer (ref. 8). A statistical analysis was performed on the decimal logarithm of the values obtained of K_{rp}, K_{sf} and K_{ads}, measured after approximate steady state was obtained at the pore bottom, with the following results:

	$log_{10} K_{rp}$	$log_{10} K_{sf}$	$log_{10} K_{ads}$
mean	4.90	0.454	0.286
standard deviation	1.37	1.72	0.611

The adsorption theory fits the results much better than the other two. Its associated standard deviation corresponds to a spread by a factor of 4, as compared with factors of 23 and 52 for the localized acidification and salt film theories, respectively.

The value of 4.90 for the average of $log_{10} K_{rp}$ would correspond to a supersaturation ratio for $Fe(OH)_2$ of approximately 100,000, which seems unrealistic. Furthermore, within each set of experiments there was a systematic variation of K_{rp} with chloride ion concentration, explaining the wide range of values obtained and the correspondingly large standard deviation.

Of the two other critieria, K_{ads} exhibited the smaller standard deviation, and does not vary appreciably with chloride ion concentration, unlike K_{sf}. This can be seen in Fig. 4, showing the variation of K_{ads} and K_{sf} for the experimental conditions of Janik-Czachor (ref. 17). The bars shown correspond to the ranges of pitting potentials reported in that work. Fig. 5 shows the

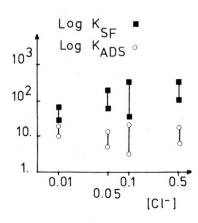 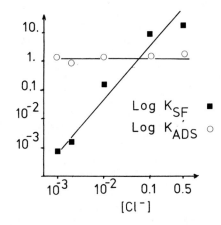

Fig. 4. Dependence of salt film and adsorption parameters on experimental conditions of Janik-Czachor (ref. 17).

Fig. 5. Dependence of salt film and adsorption parameters on experimental conditions of Heusler and Fischer (ref.8).

K_{sf} and K_{ads} values obtained from simulating the conditions of the pitting studies of Heusler and Fischer (ref. 8). Here it is clear that K_{ads} is a much more consistent parameter over a wide range of pitting potentials corresponding to different chloride ion concentrations.

CONCLUSIONS

From the values obtained for the parameter K_{rp} at the metal surface at the bottom of a defect in a passivating film, it must be concluded that in the real situation occuring in the short times (less than 1 µs) of these simulations the equilibrium between Fe^{2+} and its complexes with OH^- is not realized.

Of the three pitting theories considered, the adsorption of aggressive ion (Cl^-) critierion appears to give the best interpretation of the experimental data. This indicates that a necessary condition for the development of the flawed and breached film into a pit is the presence of a critical minimum coverage of adsorbed Cl^- on the metal surface at the flaw bottom. This is consistent with the view of Strehblow and Titze (ref. 11).

A precise definition of an adsorption criterion, including a specific critical concentration of Cl^-, must await the establishment of better kinetic data on the adsorption of chloride ions on iron surfaces at potentials appropriate to the local potential difference at the bared surface.

A more detailed description of the simulation program and its application to a range of pitting situations can be found elsewhere (ref. 19).

REFERENCES

1 V.M. Novakovski and V.M. Sorokina, Zaschita Metallov, 2 (1966) 416.
2 J.A. Richardson and G.C. Wood, Corrosion Science 10 (1970) 313.
3 G. Okamoto, Corrosion Science, 13 (1973) 471.
4 K. Videm, in Proceedings of the 5th. International Congress on Metallic Corrosion, NACE, Houston, 1974, p. 264.
5 K.J. Vetter and H.-H. Strehblow, Ber. Bunsenges. physik. Chemie, 74 (1970) 1024.
6 J.R. Ambrose and J. Kruger, Report NBSIR 74-583, National Bureau of Standards, U.S. Department of Commerce, Washington, D.C., 1974.
7 H.W. Pickering and R.P. Frankenthal, J. Electrochem. Soc., 119 (1972) 1297.
8 K.E. Heusler and L. Fischer, Werkstoffe u. Korros., 27 (1976) 551.
9 Y. Hisamatsu, in R.W. Staehle and H. Okada, eds., "Passivity and its Breakdown on Iron and Iron-based Alloys", NACE, Houston, 1976, p. 99.
10 J.R. Galvele, J. Electrochem. Soc., 123 (1976) 464.
11 H.-H. Strehblow and B. Titze, Corrosion Science, 17 (1977) 461.
12 B. MacDougall, J. Electrochem. Soc., 126 (1979) 919.
13 J.S. Newman, in "Electrochemical Systems", Prentice Hall, 1973, p. 228.
14 G.T. Burstein and D.H. Davies, J. Electrochem. Soc. 128 (1981) 33.
15 G.T. Burstein and D.H. Davies, Corrosion Science, 20 (1980) 1143.
16 A.C. Hindmarsh, "Livermore Solver for Ordinary Differential Equations", Lawrence Livermore Laboratory, Livermore, California, 94550, U.S.A.
17 M. Janik-Czachor, British Corrosion J., 6 (1971) 57.
18 S. Tajima and M. Ogata, Electrochim. Acta, 13 (1968) 1848.
19 P. Zaya, PhD Dissertation, McMaster University, Hamilton, Ontario, Canada, 1983.

Passivity of Metals and Semiconductors, edited by M. Froment
Elsevier Science Publishers B.V., Amsterdam — Printed in The Netherlands

STOCHASTIC ASPECTS OF MECHANICAL AND CHEMICAL BREAKDOWN OF PASSIVITY

C. GABRIELLI[1], F. HUET[1], M. KEDDAM[1], R. OLTRA[2] and C. PALLOTTA[1*]

[1] Groupe de Recherche n° 4 du CNRS "Physique des Liquides et Electrochimie",
associé à l'Université Pierre et Marie Curie, 4 place Jussieu
75230 PARIS Cedex 05, France.

[2] L.A. 23 "Réactivité des Solides", Professeur J.C. COLSON, 21000 DIJON, France.

ABSTRACT

Spectral analysis of the current fluctuations has been performed for two
cases of repassivation following the passive film breakdown. It has been shown
that this technique can provide information on the individual repassivation
transient and its interaction with the surrounding passive surface in the case
of mechanical depassivation. For pitting corrosion, current noise measurement
during pit growth reveals a substructure linked to micropits inside the principal one.

INTRODUCTION

Kinetics of repassivation following the film breakdown is regarded as determining for the corrosion resistance of passive metals against more or less
localized attacks.

Repassivation kinetics is tentatively estimated from the current-time transient immediately following the film breakdown. Two completely opposite
situations are usually investigated :

- single breakdown on a restricted area : i.e. single scratch,
- continuous depassivation by mechanical and chemical damages to the film...

The steady state reached during continuous depassivation is generally assumed to
arise from a large number of single depassivation-repassivation events occuring
at random.

The present work is aimed to characterize the breakdown and healing processes
over a wide range of experimental conditions and to investigate in which extent
their kinetics depends on the nature and the size of the film damages. An
approach by electrochemical noise measurement has been previously carried out
in the case of pitting (ref. 1).

---

[*] Fellow of the Consejo Nacional de Investigaciones Cientificas y Tecnicas,
Republica Argentina.

EXPERIMENTAL

As a rule, no experiment continuously covers with the same device the whole range from single independent to largely overlapping depassivation events. Two types of experiment were performed.

Mechanical abrasion of the passive surface is obtained by an impinging jet of SiC particles in suspension in the electrolyte (ref. 2). Particle impacts at the surface occur at random in time and space. The energy of an elementary impact can be controlled by varying the fluid velocity or the mass of a particle (Velocity of the order of 1 - 2 m s^{-1} and particle size 100 - 500 μm). The visible damage to the surface is restricted to small spots and the underlying plastic deformation is probably far less than in conventional single scratch or abrasion experiments. The average frequency of impacts on the target electrode is most easily fixed by the concentration of the particles in suspension in H_2SO_4,1M solution. This concentration is estimated by means of a phototransistor which measures the optical absorption of light across the pipe.

Chemical breakdown is triggered, as usual, by addition of Cl^- (3.10^{-3} M).

Cross section (0.2 cm^2) of Johnson-Matthey iron rods are used as working electrodes.

Low noise potentiostats are used. FFT spectrum analysis of the electrochemical current noise is performed by using a Hewlett Packard 5451 C Fourier analyser (ref. 3). The spectra given in this paper are characteristic of the electrochemical system and independent of the instrumentation noise except in the case of iron in chloride free solution where the potentiostat noise is detected in high frequencies (see fig. 7).

RESULTS

Mechanical breakdown

The measurement of the cross-spectrum of the current fluctuations during abrasion and of the fluctuations of the particle concentration obtained from optical absorption shows that the 1 Hz cut-off frequency is due to the fluctuations of the abrading particle density. If the impact times are assumed to follow a doubly stochastic Poisson law whose rate has a mean $\overline{\lambda}$ and a power spectral density(p.s.d.)$\psi_\lambda(f)$, the p.s.d. of the current under abrasion, given in fig. 1 can be written

$$\psi_I(f) = [\overline{\lambda} + \psi_\lambda(f)]|I(f)|^2$$

where I(f) is the Fourier transform of the elementary current transient due to one impact.

The spectrum of the current corrected for λ fluctuations by taking into

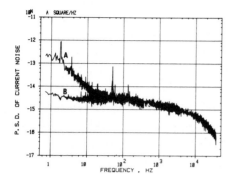

Fig. 1. P.S.D. of current fluctuations
[0.3 V/SSE - particle size=100 µm-mean
current = 22 µA].

A . measured spectrum
B . spectrum corrected for λ fluctuations.

Cut-off frequencies at 3 dB attenuation
are about : 1 Hz and 5 kHz.

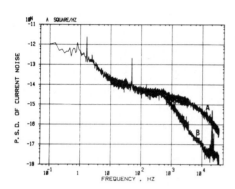

Fig. 2. P.S.D. of current fluctuations
[0.3 V/SSE - particle size 100 µm-mean
current = 55 µA] : effect of the
electrolyte resistance.

A . R_e = 0 Ω
B . R_e = 50 Ω

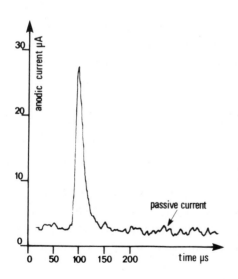

Fig. 3. a) Current transient of an
elementary impact.

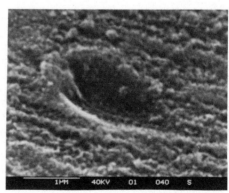

Fig. 3. b) SEM of the mechanical damage
(SiC particle size 100 µm).

account the optical response (fig. 1) shows only one cut-off in the high frequency range which can be related to the repassivation kinetics. Its value, of
the order of 1 kHz, demonstrates that the repassivation time is short. However
when the electrolyte resistance R_e is artificially increased , by adding an
extra resistance in series with the working electrode, the cut-off frequency is
decreased (fig. 2), though no effect is detected on the steady-state mean
current I. Hence the time constant of the current transient is linked to the
electrochemical characteristics of the whole surface, and is equal to R_eC where
C is the high frequency capacity. The actual repassivation time is certainly
shorter than this value. Moreover, following an impact which leads to a local
breakdown of the passive film, all the mechanically unperturbed surface parti-
cipates to the film reconstruction.

The increase ΔI of the mean current, due to abrasion is measured. As $\Delta I = \overline{\lambda}Q$,
Q, the average electric charge linked to the elementary current transient can be
evaluated when $\overline{\lambda}$ is known from optical data. For a typical experiment (particle
size : 100 µm, passive current 2 µA, abrasive current $\Delta I = 20$ µA, $\overline{\lambda} = 23000$
impacts per second, p.s.d. $\psi_I = 2.0 \ 10^{-15}$ A^2/Hz) Q = 0.9 nC which is far
larger than the double layer charge for an 1 μm^2 impact (fig. 3). This charge is
due to both the reconstruction of the passive film and the intervening metal
dissolution.

For a Poisson process, the low frequency limit of the p.s.d. equals $2\overline{\lambda}Q'^2$
Given the same data Q' = 0.2 nC which is surprisingly lower than Q. This discrepancy can be explained either by a hidden time constant in a lower frequency
range than the measured one or by a non linear effect involved in the elementary
current transients.The elementary response of a depassivation-repassivation
event is then difficult to isolate, because the mechanically unperturbed surface
can contribute by a global change of the steady state passive current.

Chemical breakdown

The current noise and the mean passive current were recorded continuously
from the chloride addition. The current time recording (fig. 4) shows six principal current peaks (the charge associated with the first one is of about 25 µC
related to a 10 µm hemispherical pit radius) in accordance with the surface
photograph which exhibits six principal pits. The exponential like current
growth is followed by a steep repassivation. This active/passive switch-off is
clearly related to the multisteady state current-voltage curve. As already shown
(ref. 4) a slight change in the dissolution regime can make the pit unstable and
trigger its repassivation.

Whenever the falling current returns to its initial value the noise spectrum
coincides again with that of a passivated electrode indicating that pit repassi-

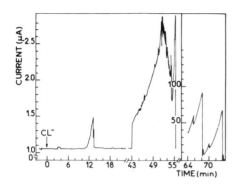

Fig. 4. Current-time dependence after chloride addition (3.10^{-3} M). Fe in 0.5 M H_2SO_4, T = 25°C, E = 0.2 V/SSE.

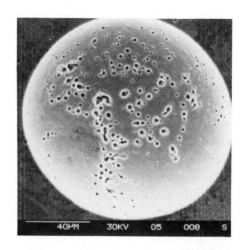

Fig. 5. Large hemispherical pit with secondary deep pits of polygonal shape on its surface.

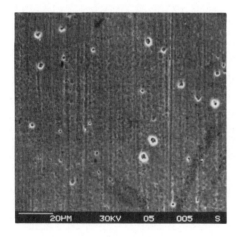

Fig. 6 . Electrode surface between large pits.

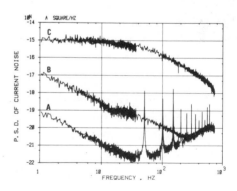

Fig. 7. Current noise spectrum at different pit activities. A) Fe in 0.5 M H_2SO_4, E = 0.2 V after 1 hour of passivation, J = 5.5 $\mu A.cm^{-2}$. B) After chloride addition (3.10^{-3}M), J = 150 $\mu A.cm^{-2}$. C) After chloride addition, J = 30 $\mu A.cm^{-2}$. Cut-off frequency at 3 dB attenuation is 60 Hz.

298

vation is achieved. The current noise measured during the current growth reveals sometimes a great pitting activity even if the d.c. current level is not too different of the passive value.

A large amount of small secondary pits of polygonal shapes can be observed inside the large pits (r > 10 μm) (fig. 5, 6). It should be noticed that the density of the small pits is much greater inside the large pits than on the remaining electrode surface. This may be related to the strongest agressivity of the solution inside the principal pit. Hence the secondary pits seem responsible for the current fluctuations detected from the noise spectrum (fig. 7). As the pitting activity increases the noise increases. Above some 1/f-like noise level, a cut-off frequency (60 Hz) and a low frequency plateau are detected (in these conditions the pit radius can reach 400 μm) which are undoubtly characteristic of the secondary pits. Large pits grow by coalescence of the secondary ones and by general attack. According to noise measurements the current of a pitting electrode can be regarded as a global measure of both general dissolution of the pit surface in a highly agressive electrolyte, and transient behaviour of secondary pits.

CONCLUSION

The two examples of noise analysis illustrate how this technique can bring new insights in localized corrosion studies. In those processes, deterministic relaxation techniques, e.g. impedances, are very difficult to use and the d.c. current provides only too overall information. Hence an analysis of the stochastic behaviour is a far more efficient approach.

In the case of mechanical depassivation spectral analysis of current fluctuations can provide information on the individual repassivation transient and its interaction with the surrounding passive surface. In the case of pitting corrosion, current noise measurement during pit growth reveals a substructure linked to micropits inside the principal one.

REFERENCES

1 U. Bertocci, J. Kruger, Surface Science 101, 608 (1980).
2 M. Keddam, R. Oltra, J.C. Colson and A. Desestret, Corrosion Science, 23, 4, p. 441-451 (1983).
3 C. Cachet, C. Gabrielli, F. Huet, M. Keddam and R. Wiart, Electrochim. Acta, in press.
4 I. Epelboin, C. Gabrielli, M. Keddam and H. Takenouti, Z. Phys. Chem. N.F.98, pp. 215-232, (1975).

Passivity of Metals and Semiconductors, edited by M. Froment
Elsevier Science Publishers B.V., Amsterdam — Printed in The Netherlands

PITTING OF ZIRCALOY 4 : STATISTICAL ANALYSIS OF INDUCTION TIMES.

G. MANKOWSKI, P. EYGAZIER, Y. ROQUES, G. CHATAINIER and F. DABOSI
Laboratoire de Métallurgie Physique, E.R.A. 263 du C.N.R.S.
118 Route de Narbonne - 31077 TOULOUSE CEDEX

ABSTRACT

Induction time for pit generation of a zirconium alloy in chloride containing aqueous solution has been analyzed from a statistical point of view. A multi-channel pitting detector was developped which allows the measurement of induction times for ten specimens simultaneously. The random variation of induction time is well described by a log-normal distribution. The median log t_{50} and the standard deviation σ of this distribution may be used to point out the influence of various parameters on pitting corrosion. The effects of applied potential and of prepolarization time are clearly shown. From all our experiments, it seems that log t_{50} and σ are not independant parameters.

INTRODUCTION

Frequently, studies about pitting corrosion concern the determination of critical potentials for pit generation and protection (or repassivation). However, while the protection potential is satisfactorily reproducible, the potential for pit generation often is widely scattered (sometimes within a few hundred millivolts). Investigations have been performed by potentiostatically measuring the induction time for pit generation at different potentials. However, induction time data also show a large variability so that it could be difficult to precisely define the potential influence on induction time as well as the effects of other parameters. Shibata and Takeyama (ref. 1-2) have shown the advantages of statistical studies of pit generation on stainless steels. For such studies two methods may be used :

- determination of the distribution of induction time for pit generation in
  potentiostatic conditions,
- determination of the distribution of potential for pit generation in potentiodynamic conditions.

The results reported here for a zirconium alloy in chloride-containing aqueous solution have been obtained by the first method.

EXPERIMENTAL

The material used in this investigation is a commercial Zircaloy-4 alloy with the following composition : 1.59 % Sn, 0.19 % Fe, 0.11 % Cr. It was supplied

300

in the form of a 25 mm thick plate in the beta-quenched condition.

Ten specimens machined into 10 mm diameter cylinders were embedded in epoxy resin on the same holder. This configuration enable a more rapid and more reproducible polishing of the specimens.

The tests were conducted in aerated NaCl 5% aqueous solutions, at room temperature, under potentiostatic conditions.

Before testing, the specimens were wet ground with 1000 grit SiC paper and passivated at -100 mV/SCE. Then, the potential was instantaneously shifted to the testing potential.

Since statistics require a large number of data, a multichannel pitting detector was developped which can simultaneously measure induction time for 10 specimens using one potentiostat. As pitting results in a sudden increase of anodic current, each channel of the detector can register the time (by stopping a time counter) when the current of one specimen reaches a threshold value. At the same time the corresponding specimen is disconnected from the polarization circuit. The passive current being about 0.2 μA/cm^2, the chosen value for the current threshold for each specimen is 20 μA. The total polarization current of the ten specimens is plotted against the time, as can be seen in Figure 1. For this reason, the sum of individual currents may exceed the threshold value, when a large number of specimens is still connected.

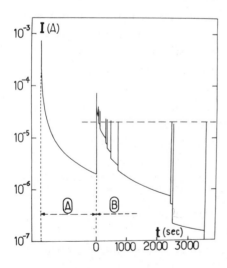

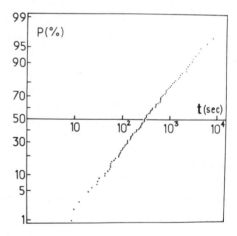

Fig. 1 - Typical shape of the total current plotted against time;
A : prepolarization
B : detection of pitting

Fig. 2 - Example of an induction time distribution (Prepolarization : 30 min at -100 mV/SCE ; pitting potential : +200 mV/SCE).

RESULTS AND DISCUSSION

Induction times for pit generation are widely scattered, from few seconds to
several hours. A large number of data is then necessary for obtaining a good
definition of the observed distributions. The distributions reported here result
from 100 measurements of induction time under the same experimental conditions.
The pitting probability P is calculated as n/N where N is the total number of
specimens (N = 100) and n is the number of specimens on which pitting occured
at time t.

Data plotted on a log-normal probability paper give quite straight lines.
An example of such a distribution is shown in Figure 2. Therefore, random variation of induction time for pit generation seems to obey a log-normal distribution :

$$P = \frac{1}{\sqrt{2\pi}\,\sigma} \int_0^t \frac{1}{t} \exp\left[-\frac{(\log t - \log t_{50})^2}{2\,\sigma^2}\right] dt$$

where σ is the standard deviation and t_{50} is the time at P = 0.5.

On figure 3, σ is plotted against log t_{50} for various distributions obtained
from different experimental conditions. Rectangles represent the error margins
of σ and log t_{50} for 95 % confident limits. It clearly appears that the two
parameters are correlated. However, the dispersion of the points don't allow us
to determine a corresponding mathematical relation. Subsequently, only the parameter log t_{50} will be used to describe the observed distributions.

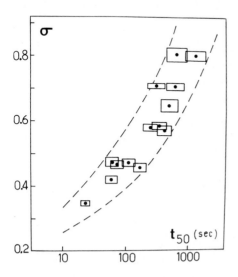

Fig. 3 - Relation between the standard
deviation σ and the median log t_{50}.

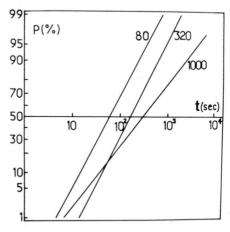

Fig. 4 - Influence of surface roughness on the induction time distribution. (Prepolarization : 30 min at
- 100 mV/SCE ; pitting potential :
+ 200 mV/SCE.

Figure 4 shows the distributions of induction time for specimens polished by wet grounding with different grit SiC papers (80, 320 and 1000). Only the regression lines are represented. One can notice that increasing the surface roughness enhances the pit generation by decreasing the induction time. Moreover, as can be seen in Table 1, the correlation coefficient is much better and the relative error margin of σ is smaller for polishing with 1000 grit SiC paper than with 80 or 320 papers. So, the induction time better conforms to the log-normal distribution for the more finely polished specimens. This may result from a better reproducibility of the surface condition. Subsequently, all the experiments are then performed with specimens polished with 1000 grit SiC paper.

TABLE 1

Influence of polishing on distribution characteristics.

Polishing SiC paper number	Correlation coefficient	$\sigma \pm \Delta\sigma$	$\Delta\sigma/\sigma$ (%)
80	0.9894	0.475 ± 0.014	2.9
320	0.9881	0.458 ± 0.014	3.1
1000	0.9985	0.709 ± 0.008	1.1

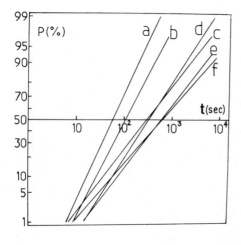

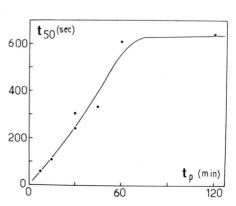

Fig. 5 - Effect of prepolarization time at - 100 mV/SCE on the induction time distribution : a : 7.5 min ; b : 15 min ; c : 30 min ; d : 45 min ; e : 60 min ; and f : 120 min.
Pitting potential : + 200 mV/SCE.

Fig. 6 - Variation of t_{50} with prepolarization time t_p : Pitting potential : + 200 mV/SCE.

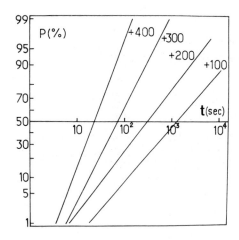
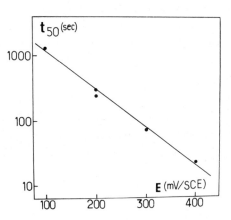

Fig. 7 - Effect of applied potential
(mV/SCE) on the induction time distribution. Prepolarization : 30 min
at - 100 mV/SCE.

Fig. 8 - Linear relationship of
log t_{50} with applied potential.

The effect of prepolarization time at - 100 mV/SCE is illustrated on Figures
5 and 6. Figure 5 shows the distributions of induction time for prepolarization
time from 7.5 to 120 minutes. Figure 6 represents the change in t_{50} with increasing prepolarization time. It clearly appears that the pitting susceptibility decreases with increasing time of prepolarization until about 1 hour where
the passive film get its highest protective ability.

The distribution curves of induction time for different applied potentials
(above the protection potential of + 75 mV/SCE) are plotted on figure 7. It
appears that the potential has a great influence on induction time. Moreover,
the median log t_{50} seems linearly dependant on potential as can be seen on figure 8. Other investigators also observed that induction time is an exponential
function of the applied potential (ref. 3, 4, 5), a behaviour which shows the
electrochemical nature of the pitting process.

CONCLUSION

Pitting corrosion data of Zircaloy-4 in chloride-containing aqueous solution
are widely scattered. Induction time for pit generation may vary from few seconds
to several hours in the same experimental conditions. This justifies our statistical study of the pitting process.

Induction time data obey a log-normal distribution where the median and the
standard deviation seem to be correlated.

The median log t_{50} increases when :
- the surface roughness decreases,
- the prepolarization time increases up to about 1 hour,
- the applied potential gets closer to the protection potential.

REFERENCES

1. T. SHIBATA and T. TAKEYAMA, Corrosion, 33, 7, (1977), p. 243.
2. T. SHIBATA and T. TAKEYAMA, Proceedings 8th ICMC, Mainz (1981), p. 146.
3. K.E. HEUSLER, L. FISCHER, Werks und Korr, 27, (1976) p. 551.
4. M. JANIK-CZACHOR, Werks. und Korr., 30, (1979) p. 255.
5. L.F. LIN, C.Y. CHAO and D.D. MACDONALD, J. Electrochem. Soc., 128 (1981) p. 1194.

Passivity of Metals and Semiconductors, edited by M. Froment
Elsevier Science Publishers B.V., Amsterdam — Printed in The Netherlands

AN ELECTROCHEMICAL STUDY OF AMORPHOUS ION IMPLANTED STAINLESS STEELS

C. R. CLAYTON[1], Y-F. WANG[1], AND G. K. HUBLER[2]

[1] Department of Materials Science and Engineering, State University of New York, Stony Brook, N.Y. 11794, USA

[2] Naval Research Laboratory, Washington, D.C. 20375, USA

ABSTRACT

304 and 316 stainless steel was implanted with phosphorous at 40 keV or with boron at 25 keV to a fluence of 1×10^{17} to form amorphous surface alloys. The modification of the corrosion properties of the steels was monitored by dynamic polarizaton.

INTRODUCTION

It has been shown by Hashimoto (ref.1) that the addition of metalloid elements is necessary to obtain an amorphous structure in alloys formed by rapid quenching techniques. Among those metalloid elements, phosphorous and boron are important to ensure good corrosion resistance of amorphous iron-based alloys containing chromium. Recently ion implantation has been used extensively to modify the properties of the surface layers of pure metals and alloys. Implantation is capable of producing a wide range of metastable surface alloys. With proper selection of implant element and implantation parameters, an amorphous surface alloy may be formed (ref.2,3,4). In this work, 304 and 316 stainless steel (ss) have been implanted with phosphorous at 40 keV to a fluence of 1×10^{17} ions cm^{-2} or with boron at 25 keV to a fluence of 1×10^{17} ions cm^{-2}. In each case, this resulted in the formation of an amorphous surface alloy (ref.2,3,5). Combined implantation of B and P was also carried out. Anodic polarization studies were carried out to determine the effects of P and B implantation on the active-passive behavior and pitting resistance of 304 and 316 ss in deaerated solutions of 0.5M H_2SO_4 and 0.5M H_2SO_4 + 0.5M NaCl.

EXPERIMENTAL

The chemical composition of 304 and 316 ss is given below. The depth profiles of P and B in 304 and 316 ss were obtained by Auger Spectroscopy with 6keV argon or 2keV xenon ion sputtering. The distribution of P and B in 304 and 316 ss were very similar. The distributions of P and B in the 304 ss samples are shown in figures 1a and 1b. It is seen from the B profile, figure 1b, that a slight

306

enrichment of Fe and Cr occurs within the first 20 nm. This is thought to be due
to radiation enhanced diffusion rather than selective sputtering.

TABLE 1.

Composition of 304 and 316 stainless steels

Steel	% Cr	% Ni	% Mn	% Si	% Mo	% C	% N	% S	% P	% Fe
304	18.18	8.48	1.75	0.5	0.36	0.051	0.05	0.005	0.028	balance
316	17.25	10.82	2.0	0.6	2.28	0.05	0.03	0.030	0.032	balance

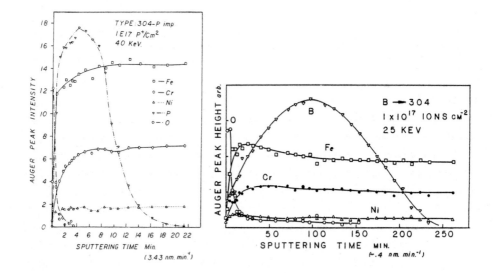

Fig. 1a.   Fig. 1b.

Fig. 1. AES composition depth profiles of P (1a.) and B (1b.) implanted 304
stainless steel.

Anodic polarization was carried out in argon deaerated solutions of 0.5M
H_2SO_4 and 0.5M H_2SO_4 + 0.5M NaCl using a conventional Greene cell. The implanted
samples were masked with epoxy resin to prevent attack on unimplanted surfaces.
All potentials were recorded relative to a saturated calomel electrode (SCE).
The samples were cathodically treated at $1mA/cm^2$ (150–200 mV below Eocp) for 30
minutes to remove the air formed film on the surfaces. Samples were than anodic
polarized with a scanning rate of 1mV/sec.

RESULTS AND DISCUSSION

Anodic Polarization in 0.5M H$_2$SO$_4$

The anodic polarization curves of implanted and unimplanted 304 and 316 ss in 0.5M H$_2$SO$_4$ are given in figure 2 and figure 3 with the corresponding parameters given in Table 2. All the implanted steels show a reduction in the critical current density (Ic). In the case of the 304 ss samples, the passivation potential (Ep) shifted to more negative values and a small decrease in the transpassive potential (E$_T$) is observed. The elevation of the slope of the transpassive curve indicated that P-implantation reduces the rate of transpassive dissolution. In the passive potential region Phosphorous tended to form a more stable film as indicated by a lower passive current density (Ipp). Boron implanted steel showed a higher current density in the same potential region. A second current maximum (Ep') was observed on each of the implanted steels. The polarization behaviour of the P and B multi-implanted steel is midway between that of both P and B implanted steel. Therefore there appears to be no synergistic effect on the anodic behaviour of 304 ss from mixing P and B.

P-implantation of 316 ss results in the complete removal of the active region. A single anodic maximum is observed at +33mV. The transpassive potential and the slope of the transpassive branch is slightly increased. B implantation provides no evident improvement in the anodic characteristics of 316 ss, but instead leads to a slight positive increase in the passivation potential.

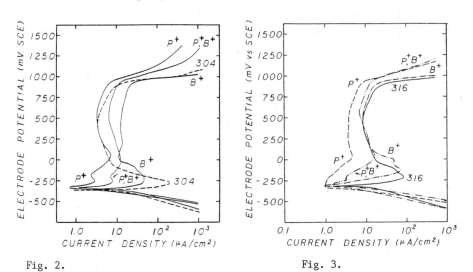

Fig. 2.   Fig. 3.

Fig. 2. Anodic polarization curves of unimplanted and implanted 304 ss in 0.5M H$_2$SO$_4$ solution.
Fig. 3. Anodic polarization curves of unimplanted and implanted 316 ss in 0.5M H$_2$SO$_4$.

TABLE 2

Anodic polarization parameters determined in deaerated 0.5M H_2SO_4

Steel	Potential (mV vs SCE)				Current Density ($\mu A\ cm^{-2}$)		
	Eocp	Ep	Ep'	E_T	Ic	Ic'	Ipp
304	−370	−225	---	+945	200.0	---	3.4
304−P	−330	−250	+50	+930	4.2	7.1	3.3
304−B	−375	−263	+200	+910	42.0	12.5	11.0
304−P,B	−330	−250	0	+933	8.2	12.4	6.6
316	−314	−206	---	+904	84.0	---	8.4
316−P	−325	---	+33	+900	---	4.5	3.7
316−B	−305	−125	−5	+935	70.0	50.0	8.4
316−B,P	−282	−245	32	+962	3.2	16.5	7.2

Anodic polarization in 0.5M H_2SO_4 + 0.5M NaCl

The anodic polarization curves for implanted and unimplanted 304 and 316 ss in the acid/chloride solutions are presented in figures 4 and 5. The corresponding parameters are given in Table 3. The beneficial effect of P-implantation on the active-passive transition is repeated in the acid/chloride solution as is the lowering of the critical current density for 304 passivation. As before, B increased the passivation potential both for 304 and 316 ss in the positive direction. P-implantation appears to have little beneficial effect upon the pitting resistance of 304 and 316 ss, while B significantly raises the pitting potential (Eb) of both steels. The multiple P and B implantation treatment combines the beneficial effect of P on the active-passive transition, and the beneficial effect of B on the breakdown potential (Eb).

It is interesting to note that Mo is added to stainless steels to improve pitting resistance. However, it is well known that Mo also tends to increase the activity of the stainless steel slightly. Therefore Mo and B appear to have some similar characteristics. At this stage the exact effect Mo and B may have on the properties of the passive film and on the metal layer interfacing the film is not known. A more comprehensive study to include the composition analysis of the passive films and the underlying alloy of both unimplanted and implanted stainless steel is currently underway and will be reported elsewhere.

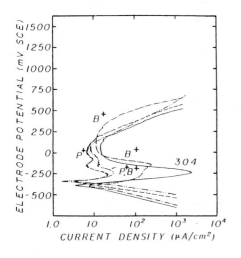 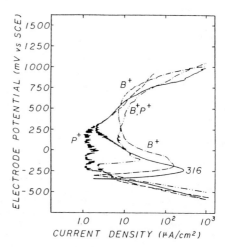

Fig. 4.   Fig. 5.

Fig. 4. Anodic polarization curves of unimplanted and implanted 304 ss
in 0.5M H_2SO_4 + 0.5M NaCl.
Fig. 5. Anodic polarization curves of unimplanted and implanted 316 ss
in 0.5M H_2SO_4 + 0.5M NaCl.

TABLE 3

Anodic polarization parameters determined in deaerated 0.5M H_2SO_4+ 0.5M NaCl.

Steel	Potential (mV vs SCE)				Current Density (μA cm^{-2})		
	Eocp	Ep	Ep'	Eb	Ic	Ic'	Ipp
304	−390	−243	----	+150	2600	----	15.0
304-P	−345	−263	−80	+160	23	9.6	6.5
304-B	−375	−145	----	+350	195	----	11.5
304-P,B	−342	−237	−80	+110	30	16.0	8.3
316	−340	−236	----	+305	280	----	2.2
316-P	−250	−135	0	+225	1.6	2.2	1.3
316-B	−295	−180	−75	+692	160	120.0	7.0
316-P,B	−250	−112	----	+640	25	25.0	7.1

CONCLUSIONS

* P-implantation improved the active-passive behavior of 304 and 316 ss in
  solutions of 0.5M H_2SO_4 and 0.5M H_2SO_4 + 0.5M NaCl.
* B-implantation had no beneficial effect on the active-passive behavior of 304
  and 316 ss in both test electrolytes.
* B significantly increased the pitting resistance of 304 and 316 ss in the 0.5
  H_2SO_4 + 0.5M NaCl solution.

* The beneficial effects of P and B were combined in the multiple implantation treatment.

ACKNOWLEDGEMENTS

This work was support by the U.S Office of Naval Research, Arlington, Virginia, under the contract # N0001477C0424.

REFERENCE

1 K. Hashimoto, M. Naka, J. Nugichi, K. Asami, and T. Masumoto. Passivity of Metals, Ed. R.P. Frankenthal and J. Kruger (Proc. of Fourth Int. Symp. on Passivity, 1977) p. 156 The Electrochem. Soc. Inc., Princeton, N.J. (1978)
2 C. R. Clayton, K. G. K. Doss, Y-F Wang, J. B. Warren and G. K. Hubler. Ion Implantation Into Metal, Ed. V. Ashworth et al. Pergamon Press, Oxford and New York. 1982.
3 C. R. Clayton, K. G. K. Doss, H. Herman, S. Prasad, Y-F Wang, J. K. Hirvonen and G. K. Hubler. Ion Implantation Metallurgy, Ed. C. M. Preece and J. K. Hirvonen, TMS-AIME 1980
4 C. R. Clayton, Nuclear Instruments and Methods 182/183, p.865-873 (1981)
5 Chen Qing-Ming et al., Proc. IBMM-82 Grenoble, France, September, 1982.

Passivity of Metals and Semiconductors, edited by M. Froment
Elsevier Science Publishers B.V., Amsterdam — Printed in The Netherlands

ELECTROCHEMICAL PROPERTIES AND PASSIVATION OF AMORPHOUS $Fe_{80}P_{20}$ ALLOY[*]

P. CADET, M. KEDDAM and H. TAKENOUTI

Groupe de Recherche n° 4 du CNRS, "Physique des Liquides et Electrochimie",
associé à l'Université Pierre et Marie Curie, 4 place Jussieu,
75230 PARIS Cedex 05 (France)

ABSTRACT

The corrosion resistance of iron base amorphous alloys is markedly improved
by the simultaneous addition of P and Cr. Their kinetic effects are, however,
still poorly understood. In this paper, the influence of P on the anodic behaviour of amorphous iron - phosphorus alloy in an acidic sulfate solution is studied. In the active range, the addition of P to Fe decreases drastically the
dissolution rate. The steady - state polarization curve indicates that passivation takes place through a potential dependent process. On the contrary, cyclic
voltammograms exhibit two current maxima, both leading to the formation of passive species. One is identifical with that found by the steady - state polarization and is easily reduced. Another formed through the second current peak improves the stability of passivity. The charges consumed to the reduction of
these species suggest a thick passive film.

INTRODUCTION

The addition of Cr and P is highly beneficial for the corrosion resistance
of Fe base amorphous alloys (ref.1). That is, the chemical composition of an
amorphous alloy is more decisive   than the desordered structure in its electrochemical behaviour. In spite of this obvious experimental fact, few papers
were devoted to the electrode kinetics of amorphous alloys emphasizing the chemical influence of metalloid elements. The aim of this paper is to characterize the influence of P on the anodic behaviour of $Fe_{80}P_{20}$ (the subscripts indicate atomic percentage) by means of several electrode kinetic methods. Results
will be compared with those obtained with $Fe_{80}B_{20}$.

EXPERIMENTAL

Amorphous ribbons obtained by melt-spinning were used. Ribbons were 0.8 to
3 mm in width and 10 to 15 μm in thickness. Two types of electrodes were prepared ; a ribbon electrode,one or both sides of which   were in contact with
electrolyte, and a   rotating electrode consisting   of the cross-section
of fifty pieces of ribbon bundled together and moulded in an epoxy-resin. The

---

[*] This work constitutes a part of P.C's thesis which will be submitted to
l'Université Pierre et Marie Curie, Paris.

electrolyte was $1M-H_2SO_4$. The potential was measured with Hg, Hg_2SO_4 in sat. K_2SO_4 (SSE). Electrochemical measurements were performed with a potentiostat and a frequency response analyser (Schlumberger, Solartron electrochemical interface 1186 and TFA 1250).

RESULTS

Active dissolution : comparison between $Fe_{80}P_{20}$ and $Fe_{80}B_{20}$

It has been shown previously that Fe dissolves as Fe^{2+} whereas P is oxidized into P^{5+} in the active range (ref.2). This result agrees with the presence of PO_4^{3-} detected by XPS in the passive film (ref.3). With $Fe_{80}B_{20}$, similar results were obtained : it dissolves as Fe^{2+} and B^{3+} (ref.4). The XPS analysis of Fe - Ni - Cr - P - B detected however atomic B on the surface (ref.5). This discrepancy may be explained since no film is formed in the active state. XPS is applied to the solid surface, and cannot detect B^{3+} dissolved in the solution. These results show that the metalloid elements are electrochemically active in the solution used.

When the polarization curves of these two alloys are compared with Fe, it is revealed that B addition changes little the dissolution rate of Fe. Besides, the electrode impedances of $Fe_{78}B_{20}Mo_2$ (Allied Chemical), are similar to those of Fe specimen (ref.6). On the contrary, the dissolution rate of Fe is drastically decreased by the presence of P, and the electrode impedances are also markedly different (ref.2). The dissolution mechanism of Fe itself is deeply modified by the presence of P.

Passivation of $Fe_{80}P_{20}$

In contrast with Fe whose active-passive transition exhibits multi steady-states, (ref.7), passivation of $Fe_{80}P_{20}$ takes place following a bell-shaped polarization curve as illustrated Fig. 1 - a. If the electrode impedance is measured in the passivation range (see Fig. 1 - A), the potential dependent film - forming process gives rise to the negative real part at low frequencies. The associated frequency (corresponding to the maximum of imaginary component) close to 0.5 Hz is somewhat higher than in the case of Fe. These experimental observations exclude the possibility of a dissolution-precipitation mechanism proposed in the litterature (ref.1), and is in favour of a process similar to the passivation of Fe.

Cyclic voltammograms (see Fig. 2) exhibit two current maxima in the anodic sweep : peak - 1 located near -0.15 V depends little on the sweep rate, and is observed at the steady-state (Fig. 1-a). On the contrary, peak - 2, located by 0.6 V depends on the sweep rate. The charge under this maximum (ca 0.3 $C.cm^{-2}$) is three orders of magnitude greater than that involved in a monomolecular layer.

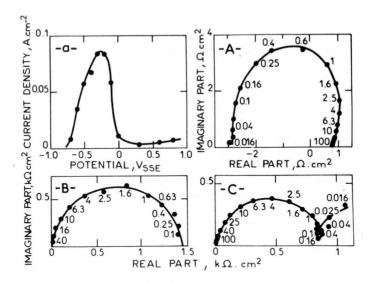

Fig. 1. $Fe_{80}P_{20}$ in $1M - H_2SO_4$ at 25°C. (a) Steady-state polarization curve with rotating electrode and impedance diagrams with ribbon electrode at (A) -0.05 (B) 0.4 and (C) 0.6 V_{SSE}.

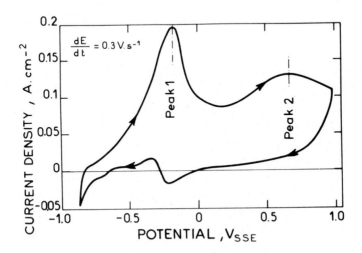

Fig. 2. Cyclic voltammogram (1st plot) with ribbon electrode showing two current maxima. Conditions see Fig. 1.

If the potential sweep is stopped near the peak - 2, the current I decreases towards the steady-state value I_p. Fig. 3 shows the time dependence of log $(I - I_p)$. This curve reveals clearly two time constants   of about 6   and 75 s. The fast current decrease is due to the oxidation process corresponding to peak - 2, and the   slow   current decrease to the completion of the passivation process similar to Fe (ref.8).

Stability of the passive film

The corrosion resistance of passive materials is closely related to the stability of the passive film. For this reason, the latter is evaluated from an open - circuit potential decay curve (Flade's experiments). The passive film is also reduced by imposing a constant cathodic current.

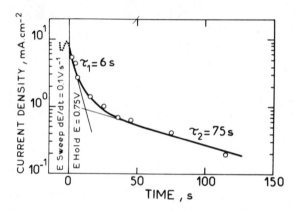

Fig. 3. Current $(I - I_p)$ decrease at the potential hold following to the potential sweep (marked by dots), conditions see Fig. 2.

Fig. 4 - A illustrates the chronopotentiometric curve exhibiting two potential arrests, hence two time intervals T_1 and T_2 are to be defined. At the open-circuit $(I_c = 0)$ $T_2 \ll T_1$, the stability of the passive film is evaluated through T_1. The results obtained at various hold-potentials are shown in Fig.4-B. A steep increase of T_1 is observed for potentials beyond peak - 2.

The influence of peak - 2 on the passivation process is also supported by impedance data obtained on the cathodic and   anodic sides of this peak - 2 (Fig.1 - B and C). The low frequency branch (f < 0.1 Hz) observed in Fig. 1 - C is related to the film formation process (ref.9). On the contrary, such a branch is hardly seen in Fig. 1 - B. With $Fe_{78}B_{20}Mo_2$ specimen, diagrams such as 1 - C were observed even at the beginning of the passivity like in the case of Fe (ref.10).

The amount of passivating species is estimated by cathodic reduction. Fig. 5 illustrates the change of T_1 and T_2 with respect to hold-potential. It can be

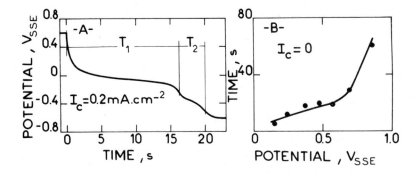

Fig. 4. (A) Chronopotentiometry with ribbon electrode. The electrode is held at 0.6 V during 60 s, then the passive film is reduced at $I_c = 0.2 \, mA.cm^{-2}$. (B) T_1 with respect to hold potential. $I_c = 0$ (open-circuit).

seen that T_1 is almost constant. The charge associated is about 25 $mC.cm^{-2}$ regardless of I_c. T_2 increases with potential and reaches asymptotically to 4 s, that is, 20 $mC.cm^{-2}$. However, the charge thus determined depends on I_c value, and the self-reduction cannot be neglected in this estimation. Consequently, the charge implied in the reduction of species is probably under estimated. It is worth noting that the charge evaluated by Fig. 6 is less than one tenth of that measured under peak - 2. If the same experiments are performed with shorter periods of potential hold, T_2 decreases whereas T_1 remains constant. T_2 species is hence related to the oxidizing process of peak - 2 and to the current decrease τ_1.

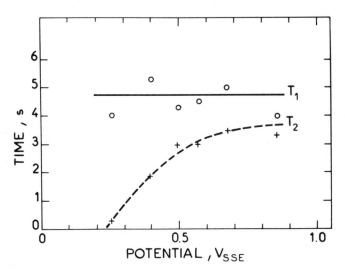

Fig. 5. T_1 and T_2 under cathodic reduction. $I_c = 5 \, mA.cm^{-2}$

The oxidation process corresponding to peak-2 improves considerably the stability of the passive film. The passive film is clearly constituted by two species, its thickness estimated from the charge involved (25 mC) is at least 100 times greater than a monolayer.

CONCLUSION

By comparing the results obtained with $Fe_{80}P_{20}$ and $Fe_{80}B_{20}$, it was shown that the anodic behaviour of these amorphous alloys depends much on the metalloid added.   On the contrary, the amorphous state itself has rather little effect. The addition of P decreases drastically the active dissolution rate of Fe in an acidified sulfate medium. The dissolution mechanism is also deeply modified by P whereas B changes scarcely either dissolution rate or its mechanism. The passivation process, on the contrary, takes place through a process similar to Fe with potential dependent film formation. In the passivity range, P changes heavily the electrode behaviour. The passive film is thick in comparison to crystalline Fe and is formed by two species. Several important aspects of the contribution of P to the kinetics of electrode processes were thus characterized in this paper.

ACKNOWLEDGEMENT

Authors wish to thank Drs. MASUMOTO and HASHIMOTO (Tohoku University, Sendai) and Dr. BIGOT (CECM, CNRS - Vitry) for kindly providing amorphous specimens, and the DGRST for a financial support of this work under contract n° 80. P.0552.

REFERENCES

1 K. Hashimoto, Suppl. Sci. Rep. Res. Inst. Tohoku Univ. Serie A, 26 (1978) 233 - 250, and 28 (1980) 201 - 216.
2 P. Cadet, M. Keddam and H. Takenouti, 4th Intern. Conf. Rapidly Quenched Metals, Sendai, 1981, 1447 - 1451.
3 K. Hashimoto, K. Asami, M. Naka and T. Masumoto, Corros. Sci., 19 (1979) 857 - 867.
4 P. Cadet, M. Keddam and H. Takenouti, C.R. Acad. Sci., Paris, t294, Serie II (1982) 509 - 512.
5 D.R. Baer and M.T. Thomas, J. Vac. Sci. Technol., 18 (1981) 722 - 726.
6 O.R. Mattos, Thesis "Sur une étude systématique des mecanismes de dissolution anodique du fer par analyse de l'impédance. Extension aux alliages fer-chrome", Paris, Juin 1981.
7 I. Epelboin, C. Gabrielli, M. Keddam and H. Takenouti, Z. Phys. Chem., N.F., 98 (1975) 215 - 232.
8 M. Baddi, C. Gabrielli, M. Keddam and H. Takenouti, 4th International Symposium on Passivity, Airlie, in R.P. Frankenthal and J. Kruger (Eds) "Passivity of Metals", The electrochemical society softbound proceeding series. Princeton, N.J., 1978, 625 pp.
9 M. Keddam, J.F. Lizée, C. Pallotta and H. Takenouti, This proceeding
10 M. Keddam, P. Mirebeau and H. Takenouti, Fall Meeting of the Electrochemical Society, Los Angeles, Ca, 1979, Abstract 675 pp.

Passivity of Metals and Semiconductors, edited by M. Froment
Elsevier Science Publishers B.V., Amsterdam — Printed in The Netherlands

EFFECT OF CHROMIUM ON THE PASSIVATION OF AMORPHOUS ALLOY Fe-Ni-B-P-xCr. EFFECT OF A HEAT TREATMENT

J.CROUSIER, K.BELMOKRE, Y.MASSIANI, J-P.CROUSIER

Equipe Corrosion. Laboratoire Chimie des Matériaux.

Université de Provence. 13331 Marseille Cedex 3 (France)

ABSTRACT

The effect of chromium addition on the passivation of amorphous alloys has been studied in 0.1 N H_2SO_4 and 3 % NaCl water solutions. In H_2SO_4 solution, the positive effect of chromium is related to its concentration in the alloy. In NaCl solution, from 5 at.% of chromium, the alloys show spontaneous passivation and an increase in chromium content does not improve it. Samples with 15 at.% chromium have been heated in argon atmosphere ; the polarization curves clearly indicate a different behavior depending on temperature. Particularly, formation of pits can be noticed when alloys are no longer amorphous.

INTRODUCTION

The effect of chromium on the corrosion resistance of amorphous iron-base alloys with C-P, B-Si, B-C and B-P as additionnal elements as well as amorphous nickel-base alloys has been studied by Hashimoto and Coll.(1,2).

The present work aims to study the effect of nickel, molybdenum, and phosphorus on the corrosion resistance of amorphous Fe-B alloy as well as effect of chromium on the passivation of amorphous Fe-Ni alloy with B-P in 0.1 N H_2SO_4 and 3 % NaCl water solutions. The effect of a preliminary heat treatment on passivation of a chromium rich alloy in NaCl solution has been studied.

EXPERIMENTAL PROCEDURES

Starting from crystalline alloys obtained by fusion of a mixture of powders of constitutive elements, amorphous alloys have been prepared by melt-spinning. The nominal compositions of amorphous alloys containing chromium (in at. %) are :

Fe	Ni	B	P	Cr	
46.16	34.71	3.89	15.23	0	Cr Ø
38.93	37.20	3.24	15.33	5.30	Cr 5
35.74	33.94	4.99	15.25	10.09	Cr 10
29.20	33.30	3.90	20.20	13.40	Cr 14

318

The amorphous structure of the alloys has been controled by differential scanning calorimetry (D.S.C.) and by X ray diffraction. The samples were not polished. The anodic polarization curves have been recorded under open air condition by a potentiodynamic method starting from corrosion potential E_{cor}.

The samples of an alloy 30Fe-35Ni-6B-14P-15Cr (Allied Chemical), indicated by 15Cr-A, have been polished and heated 24 hours in argon. The heat treatment goes from 92°C to 400°C. After each experiment, the amorphicity of the sample was controled. In order to crystallise, one of the sample has been Heated at 750°C during 24 hours.

RESULTS

0.1 N H$_2$SO$_4$ Solution

The results are obtained from the shinny side of the alloy. Amorphous Fe-B-Ni, Fe-B-Mo, Fe-B-Ni-Mo alloys show poor corrosion resistance. Addition of phosphorus tends to passivate the Fe-B-Ni alloy. The potential intensity curves of the four alloys containing chromium are shown on fig.1. Starting from the 10 at. % chromium, the critical current density is very weak and passivation is spontaneous for 14 % at.

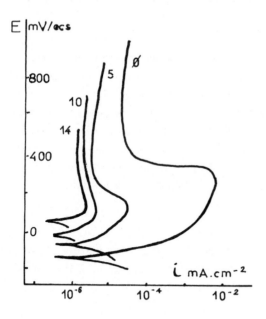

Fig. 1 . Anodic polarization curves of amorphous Fe-Ni-B-P-xCr in 0.1 N H$_2$SO$_4$ with x = 0;5;10;14.

Comparison of the polarization curves with the commun passivable alloys ones (3), shows that the passivation effect of chromium is much more important when this element is included in an amorphous alloy.

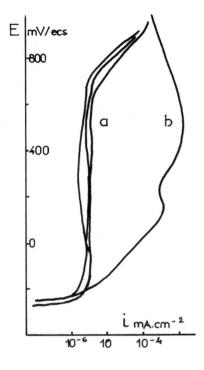

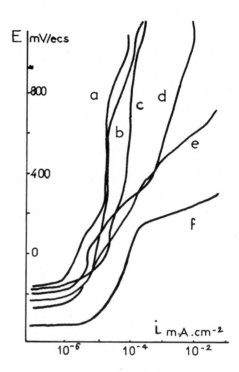

Fig. 2. Anodic polarization curves of amorphous Fe-Ni-B-P-xCr in 3 % NaCl
(a) : 5;10;14 at % Cr
(b) : 0 at % Cr.

Fig. 3. Anodic polarization curves of amorphous Fe-Ni-B-P-xCr
(a) : without heat treatment
After heat treatment :
(b) : 92°C - (c) : 300°C -
(d) : 350°C- (e) : 400°C -
(f) : 750°C.

3 % NaCl Solution

Comparison of the potential-intensity curves (fig.2) of Cr ∅ with those of Cr 5, Cr 10, Cr 14, shows the positive effect of chromium and no increase of concentration effect of this element . Indeed, from 5 at.% of chromium the alloy is spontaneously passived and an increase of chromium content has no effect. Pitting corrosion is never observed.

Influence of heat treatment

Fig. 3 shows the polarization curves of preheated samples. The curves at 150, 220 and 250°C are not shown in order to clarify the figure : they might be drawn between those of 92°C and 300°C. Curve of the sample which has been heated at 750°C during 24 hours is reported; we have verified that this sample is no longer amorphous. The polarization curve shows pitting corrosion.

CONCLUSION

This study has confirmed the importance of chromium in passivation of amorphous Fe-Ni alloys. In NaCl solution a small amount of this element is sufficient to obtain spontaneous passivation : moreover this caracteristic can be kept well over ambiant temperature without finding pitting corrosion.

REFERENCES
1 K.Hashimoto, Sci. Rep. Res. Inst. Tohoku University A, (1978), 233.
2 K.Hashimoto, Sci. Rep. Res. Inst. Tohoku University A, (1980), 201.
3 J.Crousier, Y.Massiani, J-P.Crousier, Materials Chemistry, 7, 5, (1982),587.

Passivity of Metals and Semiconductors, edited by M. Froment
Elsevier Science Publishers B.V., Amsterdam — Printed in The Netherlands

THE EFFECT OF PHOSPHORUS ON THE CORROSION RESISTANCE OF AMORPHOUS COPPER-ZIRCONIUM ALLOYS

T.D. BURLEIGH[1] and R.M. LATANISION[2]

[1]Mat. Sci. and Eng. Dept., Room 8-204, MIT, Cambridge, MA 02139 USA
[2]Mat. Sci. and Eng. Dept., Room 8-202, MIT, Cambridge, MA 02139 USA

ABSTRACT

Cu-40Zr and Cu-42Zr-2P amorphous metal alloys were prepared by melt-spinning. The corrosion properties of the two alloys were compared, and it has been determined that phosphorus has a beneficial effect on improving the corrosion resistance of the alloy.

In comparing Cu-42Zr-2P with Cu-40Zr, the results show that the kinetics of corrosion of the Cu-42Zr-2P are slower in 1N sulfuric acid, in 3.5% NaCl, and also in 1N HCl. The presumably zirconia film on the Cu-42Zr-2P alloy contains phosphorus in the +5 valence state and does not increase in thickness with time. In anodization, the Cu-42Zr-2P is not photosensitive while the Cu-40Zr does oxidize faster in the light than in the dark. It is hypothesized that the phosphorus alters the zirconia film, possibly limiting the oxygen diffusion or the electrical conductivity, and also changing the space charge distribution of the zirconia interfaces.

INTRODUCTION

Phosphorus has been shown to have a beneficial effect on reducing the corrosion rate of various metal alloys. The 1600-year old iron pillar at Delhi, India, has a high phosphorus content (0.25% P) (ref. 1), as do the weathering steels (0.07-0.15% P) (ref. 2). Previous researchers have also shown that phosphorus containing amorphous alloys, which also contain film formers (Cr or Ti), have improved resistance to corrosion compared to amorphous alloys melt-spun with the additions of the other metalloids; C, B, or Si (ref. 3,4).

This research was undertaken so as to determine the mechanism by which phosphorus improves the corrosion resistance. Two amorphous alloys were manufactured and compared under identical electrochemical conditions. The amorphous copper-zirconium alloy system was chosen for experimentation because the amorphous structure eliminates the complications due to crystalline defects (e.g., grain boundaries, precipitates). The Cu-Zr system is also a natural glass former, and the zirconium is a natural film former. The zirconium forms zirconia (ZrO_2) (see Fig. 1), which is normally a protective surface oxide. Two alloys Cu-40Zr, and Cu-42Zr-2P (numbers are atomic percentages) were melt-spun, and tested in various aqueous media.

If the role of the phosphorus in the Cu-Zr system is to stimulate rapid initial dissolution of the unfilmed surface, prior to the passive film formation, as postulated by Hashimoto in the iron-based alloys (ref. 5), then there should be little difference in the corrosion between the Cu-40Zr and the Cu-42Zr-2P. The mechanism by which the phosphorus improves the corrosion resistance has

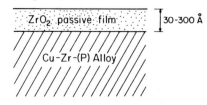

Fig. 1. The passive film on the alloy.

not been proven in this work, but the effect of the phosphorus has been demonstrated.

EXPERIMENTAL

The alloys were melt-spun at MIT, starting with Cu (99.999%), Zr (99.9%), and Cu_3P (99.5%). The original composition of the phosphorus containing alloy was Cu-37Zr-5P, but due to formation of zirconium phosphide dendrites, and to phase separation, the final composition of the melt-spun ribbon was Cu-42Zr-2P. The final ribbon compositions were determined by electron microprobe X-ray analysis.

The electrolyte selected was 1N H_2SO_4, deaerated by bubbling nitrogen gas for thirty minutes, and continued bubbling throughout the tests. The potentiodynamic polarization experiments (Figs. 2-4) were run at a slow scan of 0.1 mV/sec., starting at the open circuit potential, and going in the positive direction. The surfaces were prepared by polishing with 600 grit silicon carbide.

The photo-corrosion current (Fig. 5) was determined by first immersing the sample in solution at +1.0 V (SCE) for thirty minutes, and then illuminating the sample with two 500 watt photographic flood lamps (General Electric, model EBW-No.B2).

The oxide thickness in Figure 6 was estimated by the color of the surface. As an oxide grows thicker, it reflects differing colors, depending on its optical thickness (optical thickness equals refractive index times the actual thickness). Therefore, by comparing the color of the oxide film, with the color on a known silica step gauge (ref. 6), the thickness may be calculated.

The Auger Electron Spectroscopy and the X-ray Photospectroscopy (ESCA) studies were performed at the Surface Analytical Lab at MIT. The incident sputtering current in Figures 7 and 8 is 4 microamp/cm^2.

RESULTS AND DISCUSSION

The potentiodynamic polarization experiment results are displayed in
Figures 2 and 3. The alloys are described as glassy to denote possible short
range order (i.e., not completely random structure). Both the Cu-40Zr (Fig. 2)
and the Cu-42Zr-2P (Fig. 3), though having poor reproducibility, show passive
regions with passive currents on the order of microamps/cm^2. The Cu-42Zr-2P
shows a lower passive region current than does the Cu-40Zr. Pure zirconium
(Fig. 4) is included as a comparison. Its results are more reproducible,
and it also demonstrates a passive corrosion current, similar in magnitude
to the currents for Cu-40Zr and Cu-42Zr-2P.

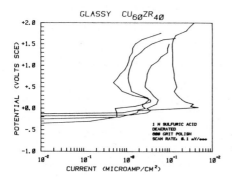

Fig. 2. Potentiodynamic polarization of glassy $Cu_{60}Zr_{40}$.

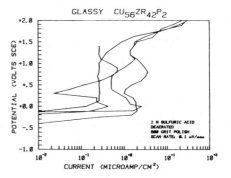

Fig. 3. Potentiodynamic polarization of glassy $Cu_{56}Zr_{42}P_2$.

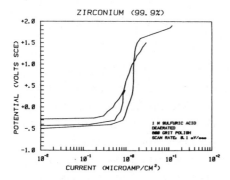

Fig. 4. Potentiodynamic polarization of pure Zr (99.9%).

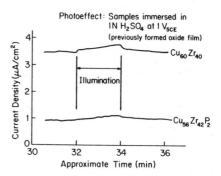

Fig. 5. Photocorrosion of the
alloys.

Illumination by the photographic lamps had no effect on the Cu-42Zr-2P, while causing a 4% increase in the current of the Cu-40Zr (Fig. 5). The slight up-ramping in current is due to the temperature increase under the hot flood lamps. The Cu-40Zr being photosensitive, while the Cu-42Zr-2P is not, suggests that the phosphorus alters the space charge distribution in the metal/metal oxide interface.

Weight loss experiments are shown in Table 1. Chloride solutions are very aggressive towards these alloys. However, as demonstrated in Table 1, the phosphorous significantly reduces the weight loss by a factor of five for the 3.5% NaCl solution, and by a factor of fifty for the 1N HCl.

In simple immersion tests in 1N H_2SO_4, the color of the Cu-40Zr sample was noticed to change over several days, indicating the thickening of the film. The Cu-42Zr-2P maintained a constant golden color, indicating no change in the oxide thickness. The colors corresponded to oxide thicknesses, which were estimated with a silica step gauge (ref. 6). The results are shown in Fig. 6.

TABLE 1: Weight loss in chloride solution.

Weight Loss Experiments$(\frac{mm}{yr})$

Corrosive Solution	Amorphous Alloy Composition	
	$Cu_{60}Zr_{40}$	$Cu_{56}Zr_{42}P_2$
3.5% NaCl	0.14	0.03
1N HCl	2.61	0.05

20 day test at room temperature.

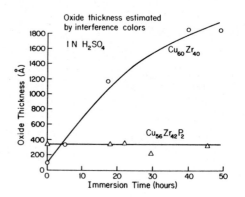

Fig. 6. Increase in oxide thickness.

The two alloys were also compared using the Auger Electron Spectrometer. The surface profiles are shown in Figures 7 and 8. The oxides were formed by anodizing at 1.0 V (SCE) for 1 hour in 1N H_2SO_4. The surface film can be identified as ZrO_2. The Cu-40Zr passive film (Fig. 7) can be seen to be slightly thicker (about 15%) than the Cu-42Zr-2P film (Fig. 8). The alloy/oxide interface is less distinct for the Cu-40Zr, indicating that it is not as protective. It is also noticeable that the Auger spectra identifies approximately 20% phosphorus in an alloy which should only have 2%. This discrepancy is believed to be caused by an inaccurate atomic composition calculation, as well as differing sputtering rates for the phosphorus as compared to the copper or zirconium. Identical tests with the ESCA identified only 10% phos-

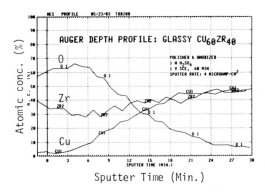

Fig. 7. Composition vs. Depth of $Cu_{60}Zr_{40}$ surface.

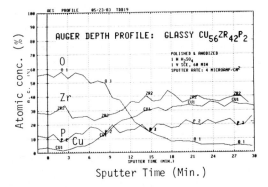

Fig. 8. Composition vs. Depth of $Cu_{56}Zr_{42}P_2$ surface.

phorus in the base alloy (still high). Due to the peak positions, phosphorus is definitely present, however, there is a need to study the quantitative aspects more fully.

The ESCA was used to determine the valence of the phosphorus. In Figure 9 is the expanded ESCA phosphorus peak. Its position is at -137 eV, which corresponds to phosphorus of a valence of +5.

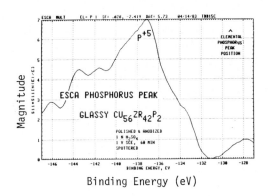

Fig. 9. Energy shift of phosphorus peak.

CONCLUSIONS

Potentiodynamic polarization studies, photoeffect observations, weight loss and immersion tests, and Auger analysis demonstrate that phosphorus improves the corrosion resistance of the copper-zirconium alloy tested. Phosphorus with a valence of +5 has been shown by Auger and ESCA to be present in the oxide film. The phosphorus might alter the ZrO_2 in the following manners:

1. Reduce the oxygen vacancy concentration and therefore reduce the oxygen diffusion rate (ZrO_2 grows by oxygen diffusion (ref. 7)).

2. Alter the semiconductor properties of the ZrO_2 film:
   a) reduce the electrical conductivity
   b) change the space charge distribution

3. Alter the ZrO_2 crystalline or non-crystalline structure.

Some unanswered questions are the following: Where is the P^{+5} in the ZrO_2 lattice? Why is there a difference in the photoelectric effects between the two alloys, with and without phosphorus? In any event, it seems clear that the role of phosphorus in Cu-Zr alloys is different than that proposed by Hashimoto in iron-based glasses.

ACKNOWLEDGEMENT

This work was funded by the National Science Foundation under Grant #DMR 81-06233. The authors wish to express thanks to members of the MIT Corrosion Laboratory and Surface Analytical Facility for their help and input into the program.

REFERENCES

1 G. Wranglen, Corrosion Sci., 10, p. 761-770 (1970).
2 H.E. McGannon, ed., The Making, Shaping, and Treating of Steel, 9th Edition. U.S. Steel,pp. 1149-1154 (1971).
3 T. Masumoto, K. Hashimoto, Ann. Rev. Mater. Sci., pp. 215-233 (1978).
4 M. Naka, K. Hashimoto, T. Masumoto, Sci. Rep. RITU, A-Vol., 27, No. 2, (March 1979).
5 K. Hashimoto, Supplement to Sci. Rep. RITU, pp. 201-216, (March 1980).
6 K.B. Blodgett, Step Gauge for Measuring Thickness of Thin Films, U.S. Patent #2,587,282 (Feb 26, 1952).
7 J.L. Whitton, J. Electrochem. Soc., 115, pp. 58-61, (1968).

Passivity of Metals and Semiconductors, edited by M. Froment
Elsevier Science Publishers B.V., Amsterdam — Printed in The Netherlands

PASSIVATION OF IRON, NICKEL AND COBALT IN CONCENTRATED NITRIC ACID SOLUTIONS

E. STUPNISEK-LISAC[1], M. KARSULIN[1] and H. TAKENOUTI[2]

[1]Institute of Physical Chemistry, Department of Technology, University of
Zagreb, Marulicev trg 20, 41000 Zagreb (Yougoslavia)
[2]Groupe de Recherche n° 4 du CNRS "Physique des Liquides et Electrochimie",
associé à l'Université Pierre et Marie Curie, 4 place Jussieu,
75230 Paris Cedex 05 (France)

ABSTRACT

The iron group metals, Fe, Ni and Co have been known for a long time to be passivated chemically in a concentrated nitric acid. However, few works were performed elucidate the passivation process by means of recent electrochemical techniques. The polarization curves of these interfaces were obtained completely by means of a regulating device having a negative output impedance. Then the anodic and the cathodic contributions in the overall current were determined by the weight-loss measurements and the Faraday's law. Electrode impedance measurements were performed to determine why the potentiostatic regulation is not able to control completely the electrode interface. The addition of paranitroaniline in the electrolyte illustrates the catalytic nature of nitrate reduction. This reaction is of importance for the spontaneous passivation, because it covers the anodic current of metal dissolution leading the metal to the passive state.

INTRODUCTION

It is known since the days of Faraday and Schönbein that Fe passivates spontaneously in   concentrated nitric acid. In spite of their outstanding importance, electrochemical behaviours of Fe, Ni and Co electrodes in a concentrated nitric medium are scarcely investigated by means of modern electrochemical techniques. The aim of this paper is to present several new results, in particular polarization curves exhibiting a multiplicity of steady-states (MSS) at a given potential (ref.1) and also electrode impedance diagrams. The MSS are closely related to localized corrosion (ref.2). The electrode behaviour of these interfaces will be compared in relation to the restoration of the passive state following a coulostatic activation.

EXPERIMENTS

Electrodes were prepared of cylinder rod of Fe, Ni, and Co (Johnson-Matthey), the diameter of which was 5 mm, without any heat treatment. Only the cross-section of rod was allowed to be in contact with electrolyte, forming a rotating disk electrode. Remaining surfaces of metal specimen were covered by allylic

resin loaded with glass-fibre or by a thermal retracting olefinic sheath. The
latter showed a high   stability even in   fuming nitric acid. These
electrodes were immersed in   aqueous HNO_3 solution, the concentration of
which was varied from 8 to 24 M according to the metal investigated. The working electrode was separated from the counter-electrode by a sintered-glass disk,
65 mm in diameter.

RESULTS AND DISCUSSIONS
Iron

    Fig. 1-a shows the steady-state polarization curves of Fe-12 M HNO_3 for three
different rotation speeds (Ω) of the disk electrode. In this figure, two steep
transitions of current are observed, the potential of which depends on the rotation speed for anodic voltage sweep and independent of it for cathodic sweep.
In other words, the curves show a hysteresis cycle. Steep transitions of current and the hysteresis cycle indicate that a potentiostatic regulation is
badly adapted to polarize this electrode interface. Stability of the interac-
ting system interface-regulation is not fulfilled in some bounded potential
ranges. Fig. 1-b illustrates the curves obtained by means of a regulating device having a negative output impedance (NOI) (ref.1). These curves are very different from those shown in Fig. 1-a. That is, the true interface characteristics

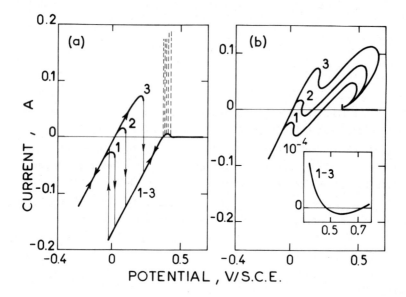

Fig. 1. Steady-state polarization curves of Fe in 12M HNO_3, 25°C desoxygenated
by Ar. Curves (1) rotation speed Ω = 400 (2) Ω = 900 and (3) Ω = 1600 rpm
(a) curves obtained under potentiostatic regulation (b) under NOI (ref.3)

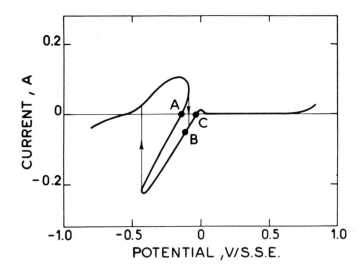

Fig. 2. Steady-state polarization curve, Ni in 8M HNO_3, 25°C Ω = 400 rpm, obtained by NOI. Lines with arrow show steep current transitions observed under potentiostatic regulation.

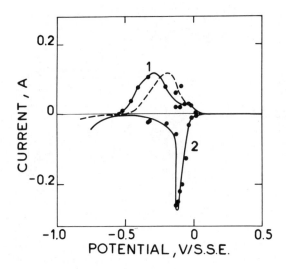

Fig. 3. Anodic (1) and cathodic (2) contributions on the overall current. Potential is corrected for ohmic drop. Experimental conditions as in Fig. 2. The broken line is the polarization curve in the presence of PNA.

330

shown in Fig. 1-b can be hardly foreseen by the potentiostatic curves. In particular, the large anodic loop (0.2 to 0.6 V) is completely ignored by potentiostatic regulation, whereas the cathodic branch located between 0 and -0.5 V
is an experimental artefact due to the unappropriate regulation. The insert in
Fig. 1-b illustrates the polarization curve in the passive range with enlarged
scale in current.

The polarization curve for Ω = 900 rpm intersects five times the voltage
axis. The reason why only two outmost points were stable under open-circuit
conditions was examined previously on the basis of electrode impedance and
Nyquist stability criterium (ref.3).

Nickel

Fig. 2 shows the polarization curve of Ni - 8M HNO_3. MSS can be seen between -0.4 and -0.1 V with respect to saturated mercurous sulfate reference
electrode (SSE). Under potentiostatic regulation a hysteresis loop was observed. The steep current changes are indicated by lines with arrow on this figure. Compared with the case of Fe, the difference between the two regulation
modes is considerably minor.

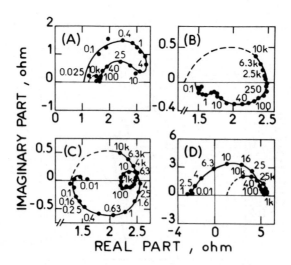

Fig. 4. Impedance diagrams, Ni - 8M HNO_3. Diagrams A to C corresponds the
points marked A to C in Fig. 2. Diagram D in the presence of PNA E = 0 V,
I = 25 mA.

The current observed includes both the anodic dissolution of metal and
the cathodic reduction of nitrate. These two contribution was separated on
the basis of Faraday low with weight loss measurements.

Results are illustrated in Fig. 3. In this figure, the potential is corrected for ohmic drop due to the electrolyte resistance (1.3 ohm). It may worth noting that once this correction is made, MSS no longer exists in contrast also with the case of Fe (ref.3). The anodic curve (1) shows a bell-shaped polarization curve with a shoulder near -0.1 V where the reduction reaction is observed. The cathodic curve reveals two potential ranges for the reduction reaction : one with a narrow peak near -0.12 V and the other for E < -0.5 V. The reason why the cathodic reaction is suddenly hindered around -0.12 V is not yet under-stood and deserves further investigations.

Electrode impedances were measured at various polarization points. A few of them are shown in Fig. 4. Diagrams A to C correspond to the points labelled in Fig. 2. In   these diagrams, the electrolyte resistance, high frequency limit of the impedance, and the polarization resistance, low frequency limit, are clo-se to each other corresponding to a steep polarization curve once the ohmic drop is corrected. Diagram A indicates that this state is unstable under both potentiostatic and galvanostatic regulations. Only NOI allows to obtain such a characteristics in agreement with the stability theory (ref.4). But, it can be concluded that the electrolyte resistance is the main cause of the failure of stability under a potentiostatic regulation.

When an excess of paranitroaniline (PNA) is added into the solution, nitri-te ions present spontaneously in HNO_3 solution, are destroyed. The polariza-tion curve thus obtained is shown by dashed line in Fig. 3, the potential is corrected for ohmic drop. The cathodic reduction observed near -0.12 V is com-pletely suppressed by the addition of this aniline. The catalytic effect of nitrite on nitrate reduction is thus established as in the case of Fe electro-de (ref.3). The active dissolution of Ni is then very similar to that determi-ned by weight-loss measurements (curves) though potentials are shifted by about 0.1 V towards cathodic direction for E < -0.05 V. In other words, there is no coupling between Ni dissolution and nitrate reduction. This contrasts also with the case of Fe passivation in nitric acid (ref.3).

Fig. 4-D shows the impedance diagram at 0 V in the presence of PNA. If this diagram is compared to Diagram C, it can be concluded that impedance below 4 Hz in Diagram C is essentially related to the nitrate reduction process. The close values of the electrolyte and the polarization resistances can then be linked to the catalytic nature of this latter process.

Cobalt

Unlike   Fe and Ni, Co is known as a poorly passivating metal (ref.5,6). In Fig. 5-a, the polarization curves of Co in 24M HNO_3 are shown. If the electrode surface is facing downward, the dissolution rate at zero over-

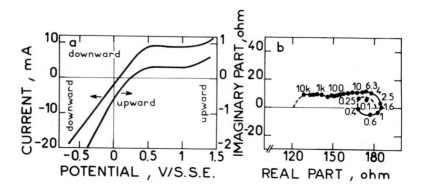

Fig. 5. Co in 24M HNO$_3$ (a) Polarization curve with the electrode facing downward (Ω = 400 rpm) and upward (stationary electrode) (b) Impedance diagram at zero overall current. Electrode facing upward and stationary.

all current, that is at the open-circuit corrosion potential, is evaluated to about 1 A cm^{-2} as evaluated by weight-loss measurement. Co is active and is considered to go into solution as Co^{2+}. If the electrode surface is facing upward, the current at a constant potential decreases markedly and the corro-sion potential is shifted by about 0.2 V in the anodic direction. The metal is considered to be passivated. However, both polarization curves are very similar in shape. Furthermore, as seen in Fig. 5-b, for the upward electrode the elec-trolyte resistance is equal to 120 ohms, whereas for the downward electrode this resistance is only about 25 ohms. It is also observed that for the upward elec-trode, its surface is covered by a raspberry coloured layer. That is, the for-mation of a compact cobalt nitrate salt resulting from the concentrated medium, protects the metal surface against active dissolution. This differs essentially from the passivation of Fe and Ni which is due to an extremely thin oxide-like passive layer.

Coulostatic activation

In 12M HNO$_3$, both Fe and Ni passivate spontaneously. If a cathodic coulosta-tic impulse is applied to those electrodes, the metals first activate, and then recover their passive state marking a potential arrest. These phenomena are il-lustrated on Fig. 6. Simultaneously, the intensity of the light reflected at the metal surface is shown below each potential-time curve. The darkening of the reflected light indicates the formation of a nitroso-complex at the metal surface (ref.7). It is also noticed that, because of strong catalytic effect of Fe^{2+} on the reduction of nitrate[3], the passivation of Fe is far faster than that of Ni.

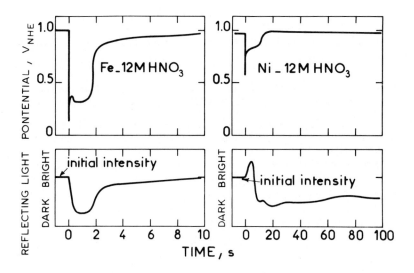

Fig. 6. The potential change after a coulostatic activation for Fe and Ni in 12M HNO₃. Below each figure, the change of reflected light intensity showing the synchronous formation of nitroso-complex.

CONCLUSION

The polarization curves obtained with a potentiostat for Fe and Ni electrode show both a hysteresis loop and the reduction of nitrate with a catalytic effect of nitrite on (nearly) passive metal surface. If the curves were obtained with a regulating device having a negative output impedance, a continuous change from active to passive state can be observed. On Fe electrode, the latter shows that the reduction of nitrate on passive Fe is due to the incapacity of the potentiostat to regulate suitably the electrode interface. On the other hand, on Ni, that process is always present, the hysteresis cycle is due to the ohmic drop across the electrolyte resistance which transforms a bell-shaped curve into a multi-steady-state polarization curve. Co is difficult to passivate in nitric acid. In this case, it seems that the so-called passivation observed in fuming nitric acids is different in nature from that of other metals of the iron group.

334

REFERENCES

1 I. Epelboin, C. Gabrielli, M. Keddam, J.C. Lestrade and H. Takenouti,
   J. Electrochem. Soc., 119, 1632 (1972).
2 I. Epelboin, C. Gabrielli, M. Keddam and H. Takenouti, Z. Phys. Chem. N.F.,
   98, 215 (1975).
3 C. Gabrielli, M. Keddam, E. Stupnisek-Lisac and H. Takenouti, Electrochim.
   Acta, 21, 757 (1976).
4 I. Epelboin, C. Gabrielli, M. Keddam and H. Takenouti, in Comprehensive
   Treatise of Electrochemistry, edited by Bockris, Conway, Yeager, and White,
   pp. 151 - 192, Plenum Press, New York, (1981).
5 I. Epelboin, C. Gabrielli and Ph. Morel, Electrochim. Acta, 18, 509 (1973).
6 E. Stupnisek-Lisac and M. Karsulin, 33rd Meeting of ISE, Ext. Abst. p. 387,
   Lyon, Septembre 1982.
7 E. Stupnisek-Lisac and M. Karsulin, 3rd International Symposium on Passivity,
   Dresden, (1975).

Passivity of Metals and Semiconductors, edited by M. Froment
Elsevier Science Publishers B.V., Amsterdam — Printed in The Netherlands

NICKEL PASSIVATION-DEPASSIVATION IN SULPHURIC ACID : INFLUENCE OF THE GRAIN SIZE

M. CID and M.C. PETIT
Laboratoire de Mécanique Physique, E.R.A. CNRS N° 769
Université de Bordeaux I, 351 cours de la Libération,
33405 TALENCE CEDEX - France -

ABSTRACT

The aim of our study is to investigate the influence of the grain size on the passivation-depassivation phenomenon, for nickel in normal sulphuric acid. Up to a value of 700 mV vs S.S.E. an instantaneous depassivation leads to an I(t) response which tends towards a stable value of the current, whatever the electrode structure might be. Corrosion is essentially of an intergranular type. For higher values of potential, electrode response during depassivation is quite different. Current intensity never tends towards a stable value whatever the grain size might be, this leads to different corrosion facies.

I. INTRODUCTION

Many parameters such as metal composition and structure, as well as previous heat treatment, have an influence on passivation-depassivation phenomena, and, subsequently, on the ability to intergranular corrosion[1]. Indeed, within transpassive range, nickel in sulphuric acid shows an intergranular corrosion phenomenon, although passive layer is not entirely destroyed[2]. So, one can observe selective dissolution of grain boundaries, in spite of a more or less protective layer. But that dissolution leads to different attack facies, depending on depassivation potential, as well as on grain size of the metal. That will lead to a specific I(t) response in each case[3]. We will study two parameters simultaneously : depassivation potential, and structure of the metal. Evolution of current intensity, after depassivation, versus time and resulting attack facies will be correlated.

II. EXPERIMENTAL CONDITIONS

In this study, we compared experimental results obtained with two electrodes, of identical chemical composition, but of different metallographic structures. The two kinds of nickel samples (cylindrical electrodes)will be :
- Johnson polycristalline nickel electrode, without any previous heat treatment, showing very irregular, small size grains,

- Johnson polycristalline nickel electrode, with 1000°C, sixteen hours
  pretreatment, then water-quenched, to cancel strains introduced by
  tooling ; this kind of sample presents regular mid-sized grains.

Experimental setting is of the potentiostatic type, using a rotating disk
electrode. The reagent is a normal sulphuric acid solution, set to a constant
temperature of (25 ± 1)°C. Samples, all presenting the same value of their active section, undergo the same surface treatment, which consists of a 1/4μ Boron
nitride diamond paste polishing, followed by a passivation step at 0 mV vs S.S.E.
until current intensity is constant.

So passived electrode is then shortly brought to the desired value of potential (i.e. 700 mV, and 1000 mV vs S.S.E.) and thus undergoes a more or less
fast and complete depassivation.

As experimental results are compared for different grain size samples, we
will talk about current intensity, rather than current density. Selecting I(t)
curves is of interest, because, in such a way, we will be able to correlate curve shapes, and corresponding attack facies[4].

The microphotographies we present here, observed with a scanning electron microscope, all correspond to the same value of the quantity of electricity passed (i.e. ≃ 1 Coulomb), and subsequently to the same quantity of dissolved metal.

III. EXPERIMENTAL RESULTS

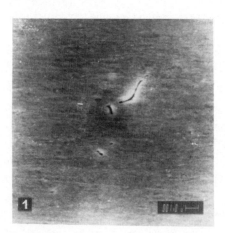

Microphotography 1. Heat treated electrode, 0 mV vs S.S.E.

Both parameters, potential and structure, are simultaneously studied. Electrodes which have been passived at a potential of 0 mV vs S.S.E. do not show
any visible attack of the grain boundaries and matrix, in the above experimental
conditions. However, at such a potential, as 1 Coulomb quantity of electricity
passed leads to a pitting phenomenon which may reveal the existence of not very
clearly marked grain boundaries (microphotography 1).

1°/ Depassivation at a potential value of 700 mV vs S.S.E.

Short depassivation of the electrode leads to I(t) curves which tend to-wards stable values of current (figure 1).

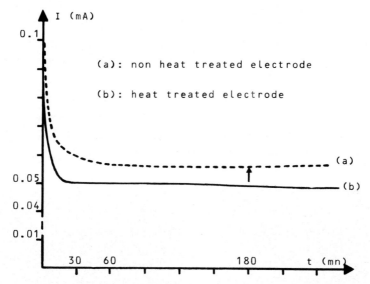

Fig.1. Evolution of current of two kinds of electrodes at 700 mV vs S.S.E. after depassivation

Attack facies (cf. microphotographies 2 and 3) are of the same type.

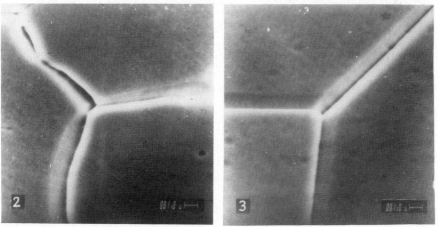

Microphotographies 2. Non heat treated electrode, 700 mV vs S.S.E.
3. Heat treated electrode, 700 mV vs S.S.E.

Corrosion is essentially of the intergranular type, with clearly marked joints. However, one may observe that current shows a very light drift after three hours for irregular small grain size electrodes. This comes out with intergranular grooves which are not so clearly marked. But, on the other hand, for regular grain size samples, grain boundaries show a regular opening.

However, at such a value of depassivation potential, the influence of structure is not preponderant. Sample attack remains well located, and grain is as a whole passived.

2°/ Depassivation at a potential value of 1000 mV vs S.S.E.

At such a potential, current due to dissolution becomes higher in the hump, that leading to the simultaneous attack of grains and grain boundaries, the latter loosing their edges (microphotographies 4 and 5).

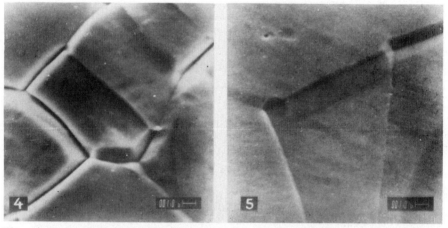

Microphotographies 4. Non heat treated electrode, 1000 mV vs S.S.E.
5. Heat treated electrode, 1000 mV vs S.S.E.

One can observe, on I(t) curves, a current drift whatever the structure might be (figure 2).

However, drift comes out earlier and is of greater importance when sample is of the small grain size type, (fig. 2a). This is quite clear when looking at the attack facies : irregular small grain size samples present a corrosion of a more heterogeneous type (microphotography 4) than regular grain size samples do, (microphotography 5).

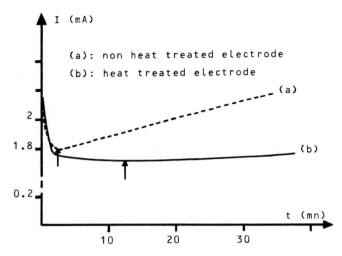

Fig.2. Evolution of current of two kinds of electrodes at 1000 mV vs S.S.E.
after depassivation

IV. CONCLUSION

For light values of depassivation current, tending towards stable values
(under 700 mV vs S.S.E.), attack is essentially of the intergranular type. The
influence of sample structure comes out chiefly with a more or less clearly
marked opening of grain boundaries.

For higher values of depassivation current, with constant current drift,
(above 700 mV vs S.S.E.), simultaneous attack of both grains and grain boundaries can be observed. Corrosion is clearly more heterogeneous with the irregular small grain size type of structure, than with the other one.

REFERENCES

1   P.L. Bonora, R. Fratesi and G. Roventi, Materials Chemistry, 6, 111-116,
     1981.
2   L. Beaunier and M. Froment, C.R.A.S., série C, 279, 91, 1974.
3   S. Ruchaud-Corretja, thesis Bordeaux I, 1982.
4   M. Cid, S. Ruchaud-Corretja and M.C. Petit, 33rd Meeting of I.S.E., Lyon,
     242-244, 1982.

THE PASSIVATION-DEPASSIVATION BEHAVIOUR OF GOLD IN A HCl-GLYCEROL SOLUTION AND
ITS APPLICATION TO ELECTROPOLISHING

J. VERLINDEN, J.P. CELIS and J.R. ROOS

Departement Metaalkunde, K.U.Leuven - B-3030 Heverlee (Belgium)

The anodic behaviour of gold in aqueous HCl-solutions has been discussed by
several investigators.[1-4] In a certain potential range, periodic current oscillations are observed. These oscillations are attributed to the competition
between two processes :

- the formation of Au (III) oxide on the electrode surface (passivation)
- the redissolution of this oxide by the Cl$^-$ ions in solution (depassivation)

This phenomenon has been used in our laboratory for electropolishing and electro-
thinning of thin gold electrodeposits for TEM examination. As the gold dissolution rate in aqueous HCl-solutions was too high for accurate control of the
thinning process, a 25% HCl - 75% glycerol solution was selected. The occurrence of passivation-depassivation phenomena in such a solution is reported in
the present work.

Gold specimens were galvanostatically prepared on electrolytically polished
brass electrodes from an additive free cyanide bath (pH 7) at current densities
of 0.5, 3.5 and 6 A/dm^2. A rotating electrode set-up was used to obtain deposits under reproducible mass-transport conditions. In this way, deposits with
constant composition but different morphologies were obtained. (see fig. 1a and
2a for the two extreme morphology types in the as-plated condition).

Anodic polarisation curves of these deposits, recorded in a 25% HCl - 75%
glycerol solution at room temperature, are shown in fig. 3. The current drop
recorded at a potential of about + 1.52 V vs SCE is due to the formation of a
passive Au (III) oxide film on the specimen surface[1-3]. This "passivation potential" seems to be independent of the structure and the morphology of the
specimen. The magnitude of the current drop, however, decreases with increasing surface roughness of the specimen surface.

At constant potential close to this "passivation potential", periodic current
oscillations are observed similar to these mentioned in literature for aqueous
HCl-solutions. These oscillations are shown schematically in fig. 4, together
with a definition of some characteristic parameters. The general shape of
these oscillations, and consequently the magnitude of the defined parameters,
seem to be dependent on the value of the anodic potential, as shown in fig. 5.

342

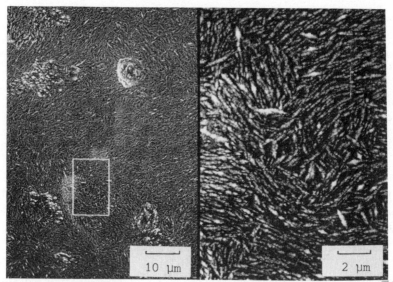

Fig. 1a. SEM-topography of as plated gold deposits prepared at 0.5 A/dm^2 from an additive free gold cyanide bath.

Fig. 1b. SEM-topography of potentiostatic electropolished specimen shown in fig. 1a.

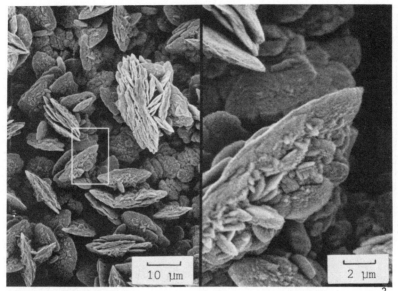

Fig. 2a. SEM-topography of as plated gold deposits prepared at 6 A/dm^2 from an additive free gold cyanide solution.

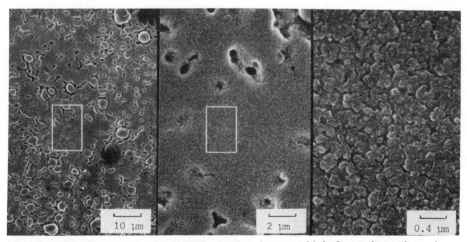

Fig. 2b. SEM-topography of potentiostatic electropolished specimen shown in fig. 2a.

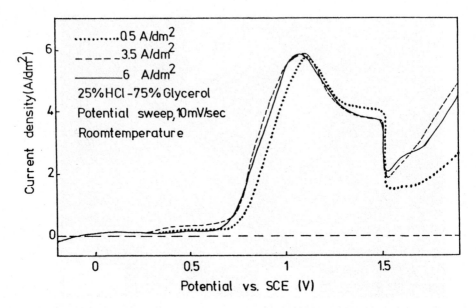

Fig. 3. Anodic polarisation curves of gold electrodeposits in a 25 % HCl – 75 % glycerol solution. The electrodeposits used were obtained in an additive free gold cyanide bath, pH 7 at 0.5, 3.5 and 6 A/dm².

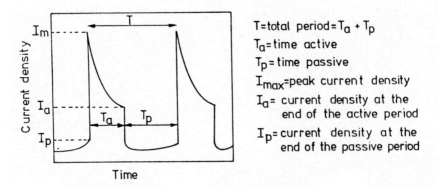

Fig. 4. General shape of the periodic current oscillations observed for gold under potentiostatic control in HCl-solutions, and definition of some characteristic parameters.

345

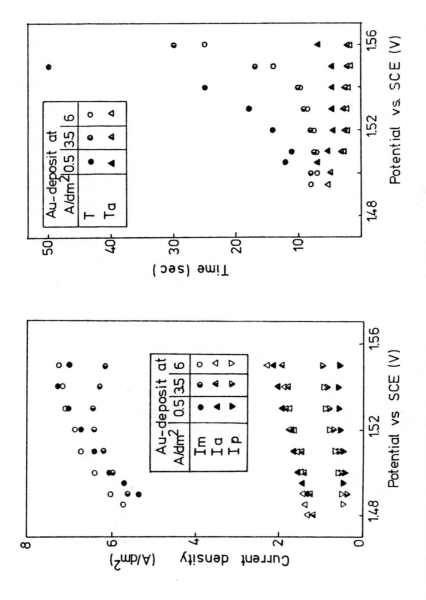

Fig. 5. Some characteristic parameters of the periodic current oscillations (see fig. 4) plotted versus anodic potential, for gold electrodeposits in a 25% HCl – 75% glycerol solution.

The structure and morphology of the specimen, however, have only a slight influence on the shape of the current oscillations.

In order to evaluate the quality of this treatment for electropolishing gold surfaces, the gold specimens deposited at 0.5 and 6 A/dm^2 were held at a potential of about + 1.52 V vs SCE in the HCl-glycerol solution mentioned above. Electropolished specimen surfaces were analysed by SEM. Figures 1b and 2b gives an excellent evidence of the quality of the suggested electropolishing technique. Moreover the results obtained are independent of the initial surface topography of the specimens.

The small dissolution rate of gold in the 25% HCl - 75% glycerol solution allows a potentiostatic electropolishing technique to be effective for the thinning of rough gold specimens for TEM-examination. An even better control of the thinning process was obtained by applying a cyclic potential sweep between two potential values, one in the active and one in the passive range. In this way the frequency of the oscillations can be chosen and the ratio active to passive time is controlled. A higher frequency results in a better electropolish in accordance with an earlier report by Peck[5]. A short active period coupled with a long passive period, determined by the chosen sweep rate, slows down the dissolution rate. This results in a better controllability of the thinning process.

REFERENCES

1 Podesta, Piatti and Arvia, Electrochim. Acta, (24) 633-638 (1979)
2 Heumann and Panesar, Zeitschr. Phys. Chem., 229 84-97 (1965)
3 Frankenthal and Siconolfi, J. Electrochem. Soc., 129 (6) 1129-1195 (1982)
4 Just and Landsberg, Electrochim. Acta, (9) 817-826 (1964)
5 Peck and Nakahara, Metallography, 11 347-354 (1978)

Passivity of Metals and Semiconductors, edited by M. Froment
Elsevier Science Publishers B.V., Amsterdam — Printed in The Netherlands

CREVICE CORROSION TEST FOR STAINLESS STEELS IN CHLORIDE SOLUTIONS

R.O.Müller
BBC, Dept. ZLC, CH-5401 Baden, Switzerland

ABSTRACT

A new method is presented to measure the crevice corrosion resistance of stainless steels. Depending on the temperature a potential-chloride diagram is obtained where the region of stability is visible. If the free corrosion potential is combined with the diagram, limiting chloride values can be extracted.

INTRODUCTION

In spite of their high corrosion resistance, stainless steels may suffer crevice corrosion in chloride solutions. Therefore it is very important to have a test method to measure their crevice corrosion resistance (ref.1). One method consists in bolting a grooved delrin washer on a test specimen and exposing it to ambient conditions (ref.2). Although the results are realistic, the test is not convenient, as its duration is too long.

A well known accelerated procedure (ref.3) is the $FeCl_3$-test which allows to compare different materials in a short time. Its limitation lies in the difficulty to extrapolate the results to real situations.

That is why we introduced a new method.

METHOD

Critical crevice corrosion potential (U_s)

The sample is located at the bottom of a container in stagnant electrolyte solution. Diameter of the steel disc: 42 to 46 mm, free surface, filled with electrolyte: 30 ÷ 36 mm. To form defined crevices, several layers of glass beads are deposited on the sample which is connected to a potentiostat (Fig.1). About 200 glass beads with a diameter of 4 ÷ 4,5 mm were used, giving approximately a covering with three layers.

The connection with the reference electrode is made by a luggin capillary at the surface of the sample. The counter electrode is located in the upper electrolyte compartment.

The electrolyte above the sample is kept stagnant by a porous membrane filter. Above it, the solution is recirculated and under thermostatic control.

For a test the potential is set at a given level and the current is observed in function of the time. If the current stays below 3 $\mu A/cm^2$ (30 mA/m^2) for

348

more than 12 h, the potential is increased until, after a certain time, the current starts to increase. This indicates the onset of crevice corrosion. The potential is then lowered by 10 ÷ 50 mV and the current-time behaviour again observed. The critical potential for crevice corrosion (U_s) is the highest potential at which the crevice sample just repassivates. This value depends in sodium chloride solutions on the chloride concentration and the temperature. (All potential values are reduced to the normal hydrogen scale). The critical potentials, measured in solutions with different chloride content are presented in chloride-potential diagrams. All critical potentials of a given alloy at a certain temperature can be connected by a line which divides the field in areas where crevice corrosion will occur or not (see Fig. 2-9).

The lowest potential in every U_s-curve, marked x (critical limiting potential U_G), was measured in a cell where the glass beads have been omitted and the filter membrane replaced by an anion exchange membrane (ref.4). Highly concentrated and acidic solutions of metal chlorides occured in contact with the stainless steel sample.

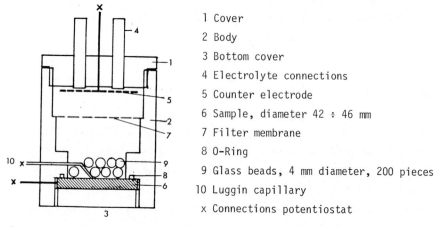

1 Cover
2 Body
3 Bottom cover
4 Electrolyte connections
5 Counter electrode
6 Sample, diameter 42 ÷ 46 mm
7 Filter membrane
8 O-Ring
9 Glass beads, 4 mm diameter, 200 pieces
10 Luggin capillary
x Connections potentiostat

Fig. 1. Crevice corrosion test cell filled with electrolyte.

Free corrosion potential

To evaluate the stability of a given material against crevice corrosion, the free corrosion potential (U_R) must also be known. For this test it is important to choose conditions, where all the surface of a sample is in the passive state. Samples in the form of rods of sheets were mechanically cleaned, pickled in a mixture of hydrofluoric and nitric acid and passivated in nitric acid.

Afterwards they were dipped in diluted (1 mg/kg), stirred sodium chloride solutions at a defined temperature.

Starting with a pH of 8, adjusted by the addition of sodium hydroxide, the potential was recorded in function of the time for 8-12 h. Then the pH was decreased by 1 unit with addition of hydrochloric acid and the procedure repeated. Most stainless steels behaved like a pH-electrode for a broad range of pH-values. Some samples suffered from pitting attacks at pH-values of 4 and below 4. This was indicated by a lower free corrosion potential U_R.
Tests made at other chloride contents at pH=7 in the range of 1000 mg/l down to < 1 µg/l showed that the chloride content had little influence on the free corrosion potential, provided that the sample stayed completely passive. The oxygen content at a given pH had the effect that the free corrosion potential in oxygen-free solutions was about 50 - 100 mV lower (4).

Evaluation of the stability against crevice corrosion

If the critical potentials for crevice corrosion are combined with the free corrosion potentials, one can predict the corrosion behaviour in respect to crevice corrosion. It results an upper chloride limit for a given temperature below which the material will be stable to crevice (and of course pitting) corrosion. The following diagrams show the behaviour of different stainless steel types at 25 and 60 °C. Above the marked line, crevice corrosion will take place. In fig.2 e.g. the free corrosion potential of pH=8 (■) is intersecting the line of the critical potential at a chloride content of 1,1 g/l. At higher chloride values crevice corrosion will occur, at lower not. For the same steel the free corrosion potential in a solution of pH 5 (◊) will be higher and intersect the critical corrosion limit already at chloride values of 0,1 g/l.

DISCUSSION

It can be seen from the diagrams (Fig. 2-9) that crevice corrosion will occur at far lower potentials than pitting corrosion, especially in diluted chloride solutions. It is also distinctly visible that the pH of the medium has a pronounced effect on the stability, especially by its action on the free corrosion potential of the passive surface. That makes a material tolerate an order of magnitude higher chloride concentrations when changing the pH from 6 to 8. In the fig. 6 and 7 for 1.4539, (X2 NiCrMoCu25205) it can be seen that the critical potential for crevice corrosion differs drastically with temperature, whereas the free corrosion potentials stay relatively constant. This means that the material is stable to higher chloride concentrations at 20 °C than at 60°C. On the same two figures one finds the critical potentials for pitting. They are much higher than the ones for crevice corrosion. In the field between the critical potentials for crevice and pitting corrosion only attacks by crevice corrosion will occur. Such attacks are the reason for the opinion that stainless steels will only be stable when crevices are absolutely excluded.

The tests performed have only delt with chloride solutions, which are technically very important, if one thinks at the broad application in sea water and many other salt solutions. But the test can easily be applied to other electrolytes and will give important answers to many questions.
The advantage of the presented method over classical exposure tests (ref.2) is the fact that the presented method gives more accurate results in much shorter times. In addition, due to the separate evaluation of the anodic reaction in the crevice and the cathodic reaction on the free surface in the passive state the influence of different parameters, such as pH and temperature can be evaluated separately. This allows a better understanding of the mechanism of crevice corrosion. The $FeCl_3$-test (3) is only intended to compare different materials quickly. But as only one parameter, the temperature is varied, all other influences, such as pH or oxygen content of the solution cannot be investigated.

CONCLUSIONS

In our opinion, crevice corrosion is in technical applications more important than pitting, as it is nearly impossible to keep surfaces clean and free from crevices. The diagrams show, that it is possible to choose materials which will be stable against crevice corrosion under the given conditions. This allows us to design with crevices which is very convenient and often less expensive. The presented method allows to test new alloys quickly and accurately in respect to crevice corrosion and thus makes materials selection safer and faster.

REFERENCES

1 F.P. Ijsseling, Br.Corros. J., 15, 2, 51-69 (1980)
2 D.B. Anderson, ASTM Spec. Techn. Publ. 576, 1974, p. 231.
3 ASTM G 48-76.
4 R.Scheidegger, R.O.Müller, Werkst. u. Korr. 31, 387-393 (1980).

APPENDIX

Fig. 2-9. Chloride-potential diagrams.
Below the line the alloy is stable against crevice (resp. pitting)corrosion, above not.

○\●	Critical potential for crevice corrosion (U_s)
○	Repassivation ● Crevice corrosion
–▲–	Critical potential for pitting corrosion (U_L)
◇□■	Free corrosion potentials in solutions with a chloride content of 1 mg/l and pH= 5, 7, 8.
X	Critical limiting potential U_G

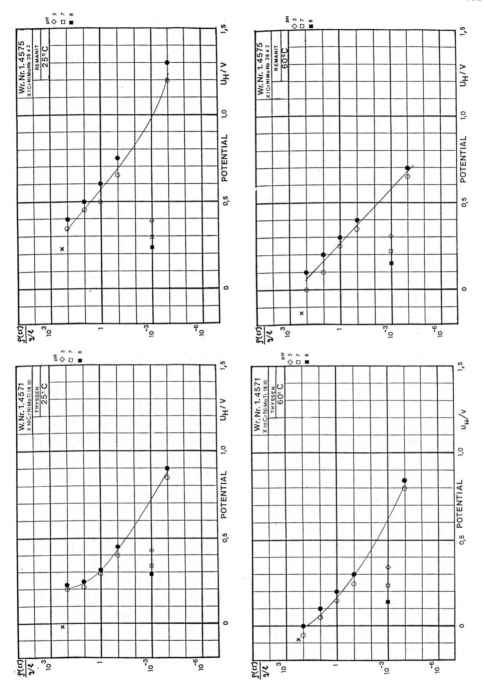

352

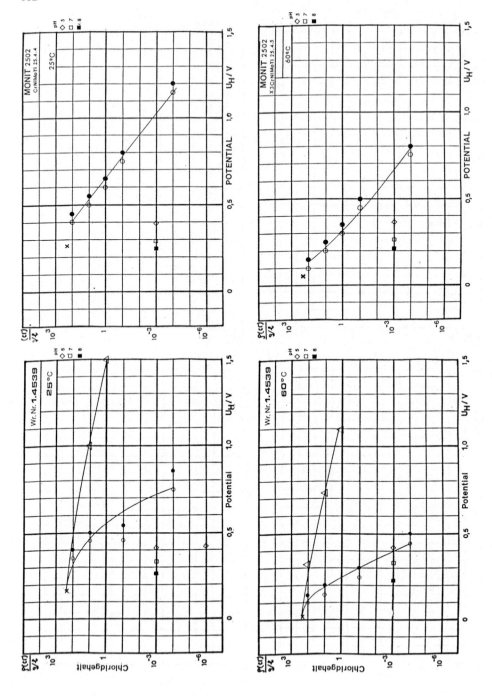

Passivity of Metals and Semiconductors, edited by M. Froment
Elsevier Science Publishers B.V., Amsterdam — Printed in The Netherlands

353

INITIATION AND INHIBITION OF PITTING CORROSION ON NICKEL

S.M. ABD EL HALEEM, M.G.A, KHEDR, A.A. ABDEL FATTAH AND H. MABROK

FACULTY OF EDUCATION, ZAGAZIG UNIVERSITY, ZAGAZIG (EGYPT)

ABSTRACT

Up to a certain concentration, Cl^- ions have no influence on the passive state of nickel in HNO_3 and NaOH media. However, higher aggressive ion concentrations cause destruction of passivity and initiation of visible pits. The change in integrated anodic charge, Δq_a, is taken as a measure of the extent of pitting. The latter varies linearly with $\log C_{Cl^-}$. Pitting corrosion potential varies with $\log C_{Cl^-}$ according to segmoidal curves. Organic amines inhibit the effect of Cl^- ion by influencing both anodic and cathodic partial reactions.

INTRODUCTION

Pitting of Ni and Ni alloys occurs when passivity breaks down at local points on surfaces exposed to corrosive environment containing Cl^- or other aggressive anions (1,2). At these points anodic dissolution proceeds whilst the major part of surface remains passive. Pitting of Ni has been shown to develop prefrentially at grain boundaries and imperfections in the surface (1). The effect of solution composition, pH, temperature and alloying elements on the pitting of Ni has been studied by several authors (1,3,4).

In the present study the technique of potentiodynamic polarization was used to investigate the pitting corrosion of Ni in HNO_3 and NaOH solutions, Trials were made to inhibit such type of attack using some organic amines.

EXPERIMENTAL

Spectroscopically pure Ni (Johnson & Matthey, England) was used as the test material throughout the present work in the form of short rods with total exposed area of $0,5cm^2$. Pretreatment of electrodes including polishing with 0-,oo- and 000- grade emery papers, degreasing with acetone and finally washing with twice-distilled water. Complete wetting of the surface was taken as indication of its cleanliness. Chemicals used were of A.R, quality. All solutions were prepared using twice-distilled water and no trials were made to dearate them. The electrolytic cell was all pyrex and described elsewhere (5).

354

Potentiodynamic polarization curves were performed using a
Wenking Standard potentioscan Type POS 73. Potential current curves
were recorded on X-Y recorder Type Advance AR 2000. All potentials
were measured relative to saturated Calomel half cell (S.C.E.) at
$25 \pm 1^{\circ}$ C.

RESULTS AND DISCUSSIONS

Figs. 1 and 2 show the cyclic voltammograms (CVs) of the nickel
electrode in 0.1M solutions of HNO_3 and NaOH devoid of-and containing increasing additions of Cl^- ions at a scanning rate of 10 and
25mv/sec, respectively. In Cl^--free HNO_3 solution, the CVs are
characterized by a well defined anodic dissolution peak and an
extending passive region before O_2-evolution commences. A model was
proposed (6) involving participation of both NO_3^- and OH^- ions, in
which all reactions but dissolution involve adsorption on a primary
adsorbed dipoles like $Ni-NO_3$ or $Ni-OH$, and never direct adsorption
on Ni. Passivation occurs by adsorption of OH^- on the same "dipole".

In NaOH solution, Fig.2, the CVs are characterized by an extended
stable passive region which is attributed to chemical or electrochemical transformation of $Ni(OH)_2$ into more difficult electroredu-
ced NiO (7,8).

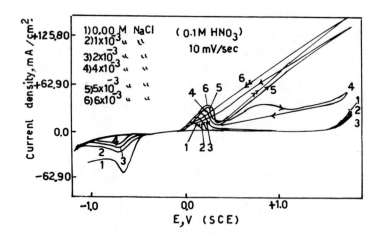

Fig.1. Cyclicvoltammograms of Ni in 0.1M HNO_3 in presence of
increasing Cl^- ion concentrations.

Inspection of Figs.1 and 2 reveals the following conclusions to be drawn:

i) The dissolution current density flowing in the active region of Ni in HNO_3 increases with increasing concentration of Cl^- ions due to adsorption of these ions on the bare metal surface (9).

ii) Presence of Cl^- ion up to a certain concentration has practically no influence on the dissolution kinetics of passive films in both media. However, higher Cl^- ion contents cause the currents to increase at potentials more active the higher the Cl^- ion concentration due to destruction of passivity and initiation of pitting. When the potential of Ni is reversed into cathodic direction, the flowing currents decrease gradually with potential while

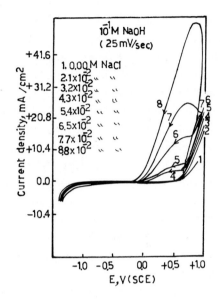

Fig.2. Cyclic voltammgrams of Ni in 0.1M NaOH in presence of increasing Cl^- ion concentration.

remaining always in the positive branch of the CVs due to continuous propagation of the formed pits. The change in anodic charge amount, Δq_a, integrated in presence and absence of Cl^- is is taken as a measure of extent of pitting. Δq_a varies with Cl^- ion concentrations according to:

$$\log \Delta q_a = a + b \log Cl^- \qquad (1)$$

where "a" and "b" are constants.

Figs.3 and 4 show potentiodynamic polarization curves of Ni in 0.01M solutions of HNO_3 and NaOH, respectively, at very slow scanning rate. Increasing Cl^- ion content causes the current flowing in the passive region in both media to increase markedly at some critical potentials due to destruction of passivity and initiation of visible pits. The critical pitting potential, $E_{pitting}$, varies with

356

logC$_{Cl}$- according to segmoidal S-shaped curves, Fig.5. These curves
are explained on the basis of formation of passivitable, limiting
active and continuously propagated pits depending on the range of
Cl$^-$ ion concentration (10).

Initiation of pitting corrosion of Ni in HNO$_3$ solution is
assumed to occur through competitive adsorption of Cl$^-$ ion with
both OH$^-$ and NO$_3^-$ ions for adsorption sites of metal surface. When
Cl$^-$ ion reaches a concentration in the electrical double layer, as
the potential becomes more positive, sufficient to overcome that
of OH$^-$ and NO$_3^-$, it succeeds in destroying passivity and initiating
pitting corrosion. However, in NaOH solution, Cl$^-$ ions are assumed
to adsorb well on passiviting oxide film, followed by their penetration through oxide under influence of the electrical field across
film/solution interface. When the latter field reaches a critical
value, corresponding to pitting potential, Cl$^-$ ions undermine the
oxide film either by vacancy condensation at metal interface, or
by releasing cations rapidly at solution interface, so that in

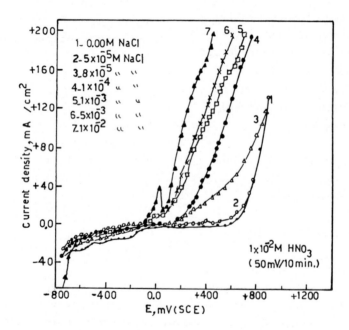

Fig.3. Potentiodynamic anodic polarization curves of Ni in
presence of increasing Cl$^-$ ion concentrations,

either case pitting proceeds.

Addition of mono-methyl-, dimethyl- and monoethyl amines to NaOH containing Cl^- ions causes a shift of pitting potential into noble direction indicating increased resistance to pitting, Fig.6. Urea behaves bifunctionally inhibits pitting at low concentrations and promotes it at higher additions. However, hydrazine cannot be used as a pitting corrosion inhibitor for nickel in alkali media.

Inhibition of pitting corrosion of Ni by organic amines in NaOH solution could be atributed to:
i) Competitive adsorption between OH^- and Cl^- ions for adsorption sites on the metal surface, and
ii) Adsorption of cations of the form RNH_3^+ on cathodic areas of metal surface.

The increased tendency of amines inhibitors of pitting of Ni in the order monomethyl $<$ monoethyl $<$ dimethyl is associated with the change of electron density on the nitrogen atom of amine, on one hand, and to their

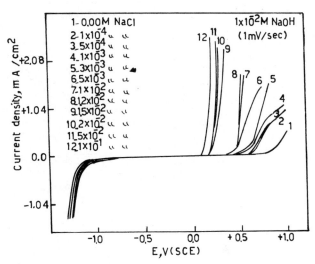

Fig.4. Potentiodynamic anodic polarization of Ni in 0.01M NaOH in presence of increasing Cl^- ion concentrations.

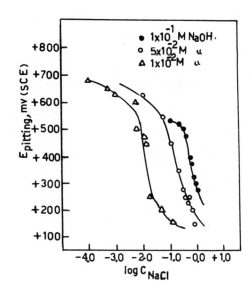

Fig.5. Variation of pitting potential with $\log C_{Cl^-}$ in 0.1M NaOH.

358

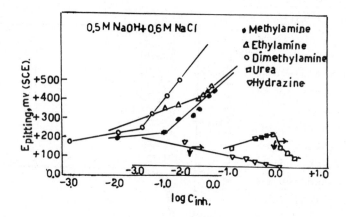

Fig.6. Variation of pitting potential
with logC$_{inh}$ in 0.5M NaOH + 0.6M NaCl.

capacity to adsorb on Ni surface, on the other.

REFERENCES

1   T. Tokuda and M.B. Ives, Corros. Sci., 11 (1971) 297.
2   K. Schwab and R. Raeglia, Werkst. U. Korros., 5 (1962) 281.
3   R. Gressmann, Corros. Sci., 8 (1968) 325.
4   J. Postethwait, Electrochim. Acta, 12 (1967) 333.
5   A.M. Shams El Din and F.M. Abd El Wahab, Electrochim. Acta,
    9 (1964) 113.
6   H. Mabrok, M.Sc. Thesis, Zagazig University (1982).
7   D.E. Davies and Barker, Corrosion, 20 (1964) 47t.
8   R.S. Schrebler, J.R. Vilche and Ariva, Corros. Sci., 18
    (1978) 762.
9   S.M. Abd El Haleem, Werkst. U. Korros., 30 (1979) 631.
10 S.M. Abd El Haleem and A. Abd El Aal, Corros. Prev. &
    Control, 29 (1982) 13.

Passivity of Metals and Semiconductors, edited by M. Froment
Elsevier Science Publishers B.V., Amsterdam — Printed in The Netherlands

INITIAL STAGES OF FILM BREAKDOWN ON PASSIVE AUSTENITIC STAINLESS
STEEL IN 3% SODIUM CHLORIDE

M.G.S. FERREIRA* and J.L. DAWSON

Corrosion and Protection Centre, UMIST, Manchester, M60 1QD (England)

*Present address:

Laboratorio de Electroquimica, Instituto Superior Tecnico, Ave. Rovisco

Pais, 1096 Lisboa Codex (Portugal)

ABSTRACT

The paper describes impedance data obtained on 316 stainless steel during pit initiation studies. The complex impedance response can be modelled using an equivalent ciruit incorporating two adsorbed intermediates. This may represent either electrocrystallisation phenomena or the adsorption of species associated with dissolution and repassivation processes.

The modelling data is discussed in terms of film breakdown, metal dissolution and film formation. It is suggested that pit initiation is first a process of film rupture due to adsorption of chloride ions on the film. Above the pitting potential the repassivation is then inhibited by a chloro-complex intermediate.

INTRODUCTION

The study of passivation and its breakdown is one of the more challenging subject areas of corrosion science, improved information of the phenomenon would have implications for both practical and theoretical studies. Surface analysis based on physical measurements are useful for studies of metals with a thick film but are less satisfactory for investigating the thin films formed on stainless steels in service conditions.

Electrochemical techniques provide some of the few methods which can provide in-situ mechanistic information on the processes involved in the corrosion and corrosion protection reactions. Impedance measurements are possibly the most useful of the deterministic methods since the data obtained is quantitative and allows the development of modelling studies.

In a previous paper (ref. 1) we presented impedance data obtained during film growth and in the passive region of 316 stainless steel in 3% sodium chloride solution. In the passive region the complex plane is essentially capacitive but there is also a slow charge transfer and a diffusion process occurring through

the film. The present paper is concerned with the initial stages of film breakdown and shows the impedance response from 316 stainless steel during pit
initiation.

EXPERIMENTAL

The stainless steel specimens were polarized potentiostatically to the required potential using a conventional three electrode arrangement incorporating
a platinum foil counter electrode and saturated calomel reference electrode
(ref. 1,2). Impedance measurements were then obtained using a Solartron 1172
frequency response analyser and the data was interpreted using a computer program developed by the UMIST Electrochemical Group (ref. 3).

The AISI 316 stainless steel specimens were machined from bar stock to form
inverted truncated cones with a height of 6 mm, a lower diameter of 5 mm and an
upper diameter of 10 mm. The electrode hung from a compression fitting was then
partially immersed in the deaerated 3% sodium chloride test solution to give an
immersed area of 60 mm^2. This arrangement prevented pit initiation occurring
at any sharp corner and provided a crevice free electrode, the use of a nitrogen
atmosphere within the cell also eliminated waterline attack. The pH of the 3%
sodium chloride solution was adjusted to 6.5 prior to the experiment.

RESULTS AND INTERPRETATION

Impedance measurements on 316 stainless steel at +55 mV vs SCE, i.e. when an
increase of current was detected, are shown in figure 1. The major feature is
the undercut appearance of the low frequency data as presented in the complex
plane plot, real impedance, Z^1, vs imaginary impedance, $-jZ^{11}$; this was also
evidenced in the Bode plot, log frequency vs log impedance, where the impedance
decreased at the low frequences.

Accurate values of the charge transfer resistance, R_{ct}, and the Warburg
diffusion coefficient, σ, were difficult to determine since replotting of the
data, as recommended by previous workers (ref. 4,5), e.g. $\sqrt{\omega}$ vs $\sqrt{\omega}/Y^1$ to obtain
R_{ct} and σ, gave distorted straight lines. An alternative approach was therefore
employed, this involved a simulation of the complex plane plot by means of an
appropriate equivalent circuit and a computer program (ref. 3). The equivalent
circuit and their kinetic parameters were developed from previous analysis as
follows.

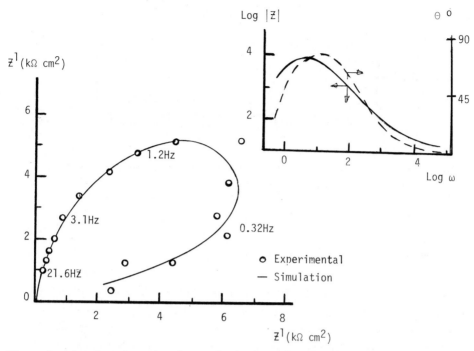

Figure 1. Complex plane impedance plot of 316 stainless steel in 3%
NaCl polarized at 55 mV vs S.C.E. Inset is the Bode Plot.

A theoretical modelling study by Haruyama (ref. 6) suggests that two electrochemical reactions, one involving charge transfer and diffusion and the second
involving only charge transfer, can interact to give a distorted impedance response similar to that shown in figure 1. However, our modelling of pit initiation suggests that the mechanism of pitting involves adsorbed intermediates
and not just a relatively simple charge transfer/diffusion interaction. As a
first approximation we assumed that the current fluctuations observed in the
pitting region were due to competition between active dissolution and repassivation processes. Although this approach is an oversimplification, since such
fluctuations or noise are more likely to be the result of electrocrystallization
processes (refs. 7 and 8), it nevertheless indicated that an equivalent circuit
incorporating adsorption pseudo-capacitances and resistances could perhaps be
used to simulate the experimental impedance response.

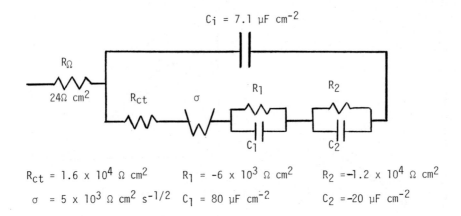

$$C_i = 7.1 \ \mu F \ cm^{-2}$$

R_Ω

$24\Omega \ cm^2$ R_{ct} σ R_1 R_2

C_1 C_2

$R_{ct} = 1.6 \times 10^4 \ \Omega \ cm^2$ $R_1 = -6 \times 10^3 \ \Omega \ cm^2$ $R_2 = -1.2 \times 10^4 \ \Omega \ cm^2$

$\sigma = 5 \times 10^3 \ \Omega \ cm^2 \ s^{-1/2}$ $C_1 = 80 \ \mu F \ cm^{-2}$ $C_2 = -20 \ \mu F \ cm^{-2}$

Figure 2. Equivalent circuit used in the simulation

Figure 2 shows the model which provided the best fit to the experimentally obtained data. The interfacial capacity, $C_i = 7.1 \ \mu F.cm^{-2}$, was estimated from a plot of imaginary part of the admittance, Y^{11}, versus angular frequency, ω. An allowance was made for time constant dispersion giving a depression angle, β, of 10° to the complex plane plot and a diffusion factor (ref. 9), $K = 0.6 \ s^{\frac{1}{2}}$, was used in the modelling where $K = \delta(2/D)^{\frac{1}{2}}$ and δ = the diffusion layer, i.e. the film thickness and D = the diffusion coefficient. Values of charge transfer R_{ct}, Warburg diffusion coefficient, σ, and the adsorption capacitances C_1, C_2, and resistances, R_1, R_2, were modified to provide the simulated curve shown in figure 1. Other equivalent circuit simulations neglecting either the second pseudo-capacitance and resistance and/or the charge transfer and Warburg diffusion failed to produce a satisfactory fit to the experimental data.

A similar equivalent circuit, describing two electrochemical reactions involving two adosrbed species, has been proposed by Epelboin et al (ref. 10) but with the assumption that the resistances for the two relaxation processes were both positive; this gives rise to a different complex plane plot. A theoretical treatment for the impedance response from metals in the passive and transpassive region has been presented by Armstrong et al (ref. 11-14) but their models were only concerned with one second semi-circle or adsorbed intermediate. There the impedance response with $+R_a +C_a$ or $+R_a -C_a$ is for active dissolution, for example of chromium (ref. 11) and nickel (ref. 12), whilst a negative resistance, $-R_a +C_a$, is obtained for the active passive transition (ref. 13,14). Essentially R_a will be negative if the velocity of the electrochemical step which produces the intermediate is slower than the following step, i.e. for $A \xrightarrow{\nu_1} B_{ads} \xrightarrow{\nu_2} C$, $\left(\frac{\delta\nu_1}{\delta E}\right)_\Gamma < \left(\frac{\delta\nu_2}{\delta E}\right)_\Gamma$. An important conclusion of their work was that the adsorbed intermediates or adsorbed oxygen species are responsible for inhibiting the metal dissolution reaction but no indication of the chemical nature of these species was given.

Electrochemical noise analysis based on the measurement of the corrosion potential fluctuations have also been used to study the pitting of stainless steel, (ref. 15). The potential-time records are essentially a series of transients each characterised by a rapid potential drop followed by a relatively slow exponential recovery. These fluctuations can be transformed into a noise spectra of small amplitude and low frequency which is typical of stochastic electrocrystallisation processes with large time constants. Electrocrystallisation processes may also be simulated in an equivalent circuit by RC combinations with large time constants similar to those in Figure 2.

The lower frequency impedance phenomena observed during pit initiation could therefore be interpreted in terms of adsorption of intermediates or electrocrystallisation phenomena. At potentials above the pitting potential the complex plane is either a charge transfer semi-circle with a capacitive loop due to adsorption or a simple charge transfer process, particularly after prolonged exposure (ref. 15). There the processes are more concerned with pit propogation rather than pit initiation.

DISCUSSION

Any proposed model which describes the passive film and breakdown leading to the pit initiation process must also take into account the known kinetics and surface analytical data. Results from auger spectroscopy obtained during the removal of passive films on stainless steel (ref. 16, 2) suggest that chloride ions do not penetrate to the metal-film interface. Electrodes scratched in the presence of chloride ions appear to form an $(-M-OH)_{ads}$ species and not $(-M-Cl)_{ads}$ (ref. 17). This is in accordance with the well established Bockris mechanisms for metal dissolution (ref. 18) where the Tafel parameters are consistant with the schemes proposed by Keddam, Mattos and Takenouti in their more recent impedance studies on iron, who also showed that a second monovalent adsorbed species and a divalent species are involved in prepassivation and passivation processes (ref. 19).

In the case of austenitic stainless steel it is the second monovalent species which is important in establishing passivation; we therefore propose the possibility of the species $(M-O-OH)_{ads}$, based on the perhydroxyl group present during oxygen reduction. Alternatively the passivating species could be considered as $(O-M-OH)_{ads}$, $(M-O-M-OH)_{ads}$, $(O-M\genfrac{<}{}{0pt}{}{OH}{OH})_{ads}$ or $(M-O-M\genfrac{<}{}{0pt}{}{OH}{OH})$ ads which are analogous to those proposed by Lorenz (ref. 20) who considers that $(M(OH)_2)_{ads}$ is not passivating. Film formation is therefore seen as a place exchange type mechanism as proposed by Sato (ref. 21). It is also suggested that the chloride ions inhibit the film formation process and assist the metal dissolution by a similar mechanism as suggested by Lorenz (ref. 22). Our preferred scheme to show the influence of chloride in preventing repassivation is :-

$$M(O-M-OH)_{ads} + Cl^- \longrightarrow M(O-M-Cl)_{ads} \longrightarrow (MOH)_{ads} + M^{++} + Cl^-$$

These processes are shown schematically in figure 3 together with the film dissolution process, which involves chloride adsorption and complexing with the outer film, $-O-M<^{OH}_{OH}$.

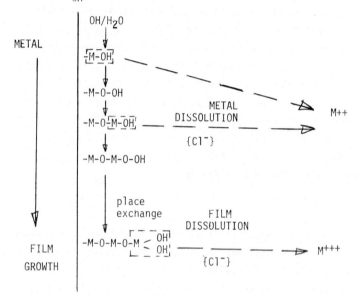

Figure 3.   Model showing film growth and dissolution.

This proposal is consistant with our earlier data and equivalent circuit model of the passivation region on 316 stainless steel, where charge transfer and diffusion occurs through the film, (ref. 1, 2). However the low frequency phenomena, observed as electrochemical noise i.e. scatter of the data points and undercutting of the complex plane, figure 2, is difficult to model unless a stochastic process of mechanical breakdown of the film is invoked. The adsorption of chloride ions on the film induces film cracking as proposed by Sato in his electrostrictive mechanism (ref. 23). This film rupture occurs both below and above the pitting potential, as evidenced by the noise measurements, and is a stochastic process. This may be modelled as an electro-crystallation phenomenon, as suggested by Bignold and Fleischmann (ref. 24) for active dissolution of iron, but in the case of pitting corrosion the initiation stage is the result of mechanical film rupture induced by chloride adsorption on the film.

CONCLUSIONS

Impedance data obtained on 316 stainless steel during pit initiation indicates that a number of processes are occurring simultaneously. The frequency response can be interpreted in terms of charge transfer and diffusion through the film and a superimposed noise source which distorts the lower frequency data. The electrochemical noise is seen as being the result of film rupture which is due to the electrostrictive mechanism following adsorption of chloride ions on the film. This stochastic process of film rupture and repassivation occurs both above and below the pitting potential.

Above the pitting potential it is possible for further chloride adsorption and complexing to occur but in this case with the adsorbed metal oxy/hydroxy species. These species would normally be the precursers to the production of the passive film but at these higher potentials the chloride is able to inhibit this mechanism.

The impedance data also indicates the presence of adsorbed intermediates; in the case of austentic stainless steels the important prepassivating species is suggested as $(M-O-OH)_{ads}$ whilst for iron or ferritic steels a species such as $(O-Fe-OH)_{ads}$ could play a more important role in the passivation process.

ACKNOWLEDGEMENTS

The authors would like to thank Professor G. C. Wood of the Corrosion and Protection Centre for provision of the research facilities and the Instituto Nacional de Investigacas Cientifica, Portugal for financial support for M.G.S. Ferreira.

REFERENCES

1. M.G.S. Ferreira and J.L. Dawson, "Electrochemical Studies of the Passive Film on 316 Stainless Steel in Chloride Media", paper presented at 150th Meeting of the Electrochem. Soc., Denver, October 1981 and submitted to the J. Electrochem. Soc.
2. M.G.S. Ferreira, Ph.D. Thesis, Manchester (1981).
3. IMPED Series of programs, available through Corrosion and Protection Centre Industrial Services, UMIST.
4. M. Sluyters-Rebach and J.H. Sluyters, Electroanalytical Chemistry, Vol. 4, Chpt. 1. Pub. Marcel Dekker, N.Y., 1969.

5. J.H. Sluyters, Rec. Trav. Chim. 79 (1960), 1092.

6. S. Haruyama, Proceedings of 5th Int. Cong. Metal Corros., Tokyo, 1972,
    Pub.-Nace, Houston 1974.

7. G. Blanc, I. Epelboin, C. Gabrielli and M. Keddam. J. Electroanal. Chem.
    62 (1975) 59.

8. G. Blanc, C. Gabrielli, M. Ksouri and R. Wiart, Electrochim. Acta. 23
    (1978) 337.

9. J.L. Dawson and D.G. John, J. Electroanal. Chem. 110 (1980) 37.

10. I. Epelboin, M. Keddam and J.C. Lestrade, Farad. Discs. Chem. Soc 56 (1974)
    264.

11. R.D. Armstrong, M. Henderson and H.R. Thirk. J. Electroanal. Chem. 35
    (1972) 119.

12. R.D. Armstrong and M. Henderson, J. Electroanal. Chem. 39 (1972) 222.

13. R.D. Armstrong. J. Electroanal. Chem. 34 (1972) 387.

14. R.D. Armstrong and K. Edmondson. Electrochimica Acta. 18 (1973) 937.

15. J.L. Dawson and M.G.S Ferreira, submitted to Corr. Sci., 1983.

16. Z. Szklarska-Smialowska and M. Janik-Czachor. Corros. Sci. 16 (1976) 644.

17. G.T. Burnstein and D.H. Davies. Corr. Sci. 20 (1980) 1143.

18. J. O'Bochris, D. Drazic and A.R. Despic. Electrochimica Acta., 4 (1961) 325

19. M. Keddam, O.R. Mattos and H. Takenouti, J. Electrochem. Soc. 128 (1981)
    257.

20. J. Bessone, L. Karakaya, P. Lorbeer and W.J. Lorenz, Electrochimica Acta.,
    22 (1977) 1147.

21. N. Sato and M. Cohen, J. Electrochem.Soc., 111, (1964) 512

22. A.A. El Miligy, D. Geana and W.J. Lorenz, Electrochimica Acta, 20, (1975)
    273.

23. N. Sato, Electrochimica Acta., 16 (1971) 1683.

24. G.J. Bignold and M. Fleischmann, Electrochimica Acta, 19, (1974) 363.

Passivity of Metals and Semiconductors, edited by M. Froment
Elsevier Science Publishers B.V., Amsterdam — Printed in The Netherlands

REPASSIVATION POTENTIAL OF CORROSION PITS IN STAINLESS STEEL

T. HAKKARAINEN

Technical Research Centre of Finland (VTT), Metallurgy Laboratory,
Metallimiehenkuja 4, SF-02150 ESPOO 15, Finland

ABSTRACT

The relation of the "true" repassivation potential, as measured close to the dissolving metal surface inside model pits in stainless steel (AISI 304 or AISI 316), to the "apparent" repassivation potential, as measured from the bulk solution, in NaCl solutions has been investigated. The true repassivation potential was several hundreds of millivolts lower than the apparent repassivation potential, and only about 100 mV higher than the free corrosion potential of the steel in a concentrated metal chloride solution expected to exist in a growing pit. In the vicinity of the true repassivation potential the dissolution rate decreased rapidly with decreasing concentration of the pit solution. It is concluded that the repassivation of pits simply involves the replacement of the concentrated solution in a growing pit by a dilute one. At higher potentials the rapid dissolution inside the pit is able to keep the concentration sufficiently high.

INTRODUCTION

It is generally agreed that the solution inside growing corrosion pits in austenitic stainless steels in chloride solutions is a concentrated metal chloride solution with a low pH value (ref.1-6). There are reports indicating that the electrode potential of the dissolving metal surface inside a pit is considerably lower than that of the metal surface outside the pits (ref.7). On the other hand, it has been reported that these two potentials could be quite close to each other (ref.8).

In this study, the potential relationships inside and outside model pits during pit growth and repassivation as well as some factors contributing to the stability of pit growth have been investigated.

EXPERIMENTAL

Test materials

Commercial austenitic stainless steels of types AISI 304 and AISI 316 were used in the experiments. The chemical compositions of the steels are given in table 1.

TABLE 1

Chemical compositions of the steels, wt-pct.

Steel	Cr	Ni	Mo	C	Si	Mn	S	P	Fe
AISI 304	18.5	9.7	0.1	0.04	0.54	2.0	0.016	0.033	Bal
AISI 316	16.5	11.2	2.6	0.05	0.45	1.72	0.006	0.032	Bal

Pit solutions

Concentrated metal chloride solutions, simulating those in growing natural corrosion pits, were prepared by dissolving about 50 g of the appropriate stainless steel into 200 ml of 10 M HCl. This results in a metal chloride solution that contains the same proportions of metal cations as the steel and is saturated at room temperature. The pH of such a solution is about 0.5 and the density about 1.5 g/cm^3 (ref.9). Less concentrated metal chloride solutions were prepared from the saturated ones by adding distilled water.

Test procedures

Model pits. To study the electrode potential of the dissolving metal surface at pit growth and repassivation, macroscopic model pits were used (see Fig. 1). The specimen surface was covered with a multiple adhesive tape that had a circular hole with a diameter of 6 mm. The pit depth (thickness of the tape layer) was typically 1.5 mm. A reference electrode (saturated calomel electrode) for potential control was placed outside the pit (R1, in Fig. 1). The potential very close to the metal surface inside the model pit was monitored with another reference electrode (R2, in Fig. 1).

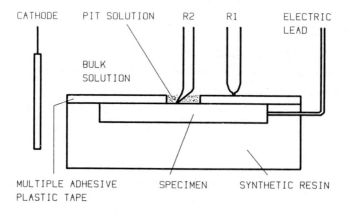

Fig. 1. Sketch of the model pit configuration. R1 is reference electrode for potential control, R2 is electrolytic connection to another reference electrode.

The test procedure was as follows. A beaker with the specimen holder in it was filled with the "bulk" solution, i.e. 0.5 M or 0.1 M NaCl, and the reference electrodes were placed into their proper positions. Next the electrode potential with respect to reference electrode R1 was adjusted to a value, where pits are expected to grow stably. Then the model pit was filled with "pit solution" by adding into the pit with a pipette about 0.1 ml of concentrated metal chloride solution, degree of saturation 0.8 to 1. After both the electric current and the electrode potential of the pit bottom were stabilized, in about 20 to 30 min, the control potential was gradually lowered at a constant rate. The readings of both reference electrodes as well as the electric current through the model pit were recorded as a function of time.

Anodic polarization. To study the effect of pit solution concentration on the anodic behaviour of the steels, anodic polarization curves starting at the free corrosion potentials were recorded using pit solutions diluted to various degrees. The rate of potential increase was 30 mV/min.

Real pits. Real pits were produced by polarizing the steel surfaces anodically in a NaCl solution until the current increased sharply. When the current reached the value 1 mA (0.45 mA/cm^2) the potential scan direction was reversed. A sharp decrease of the current to very low values was observed when the pits were repassivated.

RESULTS
Model pits

Typical results of repassivation studies on model pits are shown in Figs. 2 and 3. The initial transients after the addition of concentrated pit solution are not shown. Addition of pit solution with saturation degrees of 0.8 or 1 gave differencies in the initial transients, but resulted in similar stationary values of pit potential and current density at the constant control potential.

In the experiments, the steel was uniformly attacked over the whole exposed area. Some crevice corrosion under the tape at the edges of the pit bottom area was, however, observed after each experiment on examining the specimens.

Anodic polarization

Anodic polarization curves in pit solutions of various concentrations in the potential region that was observed inside model pits (see Figs. 2 and 3) are shown in Figs. 4 and 5. At the higher degrees of saturation, 0.8 or 1, reversing the potential scan direction at about 50 mA/cm^2 resulted in current

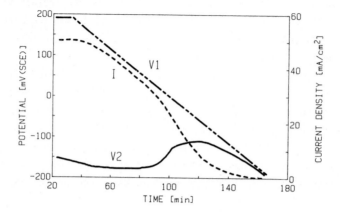

Fig. 2. Repassivation behaviour of a model pit of AISI 304 steel in 0.5 M
NaCl solution. Pit depth 1.5 mm, control potential scan rate -3 mV/min.
V1 is the control potential measured outside the pit, V2 is the electrode
potential of the pit bottom, and I is the current density.

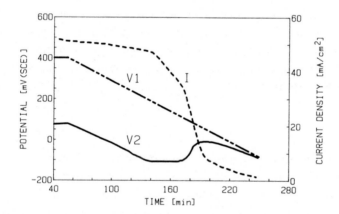

Fig. 3. Repassivation behaviour of a model pit of AISI 316 steel in 0.5 M
NaCl solution. Pit depth 1.5 mm, control potential scan rate -3 mV/min.
Curve labels as in Fig. 2.

densities at decreasing potential values close to those at increasing potential
values. At the lower degrees of saturation, 0.6 or 0.4, on the other hand,
a reversal of the potential scan direction resulted in considerably higher
current densities at decreasing potential values than were observed at increasing potential values. The latter behaviour is an indication of separate pits
or crevice corrosion on the specimen. These specimens were also found to have
been attacked inhomogeneously, whereas the corrosion attack on the specimens
in solutions of high degrees of saturation was uniform.

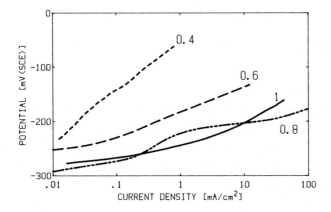

Fig. 4. Anodic polarization curves of AISI 304 steel in pit solutions of various degrees of saturation. The figures at the curves indicate the degree of saturation. Potential scan rate 30 mV/min.

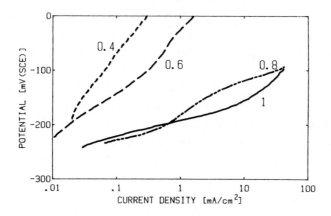

Fig. 5. Anodic polarization curves of AISI 316 steel in pit solutions of various degrees of saturation. The figures at the curves indicate the degree of saturation. Potential scan rate 30 mV/min.

Real pits

The repassivation potential of real pits, as measured at the steel surface outside the pits, was found to be 0 to 50 mV (SCE) for AISI 304 steel in 0.5 M NaCl, 100 to 150 mV (SCE) for AISI 304 steel in 0.1 M NaCl, and 250 to 300 mV (SCE) for AISI 316 steel in 0.5 M NaCl. The pit depths in these experiments varied, depending on their time of growth, between 0.05 and 0.3 mm.

DISCUSSION

The results of the repassivation experiments on model pits (see Figs. 2 & 3) suggest that during stable pit growth the electrode potential inside the pit is several hundreds of millivolts lower than that measured outside the pit. At high values of the control potential, the pit current may become saturated. In such a case the "inside" potential of the pit changes in a similar way as the control potential (see Fig. 3, between 60 and 140 min). Below a certain potential, however, the current density in the solution inside the pit becomes very strongly dependent on the electrode potential of the dissolving surface (see Figs. 4 and 5). In Figs. 2 and 3 this region is characterized by a decrease in the current with decreasing control potential, while the inside potential of the pit remains approximately constant.

When the current density inside the pit has decreased to a critical value, its rate of change becomes higher and the inside potential of the pit begins to rise towards the control potential. This is the start of repassivation, in Fig. 2 at about 85 min and in Fig. 3 at about 170 min.

It is concluded that at stable pit growth the dissolution rate of the steel is high enough to keep the solution inside the pit concentrated. As the potential is lowered, the dissolution rate slowly decreases, until below a certain "repassivation potential" the rate of dilution of the pit solution exceeds that of metal dissolution.

At repassivation, the current density is expected to fall off and the inside potential of the pit to approach the control potential at an ever increasing rate. This is what happens at first in the model pit experiments. In these experiments, however, crevice corrosion at the edges of the exposed surface intervenes. Thus, there is a point of discontinuity in the pit potential curves (V2) and a point of inflection in the current density curves (I) at about 102 min in Fig. 2 and at about 183 min in Fig. 3.

REFERENCES

1 S. Szklarska-Smialowska, in Localized Corrosion, NACE-3, Houston, p. 312.
2 T. Suzuki, M. Yamabe and Y. Kitamura, Corrosion 29 (1973), p. 18.
3 L. Gainer and G. Wallwork, Rev. Coatings and Corr., III (1978), p. 49.
4 T.R. Beck and S.G. Chan, Corrosion 37 (1981), p. 665.
5 N. Lukomski and K. Bohnenkamp, Werkstoffe u. Korrosion 30 (1979), p. 482.
6 J. Mankowski and Z. Szklarska-Smialowska, Corr. Sci. 15 (1975), p. 493.
7 W.H. Pickering and R.P.J. Frankenthal, J. Electrochem. Soc. 119 (1972), p. 1297.
8 D.A. Jones and B.E. Wilde, Corr. Sci. 18 (1978), p. 631.
9 T. Hakkarainen, 8th Int. Cong. on Metallic Corrosion - Proceedings, DECHEMA 1981, p. 157.

Passivity of Metals and Semiconductors, edited by M. Froment
Elsevier Science Publishers B.V., Amsterdam — Printed in The Netherlands

STUDY OF THE PASSIVE BEHAVIOUR OF A 304L STAINLESS STEEL IN 1N H_2SO_4 BY A METHOD OF CORROSION UNDER FRICTION.

by : D. BOUTARD, J. GALLAND.
Laboratoire Corrosion et Fragilisation par l'Hydrogène.
Ecole Centrale des Arts et Manufactures. 92290 Châtenay-Malabry (FRANCE).

ABSTRACT
        The electrochemical behaviour of a 304L stainless steel is studied under friction conditions. The polarisation curve appears to be a function of the mechanical parameters as much as the electrochemical ones.
        For corrosion alone the existence and moreover the stability of the passive layer seem to obliterate all evolutions in electrochemical kinetics or process which may occur in the passive region of the polarisation curve.
        As for the studied 304L stainless steel a second mechanism seems to arise, its influence being far from negligible long before the transpassive branch.

INTRODUCTION

        A metal in aqueous medium displays a passive behaviour if the current through the solid-liquid interface remains very low and quite steady over a wide interval of potential, i.e. the metal surface is protected by a passive layer which stability determines the extent of the passive region. But in such conditions its existence is liable to obliterate any change in electrochemical kinetics or process.

        The experimental device used for tests under friction is designed to continuously renew the metallic surface : this induces a mechanical destruction of a fraction of the passive layer. So the measured current depends on :

i)   the kinetics of electrochemical reactions occuring on the renewed surface which are function of the sole potential,

ii)   the rate of renewal which is determined by friction parameters.

EXPERIMENTAL

The tribometer

        Tests of corrosion under friction are performed using a special tribometer allowing both electrochemical measures (such as potential and current) and friction under controlled parameters (contact pressure and relative speed) (ref.1).

374

Its working cell is shown in figure 1. The surface of the metallic sample
is a horizontal plane ring (see figure 1b) on which two fingers are pressed
under a given pressure p. They are driven into a circular movement by the
rotating head and their speed may be maintained at a given value v (expressed
hereafter as s^{-1}). Both fingers are made out of alumina so that being electrochemically inert their sole intervention is a mechanical one.

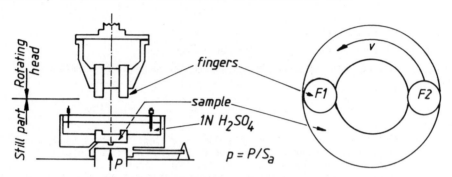

Figure 1 : Working cell of the tribometer and shape of the samples used for
tests of corrosion under friction.

The instant renewed surface may be written as :

$$S_{rcf} = S_c(v,p) \quad \text{or} \quad R_{cf} = S_c(v,p)/S_a \qquad (1)$$

where $S_c(v,p)$ is the actual contact area (calculable from mechanical properties
of both materials), S_a the apparent surface and R_{cf} the instant renewal rate.
S_{rcf} as well as R_{cf} are independent of the potential.

The polarisation curves are obtained by controlling the potential of the
metallic sample and measuring the resulting current.

The material

The 304L stainless steel we studied has the following composition :

Ni	Cr	Mn	Mo	Cu	Mg	Si	V	Nb	C	S	P
9.70	18.31	1.78	0.33	0.20	0.002	0.26	0.026	0.03	0.028	0.019	0.022

Composition of the stainless steel (given in weight %)

This alloy is homogeneous without any segregation on the grain boundaries.
When it is in aqueous medium such as 1N H_2SO_4, the polarisation curve displays a
passive behaviour over a wide range of potential. The passive layer is very efficient : the intensity remains around 10 uAxcm^{-2} (*) between -0.50 and +0.50 V/$_{SSE}$
(+0,17 and 1.17 V/$_{NHE}$), i.e. 1V wide.

(*) We will express the measures as mA.cm^{-2} but refer to them as intensities :
the term "density" is to avoid here because the metallic surface is not
electrochemically homogeneous under friction conditions.

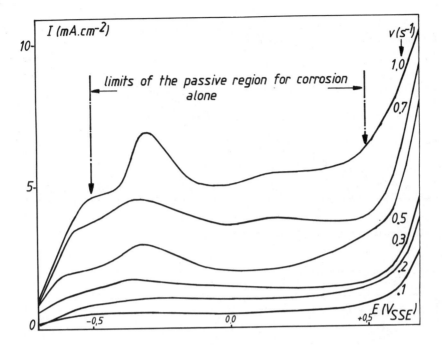

Figure 2 : Polarisation curves under friction (these curves are represented
after correction of the ohmic drop).

RESULTS

Tests performed under friction conditions were achieved with a constant
pressure equal to 2.6 MPa. The main parameter is the finger speed v.

The polarisation curves depend on the friction parameters (see figure 2)
but the electrochemical behaviour is independent of the previous "story" of the
interface, i.e. the influence of the superficial plastic deformation is
negligible when compared to the renewal effect. Intensities are at least two
orders of magnitude larger than for corrosion alone and moreover they increase
with v for a given value of the potential.

The shape of the curve is different too from classical one. We must enhance
here the anodic shift of the potential of the maximum intensity. The passive
"plateau" tends to narrow : a minimum value appears around 0.0 V/SSE when the
speed v increases.

Equipotential variations of current against speed v seems to be linear
whatever the potential is. One example of these evolutions is shown in figure 3.
Equation of these straight lines may be written as :

$$I(E,v) = B(E) + a(E) \, v \qquad (2)$$

Figure 3 displays the evolution for 0.0 V/SSE which is in the passive region
(see figure 2, for limits of passivity). In such cases the I-intercept b(E) is

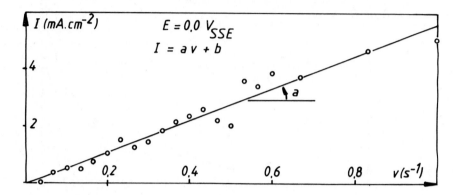

Figure 3 : Equipotential evolution of intensity against v.

Figure 4 : Variation of a(E).

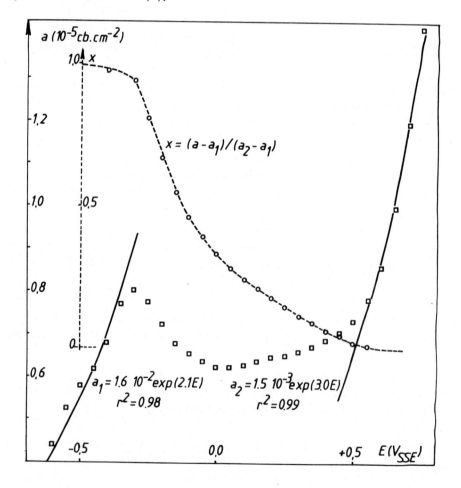

nearly null which is consistent with b(E) being the value for v = 0 (corrosion alone). b(E) increases for E   +0.55 V/$_{SSE}$, i.e. when the interface is in transpassive conditions.

As for the slope a(E) its evolution with potential is plotted in figure 4. The sensitivity of current to the speed v increases first (up to -0.35 V/$_{SSE}$) then decreases (the minimum being round 0.0 V/$_{SSE}$) to increase slowly (up to +0.5 V/$_{SSE}$), then sharply (above +0.55 V/$_{SSE}$).

It is possible to adjust two exponential variations of a(E) by the least square method : a_1 and a_2 are given on the figure (r^2 being the correlation coefficient).

INTERPRETATION.

If we suppose the electrochemical behaviour to be schematically the result of one adsorption process we may write as :

$$X_0 \underset{K_1'}{\overset{K_1}{\rightleftharpoons}} X_1 + e^- \qquad (3)$$

where K_1 and K'_1 are respectively exponential increasing and decreasing variations of potential E. We may infer from (3) that the degree of coverage θ_1 of X_0 by X_1 is such :

$$\frac{d\theta_1}{dt} = K_1 (1 - \theta_1) - K'_1 \theta_1 \qquad (4)$$

and the associated current density J equal to :

$$J = F [K_1 (1 - \theta_1) - K'_1 \theta_1] \qquad (5)$$

For corrosion alone the stationary value of θ_1, θ_{1c}, is given figure 5.

The finger sweep is then suppose to wipe away a fraction γ of θ_1, independent of the potential and calculable from the mechanical properties of the metal. This fraction may actually be approximated by :

$$\gamma = S_c(v, p)/S_a \qquad (6)$$

Where S_a is the apparent contact surface.

So the evolution of θ_1, between the fingers F_1 and F_2 may be represented as

Figure 5 : θ_1-evolution between F_1 and F_2.

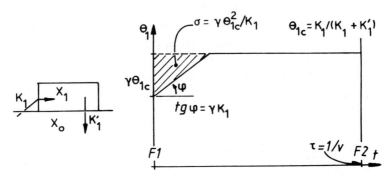

shown in figure 5, providing the electrochemical rates are much greater than the finger sweep v. In such conditions, the mean value of θ, during corrosion under friction, is :

$$\theta_{1cf} = v \left(\frac{\theta_{1c}}{v} - \sigma \right) = \theta_{1c} - \frac{v\gamma}{2} \frac{(\theta_{1c})^2}{K_1} \tag{7}$$

leading to mean value of current (the measured value) :

$$J_{cf} = J_c + F \frac{v\gamma}{2} \theta_{1c} \tag{8}$$

So the slope a(E), being equal to $F \frac{\gamma}{2} \theta_{1c}$, equipotentialy constant, means :
i) γ is independent of the speed v;
ii) the increase of a(E) with E is due to the evolution of θ_{1c}.
a(E) may be approximated by an exponential as long as θ_{1c} is, that is to say as long as the electrochemical process leads to the building of the coverage by X_1.

But as equation (8) is written, a sole process would give a constant value to a(E) when the potential increases. Experimental results shows a contradictory variation. So the decrease must be assigned to the rise of a second process which concurs the first one before overwhelming it, as the potential grows more anodic.

As the second exponential increase of a(E) occurs in the transpassive region it may be ascribed to the secondary passivity process which appears in the polarisation curve for corrosion alone as well as the first one to the building of the passive layer.

The ratio x we defined in figure 4 gives a hint of the relative importance of these two processes. Its evolution against E shows that the process is far from negligible long before the transpassive region, at least 0,5 V below the value which may be determined from the classical polarisation curve.

The a(E) values may be taken from another point of view as a measure of the stability of the passive behaviour of the metal in the aqueous medium. The higher a(E), the lower the stability of the passive layer is.

Thus it seems there is an optimal value of potential which leads to a maximum stability : around 0,0 V_{SSE}. Below that point the destruction of the passive layer retrieves mainly the first process ; above it, the renewal of the metallic surface issues mostly in the secondary passivity (transpassive one).

Anyway the influence of the latter arises at least 0,5 V lower than for corrosion alone ; this gap being mainly due to the stability of the passive layer which obliterates the shift from one mechanism to the other as long as the potential value allows the existence of passivity.

REFERENCE
(1) D. BOUTARD-GABILLET, Thèse d'Etat Univ. P. et M. Curie, Paris (1983).

Passivity of Metals and Semiconductors, edited by M. Froment
Elsevier Science Publishers B.V., Amsterdam — Printed in The Netherlands

BREAKDOWN OF PASSIVITY OF IRON AND NICKEL BY FLUORIDE

H.-H. STREHBLOW[1] and B. P. LÖCHEL[2]

[1]Institut für Physikalische Chemie, Universität Düsseldorf, 4000 Düsseldorf
(Germany)

[2]Institut für Physikalische Chemie, Freie Universität Berlin, 1000 Berlin 33
(Germany)

ABSTRACT

Breakdown of passivity by fluoride has been studied for iron and nickel by
electrochemical methods including rotating ring disc studies and by x-ray photo-
electronspectroscopy (XPS) and ion scattering spectroscopy (ISS). HF catalyses
the metal ion transfer from the oxide into the electrolyte and leads to a thinning of the passive layer. For iron a final complete breakdown of passivity is
observed with a related Fe^{2+} production. Nickel forms a NiF_2 layer at higher
potentials leading to a different dissolution mechanism. The mechanisms of the
catalytical activity of HF on the passive corrosion current density of both
metals is discussed.

INTRODUCTION

Breakdown of passivity of metals by aggressive anions like halides is of
technical importance and current interest for corrosion research. It may occur
locally or may affect the total surface depending on the kind and concentration
of the anions, the pH and the metal. Many work has been devoted to the growth
of corrosion pits and its chemical and electrochemical conditions. However, the
processes leading to the breakdown of the passive layer are still in discussion
and need further examination. For a well prepassivated iron specimen the so
called adsorption mechanism (ref.1) seems to be effective involving a catalytically increased transfer of the Fe^{3+}-ions from the oxide into the electrolyte
by the action of chloride (ref.2). For non stationary conditions of the passive
layer the film breaking mechanism seems to cause pit nucleation most effectively (ref.1,3). A further elucidation of this problem by a combination of electrochemical and surface analytical methods seems promising. The application of the
powerful spectroscopical methods working in the vacuum like x-ray photoelectron
spectroscopy (XPS) or ion scattering spectroscopy (ISS) requires however a situation where the total surface or at least its major part is affected. Therefore we studied the attack of passive iron and nickel by fluoride.

ELECTROCHEMICAL STUDIES

Fluoride causes local and general breakdown of passivity on iron, depending on the pH of the solution. In weakly acidic and alkaline solutions, pitting is found whereas in strongly acidic electrolytes a general attack is observed (ref.4). The homogeneous process has been studied in 0.1 and 1 M $HClO_4$ and H_2SO_4 with a rotating Pt-split ring Fe disc electrode. Three stages are observed, which are described in the following.

1. The Fe^{3+} transfer into the electrolyte and the related disc current are increasing with 0.1 M HF from i_D < 1 μA/cm² up to ca 1 mA/cm². The dissolved Fe^{3+} cannot be seen at a Pt ring by its reduction because of the formation of a highly stable complex $|FeF_5OH|^{3-}$. The presence of a barrier type passive layer is to be seen by a linear increase or decrease of the electrode potential with time when the electrode is switched from potentiostatic to galvanostatic conditions with current densities larger or smaller than the previous potentiostatic value (ref.5). The current density is potential independent and is increasing with the HF concentration within the acidic electrolyte with a first order dependence. All these findings suggest a catalytic activity of HF for the Fe^{3+} dissolution at the oxide-electrolyte interface, resulting in a thinning of the passive layer (ref.5).

2. After ca. 0.5 to 1 hour current oscillations are observed at the disc, accompanied by a simultaneous change of the ring current detecting Fe^{2+} ions by their oxidation. The oscillating Fe^{2+} production demonstrates the temporary exposure of bare metal surface, e. g. breakdown and repair of the passive layer.

3. Finally a complete and permanent breakdown of passivity is obtained with a further steep increase of the current density and the Fe^{2+} detecting ring current.

The dissolution of nickel is also increased by the presence of HF however no localized corrosion is obtained for all pH values (ref.4). The current density potential curve shows a broad maximum in the range of passive behaviour (fig. 2). The formation of a black visible film is observed at the current maximum (ε = 1.15 V) which disappears only for a sufficient stirring of the electrolyte, i. e. for > 400 rpm of a rotating Ni disc. At more positive and negative potentials, no visible film is obtained. At the minimum I at less positive potentials (ε = 0.5V) the current density increases with the 1.6 order of HF, whereas at minimum II (ε = 1.4 V) a decrease with the -0.8 order is obtained. At the maximum it is nearly HF independent. This completely different behaviour suggests two distinct mechanisms being responsible for the dissolution of nickel in the two potential ranges.

SURFACE ANALYTICAL STUDIES

To explain the properties of the two metals in HF containing acidic solutions XPS and ISS analysis of prepassivated specimen were performed, which have been attacked by HF under potentiostatic conditions. Iron does not show any F1S signal. However, the presence of oxide and hydroxide is indicated by the O1S signal with an O^{2-} (E_B = 530.0 eV) and an OH^- peak (E_B = 531.4 eV). The presence of the passive layer causes an attenuation of the Fe2P3/2 metal signal (E_B = 706.8 eV) and an Fe2P3/2 oxide signal at higher binding energy (E_B = 710.9 eV). It is generally accepted that the hydroxide is in the outer position (ref.6). Tilting the specimen increases the OH shoulder of the O1S signal which is caused by a virtual thickening of the layer when the electrons do not leave the surface perpendicularly to be accepted by the analyser. The resulting model for an outer hydroxide and an inner oxide layer is found for passive iron with and without HF exposure.

The total thickness of the passive layer δ may be calculated by the attenuation of the intensity I_{Me}^{Me} of the Fe2P3/2 metal peak after the following equation:

$$I_{Me}^{Me} = I_{Me,\infty}^{Me} \; \exp(-\delta/\lambda_{ox}^{Me}) \tag{1}$$

$I_{Me,\infty}^{Me}$ is the intensity for a clean oxide free specimen which may be obtained by argon ion sputtering; λ_{ox}^{Me} is the escape depth for the electrons from an oxide matrix which may be calculated for kinetic energies $E_{Kin} > 150$ eV approximately by the following relation (ref.7):

$$\lambda_{ox}^{Me} = B \; E_{Kin}^{1/2} \tag{2}$$

The subdivision of δ in the hydroxide x and the oxide part y is obtained with the unfolded two O1S signals. Their peak ratio yields:

$$\frac{I_{OH}^{0}}{I_{ox}^{0}} = \frac{D_{OH}^{0} \left[1-\exp(-x/\lambda_{OH}^{0})\right]}{D_{ox} \exp(-x/\lambda_{OH}^{0}) \left[1-\exp(-y/\lambda_{ox}^{0})\right]} \tag{3}$$

With $\lambda_{ox}^{0} = \lambda_{OH}^{0} = \lambda^{0}$ one obtains:

$$x = -\lambda^{0} \; \ln\left[\frac{I_{ox}^{0} \; D_{OH}^{0} + I_{OH}^{0} \; D_{ox}^{0} \; \exp(-\delta/\lambda^{0})}{I_{OH}^{0} \; D_{ox}^{0} + I_{ox}^{0} \; D_{OH}^{0}}\right] \tag{4}$$

which, in turn, permits to calculate y after $\delta = x + y$

Comparing the results of Fig. (1) with the thickness x and y of an only passivated specimen shows a pronounced thinning of the oxide part of about 50% by the action of HF.

382

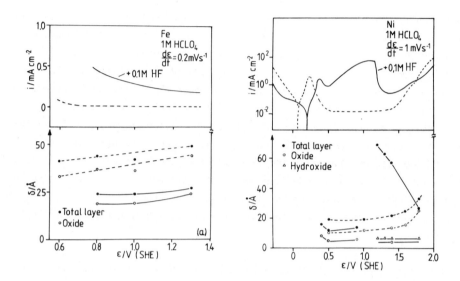

Figs. 1 and 2: Potentiodynamic current density potential curves (upper part) and layer thicknesses. Specimen electropolished and 2h prepassivated at indicated potential (**dashed line---**) and additionally 30min exposed to 0.1 M HF (**solid line ——**). Fig. 1 iron, Fig. 2 nickel

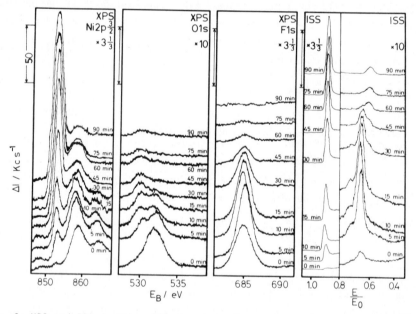

Fig. 3: XPS and ISS depth profile by argon ion sputtering of nickel prepassivated 2h at ε = 1.4V(SHE) in 1 M $HClO_4$ and 30min exposed to 1 M $HClO_4$ + 0.1 M HF at ε = 1.40 V

Nickel shows similar results by XPS for minimum I (Fig. 2). Again no F^- is
found within the layer and the oxide part is thinned. The outer position of the
hydroxide is recognized in addition by the first disappearance of the OH-shoul-
der of the O1S signal during argon sputtering. It is known from literature (ref.
8) and own tests that $Ni(OH)_2$ is not decomposed, i. e. dehydrated or reduced by
preferential sputtering. At the minimum II a large F1S signal is found, demonstrating the presence of a fluoride film (Fig. 3). After the ISS sputter profile
a rest of the oxide is found by an O-peak at smaller E/E_o values, when F^- is
already removed totally as may be seen also by the disappearing F1S signal. The
oxide thickness reaches its reduced value within the first minute. The fluoride
layer is still increasing up to several min. For this layer a more complicated
triple structure is assumed with an inner oxide covered by the fluoride and
followed finally by the hydroxide. To subdivide the total thickness into the
three parts of the hydroxide x the fluoride y and the oxide z, the F1S peak is
used additionally for the calculation. The following equations are applied for
the evaluation:

$$x = -\lambda \ln(\frac{b/c + f(h \exp(-\delta/\lambda)}{a/c - g/h}) \qquad (5)$$

$$y = -\lambda \ln(g/h + f/h \exp -(\delta-x)/\lambda) \qquad (6)$$

$$\lambda = \lambda^0 = \lambda^F \qquad\qquad \delta = x + y + z$$

$$a = I_{OH}^0 \ \sigma_F \ D_F^F + I_F \ \sigma_0 \ D_{OH}^0 \qquad\qquad f = I_F \ \sigma_0 \ D_{ox}^0$$

$$b = I_F \ \sigma_0 \ D_{OH}^0 \qquad\qquad g = I_{ox}^0 \ \sigma_F \ D_F$$

$$c = I_{OH}^0 \ \sigma_F \ D_F^F \qquad\qquad h = I_F \ \sigma_0 \ D_{ox} + I_{ox}^0 \ \sigma_F \ D_F$$

σ_0 and σ_F are the photoionisation crosssections of the O1S and F1S signal and
I and D the XPS signal intensities and atomic densities for the two elements in
the different matrix. The layer structures and the related current densities for
iron and nickel are both given in Figs. 1 and 2. The composition NiF_2 for the
surface layer in the potential range of maximum II and above was determined by
quantitative evaluation of the F1S and Ni2P3/2 signal and comparing their ratio
to that of pressed NiF_2 powder standards.

The surface analytical results might be affected by modifications during the
specimen's transfer from the electrolyte into the UHV chamber and by the action
of the X-rays or the ion beam. A change of hydroxide into oxide or Ni-metal,
however, seems unreasonable as this is even not observed during heavy ion bombardment (ref. 8). A hydroxide overlayer has been found even by in situ
                   (ref.9)
methods./The thinning of the oxide and the formation of the fluoride layer

by HF does occur during the electrochemical pretreatment as this is proceeding systematically with the time of HF exposure (ref. 10). Therefore, a major change of the surface layers after sample preparation may be excluded.

MECHANISMS OF PASSIVITY BREAKDOWN

The combined electrochemical and surface analytical results suggest a surface catalytical mechanism for the increased dissolution of passive iron by HF. A surface complex is formed by a fast reaction (eq. 7a), followed by the rate determining, but enhanced ion transfer step (eq. 7b) and a fast further complexing within the electrolyte (ref. 5).

$$Fe^{3+}_{ox} + HF_{aq} \rightleftharpoons FeF^{2+}_{ad} + H^{+}_{aq} \tag{7a}$$

$$FeF^{2+}_{ad} \longrightarrow FeF^{2+}_{aq} \tag{7b}$$

$$FeF^{2+}_{aq} + 4\ HF_{aq} + H_2O \longrightarrow FeF_5OH^{3-}_{aq} + 5\ H^{+}_{aq} \tag{7c}$$

With the potential drop $\varepsilon_{2,3,s}$ at the oxide-electrolyte interface follows for the stationary corrosion current density $i_{c,s}$ for rate determining reaction (equ. 7b).

$$i_{c,s} = 3kF \left[FeF^{2+}_{ad} \right] exp(- \frac{2\alpha F}{RT} \varepsilon_{2,3,s}) \tag{8}$$

Introducing the electrochemical equilibrium (9) for equ. (7a), and the pH dependence of $\varepsilon_{2,3,s}$ determined by the equilibrium of the 0^{2-} formation after equ. (10).

$$\Theta = \frac{\left[FeF^{2+}_{ad} \right]}{\left[Fe^{3+}_{ox} \right]} \frac{\left[HF_{aq} \right]}{\left[H^{+}_{aq} \right]} exp(- \frac{\Delta G^0}{RT}) \ exp\ (\frac{\gamma F \varepsilon_{2,3,s}}{RT}) = a \left[HF_{aq} \right] \tag{9}$$

γ = electrosorption valency

$$H_2O_{aq} \rightleftharpoons 0^{2-}_{ox} + 2\ H^{+}_{aq} \tag{10.1}$$

$$\varepsilon_{2,3,s} = \varepsilon^0_{2,3,s} - 0.059\ pH \tag{10.2}$$

yields $i_{c,s} = i^0_{c,s} \left[HF_{aq} \right] exp \left[2.303\ pH\ (1-2\alpha-\gamma) \right]$ (11)

$$i^0_{c,s} = 3kF \left[Fe^{3+}_{ox} \right] exp(- \frac{\Delta G^0}{RT}) \ exp\ (\frac{F(2\alpha + \gamma)}{RT} \varepsilon^0_{2,3,s}) \tag{11.1}$$

Equ. (11) shows the first order dependence of the Fe^{3+} dissolution on the Hf-
concentration as found experimentally. The pH dependence is difficult to con-
troll as a consequence of localized corrosion phenomena for higher pH (ref. 4).

Similar relations hold for the HF catalysed Ni^{2+} dissolution at minimum I
(ref. 10). The reaction order -0.76 at minimum II may be explained by the result
that the rate determining Ni_1^{2+} ion transfer occurs from a fluoride layer (Ni_1^{2+})
into the electrolyte. The potential drop $\varepsilon_{2,3,s}$ is now determined by the equilibrium of the F^- formation at the fluoride-electrolyte interface resulting in
the rate equation.

$$i_{c,s} = i_{c,s}^{\circ} \left[HF\right]^{-2\alpha} \exp(-4.606\ \alpha pH)$$

$$i_{c,s}^{\circ} = 2\ kF\left[Ni_1^{2+}\right] \exp(\tfrac{2\alpha F}{RT}\ \varepsilon_{2,3}^{\circ})$$

With the experimental reaction order -0.79 one obtains the reasonable charge
transfer coefficient $\alpha = 0.4$.

CONCLUSION

The combined electrochemical and surface analytical examination of the fluoride attack on passive iron and nickel yield a surface catalytically enhanced
mechanism for the transfer of the metal ions from the oxide into the electrolyte.
This process leads to a thinning of the barrier type oxide, whereas the hydroxide overlayer remains unaffected. In the case of iron this intermediate stage
is terminated by a final complete breakdown of passivity with a further increase
of the dissolution current. The production of Fe^{2+} instead of Fe^{3+} indicates the
free corrosion of an "active" metal surface. It is reasonable that similar events
occur locally during the nucleation of corrosion pits by other halides. This has
been confirmed by studies with rotating ring disc electrodes for Fe- and Fe-Cr
alloys in chloride containing electrolytes (ref. 2). A high concentration of
chloride (>1 M) may prevent the formation of a passive layer on iron and nickel
completely (ref. 11). Similarly the accumulation of halides within a pit will
stabilize its growth. In this sense, fluoride may be seen as a model system for
localized corrosion. The results for nickel at higher potentials at minimum II
are complicated by the formation of a fluoride layer which lowers the metal dissolution to some extent.

REFERENCES

1 H.-H. Strehblow, Werkstoffe und Korrosion 27 (1976) 792
2 K. E. Heusler, L. Fischer, Werkstoffe und Korrosion 27 (1976) 697
3 B. P. Löchel, H.-H. Strehblow, Werkstoffe und Korrosion 31 (1980) 353
4 H.-H. Strehblow, B. Titze, B. P. Löchel, Corrosion Science 19 (1979) 1047

5   B. P. Löchel, H.-H. Strehblow, Electrochim. Acta (1983), in press
6   N. Sato, Denki Kagaku 46 (1978) 584
7   M. P. Seah, W. A. Dench, Surf. Interf. Anal. 1 (1979) 2
8   K. S. Kim, N. Winograd, Surf. Sci. 43 (1974) 625
9   N. Sato, K. Kudo, R. Nishimura, J. Electrochem. Soc. 123 (1976) 1419
10   B. P. Löchel, H.-H. Strehblow, J. Electrochem. Soc., submitted
11 H.-H. Strehblow, J. Wenners, Electrochim. Acta 22 (1977) 421

Passivity of Metals and Semiconductors, edited by M. Froment
Elsevier Science Publishers B.V., Amsterdam — Printed in The Netherlands

ELECTROCHEMICAL BEHAVIOUR OF MILD STEEL IN SULPHIDE AND CHLORIDE CONTAINING
SOLUTIONS

C.A. ACOSTA, R.C. SALVAREZZA, H.A. VIDELA and A.J. ARVIA

Instituto de Investigaciones Fisicoquímicas Teóricas y Aplicadas - INIFTA. Casilla de Correo 16, Sucursal 4, 1900 La Plata, Argentina.

ABSTRACT

The localized corrosion of mild steel in alkaline and neutral buffered solutions containing sodium sulphide is studied. At a constant pH, breakdown potential decreases linearly with the concentration of sodium sulphide. The logarithm of the induction time for pit initiation decreases linearly with the reciprocal of the applied potential. The results suggest that two dimensional ferrous sulphide islands are formed before pitting. The addition of chlorides into the sulphide containing solutions prevents the oxide growth and consequently the electrode repassivation.

INTRODUCTION

The corrosion of iron and steel by sulphide species affects different industrial activities such as the paper and pulp industry, extractive petroleum industry, heavy water production by the Gidler process. It is known that iron and steel exposed to sulphides develop an initial film of mackinawite of poorly protective characteristics which is transformed later in a more stable iron sulphide through different chemical or electrochemical steps[1,2]. However, there is scarce information on the initial reaction leading to mackinawite formation. The purpose of this work is to study the first steps on steel corrosion by sulphide in neutral buffered and alkaline solutions using conventional electrochemical techniques complemented with scanning electron microscopy.

EXPERIMENTAL

The electrochemical measurements were made in a Pyrex glass cell at $25\pm0.1°C$. Working electrodes were SAE 1020 mild steel discs embedded in Araldite with a working area of 0.2 cm^2. A platinum wire was used as counter-electrode. Potentials are referred to the saturated calomel electrode (SCE). The reference electrode (Radiometer K-701) was designed to avoid the diffusion of chloride ions into the electrolyte.

The electrolyte solutions were prepared with bi-distilled water previously purged with purified N_2 during 3 hours. The solutions used included xM Na_2S and xM Na_2S in a KH_2PO_4 0.1 M + $Na_2B_4O_7$ 0.05 M buffer pH 8.00 ($10^{-3} < x < 3.0$ 10^{-2})

The breakdown potential (E_b) was determined by the potentiostatic method. Prior to the experiments, the working electrode was held at -1.20 V in the neutral buffered solution and -1.40 V in the alkaline solution during 2 min to electroreduce any possible surface species present on the specimen.

RESULTS

Anodic polarization curves of 1020 SAE steel in alkaline and neutral buffered solutions containing Na_2S show the active metal dissolution, the active-passive transition and the passive regions (Fig. 1). In the three regions, the current

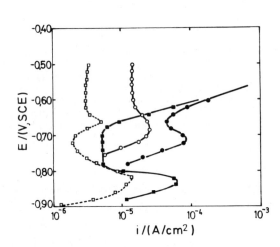

Fig. 1. Anodic polarization curves of 1020 SAE steel at 0.02 V/5 min in (□) NaOH 1.0×10^{-2} M adjusted with acetic acid to pH 11.40; (■) Na_2S 10^{-2} M adjusted with acetic acid to pH 11.40; (o) phosphate-borate solution pH 8.00 (●) phosphate-borate solution + Na_2S 2.5×10^{-2} M pH 8.00.

observed in the sulphide containing solutions is greater than in the absence of the latter. In sulphide solutions, a considerable current increase appears when the potential applied to the metal exceeds E_b. At potential values more anodic than E_b spots of ferrous sulphide which are related to isolated pits are observed on the metal surface.

E_b values were obtained in neutral buffered solutions containing different concentrations of Na_2S. At constant pH, E_b depends on the Na_2S concentration according to (Fig. 2).

$$E_b = a - b \log c_{Na_2S} \tag{1}$$

where a and b are constants. A similar relation was found in the neutral buffered solution containing NaCl (Fig. 2). In both cases b was close to 0.10 V/decade. At a constant Na_2S concentration, in alkaline solutions E_b shifts towards more cathodic values as the pH decreases (Fig. 3). This effect was also observed in neutral buffered solutions containing sulphide but then E_b decrease is smaller than the observed in alkaline solutions. E_b dependence with Na_2S concentration and pH suggest that breakdown process is related to SH^-/OH^- concen-

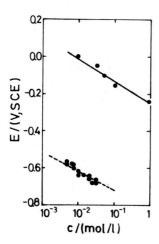

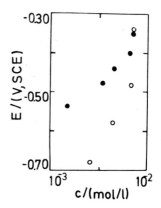

Fig. 3. E_b vs Na₂S concentration plot (●), E_b vs OH⁻ concentration plot in Na₂S 8.0×10^{-3} M (the OH⁻ concentration was adjusted with acetic acid) (o)

Fig. 2. Breakdown potential vs. concentration plot (—) NaCl, (--) Na₂S. Electrolyte: phosphate-borate solution.

tration ratio (r). In Na₂S solutions, due to the total hydrolysis of sulphide anions, r is equal to 1 but as the Na₂S concentration increases E_b shifts towards more positive values (Fig. 3). At Na₂S concentrations > 10^{-2} M pitting is inhibited. This was also observed in neutral buffered solutions containing sulphide when the applied potential exceeds -0.60 V. The value of r correlates also with pit density and morphology. In Na₂S 10^{-2} M, when r → ∞ a large number of ferrous sulphide nuclei related to small and shallow pits is observed but when r → 1 only few ferrous sulphide nuclei related to large and deep pits are formed. Besides at a constant r, both dilute and concentrate solutions show behaviour which is similar to that for r → ∞ and r → 1 respectively.

To obtain information on the species involved in the pitting inhibition, the electrode potential was held at a potential value more anodic than E_b in the neutral buffered solution containing sulphide. After certain time of anodization potentiodynamic polarization curves in cathodic direction were made. The curves were compared with those corresponding to the buffer solution without sulphide (Fig. 4). In this case, a broad cathodic current peak at -0.50 V corresponding to the electroreduction of the oxide film can be seen[3]. Conversely, in sulphide containing solutions a large anodic peak is observed at -0.50 V. Besides two cathodic peaks appear at -0.68 and -1.05 V. These results indicate that the pitting inhibition is due to the oxide formation rather than to the sulphide species formed during pitting.

Current-time transients were recorded in alkaline sulphide solutions stepping

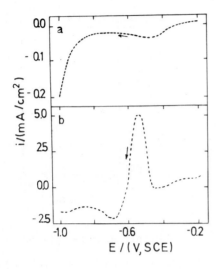

Fig. 4. Potentiodynamic polarization curves at 0.02 V/sec in: (a) phosphate-borate solution; (b) phosphate-borate solution containing 2.5×10^{-2} M Na2S. The electrode was held at -0.20 V during 30 sec before the negative scan.

the potential from -0.74 V to potential values more positive than E_b (Fig. 5). After an induction time (t_i), the current increases according to the relation:

$$I = I_p + K (t - t_i)^{b'} \tag{2}$$

where I_p is the background current and K and b' are constants. Log ($I-I_p$) vs. ($t-t_i$) plots show b values close to 1. t_i and the applied potential fit the relation (Fig. 6):

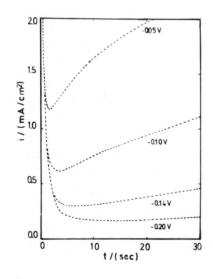

Fig. 5. Current-time transients recorded in 8.0×10^{-3} M Na2S. The electrode was held during 5 min at -0.74 V before the potential step.

$$\log t_i/t_0 = K'(E - E_b)^{-1} \qquad\qquad (3)$$

where t_0 and K' are constants.

After the initial rise, the current decreases due to the ferrous sulphide precipitation, although it is still sufficiently large due to the poor protective characteristics of mackinawite. Current-time transients recorded in phosphate-borate solution containing Na_2S also obey equation (2) and (3)[4].

Current-time transients recorded in the neutral buffered solutions containing sodium sulphide at potential values higher than -0.60 V with the addition of NaCl show that the chloride anion favorous pitting (Fig. 7). Chloride ions promote the formation of pits larger and deeper than those observed in the solutions containing only sulphide anions. At a constant potential, the anodic current increases with the concentration of sodium chloride.

DISCUSSION

The electrochemical behaviour of 1020 SAE steel at low potentials in neutral and alkaline solutions containing sulphide species shows two different processes that can be explained through a complex competitive surface reaction mechanism between the OH^- and SH^- anions and the H_2O and H_2S molecules. The OH^- and H_2O species are related to the initial metal passivation by a $Fe(OH)_2$ film. Otherwise the HS^- and H_2S species are related to the formation of a poorly pro-

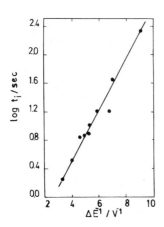

Fig. 6. Log t_i vs $1/E-E_b$. Electrolyte: 8.0×10^{-3} M Na_2S.

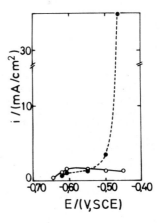

Fig. 7. Current density (read from the current-transients at t = 10 min) vs. potential relationship in phosphate-borate solution containing (o) Na_2S 2.5×10^{-2} M, ($\bullet$) Na_2Sx10^{-2} M + 1 M NaCl.

tective ferrous sulphide film and the simultaneous electrodissolution of iron through the $SHFe^+$ species[5]. The initial step in the mackinawite formation involves the specific adsorption of HS^- and H_2S replacing OH^- and H_2O. The competitive mechanism explains the dependence of E_b and pit density on r. However, the adsorption rate of aggressive species appears to be smaller than that of passivating species as can be concluded from the E_b dependence at constant r.

For a constant number of adsorption sites, the increase in concentrations of both aggressive and passivating species must result in an increase of the sites occupied by the species with higher adsorption rates (OH^- and H_2O) leading to passivation.

On the basis that equation (3) is fullfilled, it can be assumed that the electrochemical formation of mackinawite occurs through a nucleation and growth mechanism involving two dimensional ferrous sulphide nuclei formation and growth on the metal before pitting[6,7].

Pitting inhibition occurs in the potential region where the oxide film is formed. This suggests the oxide growth underneath ferrous sulphide nuclei as it is explained for the oxide growth on iron covered by salt layers[8]. The presence of chloride into the sulphide containing solutions prevents the oxide growth and, consequently, the electrode repassivation through the OH^- and H_2O displacement from the metal. This explains the enhancement of aggressiveness of chlorides and sulphide mixtures for iron and mild steel[9].

REFERENCES

1 D.W. Shoesmith, M.G. Baley and B. Ikeda, Electrochim.Acta 23, 1329 (1978).
2 A.J. Wikjord, T.E. Rummery, F.E. Doern and D.G. Owen, Corros.Sci. 20, 651 (1980).
3 K. Ogura and K. Sato, Electrochim.Acta 25, 857 (1980).
4 R.C. Salvarezza, H.A. Videla and A.J. Arvía, Corros.Sci. (in press).
5 R.C. Salvarezza, H.A. Videla and A.J. Arvía, Corros.Sci. 22, (9), 815 (1982).
6 T. Erdey-Gruz and M. Volmer, Z.Phys.Chem. A. 157, 165 (1931).
7 K.E. Heusler and L. Fischer, Werkst.Korros. 27, 697 (1976).
8 T.R. Beck, J.Electrochem.Soc. 129 (11), 2417 (1982).
9 Z.A. Foroulis, Werkst.Korros. 31, 463 (1980).

THE ROLE OF MOLYBDENUM IN THE INHIBITION OF PASSIVITY BREAKDOWN IN
NEUTRAL CHLORIDE SOLUTIONS UP TO 250°C

P. VANSLEMBROUCK[1], W. BOGAERTS[1] and A. VAN HAUTE[1]
[1]University of Leuven, Institute of Industrial Chemistry,
de Croylaan 2, B-3030 Heverlee (Belgium)

ABSTRACT

Several austenitic stainless steels and nickel alloys have been tested for
their resistance against localized corrosion in a 0.1 M NaCl solution up to
250°C.
The molybdenum containing alloys proof to have a superior resistance against
pitting attack up to 200°C. Above 200°C, for all the alloys tested the pitting
attack seems to degenerate into different kinds of localized or general attack
at rather high potentials.
The beneficial action of Mo against pitting corrosion in a neutral 0.1 M NaCl
solution can be attributed to its presence as MoO_4^{2-} (or $HMoO_4^-$) in the passive
film as indicated by the E-pH diagrams for Mo (ref. 1) and not to an insoluble
Mo-oxide.
The superior resistance of Mo containing alloys against active corrosion in
acid solutions on the other hand can be attributed to the presence of an
insoluble Mo-oxide (MoO_2) as again is indicated by the E-pH diagrams.

INTRODUCTION

For many years there has been a dispute among scientists about the improved
resistance of passive films on Mo containing alloys against acid solutions and
solutions containing chlorides, as well as to the question if Mo continues its
beneficial action above approximately 50°C (ref. 2).

Our experimental data, obtained from anodic polarisation measurements on
several austenitic stainless steels, nickel alloys and pure Mo, and from Auger
and Rutherford backscattering spectrometry (RBS) analyses performed on samples
passivated at a fixed potential close to the pitting potential, combined with
the E-pH diagram for Mo, provide some definite answers to the first question
as they do answer the second without any restrictions.

EXPERIMENTAL

Slow scan (200 mV/hr) anodic polarisation measurements have been carried
out in a static 1.5 l autoclave made of SS AISI 316 with PTFE lining. The
nominal working pressure throughout the tests was about 50 bar. The test
vessel contains a conventional three electrode electrochemical measuring
assembly with an external Ag/AgCl reference electrode.

Test solutions were deaerated by purging with pure N_2 for 15 min. All probes had a surface finish on emerypaper 600 followed by polishing down to 2 μ.

TABLE 1. Composition of the alloys tested.

Mass %	C	Fe	Cr	Ni	Mo	Cu	Mn	Ti	Nb
AISI 304	0.06	Bal.	18	9					
Haynes 600	0.05	9.4	14.8	74.4		0.11	0.41		
Incoloy 800H	0.08	45.6	20.25	31.6		0.53	0.77	0.46	
AISI 316	0.06	Bal.	17.5	12	2.8				
Haynes 825	0.30	27.3	21.66	44.4	3.05	1.82	0.29	0.84	
UDH 904L	0.02	Bal.	20	25	4.5	1.5	1.7		
Inconel 625	0.1	5.0	21.5	Bal.	9			0.4	3.6

RESULTS AND DISCUSSION

The potentials measured versus the external Ag/AgCl reference electrode have been recalculated with the method of Bogaerts (ref. 3) accounting for the thermal liquid junction potential. All results shown are given versus the NHE at 25°C unless explicitly indicated otherwise.

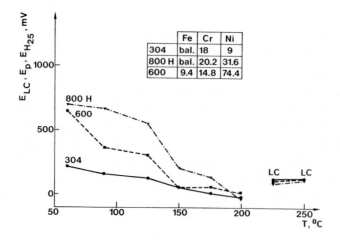

Fig. 1. Pitting potentials (E_p) of SS AISI 304, Haynes alloy 600 and incoloy 800 H in 0.1 M NaCl.

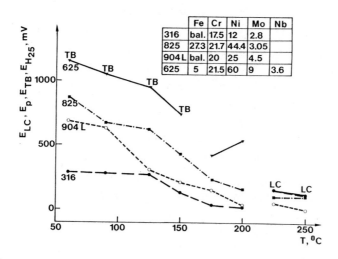

Fig. 2. Pitting potentials (E_p) and transpassive breakthrough potentials (E_{TB}) of SS AISI 316, UDH 904L, Haynes alloy 825 and inconel 625 in 0.1 M NaCl.

When we compare the pitting (E_p) and transpassive breakthrough (E_{TB}) potentials of the Mo containing alloys in fig. 2 with the non Mo containing ones
in fig. 1 it is clear that Mo has a strong inhibitive action up to 200°C.
Fig. 2 also points out the strong synergetic effect of both a high Ni and Mo
content (Alloys 625 and 825). When only the Ni content (alloys 600 and 800 H)
or Mo content (alloy 904L) is raised, the beneficial effect is far less,
especially in the temperature range between 150°C and 200°C. At temperatures
above 200°C the pitting phenomenon degenerates into different kinds of localized
attack (especially around 250°C) at high potentials.

Auger and RBS analyses on samples of the Mo containing alloys, carefully
passivated at 90°C and 150°C at potentials slightly beneath their pitting
potentials show a clear depletion of Mo in the passive film. This depletion
decreases the more the metal-oxide interface is approached. The metal just
beneath the oxide film has the nominal composition. A typical result is shown
in fig. 3.

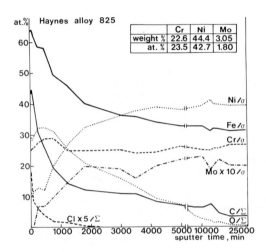

Fig. 3.   Auger depth profile of Haynes alloy 825 left at its corrosion potential
(increasing from +200 mV up to +400 mV vs. NHE 150°C) for 6 days at
150°C in an oxygenated (± 100 ppm O_2) neutral 0.1 M NaCl solution.
σ =   [Fe] + [Ni] + [Cr] + [Mo]
Σ = σ + [O] + [C] + [Cl]

The E-pH diagrams of Mo (ref. 1) at temperatures from 60°C up to 250°C predict Mo to be stable in the (soluble) +VI state at potentials even far beneath
the pitting potentials of alloy 316.   These thermodynamic predictions are confirmed by our measurements of the transpassive breakthrough potentials of pure
Mo in neutral 0.1 M NaCl solutions in the same temperature range.

All these results seem to indicate that the inhibitive action of Mo should
be ascribed to the adsorption of MoO_4^{2-} at weak spots, preventing an accumulation of Cl^-, rather than a strengthening of the passive film as a whole.   First
results from a research program still in progress show MoO_4^{2-} (or $H\,MoO_4^-$) to be a
very strong inhibitor against pitting corrosion at temperatures up to 200°C when
added to a 0.1 M NaCl solution, which again supports our views just mentioned.

The improved resistance of alloys containing Mo against acid solutions on
the other hand can be attributed to the formation of the inert MoO_2 in acid environments.   Striking evidence is shown in figure 5.
Although Haynes alloy 600 contains Mo only as a trace element, an enormous
amount (up to 50 % of the metal atoms!) of Mo is present in the oxide film after
being corroded for 3 hrs. in an acid (pH = 1.4 at 25°C) chloride solution at
150°C.   Similar results were obtained with AISI 304 and incoloy 900 which also

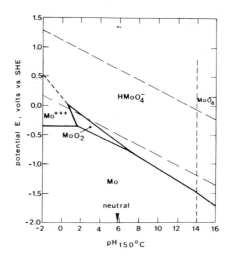

Fig. 4. Potential-pH diagram for the molybdenum-water system at 150°C (ref. 1).

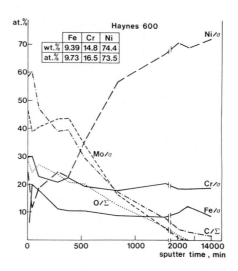

Fig. 5. Auger depth profile of inconel 600 left at its corrosion potential
(-260 vs. NHE 150°C) for 3 hrs in an acid 0.1 M NaCl solution
(pH = 1.4 at 25°C) at 150°C.

contain Mo in only fractions of a percentage. Quite surprisingly the A.E.S. and R.B.S. analyses showed again a Mo depletion for the molybdenum containing alloys which had been given the same treatment. Yet everything becomes clear when the corrosion potentials are projected in the E-pH diagram for Mo at 150°C. The corrosion potentials of the enriched alloys are all situated in the stability region of MoO_2 while the Mo depleted alloys all corroded at much higher potentials, which are situated in the stability region of the soluble $HMoO_4^-$. The reason that the enrichment of Mo in the alloys 304, 600 and 800 take such huge proportions can be found in the E-pH diagrams for Fe, Ni and Cr which indicate that these elements are all in a soluble state under the given conditions of pH, potential and temperature.

REFERENCES

1 J.B. Lee, Corrosion, 37 (1981) 467-480.
2 M.A. Streicher, Journal of the Electrochemical Society, 103 (1956) 375-390.
3 W. Bogaerts, Ph.D.Thesis, Leuven University (1981).

Passivity of Metals and Semiconductors, edited by M. Froment
Elsevier Science Publishers B.V., Amsterdam — Printed in The Netherlands

ON THE INFLUENCE OF BOTH PASSIVE FILM COMPOSITION AND OF NON-METALLIC INCLUSIONS
ON THE INITIATION OF LOCALIZED CORROSION OF STAINLESS STEEL

G. HULTQUIST, S. ZAKIPOUR AND C. LEYGRAF*
Department of Physical Chemistry, The Royal Institute of Technology, S-100 44
Stockholm 70 (Sweden)

ABSTRACT
   The role of both passive film composition and of certain non-metallic inclusions on the resistance to initiation of localized corrosion has been elucidated by comparing information from surface analysis and from electrochemical
measurements of some stainless steels after different surface treatments. With
this approach it has been possible to separate the effects of both passive film
composition and of non-metallic inclusions. Experimental evidence is presented which unambiguously shows that both effects are operating simultaneously
and must be considered when optimizing stainless steel with respect to resistance to localized corrosion.

INTRODUCTION
   In the literature on passivity breakdown of stainless steel, several mechanisms have been proposed for the initiation of localized corrosion, which
have been summarized in recent reviews (refs. 1 and 2). Somewhat simplified
the mechanisms proposed can be divided into two categories, which are of different nature. One approach emphasizes the role of the passive film itself and
postulates changes in defect structure and in passive film composition caused
by aggressive anion migration through the passive film lattice or adsorption of
aggressive ions on the passive film surface - eventually leading to local removal of the film and the initiation of localized corrosion attacks. The other
approach emphasizes the macroscopic heterogeneity of stainless steels. The
presence of second-phase particles such as non-metallic inclusions leads to
non-uniform dissolution and the formation of flaws or pits - eventually resulting in local acidification and in pit propagation. Because of the opposite
nature of these two types of models, experimental evidence shown in the literature is with almost no exclusion in favour or against either of these models.
   In this work we report experimental evidence showing that both types of
mechanisms can operate simultaneously on the same stainless steel surface.

---

*Present address: Swedish Corrosion Institute, Box 5607, S-114 86 Stockholm
(Sweden).

This has been done by comparing information from surface - and from electrochemical characterization of two stainless steels, which have been subjected to
various thermal- or chemical surface treatments.

METHODS

The results are based on two different types of commercially available
stainless steels (austenitic Fe18Cr13Ni3Mo and ferritic Fe18Cr2Mo). The thermal treatments consist of low temperature annealings (200-300 C) of the stainless steel in dilute oxygen atmospheres (ref. 3). These treatments aim at
changing the passive film composition but leaving non-metallic inclusions intact. The chemical treatments consist of immersion of the stainless steel in
pickling (HNO_3-HF) solutions. These treatments aim at changing the number of
non-metallic inclusions. Surface characterization before and after surface
treatments has been performed of both passive film composition (by Auger
Electron Spectroscopy and ESCA) and of the number and composition of non-metallic inclusions (by Scanning Electron Microscope Image Analysis combined with
particle specific X-ray analysis (ref. 4)). The same stainless steel surfaces
have also been electrochemically characterized giving a measure of the corrosion resistance in terms of critical potentials for the initiation of crevice
and of pitting corrosion respectively, see (ref. 3) for experimental details.

RESULTS

In Fig. 1 the passive film composition of both stainless steels after various thermal treatments, giving various chromium surface contents, is plotted
against the corresponding critical potential for the initiation of crevice corrosion. In the potential region between 100 and 700 mV (SCE) the ferritic
Fe18Cr2Mo alloy exhibits a monotonous increase in corrosion resistance with
chromium content of the passive film. The austenitic Fe18Cr13Ni3Mo alloy, on
the other hand, only undergoes a slight increase in corrosion resistance in
spite of a considerable increase in chromium film content. The reason is the
presence of a much larger amount of (Mn,S)-inclusions in the austenitic stainless steel (see TABLE 1) which are not eliminated during the thermal treatment and which are known to act as initiation sites for localized corrosion
(ref. 5). Hence, changes in passive film composition of the high purity ferritic alloy have a much more pronounced influence on the resistance to localized
corrosion than corresponding changes in passive film composition of the austenitic alloy, where only a limited improvement is obtained.

The results of Fig. 1 suggest that a chemical treatment of the austenitic
steel should result in a considerable increase in corrosion resistance if a
reduction can be achieved of the number of detrimental sulphide inclusions at

the surface. This is shown to be the case, as illustrated in TABLE 1, which gives the total number of non-metallic inclusions for both stainless steels as well as the number of inclusions containing both Mn and S after different sur- face treatments. Upon mechanical polishing, the number of (Mn,S)-inclusions is considerably higher in the austenitic Fe18Cr13Ni3Mo surface as compared to the ferritic Fe18Cr2Mo surface. Upon pickling both surfaces in a (15% HNO_3 + 5% HF)-solution a marked reduction in (Mn,S)-inclusions is observed in the Fe18Cr13Ni3Mo surface and a slight reduction in the Fe18Cr2Mo surface.

As inferred from TABLE 1 the pickling treatment also results in a modera- te increase in chromium film content, as measured by Auger Electron Spectrosco- py. The reduction in (Mn,S)-inclusions and the increase in chromium film con- tent upon pickling both contribute to the observed increase in critical poten- tial for initiation of crevice corrosion (TABLE 1). This critical potential is further increased after a subsequent thermal treatment resulting in an ad- ditional increase in chromium film content and in no measurable change in the amount of (Mn,S)-inclusions.

TABLE 1

Surface analysis data and resistance to initiation of crevice corrosion of Fe18Cr13Ni3Mo and of Fe18Cr2Mo after different surface treatments.

Steel	Surface treatment	Total num- ber of non- metallic inclusions (mm^{-2})	Number of (Mn,S)- inclu- sions (mm^{-2})	Chromium film con- tent, $Cr/\Sigma Me$ (atom%)	Critical potential for ini- tiation of crevice cor- rosion (mV,SCE)
Fe18Cr13Ni3Mo	Mech. polished	540	320	31±2	100±25
	Mech. polished + pickled (HNO_3-HF)		45	44±3	450±50
	Mech. polished + pickled (HNO_3-HF) + thermally pas- sivated		45	57±3	> 700
Fe18Cr2Mo	Mech. polished	260	10	32±2	50±25
	Mech. polished + pickled (HNO_3-HF)		5	45±3	300±50
	Mech. polished + pickled (HNO_3-HF) + thermally pas- sivated		5	75±5	> 700

402

Fig. 2 shows the critical potential for both pitting and crevice corrosion plotted against the chromium film content after various surface treatments. As expected, the critical potential for crevice corrosion exhibits
significantly lower values than for pitting corrosion. The results were obtained after immersion at room temperature of a polished Fe18Cr13Ni3Mo surface
in a pickling solution (15% HNO_3 + 5% HF) and in a passivating solution (20%
HNO_3) respectively. Both chemical treatments result in approximately the same
amount of (Mn,S)-inclusions. Furthermore, the passivation treatment causes a
higher chromium film content than the pickling treatment, from which follows
a higher resistance against initiation of pitting as well as of crevice corrosion (Fig. 2).

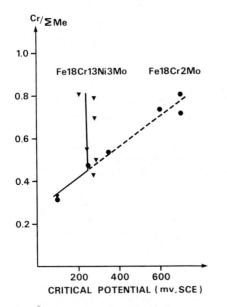

Fig. 1. Relation between chromium film content and critical
potential for initiation of
crevice corrosion (mV,SCE) of
Fe18Cr13Ni3Mo and of Fe18Cr2Mo
after different thermal treatments.

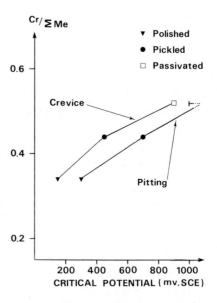

Fig. 2. Relation between chromium film content and critical
potential for initiation of
crevice and pitting corrosion
(mV,SCE) of Fe18Cr13Ni3Mo after different surface treatments.

Changes in passive film composition after pickling the steel surfaces were
also characterized by means of ESCA measurements. These measurement illustrate that pickling the Fe18Cr13Ni3Mo surface causes an increase in the amount of
oxidized chromium and a simultaneous reduction in the amount of oxidized iron

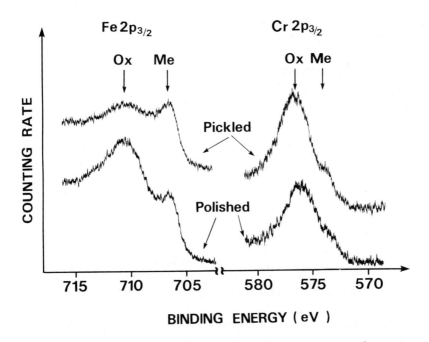

Fig. 3. Fe and Cr main peaks from ESCA measurements of Fe18Cr13Ni3Mo surfaces after mechanical polishing and after sub-sequent pickling treatment in a (15% HNO_3-5% HF) solution at room temperature.

(Fig. 3). Similar ESCA results are obtained after pickling the Fe18Cr2Mo surface and after thermal treatments of both steel surfaces. This behaviour is a consequence of both the selective oxidation behaviour of chromium in iron-chromium alloys under a broad range of oxidation conditions and of selective dissolution of iron in many liquid environments.

In Fig. 4 are plotted the number of (Mn,S)-inclusions as a function of bulk manganese content of three polished stainless steel surfaces. From this figure follows that the number of detrimental (Mn,S)-inclusions can be easily reduced by reducing the bulk manganese content. These results are in accor-dance with earlier findings that a low manganese content in various austenitic stainless steels significantly increases the resistance to initiation of pit-ting and crevice corrosion (ref. 6).

DISCUSSION

Results have been presented in this work which unambiguously show the si-multaneous effects of both chromium film content and amount of non-metallic inclusions in the surface on the resistance to initiation of localized corro-sion. These results can qualitatively be summarized as follows: Localized corrosion is a statistical process with a variety of possible initiation si-

404

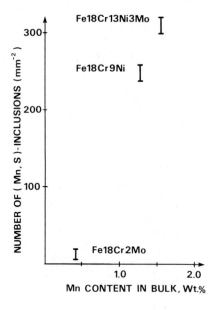

Fig. 4 Number of (Mn,S)-inclusions against manganese bulk content of some stainless steels.

tes. The most probable type of initiation site depends on the surface condition of the stainless steel. After mechanical polishing, initiation sites are expected to be found in both the passive film and next to certain non-metallic inclusions. Upon thermal passivation treatment an increase in chromium film content is observed resulting in an increase in Cr-O or Cr-OH bonds which hinder the migration of aggressive ions through the film or the dissolution of cationic film constituents. This makes non-metallic inclusions as more probable initiation sites. The highest corrosion resistance is obtained after a combination of chemical and thermal treatments resulting in surfaces with only a small number of sulphide inclusions and in a high chromium film content.

Although not discussed here, it should be added that other properties of stainless steel also are known to influence the resistance to localized corrosion. Among these properties should be mentioned the molybdenum bulk content, crystallinity of the steel matrix and possible chromium depletion of the alloy region next to the protective film.

REFERENCES

1. J.R. Galvele, in "Passivity of Metals", Ed. R.P. Frankenthal & J. Kruger. The Electrochemical Society (1978), p. 285.
2. M. Janik-Czachor, G.C. Wood and G.E. Thompson, Br. Corros. J. 15 (1980) 154.
3. G. Hultquist and C. Leygraf, Mater. Sci. Eng. 42 (1980) 199.
4. T. Werlefors and S. Ekelund, Scand. J. Metall. 7 (1978) 60.
5. G. Wranglén, Corros. Sci. 14 (1974) 331.
6. T. Sydberger, Proc. 8th Scand. Corr. Congr., Helsinki, Finland (1978), Vol. 1, p. 357.

Passivity of Metals and Semiconductors, edited by M. Froment
Elsevier Science Publishers B.V., Amsterdam — Printed in The Netherlands

THE EFFECTS OF ALLOYING ON THE RESISTANCE OF FERRITIC STAINLESS STEELS TO LOCALIZED CORROSION IN Cl⁻ SOLUTIONS

W.R. CIESLAK and D.J. DUQUETTE

Rensselaer Polytechnic Institute, Troy, New York   12181

ABSTRACT

The pitting resistance of Fe-Cr and Fe-Cr-Mo alloys has been correlated with characteristics of the passive films as analyzed by Auger Electron Spectroscopy. Increased film protectivity as a result of increased Cr in the alloy can be directly attributed to film Cr-enrichment and decreased thickness. Increased Mo in the alloy or passivation at noble potentials, while exerting beneficial effects on passive film resistance to breakdown, do little to change the macrocharacteristics of the film. Rather, it is suggested that the roles of alloying and/or film stability are related to the susceptibility and distribution of weak points of the film. In solutions which cause pitting, chloride is not generally incorporated into the film, suggesting that the role of halides is to interact with weak points of the film at the solution/film interface.

INTRODUCTION

The beneficial effects of alloyed Cr and Mo on the resistance of stainless steels to pitting breakdown are well-documented (ref. 1-3). However, the relationship of passive film composition and structure to breakdown is less straightforward. Electron analytical techniques provide a means of determining, at least qualitatively, the compositional profiles of passive films. Several investigators have noted lower or equal Mo concentrations in passive films compared to the Mo concentration in the steels, and no effect of Mo on overall film composition or thickness (ref. 4-10).

The approach used in this investigation has been to vary alloy composition and passivation treatment and to determine induction times to pit initiation. Measurements of film composition and relative thicknesses have been determined by Auger Electron Spectroscopy (AES) analysis.

EXPERIMENTAL PROCEDURE

The chemical analyses of the experimental alloys used in this investigation are shown in Table 1. Rectangular sheet specimens, 1.3 cm X 0.8 cm, were cut from the 1.5 mm thick as-rolled plates. Specimens were annealed for 10 minutes at 1040°C and water-quenched to produce a fully-recrystallized single phase ferritic microstructure.

TABLE 1

Compositions of Ferritic Stainless Steels (wt%)

Cr	Mo	C	S	P	N	Fe
17.95	0.010	0.002	0.004	0.004	0.0034	BAL.
17.94	1.98	0.001	0.004	0.004	0.0021	BAL.
18.00	4.91	0.002	0.004	0.003	0.0022	BAL.
27.65	0.10	0.002	0.003	0.004	0.0026	BAL.
27.87	1.99	0.002	0.003	0.005	0.0022	BAL.
28.05	4.92	0.002	0.003	0.004	0.0027	BAL.

All solutions were prepared from deionized water of conductivity 1.0 to
1.2 μMHO at 25°C. Deaeration was accomplished by purging the test cell with
pre-purified argon for a minimum of 4 hours prior to specimen introduction and
by continuous purging throughout the test duration. Specimens were wet ground
on 600 grit SiC paper immediately prior to testing.

Passive films for pitting induction time measurements and AES analysis
were formed by open-circuit and anodic passivation treatments for 2 hours in
80°C solutions. Following passivation, samples for AES were thoroughly rinsed
with distilled water, dried in a stream of flowing Argon and placed in the AES
chamber. The chamber was pumped down overnight and analysis performed the
following day.

The spectrometer was operated at a primary energy of 2.5 KV with a peak-
to-peak modulation of 3V. The primary beam was approximately 0.5 mm in diameter. Automated depth profiling was performed using an Argon ion gun at a
pressure of 5×10^{-5} Torr and emission current of 25 mA at 5 KV. The rastered
beam covered an area of approximately 1 cm^2.

For each specimen, AES spectra were continuously monitored during depth
profiling. Five channels were used for monitoring the depth profile, and
these were set to detect the following elements and their corresponding energy
peaks, (eV): Fe(651), Cr(529), O(518), Mo/Cl(186/181), C(273). Mo is also
identified by a peak at 221 eV. This peak was used to differentiate Mo from
Cl.

RESULTS AND DISCUSSION

The AES depth profile for an 18% Cr-0% Mo control sample (dried immediately following surface grinding instead of exposed to solution) is shown in
Figure 1. The depth at which the carbon surface contamination had decreased
75% from its initial value was arbitrarily taken as the outer surface of the
oxide. The largest signal from the substrate, Fe, was used to determine the
approximate oxide/metal interface. Each unit of thickness corresponds to the
time interval for removal of approximately 3 monolayers of NiO from a pure Ni

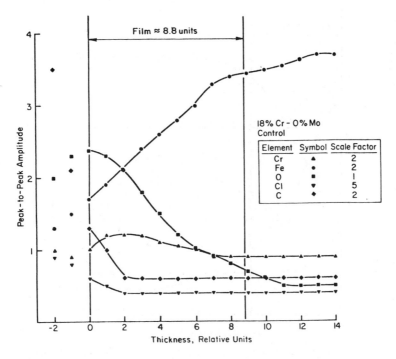

Fig. 1. AES depth profile of 18% Cr-0% Mo control sample.

calibration sample. The major distinguishing feature of the depth profiles of control samples is that the Cr-profile is relatively flat and the signal is weaker than that of Fe throughout the film.

The AES depth profile for a 28% Cr-5% Mo sample passivated in deionized water is shown in Figure 2. The enrichment of Cr relative to Fe in the film is very evident. Cl is only present at the film surface and disappears rapidly during ion-milling. According to these results, it is certain that no Mo is detected in the outermost portion of the film. It is not possible to ascertain whether any Mo is detected from within the film, since the gradual rise of the Mo-signal during ion-milling might be attributed to a gradual increase in the proportion of the signal escaping from within the bulk alloy.

Figure 3 summarizes the results of the AES analyses. The control preparation and all passivation treatments resulted in films of comparable thickness, independent of Mo, solution concentration, or passivation potential (for the range -650 to 0 mV vs. SCE). The thickness is a function only of Cr-content and is less for the 28% Cr alloys than for the 18% Cr alloys. Cr-enrichment in the films is calculated as the ratio of maximum film Cr/Fe: final Cr/Fe. The Cr-enrichment ratio is also insensitive to solution chloride concentration and

408

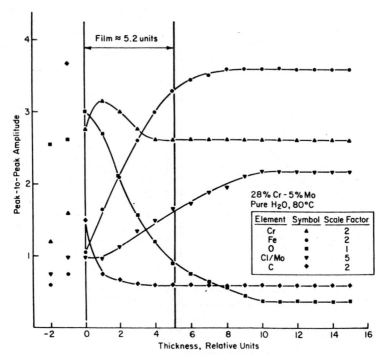

Fig. 2. AES depth profile of 28% Cr-5% Mo passivated in deionized water.

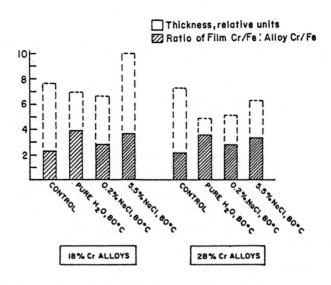

Fig. 3. Cr-enrichment ratios and relative thicknesses of films, averaged for all Mo-contents.

to passivation potential. However, the ratio is fairly consistently reduced
as a result of alloyed Mo (see Fig. 4).

Although little influence of alloyed Mo and of passivation potential on
passive film properties can be detected by AES, both variables strongly increase pitting induction times (ref. 11). Along with the increasing induction
times (more protective films), films are broken by pits at fewer sites. No
microstructural features (such as inclusions or grain boundaries) act as preferred sites for pit initiation. Therefore, it is reasonable to assume that
both Mo and potential act to alter the distribution and susceptibility of weak
points in the passive film. A possible mechanism has been suggested by Okamoto
and co-workers, related to the structure and quantity of the water incorporated
in the passive film (ref. 12,13).

Passivation treatments noble to the critical pitting potentials resulted
in film incorporation of Cl⁻, greatly increased thickness and decreased Cr-
enrichment (see Fig. 5),although analysed areas had not pitted. The changes
observed are likely to be due to a corrosion product film rather than Cl⁻ penetration of the original passive film. The Cl⁻ signal intensity decreases with
depth into the film and cannot be interpreted to indicate Cl⁻ penetration
through to the metal/oxide interface.

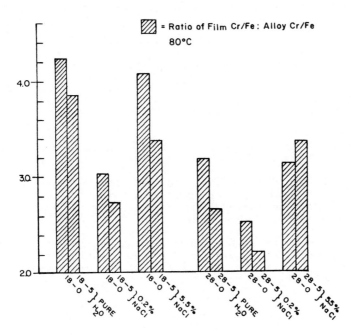

Fig. 4. Cr-enrichment ratios as a function of alloy Mo-content.

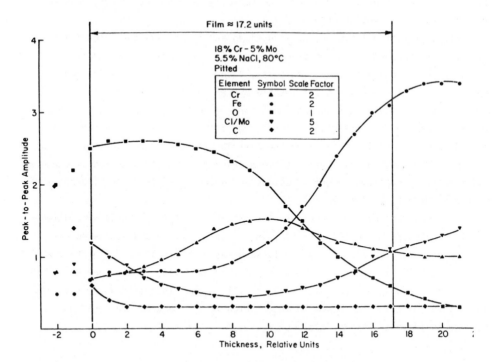

Fig. 5.   AES depth profile of 18% Cr-5% Mo sample, passivated in 5.5% NaCl.
The passivation potential was noble to the critical pitting potential
of the alloy.

CONCLUSIONS

1. The distinctly different appearance of the depth profiles of the control
    samples from that of the passivated samples shows that measurable effects
    of solution exposure were retained through the handling and analysis pro-
    cedure.

2. The beneficial effects of alloyed Cr are clearly related to increased Cr-
    enrichment and decreased thickness of the passive film.

3. The beneficial effects of alloyed Mo and of passivation potential are re-
    lated to passive film properties in a more subtle manner. It is suggested
    that the distribution and susceptibility of weak points in the passive
    film might be altered by these variables.

4. Unattacked areas of pitted samples can be distinguished from non-pitted
    samples by increased film thickness and decreased Cr-enrichment. Although
    Cl⁻ is present as more than a surface contaminant, the evidence does not
    support general film penetration by Cl⁻.

ACKNOWLEDGEMENT

The authors would like to acknowledge the financial support provided by the U.S. Department of Energy under Contract Number EY-76-5-02-2462*000 and by a Fannie and John Hertz Foundation Fellowship granted to one of the authors (WRC). Appreciation is expressed to Dr. John B. Hudson for conducting the AES analyses and the Climax Molybdenum Company for providing the alloys.

REFERENCES

1. Ja. M. Kolotyrkin, Corrosion, 19 (1963) 261t-268t.
2. Z. Szklarska-Smialowska, Corrosion, 27 (1971) 223-233.
3. M.A. Streicher, Corrosion, 30 (1974) 77-91.
4. A.E. Yaniv, J.B. Lumsden and R.W. Staehle, J. Electrochem. Soc., 124 (1977) 490-496.
5. M. da Cunha Belo, B. Rondot, F. Pons, J. LeHericy and J.P. Langeron, J. Electrochem. Soc., 124 (1977) 1317-1324.
6. K. Sugimoto and Y. Sawada, Corr. Sci., 17 (1977) 425-445.
7. H. Ogawa, H. Omata, I. Itoh and H. Okada, Corrosion, 34 (1978) 52-60.
8. K. Hashimoto, K. Asami and K. Teramoto, Corr. Sci., 19 (1979) 3-14.
9. K. Hashimoto and K. Asami, Corr. Sci., 19 (1979) 251-260.
10. R. Berneron, J.C. Charbonnier, R. Namder-Irani and J. Manenc, Corr. Sci., 20 (1980) 899-907.
11. W.R. Cieslak, Ph.D. Dissertation, Rensselaer Polytechnic Institute, May 1983.
12. G. Okamoto, Corr. Sci., 13 (1973) 471-489.
13. H. Saito, T. Shibata and G. Okamoto, Corr. Sci., 19 (1979) 693-708.

EFFECT OF Al ON THE ELECTROCHEMICAL BEHAVIOUR OF TERNARY BRASSES

F. TERWINGHE, J.P. CELIS and J.R. ROOS

Departement Metaalkunde, K.U.Leuven, de Croylaan 2, B-3030 Heverlee (Belgium)

ABSTRACT

As part of a research on the corrosion behaviour of beta and martensitic
Cu-Zn-Al alloys, the electrochemical behaviour of such alloys possessing shape
memory properties has been studied. Since shape memory alloys will be used
under periodically varying tensile stresses, the problem of their stress corrosion susceptibility has been put forward. SCC-mechanisms for α-brasses as described in the literature emphasize the role of a passivating surface layer.
Electrochemical measurements as potentiokinetic polarisation and open circuit
restpotential measurements, have been performed in order to determine the presence of such a passivating surface layer on Cu-Zn-Al alloys in tap water and in
nitrite solutions. Our results confirm the existence of such a passivating surface film but it is also demonstrated that the stability of this film is potential dependent so that an appreciable SCC-susceptibility is expected.

INTRODUCTION

Beta (β) and martensite (β') Cu-Zn-Al alloys became recently attractive engineering materials because of their shape memory properties. A phase transformation (beta $\rightleftarrows$ martensite) induced either thermally or mechanically causes
a macroscopic shape change of these materials. Since in engineering applications such alloys will frequently be used under external mechanical loading, information on the stress corrosion cracking (SCC) susceptibility is required.

SCC-susceptibility of alpha (α) brasses is a well-known phenomenon and depends on the stress level and on the type of anion present in the surrounding
medium (ref. 1, 2). Largest SCC-susceptibility for brasses occurs in NO_2^--sol-
utions, in decreasing order followed by NO_3^-, SO_4^{--} and Cl^-. The general mechanism of SCC has been explained by different researchers (ref. 3, 4, 5) and is
fundamentally based on the presence of a passivating surface layer. SCC-crack
initiation occurs when a local disruption of the passivating layer is created under the influence of tensile stresses. Extremely fast crack propagation will occur under electrochemical conditions where the crack tip remains active and the
crack walls passivate. Otherwise the crack tip will be blunted and pitting occurs.   As part of our research on the corrosion behaviour of beta and
martensitic Al-brasses, electrochemical measurements have been performed to determine to what extent aluminium creates passivating layers on such Al-brasses.

414

EXPERIMENTAL

Potentiodynamic polarisation curves and open circuit rest potentials have
been recorded at roomtemperature using a PAR potentiostat model 173 and a PAR
function generator model 175. The three-electrodes cell consists of a Pt-coun-
ter electrode and a saturated calomel electrode as reference electrode. As
working electrodes, specimens cut out of hot rolled sheets were used after grinding with 1200 grit paper and rinsing with ethanol. All materials used (Cu;
Cu-Zn 30 and Cu-Zn-Al with different Zn and Al contents) were cast from 99,9 %
pure base metals. Since up to now no information on the SCC-susceptibility of
β and β' Cu-Zn-Al is available from the literature, two types of test solutions
have been selected : tap water (pH 7.1, Cl^- ~ 25 ppm) representative for normal
use in engineering conditions and neutral $NaNO_2$ solutions known as one of the
most aggressive media for SCC of α-brasses. During testing the solutions were
not stirred.

In figures 1 and 2 potentiodynamic polarisation curves recorded in tap water
and in a 1 M $NaNO_2$ solution respectively are shown. At increasing aluminium
content more pronounced active to passive transitions under anodic polarisation
are determined. Especially in a $NaNO_2$ solution the passivating effect in the
anodic region is very pronounced and can reach several orders of magnitude.
Moreover in this solution Cu-Zn-Al alloys exhibit two different open circuit
restpotentials depending on the previously applied potential, namely points A
and E as shown in fig. 3 on a polarisation curve recorded in a 0.1 M $NaNO_2$ solution. Once a Cu-Zn-Al specimen is polarized above potential C, the open circuit restpotential determined after interruption of this polarisation corresponds
to potential E. Holding the polarization at potential C the anodic current decreases progressively until it reaches the value obtained in the potentiokinetic
reverse polarization scan as shown in figure 3. After a short subsequent cathodic polarisation (20 seconds at - 700 mV vs. SCE) the open circuit rest potential is displaced to potential A, corresponding to the open circuit rest potential of a freshly polished and immersed specimen.

The time variation of the open circuit rest potential of Cu, Cu-Zn and
Cu-Zn-Al specimens in a 0.1 M $NaNO_2$ solution is shown in fig. 4. In contrast to
Cu and Cu-Zn 30, the open circuit potential of Cu-Zn-Al alloys shows large potential variations occurring over rather long time intervals. From fig. 3 and 4 it
seems that these potential variations are limited between the Flade potential
(point B) and the potential corresponding to the onset of the passive region
(point C). The restpotential of the Cu-Zn-Al alloy moves towards the corresponding value of potential E in fig. 3. After three weeks of immersion the restpotential was still situated at this value, but even short cathodic polarization
at - 700 mV vs SCE shifted once again the restpotential to the value of potential

A in fig. 3. Polarization curves in NaNO$_2$ solutions were also recorded under applied tensile stresses. The specimen was electrically isolated from the tensile apparatus. From fig. 5 obtained for a stressed Al-brass at 52 % of the ultimate tensile strength, it can easily be concluded that there is no difference in the electrochemical behaviour of stressed and unstressed specimens.

From visual and light microscopic inspection of beta Cu-Zn-Al specimens after polarization tests described in figure 2 no colour change or apparent corrosion could be detected on the specimen surface. On the contrary Cu and Cu-Zn specimens seemed to be covered with a thin coloured surface layer. Comparison of AES-measurements on a freshly ground Cu-Zn-Al specimen (7,7 wt % Al) and on a specimen of the same alloy polarised at potential D for 2 hours did not show any representative change in the depth profile of Zn and Al; this confirms the passivation behaviour of the Cu-Zn-Al specimen in NaNO$_2$ solutions.

DISCUSSION

From electrochemical measurements in tap water and in nitrite solutions it seems thus that the addition of aluminium to brasses enables the formation of a passivating surface layer. However the stability of these layers in nitrite solutions seems to be potential dependent. Large fluctuations in the open circuit potential of Cu-Zn-Al alloys and the easy shifting of the open circuit potential by cathodic polarisation confirm the low passivation stability of the surface layer. It should be noticed that according to the literature (ref. 6) the addition of sodium nitrate and nitrite considerably increases the pitting potential of aluminium in chloride solutions. It seems that a localized corrosion process can not be responsible for the shifting of the rest potential, but rather a reduction process of a very thin passivation layer. Nevertheless this surface layer promotes a rather important passivation. But since the electrochemical breakdown of this layer is very easy, it may therefore be expected that under the influence of varying tensile stresses (as will occur in Cu-Zn-Al shape memory alloys) a local mechanical disruption of this passivating layer can be expected. Based on the SCC-mechanisms proposed in literature it may thus be expected that Cu-Zn-Al will show a large stress corrosion cracking susceptibility in nitrite solutions.

Specific stress corrosion cracking tests based on the slow strain rate technique are in progress in order to determine the SCC-susceptibility of Cu-Zn-Al in different aggressive solutions.

REFERENCES

1. E. Mattsson, Brit.Corr.Journal, 15 (1) 6-13 (1980)
2. A. Kawashima, A.K. Agrawal, R.W. Staehle, A.S.T.M. STP 665, 266-278 (1979)

416

3. E.N. Pugh, in "The Theories of Stress Corrosion Cracking in Alloys", NATO
   Report, 418-441 (1971)

4. J.C. Scully, in "The Theories of Stress Corrosion Cracking in Alloys", NATO
   Report, 1-15 (1971)

5. H.J. Engell, in "The Theories of Stress Corrosion Cracking in Alloys", NATO
   Report, 87-104 (1971)

6. M. Koudelkova, J. Augustynski, J. Electrochem. Soc., 126 (10) 1659-1661 (1979).

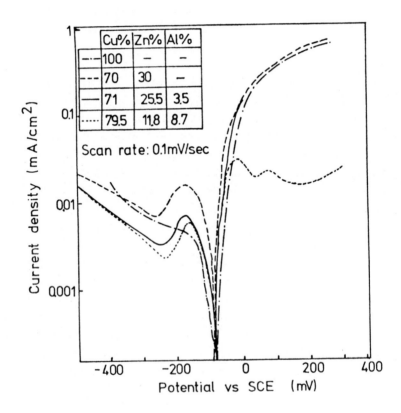

Fig. 1 Potentiodynamic polarisation curves recorded in tap water at room
temperature (pH 7.1).

417

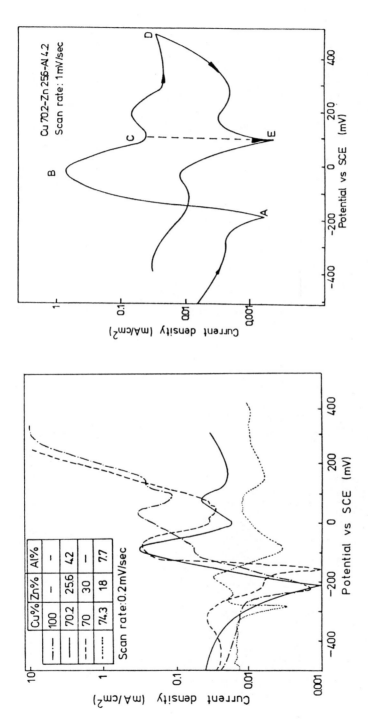

Fig. 3. Potentiodynamic polarisation curve recorded in a 0.1 M NaNO₂-solution at roomtemperature (pH 6.3).

Fig. 2. Potentiodynamic polarisation curves recorded in a 1 M NaNO₂-solution at roomtemperature (pH 8 to 9).

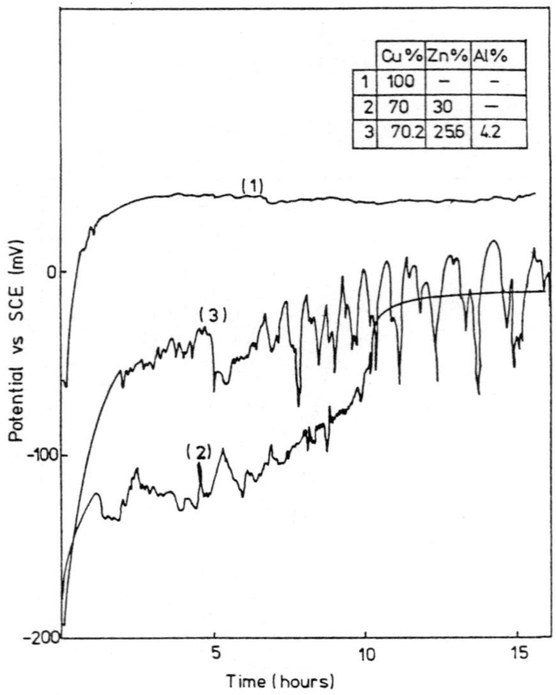

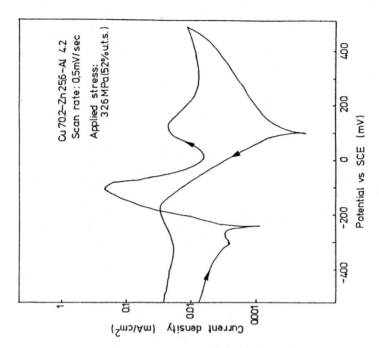

Fig. 5. Potentiodynamic polarization curve recorded in a 0.1 M NaNO$_2$-solution at roomtemperature for a mechanically stressed Cu-Zn-Al specimen.

Fig. 4. Open circuit restpotential vs time recorded in a 0.1 M NaNO$_2$-solution at roomtemperature (pH 6.3 to 7).

Passivity of Metals and Semiconductors, edited by M. Froment
Elsevier Science Publishers B.V., Amsterdam — Printed in The Netherlands

ROUND TABLE DISCUSSION

AMORPHOUS METALS - BREAKDOWN AND REPASSIVATION

N. SATO : Discussion leader suggested to divide the session into 3 parts :
amorphous alloys, passivity breakdown, stabilization of the localized corrosion.
In fact, no clear distinction was made in the discussion between breakdown and
localized corrosion.

AMORPHOUS METALS

Does the nitrogen has an effect on the corrosion of amorphous as in the case of
stainless steels ?

K. HASHIMOTO : One can predict a similar role as that of P, we have no proof of
that.

S. SMIALOWSKA : What is the behaviour of amorphous with respect to H embrittlement ?

H. TAKENOUTI : Increased rate of corrosion in the active state is due to enhanced H evolution.

K. HASHIMOTO : They have a large free volume and absorb easily H but reducing
their corrosion rate by Cr addition reduces the H penetration.

J. KRUGER : It is to say a very good passive layer reduces H embrittlement.

P.L.BONORA :Cathodic behaviour is very peculiar, H^+ reductions is more catalyzed
but the mechanism is the same. Only the recombination rate is accelerated.

PASSIVITY BREAKDOWN AND STABILIZATION OF THE LOCALIZED CORROSION

E. IRENE : We have to consider from a physical point of view the contribution of
asperities in the breakdown of passive films.

<u>H.S. ISAACS</u> : Non uniform thickness or asperities will lead to non uniform electrical field and breakdown.

<u>J. KRUGER</u> : We ignored the problem, we must look at it in details.

<u>R.P. FRANKENTHAL</u> : The problem is related to edge effects in electroplating.

<u>J. YAHALOM</u> : We also have not to forget that weak points repassivate below the pitting potential after pre-treatment in chloride.

<u>W. SCHULTZE</u> : In response to Dr. Irene's intervention, we integrate over the whole surface. With dyes a constant film thickness is expected at thicknesses of the order of 3 nm. We can check that by noble metals cathodic deposition but the anodic counterpart is impossible even though correlation with anodic breakdown must exist.

<u>N. SATO</u> : Emphasized that weak points are probably increased by Cl^-.

<u>D.E. WILLIAMS</u> : It would be possible to build a model in relation with roughness but the main aspect is the local fluctuation of concentration.

<u>H.H. STREHBLOW</u> : In response to Dr. Williams, actually what you measure is a <u>consequence</u> of local breakdown.

<u>B. MAC DOUGALL</u> : All these sorts of things play a role difficult to estimate. I found very interesting the poster on ZIRCALLOY-4 by the group of Toulouse. They showed beautiful plots of statistical analysis and the influence of pre-treatment. This is a promising approach.

<u>M. JANIK-CZACHOR</u> : We have to distinguish between breakdown and growth. The poster by Dr. Strehblow and our results show an effect of aggressive anions. Film is easier to sputter out after exposure to aggressive ions such as Cl^-, Fe^{3+} production increased linearly with the concentration of aggressive ions.

<u>N. SATO</u> : Some properties of the film may be changed.

<u>J. KRUGER</u> : No problems, statistics plays a role but we have not to forget metallurgy, for instance density of pits depends on crystallographic orientation, dislocations, inclusions ...

H.S. ISAACS : From a physical point of view, if one calculate these effects on the current variation across the film they are very small as proved by the very high impedance.

B. CAHAN : I agree with the statistical approach but we have to deal with the fundamentals. I wonder of how it was possible by connective details to establish Cl⁻ as a uniformly corroding species regardless of the system.

W. SCHULTZE : Answer and comment to Pr. Cahan's intervention.
I am not very sure we can average over many different metals. As for the poster dealing with ZIRCALLOY-4 it seems that in fact the pits growth depends on the thickness of initial film.

M. JANIK-CZACHOR : There are difficulties from the technical point of view to remove S down to a very low level. Why the aggressive anions accumulate ? In fact Cl⁻ anions do not exist in the solid state oxide and must be adsorbed on the film surface.

B. BAROUX : We should better study the real passive films on real metals with their flaws and defects instead of highly pure metals.

P. MARCUS : What do you mean by localised events ? A single monolayer of adsorbed atoms can change everything in the passive behaviour of metals.

R.C. CLAYTON : Inclusions can play an important role but are not necessarily the controlling process in pitting resistance. The nature of the passive film and the role of water as accounted for by the Okamoto's model. We have evidence that Cl^- may be incorporated in the film on 304 stainless steel as $FeOCl$ and β-$FeOOH$. The bonding of Cl^- in $FeOCl$ is expected to be stronger than for β-$FeOOH$. In β-$FeOOH$, Cl^- ions can migrate under the electric field toward the underlying hydrated passive layer ($\leqslant 10$ Å). Cl^- incorporation and mobility should be considered in studies of the initiation of pitting. Structure and bound water are modified from γ, $CrOOH$ to $Cr(OH)_3$ by exposure to Cl^-.

R.P. FRANKENTHAL : Two points - It was possible in the sixties to suppress pitting on Al by decreasing the density of defect to a very low level.
                              - We have to account for the mechanical effects
on repassivation.

B. CAHAN : When we prepared samples by high vacuum deposition impurities have no time to segregate, there is no corrosion in the laboratory environment and the samples are still bright after months.

D.E. WILLIAMS : It can also be worth studying the mechanism of dissolution of oxide.

GENERAL COMMENT FROM J. SCULLY

Perhaps it is necessary to distinguish between valve metals-Al- on which films are relatively thick and which can be grown by anodizing and Fe-Cr and general transition element alloys which have much thinner films about which less is known. One might conclude that passivity has several causes and for that reason there is likely to be more than one mechanism of breakdown.

FINAL COMMENT FROM N. SATO, DISCUSSION LEADER

1) Conceptually, it seems to be of importance to distinguish the local film-breakdown and the stability of local dissolution, as pointed out by Dr. Scully. The film breakdown is a phenomenon that occurs in the existing film itself, while the stable pit or crevice propagation and repassivation are determined by the interaction between the film-free metal surface and the local solution composition.

2) As to the breakdown of very thin passive films, such as formed on transition metals, it is statistically reasonable to assume that there occur local fluctuations with respect to film thickness and defect concentration in the film. If the film is extremely thin with several atomic layers, the maximum amplitude of thickness fluctuation would produce with a certain probability a local site where the film thickness is zero. The magnitude of the probability for such an event to occur will depend on a number of factors such as the average film thickness, the composition of the film, the homogeneity and composition of the substrate metals, and the concentration of aggressive ions in the solution. As a matter of cause, pre-existing defects on the substrate metal surface and in the film such as sulphide inclusions will be the main cause to increase the probability of local film breakdown, but even if the substrate metal and the film are homogeneous, there is a probability for local film breakdown to occur, though its probability may be negligibly small. The fundamental problem to be studied at the moment is how aggressive anions such as chloride do increase the probability of local film breakdown. There have been a number of approaches and theories dealing with this problem, but no general consensus has yet been established.

3) Turning to the stability of localization of rapid metal dissolution at a specific site or location on the metal surface, I look this phenomenon as a kind of macroscopic pattern formation from the macroscopically homogeneous metal surface to the coexistence of two different macroscopic states of the metal

surface. It comes from the non-equilibrium thermodynamics that such a transformation will occur when the localization of some of environmental factors such as the localized solution composition exceeds over a critical composition and that this localization is maintained stable as long as the energy supplied to this system is sufficient to keep the local solution composition above the critical value. It seems to me, therefore, that it is essentially important to elucidate the physical meaning of the critical solution composition for stable local corrosion from the fundamental standpoint of view and to determine this critical solution composition as functions of various factors concerning metal composition and bulk solution composition from the practical standpoint of view. It is also important to clarify the localized mass transport and solution chemistry which lead to the local change of solution composition under various geometrical and environmental conditions.

Passivity of Metals and Semiconductors, edited by M. Froment
Elsevier Science Publishers B.V., Amsterdam — Printed in The Netherlands

MECHANISMS OF PASSIVATION AND PROTECTION OF SEMICONDUCTORS IN SOLAR CELLS

S. ROY MORRISON

Simon Fraser Energy Research Institute, Simon Fraser University

Burnaby, B.C. Canada V5A 1S6

ABSTRACT

Models for photoinduced corrosion of semiconductors by holes are described and compared with experiment. The evidence is reviewed that the rate is second order in the hole concentration at the surface, and that the rate limiting step involves corrosion at surface steps, generated by dislocations. Experimental observation of the influence of such flaws as dislocations and non-conducting surface films on the corrosion are described and fitted into the models. The use of stabilizing agents in solution to protect the semiconductors is discussed in detail; the use of polymer films, metal deposits, and semiconductors with non-bonding valence bands are briefly reviewed.

INTRODUCTION

For almost a decade now, because of the pressure of the energy crisis, there has been a substantial interest in the development of improved, low cost solar cells. One approach to solar cells has been the photoelectrochemical (PEC) solar cell (ref. 1). Here a semiconductor is covered with a layer of solution, a counter electrode is inserted, and the semiconductor is illuminated. Such PEC cells in principle have advantages over the older solid state solar cells; for example in principle they are less sensitive to crystalline defects.

However, there have been several problems with PEC cells. For one thing, the efficiency is usually not as good as the efficiency with a solid state solar cell. Also it is necessary in general to put ions in solution which are highly coloured and block a significant portion of the light from reaching the semiconductor, even though the thickness of the solution is made very small. In addition, such solar cells are usually rather heavy per unit area compared to the solid state solar cells. However, the worst problem that the workers on electrochemical solar cells have had to face is the photoinduced corrosion of the semiconductor. Such corrosion leads to serious instability of the PEC cell due to resulting films or dissolution of the material. And because the life of the solar cell should be greater than ten years, even a low corrosion rate is unacceptable. Thus a significant amount of effort has gone into studies of the photoinduced corrosion of solar cells, particularly corrosion by photoproduced holes.

426

The knowledge gained in solar cell corrosion studies is not only important for solar cells. A great deal has been and is being learned about general electrode reactions at the semiconductor electrode, knowledge of more general interest. In our laboratory, for example, we are moving away from PEC cell work, but are developing a program to study electrocatalysis and photocatalysis on semiconductors, and this program depends on exactly the models that have been developed during the past years of studies on PEC cells.

CORROSION BY HOLES

   A hole is an unoccupied valence bond. In Figure 1 we show a schematic of a crystal with covalent bonding, bonding by electron pair bonds, and show one of the electrons missing. This missing electron represents a hole. The hole can move freely through the crystal by successive electron jumps. However to really understand the chemical behaviour of the hole we must use the band model of the semiconductor. In Figure 2 we show such a band model, with the energy levels associated with the valence electrons shown in Figure 1 forming the valence band and the next excited state forming the conduction band. Photo-excitation of a valence electron produces both a hole in the valence band and an electron in the conduction band. Other features indicated in Figure 2 will be discussed below.

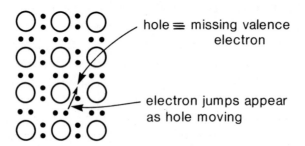

Figure 1:   Hole: an unoccupied valence bond.

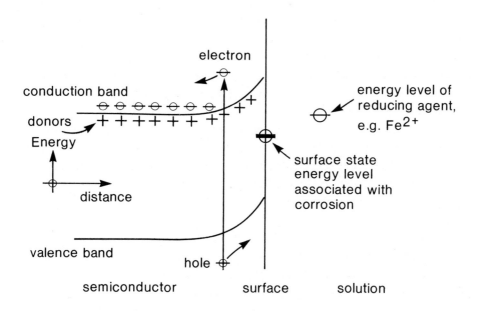

Figure 2:   Band model of the n-type semiconductor/solution interface. Illustrates band bending as electrons from the semiconductor are transferred to the solution. Illustrates photogeneration of an electron-hole pair, with the hole moving to the surface. Illustrates occupied, localized, energy levels that can capture the hole -- viz. the level gives up its electron to the unoccupied valence bond.

Holes, either present due to impurities or photoproduced by exciting an electron to the conduction band, can move to the surface and cause corrosion. Assume two mobile holes manage to meet at the same bond. Assume also that the bond is a "back bond" of a surface atom, one of the bonds holding the atom to the crystal. Two holes on the same bond, of course, means that there are no bonding electrons left and the back bond is completely broken. The now positively charged surface ion is highly susceptible to a hydroxylation process. Such a hydrolysis will probably be irreversible; restoration of the perfect crystal by capture of two electrons on the now hydrolyzed ion will be unlikely.

Now the high concentration of holes needed for this corrosion process can be common at the surface of the semiconductor. The reason for this is illustrated in Figure 2. The band picture of Figure 2 represents an n-type semiconductor with ionized impurities present which have donated electrons to the conduction band. Some of these electrons that are near the surface will move to the surface and be captured either by ions in solution or by surface-localized energy levels. This leads to a double layer, an electric field arising due to the negative charge on the surface and the positive charge associated with the ions. The "bending of the bands" near the surface shown in Figure 2 represents this double layer. The band bending is commonly called a Schottky barrier. The electric field in the Schottky barrier region as shown repels electrons from the surface and attracts holes. Thus the holes tend to become concentrated at the surface and can easily become concentrated to the point of having two holes on a single back bond.

The most common method of stabilization of a semiconductor in solution is by the use of so-called stabilizing agents. In the case of stabilization against corrosion by holes, the stabilizing agents are reducing agents in solution that will capture the holes (inject an electron into the valence band of the semiconductor), thus competing with corrosion for the available photoproduced holes.

LOCALIZED ENERGY LEVELS

At the semiconductor/solution interface adsorption, usually of water or oxygen, is almost inevitable. Dangling bonds, unoccupied orbitals of the surface atoms, will bond energetically with ligands. The presence of the dangling bonds (coordination sites) and the adsorbed water will have an inductive effect on the energy levels of the back bonds. Thus as indicated in Figure 2 we can have localized energy levels associated with corrosion (the energy levels of the back bonds) different from the valence band energy. Another set of localized surface states are the energy levels of the dangling bonds themselves (the energy levels of the bonds between the surface atoms of the crystal and the water ligands). Hole capture on the dangling bonds will not, in general, lead to corrosion. The dangling bonds are, however, generally considered to be the source of what are called recombination centres, centres which alternately capture holes and capture electrons leading to de-excitation of the excited electron.

Other localized energy levels that we must consider are the energy levels associated with ions in solution, as also shown in Figure 2. We show an energy level that we associate with a reducing agent in solution, a species able to capture holes. By hole capture of course we mean that an electron from the reducing agent moves into the semiconductor valence band and occupies the unoccupied level that is the hole. Now it is known that energy levels in solution fluctuate widely with time leading effectively to a distribution of energy levels (ref. 2). We indicate a single level for the ion which is the highest effective level in this distribution. It can easily be shown (ref. 1) that the energy level as defined is closely related for simple cases to the standard electrode potential of the reducing agent/oxidizing agent couple. For our discussion, this highest level is the important parameter. A hole coming to the surface at any energy level below the level shown can be transferred to the reducing agent, viz, the reducing agent will be oxidized by any such hole.

EXPERIMENTAL TECHNIQUES

Corrosion of the semiconductor is normally observed as a competition with other reactions. In particular the other reactions are usually either oxidation of a reducing agent in solution or electron-hole recombination through a surface state. Techniques often used are the rotating ring disc electrode (RRDE), current doubling, and (when a film is formed) the decay of current flow at constant voltage.

The principle of the RRDE is well known (ref. 3). Corrosion products or oxidized products formed at the disc (the semiconductor) can be detected electrochemically at the surrounding ring. Denoting the total hole current through the semiconductor as J_p and the current to the stabilizing agent as J_s (as measured by the current to the ring), we define S as a stabilization efficiency of the reducing agent:

$$S = J_s/J_p \tag{1}$$

Thus if all the holes go to the reducing agent the stabilization efficiency S is unity; if none of the holes go to the reducing agent the stabilization efficiency S is 0.

Current doubling (ref. 4) is a process where for each hole coming to the surface two electrons flow through the external circuit. An example, formate, is given in Equations 2 and 3. A photoproduced hole comes to the surface, oxidizes the formate ion to form a radical, perhaps the COO^- radical (ref. 5). The radical, being unstable, has a tendency to decompose, and does so by injecting an electron into the conduction band of the semiconductor.

$$2h^+ + HCOO^- \longrightarrow 2H^+ + 2COO^- \tag{2}$$

$$2COO^- \longrightarrow 2CO_2 + 2e^- \tag{3}$$

The injection of the electron by Equation 3 increases the current correspondingly.

THERMODYNAMIC MODEL OF CORROSION

Gerischer (ref. 6) has suggested a simple method of determining whether corrosion by holes of semiconductors should occur. An analogous derivation for corrosion by electron flow to the surface was also given, but in this discussion we are more interested in corrosion by holes. For clarity we shall present

the analysis of a particularly simple case. For the more general formulation
see Gerischer's paper or reviews thereof (ref. 1). We can represent the corrosion of a metal oxide by holes in a particularly useful way if we consider
that the holes are generated by the reduction of hydrogen rather than by light.
Ignoring solvation steps and anion effects for simplicity we can write:

$$2H^+ \longrightarrow H_2 + 2h^+ \tag{4}$$

$$MO + 2h^+ \longrightarrow M^2 + \frac{1}{2} O_2 \tag{5}$$

or:

$$2H^+ + MO \longrightarrow M^{2+} + H_2 + \frac{1}{2} O_2 \tag{6}$$

Equation 4 describes the transfer of electrons from the valence band of the
semiconductor to protons, where h^+ represents a hole. Equation 5 describes the
oxidation process of interest where the holes self-oxidize the semiconductor MO
to give M^{2+} in solution plus oxygen gas. In Equation 6 we show the sum of the
two equations eliminating the step involving holes.

First consider the case where the electrochemical potential of the holes
(effectively the energy of the valence band) is the same as the electrochemical
potential at the hydrogen reference electrode, (the electrochemical potential of
the hydrogen/proton couple under standard conditions). Then the free energy
change in (4) is zero, and the free energy change in (5) is the same as that of
(6). The free energy change in (6) is calculable from handbook values. Now
consider the case where the valence band of the semiconductor is lower than the
energy of the hydrogen reference electrode, say by E_v, (the holes are stronger
oxidizing agents than protons by the energy E_v). In this case we simply sub-
tract E_v from the free energy change of (6) to get the free energy change of (5).
Thus to get the free energy for corrosion by holes we analyze Equation 6 or its
equivalent and we make an experimental measurement of the location of the valence band edge of the semiconductor relative to the hydrogen reference electrode.
These two values determine whether a given hole corrosion reaction (Equation 5)
should be thermodynamically favourable.

It turns out that for most semiconductors with band gaps between one and two
electron volts, such as are of interest in solar energy conversion, self-
oxidation by holes is exothermic. In fact all the semiconductors analysed by
Gerischer (refs. 1, 6) using the above procedure were susceptible to self-
corrosion by holes.

A few wide band gap semiconductors, primarily titanium dioxide and the titanates do not photocorrode noticeably in aqueous solution (unless sulphate ions
are present). This could be either because the kinetics are too slow or be-

cause water oxidation by the holes is favoured over self-corrosion of the semi-conductor.

KINETIC MODEL FOR CORROSION PREVENTION BY STABILIZING AGENTS

In Section 2 we indicated a possible physical model for photocorrosion, where the capture of two holes at a particular bond was an irreversible corrosion process. Capture of a single hole at a two-electron bond was implicitly considered a reversible process. We will show (from refs. 7 and 8) that the mathematical model resulting from this hypothesized mechanism is in agreement with experimental results. Other possibile models will be discussed later. Consider the reactions:

$$A:B + h^+ \underset{k_{-1}}{\overset{k_1}{\rightleftharpoons}} A \cdot B \tag{7}$$

$$A \cdot B + h^+ \xrightarrow{k_2} A + B \tag{8}$$

$$A \cdot B + R \xrightarrow{k_3'} R^+ + A:B \tag{9}$$

$$R + h^+ \xrightarrow{k_3''} R^+ \tag{10}$$

Reactions 7 and 8 represent the corrosion reaction of the semiconductor AB, where A:B represents A and B bonded by a two-electron back bond. As discussed above, we assume the first hole capture (Equation 7) is reversible and the second hole capture (Equation 8) leads to an irreversible dissociation of the bond. Reactions 9 and 10 represent the capture of holes by the reducing agent in the solution, either indirectly through the intermediate (Equation 9) or directly from the valence band (Equation 10). An exact kinetic analysis of these equations in terms of J_S and S (Equation 1) has been described (ref. 8). Analysis of Equations 7, 8 and 9 yields:

$$1/S = 1 + J_S(2k_2/k_3'^2[R]^2)(k_{-1} + k_3'[R])(k_1 - k_2 J_S[R])^{-1} \tag{11}$$

Analysis of Equations 7, 8 and 10 yields:

$$1/S = 1 + J_S(2k_2 k_1/k_3''^2[R]^2)(k_{-1} + k_2 J_S/k_3''[R])^{-1} \tag{12}$$

Thus we see from both of these expressions if we plot $1/S$ vs J_S at low J_S we should obtain a straight line.

432

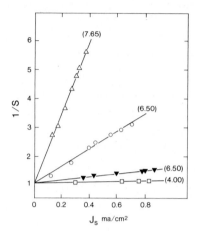

Figure 3: RRDE measurements of corrosion on GaAs with Ferrous EDTA (open symbols) or Ferrous DTPA (solid symbols), 0.02M as the stabilizing agent. The numbers represent pH, showing the effectiveness of the stabilizing agent depends on pH.

In Figure 3 results are shown indicating the linear relationship observed at times between S^{-1} and J_s (Equations 11 and 12). These are RRDE measurements on gallium arsenide with various complexes of iron as the stabilizing agent. Gomes and his co-workers have made similar observations on GaP and InP which, when plotted according to Equations 11 or 12, appear to extend into the non-linear region (ref. 8). Such results show both the power of the RRDE measurements and the agreement of the measured corrosion with the simple theory.

Gomes et.al. (ref. 9) has suggested, independantly, the above model plus several alternatives. A few of these alternatives predict a linear relation between J_3 and 1/S. These alternative models cannot be rejected entirely but arguments have been presented (ref. 8) to suggest that the simple model suggested above is the most probable.

More details regarding the corrosion mechanism can be extracted from such data if the above model is assumed correct and the experimental data is used to identify the rate constants for the various equations. Because by the model Equation 7 must be reversible, it is suggested (ref. 7) that the energy level associated with Equation 7 must be near or below the valence band edge. By analysing the rate constants under this restriction, a density of active corrosion sites is estimated (ref. 7) that corresponds reasonably to the expected density of steps on the surface of a crystal. The estimated density is too low to reflect the total density of surface atoms, too high to reflect the density of dislocations. Thus, the results suggest corrosion at surface steps as rate-limiting in photoinduced corrosion.

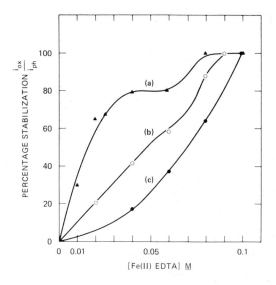

Figure 4:   Percent stabilization of n-GaAs versus Fe(II) EDTA concentration in
0.1 M EDTA aqueous solutions, intermediate pH-range.
Curve (a), surface undamaged; curves (b) and (c), surface damaged.

Figure 4 shows corrosion as measured (ref. 10) by RRDE on gallium arsenide as
a function of the surface damage on gallium arsenide. The measurements show
clearly the increased tendency toward corrosion when the surface is mechanically
damaged, in other words when dislocations are present on the surface. This observation is consistent with the conclusion that surface steps are the sites
for photo-induced corrosion on this semiconductor because the step density will
be closely related to the dislocation density.

OTHER OBSERVATIONS OF PHOTOINDUCED CORROSION OF SEMICONDUCTORS
    The effect of a film as the corrosion product has been studied on silicon
(ref. 11) by the simple procedure of observing the decay of the photocurrent
(corrosion current) as the film thickens. The analysis of this effect, although
qualitative, led to the conclusion that the stabilizing efficiency may be higher
with a thin film present. This conclusion was later confirmed (ref. 12) by combining RRDE with ellipsometer measurements of oxide thickness. The results from
ref. 12 of stabilization efficiency as oxide thickness are shown in Figure 5.

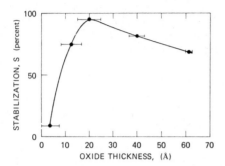

Figure 5:   Effect of oxide thickness on stabilization of n-silicon by ferrocyanide (0.1 molar).

   This improved stability observed with a thin oxide layer, when using ferrocyanide as the stabilizing agent for silicon, is explained as follows. Ferrocyanide should be a poor stabilizing agent when there is no oxide, because the
energy level of ferrocyanide is close to the energy level of the silicon valence
band. Thus hole capture is not strongly favoured energetically. However, the
presence of a very thin oxide and the resulting electric field across the oxide
during illumination raises the energy level of the ferrocyanide well above the
valence band. This permits hole capture by the ferrocyanide. As the oxide layer continues to grow so that holes must tunnel through a thick oxide to reach
the ferrocyanide, the stability slowly decreases.

   It is possible to have corrosion of the semiconductor even when the stabilizing agent is also being oxidized. This is best illustrated by the results of
Fujishima et.al. (ref. 13). Using the current doubling technique (refs. 1 and 4)
they studied the photoinduced corrosion of zinc-oxide, with formate as the reducing agent, the stabilizing agent. Current doubling is observed, indicating
the capture of holes by the formate, in a reaction sequence similar to Eqs. 1
and 2. Fujishima et.al. tested the stability further by examining the product
yield, using a RRDE, simultaneously with the current doubling measurements.
They found, surprisingly, that not only did each hole coming from the surface
current double, but also zinc ions were produced in solution, indicating efficient corrosion. The conclusion was that the reaction process must occur as follows:

$$2h^+ + ZnO \longrightarrow Zn^{++} + O^* \tag{13}$$

$$O^* + 2HCOO^- \longrightarrow 2COO^- + H_2O \tag{14}$$

$$2COO^- \longrightarrow 2e^- + 2CO_2 \tag{15}$$

Thus, in some cases the observation of the oxidized product of the stabilizing agent may not definitely indicate a complete lack of corrosion.

OTHER TECHNIQUES FOR STABILIZATION

Other techniques for stabilization of photo-electrochemical solar cells have been used. These have not been emphasized here because the models are less interesting from a fundamental point of view. Examples of such mechanisms are the use of polymer coatings, the use of metal coatings on the semiconductor and, finally the use of semiconductors that are layer compounds with non-bonding valence bands.

Electrically conducting polymers and thick layers of noble metals on the surface of semiconductors are useful (ref. 14), but for many of these cases it may be that the potential advantages of the electrochemical solar cells are lost. In both cases the Schottky barrier appears now between the semiconductor and the conductive coating and the device acts as a solid state Schottky barrier solar cell (ref. 15). The solution simply makes an electrical contact to the metal or to the polymer.

On the other hand, very small metal deposits, layers of thickness the order of just a few angstroms or less (if evenly dispersed) have been used in attempts to stabilize PEC cells. In studies in our laboratory on silicon (ref. 16) we have found such thin layers are effective, but it is clear that in between the crystallites of the deposited metal an oxide grows. The stability arises because the photoproduced holes can still flow through the metal contact to the solution. So if the deposits form crystallites, such "thin" metal deposits may act as thick metals and, to some extent, lose again the advantages of the electrochemical solar cell. However, there is some evidence that in some cases deposits of metals in thin layers on semiconductors do deposit in a form approaching a simple monolayer. In some cases there is evidence that the dominant effect of such metal adsorbates is to lower surface recombination velocity and thus increasing the efficiency of the solar cell.

Finally, the use of layer compounds or compounds with non-bonding valence bands is of great interest. If the valence band orbitals are non-bonding, the presence of a hole does not break a back bond. Such studies were initiated by Trivitch (ref. 17) and have been pursued in many laboratories, including our own. Layer compounds with non-bonding valence bands, such as MoS_2 are found to be highly stable against attack by holes on the basal plane, but unfortunately are found unstable to attack on an edge plane. Thus studies on broad single crystals of such materials show a significant stability, whereas studies on thin film deposits are less promising. Studies designed to deposit coatings on the edge planes and thus eliminate them from the area of holes transferred to the

surface are in progress. Another case where non-bonding valence bands are considered to provide improved stability is the case of cadmium selenide. In this
case the corrosion is allowed to proceed and a selenium layer is produced on the
surface. Selenium has a valence band associated with non-bonding orbitals, and
the valence band is suitably located to conduct photoproduced holes between the
cadmium selenide and the solution. Under these conditions a significant improvement in stability is found (ref. 18) together with a high solar cell conversion
efficiency.

CONCLUSIONS

As described in the Introduction, the most interesting aspect of studies of
passivation of electrochemical solar cells is the basic knowledge that has been
and will be generated about corrosion processes and about electrode processes.
Semiconductor electrodes permit novel experimental methods -- for example, with
semiconductors one can easily change the concentration of reactants (holes or
electrons) in many instructive ways. Such versatility in experimental techniques
will hopefully lead to rapid and significant improvement in models of electron
transfer processes, including the role of the solid in electrochemical corrosion.

REFERENCES

1　S.R. Morrison, Electrochemistry at Semiconductor and Oxidized Metal Elec-
　　trodes, Plenum, New York, 1981.
2　H. Gerischer, Z. Phys. Chem. N.F. $\underline{27}$ (1961) 48.
3　Yu V. Pleskov and V. Yu Filinovskii, The Rotating Disc Electrode, Consultants
　　Bureau, New York, 1976.
4　S.R. Morrison and T. Freund, J. Chem. Phys. $\underline{47}$ (1967) 1543.
5　J.R. Harbour and M.L. Hair, J. Phys. Chem. $\underline{83}$ (1979) 652.
6　H. Gerischer, J. Electroanal. Chem. $\underline{82}$ (1977) 133.
7　K.W. Frese, Jr., M.J. Madou and S.R. Morrison, J. Electrochem. Soc. $\underline{128}$ (1981)
　　1527.
8　S.R. Morrison, in Photoelectrochemistry, Fundamental Processes and Measure-
　　ment Techniques, eds. W.W. Wallace, A.J. Nozik, S.K. Deb, and R.H. Wilson,
　　Electrochemical Society Inc. 1982.
9　W.P. Gomes, F. Van Overmeire, D. Vanmaekelbergh, F. Vanden Kerchove and F.
　　Cardon, in Photoeffects at Semiconductor-Electrolyte Interfaces, ed. A.J.
　　Nozik ACS Symposium Series 146, 1981.
10 K.W. Frese, Jr., M.J. Madou and S.R. Morrison, J. Phys. Chem. $\underline{84}$ (1980) 3172.
11 M.J. Madou, K.W. Frese, Jr. and S.R. Morrison, J. Phys. Chem. $\underline{84}$ (1980) 3423.
12 B.H. Loo, K.W. Frese, Jr., and S.R. Morrison, Appl. Surf. Sci. $\underline{8}$ (1981) 290.
13 A. Fujishima, T. Kato, E. Malkawa and K. Honda, Bull. Chem. Soc. Jpn. $\underline{54}$
　　(1981) 1671.
14 for example: R. Noufi, A.J. Nozik, J. White and L.F. Warren, J. Electrochem.
　　Soc. $\underline{129}$ (1982) 2261.
15 for example: A.J. McEvoy and W. Gessler, J. Appl. Phys. $\underline{53}$ (1982) 1251.
16 M. Matsumura, to be published.
17 H. Tributsch, Ber. Bunsenges, Phys. Chem. $\underline{81}$ (1977) 361.
18 K.W. Frese, Jr., Appl. Phys. Lett. $\underline{40}$ (1982) 275.

Passivity of Metals and Semiconductors, edited by M. Froment
Elsevier Science Publishers B.V., Amsterdam — Printed in The Netherlands

PLASMA OXIDATION OF SEMICONDUCTOR AND METAL SURFACES

R. P. H. CHANG

Bell Laboratories, Murray Hill, New Jersey 07974

ABSTRACT

Plasma oxidation is a process whereby surfaces are oxidized in an oxygen gas discharge via active neutral and charged oxygen species. In recent years it has attracted much interest in the microelectronics industry where low temperature, dry processes are widely used for fabricating VLSI devices. For example, oxides can be used for surface passivation, device isolation, charge storage, optical wave guides, and diffusion inhibition, to mention just a few. A brief overview of the current understanding of plasma oxidation of Si, compound semiconductors, and metals is presented. Using analytical instrumentations such as ion, electron, and photon spectroscopy much information have been gathered on the oxide films and their interfacial properties. Discussions on oxidation mechanisms pertaining to plasma-surface interaction, bulk oxygen species transport, and oxide-substrate interfacial reactions are given with illustrations from different oxide systems. In the area of practical applications, examples are given to show how plasma oxides can be used in micro-electronic device fabrication. Plasma oxidation mechanisms are not yet fully understood in terms of the physical and electrical properties of the oxides. Areas of potential future research and development are also discussed.

INTRODUCTION

Plasma oxidation is a low temperature vacuum process for the growth of native oxides on metals and semiconductors. For instance, it can oxidize silicon wafers at a reasonable rate with substrate temperatures below 600°C, thus reducing wafer warpage, dopant diffusion, and the generation of defects such as stacking faults. Unlike thermal oxidation, it is an anisotropic process which minimizes lateral oxidation beneath the mask (the so-called bird's beak effect). It is also compatible with other dry processes such as plasma etching, and ion beam sputtering. Due to their low evaporation temperatures, compound semiconductors also require low temperature surface preparation techniques. Plasma oxidation has been used extensively in the Josephson junction technology where oxides of precise thickness and high tolerance to temperature recycling have to be formed. In this paper a brief overview of the current understanding of plasma oxidation of Si, compound semiconductors, and metals is given. Due to space limitation only selected current topics are discussed. For example, theoretical models for plasma oxidation (see ref. 1) are omitted. The references given are by no means complete or exhaustive. The reader is encouraged to study the references for details. A simple physical picture for plasma oxidation is included for the sake of completeness. Discussions on oxidation mechanisms pertaining to plasma-surface interaction, bulk oxygen species transport, and oxide-substrate interfacial reactions are given with illustrations from Si, compound semiconductors, and metallic surfaces. Finally, possible future areas of research are given in the concluding section.

THE PLASMA OXIDATION PROCESS

When a substrate such as silicon is placed in an oxygen plasma the surface is usually charged negatively with respect to the plasma as a result of higher electron mobility. Unless the Si surface is biased with respect to the plasma potential, it is continuously bombarded by both positively and negatively charged plasma particles as well as neutral oxygen species. As a consequence of these interactions the surface is oxidized to few tens of angstroms due to low energy implant, diffusion, and chemical reactions. Further growth is made by positively biasing the substrate with respect to the plasma to collect negatively charged oxygen species. Oxides can be grown in this way (anodization) to thicknesses exceeding 1 μm. The rate of oxide growth depends strongly on the O_x^- flux available at the Si surface, and the rate of transfer (drift and diffusion) of these active charged species across the oxide to the Si-SiO$_2$ interface for the formation of native oxide. Thus the important parameter for oxide growth at low temperatures is the strength of the electric field impressed across the oxide. Of course, increasing the substrate temperature will also increase the oxidation rate via diffusion. Although oxygen plasmas do possess a small fraction of negatively charged ions, it has been determined that the main source of negative oxygen species is created on the surface of the substrate as a consequence of electron capture by adsorbed oxygen atoms or molecules (ref.2). Thus the plasma electron density and its energy distribution are important parameters for the optimization of the surface electron capturing cross-section. Unlike the case of wet chemical anodization, when the substrate is biased to collect negatively charged species the current is largely composed of electrons (>95%) and therefore the current efficiency is very low (only a few percent at best). This problem can be overcome, however, by using an electron filter such as a thin film of ZrO$_2$ on Si. Recently, it has been shown that by using such a filter, the oxidation rate can be increased by an order of magnitude as a result of blocking the electron flow and thus increasing the electric field (to $\approx 10^7$ V/cm) in the oxide (ref.3).

In the case of compound semiconductors, it has been found that both the oxygen ions and the substrate ions move in the oxide under the applied electric field (ref.4). This greatly complicates the model for oxidation. But it has been shown recently that even under such complex situation a single moving species model can accurately describe the gross features of the oxidation (ref.5). The basic difference in the oxidation mechanism between Si and compound semiconductors is that the different elements in the compound oxidize with different rates.

Oxygen plasmas have also been used for growing very thin tunnel barrier oxides (20-80Å) for the Josephson junction technology. It has been demonstrated that by controlling the rf power and the oxygen pressure, a predetermined oxide thickness can be easily achieved. This is accomplished by noting the following facts: By placing the sample on the active rf electrode both sputtering and oxidation take place simultaneously on the sample surface. Increasing the rf power and decreasing the oxygen gas pressure will increase the ion density and increase the sample surface potential, thus increasing the surface sputtering by ions. Lowering the rf power and increasing the pressure will have the opposite effect. However, if the substrate is not biased to collect the negative oxygen species, the oxide film thickness is ultimately determined by diffusion processes in the oxide film. By controlling these parameters useful thin oxide films have been fabricated for tunnel junctions on many metallic substrates. Thus plasma-surface interaction plays an important role in the formation of useful thin oxide films for tunnel junctions.

PLASMA OXIDATION OF SILICON

Plasma oxidation of Si was first carried out successfully by Ligenza in 1965 using a microwave discharge (ref.6). Since then there has been numerous publications using various kinds of discharges (from DC to microwave) and a wide range of operating parameters (ref.7). For instance, oxygen gas pressure used in these experiments varied from 10^{-3} to a few Torr. The substrate temperature varied from 40°C to 600°C, while the sample can be either biased or floating with respect to the oxygen plasma. The growth rates varied from linear to parabolic or nonlinear, indicating that the SiO_2 could be formed by a combination of sputter deposition and complex growth mechanisms (such as space charge effects). In this section, discussions will focus on some of the most recent experiments with emphasis on oxide properties, oxidation mechanism, and applications.

Recently Ray and Reisman have reported growth of oxide on 57 mm diameter silicon wafers in an electrodeless plasma with O_2 pressure $> 10m$ Torr (ref.8). In this case the plasma was excited by an external rf coil at one end of a quartz tube and the charged species diffused into the region where the tube was heated by an oven. The substrate temperature could be varied in the range of 300-600°C and the rf (0.5-3 MHz) power was 1-7 kW. By placing the silicon wafer close (1-2 cm) to the rf coil, these authors observed oxide growth on the surface facing away from the dense plasma (which is most intense under the rf coil). Much thinner oxide was found on the surface facing the plasma. In this geometry it is most likely that the plasmas on either side of the wafer have different potentials, thus setting an electric field across the Si substrate and causing the oxide to grow. This electrodeless scheme is necessary for maintaining high oxygen purity in the oxidation chamber, and it is also attractive for scaling up the machine. A typical parabolic like growth rate is observed with an initial rate of about 1000Å/hr. The physical properties of these oxides are very similar to those of thermal oxides, but with added benefits. For example, no dopant diffusion and oxidation-induced defects were observed.

More recently, Ho and Sugano have reported the growth of high quality SiO_2 in a low frequency (420 kHz) rf discharge for the fabrication of devices and ICs (ref.9). They have demonstrated that 1 μm thick SiO_2 films (for isolation purposes) can be grown in less than an hour in an oxygen plasma at a pressure of 0.2 Torr and a temperature of 600°C, by biasing the substrate with respect to the plasma. Using a double-masked Al_2O_3/SiO_2 structure it was found that the bird's beak effect was suppressed. This implies a tremendous amount of Si real estate savings for device architecture. The anisotropic growth of SiO_2 in plasma is one of the big advantages of this technique. Table 1 gives a summary of properties of SiO_2 films grown by these authors. Aside from the higher density of interfacial states (by a factor of 10), the plasma grown oxides again have identical properties as those formed thermally. The only reported way to reduce to the surface state density at present is to anneal the sample in a chlorine containing ambient at a temperature of 1000°C. It is hoped that with further research and development, such high temperature annealing would not be necessary! By using this plasma oxidation technique, Ho and Sugano have fabricated p-channel MOSFETs with a threshold voltage of -1.7 V and a mobility of 225 cm^2/V sec. They have also fabricated MOSFETs on SOS (silicon on sapphire) as well as neutron-irradiated silicon substrates.

To reduce the interface state density it is desirable to first understand the oxidation mechanism in some detail (e.g., how do the species and defects migrate). Perriere *et al.* have performed low temperature ^{16}O plasma anodization of silicon through thin ^{18}O-enriched ZrO_2 layers (ref.10).

TABLE I

Summary of the Properties of SiO$_2$ Films (After Ho and Sugano)

Property	Plasma-grown SiO$_2$	Thermally grown SiO$_2$
Refractive index	1.46	1.46
Etch rate[a] ($\overset{\circ}{A}s^{-1}$)	2	2
Breakdown strength (V cm^{-1})	(6-7) × 10^6	5 × 10^6
IR absorption (μm)	9.3, 12.4	9.3, 12.4
Dielectric constant	3.5-4.0	3.5-4.0
Resistivity (Ω cm)	(0.1-5) × 10^{16}	(0.1-5) × 10^{16}
Density of interface states (cm^{-2})	$\approx$10^{12}	$\approx$10^{11}

[a]Pliskin etch (48% HF(aq) plus 1 M Cr$_2$O$_3$(aq), 1:1 by volume).

Measurements of the overall oxygen (^{16}O and ^{18}O) and cation contents as well as their distribution in the films were carried out using combined Rutherford backscattering and nuclear microanalysis techniques. Their data show (1) the conservation of the order of oxygen atoms during SiO$_2$ growth and (2) a partial exchange of ^{18}O present in the Zr^{18}O$_2$ with the ^{16}O in the plasma. These results show that the microscopic mechanism of oxygen transport during low temperature plasma anodization involves a short-range oxygen migration, in contrast with the case of high temperature thermal oxidation. Analysis of the ^{18}O depth profiles in the Zr^{18}O$_2$ shows that the exchange phenomenon is not only at the plasma-oxide interface but it extends to a depth of about 200Å. This strongly suggests that the formation of positive oxygen ions in the Zr^{18}O$_2$ is due to high energy (about 30 eV) plasma electron bombardment. Further studies on the oxidation mechanism are clearly necessary.

In this section we have seen that due to the continuous effort of many research groups around the world the quality of plasma SiO$_2$ has become near identical to that of thermal SiO$_2$. Because of the inherent advantages of this technique over thermal oxidation, it should become useful in future VLSI technology.

OXIDATION OF COMPOUND SEMICONDUCTORS

Formation of high quality insulators on compound semiconductors is an important step in the fabrication of integrated circuits with these materials. Due to their low evaporation temperatures, growth of native oxides on compound semiconductors by conventional thermal oxidation has not been successful. However, during the past several years studies of plasma grown oxides on various compound semiconductors (e.g., InP, GaAs, InGaAs, HgCdTe, etc.) have been carried out. In this section, oxides of GaAs will be discussed in some detail simply because it has been studied the most.

A chemically polished GaAs substrate is usually held in a temperature controlled sample holder which is then placed in an oxygen plasma and biased positively with respect to the discharge potential to collect active species for oxide growth. Various plasma systems (from simple D.C. discharge to magnetically confined plasmas) have been used in these studies. They all produced similar overall results, except for the oxidation efficiency which can affect the quality of the oxides. The rate of film growth was found to increase with the substrate temperature and the applied electric field in the oxide. An optimum growth rate of 50 Å/sec can be easily achieved by carefully adjusting the plasma parameters.

The properties of plasma grown oxides have been studied in detail. (See Ref. 11 and also references therein.) In particular, combination of Rutherford backscattering and ion induced X-ray measurements show that the bulk of the oxide is composed of Ga_2O_3 and As_2O_3. Quantitative Auger depth profile of a typical oxide (of about 1600Å thick) shows that the oxide is very uniform in composition and it possesses a very sharp interface (on the order of 20Å). However, if one expands the interface region and examines it more carefully, one would find that there are excess amounts of arsenic over gallium in this region (i.e., As/Ga > 1). This is a consequence of the thermodynamics and kinetics in the oxidation process. Although this amount of elemental As is only of the order of one monolayer, it turns out that it produces deleterious effect on the electrical properties of the film. Upon thermal annealing in inert gases (e.g., N_2, Ar, or He) the elemental As forms metallic domains at the interface which cause large hysteresis (over 1 Volt) in the C-V curve (ref.12). On the other hand annealing in pure hydrogen does seem to remove the elemental As but it also reduces the bulk oxide. Oxides annealed in hydrogen do not possess large hysteresis but tend to be quite leaky due to the reduction in density (about 30%).

These important results encouraged Chang et al. to come up with a processing technique whereby the elemental arsenic layer could be removed or transformed into another compound during oxidation (ref.13). It is also well known that oxides on GaAs tend to pin the Fermi level at the semiconductor surface. Thus by adding a new layer between the oxide and the GaAs surface might overcome this problem. The key to the success of such an endeavor was the identification of the proper reactant. The desirable properties of the reactant are as follows: (1) It should be a gas which can form a plasma that can coexist with the oxygen plasma without reacting with it. (2) The reactant should react more readily with arsenic than with gallium. (3) This added reaction should not compete with the formation of arsenic oxide. An element that satisfies these conditions is fluorine. Therefore, oxides were grown using a fluorinated oxygen (e.g., CF_4 in O_2) discharge. Chemical depth profile through an as grown oxide, obtained using Auger analysis, shows that the amount of fluorine is almost 6 atm. %, and that the oxide is essentially stoichiometric (except for the presence of fluorine), i.e., the Ga/As concentration ratio is nearly unity all the way through the interface region. What has been formed at the interface is a layer of oxyfluoride film (ref.14). Ellipsometric measurements show that oxides with fluorine are more transparent and that the absorption edge is shifted towards higher energies indicating the removing of elemental arsenic (ref.15).

A typical high-frequency (1 MHz) C-V curve shows essentially no hysteresis (in contrast to the case without fluorine) indicating that the selective oxidation process has removed most of the trap states. The C-V curve goes into "inversion" at negative bias, indicating that the oxide does not leak. (See ref. 16 for more details.) Thus it has been shown that the interface properties of GaAs-oxide can be modified by a selective plasma processing technique. More recent results have also confirmed these observations (ref.17).

Following the favorable results of plasma oxides grown on GaAs, oxygen plasmas have also been used successfully to grow native oxides on InP, InGaAs, and HgCdTe. Using the native oxides as gate dielectric, field effect transistors have also been fabricated on these materials (ref.18).

THIN METAL OXIDE FILMS

The formation of thin metal oxide films in rf plasmas have been studied extensively during the past decade. Oxide layers less than 50Å in thickness are used in the fabrication of tunnel barriers for Josephson junction devices. Oxides with good electrical properties (such as low leakage currents, etc.) and high tolerance to temperature recycling are needed in this technology. A technique which combines plasma oxidation, sputter etching, and sputter deposition has been successfully used to form thin oxide films on metal surfaces (e.g., Nb, Pb, Pb alloys, etc.). In this section a brief discussion is given on how this technique is applied, and what are the physical mechanisms for thin film growth.

The sample to be oxidized is affixed to a temperature controlled rf cathode of an oxygen discharge. In this manner the sample can be oxidized by chemically active neutral (as well as negatively charged) species and sputtered by positively charged ions. The discharge can be sustained by driving the cathode at 13.5 MHz. This results in the usual ion sheath near the cathode surface which has a peak negative potential as large as the peak amplitude of the applied potential. Sputtering occurs when positive ions from the sheath are accelerated toward the sample by the negative potential. Negative ions for oxide growth are either extracted from the plasma or created at the oxide surface by electron capture. The negative bias results from oxide surface charging provide an additional driving force for diffusion of cations through the oxide. The oxidation rate is limited by the diffusion process, thus it decreases with increasing oxide thickness. However, the sputtering rate is independent of the oxide thickness. If the oxidation rate is initially higher than the sputtering rate, the oxide thickness will increase until the rate of oxidation becomes equal to the sputtering rate, at which time a steady state oxide thickness is attained. Assuming that these two processes are additive, and by judiciously adjusting the rf power density, substrate temperature, and the gas mixtures (such as Ar/O_2 ratio), desired oxide thicknesses can be obtained. By taking a logarithmic oxidation law and a constant sputtering rate, Greiner has studied the oxide growth as a function of the oxygen gas pressure and the peak cathode voltage using an in situ ellipsometer (ref.19). His findings confirm that a stead-state oxide thickness can be obtained. He also found that when a steady-state oxide thickness was reached, further variation of the oxide thickness with discharge exposure time was in accordance with what could be expected from oxide formation as a result of additive oxidation and sputtering rates.

Recently Baker et al. have done a series of studies on the sputter-oxidation of Pb alloy films in a rf system similar to that described above (ref.20). Due to the preferential oxidation and sputtering rates of the elements in the alloy, complex oxide layers were formed.

DISCUSSIONS AND CONCLUSIONS

A brief review of the present status on plasma oxidation of semiconductors and metals has been given. It is seen that the plasma oxidation has the following attractive properties: (a) it is a vacuum process, (b) with low substrate temperature ($0°$-$600°C$) during oxidation, (c) oxides grow anisotropically (in the direction of the applied electric field), (d) the process can be controlled precisely by pulsing the plasma, (e) ion sputter etching can be incorporated during oxidation to tailor the composition and the thickness of the oxide, (f) by doping the oxygen gas with halogens, the oxide properties, especially the interface can be modified (ref.13). Using these intrinsic properties of plasma oxidation, oxides have been formed on Si, compound semiconductors, and metal surfaces for the fabrication of microelectronic devices. Although devices such as MOSFETs have been made using plasma oxides, the fundamental mechanisms involved in the oxidation are not fully understood. For example, it is not clear what fraction of the active species is generated on the substrate surface for oxidation, and how does the flux of active species depend on the plasma parameters. For the plasma oxides to be useful, no high temperature annealing step should be needed to reduce interfacial traps. Thus it is essential to understand how the interface is formed and how are the defects propagated during oxidation. To answer these questions, controlled experiments are needed to separate the various parameters. With the help of analytical instruments, it should be possible to pinpoint the cause for interfacial defects. This seems to be the approach that is being taken by research scientists in the field.

REFERENCES

1 A. T. Fromhold, Jr., Thin Solid Films, **95**, 297 (1982) and references therein; V. Lakunov, V. Parkhutik, and E. Tkharev, J. of Crystal Growth **45**, 399 (1978).
2 K. Ando and K. Matsumura, Thin Solid Films, **52**, 173 (1978).
3 S. Gourrier, P. Dimitriou, J. B. Theeten, J. Perriere, J. Siejka, and M. Croset, Appl. Phys. Lett., **38**, 33 (1981); R. P. H. Chang, J. Siejka, J. Perrier, and M. Croset, Proc. Electrochemical Soc. Meeting, Denver, CO, Oct. 1981.
4 R. P. H. Chang, Thin Solid Films, **56**, 89 (1979).
5 S. Gourrier and M. Bacal, J. of Plasma Chem. and Plasma Process., **1**, 217 (1981).
6 J. R. Ligenza, J. Appl. Phys., **36**, 2703 (1965).
7 J. Kraitchman, ibid, **38**, 4323 (1967); L. Bardos, G. Loncar, I. Stoll, J. Musil and F. Zacek, J. Phys. D., Appl. Phys., **8**, L195 (1975); J. Musil, F. Zacek, L. Bardos, G. Loncar and R. Dragila, ibid, **12**, L61, (1979); R. Dragila, L. Bardos and G. Loncar, Thin Solid Films, **34**, 115 (1976); E. R. Skelt and G. M. Howells, Surface Science, **7**, 490 (1967); D. L. Pulfrey, F. G. M. Hathorn and L. Young, J. Electrochem. Soc., **120**, 1529 (1973); M. A. Copeland and R. Pappu, Appl. Phys. Lett., **19**, 199, (1971); R. B. Beck, M. Patyra, J. Ruzyllo and A. Zakubowski, Thin Solid Films **67**, 261 (1980); J. R. Ligenza and M. Kuhn, Solid State Technology, p. 33, (1970); R. P. H. Chang, C. C. Chang, and S. Darack, Appl. Phys. Lett. **36**, 999 (1980).
8 A. K. Ray and A. Reisman, J. Electrochem. Soc., **128**, 2424, 2460, 2466, (1981).
9 V. Q. Ho and T. Sugano, Thin Solid Films, **95**, 315 (1982).
10 J. Perrier, J. Siejka, and R. P. H. Chang, Thin Solid Films, **95**, 309 (1982).
11 R. P. H. Chang, A. J. Polak, D. L. Allara, C. C. Chang, and W. A. Lanford, J. Vac. Sci. Technol. **16**, 888 (1979), and references therein.
12 R. P. H. Chang, T. T. Sheng, C. C. Chang, and J. J. Coleman, Appl. Phys. Lett. **33**, 341 (1978).
13 R. P. H. Chang, J. J. Coleman, A. J. Polack, L. C. Feldman, and C. C. Chang, Appl. Phys. Lett. **34**, 237 (1979).
14 R. K. Ahrenkiel, L. L. Kazmerski, O. Jamjourn, P. E. Russell, P. J. Ireland, and R. S. Wagner, Thin Solid Films, **95**, 327 (1982).
15 S. Gourrier, A. Mircea, J. B. Theeten, M. Bacal, Conf. Proc. — Int. Symp. Plasma Chem., **1**, 181 (1979).

444

16   R. P. H. Chang, Jap. J. of Appl. Phys. **19**, 483 (1980).
17   R. K. Ahrenkiel, R. S. Wagner, S. Pattillo, D. Dunlavy, T. Jervis, L. L. Kazmerski, P. J. Ireland, Appl. Phys. Lett. **40**, 700 (1982).
18   K. Kanazawa and H. Matsunami, Jpn. J. Appl. Phys. **20**, L211 (1981); T. Mimura, N. Yokoyama, M. Fukuta, Fujitsu Sci. Tech. J., **14**, 45 (1978); B. Tell, R. E. Nahory, R. F. Leheny, J. C. DeWinter, Appl. Phys. Lett. **39**, 744 (1981); Y. Nemirovsky, R. Goshen, Appl. Phys. Lett. **37**, 813 (1980).
19   J. H. Greiner, J. Appl. Phys. **45**, 32 (1974) and references therein.
20   J. M. Baker, C. J. Kircher, and J. W. Matthews, IBM J. of R&D, **24**, 223 (1980) and references therein.

Passivity of Metals and Semiconductors, edited by M. Froment
Elsevier Science Publishers B.V., Amsterdam — Printed in The Netherlands

THERMAL OXIDATION OF NIOBIUM NITRIDE FILMS

R.P. FRANKENTHAL, D.J. SICONOLFI, W.R. SINCLAIR, and D.D. BACON
Bell Laboratories, Murray Hill, NJ 07974, USA

ABSTRACT

The oxidation of cubic δ—NbN has been studied by Auger electron spectroscopy at temperatures from ambient to 400°C. Below 200°C, oxidation is limited to 1 to 2 monolayers. The reaction product is most likely a niobium oxynitride. Above 250°C, oxidation is not limited to the surface region. The reaction product at these higher temperatures initially is a suboxide of Nb_2O_5 but becomes Nb_2O_5 as the reaction proceeds further.

INTRODUCTION

In some recent studies, superconducting niobium nitride has been used as the base electrode in Josephson tunnel junction devices (ref. 1-4). Several oxides, including niobium oxide, have been proposed as the barrier layer on NbN to provide tunnel diodes. Further, it has been shown that the air oxidation of niobium nitride at temperatures from ambient to 90°C produces suitable junctions (ref. 4). There have, however, been only a few studies of the oxidation of niobium nitride, and all at temperatures above 400°C (ref. 5-7). At these temperatures, bulk oxidation is rapid, and the thin films required for a good junction cannot be formed reliably. Therefore, we have undertaken a study of the oxidation of niobium nitride thin films in air at temperatures below 400°C and present a summary of some of the results here. Full details and additional data will be presented elsewhere (ref. 8).

EXPERIMENTAL

Films of cubic NbN, also known as δ—$NbN_{0.9}$, 200-600 nm thick, were deposited on silicon wafers by dc reactive magnetron sputtering of niobium in a $15\%N_2$ — $85\%Ar$ gas mixture. The films have an average grain size of 5 nm and a superconducting transition temperature of 14 K. The thickness of the films was calculated from the weight gain during deposition, assuming a density of 8.4 g/cm^3 for NbN. Films of Nb_2O_5 about 20 nm in thickness were deposited on silicon wafers by dc reactive magnetron sputtering of niobium in oxygen. Their thickness was calculated from the weight gain assuming a density of 4.47 g/cm^3 for the oxide.

The niobium nitride films were oxidized in air at temperatures from 20°C to 400°C for various times ranging from 30 minutes to 64 hours.

The oxidized films were analyzed by Auger electron spectroscopy in the derivative mode with a single pass cylindrical mirror analyzer. The electron beam energy and current were 5 kV and 10 μA, respectively, and the modulation amplitude was 2 eV. Depth profiles through the films were obtained by argon ion sputter etching. The sputter rates, calculated from the time to sputter through the films and their known thicknesses, were 5.7 and 11 nm/min for the NbN and Nb_2O_5 films, respectively.

RESULTS and DISCUSSION

Electron stimulated desorption and reduction

After rapid electron-stimulated desorption of adsorbed oxygen and nitrogen, the O peak intensity decays slowly until it reaches a steady-state. For very thin oxide films, i.e., those formed below 100°C, the N peak also decays slowly to its steady state. However, for thick oxide films, the N peak reaches its steady state value with the initial desorption. The slow decrease in the O peak intensity most likely is due to some reduction of the reaction product by the electron beam. The observation that, for thin films, the N peak also decreases in intensity is an indication that the product contains some N and is probably a niobium oxynitride. More evidence for this will be presented below.

Ion sputter profiles

Nitrogen depth profiles are shown in Figure 1 for a thin and for a thick oxide film. Each profile shows the intensity of the N peak, I_N, normalized against the intensity, $I_{Nb'}$, of the Nb peak from the bulk NbN film.

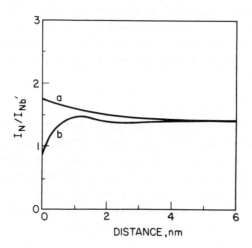

Fig. 1. Nitrogen depth profiles of products formed by oxidation at (a) 20°C for 24 hr and (b) 270°C for 16 hr.

For thin oxide films, the N peak decreases slowly until it reaches the steady state corresponding to bulk NbN. For thick films, the N peak first increases in intensity, goes through a small maximum, and then decreases to the steady state. For the thick oxide, the initial increase in the N intensity is due to the thinning of the oxide with sputtering. The decrease after the maximum is probably caused by some preferential sputtering of N from NbN. For the thin oxide, the initial decrease in the N intensity again indicates that there is N in the outer layer of the surface. This means either that the oxidation product is discontinuous, that is, N from NbN is exposed at holes, or that it contains both O and N. Since it is known that these films are good junctions and that they can be made reproducibly (ref. 4), the film cannot be discontinuous. Thus, the thin oxide film must contain N and is, most likely, an oxynitride. This conclusion is confirmed by the observation that the N peak changes shape as the oxidation product is etched to expose the NbN film, which implies that the N peak from the oxidized specimen comes from a species other than NbN.

The oxygen depth profiles, plots of $I_O/I_{Nb'}$ against sputter time, are shown in Figures 2-4 as a function of temperature and time of oxidation. At 20°C and 95°C, assuming a sputter rate of about 6 nm/min, the oxide is limited to a depth of less than 0.5 nm, that is,

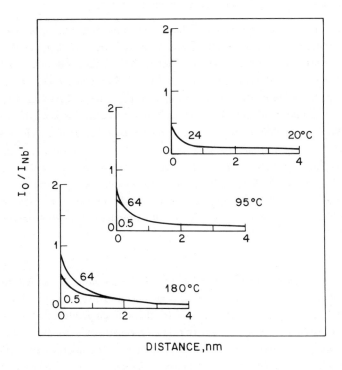

DISTANCE, nm

Fig. 2. Oxygen depth profiles of products formed by oxidation. Temperature and time of oxidation in hours shown on figure.

to the first monolayer (Fig. 2). It does not grow thicker with further exposure to air. As noted above, the film formed at these temperatures is probably an oxynitride. At 95°C, the initial O intensity is greater than that observed at 20°C, indicating that the composition of the oxynitride varies with the oxidation conditions. Only after extensive (64 hr) oxidation at 180°C is the oxidation product detected in the second molecular layer (Fig. 2). At 270°C and 400°C, assuming the sputter rate for Nb_2O_5, oxidation extends into the bulk of the NbN film (Figs. 3 and 4), and the reaction appears to obey a linear rate law. This is consistent with studies of the oxidation of bulk and film NbN at temperatures above 425°C (ref. 5-7). The rate constants, determined from the linear growth plots at 270°C and at 400°C, obey the equation defined by the apparent activation energy and pre-exponential term calculated by Gallagher et al. (ref. 7) from thermogravimetric data obtained at temperatures above 425°C. Only at temperatures above 400°C can the oxidation reaction be followed by non-surface sensitive techniques. At lower temperatures, a surface sensitive technique, such as Auger electron spectroscopy, must be used.

At both 180 and 270°C, the shape of the N peak indicates the presence of an oxynitride in the oxidation product, although it is not possible to determine whether the oxynitride exists throughout the product or only at the NbN interface. At 400°C, the N peak from the surface is too small to analyze its shape. However, as the NbN substrate is approached, the peak shape again corresponds to the oxynitride.

To understand better the composition of the oxidation product, the shape of its Nb peak for each temperature and time of oxidation was compared with the shape of the Nb peaks from the NbN substrate and from Nb_2O_5. Below 200°C the oxidation product is not Nb_2O_5. The shape of the Nb peak lies between those of NbN and Nb_2O_5 and is consistent with an oxynitride of niobium. At 270°C, the film formed after a short period of oxidation appears to be a suboxide of Nb_2O_5. As oxidation proceeds further, the suboxide is converted to Nb_2O_5. At 400°C, the reaction product contains Nb_2O_5. However, at the air/oxide interface, there appears to be a thin film of the oxynitride. As one sputters through the oxidation product to the NbN substrate, the shape of the Nb peak slowly changes to that for NbN, the final NbN shape being obtained when the intensity of the O peak has reached its background level. The results indicate that the oxidation reaction occurs by the diffusion of oxygen through the reaction product to the NbN interface.

CONCLUSIONS

At temperatures below 100°C, oxidation of NbN thin films is limited to the first monolayer. Only at temperatures above 250°C is significant oxidation of the bulk NbN film observed. Based on the present data, it appears that at temperatures above 400°C the oxidation product is Nb_2O_5. For the very thin films formed at temperatures below 100°C, the reaction product is probably an oxynitride of niobium. At intermediate temperatures for films more than 2-3 monolayers in thickness, the product is a suboxide of Nb_2O_5. However, there appears to be some oxynitride at the oxide/NbN interface.

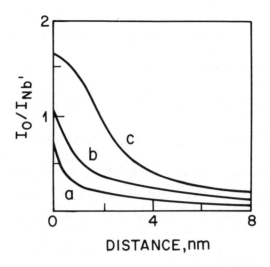

Fig. 3 Oxygen depth profiles of products formed by oxidation at 270°C. Time of oxidation: (a) 0.5 hr, (b) 16 hr, (c) 30 hr.

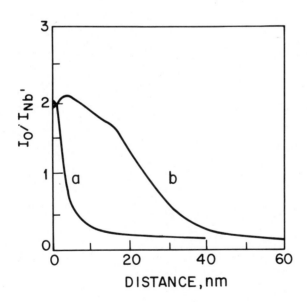

Fig. 4 Oxygen depth profiles of products formed by oxidation at 400°C. Time of oxidation: (a) 0.5 hr, (b) 2 hr.

REFERENCES

1. F. Shinoki, S. Takada, S. Kosaka, and H. Hayakawa, Japan. J. Appl. Phys., 19, 591 (1980).

2. S. Kosaka, F. Shinoki, S. Takada, and H. Hayakawa, IEEE Trans. Mag., MAG-17, 314 (1981).

3. A. Snoji, F. Shinoki, S. Kosaka, and H. Hayakawa, Japan. J. Appl. Phys., 20, L587 (1981).

4. R. B. van Dover, D. D. Bacon, and W. R. Sinclair, Appl. Phys. Lett., 41, 764 (1982).

5. P. Lefort, J. Desmaison, and M. Billy, Mat. Res. Bull., 14, 479 (1979).

6. P. K. Gallagher and W. R. Sinclair, Israel J. Chem., 22, 222 (1982).

7. P. K. Gallagher, W. R. Sinclair, D. D. Bacon, and G. W. Kammlott, J. Electrochem. Soc., in press.

8. R. P. Frankenthal, D. J. Siconolfi, W. R. Sinclair, and D. D. Bacon, J. Electrochem. Soc., in press.

Passivity of Metals and Semiconductors, edited by M. Froment
Elsevier Science Publishers B.V., Amsterdam — Printed in The Netherlands   451

COMPARATIVE STUDY BY XPS AND LEED OF NITRIDATION PROCESSES ON
SILICON (111).

C.MAILLOT, H.ROULET and G.DUFOUR
Laboratoire de Chimie Physique (LA 176), Université Pierre et
Marie Curie, 75231, PARIS CEDEX 05, FRANCE.

ABSTRACT

Two thermal, low pressure, nitridation processes were achieved
on silicon (111) and studied in situ by XPS and LEED.The results
show two different growth rates but the same evolution of the
electronic and surface cristallographic structure. XPS study
permitted to distinguish two stages for the growth in both cases.
For thicknesses less than 1 nm a split N 1s signal was interpreted
by the presence of a mixture of intermediate nitrides at the
interface. For films thicker than 1 nm the single N 1s and Si 2p
peaks were attributed to stoichiometric Si_3N_4. Films up to 4 nm
thick have been prepared in that way.

INTRODUCTION

Recent progress in microelectronics have underlined the need for
very thin insulators (thickness (T) $\leqslant$ 20 nm) for very large scale
integration (VLSI) devices.

SiO_2 is the most commonly used insulator but presents some drawbacks as for instance poor masking against dopant or impurity diffusion and interface trap densities inducing a shift in the
threshold voltage. Using a thermally grown nitride as insulator
instead of silicon dioxide could eliminate these disadvantages.

The main problem for the use of nitride films seems to be the
lack of understanding concerning the interface Si_3N_4/Si. The
thermal growth of Si_3N_4 on silicon has been studied by different
methods. For example Heckingbottom and Wood (1), Schrott and Fain
(2, 3) have applied Auger electron spectroscopy (AES) and low
energy electron diffraction (LEED) to study ultra-thin nitride
films. When argon ion sputtering is used to investigate the
interface some artifacts can complicate the interpretation of the
spectra. We present in this work results of X-ray photoelectron
spectroscopy (XPS) and LEED, used to compare the two nitridation
processes.

EXPERIMENTAL

Ultra-thin nitride films were prepared on n-type, polished, (111) oriented silicon with a resistivity of about 5Ω. cm in an ultra high vacuum chamber. The samples were at first cleaned by sputter etching with argon ions followed by annealing at 950°C for about 10 minutes. The pressure was better than 5.10^{-10} Torr during annealing. The starting surface exhibits a sharp 7x7 LEED pattern and no impurities were detectable by XPS.

Two nitridation processes were carried out on the reconstructed surface Si (111) 7x7. One treatment was achieved by entering pure ammonia gas at constant pressure of about 10^{-6} Torr, the other one by dissociation of nitrogen gas at the heated filament of our ion gun. Instead of the unknown nitrogen ions and atoms arrival rates on the sample we give the constant N_2 pressure (5.10^{-5} Torr), the emission current (25mA) and the ion acceletating voltage (300V). The surface was then characterized by LEED and transferred under ultra high vacuum to the electron spectrometer. Using Al K$\alpha_{1,2}$ (1486.6 eV) monochromatized radiation we scanned the regions of the Si 2p and N 1s levels. The total instrumental broadening was less than 0.5 eV.

We deduced the relative thicknesses of nitride films from the ratio of the integrated substrate Si 2p signal through the film and from the nude substrate. Our measurements were then calibrated by the nuclear reaction microanalysis method. To this end, absolute equivalent thickness of the thickest sample that we prepared with ammonia was evaluated from the amount of ^{14}N obtained from the nuclear reaction ^{14}N(d,α)^{12}C induced by deutons beam from a 2MeV Van de Graaf accelerator (4) by comparison with a LPCVD Si_3N_4/Si film whose thickness was known, assuming the same density for both our sample and the LPCVD film. Furthermore using these results we estimated the photoelectron mean free-path through the overlying film : λ(1385 eV) = 3.2 nm. This value is close to the estimate 3.6 nm derived by Taylor et al. (5) from their XPS results and to the average of the available estimates 3 nm selected by Wurzbach and Grunthaner (6).

RESULTS

Figure 1 shows that the growth is much quicker using ammonia than using nitrogen.

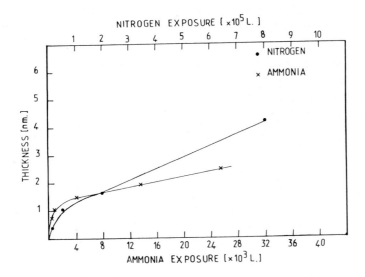

Fig. 1. Variation of the equivalent thickness (T) of the nitride films versus exposure for nitridation with ammonia and nitrogen. The exposures are given in Langmuir (10^{-6} Torr . 1s) and the thicknesses in nm.

The LEED patterns evolution with exposure is the same for ammonia and nitrogen treatments and is in good agreement with previous studies by Heckingbottom and Wood (1), Schrott and Fain (2). For the first step of the reaction (T$\leqslant$0.5 nm) quadruplet spots in addition with bright Si (111) 7x7 spots are observed. Si (111) 7x7 spots disappear when the film is thicker than 1 nm. When the reaction continues the quadruplet pattern becomes more and more diffuse. For a thickness larger than 2.5 nm the LEED pattern is completly diffuse.

XPS results evidence that the nitride electronic structure is the same for both nitridation processes. For a same thickness the spectra are similar either with ammonia or with nitrogen. The N 1s region exhibits a doublet structure until the thickness is above 1 nm (figure 2). These two peaks are separated by 1 eV at the beginning of the reaction but this energy difference becomes smaller when the thickness of the nitride film is increased. For thicknesses larger than 1 nm, the high binding energy side N 1s

peak disappears. The Si 2p region always shows the Si 2p signal
from the substrate. For the very thin films (T ≤ 0.5 nm) a broadening of the Si 2p peak from the substrate can be observed as well
as a small structure on its high binding energy side. This broadening disappears when the exposure increases and the Si 2p region
displays a single Si 2p signal from the nitride separated by
2.6 eV from the substrate.

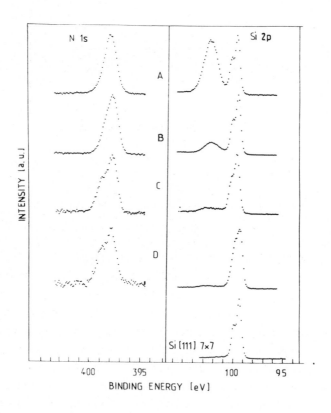

Fig. 2. Si 2p and N 1s spectra of the nitride films for equivalent
thicknesses of 4 nm (A), 2 nm (B), 1 nm (C), 0.5 nm (D) and
Si 2p signal from nude substrate.

DISCUSSION

It was proposed (7) that thermal nitridation is limited by the
diffusion of reactive species through the nitride film itself.
Our results are consistent with this argument (Figure 1).

The N 1s doublet structure is explained by two chemically dif-
-ferent nitrogen species. When the film is thicker than 1 nm the
spectra display only a single N 1s peak and a single nitride Si 2p
signal shifted by 2.6 eV from the Si 2p peak from the substrate.
Furthermore the energy difference between the single N 1s peak and
the Si 2p signal from the nitride is 295.8 eV, in very good
agreement with a measurement on stoichiometric LPCVD Si_3N_4 (8).
The integrated intensities were converted to the atomic ratio of
nitrogen/silicon. They were corrected for relative photoionisation
cross section (9), mean free-paths (10) and experimental transmission (11). This procedure gives a ratio nitrogen/silicon very close to the stoichiometric Si_3N_4 (1.33).

For the thinnest films (T$\leqslant$0.5 nm) we can interpret the broadening of the Si 2p substrate peak by the presence of intermediate
nitrides as for instance claimed by Schrott and Fain (2). Indeed
in bulk Si_3N_4, each silicon is bound to four nitrogen atoms arranged in a tetrahedron arrangement. In intermediate nitrides one,
two, or three nitrogen atoms are replaced by silicon atoms. In
similar studies on the SiO_2/Si interface Grunthaner et al. (12)
give the energy shift values corresponding to the three intermediate oxides. If we extrapolate these results to nitrides we find
1.5 eV for silicon bound to three nitrogen atoms, 0.8 eV for silicon bound to two nitrogens, 0.3 eV for silicon bound to a single
nitrogen. The presence of such intermediate nitrides explains the
broadening of the Si 2p substrate peak and the split of the N 1s
signal. When the thickness of the nitride film increases
(0.5$<$T$\leqslant$1nm) we notice a narrowing of the Si 2p substrate peak.
The electronic structure of the interface seems to change during
the reaction.

Thus, two different nitridation processes studied by a combination of XPS and LEED show two different growth rates but the same
electronic structure and the same cristallographic structure of
the surface. The results, for both treatments, reveal the existence of a two-stage process for the nitridation of silicon. The
first stage is characterized by a mixture of chemically different
species whereas in the second stage, for films thicker than 1 nm,
we only observe stoichiometric Si_3N_4.

456

ACKNOWLEDGMENTS

The authors are grateful to B.Agius, S.Rigo and F.Rochet for the thickness measurement by nuclear reaction microanalysis.

Financial support was provided in part by Centre de Spectrochimie (Université Pierre et Marie Curie).

REFERENCES

1 R.Heckingbottom and P.R.Wood, Surf. Sci., 36 (1973) 594.
2 J.F.Delord, A.G.Schrott and S.C.Fain Jr, J.Vac. Sc. Tech.,17 (1980) 517.
3 A.G.Schrott and S.C.Fain, Surf. Sci., 123 (1982) 204.
4 G.Amsel, J.P.Nadai, E.d'Artemar, D.David, E.Girard and J.Moulin, Nucl. Instrum. Meth., 92 (1971) 481.
5 J.A.Taylor, G.M.Lancaster, A Ignatiev and J.W.Rabalais, J·Chem. Phys., 68 (1978) 1776.
6 J.A. Wurzbach and F.J.Grunthaner, J. Electrochem. Soc.,130 (1983) 691.
7 D.S.Thomson, P.L.Pratt,The structure of silicon nitride, Science of Ceramics 3, ed. G.H.Stewart, Academic Press, 1967.
8 C.Maillot, Thèse de 3 eme cycle (Université Pierre et Marie Curie Curie), to be published
9 J.H.Scofield, Lawrence-Livermore Laboratory Report, No. UCRL. 51326 (1973) and J.Electron. Spectr. Rela. Phenom., 8 (1976)129.
10 D.R.Penn, J.Electron. Spectr. Rela. Phenom., 9 (1976) 29.
11 P.C.Kemeny, A.D.Mc Lacklam, F.L.Battye, R.T.Poole, R.C.G.Leckey, J.Liesegang and J.G.Jenkin, Rev. Sci. Instrum., 44 (1973) 1197.
12 F.J.Grunthaner, P.J.Grunthaner, R.P.Vasquez, B.F.Lewis, J.Maserjian and A.Madhukar, Phys. Rev. Lett., 43(1979) 1693.

Passivity of Metals and Semiconductors, edited by M. Froment
Elsevier Science Publishers B.V., Amsterdam — Printed in The Netherlands   457

STRUCTURE AND GROWTH KINETICS OF SiO$_2$ ULTRA THIN FILM ON Si(111) SURFACE[*]

J. DERRIEN[1], F. RINGEISEN[1] and M. COMMANDRE[2][**]

[1]I.S.E.A., Université de Haute-Alsace-4 rue des Frères Lumière-68093 Mulhouse
Cédex, France

[2]ERA CNRS n° 899, Département de Physique, Faculté des Sciences de Luminy,
Case 901 - 13288 Marseille Cédex 9, France.

ABSTRACT

The structure and the growth of very thin films of silicon oxide on top of
atomically clean silicon surfaces have been investigated with surface techniques
such as AES, ELS, LEED and XPS under UHV conditions. A transition layer was
found to form at the Si-SiO$_2$ interface. The growth kinetics could be explained
by a phenomenological model based on oxygen diffusion through the oxide layer
under the presence of a surface electric field. Finally the Si oxidation has
also been examined with XPS technique when some metallic impurities were evaporated onto the Si surface prior to oxygen exposure. An enhanced reactivity was
observed and attributed to the covalent sp^3 bond disruption which was provoked
by metallic impurities.

INTRODUCTION

The study of the initial stages of silicon oxidation is very fascinating for
both fundamental and technological reasons. Indeed it can shed light on the
nature and transport mechanism of oxidizing species that in thin or thick film
could be quite different. It allows also a control over ultra thin film
(< 50 Å) silicon oxide which is a vital part in present semiconductor device
technology (MOS), particularly for those operating with carrier tunneling
through the insulator such as solar cells and switching devices.

The preparation of ultra thin and pure oxide films by conventional methods
appears to be a rather difficult task due to : i) the presence of a native and
unknown composition oxide film at the semiconductor surface, extending sometimes to $\gtrsim$ 10 Å ; ii) the very rapid growth rate of the early stages of thermal
oxidation in the oxygen range usually used (several Torr up to atmospheric
pressure).

To avoid these difficulties, we have studied the growth process of ultra

---

[*] Work supported by "Groupement du Circuit Intégré au Silicium".

[**] Now at Ecole Nationale Supérieure de Physique, Rue H. Poincaré -
13397 Marseille Cédex 13.

thin oxide layers by thermally oxidizing atomically clean silicon surfaces under well controlled low oxygen ambient pressures (10^{-4} to 1 Torr). The sample cleanliness was achieved under ultra-high vacuum conditions (UHV) in order to get an uncontaminated surface, and the low oxygen pressure during oxidation permitted initial slow growth rate.

The paper is organized as follows : we very briefly discuss experimental details in section 2. Section 3 deals with the SiO_2-Si interface formation. Oxidation kinetics are discussed in section 4 and finally in section 5 we show the enhanced reactivity of Si atoms with oxygen when they are out of their usual sp^3 configuration.

EXPERIMENTAL

Experiments have been conducted in two UHV vessels. The first one is equipped with Auger electron spectroscopy (AES), energy loss spectroscopy (ELS) and low energy electron diffraction (LEED), sputter ion gun, gas entries and metal evaporators. The other one is an E.S.C.A. apparatus with nearly the same facilities, allowing X-ray photoemission measurements (XPS).

LEED, AES, ELS, XPS were used to characterizing the SiO_2-Si interface during its formation in situ. AES and/or XPS allowed also to plot oxidation kinetics of thin oxide films basing on the intensity of the elemental Si and oxidized Si peaks. More details can be found in ref. 1.

THE SiO_2-Si INTERFACE STRUCTURE

AES results

Fig. 1 shows AES spectra recorded during the formation of the SiO_2-Si interface. Fig. 1a displays a clean (111) Si surface spectrum. Only a characteristic Si transition (LVV, 92 eV) is observed in the low energy range. Fig. 1b corresponds to an oxygen monolayer adsorbed on the surface. A new structure appears then at $\sim$ 83 eV (due to Si-O bonds) concurrently with the oxygen transition (KLL, 510 eV). Fig. 1c represents a slightly oxidized surface (thickness $\sim$ 5 Å). Several features, characteristic of SiO_4 tetrahedra can already be distinguished. Besides the 83 eV peak still not obscured completely by SiO_4 tetrahedra, new peaks appear at 76, 63 and 59 eV. They are assigned to transitions from SiO_4 bond energy levels in a molecular orbital scheme. Fig. 1d shows a more heavily oxidized surface (thickness $\sim$ 13 Å). A very strong SiO_2 peak takes place at $\sim$ 78 eV, although the Si substrate transition (LVV, 92 eV) is still measured. Finally, Fig. 1e displays the AES spectrum of a thick SiO_2 film on top of a Si sample (thickness $\sim$ 30 Å). Only peaks characteristic of SiO_2 are recorded. During the oxide growth, the O(KLL) peak increases as shown in Fig. 1. By following the oxide growth step by step starting initially from a clean Si

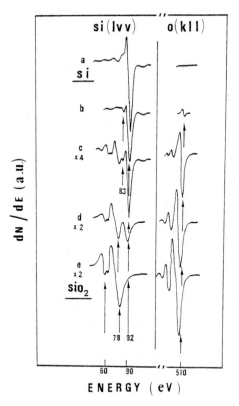

Fig. 1 .

sample, we confirmed during the interface formation that a transition oxide layer of non-stoichiometric composition appeared in between the Si substrate and the stoichiometric SiO_2 film. The thickness of this transition layer is probably less than 5 Å. During the oxide growth, LEED measurements were also monitored. Usually the 7x7 sharp spots of the clean Si(111) surface disappeared very quickly and an intense background intensity was observed. The SiO_2 film seems to be amorphous or possibly composed of very small crystallites, highly disoriented.

ELS results

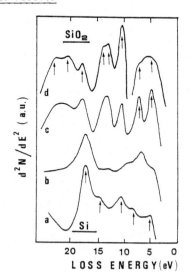

Fig. 2 .

The interface structure was also followed with ELS technique during its formation. Fig. 2a shows an ELS spectrum of a clean surface. Bulk ($\sim$ 17.5 eV) and surface ($\sim$ 10.5 eV) plasmons are observed together with an interband transition (4.9 eV). The two transitions located at 7.5 and 14.7 eV are very sensitive to the degree of surface cleanliness and are therefore assigned to surface states. Fig. 2b shows the Si surface when an oxygen monolayer has been adsorbed. Clean surface states disappear and are now replaced by several oxygen induced features (7, 8.2, 10 and 13 eV) which are attributed to Si-O bonds in a

460

molecular scheme (ref. 2). Fig. 2c corresponds to a slightly oxidized surface with SiO$_4$ tetrahedra on top of the transition layer. All the loss peaks observed in the range 10 to 23.5 eV are attributed to transitions between energy levels of SiO$_4$ tetrahedra (ref. 2). Only the two low energy (5 and 7 eV) peaks show the presence of broken or unsaturated Si-O bonds in incomplete SiO$_4$ tetrahedra. Finally, on a thick SiO$_2$ film (Fig. 2d), these two peaks broaden in a large bump usually found with perfect SiO$_2$ film. Thus ELS gives another confirmation of the interface structure. A transition layer of non stoichiometric composition is found to form at the Si-SiO$_2$ interface.

OXIDATION KINETICS

A challenging problem in a study of ultra thin SiO$_2$ layer growth is the thickness measurement. Clearly it is desirable to measure the oxide thickness in situ during growth to avoid the possibility of additional oxidation outside of the sample chamber. This problem is particularly severe when dealing with ultra thin films. We therefore utilized AES to estimate the SiO$_2$ thickness in situ. The details are already described in ref. 1. Representative set of results, at given oxygen pressure, are shown in Fig. 3. The points in these figures represent experimental measurements while the dotted lines are calculated from a phenomenological model (ref. 1). Several common features can be

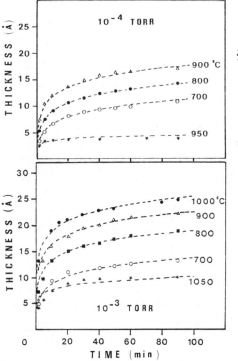

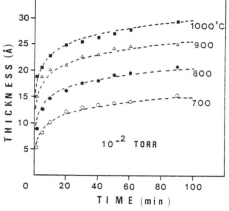

Fig. 3 - Thermal SiO$_2$ layer thickness versus oxidation time. The oxygen pressure is maintained at fixed values (10^{-4}-10^{-2} Torr) and the sample temperature in the 700°C-1000°C range.

observed in the data. First, each growth curve shows two distinct regimes, a rapid growth rate initially followed by a slow one, the value of which is very close to those found for thick layer growth (a few Å per hour). It appears that saturation occurs at this stage. Secondly, for each oxygen atmosphere, there exists a substrate temperature limit beyond which the sample can no longer be oxidized ($\sim$ 950°C at 10^{-4} Torr and $\sim$ 1050°C at 10^{-3} Torr). This behaviour is observed only with low oxygen pressures. We suggest that this limit is due to SiO_2 decomposition and its desorption from the substrate at high temperature and low partial oxygen pressure. Data can be fitted with computed plots derived from a model based on oxygen ion diffusion through the oxide layer under the presence of a surface electric field (ref. 1).

ENHANCED REACTIVITY OF Si ATOMS UPON OXYGEN EXPOSURE

A clean Si(111) surface can be easily oxidized even at room temperature if the Si atoms display their covalent sp^3 configuration disrupted. This occurs when a few metal atoms such as Au, Cu, Pd... are evaporated onto the Si surface. It is well known (ref. 3,4,5,6) that these atoms interact strongly with Si and destroy the sp^3 configuration giving rise to an intermixed surface zone where Si and metal atoms are embedded together. These Si atoms, liberated from their sp^3 configuration, behave as metal atoms and react very easily with oxygen to form a SiO_2 film even at room temperature.

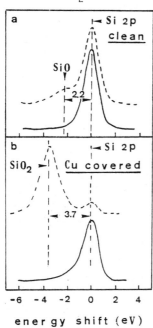

Fig. 4 .

Fig. 4 illustrates this phenomenon by means of XPS core level spectroscopy. Fig. 4a shows the Si 2p core line of a clean Si (111) surface (solid line). Its energy location is taken as origin. It has been excited by an ESCA apparatus using a Mg target as photon source. After an oxygen exposure (0.5 Torr 30 min.) at room temperature only a small shoulder is observed (dotted line). This is due to oxygen chemisorbed onto the Si surface atoms and this induces a small shift (2.2 eV) characteristic of Si-O chemisorbed bonds. Fig. 4b shows the Si surface precovered with 3 monolayers of copper (solid line). The surface Si atoms are now embedded in a metallic matrix (ref. 6). After the same oxygen exposure at room temperature (dotted line), a Si core level characteristic of SiO_2 is found at ~ 3.7 eV far from the initial Si core level. This result clearly demonstrates that Si atoms can be very easily oxidized once their sp^3 configuration has been broken.

CONCLUSION

We have shown that combination of several surface techniques allows an accurate access to the microscopic properties of the SiO_2-Si interface during its formation.

REFERENCES

1   J. Derrien and M. Commandré, Surf. Sci. 118 (1982) 32.
2   H. Lieske and R. Hezel, Thin Solid Films 61 (1979) 197.
3   A. Cros, J. Derrien and F. Salvan, Surf. Sci. 110 (1981) 352.
4   I. Abbati, G. Rossi, L. Calliari, L. Braicovich, I. Lindau and W.E. Spicer, J. Vac. Sci. Techn. 21 (1982) 409.
5   J. Derrien and F. Ringeisen, Surf. Sci. Letters, in press (1983).
6   F. Ringeisen, J. Derrien, E. Daugy, J.M. Layet, P. Mathiez and F. Salvan, J. Vac. Sci. Techn. (to be published). Proceedings of the 10th Conference on Physics and Chemistry of Semiconductor Interfaces, Santa Fe (USA), January 1983.

Passivity of Metals and Semiconductors, edited by M. Froment
Elsevier Science Publishers B.V., Amsterdam — Printed in The Netherlands

OXYGEN TRANSPORT STUDIED BY ^{18}O LABELLING IN THIN THERMAL SILICON OXIDE FILMS
IN CONNECTION WITH THEIR STRUCTURAL CHARACTERISTICS

B. AGIUS, M. FROMENT*, S. RIGO, F. ROCHET
Groupe de Physique des Solides de l'Ecole Normale Supérieure,
Université Paris VII, Tour 23, 2 place Jussieu, 75251 Paris Cedex 05, France.
*Physique des Liquides et Electrochimie, Université Pierre et Marie Curie,
Tour 22, 4 place Jussieu, 75230 Paris Cedex 05, France.

ABSTRACT

Using ^{18}O as a tracer, it was shown that during thermal growth of a silicon oxide films (at 930°C, from 10^{-1} to 100 Torr), the oxygen transport shows two aspects : one is related to the interstitial transport of O_2 through the network, the other to a step-by-step motion of network oxygen atoms. The relative contribution of the later increases as oxide films are thinner and O_2 pressures lower. For oxide films grown at 10 Torr during less than 1 hour ($d_{ox} \leqslant 4.6$ nm), a very slow chemical dissolution rate layer (~ 1.1 nm) was observed. Its existence is correlated to a structural order seen by RHEED.

INTRODUCTION

In their well known model Deal and Grove (1) put forward these two basic
assumptions : (i) silica has an open structure allowing interstitial dissolution of atoms or molecules, (ii) the oxidant is assumed to be molecular : the
molecule is not dissociated when crossing the oxide before reacting at Si-SiO$_2$
interface. This leads to linear kinetics followed by parabolic growth when
diffusion across silica film becomes rate-limiting, the corresponding constants being k_L = k C*/N$_1$ and k_p = 2 D$_{eff}$ C*/N$_1$, respectively (where C* is the
oxidant solubility, N$_1$ the number of network oxygen atoms per volume unit,
k the reaction rate at Si/SiO$_2$ interface and D$_{eff}$ the effective diffusion coefficient. When kinetics match well with the theory of Deal and Grove in the case
of water vapor oxidation,for dry oxygen this theory only applies for thicknesses greater than 20 - 30 nm. In a first work, using ^{18}O as a tracer
Rosencher et al. (2) have studied the oxygen transport across the film during
thermal oxidation. SiO$_2$ layers first grown in natural oxygen (130-300 nm) were
further grown in highly ^{18}O enriched oxygen during 8.5 h. at 930°C. The resulting SiO$_2$ films consisted of two ^{18}O rich zones, about 7 % near the SiO$_2$
surface ^{18}O FNES and $\sim$ 93 % near the Si/SiO$_2$ interface ^{18}O FNI while the
bulk ^{18}O concentration was very low (under 0.3 %).

The aim of this work is to determine : (i) the possible origins of ^{18}O fixa-
tion near the external surface of the oxide (^{18}O FNES) ; (ii) the possible

growth mechanisms of films of thicknesses ranging from 260 nm down to 5 nm, and (iii) the structural characteristics of such films.

OXYGEN TRANSPORT MECHANISMS

The mechanism of thermal oxidation of silicon in dry oxygen was studied using ^{18}O as a tracer. SiO_2 layers first grown in natural oxygen were further grown in ^{18}O enriched oxygen.

All experiments were carried out at 930°C from 10^{-1} to 100 Torr and various durations on (100) oriented silicon substrate.

The nuclear reactions $^{16}O(d,p)^{17}O$∷ and $^{18}O(p,\alpha)^{15}N$, induced by deuteron or proton beams from a 2-MeV Van de Graaff accelerator, were used to measure the number of ^{16}O (N_O^{16}) and ^{18}O (N_O^{18}) atoms/cm^2 in the oxide layers.

The ^{18}O concentration depth profiles were deduced from two different techniques : (i) analysis of the excitation curves of the $^{18}O(p,\alpha)^{15}N$ reaction near the 2-keV wide 629 keV resonance or (ii) step-by-step dissolution combined with nuclear microanalysis.

Origin of the ^{18}O fixation near the external surface

^{18}O FNES can be due either to :

a) some interfering phenomena : network oxygen exchange with water vapor traces or effects connected with the experimental procedure (exposure to air between thermal treatments and particular properties of the possibly remaining native oxide).

b) Other growth mechanism by Si transport.

c) Step by step motion of network oxygen atoms.

a) Interfering phenomena

We have measured the amounts of ^{18}O FNES. From these values we calculated minimum values for the partial pressure necessary to obtain such a quantity assuming an exchange with water vapor which follows a model of molecular water diffusion and chemical reaction (3). If, in such a model, we assume a linear dependence on the ^{18}O carrier solubility with pressure, one could then obtain the water content of the gas up to 300 ppm. This value is two orders of magnitude larger than those we actually measured.

In order to avoid possible effects related to an exposure to air and/or cooling phenomena between two subsequent growth treatments in ^{16}O and then in ^{18}O, we had oxide films grown in ^{16}O gas and then we increased the pressure in adding ^{18}O gas without removing the sample from the hot area. We still observed ^{18}O FNES.

Concerning a possible effect of the presence of the native oxide, the following experiment was carried out : one Si $^{16}O_2$ sample was partly dissolved to be sure that the native oxide, that could stay at the external surface, was removed. This sample was then submitted to the following further growths : in ^{16}O and then in adding ^{18}O gas. ^{18}O FNES was still observed.

b) Si_transport : samples with $Si^{16}O_2$ oxide films were successively submitted to the following further growths : in ^{18}O ambient and then in ^{16}O ambient during the same time and the same pressure. We noticed that the ^{18}O FNES considerably diminished. This proves that ^{18}O FNES is due to some exchange mechanism and not to silicon transport towards the external surface.

c) Step_by_step_motion_of_network_oxygen_atoms : in the case where an interfering phenomenon is negligible, the ^{18}O FNES is due to a step by step motion of network oxygen atoms. This means that the external surface of the oxide plays some particular role.

These results will be discussed in a paper recently submitted to the Journal of Electrochemical Society.

Study of growth mechanism for various thicknesses (5 - 260 nm)

For films of thicknesses higher than 50 nm, the isotopic concentration of ^{18}O FNI is equal to that of the oxidized labelled ambient. This is in favor of a transport of molecular oxygen across the film with no interaction with the network in agreement with the assumption of Deal and Grove (1). In whatever case, for a given ^{18}O thermal treatment, the amount of ^{18}O FNES is all the more important as the original thickness of the film is smaller. Even for thicknesses under 20 nm, when kinetics do not follow the linear parabolic law of thick films, we always found ^{18}O FNI. However, the amount of ^{18}O FNES can reach 50 % of the total amount of ^{18}O for $\sim$ 5 nm thickness films. The continuity in the behavior of the two ^{18}O transport mechanism through a great range of thicknesses (5 - 260 nm) suggests that a simple assumption of a modification of the mechanism because of space charge is not sufficient in the domain of very thin films.

We present in more details the results obtained for the thinnest films ($d_{ox} \leqslant 22$ nm) (Table 1).

Two groups of sample were treated. First, samples J_1, J_2, J_3, grown at 950°C in an $^{16}O_2/N_2$ mixture up to a thickness of about 19 nm, were subsequently submitted to an $^{18}O_2$ treatment at 930°C during about 1 hr, under a pressure of 100 Torr. These samples were exposed to air between these two thermal treatments. Although we approach the thickness domain where kinetics do not seem to follow the Deal-Grove model (1) any longer, we can notice that ^{18}O is still

TABLE 1

	$^{16}O_2$ TREATMENT				$^{18}O_2$ TREATMENT							
	N_o^{16} (10^{15} at/cm^2)	T (°C)	t (min)	P (Torr)	T (°C)	t (min)	P (Torr)	N_s^{18} (10^{15} at/cm^2)	N_i^{18} (10^{15} at/cm^2)	$\dfrac{N_s^{18}}{N_s^{18}+N_i^{18}}$	$N_o^{16}+N_o^{18}$ (10^{15} at/cm^2)	Total equivalent thickness (nm)
J_1	84.3	$O_2 + N_2$ mixture 950°C			930	67	100	3.3	14.7	0.18	101	22.8
J_2	84.3				930	67	100	5.7	14.1	0.28	109	24.6
J_3	84.3				930	67	100	2.9	14.3	0.17	99	22.3
K	(+)	930	30	9	930	60	9	3.2	3.0	0.52	23	5.2
L_1(*)	(+)	-	60	-	-	-	9	2.1	2.1	0.5	22.4	5
L_2(*)	(+)	-	60	-	-	-	1	1.4	0.5	0.74	21	4.7
L_3(*)	(+)	-	60	-	-	-	0.1	0.5	0.12	0.81	22.1	4.9
M	(+)	-	90	-	-	-	9	2.6	3.0	0.46	26.4	5.9
N	(+)	-	120	-	-	-	9	2.7	3.2	0.46		

Measurement of ^{16}O and ^{18}O total amounts (N_o^{16} and N_o^{18}) and of ^{18}O FNES and ^{18}O FNI amounts (N_s^{18} and N_i^{18} respectively) obtained from step-by-step dissolution combined with nuclear micro analysis for ^{16}O oxide films of original thicknesses ≤ 20 nm. (*)samples simultaneously grown in $^{16}O_2$; (+) no re-exposure to air between two thermal treatments thus no ^{16}O content measurement available.

mainly found near the interface Si/oxide but the ^{18}O FNES amount now is around 20 % of the whole ^{18}O amount.

Second, samples K, L_1, L_2, L_3, M, N grown in an Ultra High Vacuum Technology furnace in dry natural oxygen at 930°C, 9 Torr, during various durations (from 30 min to 120 min) were subsequently treated in $^{18}O_2$, with no reexposure to air, at 930°C during one hour and various pressures (from 0.1 Torr to 9 Torr) up to total thicknesses of about 5 nm. In this range of thicknesses, tempera- ture and pressure where Kamigaki and Itoh (4) needed the Mott-Cabrera model in order to explain the kinetics they found, we always found the heavy isotope located near the interface Si/oxide and near the external surface, but the ^{18}O FNES amounts are around 50 % of the whole ^{18}O content when samples are still treated in $^{18}O_2$ at 9 Torr. This ratio increases when the $^{18}O_2$ treatments are carried out under lower pressures and can reach 80 % of the whole ^{18}O amount at $p^{18}O_2$ = 0.1 Torr. It appears that N_i^{18} seems to follow a $\sim pO_2^{0.62}$ law, when N_s^{18} seems to follow a $\sim \log pO_2$ law (N_s^{18} and N_i^{18} corresponding to the ^{18}O FNES and ^{18}O FNI amounts respectively). These above results could be related to some structural modifications of the oxide film during growth. This point is investigated in the following chapter.

STRUCTURAL STUDIES

RHEED (accelerating voltage, 100 KV) and step-by-step dissolutions of oxide thermally grown in $^{18}O_2$ (at 930°C, 10 Torr) were carried out.

Dissolution

The step-by-step dissolution was performed in dilute HF (3 % volume in H_2O) at 20°C. Table 2 gives the number of oxygen atoms of the studied samples named

TABLE 2

Samples	t_{ox} (hour)	$\hat{n}^{18}O$ (10^{15} at/cm^2)	d_{ox}^{eq} (nm)
A	0.5	14.1	3.2
A' (∷)	-	8.8	2.0
A" (∷∷)	-	5.0	1.1
B	1	20.3	4.6
C	3.5	36.1	8.2
D	22.5	92.2	20.8

^{18}O content measurements of samples A, B, C, D. t_{ox} is the oxidation time, $\hat{n}^{18}O$ the ^{18}O content and d_{ox}^{eq} the equivalent thickness.
∷ A' corresponds to A dissolved during 4 s.
∷∷ A" corresponds to A dissolved during 40 s.

A, B, C, D. On figure 1 is represented the remaining ^{18}O amounts vs etching time. Sample D, although not entirely represented on this figure, shows a constant etching rate and after 120 s the remaining ^{18}O amount is less than 0.3×10^{15} atoms/cm^2 (< 0.07 nm equivalent).

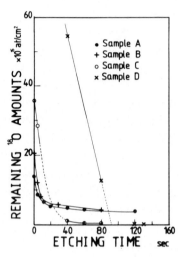

Fig. 1. Remaining ^{18}O amounts vs etching time.
● sample A
+ sample B
O sample C
X sample D

The 8.2 nm thick oxide sample, C, shows at the beginning an etching rate practically equal to that of sample D, but after 40 s the slope changes and at 80 s an ^{18}O content of 0.6×10^{15} atoms/cm^2 (∿ 0.14 nm equivalent) remains. Concerning the thinnest samples, A and B, a strong change in etching rate occurs after 8 s, the slope being divided by more than two orders of magnitude. Let us emphasize that for A and B a comparatively rather thick layer (∿ 1.1 nm equivalent thickness) shows a much slower dissolution rate, whilst at the beginning of etching the rate is similar to that of thicker oxides. We must add that a 30 minutes dissolution time is necessary to reduce the ^{18}O amount of oxygen down to 0.2×10^{15} atoms/cm^2.

RHEED observations

Figure 2-1 shows a RHEED diagram in the azimuth [011] obtained after a chemical etching of the (100) Si surface. A Kikuchi pattern clearly appears showing that the surface has a good crystallographic organization and a high degree of smoothness.

Figure 2-2 shows that the thicker oxides (more than 8 nm) have an amorphous structure.

Fig. 2-1.

Fig. 2-2.

Fig. 2-3.

Fig. 2-4.

Fig. 2. Diffraction patterns observed on bare Si (Fig. 2-1),
and samples D (Fig. 2-2), A (Fig. 2-3) and A' (Fig. 2-4).

When the thickness of the film lies between 2 nm and 4 nm, ring shaped diffraction patterns are observed (Fig. 2-3 and 2-4). The intensity and location
of the rings do not change when the sample is rotated around the axis perpendicular to the layer. This proves that the layer is isotropic in a direction
parallel to this plane, but presents a structure in a direction perpendicular
to this plane. The succession of interreticular distances d_{hkl} (Table 3) and
the intensity I of the rings, deduced from different patterns, do not correspond
to any usually known crystalline SiO_2 structure. The fact that sample A' shows

TABLE 3

d_{hkl} (nm)	0.279	0.254	0.233	0.199	0.187	0.173	0.160
I	W	S	S	W	M	M	M
d_{hkl} (nm)	0.151	0.140	0.129	0.110	0.103	0.098	0.093
I	S	S	M	W	W	W	W

Interreticular distances d_{hkl} and intensities deduced from diffraction
pattern   (S = strong, M = medium, W = weak).

the same diagram as sample A but with less bright rings indicates that this peculiar structure is present in the whole film thickness.

In conclusion, an intermediate layer (equivalent to about 1 nm) is observed (at 930°C, 10 Torr) for oxidation times less than 1 hour ; this layer disappears for oxidation times greater than 3.5 hours. The existence of this intermediate layer is correlated to a structural order observed by RHEED but this order is present throughout the entire oxide film.

REFERENCES

1 B.E. Deal and A.S. Grove, J. Appl. Phys., 36 (1965) 3770.
2 E. Rosencher, A. Straboni, S. Rigo and G. Amsel, Appl. Phys. Letters, 34 (4) (1979) 254.
3 S. Rigo, F. Rochet, B. Agius and A. Straboni, J. Electrochem. Soc., 129 (1982) 867.
4 Y. Kamigaki and Y. Itoh, J. Appl. Phys., 48 (1977) 2891.

Passivity of Metals and Semiconductors, edited by M. Froment
Elsevier Science Publishers B.V., Amsterdam — Printed in The Netherlands   473

GROWTH OF THIN OXIDE LAYERS ON SILICIDE COMPOUNDS

A. CROS

IBM, T.J. Watson Research Center, Yorktown Heights, NY 10598

Permanent address : ERA CNRS 373 - Faculté des Sciences de Luminy, Département
de Physique - 13288 Marseille Cédex 9, France.

ABSTRACT

The first stages of the room temperature oxidation of Ni, Pd and Pt silicides
have been studied by ESCA. It is suggested that the observed variations in oxide
stoichiometry and oxide thickness are related to the more or less pronounced
metallic character of the silicide surface.

The peculiar properties of silicides have produced   interesting device
applications. The existence of a superficial oxide on the silicide can be a
serious problem in the realization of an ohmic contact or it can be a basic
requirement of tunneling devices.

We present here an ESCA study of the first stages of the room temperature
oxidation of near noble metal silicides. This usually results in the formation
of a thin (10-50 Å) dielectric film,which also acts to protect the compound
from further attack.

The oxidation of Ni_2Si, $NiSi$, $NiSi_2$, Pd_4Si, Pd_2Si, $PdSi$, Pt_2Si and $PtSi$ has
been investigated and several general tendencies have been evidenced.

All silicides except Pd_4Si have been grown by thermal annealing of a metal
film ($\sim$ 1000 Å thick) deposited on a Si substrate. The formation of these
compounds has been extensively studied previously,and more details on it can be
found elsewhere [1]. Clean Pd_4Si surfaces have been obtained by scraping in
situ a Pd_4Si amorphous alloy or by annealing in situ the structure composed of
Pd and Si films successively deposited on a SiO_2 substrate. We can mention that
sputtering a Pd_2Si surface also produces an enrichment in Pd atoms and the
surface is spectroscopically indistinguishable from the two Pd_4Si surfaces
described previously.

Characterization of these surfaces has been made by ESCA by using a mono-
chromated Al Kα (1486.6 eV) source with a resolution of $\sim$ 1 eV. Oxidation was
made by air exposure at room temperature. Our results concern essentially the
initial chemisorption processes,and the oxide products which have been identified may thus be different from the ones corresponding to terminal processes

obtained after days of air exposure.

Several simple questions may be formulated : i/ which atom will oxidize, metal or silicon ? ii/ what is the oxidation state of the oxidized species? iii/ what oxide thicknesses can be obtained under a given set of well defined conditions ?

For all the compounds which have been mentioned, only Si atoms have been found to oxidize. This is probably due to the very large heat of formation of SiO_2 (ΔH = - 70 kcal/gm-atom) when compared to the one of near noble metal oxides ($\Delta H(NiO)$ = - 29.2 kcal/gm-atom, $\Delta H(PdO)$ = - 10.5 kcal/gm-atom, $\Delta H(PtO)$ = - 8.5 kcal/gm-atom).

Concerning the oxidation states of these Si atoms, we know from previous studies of the oxidation of crystalline Si, that several oxidized states can exist : they are distinguished by the tetrahedral environment of Si atoms which can be represented as $Si(Si_3O)$, $Si(Si_2O_2)$, $Si(SiO_3)$ and SiO_4 and they have been found [2] to show ESCA chemical shifts of the Si 2p core level of 0.9 eV, 1.9 eV, 2.6. eV and 3.4. eV, respectively, from pure Si towards higher binding energy. After oxidation of the silicides, we have found that all these states exist simultaneously in the oxide films. However their relative abundance depends strongly on the silicide stoichiometry. As can be seen in Fig. 1, which presents the case of Pd silicides, the oxidation of Pd_4Si surfaces gives a

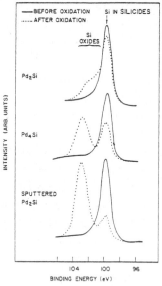

Fig.1 : Si(2p) core levels before and after the oxidation of Pd silicides (2 min, air exposure, room temperature).

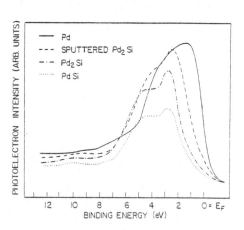

Fig.2 : Valence bands of clean Pd silicides. The spectrum of Pd_4Si is identical to the one of "sputtered Pd_2Si".

thicker and more completely oxidized oxide than the one of Pd_2Si or PdSi.
Similar behaviors have also been observed in the case of Ni and Pt silicides.
Quantitatively and qualitatively, the oxidation of the silicides is enhanced
when the metal concentration increases. This corresponds to different growth
rate of the oxide. The oxide layer on top of Pd_4Si does not change with further
air exposure. On the contrary, a more complete oxidation state of Si atoms and
a thicker oxide can be obtained in the case of Pd_2Si and PdSi, but this requires
oxidation times which are between one and two orders of magnitude larger.

These differences in the passivation of silicides can have different origins:
a first obvious explanation consists in the possible segregation of Si atoms on
top of Si rich silicides. However from the shape of the shifted Si (2p) core
levels on ESCA spectra, we conclude that the amount of segregated Si atoms
cannot exceed one monolayer. Since the oxide thicknesses are in the range
10-50 Å, we do not expect the first monolayer to play a crucial role in the
observed phenomenon. A second origin could be differences in the stability of
the compounds. Within this framework, the oxidation rate would be controlled by
the activation energy required for the motion of the atoms. The driving force
for the diffusion during oxidation is expected to be proportional to the difference between the heat of formation of SiO_2 and of the silicides. This would
explain the fact that Pd_4Si oxidizes more easily than Pd_2Si and PdSi. However,
it fails to explain the fact that Ni_2Si which is the most stable of Ni silicides,
oxidizes more easily than NiSi and $NiSi_2$. Finally, a possible explanation may be
the different metallic character of the surfaces. A common trend to all the
silicides investigated is that upon increasing Si concentration, the density of
electronic states at the Fermi level $N(E_F)$ decreases strongly. Although ESCA
does not allow a precise determination of these densities, it is clear in
Fig. 2 that the variations of $N(E_F)$ are much larger than the uncertainties in
the measurements. The shift of the metal conduction d band to higher binding
energies with increasing Si concentration is responsible for these tendencies.

We suggest here that the availability of transferable electrons in the
vicinity of E_F plays an important role in the enhancement of the oxidation rate
observed for metal rich silicides. A possible mechanism of this enhancement is
the breaking of the oxygen molecule by electron transfer from the silicide
surface.

Finally, we want to emphasize the similarities between these experiments and
some results concerning the oxidation of the Si-Au interface. It has been
reported that metallic amorphous Au-Si alloys oxidize very easily [3], while
ordered $\sqrt{3}\times\sqrt{3}$ gold structures "passivate" the Si surface [4]. These differences
have been ascribed to differences in the metallic character of the surfaces [5],
the amorphous one being clearly metallic, while the ordered one corresponds to

a more semiconducting character. These results are in qualitative agreement with what we have presented here. Both correspond to the physical concept that the oxidation rate of Si atoms is enhanced whenever they are in a more metallic environment.

REFERENCES

1 See for example : Thin films - Interdiffusion and Reactions, ed. by J.P. Poate, K.N. Tu and J.W. Mayer.
2 G. Hollinger and F.J. Himpsel, to be published.
3 A. Cros, J. Derrien and F. Salvan. Surf. Sci. 110 (1981) 471.
4 A. Cros, F. Houzay, G.M. Guichar and R. Pinchaux. Surf. Sci. Lett. 116 (1982) L 232.
5 A. Cros, to be published in Journal de Physique (June).

Passivity of Metals and Semiconductors, edited by M. Froment
Elsevier Science Publishers B.V., Amsterdam -- Printed in The Netherlands

PROPERTIES OF THE PASSIVE FILM ON IRON ELECTRODES BY CAPACITY AND PHOTOCURRENT MEASUREMENTS

Ulrich Stimming

IBM, T.J.Watson Research Center, Yorktown Heights, N.Y.10598, U.S.A.*

ABSTRACT

Passive iron electrodes were investigated in 1M $HClO_4$, 1M $NaClO_4$ - pH 8 and 1M $NaOH$ covering an pH range of almost 14 units. Capacity measurements were carried out in the frequency range ca. 3 Hz - 3 kHz. Applying a Schottky-Mott analysis to the results yields donor concentrations on the order of $10^{20} cm^{-3}$. At higher potentials, however, a frequency dependent capacity is encountered which is probably due to OH^- adsorption. Photocurrent spectra measured in all electrolytes yielded a similar shape of the spectrum. The onset of the photocurrent was between 2.0 and 2.5 eV photon energy depending on the electrolyte the passive film was formed in. Photocurrents have a pronounced transient like behaviour in neutral and basic solutions. The potential dependence of the photocurrent shows an exponential increase with additional peaks at low potentials in neutral and basic solutions after prolonged polarization of the electrode.

INTRODUCTION

The properties of the passive film on iron electrodes have been discussed for a long time but as yet there is no concensus of opinion. This is clear from the diversity of views regarding the "nature" of the passive film (ref.1,2,3).

A picture of the charge distribution in the passive film can be obtained from capacity measurements. The first results of the potential dependence of the electrode capacity were obtained by means of the potentiostatic double pulse technique (ref.4). They were interpreted in terms of the semiconducting properties of the passive film with a behaviour of a highly doped ($10^{20} cm^{-3}$) n-type semiconductor. Impedance measurements by other authors (ref.1) lead them to the assumption that the passive film is basicly of insulating properties.

Photoelectrochemistry is a good in-situ method to investigate the electronic structure of the film, however, results can be complex. Recent investigations of photocurrent behaviour (ref.3,5) confirmed the picture of a semiconducting passive film, but could explain only partial aspects of the problem. Photocurrent measurements with bulk $\alpha - Fe_2O_3$ electrodes (ref.6) demonstrated the complexity of the behaviour of iron oxide.

---

*Permanent Address: Institut fuer Physikalische Chemie, Universitaet Duesseldorf, D-4000 Duesseldorf 1, W.-Germany

To further contribute to the discussion on the properties of the passive film, investigations of the capacity behaviour and the photoelectrochemical properties were undertaken. To understand how far the electrolyte and in particular the pH plays role, measurements were carried out in 1M $HClO_4$, 1M $NaClO_4$ buffered to pH 8 by borate and 1M $NaOH$.

EXPERIMENTAL

The working electrode was made from zone refined iron (99.99% Fe), while the counter electrode was a gold foil separated by a diaphragm from the working electrode chamber. For the capacity measurements iron rods of approxamitely 2 mm diameter and 15 mm length were used and for the photocurrent measurements circular discs of approxamitely 10 mm diameter. Details of the pretreatment of the electrodes are described in (ref.4). Reference electrodes were acid or neutral mercury/mercurous sulphate or mercury/mercurous oxide electrodes, but all given potentials refer to the normal hydrogen electrode (nhe) unless otherwise stated. The electrodes were polarized by means of a fast rise potentiostat (bandwidth 1 MHz); currents were measured by a differential amplifier.

The passive film was formed at high film formation potentials ε_p, but at a value below oxygen evolution. With $HClO_4$ the highest potential was 1.5 V because of the pitting corrosion potential of ClO_4^-. Film thicknesses were on the order of 3 nm estimated from previous data (ref.4). The film formation time was 3600 s, for some photoelectrochemical measurements prolonged polarization times (up to 10^5 s) were used.

Electrode capacities were measured by superimposing a small a.c. voltage (0.5-5mV RMS) on a linear potential sweep and detecting the imaginary part of the impedance by means of a lock-in amplifier (Ithaco 393). The frequency was varied in the range of ca. 3 Hz - 3 kHz.

In the photoelectrochemical measurements, for monochromatic illumination of the electrode in the wavelength range $\lambda = 200 - 800nm$, a $900W$ Xenon lamp and a grating monochromator were used. Measurements with chopped light were carried out normally in the Hz range, and the signal was detected by means of a lock-in amplifier (Ithaco 395). Photocurrent transients were measured directly.

CAPACITY BEHAVIOUR

Capacity-potential curves, $C(\varepsilon)$, were obtained in 1M $HClO_4$, 1M $NaClO_4$ pH 8 and 1M $NaOH$ marking a pH range of nearly 14 units. The potential range was limited on one side by the film formation potential, ε_p, and, on the other, by the onset of oxide reduction. Starting at ε_p, the electrode capacity was measured during a slow potentiodynamic sweep (1-10mV/s). Curves obtained in 1M $NaOH$ are shown in Fig.1 for different frequencies. At low potentials the capacity is fairly high, on the order of $30\mu F \cdot cm^{-2}$ taking into account a roughness factor of 1.6 (ref.4). The capacity drops with increasing potential to a value of ca. $5\mu F \cdot cm^{-2}$ at $\varepsilon = 0.5V$. The magnitude of the capacity in this range depends only slightly on the frequency. At higher potentials, however, the capacity increases again to maximum values which do depend strongly on the frequency. The capacity behaviour displayed in Fig.1 is representative for the results obtained in the other electrolytes. However, the capacity curves are shifted to

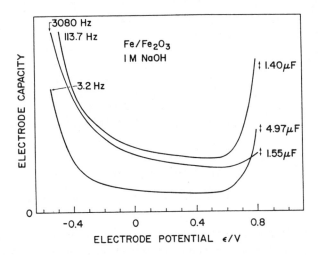

Fig.1: Capacity-potential curves in 1M NaOH for different frequencies.

higher potentials with lower pH. In perchloric acid, a capacity increase at higher potentials could not be observed because of the potential limitation due to the pitting corrosion potential of ClO_4^-.

Assuming that the solid state properties of the passive film determine the capacity behaviour and the Schottky-Mott relation (Eq.1) is applicable, informations on the donor concentration and distribution can be obtained:

$$1/C^2 = 2/eND'D_0(\varepsilon - \varepsilon_{fb} - kT/e)$$ (1)

with N, the donor concentration, D' and D_0, the dielectric constant of the film and the vacuum, respectively, ε_{fb}, the flatband potential, k, T and e which have the usual meaning.

Plots according to Eq.1 are shown in Fig.2. In fact, straight lines are observed at low potentials in all electrolytes. Using a dielectric constant D'=10, which was determined previously (ref.4), from the slope of the straight lines in Fig.2 the following donor concentrations are calculated: $1.7 \cdot 10^{20} cm^{-3}$ ($HClO_4$), $4.0 \cdot 10^{19} cm^{-3}$ ($NaClO_4$) and $2.8 \cdot 10^{20} cm^{-3}$ ($NaOH$). From the extrapolation to $1/C^2 = 0$ the flatband potential ε_{fb} can be calculated considering an additional potential drop in the Helmholtz layer of 0.1V due to the high donor concentration (ref.7,4). Then the shift on the potential scale can be related to the pH change which is roughly 60mV/pH, i.e. the

480

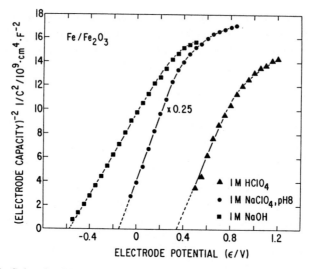

Fig.2: Schottky-Mott plots of capacity data obtained at ca. 1kHz.

flatband potential is pH independent with respect to the reversible hydrogen electrode (rhe). This confirms earlier assumptions (ref.2) that only the potential drop in the Helmholtz layer is changed with pH.

The frequency dependence of the capacity is demonstrated in Fig.3 for three selected potentials, two in the region of a linear Schottky-Mott behaviour and one at higher potential where a frequency dependence was observed. Fig.3 demonstrates that

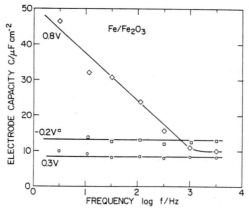

Fig.3: Frequency dependence of the capacity at different potentials.

in the entire frequency range which was investigated the capacity is almost constant at the low and medium potential, but exhibits a strong decrease with increasing frequency at high potentials. This happens at potentials just before oxygen evolution starts. Therefore, it is possible that OH^- is adsorbed at the oxide surface creating a surface charge which increases the capacity but is slower than the electronic process which determine the capacity in the Schottky-Mott regime. Further details of this discussion and especially of the frequency influence on the Schottky-Mott parameter are given elsewhere (ref.8).

PHOTOELECTROCHEMICAL PROPERTIES

Photocurrents were investigated as a function of the wavelength of the incident light, the electrode potential and the time. The wavelength dependence of the photocurrent measured at a potential slightly below the film formation potential ε_p is shown

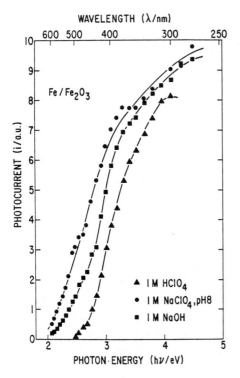

for all three electrolytes in Fig.4. The onset of the photocurrent is at 2.0-2.5 eV and the photocurrent increases continuously up to 4.5 eV. For $HClO_4$, the onset is at slightly higher photon energies. However, the profile of the spectra is basicly the same at all pH values disregarding some minor differences. An analysis of the absorption edge according to

$$i_{ph} = (h\nu - E_g)^n / h\nu \qquad (2)$$

with n=1/2 for direct and n=2 for indirect transitions, shows different exponents n depending on the electrolyte, the electrode potential and the photon energy. This behaviour does not allow for a simple explanation and will be discussed in a subsequent paper (ref.9). It is probably due to two independent processes and an amorphous structure of the passive film.

Fig.4: Photocurrent spectra in different electrolytes; in neutral and basic solution after prolonged polarization.

The potential dependence of the photocurrent is shown in Fig.5. In neutral and basic solutions, the onset of the photocurrent is close to the flatband potential found from the capacity measurements. These curves are measured after prolonged polarization of the electrode ($10^4 - 10^5$ s). A peak is observed at low potentials followed by an exponential increase at higher potentials. This exponential increase is also observed in the acid solution but with a shift on the potential scale by ca. 0.85 V compared to $NaOH$ which corresponds to 60 mV/pH. The photocurrent in the neutral and basic solution shows a transient behaviour of the anodic photocurrent at low and high potentials. At medium potentials, in the range of the minimum, however, the anodic transient changes to a stationary cathodic photocurrent. In the acid solution the photocurrent is a stationary anodic one.

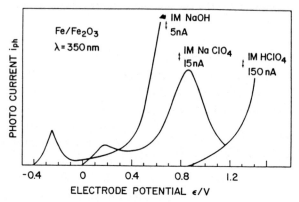

Fig.5: Potential dependence of the photocurrent at constant wavelength; in neutral and basic solution after prolonged polarization.

The results indicate, that at least in neutral and basic solutions, after prolonged polarization of the electrode, two different photoinduced processes can be distinguished. A more detailed discussion of the absorption characteristics and the potential dependence will be given elsewhere(ref.9).

CONCLUSIONS

Combining the results of the capacity and photocurrent measurements shows that the behaviour of passive iron electrodes is basicly determined by the solid state properties of the passive film. Independent of the electrolyte in the whole pH range, the characteristics of a highly doped n-type semiconductor are found. Photocurrent spectra show a continious increase of the photocurrent from ca. 2 eV to 4.5 eV. However, a more complex behaviour has to be considered with a frequency dependent capacity at high potentials and two different photoinduced processes.

REFERENCES

1 B.D. Cahan and C.T. Chen, J.Electrochem.Soc 129 (1982) 17,474,700,921
2 U. Stimming and J.W. Schultze, Electrochim.Acta 24 (1979) 859
3 S.M. Wilhelm and N. Hackerman, J.Electrochem.Soc. 128 (1981) 1668
4 U. Stimming and J.W. Schultze, Ber.Bunsenges.Phys.Chem. 80 (1976) 1297
5 L.M. Abrantes and L.M. Peter, J.Electroanal.Chem., in press
6 P. Iwanski, J.S. Curran, W. Gissler and R. Memming,
   J.Electrochem.Soc. 128 (1981) 2128
7 B. Pettinger, H.P. Schoeppel, T. Yokogama and H. Gerischer,
   Ber.Bunsenges.Phys.Chem. 78 (1974) 1024
8 U. Stimming, in preparation
9 U. Stimming, in preparation

Passivity of Metals and Semiconductors, edited by M. Froment
Elsevier Science Publishers B.V., Amsterdam — Printed in The Netherlands   483

COUPLED ELLIPSOMETRIC AND CAPACITANCE MEASUREMENTS DURING GROWTH OF ANODIC
OXIDE ONTO GaAs

P. CLECHET, J. JOSEPH, A. GAGNAIRE, D. LAMOUCHE, J.R. MARTIN and E. VERNEY
Laboratoire de Physicocnimie des Interfaces - Ecole Centrale de Lyon - BP 163
69131 ECULLY Cedex - France

ABSTRACT
The coupled utilization of in situ ellipsometric and capacitance measurements appears to be a usefull method for the studying of anodization processes.
We applied this new technic in order to examine kinetic, electrical and optical
phenomena during the growth and stripping of anodic oxide onto GaAs.

INTRODUCTION
In the field of semiconductor devices, the fabrication of MOS type structures requires low surface states densities at the oxide - semiconductor interfa-
ces.Such structures having good dielectric and interfacial properties are difficult to realize on III V compounds in particular on GaAs. The rather poor literature (compared with the Si) shows that the situation is still unclear, the
reported results being often contradictory.

Electrochemical techniques (1,2) should have advantages that overcome many of
the difficulties;

- on the one hand, the large number of controled parameters (temperature,
electrolyte, current density...) make the anodic oxidation a powerful method

- on the other hand they allow the possibility of in situ control of the
oxide and interfa ce properties.

With this aim, two experimental techniques seem to be particulary interesting:

- spectroellipsometry which provides an optical measurement of the surface
evolution (thickness, refractive indexes and to some extent structural properties of the oxide layer)

- capacitance measurements, which give valuable information on the electrical and energetical properties of the sample (space charge capacitance, density
of interface states ...).

The coupled utilisation of these two methods enables us to obtain a maximum
of information at any time during experimental procedure. We applied them to
studying the anodic growth of oxide on GaAs using a AGW type solution (4) propy-

lène glycol and aqueous 3% tartaric acid 2 : 1 v/v, the pH of the tartaric solution being ajusted either to 7.6. or to 7.2. with NH_4OH). The advantages of this solution are the good dielectric properties of the grown oxide and thickness reproducibility. Moreover, chemical dissolution of the oxide occurs in this solution in absence of any bias (this phenomenon being very useful in order to explain the growth mecanism). Yamagishi et al. have presented a study of anodization process on Ga As by in situ differential reflectance (10), but they were mostly interested in the initial stage of oxide growth. Even if this study seems to be closely related to the present paper, it differs on two important points: first, we did not specially consider the nucleation process but we mostly studied the growth conditions of thick oxide films; twice, our aim was especially here to describe the principles and some advantages of the coupled utilization of optical (ellipsometric) and electrical (capacitance) in situ measurements.

EXPERIMENTAL

For ellipsometry as for impedance, all measurements were monitored by a microcomputer. Ellipsometric measurements were made in a quartz window cell with a 70° incidence angle. Our ellipsometer had a rotating analyser which rotated at 15 Hz (5). The amplified signal of photomultiplier was sampled at a rate of 720 points per revolution and averaged for 40 revolutions. The wavelength of light (620 nm) was chosen in order to obtain the most precise measurement of the thickness of the oxide film. The analysis of experimental measurement (the ellipsometric angles ψ and Δ versus time) was performed by a minimisation procedure (6). We assumed a three phase model: in the ambient medium (electrolyte) an homogeneous oxide film of variable thickness (t) was grown onto a substrate. For a choice of the indices of ambient, oxide and substrate, we plotted a theoretical curve of ψ Δ with respect to t. By a least squares method we found the indices of oxide and substrate corresponding to the best fit of experimental values. Thus this method gives a set of indices for the oxide and substrate and the thickness of oxide for each (Δ,ψ) measurement. Impedance measurement were performed using a home made potentiostatic device in order to obtain good high frequency behaviour. DC bias, which is directly generated by the microcomputer (D/A converter), is applied to the electrode with respect to the saturated calomel electrode(SCE). A typical sinusoïdal signal of 10 mV amplitude is surimposed on the DC bias. The current response of the cell was detected by a lock-in amplifier. Every part of the measurement system was monitored via the IEEE bus. In all experiments, we used a frequency of 1000 Hz. The interface capacitance can be calculated with an accuracy of 1%.

All experiment were carried out using the same sample. Only the electrolyte was renewed between each oxidation-dissolution cycle, in order to avoid depletion and/or contamination effects. During each cycle, in situ ellipsometric and capacitance measurements were performed automatically every 30 sec. for the former and every 30 sec. to 5 min. for the latter. The anodization bias had to be interrupted before each capacitance measurement, which was performed at the rest potential (0,0 volts/SCE in our case for both the "clean" Ga As and the Oxide-GaAs systems. See the discussion). The capacitance measurements were fast enough (about 15 sec. each) not to disturb the growth and dissolution phenoma, and this assumption was verified by ellipsometric measuments. Oxidation step were discontinued when the anodization current evolution drops to a value lower than $0.02 \mu A$ cm^{-2} mn^{-1}. A simple series equivalent circuit was used for the treatment of capacitance data. At 1 kHz, the capacitance of surface states can be neglected until the surface barrier is greater than 0.3 ev(1). When, for thicker oxide films ($\approx$ 80 nm) the surface barrier may be lower than 0.3 ev, the capacitance of the oxide layer plays a dominant role so that we can neglect the other components of the system. Thus, the simple series equivalent circuit can be applied in every case.

RESULTS

A first anodization was performed at 25 volts bias, the current density being limited to 1 mA cm^{-2}. The ellipsometric parameters (ψ and Δ) were measured in situ every 30 sec. during both the consecutives stages of growth and dissolution. A simple analysis bases in the comparison of the complex dielectric constant of the initial and the final surface clearly shows that this process (growth and stripping of an oxide film) improves surface quality (7,8).

Figure 1 shows a Mott-Schottky plot obtained in the AGW electrolyte (pH=7.2) after the anodization dissolution treatment described above. Some important electrical charasteristics of the junction are thus available: we find N_d=6.7 10^{17} cm^{-3} for the doping level of the semiconductor which is in good agreement with the known properties and V_{FB}=-1.46 volts/SCE for the flat band potential, which is the expected value for a junction involving the use of an aqueous electrolyte with pH near from 7.

A second anodization-dissolution cycle was performed at 150 volts bias, with current density limited to 1 mA cm^{-2}. Figure 2 represents the changes in the ellipsometric angles (ψ and Δ) during this cycle, with the calculated thickness of the oxide film as a parameter. The parts corresponding to the growth and the dissolution are identical and are caracteristic of an homogeneous oxide. Least squares analysis leads to the complex dielectric constants of both the substrate and the film at 620 nm; we found ε_1=14.8, ε_2=1.54 for GaAs and ε_1=3.24, ε_2=0 for the oxide (values which are in good agreement with previous results (7,10).

Figure 3 shows the evolution of the total series capacitance (C) and its reciprocal (1/C) during the same cycle. One can easily see that the parts corresponding to the growth and the dissolution are nearly identical, although the

measurements dispersion is greater in the case of the growth (mainly because of
the large variations of potential and of the incertainties corresponding to the
calculated oxide thicknesses). Moreover the plot 1/C as a function of the oxide

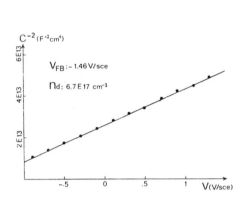

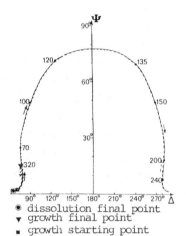

● dissolution final point
▼ growth final point
■ growth starting point

Fig.1 - Mott-Schottky plot for a clean
GaAs surface in AGW pH 7.2 electrolyte.

Fig.2 - Experimental changes of the
ellipsometric parameters during
growth and dissolution. Anodization
bias = 150 v. Initial current density limited to 1 mA cm^2. The calculated oxide thicknesses (in nm.) are
plotted.

thickness appears to be a straight line from the very begining with the starting
point corresponding exactly to the inverse of the capacity of the "clean" GaAs
surface. From now, we can conclude (and this point will be discussed with more
details below) that the potential drop in the oxide is negligible compared with
the surface barrier in the semi-conductor: we can thus write in this case:

$$\frac{1}{C} = \frac{1}{C_{sc}} + \frac{1}{C_{ox}}$$

where $\frac{1}{C_{ox}} = \frac{d}{\varepsilon_{ox}\,\varepsilon_0}$ represents the inverse of the capacity of the oxide

and $\frac{1}{C_{sc}} \simeq$ constant (5 10^6 F^{-1} cm^2) represents the inverse of the capacity of
the space charge layer in the semiconducteur.

The value of the dielectric constant of the oxide can be easily calculated:
we found $\varepsilon_{ox} = 10,5 \pm 0.5$, which is slightly greater than the one previously measured (8 at 1 MH$_z$) by HARTNAGEL et al. (3). This difference may be explained by
the quality of both the oxide itself and the GaAs/Oxide interface.

Besides, the dissolution rate of the oxide layer appears to be independant
of the thickness and of the growth conditions. On the other hand, it seems to
be highly dependant on the pH of the solution.

Furthermore, by assuming that the current efficiency is close to unity, and that the oxide is composed of As_2O_3 and Ga_2O_3 in equal parts, one can verify that the dissolution rate corresponds to the final anodization current density, within experimental errors. Table 1 shows the values of these parameters for two AGW solution with slighly different pH.

pH	dissolution rate nm/mn	final current density $\mu A/cm^2$
7.6	1.49 ± 0.05	29 ± 2
7.2	0.825 ± 0.05	14.5 ± 1

Table 1

Hence tne final oxide layer thickness is reached when the growth rate equals the dissolution rate, leading to a dynamic equilibrium.

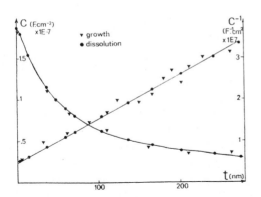

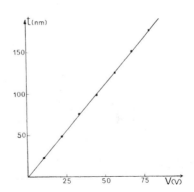

Fig.3 - Total series capacitance (C) and its reciprocal (C^{-1}) as a function of the thickness (t) of the oxide film. Anodization bias 150 v. Initial current density limited to 1 mA cm^{-2}.

Fig.4 - Thickness oxide film as a function of anodization bias.

Figure 4 represents the final oxide thickness as a function of the anodization bias, the initial current density being limited to 1 mA/cm^2. The calculated growth rate is constant all over the explored range of potential (5 v to 120 v) and is equal to 21.3 Å/V as expected from reported results (5): we may think that the growth is limited by the electric field in the oxide (about 510^6 v/cm).

At last, an anodization-dissolution cycle was carried out with an initial oxydation current density limited to 0.1 mA cm^{-2} (110 volts oxidation bias).

488

The analysis of tne ellipsometric data clearly proved that the grown oxide had the same optical properties than the previous oxide grown under greater current density. Figure 5 represents the evolution of the total series capacitance during the growth and the dissolution, as a function of oxide thickness. In contrast with the Figure 3, the part of this curve corresponding to the growth

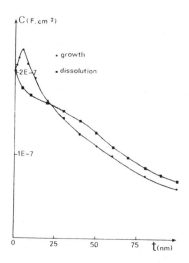

Fig.5 - The final part of the growth and the starting part of the dissolution are omitted.

presents a maximum for an oxide thickness which is equal to about 5 nm. In order to explain this phenomenon, we may invoke several assumptions, such as:

- a phenomenological effect: for this low current density, the initial stage of anodic oxidation could occur by islands nucleation (9-10) up to a thickness of about 5 nm, afterwards the growth, and consequently the dissolution, would be homogeneous;

- a energetical effect, for example charge storage in the bulk of the oxide and/or near the GaAs/oxide interface.

Further complementary investigations are needed in order to solve this problem.

DISCUSSION

This study exhibits two points which seem to be particularly interesting to discuss. The first point deals with the kinetic aspect of the growth of the oxide layer. The evolution of the ellipsometric parameters during the growth and the dissolution of an oxide film from an initialy contaminated GaAs surface allows us to think that the growth mainly occurs at the GaAs/oxide interface by in diffusion of oxygen-species through the oxide. This species would probably be hydroxyle ions which could easily drift within the oxide under the intense electric field. The oxidation reaction at the oxide/GaAs interface would lead to the formation of protons which could drift in the opposite direction under the same field. The electrochemical reaction can be written as follow:

$$AsGa + 30\ H^- + 6\ h^+ \longrightarrow 3\ H^+ + 1/2\ As_2\ O_3 + 1/2\ Ga_2\ O_3$$

This assumption could thus explain the influence of alcohol additives (glycol, methanol) in term of control and of regulation of the concentration of OH^- species (4). It can also explain that the impurities contaminating the initial GaAs surface are carried away to the outer surface of the oxide during it's growth. Furthermore, when the oxide has reached its final thickness, the dissolution and oxidation rates are equal, leading to a dynamic equilibrium which continuously renews the outer surface of the film, hence eliminating the impurities. The cleaning ability of the "oxidation-dissolution" process could be explained this way.

The second point deals with the peculiar electrical properties of the oxide. We noticed, effectively, that the rest potentials of clean GaAs and of the oxide/GaAs system in our AGW electrolyte are the same and equal to 0 volts with respect to SCE, which corresponds to an extremely deep depletion,(in other words to a large potential drop) in the semiconductor. Furthermore, and concerning with the oxide grown with the higher initial current density ($1\ mA\ cm^{-2}$), we showed that the curve representing the inverse of the total capacity (1/C) as a function of oxide thickness, is linear from the very beginning. The potential drop in the oxyde may thus be considered as negligible with respect to the drop in the semiconducteur space charge, at least under quasi-steady state conditions as it is the case for our capacitance measurement. This behaviour can be explained in term of leakage current and/or in term of the existence of an impurity band (2) in the oxide. Such a behaviour is not observed for the oxide grown under lower ($0.1\ mA\ cm^{-2}$) initial current density.

CONCLUSION

These preliminary results clearly show the interest in carrying out in situ coupled optical and electrical measurements in order to collect a maximum of independant informations. We applied this technique to studying the growth and dissolution mechanisms during the anodization of GaAs. Kinetics phenomena, electrical and optical parameters are thus available. We evidenced the cleaning ability of such a growth-dissolution cycle in AGW medium. Differences of electrical behaviour were underlined for the oxides grown at initial current density limited either to $1mA\ cm^{-2}$ or to $0.1\ mA\ cm^{-2}$. A more thorough study would be able to confirm our assumptions.

490

1 K.W. Frese Jr, S.R. Morrisson, J. Electrochem. Soc. (1979), 126 (7), 1235.
2 K.W. Frese Jr, S.R. Morrisson, J. Vac. Sci. Technol. (1980) (March,April), 17 (2).
3 H. Hasegawa, K.E. Foward, H.L. Hartnagel, Appl. Phys. Lett. (1975) 26 (10), 567.
4 H. Hasegawa, H.L. Hartnagel, J. Electrochem. Soc. (1976) 123 (5), 713.
5 D.E. Aspnes, Optic. Com. 8 (3) (1973), 222.
6 J.B. Theeten, Surf. Sci. 96 (1980), 275.
  and J.Joseph, A. GAGNAIRE, Thin Solid Film (in press).
7 D.E. Aspnes,G.P. Schwartz, G.J. Gualtieri, A.A. Studna, B. Schwartz, J. Electrochem. Soc. 128, (1981), 590.
8 S.M. Spitzer, B. Schwartz, G.D. Weigle, J. Electrochem. Soc. 122 (3), 1975, 397.
9 W.H. Makky, F. Cabrera, K.M. Geib, C.W. Wilmsen, J. Vac. Sci. Technol. 21(2), July/Aug. 1982, 417.
10 C.Yamagishi, A.Moritani, J. Nakai, J.Electrochem. Soc. (1980), 127 (1), 169.

Passivity of Metals and Semiconductors, edited by M. Froment
Elsevier Science Publishers B.V., Amsterdam — Printed in The Netherlands

STUDY BY SPECTROPHOTOELECTROCHEMISTRY (SPEC) OF THE PASSIVITY OF IRON IN
NEUTRAL SOLUTIONS

M. FROELICHER, A. HUGOT-LE GOFF, V. JOVANCICEVIC[*]
Groupe de Recherche n° 4 du C.N.R.S. "Physique des Liquides et Electrochimie",
associé à l'Université Pierre et Marie Curie,
4 place Jussieu, 75230 Paris Cedex 05, France.

ABSTRACT

The current photogenerated in an oxide polarised in an electrolytic solution
can be separated in two parts : transient, and steady-state currents. The
spectral analysis of these currents, their variations with the polarisation
potential, the shape of the transient part give informations about the nature
of the oxide. We have studied α-Fe_2O_3 and Fe_3O_4 single-crystals, and thermal
oxides on iron which are mixtures of these two modifications ; their identification is made by Raman spectroscopy. With reference to these standards, we
have studied the passive films grown on iron in buffer borate solution.

The measurement of the electrolytic currents, photogenerated in an oxide
film, can be a way to identify this film, which is particularly interesting
since it allows to work in-situ without any perturbation of the film growth
conditions. After illumination, the current has a sharp increase, reaching a
value i_{Ta}, and after a transient decay, stabilizes to a steady-state value,
i_{ssa}. When the surface is darkened, one observes an other transient current, i_{tc},
having an opposite sign and a value comparable to the value of i_{ta}. Figure 1
gives a schematic picture of these currents. The spectral distributions of i_{ssa}
and i_{ta} or i_{tc} are plotted for different polarisation potentials ; the whole of
these data is quite characteristical of a given material ; in a first approximation i_{Ta} reproduces the charge transfer spectrum of the material, i.e. the
main part of its optical absorption spectrum. This phenomenon is quite general
in the oxides and other semiconductor materials, but the main problem is the
frequent weakness of the photoresponse (particularly in the case of a thin
film), which is superposed at the d.c electrolytic current.

---

[*] Present address : CTFB, Thionville.

EXPERIMENTAL

To overcome this difficulty, we have built an high sensitivity differential amplifier, which allows to separate this weak response of the very much larger d.c current. Practically, we are able to appreciate the photocurrents having an amplitude equal to 0.5 nA. The samples are illuminated with a 450 W Xenon lamp.

All the oxides are polarised in the same solution of buffered borate (pH 8.4) and in the same range of potentials ; from - 0.5 V/SSE to 0.7 V/SSE. Of course, all the photocurrent plots given here are corrected of the response of the different elements of the optical device.

As standards, we have used α-Fe_2O_3 (doped with Zr to ensure the electric conductivity) and Fe_3O_4 single crystals. We have used also thermally oxidised iron samples at different temperatures : we can pass from a Fe_3O_4 film after heating below 200°C, to a α-Fe_2O_3 film above 400°C. At 300°C, one obtains a mixture of the two modifications, identified by Raman spectroscopy (1).

RESULTS

α-Fe_2O_3

On figure 2 is plotted the spectrum of photocurrent at 0.7 V/SSE, potential where this current is purely steady-state. At this potential, hematite cannot be electrochemically reduced, and one can consider that the "true" α-Fe_2O_3 photocurrent spectrum is obtained. Its part at low energy is classical (2 for instance). The assignment of the different electronic transitions in hematite is really controversial in the literature : in particular, taking in account the strength of the optical absorption at about 3 eV, the first charge transfer transition is often assigned at 3 eV or below this value. We agree with the results of Marusak and coll (3), which show by SCF-Xα calculations, that all the transitions up to 2.95 eV are ligand and crystal-field transitions, and assign the charge transfer transitions at : 3.31 eV and 3.94 eV ($O2p \rightarrow 2t_{2g}$), 4.82 and 5.56 eV ($O2p \rightarrow 3e_g$). In these conditions, the convolution of the crystal field transition at 2.85 eV and of the first charge-transfer transition creates a maximum in the optical absorption at 3.1 eV, which of course is not found again in the photocurrent spectrum. On the other hand, this spectrum is easily obtained by the first charge transfer transitions.

When the potential is below 0.7 V/SSE, a certain amount of the total current becomes transient, and i_{ssa} disappears at - 0.5 V/SSE. The main point is that, at each potential, i_{ssa}, t_{ta} and i_{tc} reproduce the same spectral features. When the potential decreases, the low energy part of the spectra remains unchanged but a shoulder, and then a peak, appears and growths at 4.65 eV. One knows that the wustite FeO has an absorption edge at 3.56 eV, and an absorption maximum at 4.65 eV. One can assume that this new peak is related to an electrochemical reduction of Fe_2O_3 leading to a certain amount of Fe^{2+} ions at low potentials.

We have to notice finally that the time constants of the transient currents are always very high (several seconds), that would be consistent with the idea that they are related to a chemical reaction in the bulk of the oxide (photo-activated reduction of Fe^{3+}) rather than to a classical recombination of e-h pairs after trapping in bulk or surface states.

Fe_3O_4

Whatever the polarisation potential of this oxide can be, i_{ta}, i_{tc} and i_{ssa}, have always quite different features. The transient currents reproduce the spectral features of hematite (i.e., Fe^{3+} ions) ; as one can see on figure 3, plotted at 0.2 V/SSE, i_{ssa} would be related to the presence of Fe^{2+} ions, as deduced from the optical behaviour of FeO. The time constants of the transients are always very fast. From these results, one can suppose that the presence of the Fe^{2+} sites allows the recombination of e-h pairs created on Fe^{3+} sites. The short values of the time constants are always in favor of a classical mechanism of trapping of the carriers.

Thermal oxides

The existence of two "photocurrent behaviours" characteristical of the two modifications of iron oxides is found again when these modifications are present in the form of thin layers on an iron substrate. One will be able then to extend the precedent analysis to the identification of the passive layers.

We have studied two films, having respectively thicknesses of about 300 and 900 Å, and approximatively the same composition (i.e., the same ratio between the amplitude of peaks of Fe_3O_4 and α-Fe_2O_3 in the Raman spectra). The intensities of their transient photocurrents are also in a 3 ratio, which shows that the recombinations give rise in the bulk of material (and not in surface states) and corroborate the role of Fe^{2+} sites as points of recombination.

Passive films

On figure 4, are plotted the photocurrent spectra for a passive film grown at 0.6 V/SSE, and measured at 0.4 V/SSE. They clearly display a "Fe_3O_4-type behaviour". On the figure 5, the photocurrent/potential curve for this film is plotted, the measurements being made for decreasing potentials. The film is stable down to - 0.4 V/SSE. When the photocurrents/potential curve is plotted for increasing potential, starting from a bare iron electrode, i_{ta} increases only from + 0.2 V/SSE. Then, a solid "Fe_3O_4-type" film exists between + 0.2 and + 0.6 V/SSE.

In the first part of the passive range, the photocurrents are purely steady-state and Fe^{2+} -featured (at least if the measurement is done rapidly because the spectra change with the time and a steady-state part can appear after several hours of polarisation), as seen on figure 6. But, of course, it is question-nable to suggest that FeO is the species present at the beginning of the passive

494

range. Taking in account the presence of Fe(OH)$^+$ as an intermediate of reaction, the existence of Fe(OH)$_2$ would be more easy to interpret, but this compound is not able to give a photoelectrochemical effect. Finally, we assume that the composition of the films grown in the first part of the passive range is probably really complex, nearer of a polymeric form (4), with Fe-O and Fe-(OH) bondings than of a crystallographically well defined modification. This form would be able to advance with the time toward crystalline rearrangements.

In the second part of the passive range (0.2 - 0.6 V/SSE), where the photocurrent spectra reproduce always the Fe$_3$O$_4$ shapes, one can then assume that this modification is present.

CONCLUSION

Here we have shown that the observation of the currents photogenerated during the polarisation of electrodes can be an interesting way to study the thin oxide films grown at their surfaces. In the case of the passivity of iron, we have proved, for the first time, the importance of the divalent form of Fe ions. A divalent species is present alone at the beginning of passive range, and it is not possible to find a purely trivalent film before the oxygen evolution range.

REFERENCES

1 M. Froelicher, A. Hugot-Le Goff, V. Jovancicevic, C. Pallotta, R. Dupeyrat, M. Masson, This book.
2 J. Kennedy, K. Frese, J. Electrochem. Soc.,125 (1978) 723.
3 L. Marusak, R. Messier, W. White, J. Phys. Chem. Solids, 41 (1980) 981.
4 W. O'Grady, J.O'M. Bockris, Surf. Sci.,66 (1977) 581.

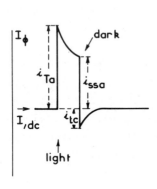

FIGURE 1

Schematic picture of
the photocurrents

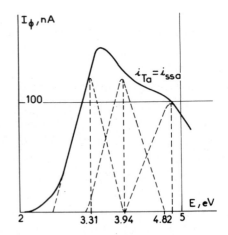

FIGURE 2

Photocurrent response of
α-Fe$_2$O$_3$ (V = 0.7 V/SSE)

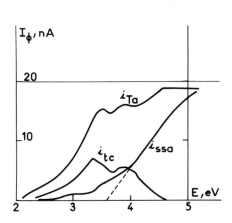

FIGURE 3
Photocurrent responses of Fe_3O_4
(V = 0.2 V/SSE)

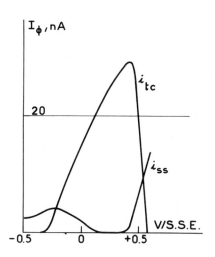

FIGURE 5
Photocurrent/voltage curve of
the film grown at 0.6 V/SSE

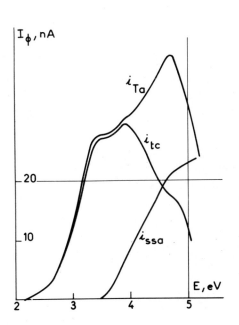

FIGURE 4
Photocurrent responses of a
film grown at the end of the
passive range : 0.6 V/SSE
(V = 0.4 V/SSE)

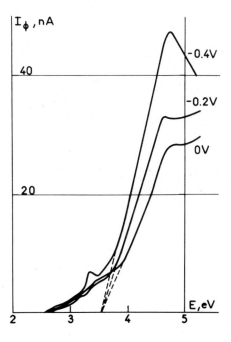

FIGURE 6

Photocurrent responses in the
first part of the passive range

Passivity of Metals and Semiconductors, edited by M. Froment
Elsevier Science Publishers B.V., Amsterdam — Printed in The Netherlands

INFLUENCE OF THE ANODIZATION CONDITIONS ON THE ELECTRONIC PROPERTIES AND
CRYSTALLOGRAPHIC STRUCTURES OF THE CORROSION LAYERS ON TUNGSTEN.
A PHOTOELECTROCHEMICAL APPROACH.

F. DI QUARTO, S. PIAZZA and C. SUNSERI
Istituto di Ingegneria Chimica, Università di Palermo, viale delle Scienze,
90128 Palermo (Italy)

ABSTRACT

The influence of morphology, composition and crystallographic structure on the photoelectrochemical behaviour of anodic oxide films on tungsten obtained in various conditions of anodization has been investigated. Different photo-current spectra and absorption edges were obtained for each type of film. Optical band gaps ranging between 2.55 eV and 3.15 eV were determined for crystalline and amorphous WO_3 films grown in different conditions. The low quantum efficiency of the anodic films must be attributed to the presence of an amorphous (a-WO_3) film which controls the transport of the injected photocarriers.

INTRODUCTION

In previous works (ref.1,2) we investigated the role of different experimental parameters on the kinetics of growth as well as on the structure and composition of anodic oxide films on tungsten. More recently (ref.3,4) we studied the semiconducting and photoelectrochemical behaviour of amorphous WO_3 (a-WO_3) anodic films in acidic solutions. In this paper we will show how the photoelectrochemical study can be used to get useful informations on the electronic structure and morphology of different corrosion layers on tungsten.

EXPERIMENTAL

Tungsten electrodes were prepared from spectrographically pure W sheets (0.1 mm thick) and rods (4 mm diameter). Different anodic films were grown in various acidic solutions at constant current density (i = 8 mA cm^{-2}) by varying both the temperature and the final cell voltage. Each film was characterized by X-ray analysis, SEM microscopy and photoelectrochemical measurements.

498

The experimental apparatus has been described elsewhere (ref.1-4). The photo-electrochemical measurements were performed at room temperature in 1N H_2SO_4 so-lution. The reference electrode was $Hg/Hg_2SO_4/1N$ H_2SO_4 (MSE).

EXPERIMENTAL RESULTS AND DISCUSSION

In Fig. 1a we report the photocurrent action spectrum of an amorphous anodic film obtained at room temperature in a galvanostatic experiment stopped before the breakdown. After correction for the lamp emission the photocurrent spectrum appears very poorly structured with a maximum at 270+20 nm. The wavelength of the maximum photocurrent was sligthly dependent on the anodization conditions and film thickness. By assuming non-direct optical transitions between the sta-tes of the valence band (V.B.) and those of the conduction band (C.B.) an opti-cal band gap E_g^{opt} = 3.05+0.1 eV is obtained from the $(I_{ph} h \nu)^{\frac{1}{2}}$ vs. hν plots as reported in Fig. 2. The use of these plots rests on the demonstrated proportio-nality between the measured photocurrent and the light absorption coefficient α (ref. 4). The above reported E_g^{opt} value was independent from the film thickness, anodizing current density and composition of electrolytic solution. It is note-

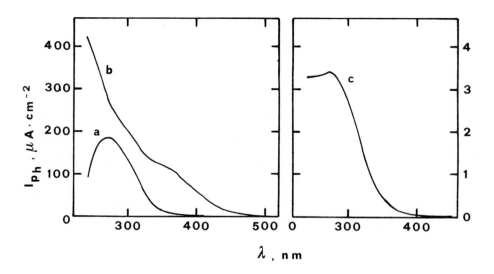

Fig. 1. Photocurrent action spectrum corrected for the lamp efficiency for dif-ferent WO_3 films. a) Amorphous film anodized in 0.1N H_3PO_4 until 70 V; b) tri-clinic crystalline WO_3 obtained by oxidation in 1N HNO_3 for 10'; c) crystalline layer of $WO_3 \cdot H_2O$ obtained by oxidation in 1N H_2SO_4 at 70 °C for 1h.

worthy that these amorphous films after crystallization under argon atmosphere at 350 °C showed a lower optical band gap (E_g^{opt} = 2.75±0.05 eV) and a large increase in the measured photocurrent at any wavelength. Moreover, the dissolution rate after crystallization was negligible both in the dark and under light.

All these findings show that the lack of a long-range order in a-WO$_3$ films has relevent influence on the elctronic structure as well as on the chemical stability of the WO$_3$ films. As for the electronic structure of amorphous WO$_3$ semiconducting films we believe that the band model proposed by Mott for amorphous semiconductors (ref.5) agrees with our experimental findings. In fact both the lower photocurrent values and the higher E_g^{opt} values of a-WO$_3$ films with respect to the crystallized ones suggest the existence of localized states near the mobility edges E_C and E_V (see Fig. 4) where the mobility of the injected carriers is very low. Moreover, the optical transitions between localized states of the V.B. and those of the C.B. do not contribute to the observed photocurrent. According to Mott (ref.5) the E_g^{opt} value defines an extrapolated rather than a real zero in the density of states. In the case of a-WO$_3$ anodic films the localized states seem to be due to the long-range lattice disorder rather than to a spcific electronic defect. Weak bonds as consequence of distorted bond-lengths

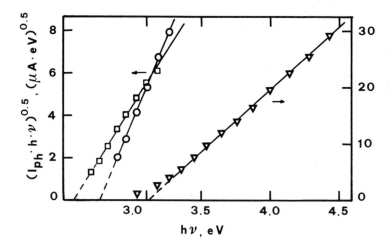

Fig. 2. Determination of the optical band gap of different WO$_3$ films. ▼ Amorphous film anodized in 0.1N H$_3$PO$_4$ until 70 V; □ triclinic crystalline WO$_3$ obtained by oxidation in 1N HNO$_3$ for 10'; ○ monoclinic crystalline WO$_3$ obtained by oxidation in 0.1 N H$_3$PO$_4$ until 70 V followed by annealing under argon atmosphere at 350 °C for 3h.

500

and/or angles can play a fundamental role in the acceleration of the dissolution rate in dark conditions. Moreover under illumination the increase of the broken bonds at the surface has a more pronounced effect for the more defective amorphous films than for the crystalline ones. This fact could explain the poorer electrochemical stability of the amorphous films with respect to the crystallized ones under otherwise identical conditions.

In Fig. 1 we also report the photocurrent spectra of anodic films grown in different conditions. In the case of films grown in HNO_3 solutions a thick porous film was obtained by keeping the anodizing voltage above the breakdown value ($\simeq$ 80 V) for about ten minutes. In this case the X-ray patterns showed a triclinic structure for the anodic WO_3 layers (ref.2). The $(I_{ph} h \nu)^{\frac{1}{2}}$ vs. $h\nu$ plots reported in Fig. 2 give an intercept to the energy axis E_g^{opt} equal to 2.55 ± 0.05 eV. The photocurrent efficiency was still very low but it could be raised by more than one order of magnitude by annealing at 350 °C under argon atmosphere. After the annealing process the photocurrent spectrum changes noticeably but the E_g^{opt} value remains constant.

In the case of anodic films grown in H_2SO_4 solution at 70 °C the X-ray patterns showed a crystalline $WO_3 \cdot H_2O$ layer. The SEM pictures showed a spongy ex-

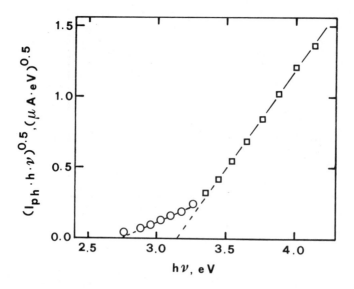

Fig. 3. Determination of the optical band gap of the crystalline layer O , and of the amorphous film underlying the crystalline layer $\square$, for $WO_3 \cdot H_2O$ film obtained by oxidation in 1N H_2SO_4 at T = 70 °C for 1h.

ternal layer having a needle-like structure. The $(I_{ph} h \nu)^{\frac{1}{2}}$ vs. $h\nu$ plot reported in Fig. 3 shows the existence of two straight line regions with intercept at ~ 2.70 eV and 3.15 eV. Also these films showed a large increase in the photocurrent values and some changes in the shape of the photocurrent spectrum as well as in the absorption edge (E_g^{opt} = 2.60±0.05 eV) after annealing under argon at 350 °C.

In order to explain the photoelectrochemical behaviour of these porous layers a schematic picture of the morphology and of the energy band scheme is reported in Fig. 4. According to this scheme the external crystalline layer determines the optical absorption edge while the underlying amorphous film controls the transport of the photogenerated carriers. However, in the case of the spongy hydrated layer a direct contribution to the measured photocurrent from the underlying amorphous film is supposed to be operating. According to this hypotesis an optical band gap of 2.70 eV has been assigned to the crystalline orthorombic $WO_3 \cdot H_2O$. The annealing process affects the photoelectrochemical behaviour of these layers by increasing noticeably the mobilities of the photocarriers in the underlying barrier film which undergoes the amorphous-crystalline modifica-

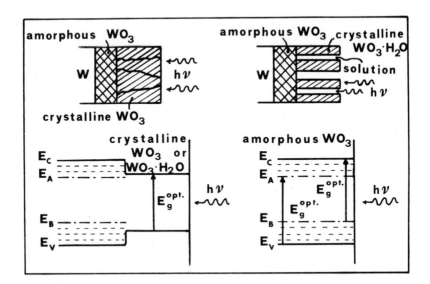

Fig. 4. Energy band schemes and schematic pictures of the morphology of corrosion layers grown on tungsten in different conditions. (E_C - E_A) and (E_B - E_V) represent regions of localized states near the mobility edges E_C and E_V respectively. The optical band gap E_g^{opt} is equal to (E_C - E_B) or (E_A - E_V) whichever is the smaller.

502

tion. This modification also affects the E_g^{opt} values if a direct contribution to the measured photocurrents comes out from the amorphous barrier layer. The decrease in the E_g^{opt} values is due to the fact that the localized states disappear after crystallization making possible the contribution of lower energy optical transitions to the photocurrent.

In the case of hydrated $WO_3 \cdot H_2O$ layer a change in the crystallographic structure as well as in the chemical composition happens during the annealing process owing to the dehydration of the external layer. Both these facts are responsible of the decreasing of the E_g^{opt} value to the measured 2.6 eV.

CONCLUSIONS

We have shown that a careful analysis of the photoelectrochemical behaviour of the corrosion layers on tungsten can give useful informations in order to characterize the chemical-physics nature of the films grown in different conditions. The existence of an amorphous barrier-type anodic WO_3 film underlying the crystalline external layers has been deduced on the basis of the photoelectrochemical response of these duplex structures (amorphous barrier film / porous crystalline layer). Different optical band gaps have been attributed to the different electronic structures as well as to the different crystallographic modifications and chemical compositions.

REFERENCES

1 F. Di Quarto, A. Di Paola and C. Sunseri, J. Electrochem. Soc. 127, 1016 (1980).
2 A. Di Paola, F. Di Quarto and C. Sunseri, Corrosion Sci. 20, 1067 (1980); ibid. 20, 1079 (1980).
3 F. Di Quarto, A. Di Paola and C. Sunseri, Electrochim. Acta 26, 1177 (1981).
4 F. Di Quarto, G. Russo, C. Sunseri and A. Di Paola, J. Chem. Soc. Faraday Trans. I 78, 3433 (1982).
5 N.F. Mott and E.A. Davis, Electronic Processes in Non-crystalline Materials, p. 249, Clarendon Press, Oxford (1971).

INFLUENCE OF ION IMPLANTATION ON THE ELECTROCHEMICAL BEHAVIOUR OF PASSIVE
TITANIUM AND HAFNIUM ELECTRODES

B. DANZFUSS[1], J. W. SCHULTZE[1], U. STIMMING[1], and O. MEYER[2]

[1] Institut für Physikalische Chemie, Universität Düsseldorf, Universitätsstr. 1
D-4000 Düsseldorf (FRG)

[2] Institut für Angewandte Kernphysik I, Kernforschungszentrum Karlsruhe,
D-7500 Karlsruhe (FRG)

ABSTRACT

Passive films of hafnium and titanium electrodes were implanted with Xe^+- and Pd^+-ions. The capacity of the insulating HfO_2-film increases due to an increase of the dielectric constant D, but the capacity remains independent of potential. The rate of electron transfer reactions is enhanced. Implantation of Pd^+ is more effective than that of Xe^+. The anodic photocurrent at HfO_2 is increased by implantation and the spectrum is shifted to lower energies. These measurements demonstrate that the implantation of foreign metal ions yields the expected enhancement of electron transfer reactions, but the lattice perturbations caused by the implantation have a similar and sometimes even stronger influence on the electrochemical behaviour.

INTRODUCTION

Passive titanium as well as passive hafnium electrodes show low electrochemical activity because of the semiconducting (E_g = 3.4 eV) and insulating (E_g = 5.1 eV) character of TiO_2 and HfO_2 passive films. Implantation of Pd^+-ions was carried out to change the electrochemical behaviour of the passive film, since additional electronic terms are introduced into the band gap. Perhaps, a metastable compound can be formed (ref. 1), or the energy bands can be deformed. On the other hand, ion implantation itself is able to induce structural transformations in the substrate like crystallisation (HfO_2), amorphization (TiO_2), and stoichiometric changes ($TiO_2 \rightarrow Ti_2O_3$ for high implantation doses) (ref. 2). This was investigated by implantation of Xe^+. After implantation, the capacity $C(\varepsilon)$, the rate of electron transfer reactions $i(\varepsilon)$, and the photocurrent $i_{ph}(\varepsilon,\lambda)$ were measured in dependence on the electrode potential ε.

EXPERIMENTAL

Titanium (99,6%) and hafnium (99,5%) were anodized galvanostatically (i = 14 mA/cm²) in 0.5 M H_2SO_4 up to an electrode potential ε = 20 V (she). The result was a film thickness d = 49 nm for HfO_2 and some 30 nm for TiO_2 (ref. 3).

Implantation was carried out with ion beams up to 220 keV, successive cycles
leading to (previously calculated) nearly uniform distributions of the implanted
element within the passive film. Implanted elements were Pd and Xe with the following implantation concentrations (= atom-% relative to metal plus oxygen):

Ti/TiO$_2$ 0.1 to 10 % Pd, and 1 % Xe
Hf/HfO$_2$:   0.1 to 10 % Pd, and 5 % Xe

XPS measurements proved the presence of Pd or Xe in the outer part (4 nm) of
the film. Rutherford backscattering was used to check the concentration profile
within the film (±10 nm) and the film thickness (±2 nm). Film sputtering was
negligible during implantation. Implanted electrodes were brighter and had a
more metallic reflectivity than the passive electrodes, which showed only the
interference colours.

RESULTS

On Hf-electrodes, the passive film is   insulating, and the capacity is
small and potential independent. Ion implantation increases the capacity parallel to the implantation concentration (Fig. 1). During the first cycles, the

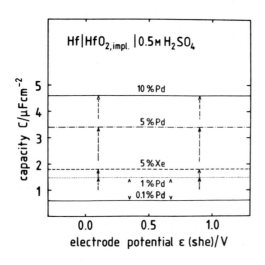

Fig. 1: Electrode capacity of
ion-implanted passive hafnium
(d = 49 nm). The arrows indicate the capacity increase
during the first experiments.

capacity shows a further increase by 30%, which is indicated by the arrows in
Fig. 1. Surprisingly, Xe$^+$ yields a similar increase as Pd$^+$. In both cases, however, the capacity remains potential independent. This means, that the energy
terms of donors or defects produced by implantation must be so low that they
don't contribute to the capacity. Therefore, the increase of capacity can be explained only by an increase of the dielectric constant D from D = 14 (ref. 4) up

to D = 70 and more. This is reasonable since some crystalline oxides (e. g. anatase) or mixed compounds (e. g. titanates) have much larger D-values than the amorphous films of TiO_2 or HfO_2. With passive Ti electrodes, a strong increase of the potential dependent capacity is observed which can be explained by an increase of D and a simultaneous increase of donor concentration.

Measurements in Fe^{2+}/Fe^{3+} containing electrolyte give information about the electronic properties of the passive film, especially whether additional electronic states have been introduced and whether they can participate in the electron transfer process.

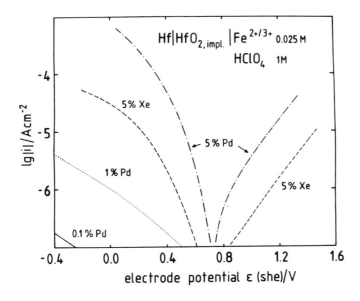

Fig. 2: Tafel plot of current densities of the Fe^{2+}/Fe^{3+} redox system at Hf/HfO_2-electrodes (d = 49 nm) implanted with different amounts of Pd and Xe.

Fig. 2 shows the Tafel plot for the implanted Hf/HfO_2 electrodes. At pure HfO_2 passive films, an electron transfer does not take place and cannot be seen in Fig. 2. But a strong catalysis can be observed for implantations exceeding 1%. Increasing the implantation dose up to 5% causes a further catalysis by a factor of 100. In any case, however, the exchange current density is much smaller than on a metal electrode. Xe^+-implantation yields a similar but smaller electron transfer catalysis. After thermal treatment at 400°C, the redox current becomes smaller by 2 orders of magnitude. This indicates, that lattice defects within the oxide produced by the implantation can contribute to the electron transfer reaction, but their influence is smaller than that of metal ions. From the slope of the Tafel-lines, $b = d\varepsilon/d \lg i$, transfer coefficients $\alpha_+ = 0.2$ and $\alpha_- = 0.1$ to 0.5 (increasing with concentration) are obtained. The result $\alpha_- > \alpha_+$

resembles those for n-type semiconductors (ref. 5).

Similar effects were observed with passive Ti. A change of the transfer co-efficient with increasing overvoltage indicates the dominating resonance tunnel-ling process.

Photocurrent spectra of pure Hf/HfO_2 show a threshold energy of 5.1 eV (ref.4) Ion implantation shifts this value to much lower energies of about 3.5 eV (Fig.3).

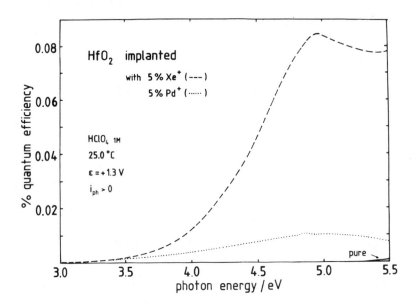

Fig. 3: Quantum efficiency η as a function of the photon energy for pure and ion implanted HfO_2 (d = 49 nm)

There exists a similarity to photocurrent spectra of anodized Pd, but the measu-rement with Xe^+-implanted HfO_2 yields the same effect with a tenfold magnifica-tion. The pronounced effect of the Xe-implantation is even more striking in the potential dependence of the photocurrent, shown in Fig. 4. For all Pd^+-implan-ted electrodes, the anodic current starts at $\varepsilon \geq 0\,V$, but for Xe^+ already at -0.3 V, and the slope $di_{ph}/d\varepsilon$ is much steeper.

There is only a small influence of concentration of Pd^+ on the photocurrent: the curve for 10% equals that of 1%.

DISCUSSION

The measurement of C, i_{ETR} and i_{ph} of the implanted oxide films show that the electronic properties are changed by Pd^+ as well as by Xe^+. For i_{ETR} only, the effect of Pd^+ exceeds that of Xe^+ clearly and indicates a chemical influence of the implanted metal ion. For C and i_{ph}, the change by Xe^+ equals or exceeds that

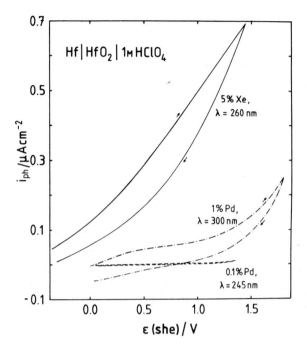

Fig. 4: Potential dependence of the photocurrent of ion implanted Hf/HfO$_2$ at
λ = 300 nm

by Pd$^+$. This means that the number of dislocations produced by implantation is
of the same order of magnitude as the number of introduced Pd$^+$-ions. It can be
assumed that they are located around the foreign atom, since the total number
of dislocations produced along the track should be much larger.

Then, the influence of implantation dose can be explained. For c = 0.1%, the
concentration of dislocations would not be sufficient to change the bulk properties of the oxide such as the dielectric constant and the photocurrent. For
electron transfer reactions, the presumed distance between dislocations
$d_{dis} \sim 1/c^{1/3}$ is too large for tunnelling. Hence, these experiments give the
same result as the blank experiment. For c = 5%, local changes of D up to 100
and even more cannot be excluded. Since d_{dis} decreases with increasing c, resonance tunnelling of electrons becomes possible. The tunnel probability

$$W \approx \exp(-d_{dis}/d_o) = \exp(-1/c^{1/3} d_o')$$

increases with c exponentially. Hence, the strong effect of c shown in Fig. 2
can be understood.

The energetic position of the dislocations cannot be determined unambiguously.
From the photocurrent spectra we know that for HfO$_2$ the excitation energy decreases to 3.5 eV, but we don't know if occupied terms near the valence band or

empty terms near the conduction band are produced, since the capacity measurements give no hint for donors or acceptors. The catalysis of electron transfer reactions via resonance states is most effective for $E \approx E_F$ (ref. 6). Hence, a position at intermediate energies is probable. Further work is necessary to clarify all the effects of ion implantation into oxide films.

ACKNOWLEDGEMENT

The support of this work by the Bundesministerium für Forschung und Technologie is gratefully acknowledged.

REFERENCES

1   G. K. Wolf, Chem. Ing. Tech. 54 (1982) 23
2   H. M. Naguib, R. Kelly, Radiation Effects 25 (1975) 1
3   J. W. Schultze, U. Stimming, J. Weise, Ber. Bunsenges. Phys. Chem. 86 (1982) 276
4   C. Bartels, J. W. Schultze, U. Stimming, M. A. Habib, Electrochim. Acta 27 (1982) 129
5   J. W. Schultze, in: Frankenthal and Kruger (Eds.) "Passivity of Metals" The Electrochemical Soc., Princeton, N.J. (1978) 78
6   W. Schmickler, ibid., 102

Passivity of Metals and Semiconductors, edited by M. Froment
Elsevier Science Publishers B.V., Amsterdam — Printed in The Netherlands

PHOTOELECTROCHEMICAL APPROACH TO PASSIVITY

Ulrich Stimming
IBM, T.J.Watson Research Center, Yorktown Heights, N.Y.10598, U.S.A.*

ABSTRACT

Most passive metals show photoeffects due to excitation processes within the electrode which cause a change in electrochemical behaviour. In most cases light induces electron-hole pair formation in the passive layer which is semiconducting or insulating. Excitation in the underlying metal followed by photoemission is observed as well. The parameters which influence wavelength and potential dependence of the photocurrent are discussed. Typical examples of passive metals which belong to transition metals, noble metals and valve metals are described. From the discussion it can be inferred that photoelectrochemistry could develop to a powerful in-situ method to characterize passive films.

INTRODUCTION

In the understanding of passivity, knowledge of the composition and the properties of the passive film plays a crucial role. Although considerable progress has been reached by the employment of methods like X-ray and electron diffraction and u.h.v. vacuum spectroscopy, results are questionable in some instances because of possible changes of hydrous or amorphous passive films during investigation. Spectroscopic methods to investigate the optical properties of the passive film have the advantage of being in-situ methods, however, they lack a direct connection to electrochemical behaviour. Photoelectrochemistry on the other hand, seems to be a promising method because excitation by light is directly connected to electrochemical parameter such as potential, current and capacity. Thus information can be obtained on the electronic structure of the electrode surface and electrochemical reactions induced by excitation processes in the electrode.

Although photoeffects have been known for a long time (ref.1), till recently, experiments and interpretations were only qualitative and the picture of photoeffects with passive metals was controversial. In the last decade, photoelectrochemistry has made considerable progress because of the recognition that solid state properties of the electrode material play a significant role. In addition, the search of materials suitable for electrochemical solar cells increased the knowledge concerning photoelectrochemistry of semiconductors. On the other hand, a systematic application of photoelectrochemistry on problems of passivity was only possible after it was recognized with passive films as well, that even films with a thickness of $d < 1$ nm exhibit solid state

---

*Permanent Address: Universitaet Duesseldorf, Institut fuer Physikalische Chemie, D-4000 Duesseldorf 1, West-Germany

properties. Photoelectrochemistry then becomes an important tool to investigate the electronic properties of the passive film and, as a consequence, to explain electrochemical reactions on passive electrodes which are influenced by these properties.

Following, some principles of photoelectrochemistry with passive metal electrodes are discussed and, for illustration, typical examples of the photoelectrochemical behaviour of passive metals are described.

THEORETICAL ASPECTS

If the electrode is illuminated by light of a suitable photon energy $h\nu$, electrons can be excited from occupied to unoccupied states within the electrode:

$$h\nu \to e^- + h^+ \tag{1}$$

The electron-hole pair formation according to Eq.(1) is the basic reaction for photoelectrochemical effects. Most commonly these effects are observed as changes in the current or the potential. Depending on the properties of the passive film, different energetic situations can be distinguished for an n-type or p-type semiconducting or an

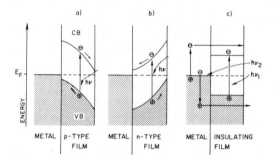

Fig.1: Schematic representation of an illuminated p-type (a) and n-type (b) passive film under depletion conditions and simultaneous excitation in an insulating film by $h\nu_1$ and photoemission by $h\nu_2$ (c).

insulating film. This is shown in Fig.1. Under the condition of a depletion layer in the semiconducting film, which is most likely to measure a photoeffect, the electron-hole pair is seperated in the electric field of the space charge layer. In case of an n-type film (Fig.1a) the hole migrates to the surface and can be transferred to an occupied (donor) state, forming an anodic photocurrent. Such an occupied state represents a reduced chemical species in the electrolyte. For a p-type film (Fig.1b) the situation is reversed. The electron migrates to the surface and reacts with an unoccupied (acceptor) state, the photocurrent is a cathodic one. In the case of an insulating film (Fig.1c) the electric field is constant throughout the film. According to the sign of the field, anodic or cathodic photocurrents are obtained. Reversing the field in the film changes the sign of the photocurrent at the point of zero field in the passive film which is the flatband potential ε_{fb}. Therefore, the flatband potential which is normally accessible from capacity data can also be obtained from photocurrent measurements. In addition to an electron-hole pair formation in the passive film, an excitation in the metal with a subsequent emission of hot electrons or holes has to be considered. If the passive film is thin enough emission into the electrolyte can occur as shown for hole emission in Fig.1c. However, internal photoemission, i.e. into the passive film, can happen as well, as is shown for electron emission in Fig.1c. Photoemission is not necessarily connected to an insulating film but can happen also with n-type and p-type films. In the following, some basic characteristics of the photocurrent are described, particularly wavelength and potential dependence.

Following Gaertner (ref.2) and Butler (ref.3) the photocurrent i_{ph} of a bulk crystalline semiconductor electrode is given by

$$i_{ph} = e\Phi I_o(1 - R)\{1-\frac{\exp{[-\alpha d_{sc}\sqrt{\varepsilon-\varepsilon_{fb}}]}}{1+\alpha L}\} \tag{2}$$

with e, the electronic charge, Φ, the quantum efficiency, I_o, the incident photon flux, R, the reflectivity of the electrode-electrolyte interface and ε the electrode potential.

The absorption coefficient α depends on the photon energy $h\nu$ according to

$$\alpha = A\frac{(h\nu-E_g)^n}{h\nu} \tag{3}$$

with n=1/2 for direct and n=2 for indirect transitions (ref.4).

The Debye length d_{sc} is given by

$$d_{sc} = \sqrt{\frac{\kappa\kappa_o kT}{e^2N}} \tag{4}$$

with κ, and κ_o, the dielectric constant of the semiconductor and the vacuum, respectively, and N, the doping concentration. k and T have the usual meaning.

L is the diffusion length of the minority carriers and is given by

$$L = \sqrt{D'\tau} \tag{5}$$

with D', the diffusion coefficient of the minority carrier and τ, the carrier lifetime .

Eq.(2) is based on the assumption that the charge transfer at the electrode-electrolyte interface is not rate determining and describes excitation in the space charge region of the actual thickness $d_{sc}\sqrt{\varepsilon-\varepsilon_{fb}}$ and in the field free diffusion region of the semiconductor. Under the condition of passive films, in most cases certain simplifications in Eq.(2) and (3) are possible because of the limited oxide thickness d. In considering a crystalline passive film at not too high absorption coefficients, $\alpha L \ll 1$ should hold. Furthermore, at high doping concentrations N with, according to Eq.(4), a small d_{sc} and, in addition, $d_{sc} < d$, $d_{sc}\sqrt{\varepsilon-\varepsilon_{fb}} \ll 1$ holds at not too high potentials. Under these assumptions and given that R is constant, Eq.(2) can be expanded to

$$i_{ph} = const \; \alpha d_{sc}\sqrt{\varepsilon - \varepsilon_{fb}} \tag{6}$$

A $i_{ph}^2(\varepsilon)$ plot should then be linear with different slopes at different photon energies. However, according to ref.5, from a linear $i_{ph}^2(\varepsilon)$ plot the validity of Eq.(6) cannot be concluded; a rate determining charge transfer might yield similar plots.

At constant potential and under the same assumptions as in Eq.(6), Eq.(2) after insert of Eq.(3) changes to

$$i_{ph}h\nu = const \; (h\nu - E_g)^n. \tag{7}$$

With Eq.(7) the band gap of the passive film and the type of transition can be obtained from photocurrent spectra.

In the case of thin films, not all of the incident light is absorbed by the passive film and is reflected at the oxide-metal interface. Assuming that the thickness of the absorbing layer in Eq.(2) is the film thickness d, Eq.(2) changes to

$$i_{ph} = e\Phi I_o(1 - e^{-\alpha d})(1 + R'e^{-\alpha d}) \tag{8}$$

with R', the reflectivity of the metal.

Consider the two limiting cases $R' = 0$ and $R' = 1$. In the first case, one obtains

$$i_{ph} = e\Phi I_o(1 - e^{-\alpha d}) \tag{9}$$

and in the second, the film thickness is doubly counted:

$$i_{ph} = e\Phi I_o(1 - e^{-2\alpha d}) \tag{10}$$

Such an effect of reflections at the oxide-metal interface has to be considered as long as $d < \alpha^{-1}$ and $R' >> 0$. If the reflectivity spectrum $R'(h\nu)$ of the metal is known, it is possible with Eq.(8) to calculate the absorption coefficient of the passive film from photocurrent spectra.

For $R' > 0$ part of the light is absorbed by the metal which can, if the photon energy is high enough, give rise to photoemission processes. The excitation most likely occurs from the Fermi level to unoccupied states or from occupied states to the Fermi level with a subsequent emission of hot electrons or holes, respectively. The quantum yield often follows a Fowler plot (ref.6, but see also discussion of the so-called 5/2 law in ref.7):

$$\Phi^{1/2} = A(h\nu - h\nu_o) \tag{11}$$

where the threshold energy $h\nu_o$ is proportional to the electrode potential

$$hv_o = e\varepsilon + const \qquad\qquad (12)$$

because the height of the barrier is directly related to the position of the Fermi level. With Eqs.(11) and (12) photoemission can be distinguished from excitation in the passive film.

As can be seen from Eq.(8), when $d < d_{sc}\sqrt{\varepsilon - \varepsilon_{fb}}$, the thickness of the absorbing layer in Eq.(2) becomes equal to the film thickness d. According to Eq.(2) then, in contrast to experience, i_{ph} should be potential independent. The reason being that Eq.(2) is derived without consideration of recombination processes in the space charge region. The execess carrier life time τ in Eq.(2) describes only recombination in the field free region. A discussion of different approaches to calculate recombination in the bulk space charge region is given in ref.8. Results are very much dependent on various assumptions made for number, energy and spatial distribution of recombination centers which are necessary to obtain analytical expressions. Commonly, the recombination depends exponentially on the electric field. The potential dependence of the photocurrent therefore changes upon the influence of recombination. Thus photocurrent-potential curves can have varying shapes with a greater potential influence than expressed in Eqs.(2) or (6).

In various cases it had been suggested that the passive film is amorphous rather than crystalline. A quantitative description of photoelectrochemistry with amorphous semiconductors is, however, not yet available. Over all in electrochemistry, there is little experience with amorphous semiconductors as electrode material. In a recent work on iron-titanium oxides some aspects of the impact of non-crystallinity of semiconductors on the electrochemical and photoelectrochemical behaviour are discussed (ref.9), but explanations must be considered tentative. The general feature of amorphous semiconductor is that a considerable number of states exist within the band gap. Their density $D(E)$ is a function of the material and its preparation conditions, as is the band gap. The band edge is no longer where $D(E)$ changes drastically as in crystalline materials but where $\mu(E)$, the mobility, changes by orders of magnitude. Band gap energies of amorphous semiconductors are found to be higher and lower compared to crystalline ones. Absorption by localized states in the mobility gap should occur at relatively low photon energies, but whether the photocurrent is measurable depends on the density of the localized states as well as on the mobility of the carrier. A determination of the type of transition as in Eq.(7) is not possible according to ref.10, since the selection rules are relaxed for amorphous semiconductors. However, in solid state devices, plots according to Eq.(7) yield exponents n between 1 and 3, with 2 being the most common. The result n=2 which corresponds to indirect transitons in a crystalline material is related to the change in density of states $D(E)$ at the band edge (ref.11). In photoelectrochemical measurements, for iron-titanium oxides of the composition $Fe_{0.5}Ti_{0.5}O_{1.75}$, n=0.5 was found for the amorphous and n=2 for the crystalline state (ref.9).

According to ref.12, the photocurrent in amorphous semiconductors can depend exponentially on the electric field. In solid state devices, a relation

$$\log i_{ph} = const \sqrt{E'} \tag{13}$$

is often found at high fields $E' > 10^4 V/cm$. This is explained by an influence of the field according to a Poole-Frenkel mechanism. However, other effects such as space charge limited currents yield similar results. For photoelectrochemical measurements, such a dependence would be a quite different one than that given by Eq.(2) or (6). For amorphous iron-titanium oxides (ref.9) a linear $\log i_{ph}(\sqrt{\varepsilon})$ relation, similar to Eq.(13) was found, which demonstrates that the $i_{ph}(\varepsilon)$ relation with amorphous semiconductors is different from Eq.(2). An unambiguous reasoning will only be possible after more experimental results and more theoretical treatment.

In electrochemical systems, a measured current always corresponds to a chemical reaction at the interface. Different possibilities of electrochemical reactions induced by a band to band excitation in the passive film are:

- electron transfer reaction with a suitable redox system in the electrolyte

$$Red + h^+ \rightarrow Ox \tag{14}$$

or

$$Ox + e^- \rightarrow Red \tag{15}$$

- electron transfer reaction with H^+ or OH^- in the electrolyte,

$$2H^+ + 2e^- \rightarrow H_2 \tag{16}$$

or

$$2OH^- + 2h^+ \rightarrow O_2 + 2H^+ \tag{17}$$

- a reaction with the film itself, like

film reduction to a lower valent oxide or to the metal

$$MeO_x + 2xe^- + 2xH^+ \rightarrow Me + xH_2O, \tag{18}$$

photocorrosion or

photoinduced film growth.

Photoemission caused by an excitation in the metal leads to a reaction of hot electrons or holes with a scavenger in the electrolyte or with water itself (ref.13).

INSTRUMENTATION

The instrumentation for photoelectrochemical measurements is relatively simple. Normally only a monochromatic light source has to be added to the usual electrochemical equipment. This can be either a polychromatic source like Xe or W lamp together with a monochromator or a tunable laser. If the photoeffects are large enough the photoresponse can be measured directly. This can be done for example with slowly chopped light (0.1 Hz), recording dark and photocurrent (or -potential) at the same time. Under similar conditions, photoresponse transients can be measured which may give useful information on the nature of the photoelectrochemical reaction or possible recombination reactions. In many cases, however, photoeffects are small and modulation techniques have to be employed. The light beam is chopped and the modulated signal is amplified with reference to the chopper frequency by a lock-in amplifier. Under stable electrochemical conditions photocurrents down to some ten pA can be detected. In most cases the light source has no flat characteristic and it is necessary to correct for the intensity change with wavelength. Calibration curves of the light source are determined either by a detector of known spectral sensitivity or by one of a flat response. The latter is usually a thermopile or a pyroelectric detector. The knowledge of the number of incident or, if possible, absorbed photons, allows for a calculation of the quantum yield.

SOME EXAMPLES

Following are some examples of systems which are typical in their photoelectro-chemical behaviour. A distinction will be made between transition, noble and valve metals. Essentially, more recent publications are considered. For older measurements, a comprehensive bibliography exists in an review article by Kuwana (ref.14).

Transition Metals

Passive transition metals are the most investigated because the oxides which form the passive film normally exhibit absorption of visible light. Recently, measurements with passive bismuth (ref.15,16,17), tungsten (ref.18), tin (ref.19), mercury (ref.20) are reported.

Passive nickel was investigated at pH 6.9 - 9.0 (ref.21). The band gap of the p-type NiO passive film was determined to 3.1 eV which is considerably lower than the value of the bulk material with 3.6 (ref.22). The flatband potential obtained from the change in sign of the photocurrent was ca. 0.1 V higher than values from capacity measurements.

Recently, several investigations of passive iron were performed (ref.21,23,24,25). The obtained picture is not conclusive. Band gap energies between 1.9 and 2.5 eV are reported. Photocurrents in neutral and alkaline solutions are anodic transient like with a stationary cathodic value (ref.25). This suggests two independent photoexcited processes. The potential dependence of the photocurrent also shows a complex behaviour.

The first more detailed photoelectrochemical measurements with copper were carried out to study the influence of inhibitors on the corrosion behaviour of copper (ref.26). While the occurrence of anodic photocurrents is still a matter of discussion, cathodic photoeffects in current and potential are observed in neutral and basic solutions after reduction of the $CuO/Cu(OH)_2$ layer. In a cyclic voltammogram, the reduction peak of the Cu_2O film is enhanced which indicates that the reduction of p-type Cu_2O is the photoinduced reaction. The spectral distribution of the photocurrent is shown in Fig.2. It displays a pronounced peak structure with several smaller peaks superimposed. This is similar to the absorption spectrum of Cu_2O (ref.27)(The band gap of the passive film is ca. 2.4 eV. Using the absorption data in ref.27 and the

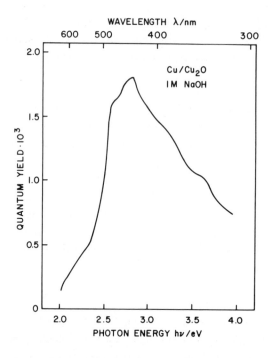

Fig.2: Photocurrent spectrum of passive copper in 1M $NaOH$ after $CuO/Cu(OH)_2$ reduction.

data in Fig.2, a maximum quantum efficiency (related to the absorbed photons) of ca. 20 % is estimated. This is a fairly high value, but it is comparable to that obtained for oxide covered gold (ref.28).

Noble Metals

Noble metals like silver (only at higher pH), gold and platinum group metals are known to form oxide films at potentials some 0.1 V below oxygen evolution. Although they are not passive metals in the sense that the oxide film prevents dissolution of the metal, the description of the behaviour of their respective thin oxide films by means of photoelectrochemistry is similar.

Photocurrents with oxide covered silver electrodes in 1M KOH were assigned to an n-type semiconducting behaviour of the Ag_2O film (ref.29). The photocurrent is anodic during Ag_2O formation and cathodic during Ag_2O reduction. AgO which is formed at higher potentials does not show direct photoeffects.

Measurements with platinum electrodes show a complex behaviour to the illumunation with u.v. light (ref.30): The so-called α-oxide shows an anodic photocurrent which decreases the oxide thickness during illumination which was controlled by simultaneous ellipsometrical measurements. With the formation of so-called β-oxide at higher potentials, the photocurrent changes to a cathodic one. It was concluded that different, n-type and p-type, oxide films are formed on platinum electrodes.

Oxide films on gold electrodes absorb visible light of an energy $h\nu > 2.8$ eV (ref.13,28). The spectra are hardly dependent on the electrode potential between 0.9

and 1.5 V (rhe). In this range the thickness of the oxide is d = 0.5-1 nm. At high photon energies a deviation is observed which is assigned to photohole emission. In Fig.3 a Fowler plot according to Eq.(11) is shown. Straight lines are obtained at

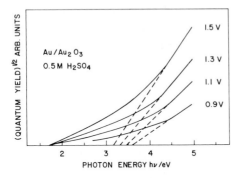

Fig.3: Fowler plot of photocurrent spectra with oxide covered gold electrodes, data are taken from ref.13.

$h\nu$ > 4.3 eV. The intercepts with the x-axis are potential dependent according to Eq.(12). The extrapolation at low photon energies leads to the common value 2.8 eV. Gold is a good example of a simultaneous excitation in the superficial oxide film and the underlying metal. It corresponds to Fig.2c with an n-type semiconducting film and a hole emission from the metal into the electrolyte.

Valve Metals

Wavelength and potential dependence of the photocurrent are reported for hafnium covered with passive films of 8 - 22 nm thickness (ref.31). The estimated band gap is 5.1 eV which is lower by 0.4 eV compared to the value of bulk HfO_2. The photocurrent increases almost linearly with ε. This changes to an exponential-like increase after Xe ion implantation of thicker passive films (ref.32). Ion implantation also changes the spectrum: the photocurrent onset is lowered to ca. 3.5 eV. This is probably due to structural changes and an increased disorder in the passive film and is discussed in detail in ref.32.

Passive titanium is a widely investigated system (ref.33-37). For passive films of a thickness d > 20 nm, the basic characteristics of TiO_2 are generally found. Measurements with thinner films, e.g. d < 10 nm, show considerable differences (ref.37). The spectrum is shifted to shorter wavelengths by ca. 0.4 eV with indirect transitions at 3.4 and direct at 3.9 eV. The potential dependence of the photocurrent also changes from a type similar to Eq.(2) to an almost linear dependence. The influence of the film thickness is remarkable: at constant electrode potential the photocurrent decreases with increasing film thickness. This was explained by a strong influence of the electric field in the film on the recombination (ref.37). In Fig.4 the change of the photocurrent with film thickness is shown, however at constant field strength E'. This shows the expected increase due to higher absorption with film thickening. The strong effect of the field

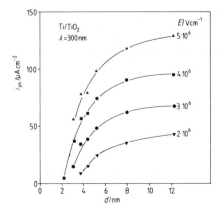

Fig.4: Photocurrent as a function of passive film thickness on titanium electrodes in 1M $HClO_4$ at different field strengths E' taken from ref.37.

on the photocurrent shows that recombination plays a mayor role. A high number of localized states are present in the band gap and the film is most likely amorphous.

Tantalum which is known to form amorphous passive films shows photoeffects upon irradiation with u.v. light (ref.38). It has been shown that the illumination stimulates oxide growth. The photogrown oxide, however, shows properties different from the original film; it is probably porous and/or strongly hydrated. No conclusive

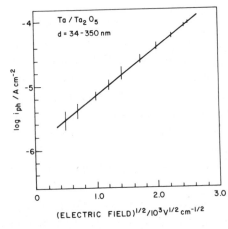

Fig.5: Field dependence of the photocurrent with passive tantalum plotted according to Eq.(13); data taken from Fig.10.03 in ref.38.

explanation is given in ref.38 but from the experimental conditions and described results it might be inferred that the photoinduced holes at the surface cause corrosion and the additional oxide is formed by a dissolution-precipitation mechanism. The photocurrent shows an overproportional change with the potential. Using the data in ref.38 and replotting it according to Eq.(13) in a $\log i_{ph}$ vs. $\sqrt{E'}$ plot yields a straight line. This is shown in Fig.5. The bars indicate that there is little variation with film thickness in the range $d = 34$-350 nm. This behaviour is in accordance with the finding of linear $\log i_{ph}(\sqrt{E'})$ plots for other amorphous oxides or passive films (ref.9,39).

CONCLUSIONS

Although the theoretical basis of photoeletrochemistry with passive metals is not yet complete, especially in the description of amorphous films, it can be assumed from the above description that photoelectrochemistry has a good potential to develop to a powerful tool in in-situ studies of thin passive films. The sensitivity is high enough to analyze films of less than 1 nm thickness. In contrast to other spectroscopic methods contributions from the underlying metal are low or easy to discriminate. Measured

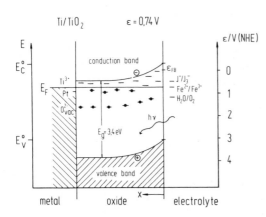

Fig.6: Model of the passive film on titanium in 1M $HClO_4$, taken from ref.40.

quantities like photocurrent or photopotential are directly related to electrochemistry. From photocurrent spectra the band gap and from the sign of the photocurrent and its potential dependence the conduction type (n- or p-type) can be concluded. These informations allow for a sketch of conduction and valence band in respect to each other and to the electrolyte. This is shown in Fig.6 for the example of passive titanium (ref.40). Such a diagram helps to interpret electrochemical behaviour which is related to the electronic properties of the passive film.

REFERENCES
1   E.Becquerel, C.R.Acad.Sci (Paris) 9 (1839) 561
2   W.W.Gaertner, Phys.Rev. 116 (1959) 84
3   M.A.Butler, J.Appl.Phys. 48 (1977) 1914
4   R.H.Bube, Photoconductivity of Solids, New York 1967
5   W.J.Albery, P.N.Bartlett, A.Hamnett and M.P.Dare-Edwards,
       J.Electrochem.Soc. 128 (1981) 1492
6   J.K.Sass and H.Gerischer in B.Feuerbacher et al. (Eds.),
       Photoemission and Electronic Properties of Surfaces, New York 1978
7   Yu.Ya.Gurevich, Yu.V.Pleskov and Z.A.Rotenberg,
       Photoelectrochemistry, New York and London 1980
8   D.Haneman and J.F.McCann, Phys.Rev.B 25 (1982) 1241
9   B.Danzfuss and U.Stimming, J.Electroanal.Chem. submitted for publication
10 N.F.Mott and E.A.Davis,
       Electronic Properties in Non-crystalline Materials, Oxford 1971
11 J.Tauc in F.Abeles (Ed.), Optical Properties of Solids, London and New York 1972
12 E.A.Davis, J.Non-cryst.Solids 4 (1970) 107
13 T.Watanabe and H.Gerischer, J.Electroanal.Chem. 122 (1981) 73

520

14 T.Kuwana in A.J.Bard (Ed.), Electroanalytical Chemistry Vol.1, New York 1966
15 D.E.Williams and G.A.Wright, Electrochim.Acta 24 (1979) 1179
16 M.Metikos-Hukovic, Electroctim.Acta 26 (1981) 989
17 L.M.Castillo and L.M.Peter, submitted for publication
18 B.Reichman and A.J.Bard, J.Electrochem.Soc. 126 (1979) 2133
19 S.Kapusta and N.Hackerman, Electrochim.Acta 25 (1980) 1001
20 M.I.Da Silva Pereira and L.M.Peter, J.Electroanal.Chem. 131 (1982) 167
21 S.M.Wilhelm and N.Hackerman, J.Electrochem.Soc. 128 (1981) 1668
22 M.P.Dare-Edwards, J.B.Goodenough, A.Hamnett and N.D.Nicholson,
     J.Chem.Soc.,FaradayTrans. 77 (1981) 643
23 M.Froelicher and A.Hugot-le Goff,
     Extended Abstracts of 32nd I.S.E. Meeting, Cavtat, Yougoslavia 1981
24 L.M.Abrantes and L.M.Peter, J.Electroanal.Chem. in press
25 U.Stimming, Thin Solid Films this volume
26 W.Paatsch, Ber.Bunsenges.Phys.Chem. 81 (1977) 645
27 S.Brahms and S.Nikitine, Solid State Communications 3 (1965) 209
28 T.Watanabe and H.Gerischer, J.Electroanal.Chem. 117 (1981) 185
29 R.S.Perkins, B.V.Tilak, B.E.Conway and H.A.Kozlowska,
     Electrochim.Acta 17 (1972) 1471
30 Yu.Ya.Vinnikov, V.A.Shepelin and V.J.Veselowskii,
     Elektrokhimiya 10 (1973) 1557
31 C.Bartels, J.W.Schultze, U.Stimming and M.A.Habib,
     Electrochim.Acta 27 (1982) 129
32 B.Danzfuss, J.W.Schultze, U.Stimming and O.Meyer, Thin Solid Films this volume
33 W.Paatsch, Ber.Bunsenges.Phys.Chem. 79 (1975) 694
34 D.Laser, M.Yamir and S.Gottesfeld, J.Electrochem.Soc. 125 (1978) 358
35 J.C.Pesant and P.Vennereau, J.Electroanal.Chem. 106 (1980) 103
36 J.F.McAleer and L.M.Peter, Faraday Discuss. 70 (1980) 67
37 J.W.Schultze, U.Stimming and J.Weise, Ber.Bunsenges.Phys.Chem. 86 (1982) 276
38 L.Young, Anodic Oxide Films, London and New York 1961
39 U.Stimming, in preparation
40 U.Stimming in F.v.Sturm (Ed.),
     Dechema Monographien Vol.90, Weinheim and New York 1981

Passivity of Metals and Semiconductors, edited by M. Froment
Elsevier Science Publishers B.V., Amsterdam — Printed in The Netherlands

PASSIVATION AND SELECTIVE ETCHING OF III-V N-TYPE SEMICONDUCTORS UNDER POTENTIAL AND PH CONTROL IN H_2O_2 MEDIA

P. CLECHET[1], E. HAROUTIOUNIAN[2], D. LAMOUCHE[1], J.R. MARTIN[1] and J.P. SANDINO[2]

[1]Laboratoire de Physicochimie des Interfaces, Ecole Centrale de Lyon, 69131 Ecully Cedex (France)

[2]RTC - La Radiotechnique-Compelec, Laboratoire de Chimie, 92153 Suresnes (France)

ABSTRACT

The corrosion rate of Ga As, Ga As Al and Ga As P monocristals under controlled potential, in aqueous H_2O_2 media, has been measured with the help of an analytical automatic system which directly gives the dissolved gallium concentration versus potential. Voltamperometric, impedance and ellipsometric measurements prove that passivation which soon occurs in neutral media under large cathodic polarisation, results from the growth of a protective amphoteric oxide layer which appears to be reversible with reference to the polarisation.

INTRODUCTION

The control of the corrosion/passivation mechanism of semiconductor materials in liquid media can lead to the development of processes which are of interest in the electronic industry: cleaning of surfaces, selective etching of heterostructures, passivating oxide layers growth... Voltamperometry, even combined with other optical or electrical method, oftenly doesn't provide enough information for such studies because of the lack of electric current which is frequently observed during semiconductor corrosion. Accordingly, the direct chemical analysis of the corrosion products appears to be a valuable complementary method to elucidate some mechanisms or to choose working conditions. An exemple of such studies, involving Ga As, Ga Al As and Ga As P, is given below.

EXPERIMENTAL METHODS

A microelectrochemical cell, with continous electrolyte flow (4 ml min^{-1}) is fitted with a classical three-electrode system. The colorimetric analysis of gallium in the solution, which results from the corrosion of the semiconductor electrode, is performed, at successive preselected potentials with an Autoanalyser II Technicon (Ga-Rhodamine complex). Conversion of gallium concentrations into current is made by assuming a six-hole corrosion process (ref.2). We use, in figures 1 and 3, $I = f(v)$ and $I_{Ga} = f(v)$ respectively for current-voltage and gallium current-voltage curves. The choice of the complexing electrolyte and of the corrosive agent has been dictated by their common use in the semiconductor industry. All potentials are expressed with respect to the saturated calomel electrode (SCE). Between two consecutive voltage steps, etching of the semiconductor wafers was carried out automatically with the help of an H_2SO_4/ H_2O/H_2O_2 (3:1:1) solution. The ellipsometer and the capacitance measurement apparatus, automatically monitored by a microcomputer, are described elsewhere (ref.1).Flat band potentials(V_{fb}) are obtained from MOTT-SCHOTTKY plots calculated by using a series equivalent circuit at 1 kHz (Fig.2). In the absence of H_2O_2, their values are perfectly reproducible and nearly frequency independent, whatever the nature of the electrolyte be. They only depend on the pH (60mV/ pH unit). In the presence of H_2O_2, the same measurements are not possible owing to gas evolution and changes in Ga As surface (corrosion). The III-V samples were <100> single cristals Si doped (Ga As, $\sim 10^{18}$ cm^{-3}, $Ga_{0.7}$ $Al_{0.3}$ As, $\sim 2.10^{17}$ cm^{-3}) or Te doped (Ga $As_{0.6}$ $P_{0.4}$, $\sim 4.10^{16}$ cm^{-3}).

DISCUSSION

As illustrated by the results obtained with Ga As at pH 2 shown in Fig.1, the corrosion of the three III-V n-type semiconductors considered here starts at a potential a little more positive than their flat-band potential (Fig.2), in a region where the faradic current is still cathodic. Then it reaches a plateau when this current disappears and, lastly, sharply increases as the avalanche breakdown raises. Thus, there is a large range of potential (about 0.6 volt) where corrosion of these n-type semiconductors proceeds in the absence of any current and where, in accordance with our above assessment, chemical analysis must supply electrochemical measurements to detect corrosion.

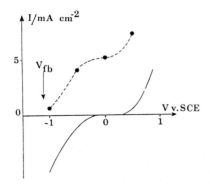

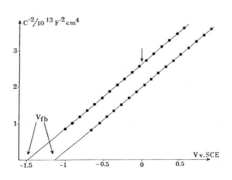

Fig.1 I_{Ga}=f(v)(--•--) and I=f(v) (——) for n-type Ga As in 1% H_2O_2, 0.1 M citric acid medium (pH 2).

Fig.2 MOTT-SCHOTTKY plot for a freshly etched "clean" surface of n-type Ga As in 0.1 M H_3PO_4 et pH 7 (-•-) and 0.1 M citric acid at pH 2 (-■-).

We may discuss the mechanism by which the corrosion of these n-type III-V semiconductors by H_2O_2 occurs, without a net exchange of electric charges between the corroded material and the corrosive solution, with reference to the results obtained with n-type Ga As at pH 2 (Fig. 1 and 2). It is well known (ref.3) that the reduction of H_2O_2 is a two-step electronic process and that the first of these two steps, which leads to an intermediate OH• radical, requires an electron of high energy (more than 0.5 volt). Owing to the value of the flat-band potential of Ga As measured at pH 2 in the absence of H_2O_2 (-1.1 volt, Fig.2) it is clear that the electrons at the bottom of the conduction band edge are energetic enough to induce this reduction. A direct proof of the feasibility of such a transfer lies in the observed current-doubling process obtained, under illumination with p-type Ga As in the same medium and in the same range of potential (ref. 1 and 10). In fact, a mechanism which would thus involve a first electrochemical charge transfer step for H_2O_2 reduction must be rejected for two obvious reasons. The first is that such a donation of free electrons from the conduction band to H_2O_2 might be compensated by an equal injection of electrons in the valence band, a process which cannot lead to the oxydation of the semiconductor. The second is that the conduction band of Ga As is depleted in this range of potential more positive than V_{fb}, even if the band edges of this material are lightly shifted downward during its corrosion. The reverse electrochemical mechanism, which would implicate holes injection in the valence band, a process which effectively can induce corrosion, coupled with electrons injection in the conduction band must also be abandonned for two reasons. The first, which concerns electrons injection in the conduction band, is the absence of high energy donor levels in the strong oxidizing solution (V redox $\sim$ 1.5 volt). The second is the lack of cathodic current observed in the dark with p-type Ga As (ref.1 and 10) which clearly reveals that holes injection in the valence band of Ga As is likewise not possible.

Thus, it appears that this kind of corrosion of n-type Ga As by H_2O_2, in the absence of current, is not an electrochemical process involving equal anodic and cathodic current densities but is more probably the result of a direct chemical reaction between Ga As and H_2O_2. This chemical corrosion is only interrupted at low potential ($V < V_{fb}$, Fig.1), when the rate of H_2O_2 reduction at the surface of Ga As by the electrons of the conduction band, which are then numerous, is high enough to prevent its corrosive effect.

The question which then arises is the origin of the bonding electrons which are given to H_2O_2 during this chemical corrosion. As these electrons must be more energetic than the bonding electrons of the bulk semiconductor (valence band, see before) and as they cannot be donated by the empty conduction band, they are likely given by occupied surface states lying in the forbidden band gap. This is supported by the high density of such states which are currently measured at the surface of Ga As (ref. 4 and 5), especially after acidic H_2O_2 treatment (ref.6). These surface states do correspond to electrons belonging to weakly bounded superficial atoms which are thus continuously renewed as the corrosion proceeds (ref.7). Refilling of these surface states by free electrons from the conduction band is thus unneeded. According to this mechanism, the first electron transfer from such a surface atoms to an adsorbed H_2O_2 molecule would give a short lived adsorbed OH• radical which, in turn, would be in a favourable situation to capture a second electron from the same partially boun-ded surface atom. The result would be a progressive removal of material accom-panied by the total reduction of H_2O_2 into hydroxyle ions.

As the pH (adjusted with NaOH) increases, this etching process begins at potential more and more negative, in accordance with the observed evolution of the flat-band potential (fig.2 and 3).

More important, it is observed that the shape of the $I_{Ga}=f(v)$ curve is stron-gly pH dependent, the stricking point being the appearance of a pronounced peak in neutral medium. This peak reveals a strong passivation of the semiconductor (fig.3).

The formation of an amphoteric oxide layer at the surface of Ga As, at high voltage, which would protect this material against a further dissolution is pro-ved by the capacitance measurements of fig.4, combined with ellipsometric obser-vations. These capacitance measurements were periodically and automatically performed after a fast shift of the voltage into a region where the faradic current is negligible (0.0 volt, Fig.4).

At - 1.4 volt, i.e. on the left side of the activation/passivation peak, the capacitance (curve 1, Fig.4) decreases to a constant value of 2.10^{-7} F cm^{-2} which corresponds to the highly reproducible value obtained at the same voltage, but in the absence of H_2O_2, during the determination of the MOTT-SCHOTTKY plot at the same pH (c.f. the arrow of Fig.2). Since it was impossible to detect the presence of an oxide by in-situ ellipsometry at this potential, we conclude that this value of the capacitance must correspond to a "quasi clean" surface, although it surely cannot be considered as completely exempt of oxygenated spe-cies.

On the contrary, at - 0.75 volt, i.e. on the right side of the peak, the series capacitance first increases, then decreases (curve 2, Fig.4), whereas in-situ ellipsometry clearly shows the growth of an oxide layer which explains the Ga As passivation at this potential.

Thus, the control of the potential in neutral H_2O_2 medium allows either to dissolve or to passive the semiconductor. As the peaks corresponding to Ga Al As and Ga As P appears on the left of the Ga As peak (Fig.3), it is moreover

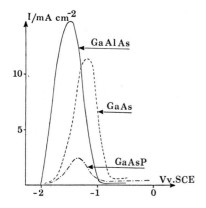

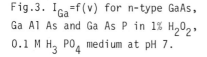

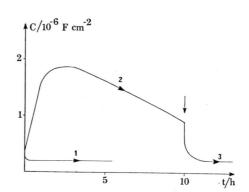

Fig.3. $I_{Ga}=f(v)$ for n-type GaAs, Ga Al As and Ga As P in 1% H_2O_2, 0.1 M H_3 PO_4 medium at pH 7.

Fig.4. Series differential capacitance measurements (1 kHz) on n-type Ga As in 1% H_2O_2, 0.1 M H_3 PO_4 medium at pH 7 at V=-1,4 volt (1),V=-0.75 volt (2) followed by a back-shift (vertical arrow) to - 1.4 volt (3).

possible, by potential control of an heterostructure, to protect the formers against corrosion while Ga As dissolves.

Preliminary experiments on Ga As/ Ga Al As heterostructure at - 1.2 v show that Ga As quickly dissolves leaving a very smooth mirrorlike Ga As Al surface. The same result is obtained for a Ga As/ Ga As P heterostructure, but the remaining Ga As P surface is less smooth. The sharpness of the Ga As wall and the extend of the undercutting have not been checked because of their known dependence on the H_2O_2/acid composition (ref.8 and 9).

The series capacitance of a Ga As wafer covered by a passivating oxide obtained at - 0.75 volt quickly turns-back to the value corresponding to a "clean" surface when it is maintained at - 1.40 volt (curve 3, Fig.4). The potential controlled growth and dissolution of the passivating layer being absolutely reversible could thus be used for cleaning semiconductor surfaces. This could be of interest for component technology.

This work has been supported by the french -DRET.

REFERENCES

1 E. Haroutiounian, J.P. Sandino, P. Cléchet, D. Lamouche and J.R. Martin, J. Electrochem. Soc. (accepted, in press) and P. Cléchet, J. Joseph, A. Gagnaire, D. Lamouche, J.R. Martin and E. Verney, this volume.
2 W.W. Harvey, J. Electrochem. Soc., 114 (1967) 473.
3 R. Memming, in A.J. Bard (ed.), Electroanalytical Chemistry, Vol. 11, M. Dek-

526

    ker Inc, 1979, pp. 1-84.

4 W.E. Spicer, J. Lindau, P. Skeath and C.Y. Lu, J. Vac. Sci. Technol., 17 (1980) 1019.

5 E. Kamieniecki and G. Cooperman, J. Vac. Sci. Technol., 19 (1981) 453.

6 A. Munoz-Yague, J. Piqueras and N. Fabre, J. Electrochem. Soc., 128 (1981), 149.

7 A. Heller in A.J. Nozik (ed.), Photoeffects at semiconductor/electrolyte interface, ACS Symp. n°146 Washington, 1981, pp. 57-77.

8 D.W. Shaw, J. Electrochem. Soc., 128 (1981) 874.

9 E. Kohn, J. Electrochem. Soc., 127 (1980) 505.

10 H. Gerischer, N. Müller and O. Haas, J. Electroanal. Chem., 119 (1981) 41.

Passivity of Metals and Semiconductors, edited by M. Froment
Elsevier Science Publishers B.V., Amsterdam — Printed in The Netherlands

ROUND TABLE DISCUSSION

SEMI-CONDUCTORS - PHOTOELECTROCHEMISTRY

A. OTTO : I would like to ask a question regarding the approach of Mme Hugot
and of U. Stimming. The optical properties of a layer may change as a function
of potential for instance due to the electrical field inducing transformation.
We must be careful at using this technique as a finger-prints method.

U. STIMMING : There is no clear indication of change of the edge with potential
for instance with Au. A threshold can appear in the higher energy region showing
a change of a factor 2 with respect to the potential. I did not consider it is
a finger-print method. At least with Cu, the spectrum of Cu_2O and of the layer
are similar.

A. HUGOT : Two cases to be separated.
1) Photo electrochemical reaction = absorption edge can be potential dependent
if a modification of the illuminated material occurs.
2) No reaction for instance TiO_2 = absorption edge not dependent of potential but
a little dependent on crystallography (0.1 - 0.2 eV).

S.R. MORRISON : What is the value of a band model for films of some Å ? A two
dimensional structure of the surface is to be considered. Don't we have different behaviours in directions parallel and normal to the surface ?

U. STIMMING : I did not claim to have a band structure. In most case we have an
amorphous material and the problem is mostly the difference between a crystalline and an amorphous layer. You have atomic effect parallel and perpendicular
to the surface but no indication on the angle of incidence.

ELECTROCHEMICAL STABILITY OF SEMI-CONDUCTORS

B. CAHAN answering to S.R. MORRISON : Some years ago, we found at small inci-
dence glancing angle, I would say a glare effect fairly important.
I would like to emphasize that electrons are physical entity and holes are not.
In the chemical sense, in the oxide where many defects create band of defects

which are mobile and induce time effects. Under illumination we observe a "battery effect" and iron is a very poor battery material. The band model is a mean of conceiving nothing more. The behaviour is dominated by chemical effects. In fact, we have a bad insulator, several orders of magnitude below the semi-conductor.

C. D'ALKAINE : I suggest that we have an insulator in the direction normal to the film and a semi-conductor parallel to the film. Can somebody answer the question ?

B. CAHAN : In many case collective excitation of electron such as plasmons in metals give resonance parallel to the surface.

W. SCHULTZE : We have to describe difference between bulk crystalline and amorphous films. Does exist differences in the chemical properties between crystalline and amorphous oxides ? We obtained results showing only little difference (some percent). In some examples gap in the film exceed gap in the oxide.

F. DI QUARTO : I would like to ask N. Mott in the case of valve metals the defects are long range disorder, what is the relation between defect and band gap ?

N. MOTT : In the case of SiO_2, there is almost no difference. All the material we talk about are glasses. There is small telling about the gap for well defined defect. H creates states in the gap.

A. HUGOT to B. CAHAN : In your semi-conductor model, how can you explain the similarity of the photoresponse of the film with that of pure magnetite ?

B. CAHAN : I have not done any measurement. Mossbauer experiments lead to Fe_3O_4 thickness of the order of 0.1 monolayer. Some of the atomic transition of Fe_3O_4 are similar to Fe_2O_3. Band gap transition may lead to conclude at Fe_3O_4 instead of Fe_2O_3.

U. STIMMING : You claimed that Fe with an electron missing is more mobile than an electron.

B. CAHAN : Apparent Fe is moving not Fe itself. H^+ moves as a hopping hole. We have a donor concentration 10^{20} - 10^{21} with extremely low mobility.

U. STIMMING : There is no difference with what I call hopping conductor.

S. GOTTESFELD : It is a matter of terminology.

R.P. FRANKENTHAL : I ask Agius's elegant poster. If you take a clean Si surface, what sort of experiment is planned to investigate the crystalline layer grown at first or before the amorphous layer develop.

B. AGIUS : It is difficult to give more informations than those shown on the poster. When we dissolve a 20 nm thick oxide film, the entire film is removed. But, if the dissolution starts from a 4 nm thick oxide, we cannot remove the whole oxide : we have always about 1 nm oxide which remains. As a matter of fact RHEED observations made by M. Froment have shown that thick oxides (20 nm) are amorphous while thinner (4 nm) are crystalline.

M. FROMENT : XPS shows a change of Si peak which can be explain by Si-O tetrahedron angle. Electron diffraction pattern does not correspond to a known Si oxide.

J. YAHALOM : I ask B. Agius on Si oxidation. What is the mechanism, molecular or atomic oxygen transport ?

B. AGIUS : Mainly molecular.

N. MOTT : The oxidation rate is proportional to O_2 pressure.

S. RIGO : I think it is molecular oxygen. Recent results show a power law $(PO_2)^{0.6 \text{ or } 0.7}$ for the linear constant rate k_L. It could be interpreted by a decomposition of O_2 in a non-equilibrium regime. At thin film the reaction changes to a higher rate regime.

N. SATO to S.R. MORRISON : You talked about Si oxidation and also protection of GaAs by layers. What is the concept used for the passivation of semi-conductors?

S.R. MORRISON : It is a problem of semantic.

N. SATO : There are many definitions in electrochemistry. What about semi-conductors ?

S.R. MORRISON : Growing a very thin film to protect the material. Have competition for holes not to corrode the material. We can also use polymers layers, inhibitors or stabilizers (more popular word).

530

E. IRENE : We dont use the word with the same sense. We are making a non reactive, stable surface in semi-conductor devices.

D.E. WILLIAMS : A. Hugot said : we have the same spectrum. We have to keep in mind that Mossbauer and Raman give the structure, photoresponses are related to minor species so that one can observe the same spectrum with quite different structures.

A. HUGOT : It is a problem of long or short range order.

D.E. WILLIAMS : If you prepare two crystals of iron oxide with different small amounts of impurities, you get the same Mossbauer or Raman, the photoresponse spectra are very different.

Passivity of Metals and Semiconductors, edited by M. Froment
Elsevier Science Publishers B.V., Amsterdam — Printed in The Netherlands

PASSIVATION AND LOCALIZED CORROSION OF STAINLESS STEELS

B. BAROUX
Ugine-Aciers Research Center 7340C Ugine France

ABSTRACT

A review of the recent studies on composition and structure of stainless steels passive layers is presented.

The passivation is described as the result of a deprotonization process occuring in an adsorbed water layer. The effects of bound water, chloride ions and alloying elements are discussed.

The various types of localized corrosion in "quasi-neutral" media are also reviewed. Their common propagation mechanism and their specific initiation mechanisms are described.   .

The difficulty to assess the "true" pitting potential is highlighted, according to the stochastic character of the initiation of pits ; a pseudo pitting potential is defined, related to the "elementary pitting probability" function. Finally the effect of non-metallic phases on the resistance to pitting corrosion of industrial stainless steels is emphasized.

1 . INTRODUCTION

Stainless steels are iron base alloys with different amounts of Chromium (more than 12 %), Ni and/or Molybdenum. Their corrosion resistance in most media is based on the existence a superfical passive layer. When the steel is well adapted to the application, this layer remains stable with respect to the chemical species present in the agressive medium and is also able to regenerate itself from the base metal in case of an accidental breakdown. These two stability requirements, however, do not exclude some local breakdowns, when certain conditions are met. This last case appears indeed a fundamental feature of the corrosion of stainless steels in neutral media. A very large body of literature has been already devoted to this topic and we shall not try in this paper to review all the theories which have been proposed. We shall rather attempt to underline few particular points which, we feel, deserve special consideration.

2 . PASSIVITY OF STAINLESS STEELS

2.1 Basic features

The corrosion resistance of stainless steels is based on the existence of a superficial passive layer with low ionic conductivity. This film act by reducing the current density of the dissolution reaction $M \longrightarrow M^{Z+} + Z\ e^-$. The stability of the passive layer is closely related to the pH of the solution and to the metal solution potential (Figure 1) ; it is also influenced by the composition of the metal and the solution.

According to UHLIG (1) and FRANKENTHAL (2), the primary cause of passivity is an adsorbed oxygen film. The formation of this film is favoured by the existence of unfilled d - electron levels in the metal. This primary film would be stable only within a few millivolts of the activation potential ; a secondary film which forms at more positive potentials grows to a thickness up to 10 to 100 A, depending on the potential and the time.

More recently (3,4,5), there has been some evidence that water and OH^- ions are present in the passive layer. Moreover, some metallic cations are inserted in the layer, giving a rather complex $M - O - OH - H_2O$ structure.

Chloride ions destroy passivity by displacing adsorbed O (or OH) at local sites. Lowering the pH has the same effect, by modifying the equilibrium of the OH^-, H^+ and H_2O species, beetween the film and the solution. It is a matter of fact that stainless steels does resist fairly well to corrosion in chloride neutral media or in strong acids free of chlorides, but that hydrochlorhydric acid is generally considered somewhat as a poison.

The composition of passive layers has been studied by AES, ESCA, SIMS, and recently by GDS (for more practical investigations). Figures 2a and 2b show the typical composition profiles at the surface of a 17 % Cr stainless steel passivated at + 200 mv or 600 mv in a formic acid solution (6). In that example, the passivation potential has no apparent influence.

The thickness of the passive layer, which can be estimated by ellipsometry, depends on the conditions in which the passive film has been formed, and also on the composition of the steel (7,8). Typical values range beetween 10 and 100 A.

2.2 Bound water, structure, and chloride ions effect

The presence of O and OH^- in the passive film is now well established. Moreover, several years ago, OKAMOTO and SHIBATA have shown that the passive film contains also bound water, in amount which depend on the film formation potential (7,9). These authors have proposed a model for the structure of the passive layer (Figures 3a and 3b), in which the passive film consist of a hydrated oxide having an amorphous or gel-like structure. According to this theory the proton included in the film is pulled out by ageing and anodic polarization. The metal ion produced by anodic dissolution forms a $(MOH)^+$ intermediate. This $(MOH)^+$ is captured by the surrounding H_2O molecules and precipitates as a solid film (Figure 3b). Freshly formed films contain a large amount of bound water, but with time their structure dehydrates. In this process $H_2O - M - H_2O$ bridges are in turn progressively replaced by $OH - M - OH$ bridge, followed by $O - M - O$ bonds as in a perfect oxyde. The initial structure $H_2O - M - H_2O$ would be most sensitive to corrosion than the deprotonized ones.

Conversely, when chloride ions are present, films containing a large amount of bound water would be more resistant. Chloride ions, indeed, can replace OH^- in the $(MOH)^+$ intermediate compound, resulting in the formation of a soluble chloride complex. The bound water may capture the dissolving metal ions and has a self-repairing action on the passive film. In this way, bound water would inhibit the breakdown of passivity due to chloride ions.

Furthermore, SAITO and al. (4) have shown that two types of bound water could exist, depending of the conditions under which the film has been formed. One would have a good self-repairing action, whereas the second would act chiefly by increasing the initial resistance to chloride attack.

Recently (10), BORCH, RANER, SOMMER and SCHATT have proposed a crystallographic model of the passivation processes, not specific to stainless steel, yet consistent with the OKAMOTO's hypothesis. Adsorbed water is assumed to form a trydimit-like structure, with some distortions favouring the implantation of metallic cations during the anodic process. This implantation is accompanied by a loss of protons, in order to preserve electrical neutrality. With pure iron, the deprotonization leads to a γ-FeOOH structure, i - e the formation of rust. If other cations with larger charges (for example : Chromium), are present, deprotonization is more complete and a nearly anhydrous passivating layer is therefore produced.

2.3 Role of the alloying elements

As it was rewieved by EPELBOIN and alias (11), when Chromium is added to iron up to 5 %, the polarization curve are similar to that of a pure iron electrode ; from 7 to 17 % Cr the curve exhibit 2 maxima, and beyond 22 % Cr they are similar to that of a pure Cr electrode. In fact, increasing the Chromium content in a Fe-Cr alloy at a given pH, could have the same effect than increasing the solution pH at a given Cr content. From a practical standpoint, a steel is considered as stainless if its Chromium content is greater that 11 - 12 %. For many applications, 17 % to 18 % Cr is needed. It has been suggested that selective dissolution of iron could enrich the surface of the Fe - Cr alloys with Chromium, provided that Chromium content is sufficient.

The very special role played by Chromium in stainless steels has been discussed by UHLIG and WULFF (12) on the base of the "electron configuration theory" : the Cr - Fe alloys are considered to retain Chromium passive properties as long as the five d- electron vacancies of atomic Chromium were not completely filled by one electron per atom donated by iron, i . e if the atom ratio Fe - Cr is less than 5.

CHARBONNIER and al. (6), investigating the surface composition of a 17 % Cr steel (Figure 2a et 2b), noticed the presence of a slight maximum in Cr concentration inside the passive layer. A similar maximum was found by TJONG and al. (5) on Fe - 9 % Cr, Fe - 12 % Cr and Fe - 18 % Cr alloys passivated in a borate buffer at different potentials. This maximum is not observed on a Fe - 3 % Cr alloy. Such a maximum has also been evidenced by DA CUNHA BELO (13) and YANIV (14). However, our own results (15) show that when the metal is aged in pure water, no maximum is observed. From an industrial point of view, the Chromium surface concentration depends very largely on the final stages in the fabrication of the sheets, bars or wires of metal.

Molybdenum content in the passive layer was also investigated by several authors. Following OLEFJORD (16) the Mo content present at the surface of a 17 Cr - 13 Ni - 2,6 Mo steel is of the same order of magnitude as in the base alloy. For a 6 % Molybdenum steel, the passive film would be enriched in Mo (17). Using ESCA measurements SUGITOMO and SAWADA (18) have found Mo^{6+} to be present, whereas other workers, on the opposite, never found any Mo in the film (6, 13, 14). CHARBONNIER and al. pointed out that detection of Mo in the film is difficult with AES, whereas ion analysis on 98 Mo^+ showed the presence of Mo at the outer surface of the layers. Some authors assume that even when Mo is not detected in the passive layer, the enrichment of Cr is dependent of the Mo content of the alloy.

Nickel does not appear to be present in the film (16, 17). Nevertheless, following OLEFJORD (17), the beneficial effect of Ni and Mo would not be related to their occurence in the passive layer but rather to their lowering effect on the dissolution rate in the active stage. The discussion seems to be open on these mecanisms.

3 . BREAKDOWN OF THE PASSIVE STATE AND LOCALIZED CORROSION

3.1 Basic features

In a perfectly homogeneous system, the passive layer will be generally either active or passive, depending on the metal/solution potential, pH value, chemical composition of the electrolyte, flaw velocity, etc...etc...We have to consider separately the case of the acid media and that of the "quasi-neutral" media. A medium will be considered as acid for stainless steel, if its pH is lower than the "depassivation pH" (pHd), and as "quasi-neutral" if its pH is greater than pHd. The depassivation pH, firstly introduced in order to characterize the resistance to crevice corrosion of stainless steel (19), is in fact a quantity of more general signification, but depends also of all the compositional parameters of the medium.

In an acid medium, general passivity or general corrosion is the usual rule for stainless steels, and the few cases of localized corrosion which have been found are always due either to heterogeneities in the system (concentrations or flow velocity gradients) or heterogeneities of the metal (such as those found in intergranular corrosion).

Conversely, in the "quasi-neutral" media, general corrosion is ruled out and therefore breakdown of passivity will be necessarily localized. As pointed out by several authors (20,21,22), localized corrosion implies a local stabilization of an acid corrosive medium, differing from the surrounding one ; this local acidification induces a local dissolution of the metal in the active state. This dissolution in turn may provide further local acidification and the process will be self-sustained. The penetration rate of this type of corrosion ought to be very high, owing to the galvanic coupling beetween a restricted anodic area (corresponding to the acidified zone) and a large cathodic area (corresponding to the surrounding passive surface).

In summary, all types of localized corrosion of stainless steels in "quasi-neutral" media can be considered as having the same propagation mechanisms, even though their initiation processes differ. The stable propagation stage is characterized by gradients in potential, pH, and contents in various chemical species. These gradients set up diffusion or convection flows, compensated by the local anodic dissolution. The outcome of these two phenomena is a dynamic steady state, resulting in a propagation of the corrosion.

Local acidification, however, is not initiated spontaneously on a passive metal and thus requires some specific conditions for its onset. These conditions vary according to the type of corrosion considered. We shall farther detail these "initiation mechanisms" for the three main form of localized corrosion, namely : stress corrosion cracking, crevice corrosion and pitting corrosion.

3.2 Protection potential

A consequence of the previous consideration is that all the types of localized corrosion of stainless steels in quasi neutral media are opened to cathodic protection. From a qualitative standpoint we may write :

$$\frac{dx}{dt} = ai\,(U,x) - bx$$

Where x is the concentration in corrosion products, U the metal/solution potential and i the dissolution current. Parameters a and b represent respectively the production in corrosion products by the anodic dissolution, and their dilution into the ambient medium by a diffusion process. In a dynamic steady state we have :

and

$$\frac{dx}{dt} = 0$$

$$\frac{d}{dx}\left(\frac{dx}{dt}\right) < 0$$

In some potentiostatic conditions we find :

$$\frac{d}{dx}\left(\frac{dx}{dt}\right) = a\frac{\partial i}{\partial x} - b$$

and the steady state condition can be written :

$$\frac{\partial i}{\partial x}\,(U,x) < \frac{b}{a}$$

Since i increases when U or x increases, there exists a potential Up (x), which we shall refer to "protection potential", below which this condition is fullfilled.

Using irreversible thermodynamic considerations, OKADA and HASINO (23) have recently used a more sophisticated approach : these authors have calculated the shift in potential beetween the bottom of an activated pit and the free passive surface ; a "protection potential" can also be defined, whose value is the free surface potential such that the pit bottom potential remains greater than the "Flade potential" inside the pit.

The "protection potential" is of a very successfull use for low alloyed steels (24) but we feel it cannot be used without caution for stainless steels. As a matter of fact, this potential is not a constant. It depends upon the various "x" concentrations, the actual stage of corrosion, and the ohmic drop in the anodic zone (19, 25). It must rather be considered as a mixed potential of a complex electrode (25). Consequently it cannot serve as a parameter characterizing the ability of the steel to undergo corrosion.

The dissolution rate i (U,x) in the anodic area depends strongly of the concentration x, but it can also be governed by the mechanism of "salt passivity" as pointed out by ROSENFELD (25). Other authors feel that a high resistance path can also take place, caused by the constriction due to an hydrogen bubble (26).

Anyway, stainless steels are made to remain uncorroded ; thus, from a practical point of view, the study of the corrosion initiation is more important than that of corrosion propagation.

3.3 Initiation mechanism for stress corrosion cracking (SCC)

Many studies have been published on SCC and various mechanisms have been put forward : see for instance (27,28,29). The most probable initiation mechanism seems to be microcreep, inducing the repeated formation of strain assisted micro-breaking in the passive layer. The onset of the SCC results from a competition beetween the formation of these "fresh" microsurfaces and the repassivation rate. Crack initiation actually depends upon a large body of parameters, and in particular the stress level, the metal solution potential (30) and the chloride content (31). For example, one can define a critical potential, which depends on the applied stress level (30). In pratice however, for free corroding surfaces, a critical stress can be defined, which give the limit beyond which the microcreep results in an inacceptable risk of cracking. Austenitic stainless steels show a very limited elastic range and are much more sensitive to SCC than ferritic stainless steels, which may often be considered as immune to this type of corrosion. MOHR and al. however, have shown (32) that this immunity is certain only at the open circuit corrosion potential. This potential would be less noble than a so called "critical cracking potential" Escc. Plastic deformation or grain coarsening, however, may shift Escc to the less noble direction without affecting the rest potential. Nickel additions to Chromium ferritic steels shift the rest potential to the more noble direction, and then may also induce S.C.C. From this point of view a close parrallelism would exist beetween austenitic and ferritic stainless steels.

DA CUNHA BELO and al. (33) have argued that the critical potential for SCC deals with the change in the chemical composition of the passive layer ; they also assume that the conductivity of the films formed beyond the critical potential are profoundly affected by plastic deformation of the metal lattice.

3.4 Initiation mechanism for crevice corrosion

The crevice corrosion initiation mechanism of stainless steels is now well known (34) and is certainly the simplest, in so far it is a slow process. The very high surface/volume ratio in a very narrow zone, and the lack of convective exchanges beetween the inside and the outside of the crevices, leads quickly to a highly concentrated solution entrapped in the crevice, even when the anodic current across the passive layer remains very low. The hydrolysis of chromium salts lowers continuously the pH, and the chloride content locally increases, so far this hydrochlorhydric acidity becomes high enough (or the pH low enough) to cause a general activation of the inner surface. As pointed out by ROSENFELD (25) crevice corrosion resistance of stainless steels is mainly determined by the ability to resist activation in acidic chloride solutions.

Following CROLET (19,34) the intrinsic resistance of a given alloy can be characterized by its "depassivation" pH which is in fact the pH below which the passive layer is not stable at the actual metal/solution potential. Other authors (35) suggest the existence of a critical passive film composition. This critical composition would be determined by the enrichment of chloride ions of the crevice solution rather than by its low pH. The very successful use of the depassivation pH in practical problems, however, shows that this quantity is one the main characteristics of a steel for crevice corrosion resistance.

3.5 Initiation mechanisms for Pitting Corrosion

It would be a wager to try to describe here all the theories which have published on the Pitting Corrosion. Fortunately, some review articles are available and very useful to find his way through this very large body of works (36 to 40). In the next paragraph, we shall put the accent on few points of particular interest for stainless steels. Hereafter, we shall only give the outlines of the main theories invoked to explain the pit initiation and to give a sense to the so called "pitting potential". The most of these theories are related to one of the following mechanisms :

a) competitive ionic adsorption : this model has been first proposed by KOLOTYRKIN (41) and UHLIG (42). Pitting potential would be that beyond which Cl^- may be adsorbed in, or on the passive layer, replacing passivating ions such as OH^-. A treshold concentration in Cl^- could be necessary for this adsorption to occur. HOAR and JACOB (43) feel that the formation a transitory chloride complex may favour the cations extraction from the passive layer.

b) ionic penetration : all these models assume that the pitting corrosion occurs when the Cl^- ions reach the inner passive layer/bare metal interface ; the observed "incubation time" would be the time of ion migration from the outer solution/ passive layer interface to this inner interface. However, the proposed mechanisms for the ionic migration differ among the authors (44,45,46).

c) <u>mechanical breakdown of the passive layer</u> : following HOAR (47) the adsorption of Cl⁻ decreases the free energy of the film and the breakdown is induced by the initial repulsion of the adsorbed ions. Following SATO (48), the film is in equilibrium beetween an electrostriction pressure due to a high electric field and the surface tension. Cl⁻ adsorption decreases the surface tension and allows the breakdown to occur.

In all these theories, the metal/solution potential acts as a "driving force" to initiate the pitting corrosion. Others authors (39,49) assume the preexistence of "nuclei" which may, or may not, develop, depending on the metal/solution potential. From this standpoint the pitting potential would be the "protection potential" of the most detrimental nuclei, just at the zero - time limit, where the pit has not already propagated. Beyond this pitting potential, the nuclei can develop in "stable" pits. The nature of these nuclei is not well known ; some of them are certainly linked to metallurgical defects, such as the non metallic inclusions, geometric flaws at the surface, whereas others may be connected to the passive layer and to the adsorption of Cl⁻ anions.

Few studies have been dealing with the relation beetween the pitting susceptibility in neutral media and the nature of the passive layer. It is very likely that the effect of bound water emphazised by SAITO for chlorides containing acid solutions (4), may also be put forward. In practice, the "passivation treatments" modify both the passive layer and the pitting resistance. However, their role is not always very clear : they may not only reinforce the passive layer, but also dissolve the pit "nuclei" (49). Our own work (50) shows that a passive layer formed in a H_2/H_2O high temperature atmosphere can be more or less resistant to pitting corrosion, depending on the pH_2O/pH_2 ratio and on a further ageing treatment in water. We are now working further on these problems.

3.6 Stochastic character of pitting corrosion

Whatever the mechanism proposed for its initiation, there is a growing agreement to consider pitting corrosion as a stochastic process, as it has been emphasized particularly by SHIBATA and TAKEYAMA (51). These authors have analysed the effect of time on the survival probability, in some potentiostatic conditions. We feel however that the stochastic hypothesis must also be applied to the localization of pits and to the effect of the sample area.

If we note ϖ dS the pitting probability of a very little area dS, the survival probability of the macroscopic surface S will be :

$$P = (1 - \varpi\, dS)^{(S/dS)} = \exp(-\varpi.\,S)$$

The "elementary pitting probability" (EPP) : ϖ, does not depend on the sample area. The EPP can be determined in potentiostatic conditions and will be a function ϖ_{PS} (E,t) of potential E and time t. It may also be obtained in the potentiokinetic mode and will be a function ϖ_{PK} (E,v), where v is the potential sweep velocity. If this sweep is slow enough, we can assume that :

$$\varpi_{PK} (E,v) = \varpi_{PS} (E,) = \varpi(E).$$

Figure 4 shows the results we obtained on a 17 % Cr - 1 % Mo steel, using a Multichannel-testing apparatus (51, 52). The variations of EPP cover a rather large range of potential and can explain the dispersion observed in the determination of pitting potential, using a singlechannel apparatus. The "genuine" pitting potential is that for which $\overline{\omega}$ (E) = 0. Unfortunately, its determination cannot be very precise, because it corresponds to a very high survival probability. The best precision is obtained when $\overline{\omega}$ is of the same order than $1/S$, where S is the sample area. From a practical standpoint, good results are obtained in considering a "pseudo pitting potential" for which $\overline{\omega}$ = 1 cm^{-2}. We feel that such a methodology could be very useful in the future.

4. EFFECT OF SOME NON METALLIC PHASES ON THE RESISTANCE TO PITTING CORROSION

4.1 Basic features

Stainless steels, and more generally metals and alloys, are often considered by fundamental scientists as something very ideal. If need be, they sometimes take into account some of the unavoidable cristallographic defects, such as dislocations, grain boundaries, etc... The metallurgists, for their part, are aware that such an approach is unfortunately oversimplified. Carbides, nitrides, sulfides, oxides, secondary phases, etc... often cause some trouble in the most sophisticated alloys. Pitting corrosion does not stand apart from this rather depressing situation. Such metallurgical defects are very likely to be at the origin of many pitting corrosion "nuclei".

4.2 Sulfides

Following SZLARSKA-SMIALOVSKA (53), the pits are not initiated randomly, but preferred sites are observed on manganese sulfides. For stainless steels, these sulfides generally surround Al and Cr oxides. WRANGLEN (54) feels that sulfide inclusions are electrochemically less noble than the adjacent passive layer. EKLUND assumes (55) that the sulfides cannot thermodynamically exist at the potentials where stainless steel is passive. BRENNERT and EKLUND suggest (56) that the attack start because a low pH and a high H_2S concentration, due to the dissolution of sulfides. WRANGLEN (54) suppose that the initiation of pitting is counteracted by the metals which forms some unsoluble sulphides ; these sulphides may either be present in the steel (as TiS), or formed during the corrosion process (as Mo-Sulphides).

That is a matter of fact that Titanium stabilized ferritic stainless steels have a higher pitting potential than Ti - free ferritic stainless steels (57). In these steels, Titanium sulfides replace manganese sulphides. These sulphides are associated with Titanium carbonitrides and, following ANDERSSON and SOLLY (58), the attack starts within sulphides, close to the carbonitride interface. In Titanium stabilized austenitic stainless steels, pitting would occur at particles of complex oxides or oxysulfide containing Titanium, but not at Titanium carbonitrides (59).

Following CROLET (49), the pitting probability decreases with the time, due to the dissolution of the sulfides "nuclei". A well adapted surface treatment, or natural ageing, may then improve the pitting resistance of the steel.

4.3 Other inclusions

Using a microprobe analyser, SZUMMER and JANIK-CZACHOR detected (60) a strong Cl⁻ accumulation at the interfaces beetween the metallic matrix and non metallic inclusions (not only sulphides).They suggest than a microcrevice take place at these interfaces.

POYET, DESESTRET and alias (61, 62) have investigated the detrimental effect of the various kinds of inclusions. The authors gave a classification of the inclusions nocivity, according to their wetability in liquid steel, their thermal expansion and their hot plasticity characteristics regarding to the matrix ones. All these parameters govern the formation of an eventual microcrevice beetween inclusion and matrix. The inclusion ability to adsorb the chloride ions has also to be considered.

Finally, we must consider that the pitting resistance is the resistance of the weakest link of the following chain : sulphides, oxides, metallic matrix.. Sulphides, when they are present in sufficient amount, are the main sites of pits. When they are not, oxides (MgO, then SiO_2, Al_2O_3, Cr_2O_3) initiate the pitting. In very pure metals, grain boundaries or dislocations should perhaps be to considered.

4.4 Carbides

It is well known than some heat treatments may induce a sensitization in steels, i.e. a carbide precipitation in such a way than chromium (or even molybdenum) depletion take place arround the carbides. Such precipitations are met in welded products in the heat affected zone. They are responsible of the so called "intergranular" corrosion, and also give rise to pitting sensitization (63) ; in that case, the pits "nuclei" are not the carbides themselves, but the depleted zone surrounding them. This sensitization can be healed by the same treatments or alloying additions (Nb, Ti,...), than those which are used to cure the intergranular corrosion tendancy.

5 . CONCLUSIONS

We have reviewed few particular points related the passivation and localized corrosion of stainless steels. We would like to stress the following ones :

- Passive layer has probably a water-like structure, more or less deprotonized. It includes some inserted metal-cations and some various anions coming from the external solution. The properties of the layer depend on the degree of deprotonization, and on the amount of bound water it contains.

    - All the types of localized corrosion in "quasi neutral" media have the same propagation mechanism even if their initiation processes differ. This mechanism implies a dissolution of active areas in a locally acidified medium, which is self supported by the hydrolysis of corrosion products. Localized corrosion may be considered as a dynamic instability, leading to a high local dissolute rate in a narrow anodic zone, surrounded by a very large cathodic zone.

    - The pitting corrosion theories are very numerous. Whatever the mechanism which will be retained, we feel that two points must be underlined :

      ◦ Pitting corrosion is a stochastic phenomenum, regarding to not only the time, but also the sample area. Rather than the "pitting potential", we propose to consider the "elementary pitting probability" ϖ (E), whose the variations cover a range of potentials. The survival probability P is then :

$$P = \exp\left(-\varpi . S\right)$$

where S is the sample area.

      . In industrial steels, initiation of pitting mainly occurs on non-metallic phases, such as sulphides or oxides, or on others preexisting flaws.

REFERENCES

1 - H.H. UHLIG. Corrosion Science (1979), vol 19, p. 777

2 - R.P. FRANKENTHAL. J. Electrochem. Soc. (June 1967), vol 114, n° 6, p 543

3 - O.J. MURPHY , J.O. BOCKRIS and T.E. POU. J. electrochem. Soc. (Septembre 1982), vol 129, n° 9, p. 2149

4 - H. SAITO, T. SHIBATA and G. OKAMOTO. Corrosion Science (1979) Vol 19, p. 693

5 - S.C. TJONG, R.W. HOFFMAN and E.B. YEAGER. J. Electrochem. Soc. (Aug.1982), vol 129, n° 8, p. 1162

6 - J.C. CHARBONNIER, PH. MAITREPIERRE, P. NOVAL and R. NAMDAR IRANI. Proc. 7th Intern. Vac. Congr and 3 rd. Intern. Conf. Solid Surfaces (Vienna 1977)

7 - G. OKAMOTO. Corrosion Science (1973), vol 13, p. 471

8 - K. SUGIMOTO and T. MATSUDA. Mat. Science and Engineering (1980), Vol 42, p. 181

9 - G. OKAMOTO and T. SHIBATA. Corrosion Science (1970), Vol 10, p. 371

10 - G. BORCH, D. RANER, J. SOMMER and W. SHATT. Protection of metals, (July 1982), p. 540
See also W. FORKER, H. WORCH, D. RANER. Werkstoffe und Korrosion (1981), Vol 32, p. 545

11 - I. EPELBOIN, M. KEDDAM, O.R. MATTOS and H. TAKENOUTI. Corrosion Science (1979), Vol 19, p. 1105

12 - H.H. UHLIG. The corrosion Handbook (editor) (1948), Section I, p. 23 to 25

542

13 - M. da CUNHA BELO, B. RONDOT, F. PONS, J. Le HERICY and J.P. LANGERON.
J. Electrochem. Soc. (Sept 1977), Vol 124, n° 9, p. 1317
14 - A.E. YANIV, J.B. LUMDSEN and R.W. STAEHLE. J. Electrochem. Soc.
(April 1977), Vol 124, n° 4, p. 490
15 - G. BLANC, B. BAROUX. Symposium CEFRACOR 1980 "Etats de Surface et
Corrosion" (1980), Compiègne - FRANCE
16 - I. OLEFJORD. Mat. Science and engineering (1980), Vol 42, p. 161
17 - I. OLEFJORD and B.O. ELFSTROM. Corrosion - NACE (January 1982),
Vol. 38, n° 1, p. 46
18 - K. SUGIMOTO and Y. SAWADA. Corrosion Science (1977), Vol 17, p. 425
19 - J.L. CROLET, J.M. DEFRANOUX, L. SERAPHIN and R. TRICOT (1974). Mém.
Sci. Rev. Mat., Vol 71, p. 797
20 - H.H. UHLIG. Metals technology (1940), p. 1150
21 - W.SCHWENK. Corrosion Science (1963), Vol 3, p. 107
22 - J.L. CROLET. Rev. Coat. Corrosion (1979), Vol 3, p. 159
23 - T. OKADA and T. HASHINO. Corrosion Science (1977), Vol 17, p. 671
24 - M. POURBAIX. Corrosion - NACE (October 1970), Vol 26, n° 10, p. 431
25 - I.L. ROSENFELD, I.S. DANILOV and R.N. ORANSKAYA. J. Electrochemical
Society, (Nov 1978), Vol 125, n° 11, p. 1729
26 - H.W. PICKERING and R.P. FRANKENTHAL. J. Electrochemical Society
(october 1972), Vol 119, n° 10, p. 1297
27 - R.H. LATANISION and R.W. STAEHLE. Conférence on fundamental aspects of
SCC. Edited by NACE (1969)
28 - W. HUBNER and B.J.E. JOHANSSON. A review of data up to 1971, by the
Aktiebolaget Atomenergi Studsvik, Nyköping, SWEDEN
29 - J.C. CHARBONNIER. Mem. Sci. Rev. Met. (Feb.1981), p. 65 and (March
1981), p. 121
30 - R.T. NEWBERG and H.H. UHLIG. J. Electrochemical Society (1973),
Vol 120, p. 1629
31 - J.E. TRUMAN. Corrosion Science (1977), Vol 17, p. 737
32 - T.W. MOHR, A.R. TROIANO and R.F. HEHEMANN. Corrosion - NACE (April
1981), Vol 37, n° 4, p. 199
33 - M. Da CUNHA BELO, J. BERGNER and B. RONDOT. Corrosion Science (1980),
Vol 21, n° 4, p. 273
34 - J.L. CROLET and J.M. DEFRANOUX. Corrosion Science (1973), Vol 13,
p. 575
35 - S. ZAKIPOUR and C. LEYGRAF. Corrosion - NACE (June 1981), Vol 37,
n° 6, p. 363
36 - S. SKLARSKA - SMIALOVSKA. Corrosion - NACE (1971), Vol 27, p. 223
37 - Y.M. KILOTYRKIN. Corrosion - NACE (1961), Vol 108, p. 209
38 - J. KRUGER. National bureau of Standards (U.S.A.) (1976),
report n° 212.110
See Also : The proceedings of USA - JAPAN Seminar on passivity and its
Breakdown on Iron base alloys, Honolulu (March 1975), p. 91 (ed. by
NACE).
39 - M. JANIK - CZACHOR, G.C. WOOD and G.E. THOMSON. British Corrosion J.
(1980), Vol 15, n° 4, p. 154
40 - J.A. PETIT and F. DABOSI. Ecole d'été de Corrosion (1982) Les Houches
FRANCE
41 - Y.M. KOLOTYRKIN. J. Electrochemical Society (1961), Vol 108, p. 209
42 - H.P. LECKIE and H.H. UHLIG. J. Electrochemical Society (1966), Vol 113,
p. 1262
43 - T.P. HOAR and W.R. JACOB. Nature (1967), Vol 216, p. 1299
44 - T.P. HOAR, D.C. MEARS and G.P. ROTWELL. Corrosion Science (1965),
Vol 5, p. 279

45 - M.H. PRYOR. Localized Corrosion (1974), Ed. by NACE

46 - C.L. Mc BEE and J. KRUGER. ibid (45)

47 - T.P. HOAR. Corrosion Science (1967), Vol 7, p. 341

48 - N. SATO. Electrochem. Acta (1971), Vol 19, p. 1683

49 - J.L. CROLET, L. SERAPHIN and R. TRICOT. Mem. Sci. Rev. Metall. (1977), Vol 74, p. 647

5C - B. BAROUX, B. SALA, J. SUZAN, J.C. JOUD. This Symposium, poster Session

51 - T. SHIBATA and T. TAKEYAMA. Corrosion - NACE (July 1977), Vol 33, n° 7, p. 243

52 - B. BAROUX, B. SALA, T.JOSSIC, J. PINARD. Symposium CEFRACOR (1983), Informatique statistique et corrosion, Paris - FRANCE

53 - Z. SZKARSKA - SMIALOWSKA. Corrosion - NACE (October 1972), Vol 28, n° 10, p. 388

54 - G. WRANGLEN. Corrosion Science (1974), Vol 14, p. 331

55 - G.S. EKLUND. J. Electrochem. Society (April 1974), Vol 121, n° 4, p. 467

56 - S. BRENNERT and G. EKLUND. Scandinavian Journal of metallurgy (1976), Vol 5, p. 16

57 - Y. BARBAZANGES, B. BAROUX, PH. KRAEMER et PH. MAITREPIERRE. Aciers Spéciaux (1980), n° 52, p. 3

58 - B.R.T. ANDERSSON and B. SOLLY. Scandinavian Journal of metallurgy (1975), Vol 4, p. 85

59 - L.I. FREIMAN, L.Y. KHARITONOVA, G.S. RASKIN, L.S. PUTSENKO and O.N. LUKINA. Zashita Metallov, (March-April 1978), Vol 14, n° 2, p. 143

60 - A. SZUMMER and M. JANIK-CZACHOR. British Corrosion Journal (1974), Vol 9, n° 4, p. 216

61 - P. POYET, P. COUCHINAVE, J. HAHN, B. SAULNIER, J.Y. BOOS. Mem. Sci. Rev. Metall. (August-Sept. 1979), p. 489

62 - P. POYET, A. DESESTRET, H. CORIOU, L. GRALL. Mem. Sci. Rev. Metall. (Feb. 1975), p 133

63 - C.R. RAREY and A.H. ARONSON. Corrosion - NACE (July 1972), Vol 28, n° 7, p. 255

544

Fig. 1 : Polarization curves
on 12 % Cr (●) and 22 % Cr
(O) steels in Na$_2$SO$_4$ (0.5 M)
acidified to pH 1.5 (-----)
             or pH 2.5 (———)

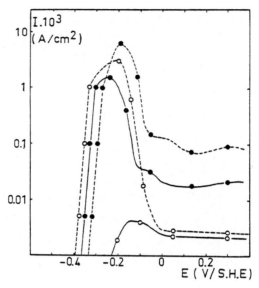

Fig. 2 : Fe - 17 Cr steel
composition profiles for
various elements in the passive film. (6)

    2 a : From A.E.S.
    ——— measurements.

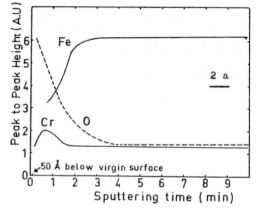

    2 b : From S.I.M.S.
    ——— measurements.

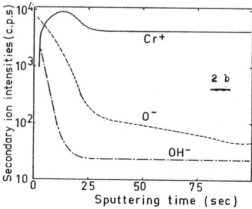

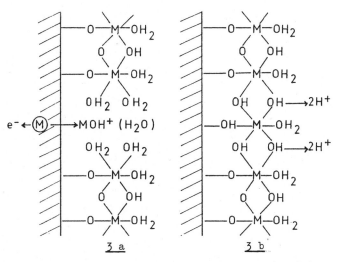

Fig. 3 : Structure of the passive layer (from | 9 |)

    3 a : freshly formed layer.

    3 b : deprotonization and bridging of an OH bond,
          to form a less hydrated film.

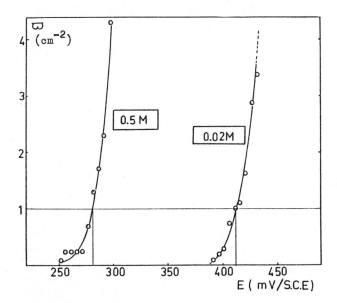

Fig. 4 : Elementary pitting probability (ϖ) in NaCl (0.02 and 0.5 M)
on a 17 % Cr + 1 % Mo steel. Test temperature 23° C, pH 6.6
potential sweep rate : 10 mV/mn ; specimen from an industrial
sheet grindéd with 000 paper, aged in air for 24 h.

Passivity of Metals and Semiconductors, edited by M. Froment
Elsevier Science Publishers B.V., Amsterdam — Printed in The Netherlands

EFFECT OF CHEMICAL HETEROGENEITY WITHIN THE METAL PHASE ON THE STABILITY OF THE PASSIVATING FILM ON IRON ALLOYS

M. JANIK-CZACHOR[1] and A. SZUMMER[2]

[1]Institute of Physical Chemistry, Polish Academy of Sciences, Warsaw (Poland)

[2]Institute of Materials Science, Technical University of Warsaw, Warsaw (Poland)

ABSTRACT

Chemical heterogeneity within the metal phase, notably non-metallic inclusions and second phase particles, can diminish the stability of passive state of iron and its alloys in various environments thus enhancing their susceptibility to localized corrosion. Effect of carbide, nitride, oxide and sulfide particles is considered with respect to their chemical and electrochemical stability, their influence on the chemical composition of the adjacent matrix and the composition and structure of the passivating film.

INTRODUCTION AND BACKGROUND

Instability of the passive state of iron and its alloys in many environments leads to localized corrosion and causes serious problems in practical applications. The primary regions of the instability of the passive film are heterogeneities within the metal matrix, like nonmetallic inclusions, second phase particles, impurity segregates at grain boundaries, etc., for the passivating film over these regions is more defective and its chemical composition is different than that over the substrate. One can eliminate these instabilities by using electrochemical methods, since various types of localized corrosion usually occur within definite potential regions. An additional, not so obvious possibility, lies in controling the level of impurities and technological additives which in turn influences the amount and composition of second phase particles and nonmetallic inclusions (mainly carbides, nitrides, oxides and sulfides). In this paper we examine the role of these phases in provoking instabilities of the passive state of iron and its alloys and discuss the ways of reducing their detrimental effect.

In the passive state metal (M) is covered by an oxide film (F) which nature is not well understood (see e.g. ref.1). The F is several nm thick, amorphous in structure, supporting high fields $\sim 10^6$ V/cm. For iron it is rather an insulator with a principal energy gap over 5eV (ref.2). At the F/So interface at least 2 parallel electrode reactions occur (ref.3):

- deposition of oxygen ions from water into the oxide film (film growth):

$$_{So}H_2O \rightleftharpoons {}_F O^{2-} + {}_{So} 2H^+ \qquad (I)$$

- dissolution of the metal ions into electrolyte:

$$\left[_M M - ze \right] \rightleftharpoons {}_F M^{z+} \dashrightarrow {}_{So} M^{z+} \qquad (II)$$

At the inner M/F interface metal is transferred via M^{z+} and electrons.

At the steady state, E=const., both the anodic current and the thickness of the passivating film are constant in time. Thus the rate of the reaction (I) is zero, and the steady state anodic c.d. is equal to the transport rate of M^{z+} through the F.

The situation is more complex for alloys. Preferential oxidation, differences in the relative concetration of components at the surface and in the bulk, and also in the F and in the matrix should be taken into account. A well known enrichment of Cr within the F (ref.4) on Cr bearing alloys, e.g., is responsible for a decrease of the rate of reaction (II) and the enhanced stability of the passive state. The occurence of particles of separate phases within the metal matrix may also affect its behaviour in many ways.

EFFECT OF CARBIDES AND NITRIDES

The detrimental effect of Cr-carbide precipitates at grain boundaries in provoking intergranular attack of Cr-bearing alloys was recognized long ago. The widely accepted view is that these precipitates reduce the Cr content in the adjacent regions and in the F thus increasing the rate of the reaction (II) and making the F unprotective there. Nitrides might have a similar effect (ref.5). Alloying with elements bounding C, notably with Ti or Nb, helps to avoid the Cr impoverishment and succesfully protects the metal from the intercrystalline attack. These are well known facts and will not be discussed here. However, Čihal et al. (ref. 6) were able to show recently that TiC and many other carbides

are unstable at high anodic potentials, thus being responsible
for intergranular corrosion of Ti stabilized steels in strongly
oxidizing media. Hence, the electrochemical instability of the
second phase particles is the factor responsible for the instability of the passive state of the material there. Reducing the C
content and not alloying with Ti will protect the metal from intergranular attack in such media.

EFFECT OF SULFIDES AND OXIDES

a) <u>solutions not containing halide ions</u>

The both kinds of inclusions are very common in steels. They
may occur separately or, more often, as oxy-sulfides. Although
the important effect of these phases on mechanical properties of
steels has been recognized long ago, their influence on the corrosion properties did not received much attention.

The close association of the inclusions with initiation of
SCC has been reported (ref.7),but,it is not clear whether this
is a result of a local depassivation at the inclusions, or concentration of stresses there resulting in a decohesion at the
inclusion/ matrix interface.

It has been estabilished that chemical dissolution of sulfide
inclusions in acidic media is a source of sulfide ions strongly
enhancing anodic dissolution of metal in the active potential
region (ref.8). In the passive region, howerer, the effect was
neglegible (ref.9,10). The inclusions were undermined or etched
out and the eventually revealed bare metal underwent repassivation[x/]. Apparently the inclusions do not affect the stationary
rate of the reaction (II). The difference in the behaviour in the
active and passive potential region of a metal with the inclusions is demonstrated in Fig. 1.

The critical c.d. for passivation increases considerably with
S content, but the c.d. within the passive state does not depend
on S level, i.e. on the amount of the sulfide inclusions within
the matrix (ref.10).

[x/] This kind of behaviour is explored in order to get rid of the
sulfide inclusions from the surface of the commercial stainless
steels, what improves their resistance against rusting (ref.11).

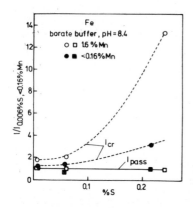

Fig.1.Effect of S content on the critical c.d. of passivation (I_{cr}) and stationary c.d. within the passive state (I_{pass}). Inclusions at: 1.6%Mn-MnS; 0.16%Mn-FeS

b) solutions containing halide ions

The situation changes dramatically in solutions containing aggressive ions, notably chlorides, and it appears that MnS inclusions are the main source of pit nucleation in steels (ref.12-19). Only occasionally pitting on oxide inclusions was reported (ref.20).

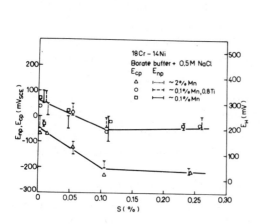

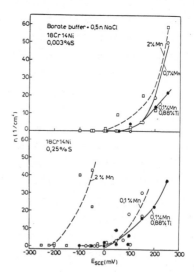

Fig.2.Effect of S,Mn and Ti content on the potential of pit nucleation and the critical potential of pitting (E_{np} and E_{cp}).
Inclusions: (....) - MnS, (---) - TiS, (——) - (Cr,Mn)S_x

Fig.3. Effect of potential on number of pits. Inclusions: 2% Mn- MnS; 0.1% Mn - (Cr,Mn)S_x; 0.8% Ti- TiS

As can be seen in Fig. 2 and 3, reducing the amount of S from 0.2% to 0.003% and Mn from 2% to 0.1%, i.e. diminishing the amount of sulfides and/or changing their chemical composition from MnS to $(Cr,Mn)S_x$, results in a distinct increase of the potential of pit nucleation and a decrease of the number of pits (ref.15). It has been suggested that the chemical dissolution of sulfides revealing the bare metal surface gives rise to pitting, since the aggressive ions prevent repassivation there. However, this concept seems unapplicable in neutral and alkaline solutions, as e.g. for the case in Fig. 2 and 3. Also the electrochemical dissolution of MnS is an unlikely reason for the enhancement of pitting since in these media at E<0.8V sulfide ions are not produced (ref.21) and the anodic c.d. is very low (see Fig.4)

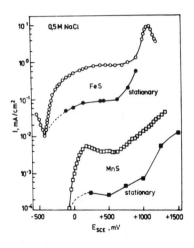

Fig.4. Anodic polarization curves of MnS and FeS: potentiodynamic; 5 min.each step, and stationary (after ref.21)

The likely explanation is that these inclusions operate as a source of discontinuities or other "weak spots" within the passivating film. As already discussed (ref.22), the aggressive ions are preferentially adsorbed at these most defective sides, and accumulate (ref.23), until the critical concentration builds up. The halides increase the rate of the reaction (II) locally, and this process is likely to occur preferentially at the defective sides. This mechanism probably operates also in free corrosion conditions since action of the sulfide inclusions as local cathodes should be excluded (ref.21).

In order to improve stability of the passive state of steels in the presence of aggressive ions an attempt was made to com-

pletely eliminate Mn containing sulfide inclusions by using supplementary alloying elements that have higher affinity to S than
Mn. Ti and rare earth elements (RE = La, Nd+Pr, Ce) were used as
the alloying elements.

c) Effect of Ti.

This element not only binds S but also is thought to improve
the ability to passivate and to repassivate when present within
the solid solution (ref.24). Low C content was kept in the alloys
under investigations in order to avoid binding of Ti by C (ref.15).
A minor effect was observed for 0.8% Ti. As can be seen in Figs.
2 and 3., the characteristic potentials of pitting were not changed and the number of pits decreased slightly. Pits were formed
at TiS inclusions (Fig.5) what suggests that these inclusions can
be a source of weak spots within the passivating film, when MnS
inclusions are absend.

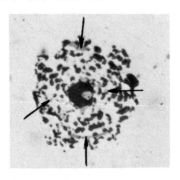

Fig.5. Early stage of a pit formed on a TiS inclusion in 18Cr-
-14Ni stainless steel. Small
pits formed around the central
one nucleate at the TiS inclusions as well (some marked by
arrows). 1500X

d) Effect of RE.

These elements are known for their beneficial effect on the mechanical properties of low alloy steels. However their influence
on the corrosion resistance of the steels appear detrimental
(ref.25). The RE sulfide inclusions are unstable in air at room
temperature (particularly these containing La and Nd) what results in formation of "flowers" of the spontaneous oxidation products (Fig.6) and thus produces holes that enhance pitting.
A dramatic increase in pit number was observed for the steels alloyed with RE and the induction period was shortened considerably
(Fig.7).

CONCLUSIONS

The experimental data available so far show that not only the
second phase particles but also nonmetallic inclusions occuring

in steel are harmful for stability of its passive state. The affecting factors are: preferential dissolution (chemical or electrochemical) of the particles, their influence on the composition
of the adjacent metal and thus of the passivating film, discontinuities or other "weak spots" within the film over the boundary
particle/matrix where the aggressive ions will adsorb preferentially. Which of these factors will prevail depends on the composition of the particle and the solution.

Reducing heterogeneity of the metal substrate and thus of the
passivating film results in an enhanced stability of the passive
state. The excellent corrosion resistance of the amorphous alloys
which are practically completely homogeneous materials is in agreement with this conclusion.

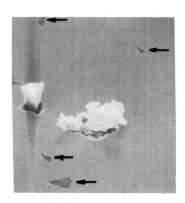 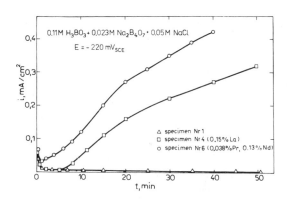

Fig.6. "Flowers" of the spontaneous oxidation products of La
rich sulfide inclusions in a low alloy steel. The stable inclusions remained unattacked (marked by arrows). 1500X
Fig.7.C.d. vs. time curves for a low alloy steel with various
additions of RE elements.

REFERENCES

1 This volume
2 B.D.Cahan, C.T.Chen, J.Electr.Soc. 129 (1982) 17,474 and 921
3 K.E.Heusler, Ber.Bunsenges. Phys.Chem. 72 (1968)1197,and
   Electr. Acta (in press)
4 R.P.Frankenthal, D.Malm, J.Electr.Soc. 123 (1976) 186; A.E.
   Yaniv, J.B.Lumsden, R.W.Staehle, in "Passivity of Iron and
   Iron Base Alloys" NACE 1975, p.72; H.Okada, H.Ogava, I.Itoh,
   ibid., p.82
5 E.G.Feldganger, L.Ya.Savkina, Zashch.Met. 11 (1975)31
6 V.Cihal, I.Kasova, V.Masarik, ibid.4(1968)355 and V.M.Knyaz-
   heva, V.Cihal, Ya.M.Kolotyrkin, ibid. 11(1975)531
7 J.G.Parker, Brit.Cor.J. 8(1973)124 and 13(1978)75; W.L.Clarke,
   G.M.Gordon, Corrosion 29(1973)1

554

8 T.P.Hoar, D.Havenhand, J.Iron.Steel.Inst. 133(1936)239
9 L.I.Freiman, Ya.M.Kolotyrkin, "Korrozija i zashchita ot korrozii" Moscow 6(1978)5 and Zashch.Met. 16(1980)714
10 Unpublished results
11 G.Hultquist, G.Leygraf, Mat.Sci.Eng. 42(1980)199
12 M.Smialowski, Z.Szklarska-Smialowska, M.Rychcik, A.Szummer, Corr.Sci. 9(1969)123
13 Z.Szklarska-Smialowska, A.Szummer, M.Janik-Czachor, Brit.Cor. J. 5(1970)159
14 M.Janik-Czachor, A.Szummer, Z.Szklarska-Smialowska ibid. 7(1972)90
15 M.Janik-Czachor, Bull.Acad.Sci.Polon.sér.sci.chim. 25(1977)561
16 P.E.Manning, D.J.Duquette, W.F.Savage, Corrosion 36(1980)313
17 G.Wranglen, Corr.Sci. 9(1969)585 and 14(1974)331
18 M.Kesten, Corrosion 32(1976)94
19 G.S.Eklund, J.Electr.Soc. 121(1974)467
20 P.Poyet, A.Desestret, H.Coriou, L.Grall, Comp.Rend Nr12(1970) 23 and Mem.Sci.Rev.Met.Nr.2(1975)133, and S.A.Glazkova, L.I. Freiman, I.Raskin, G.C.Schwarc, Zashch. Met. 8(1972)960
21 H.Keller, H.J.Grabke, Werkst.Korros. 32(1981)275 and 540
22 M.Janik-Czachor, Zashch.Met. 16(1980)265 and J.Electr.Soc. 128(1981)513C
23 A.Szummer, M.Janik-Czachor, Brit.Cor.J. 9(1974)215
24 M.Seo, Y.Matsumura, N.Sato, Proc.8th ICMC,Mainz 1981, p.108
25 A.Szummer, M.Janik-Czachor, Werkst.Korros. 33(1982)150

Passivity of Metals and Semiconductors, edited by M. Froment
Elsevier Science Publishers B.V., Amsterdam — Printed in The Netherlands

THE ROLE OF WATER IN THE KINETICS OF STAINLESS STEEL DISSOLUTION AND PASSIVATION IN ORGANIC MEDIA

B. ELSENER and H. BOEHNI

Institute of Materials Chemistry and Corrosion, Swiss Federal Institute of Technology, ETH-Hönggerberg, CH-8093 Zürich, Switzerland

ABSTRACT

The corrosion and passivation of type 304 stainless steel has been studied in deaerated ethanolic solutions containing hydrogen chloride and different amounts of water. The steady state polarization curves show that the passivation potential E_p and the critical current density for passivation i_{crit} strongly depend on the water content of the solution. Impedance measurements indicate the onset of passivation before the maximum current density. The reaction model of the passivation process proposed allows to deduce exactly the experimentally found water dependence of E_p and i_{crit}. The fundamental passivation reaction is described as a potential-dependent adsorption equilibrium where the water molecules are directly involved in the formation of the primary passivating film. This passivation reaction is continuous and reversible and begins long before i_{crit} is reached.

INTRODUCTION

Although a great deal of attention has been payed to the formation of passive films in aqueous solutions with different pH and composition, the importance of the solvent molecule water in the passivation process has not been clarified yet. Therefore the goal of this investigation in organic media (ethanol) with controlled amounts of water was to get insights in the role of water in the passivation process and to give a kinetic description for the active-passive transition.

EXPERIMENTAL

The experiments were performed with the usual electrochemical setup (all type Wenking) and additional devices for measuring the ohmic potential drop with the computer assisted interrupter technique (ref.1) (Figure 1). The electronic interrupter unit PU 1 (Meinsberg Scientific Institute GDR) (ref.2) was connected between potential control amplifier and working electrode. The potential/time transients during interruption of the current were recorded by means of a digital oscilloscope (Explorer IIIA, Nicolet Instr.Corp.) which was controlled by the microcomputer system (Sord M222). Impedance measurements were performed with a transfer frequency analyser (Solartron 1172) using the electrochemical interface ECI 1186 (Solartron) as potential control amplifier. Commercial stainless steel (SS304) was used as specimens. The electrochemical cell and the preparation of the specimens and the solutions are described elsewhere (ref.3).

RESULTS
Steady-state polarization curves

In Figure 2 the steady state polarization curves for type 304 stainless steel in 0.5N HCl-ethanol solutions with different water content are shown after correction for the measured ohmic potential drop. Two kinds of Tafel slopes can be seen: an active dissolution range close to the corrosion potential with water-dependent slopes is followed by a prepassive region with a Tafel slope of 120 mV/decade. This prepassive Tafel line is independent on the water content of the solution. The critical current density for passivation and the passivation potential both decrease with increasing water content of the solution* (Fig. 5)

$$E_p = 0.015 - 0.126 \cdot \log(C_{H_2O}) \qquad [\text{Volt}] \tag{1}$$

$$\log i_{crit} = 1.5 - 1.0 \cdot \log(C_{H_2O}) \qquad [\text{mA/cm}^2] \tag{2}$$

In solutions with low water contents (< 10 Vol.percent water) no passivation was possible under the present experimental conditions.

Impedance measurements

Impedance measurements at different points of the polarization curves are presented in Figure 3, where the results are plotted as normalized impedance Z_N (ref.4) in the complex plane

$$Z_N = (Z_{exp} - R_\Omega) \cdot I/U_N \qquad \begin{array}{l} Z_{exp} \quad \text{measured impedance} \\ I \qquad \text{stationary current} \end{array}$$

The normalization voltage U_N was taken as 1 mV.

In the active dissolution range, the impedance diagrams have the same shape with an inductive time constant and the polarization resistance $R_p \cdot I = Z_N(\omega \to 0)$ agrees well with the value calculated from the Tafel slopes. In the same way as $R_p \cdot I$ the transfer resistance $R_t \cdot I$ depends on the water content (Fig. 4). In the prepas-

---

*In order to describe the water content in a kinetically meaningful manner, the volume percent data were recalculated to give

$$C_{H_2O} = (x_{H_2O}^2/x_{H_3O^+}) \qquad \text{x mole fraction (ref.8)}$$

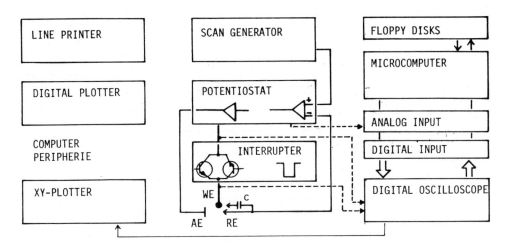

Fig. 1   Schematic diagram of the experimental setup for the computerassisted
interrupter technique

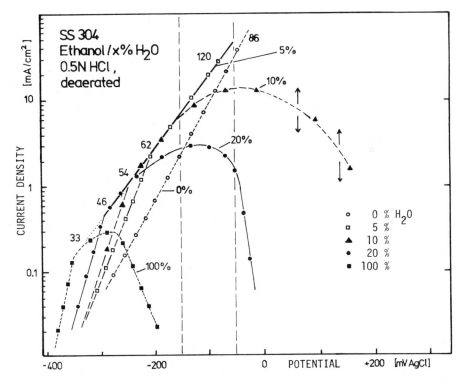

Fig. 2   Steady state polarization curves for stainless steel SS 304
in solutions with different water content
Numbers: Tafel slope mV/decade

passive region at low frequencies the inductive loop is transformed into a ca-
pactive one. The value $R_t \cdot I$ becomes independent of the water content ($R_t \cdot I=55$),
it remains constant over the whole range of primary passivation. The polarization resistance continuously increases from $R_p \cdot I=55$ to higher values. At potentials where the polarization curve deviates from the prepassive Tafel line
two capacitive loops can be observed at low frequencies. Below the passivation
potential negative real parts of the impedance are measured.

DISCUSSION

All interpretations of the behaviour of metals in the active-passive transition since the early work of Mueller (ref.5) are based on the idea of a potential dependent coverage of the electrode surface with intermediate adsorbed or
passivating $Me(OH)_n$ species. Metal dissolution and passivation reaction occur
simultaneously on the surface. A general reaction model for iron (ref.6) and
the reversible primary passivation potential (ref.7) allow the kinetic description of the passivation process:

$$Me \underset{①}{\rightleftharpoons} [Me(OH)]_{ad} \underset{②\downarrow rds \quad ③}{\rightleftharpoons} [Me(OH)_2]_{ad} \underset{④\downarrow rds \quad ⑤}{\rightleftharpoons} [Me(OH)_3]_{ad} \qquad (3)$$

$$Me(OH)^+ \qquad\qquad Me(OH)^+ \qquad\qquad Me\text{-}Oxide$$

(reaction ④: $Me + [Me(OH)_2]_{ad} \xrightarrow{rds} Me(OH)^+ + Me(OH)_{ad} + e^-$)

With the assumption of i) a Tafel law for the reaction rate of the rate deter-
ming step ④ in the prepassive range ii), the formation of the primary passivating layer according to

$$[Me(OH)_2]_{ad} + 2 H_2O \overset{⑤}{\rightleftharpoons} [Me(OH)_3]_{ad} + H_3O^+ + e^- \qquad (4)$$

and iii) Langmuir-conditions for $[Me(OH)_3]_{ad}$ (justified by a high degree of total coverage), the current density/potential relation in the active-passive
transition results in (ref.8)

$$i = \frac{K_4 \exp(\beta_v \cdot E)}{1+K_5 \cdot [C_{H_2O}] \exp(k \cdot E)} \qquad\qquad \begin{array}{l} \beta_v \quad 120 \text{ mV Tafel slope.} \\ K_i, k \text{ reaction constants} \end{array} \qquad (5)$$

At the passivation potential E_p the current density reaches its maximum
$(di/dE)_{E_p} =0$, and thus the water dependence of E_p can be decuced from the proposed model to

$$E_p = -a \cdot \log[C_{H_2O}] + b \qquad\qquad a,b \text{ constants} \qquad (6)$$

The constants in the deduced equation (6) are in good agreement with the experimental values (ref.3). Moreover the equations (5) and (6) allow to calculate

559

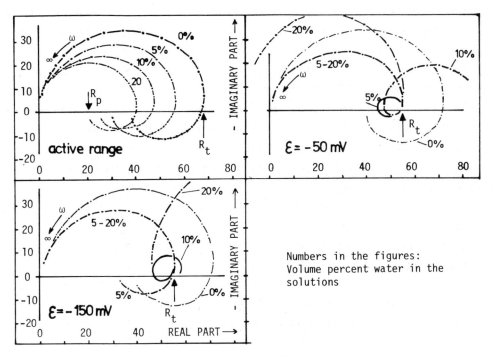

Fig. 3 Complex plane plot of the normalized impedance Z_N (Frequency range
10 kHz - 10 mHz) at different potentials
R_p polarization restistance   = $Z_N(\omega \to 0)$
R_t charge transfer resistance = $Z_f^N(\omega \to \infty)$

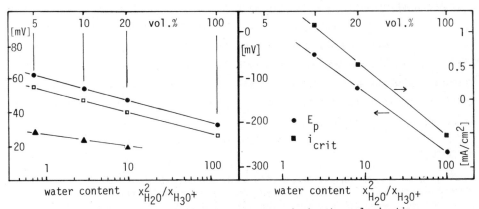

Fig. 4 Water dependence of the tafel slope b_a (-•-), the polarization
resistance $R \cdot I$ (-▲-) and the transfer resistance $R_t \cdot I$ (-□-) in the <u>active
dissolution region</u>. x = mole fraction
Fig. 5 Passivation potential E_p (•) and critical current density $\log i_{crit}$(■)
as a function of the water content

the water dependance of i_{crit}, too.

The steady-state measurements lead to the conclusion, that simultaneous with the prepassive dissolution process a highly protective primary passivating film grows on the metal surface. Considering the results of the impedance measurements in this potential region the constant charge transfer resistance $R_t \cdot I$ can be attributed to a charge transfer reaction occuring during the whole potential range of primary passivation, thus, to the water-independent dissolution reaction ④. This one-electron transfer shows the expected potential dependance of 120 mV/decade (equal to $2.3 \cdot R_t \cdot I$). The electrode impedance of the passivation process ⑤ (equation (4)) is expected to be capacitive (ref.9). Indeed a capacitive loop related to the formation of the primary passivating film can be observed long before the passivation potential E_p is reached (R_p is still > 0).

In the active dissolution range, the water dependence of both, $R_t \cdot I$ and $R_p \cdot I$, clearly show the important role of water molecules for the dissolution process. Due to the total coverage of the electrode with $[Me(OH)_2]_{ad}$ $R_t \cdot I$ becomes constant and independent of the water content of the solution. This indicates a change of the dissolution path in agreement with the steady-state measurements.

SUMMARY

The passivation of stainless steel in deaerated ethanolic solutions containing HCl is possible only above a critical water content. The passivation reaction is a potential dependent equilibrium between adsorbed Me(II) and passivating Me(III)-hydroxide film. The formation of this film starts when the electrode is totally covered with $Me(OH)_2$. Higher water contents allow the formation of the primary passivating layer at lower current densities. The passivation reaction is reversible and begins long before the critical current density is reached. The whole active-passive transition can be described by one reaction model. The water dependence of E_p and i_{crit} experimentally found can be calculated from the reaction model.

REFERENCES

1 B. Elsener and H. Böhni, Werkstoffe + Korrosion, 33 (1982) 207.
2 M. Berthold and S. Herrmann, Corrosion NACE 38 (1982) 241.
3 P. Hronsky, Werkstoffe + Korrosion, 31 (1980) 619.
4 H. Schweickert, W.J. Lorenz and H. Friedburg, J. Electrochem. Soc., 127 (1980) 1693.
5 W.A. Mueller, ibidem, 107 (1960) 157.
6 M. Keddam, O.R. Mattos and H. Takenouti, ibidem, 128 (1981) 257.
7 R.P. Frankenthal, ibidem, 114 (1967) 542.
8 B. Elsener and H. Böhni, Werkstoffe + Korrosion, 33 (1982) 213.
9 M. Keddam, O.R. Mattos and H. Takenouti, J. Electrochem. Soc., 128 (1981) 266.

Passivity of Metals and Semiconductors, edited by M. Froment
Elsevier Science Publishers B.V., Amsterdam — Printed in The Netherlands   561

QUANTITATIVE ESCA ANALYSIS OF THE PASSIVE STATE OF AN Fe - Cr ALLOY AND AN
Fe - Cr-Mo ALLOY

I. OLEFJORD and B. BROX
Department of Engineering Metals, Chalmers University of Technology,
S-412 96 Göteborg, Sweden

ABSTRACT
    Single crystals of Fe19Cr and Fe24Cr2Mo alloys were polarized to their
active and passive ranges in 0.5 M H_2SO_4 at 25°C. The compositions of the
passive film and the metallic phase below the film were analysed by the
ESCA technique.
    The analyses show that the passive film formed on both alloys consists main-
ly of Fe- and Cr-oxide. The Mo-containing alloy exhibits enrichment of Mo in
the outer part of the passive film. The valency state and the content of Mo in
the film depends on the potential to which the alloy is polarized. Hydroxide
is present in the outer part of the passive film.
    The alloying elements, Cr and Mo, are enriched on the surface at polarization
of the alloys to potentials in their active ranges. The enrichment is caused by
preferential dissolution of iron. It is suggested that the dissolution rate and
the corrosion potential of the alloy are controlled by the enriched layer. It
is also emphasized that the enrichment of the alloying elements enhances the
passivation of the alloy.

INTRODUCTION
    The corrosion properties of metals and alloys are affected by the reaction
products formed on their surfaces. In certain environmental conditions the dissolution rate of the passivatable metals is negligible. The composition and the
thickness of the passive film has been discussed for a long time. Today,
surface sensitive techniques such as ESCA and Auger have made it possible to
analyse the composition of the film.

    By using ESCA it has been shown that the oxide products formed on the surface
of austenitic and ferritic steels during their exposure in oxygen (1,2),in neutral (3,4,5) and in acid (6-11) aqueous solutions consist mainly of Fe-Cr-oxide.
Beside that, a thin layer of hydroxide is formed on the top of the oxide.
The average Cr content in the film is about twice as high as the Fe content. In
the case of Ni and Mo alloyed steels, these elements are present in the oxide
only to a very low degree. The Mo-content and the chemical state of Mo are due
to the potential (10,11).

    The earlier ESCA studies (10,11) have shown that during anodic dissolution,
the alloying elements are enriched in their metallic states in the outer atomic

layers of the metal phase. It has been suggested (11) that a contribution to the formation of the enriched layer is that the alloying elements are able to form intermetallic phases. The precipitation of the metal interphase on the surface is enhanced by selective dissolution of the base alloying component. Further, it has been stated that the dissolution, the corrosion potential and the passivatability are due to the composition of the outer metallic phase.

EXPERIMENTAL

Single crystals of the alloys Fe19.8Cr (100) and Fe24.1Cr1.9Mo (110) were polarized in 0.5 M H_2SO_4 and then analysed by the ESCA technique. The samples were pretreated by grinding, electropolishing (perchloric acid) and ion etching.

The polarization experiments were carried out in a deoxygenated and stirred electrolyte which was prepared from p.a. grade H_2SO_4 and titrated to ensure the concentration. After activation, the sample was firstly polarized to an anodic potential just above the corrosion potential. The aim was to normalize the surface composition by dissolution of outer atomic layers of the metals. Then, the potential was: set in the cathodic range (current = -100 $\mu A/cm^2$) for 10 min; swept (3 mV/s) to the holding potential either in the active or in the passive range (fig. 1); held constant for 6 min. The polarization was interrupted by pouring acetone and then methanol through the cell. The electrochemical cell is located inside a glass vessel attached to the spectrometer system (7,10,11). This allows the sample to be transported between the analyser and the cell in vacuum and protecting pure nitrogen gas. This precaution is done to minimize the influence of the atmosphere on the surface composition.

RESULTS

Figure 1 shows polarization curves recorded from the alloys. It appears that they are passivated within a wide potential range. However, noticeable differences exist between the alloys; the Mo-alloyed sample shows a higher corrosion potential, a lower maximum current in the active range and a lower passivation potential than the Fe-Cr-alloy. The arrows in the figure mark the potentials at which the samples were polarized before the surface analysis.

ESCA spectra recorded from electrochemically-treated samples are illustrated in figure 3. Because the Mo spectrum is very complex, the individual chemical states of Mo will be discussed first. The solid signals in row a, figure 2, are recorded from a Mo crystal cleaned by ion etching. The thin solid peaks represent the elemental metal states (Mo $3d_{5/2}$ and Mo $3d_{3/2}$) obtained by deconvolution of the recorded peaks. The dotted curve is the sum of the deconvoluted peaks and the background. The binding energy of metal state Mo $3d_{5/2}$ is 227.4 eV. (The spectrometer is calibrated by setting the binding energies of carbon (graphite) and metallic Ni to 284.3 eV and 852.8 eV respectively.) Ion-etched

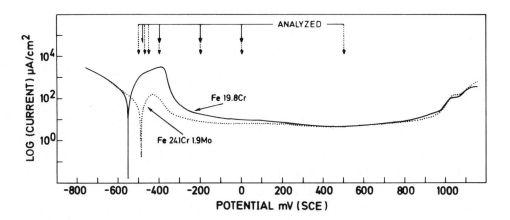

Fig. 1. The polarization diagram of Fe19.8Cr (100) and Fe24.1Cr1.9Mo (110). Sweep rate 1 mV/s. All concentrations are given in atomic percentage.

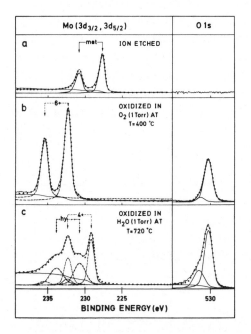

Fig. 2. ESCA spectra recorded from Mo: a) ion etched; b) oxidized in O_2; c) oxidized in H_2O vapour.

Mo samples were oxidized for 10 min. in the reaction chamber of the ESCA instrument in oxygen (P_{O_2} = 1 torr) at $400^{\circ}C$ and in H_2O-vapour (P_{H_2O}= 1 torr) at $700^{\circ}C$. The rows b and c in figure 2 show the spectra obtained. Oxidation in pure oxygen gives Mo in its six-valency state. The oxide is suggested to be MoO_3. The chemical shift (the difference in binding energies between the peaks representing the oxide and the metallic states) is 4.8 eV. From row c it appears that the oxidation in the H_2O vapour gives multichemical states of Mo. The main peak is denoted Mo^{4+} and the oxide is suggested to be MoO_2, which is the only stable

564

pure Mo-oxide beside MoO_3. The chemical shift of Mo^{4+} is 1.6 eV. It also appears that an extra signal, Mo^{hy} is located between the four- and six-valency states. The position and broadness of the extra peak indicates different coordination for the Mo atoms and it is suggested that it represents the four-valency state in an oxyhydroxide. A further indication for occurrence of hydroxide is the extra signal in the oxygen spectrum at 531.5 eV.

Figure 3 illustrates the ESCA spectra recorded from the Mo-alloyed sample after electrochemical treatments. The rows a, b and c represent the surface condition: dissolution in the active range (-450 mV (SCE)); passivation at 0 mV (SCE) and at +500 mV (SCE). The simultaneous occurrence of peaks representing the metallic and the oxide states shows that the oxide is extremely thin. The oxide present on the surface after active dissolution is formed during rinsing and transferring of the sample to the analyser. The cations in the passive film (rows b and c) are mainly Cr^{3+}. The contents of Fe^{2+} and Fe^{3+} are rather low. The figure shows that the positions of the deconvoluted Mo peaks correspond exactly to the positions of the states described in figure 2. The largest peaks are those denoted Mo^{6+} and Mo^{hy}, while the peak denoted Mo^{4+} (MoO_2) is very low. It also appears that the total intensities of the Mo signals decrease with the potential. At the highest potential almost only the six valency state exists. In the energy range of the Mo spectra a signal from sulphur bonded in SO_4^{2-} occurs. Its intensity is established from the S 2p signal. The sulphate is adsorbed on the surface of the passive film and is removed by a very slight ion

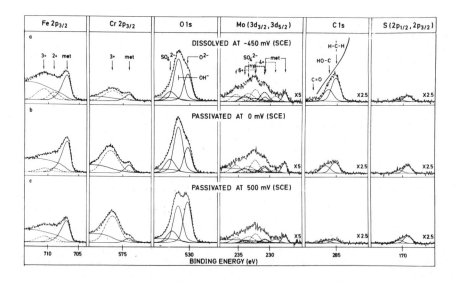

Fig. 3. ESCA spectra of the Fe24.1Cr1.9Mo (110) alloy recorded after polarization at: a) active range (-450 mV (SCE)); b) 0 mV (SCE); c) 500 mV (SCE).

etching. The carbon spectra show contributions from carbon bonded to hydrogen and OH⁻ in acetone and methanol. The compounds are adsorbed on the surface during rinsing of the sample. The oxygen signals are split into three components. The signals marked 0^{2-} and SO_4^{2-} represent oxygen in oxides (Cr_2O_3 etc.) and sulphate respectively. The "OH⁻" peak should in principle be deconvoluted into a series of signals within a narrow energy range representing oxygen in methanol and acetone. However, it appears from row c that the carbon signal is very low and therefore the contribution from these compounds to the oxygen signal is low. Thus, the main components of the OH⁻ signal will be attached to oxygen in hydroxide.

The composition of the passive film and the metal phase underneath the oxide were calculated using the expressions (11):

$$I_M^{OX} = Y_M^{OX} \, D_M^{OX} \, \lambda_M^{OX} \cdot \sin 38.5 \; (1-\exp(-a^{OX}/(\lambda_M^{OX} \cdot \sin 38.5))) \tag{1}$$

$$I_M^{met} = Y_M^{met} \, D_M^{met} \, \lambda_M^{met} \cdot \sin 38.5 \cdot \exp(-a^{OX}/(\lambda_M^{OX} \cdot \sin 38.5)) \tag{2}$$

$$C_M^{OX} = D_M^{OX}/\Sigma \, D_N^{OX} \quad \text{alt.} \quad C_M^{met} = D_M^{met}/\Sigma \, D_N^{met} \tag{3}$$

where I_M^{OX} and I_M^{met} are the measured intensities from the element M in the oxide and metal phases. The symbols represent: Y - the relative photoelectron yield (experimentally interpreted from pure metals and well-defined oxides); D - the atomic density of the element; λ - the attenuation length of the photoelectrons; 38.5 is the angle between the surface and the spectrometer axis. The values of the parameters Y and λ are shown in figure 4. The thickness of the oxide, a^{OX}, is estimated from eqn. 1 and eqn. 2 by making the assumption that the passive film consists of Cr_2O_3.

Figures 4 and 5 show the contents of the cations and the thicknesses of the reaction products formed on the surfaces of the two alloys after active dissolution and during passivation. It appears that the Cr^{3+} content in the passive films of the two alloys are the same, ~ 70%. Thus, Mo does not markedly influence the Cr content in the passive film. The three-valency state of Fe dominates at the lower potential, while for increasing potentials the contribution of Fe^{2+} becomes higher. From figure 5 it appears that the total Mo-content in the oxide is rather low. At low potential in the passive range all Mo states are represented. At the highest potential the overall contribution of Mo is lowered while at the same time almost only the six valency state exists. Thus, Mo is oxidized above a certain potential to its highest oxide state.

The thickness of the passive film is potential dependent. The thickness of the film formed on the Mo alloys varies between 8Å and 16Å in the potential range -200 mV to +500 mV (SCE). The film formed on the Fe-Cr alloy is slightly thicker.

566

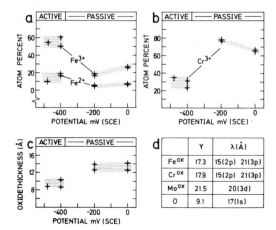

Fig. 4. a-b) The integrated cation content of the oxide product formed on Fe19.8Cr (100) alloy. c) The thickness of the oxide product. d) The parameters used in eqn. 1 to 3.

	Y	λ(Å)
Fe^{OX}	17.3	15(2p) 21(3p)
Cr^{OX}	17.9	15(2p) 21(3p)
Mo^{OX}	21.5	20(3d)
O	9.1	17(1s)

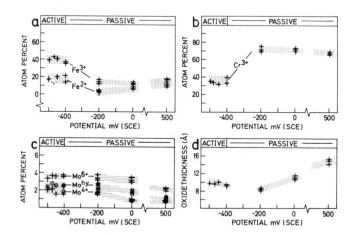

Fig. 5. a-c) The integrated cation content of the oxide product formed on Fe24.1Cr1.9Mo (110) alloy. d) The thickness of the oxide product.

The estimated composition of the passive film was made as if the elements are distributed uniformly through the layer. An attempt was performed to analyse the film in depth by successive ion etches and analyses. Spectra were recorded from three different electron energies to ensure that the ion etching would not influence conclusions about the surface composition. The measured electron levels were 2p and 3p of Cr and Fe. The attenuation lengths of the 2p and 3p electrons are 15Å and 21Å respectively. To obtain information from only a few atomic layers, Auger transitions at ~ 40 eV kinetic energy (attenuation 4Å) were recorded by the Auger spectrometer attached to the analysis system.

The figures 6a to 6d show the measured intensities in depth of the signals recorded from the Mo-alloyed sample passivated at +500 mV (SCE). It appears (fig. 6a) that sulphur and carbon are only adsorbed on the surface. The intensities of 2p, 3p and Auger-signals are shown in the figures 6b to 6d respectively. Due to the short attenuation of the Auger electrons (fig. 6d), the metal phase is not detected until 7Å of the layer is removed by etching; in the spectrum obtained by ESCA the metallic state is present before etching.

From the recordings the chemical composition in depth of the passive film was calculated by dividing the layer in a number of parallel layers. The composition of the layer at the oxide/metal interphase was obtained directly from the measured intensities. The composition of layer no. 2 was calculated from the recorded intensities and the estimated composition of layer no. 1. Then, the calculation was repeated in the same way.

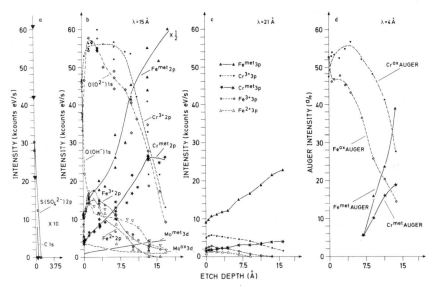

Fig. 6. Etch profiles of: a) contamination elements C and S; b) Fe 2p, Cr 2p, Mo 3d and O 1s; c) Fe 3p and Cr3p; d) Auger signals of Fe and Cr.

Fig. 7a shows the ratio, $Cr^{3+}/(Cr^{3+} + Fe^{3+} + Fe^{2+})$ calculated from the intensities of the signals representing three different attenuation lengths. It appears that the Cr^{3+} concentration in the inner atomic layers is almost 100 at.%. Further, the diagram shows that in the inner part of the passive film the agreement between the three independent measurements are rather good. On the other hand, at the outer part of the oxide, the disagreement is pronounced. It appears that the Auger-method gives an apparent high Cr^{3+} content. The interpretation

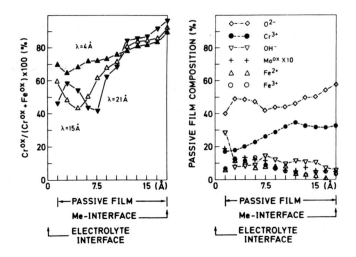

Fig. 7. a) Calculated $Cr^{3+}/(Cr^{3+} + Fe^{2+} + Fe^{3+})$ ratios.   b) The composition of the passive film in depth.

is that due to selective sputtering, Cr^{3+} ions are enriched on the surface which gives a high Cr/Fe ratio in Auger. Thus, the figure indicates that the Cr^{3+} content varies through the passive film. A further piece of evidence is that it would not have been possible to get the low values in the 2p and 3p profiles if the Cr^{3+}-ions are uniformly distributed through the passive film.

The conclusion of the above is that a composition gradient exists in the passive film. The distribution of the elements can be described approximately by the high intensity electrons (Cr 2p, Fe 2p, O 1s etc). Figure 7b shows the composition of the passive film obtained from these spectra. It appears that the ions Fe^{2+}, Fe^{3+} and Mo^{ox} are concentrated in the outer part of passive film and compensate for the lowering of the Cr^{3+} content.

DISCUSSION

This study confirms earlier results (3-11) that the passive film formed on the surface of Fe-Cr and Fe-Cr-Mo alloys consists mainly of Cr-compounds, which cation content of Cr^{3+} is about 70% (obtained from the recorded intensities before ion etching). At low potential in the passive range Mo is enriched in the passive film mainly in its four and six valency states. It has been suggested that Mo exists in a complex oxyhydroxide. At high potential (500 mV SCE) the Mo content is lowered at the same time as it is oxidized to Mo^{6+}. This occurs above the transpassive potential of pure Mo. Iron is present as Fe^{2+} and Fe^{3+}. Their contents increase with the potential and compensate for lowering in Cr^{3+} content. That is most pronounced for the Fe-Cr alloys. It is suggested that one positive effect of Mo is that it stabilizes the passive film.

The structure of the oxide products naturally can not be directly determined by the spectroscopic methods, but the ESCA measurements demonstrate that the anions are OH^- and O^{2-}. Thus, the conclusion is that oxide and hydroxide or their mixture coexist in the passive film. The profiling of the film was an attempt to get information in depth about the chemistry of the film. The interpretation has to be made very carefully because, as demonstrated by the Auger analysis, the ion etching changes the composition of the surface. From the obtained information it appears that OH^- ions are present in the underlying layers. However, the measured OH^- content is small compared to the O^{2-} content. This indicates that at least in depth the O^{2-} state is dominant. A possibility is that OH^- ions are partly transformed to O^{2-} by the etching. This reaction can not be excluded, but the OH^- content in the film is limited because summation of the positive and the negative charges of the cations and the anions respectively gives, at least for the inner layers, almost the same values.

After taking all objections into account it is concluded that the inner layer of the passive film consists of a compound with composition close to Cr_2O_3. As mentioned above, it is not possible to determine the structure, but the chemical shift of Cr^{3+} before ion etching is exactly the same (2.8 eV) as obtained from Cr_2O_3 formed at high temperature. The chemical shift of Cr^{3+} in $Cr(OH)_3$ is 3.3 eV (3). This indicates that the Cr^{3+}-ions in the passive film are mainly coordinated and bonded as Cr^{3+} ions in Cr_2O_3.

It is obvious that hydroxide ions are present in the outer part of the passive film. The analysis indicates that a continuous transition occurs from the inner oxide to the outer hydroxide layer. Thus, the layer can be characterized as oxyhydroxide. It has been suggested that Mo, at least at low potential, is bonded in an oxyhydroxide. Because the recorded OH^- signal is much larger than the total Mo signal the hydroxide compound must also contain Fe and/or Cr. This leads us to the conclusion that the positive effect of Mo in high alloyed steels is that Mo stabilizes the passive film.

It has been made clear that Mo influences the thickness of the passive film in that way that the film becomes thinner when it is alloyed with Mo. This contradicts to Sigimoto and Sawada (6) who found the opposite but their results were for austenitic stainless steels exposed to hydrochloric acid.

The description above is based on analyses after a constant passivation time. It has been shown earlier (7) that at least the Cr^{3+} content increases with the passivation time. It was stated that the change of composition is controlled by kinetic effects: the initially formed film contains a high concentration of Fe-ions due to its availability; prolonged exposure enriches the strongest oxygen-binder at the passive film/metal interface as at thermal oxidation.

The analyses of the samples polarized at the active range show enrichment of Cr and Mo in the surface products. It was argued that the oxide products were formed after interrupting the current and that the composition of the oxide reflects the composition of the metal at the interface. It can also be read from the spectra that Mo is enriched in the metal phase below the oxide film. It has been suggested (10,11) that the selective dissolution of Fe and thereby the enrichment of the alloying elements are provoked by the tendency of the alloying elements to form intermetallic compounds. The change in surface composition changes the electrochemical behaviour of the alloy; the increased corrosion potential and lowering of the peak current at passivation (fig. 1) of the Mo-alloyed sample are suggested to be due to enrichment of Mo and Cr. Further, it has been stated that the formation of the passive film is enhanced by the enrichment of the film forming elements enriched in the outer atomic layers.

CONCLUSION

The passive film formed on Fe-Cr and Fe-Cr-Mo alloys consists of an inner layer of Cr-rich oxide with the stoichiometry of Cr_2O_3 and an outer layer of hydroxide containing Fe, Mo and Cr.

The role of Mo in Fe-base Cr-alloys is that: Mo and Cr are enriched on the surface of the metal phase during anodic dissolution and thereby the passivatability of the alloy is enhanced; Mo stabilizes the passive film and contributes to the improvement of the overall corrosion resistance.

ACKNOWLEDGEMENT

The authors acknowledge the Swedish Board of Technical Development for financial support.

REFERENCES

1 I. Olefjord, Proc. 6th Scand. Corrosion Congress, Swedish Corrosion Institute, Stockholm, 1971.
2 H. Fischmeister and I. Olefjord, Monatshefte für Chemie, 102 (1971) 1486.
3 I. Olefjord and H. Fischmeister, Corrosion Science, 15 (1975) 697.
4 J.E. Castle and C.R. Clayton, Corrosion Science, 17 (1977) 7.
5 B-O. Elfström, Materials Science and Engineering, 42 (1980) 173.
6 K. Sugimoto and Y. Sawada, Corrosion Science, 17 (1977) 425.
7 I. Olefjord and B-O. Elfström, Proc. 8th Intern. Symposium on the Reactivity of Solids, ed. J. Wood et al. Plenum Press, New York, p. 791, 1977.
8 K. Hashimoto, K. Asami and K. Teramoto, Corrosion Science, 19 (1979) 3.
9 K. Hashimoto and K. Asami, Corrosion Science, 19 (1979) 251.
10 I. Olefjord, Materials Science and Engineering, 42 (1980) 161.
11 I. Olefjord and B-O. Elfström, Corrosion NACE, 38 (1982) 46.

Passivity of Metals and Semiconductors, edited by M. Froment
Elsevier Science Publishers B.V., Amsterdam — Printed in The Netherlands

MECHANISMS CONTROLLING PASSIVE FILM COMPOSITION OF Fe-Cr-Ni STEELS
IN ELEVATED TEMPERATURE SOLUTIONS

W. BOGAERTS, P. VANSLEMBROUCK, A. VAN HAUTE and M. BRABERS
University of Leuven (K.U.Leuven), Institute of Industrial Chemistry
de Croylaan 2, B-3030 Heverlee (Belgium)

ABSTRACT

Electrochemical reactions, together with the influence of electrode-potentials
and temperature on the passive film composition, have been studied for stainless
steel Type 304 (18 Cr - 8 Ni) in high-temperature alkaline solutions by combining
in situ electrochemical measurements with Auger Electron Spectroscopy analyses.

Important relative enrichments or depletions of the alloying constituents
iron, chromium and nickel have been detected in the surface layers, and the
results reveal a surprisingly strong influence of the electrochemical potential
on the chemical composition of the passive surface. This becomes particularly
pronounced at the more elevated temperatures (150 - 300 °C).

All of these composition changes can be explained in terms of some well-
defined selective dissolution reactions.

INTRODUCTION

The understanding of the passivity and corrosion behaviour of metals at
elevated temperatures becomes more and more important since operation temperatures have been raised in many industrial processes. Especially in the chemical
industries and power generating plants many attempts have been made to understand
and solve corrosion problems encountered in high temperature and high pressure
aqueous environments.

Unfortunately, numerous experimental difficulties have impeded a detailed
study of the mechanisms and electrochemical reactions controlling the composition
and structure of anodic passive films under these circumstances. Recent improve-
ments in the electrochemical measuring techniques for autoclave systems (e.g.
reference electrode concepts and their appropriate correction factors) and the
use of modern surface analytical techniques (e.g. Auger Electron Spectroscopy
with high spatial resolution) have provided powerfull means for a definite answer
to many unresolved questions.

At the present time, however, only sporadic information is available in open
literature. Some preliminairy results of Scanning Auger Microprobe investigations
on samples, previously exposed to high temperature aqueous solutions, have for
instance been published by the present authors (ref.1 and 2). Materials covered
were : stainless steel AISI 304, incoloy 800 and inconel 600. These data revealed

Cr-depletion processes in the passive film and possible Ni-enrichment phenomena
in the deeper oxide layers or at the oxide-metal interface.

The present paper reports more details about the electrochemical reactions
and the influence of electrode potentials and temperature on the passive film
composition for Type 304 (18 Cr - 8 Ni) stainless steel in alkaline chloride-
containing solutions.

EXPERIMENTAL

Anodic polarisation curves have been determined and samples have been treated
potentiostatically at different anodic potentials in an autoclave system at
temperatures ranging from 65 to 275 °C. The test vessel contains a conventional
three electrode electrochemical measuring assembly with an external Ag/AgCl
reference electrode. Accurate correction factors for the measured potential
values, taking into account the thermal diffusion phenomena, have been determined
as a function of the temperature difference (ΔT) across the cooled salt bridge
of the reference system.

Based on a large number of experiments, we have been able to show that all
the externally measured potential values (E_{meas}) could be transformed into
thermodynamically significant values referred against the standard hydrogen
electrode at the solution temperature (E_{H_T}), by means of the following equation:

$$E_{H_T} = E_{meas} + E_{correction} \qquad\qquad (1)$$

where $E_{correction}$ can be calculated from :

$$E_{correction} = E_{AgCl}(T_o) - 0.93 \times (\Delta T) \qquad , \text{ for } \Delta T < 75°C \qquad (2)$$

or

$$E_{correction} = E_{AgCl}(T_o) - \{70 + 1.20 \times (\Delta T - 75)\} , \text{ for } \Delta T > 75°C \quad (3)$$

in which $E_{AgCl}(T_o)$ is the electrochemical potential of the Ag/AgCl reference
electrode at the lower temperature of the cooled salt bridge system.

The possibility of transforming the experimentally obtained results into
thermodynamically significant data allows a meaningful use of available theoretic
high-temperature thermodynamic equilibrium data (E - pH diagrams of the
different alloying elements) for the interpretation of a number of phenomena.

After exposure to the high-temperature solution the samples were transferred
to a Scanning Auger Microprobe (SAM) with sub-micron spatial resolution. Depth-
profiles and surface distributions of different alloying or environmental
species in the surface oxide layer were determined. More details about the

adopted experimental procedure have been described elsewhere (ref. 1).

Deoxygenated test solutions, containing various amounts of Cl^- (475 - 3550 ppm) together with HCO_3^- and/or OH^- (0.1 M) were used. The chemical analysis of the alloy samples used is given in Table 1.

Fe	Cr	Ni	Mn	Si	C	S	N	P	Other
bal.	18.65	8.80	1.62	0.48	0.06	0.01	0.008	0.019	Cu: 0.10 Mo: 0.05

TABLE 1 :   Chemical composition of steel AISI 304 samples (wt %).

RESULTS and DISCUSSION

At lower temperatures (< about 100°C) only minor relative enrichment phenomena of different alloying constituents have been found in the passive potential range (Figure 1). Typical results show rather small oxide thicknesses and an initial passive film composition which is greatly influenced by previous surface treatment procedures and by the bulk composition of the alloy.

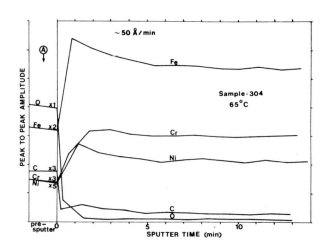

FIGURE 1 :   AES composition profile of surface film on Type 304 stainless steel sample after anodic treatment during 1000 min. at +150 mV vs Ag/AgCl (solution: 1000 ppm KCl + 0.1 M KHCO$_3$;   65°C). Surface enrichment or depletion phenomena of the alloying constituents are not very pronounced. Relative amounts of Fe, Cr and Ni at Ⓐ in the depth profile of the surface oxide are :   Fe ≈ 67 at %, Cr ≈ 24.5 at % and Ni ≈ 8.5 %.

574

At somewhat more elevated temperatures (e.g. 150°C) increasing enrichments of chromium and nickel, due to selective dissolution of iron, occur at electrode potentials in the primary passive potential range. Strong transpassive dissolution behaviour of chromium causes, however, a sharp decrease of the chromium content and a corresponding relative Ni-enrichment at higher potential values. Secondary passivation phenomena may, however, be observed at these elevated potentials in some of the test solutions (rather low Cl^- concentrations and/or higher pH), and may primarily be ascribed to a favourable anodic behaviour of nickel under these experimental conditions.

With increasing temperature, the transpassive dissolution reaction of chromium already starts at lower electrode potentials and becomes increasingly significant. Transpassive processes and secondary passivation may almost completely determine the oxide film composition of the alloy in the high temperature range. This is clearly illustrated by means of the AES depth profile in Figure 2 (Type 304 stainless steel ; 225°C ; 1000 ppm KCl + 0.1 M $KHCO_3$, pH = 8.4 ; +150 mV vs Ag/AgCl). The corresponding anodic polarisation curves at 225°C and pH 8.4 solutions are shown in Figure 3 for stainless steel 304 and for pure Fe and Ni. The different potentials at which passivation treatments and subsequent AES analyses have been carried out, are indicated at the left of the figure.

In the lower potential range (e.g. -300 mV),strong enrichments of Cr and Ni are found (Table 2). This is in agreement with the high anodic dissolution

Sputter time (min.) and corresponding current (mA)	0	5 3	30 3	90 3	145 3	+95 5	200+95 5 20
S (at %)	0	0	1.1	0	0.4	0	0
Cl	0	0	0	0	0.3	0	0
C	22.2	18.2	12.0	8.8	9.6	5.5	0
O	44.9	45.0	39.3	29.4	25.6	17.6	0
Cr	15.3	18.2	16.2	13.4	12.2	17.1	20.7
Cr/Σ =	(46.6)	(50.1)	(34.3)	(21.6)	(19.0)	(22.2)	
Fe	12.5	12.7	20.3	36.9	41.7	50.0	70.1
Fe/Σ =	(38.0)	(35.7)	(42.7)	(59.7)	(65.2)	(65.0)	
Ni	5.0	5.9	10.9	11.5	10.1	9.8	9.2
Ni/Σ =	(15.4)	(13.4)	(23.0)	(18.6)	(15.8)	(12.8)	

TABLE 2. Quantification of AES depth profiling results after exposure of Type 304 steel sample at 225°C and -300 mV (other experimental conditions are the same as in figure 1). Σ = [Fe] + [Cr] + [Ni].

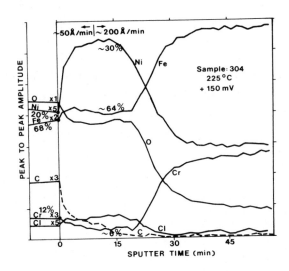

FIGURE 2 : AES composition profiles of surface film on Type 304 stainless steel sample after anodic treatment during 1000 min at +150 mV vs Ag/AgCl (solution: 1000 ppm KCl + 0.1 M KHCO$_3$; 225°C).

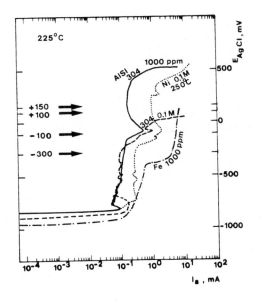

FIGURE 3 : Anodic polarisation curves of Type 304 stainless steel, Fe and Ni in x KCl + 0.1 M KHCO$_3$ at 225°C (x = 1000 ppm or 0.1 M).

current of Fe under the present experimental conditions (cf. Fig. 3 : pure Fe). At electrode potentials higher than about -150 mV transpassivity of chromium and a corresponding depletion of this element occurs (see Table 3 : surface compositions at -100 mV). This possibly results in a generalised passivity breakdown if experimental conditions are unfavourable for establishing secondary passivity. Secondary passivity is mainly governed by the behaviour of Ni (see Figure 2 : surface concentration of Ni increased to about 30 % at + 150 mV) and is favoured by lower Cl^- concentrations and higher values of the solution pH (e.g. 10.5).

Sputter time (min.) and corresponding current (mA)	0	5 10	30 10	60+50 10 30	+160 30	+90 30	+300 30	- 30
S (at %)	1.5	0	0	0	0	0	0	0
Cl	2.5	3.0	0.9	0	0	0	0	0
C	45.7	14.7	9.7	5.9	4.7	4.5	0	0
O	25.7	35.8	39.3	42.1	29.1	19.2	8.4	1.24
Cr	0	0	0	0	8.7	9.7	14.8	19.9
Cr/Σ =	(0)	(0)	(0)	(0)	(13)	(13)	(16)	(20.1)
Fe	20.8	35.3	40.8	39.1	45.9	57.3	68.4	67.9
Fe/Σ =	(85)	(76)	(81)	(75)	(69)	(75)	(75)	(68.7)
Ni	3.7	11.3	9.3	12.9	11.6	9.3	8.5	11.0
Ni/Σ =	(15)	(24)	(19)	(25)	(18)	(12)	(9)	(11.1)

TABLE 3. *Quantification of AES depth profiling results after exposure of Type 304 steel sample at 225°C and -100 mV (other experimental conditions are the same as in figure 1). Σ = [Fe] + [Cr] + [Ni].*

The situation might, however, become more complicated because of additional selective dissolution processes of Ni (e.g. at about -200 mV), resulting in an increased anodic reactivity. This is particularly apparent for alloys with increased Ni-content and of course for pure Ni-samples (Fig. 3). The combination of these processes might provide an explanation for many aspects of the corrosion behaviour of Fe-Cr-Ni alloys in the present type of solutions. The discussion of this topic is beyond the scope of this article and will be discussed more detailedly in a future communication.

CONCLUSIONS

1. Important relative enrichments or depletions of the alloying constituents iron, chromium or nickel have been detected in the passive surface layers after exposure of Fe-Cr-Ni steels to alkaline solutions at elevated temperatures.

2. These phenomena have been found to be mainly controlled by the electrochemical potential of the alloy and all of these composition changes can be

explained in terms of some well-defined selective dissolution reactions.

3. Especially the transpassive dissolution of chromium will dominate many aspects of the corrosion behaviour of Fe-Cr-Ni alloys in the present type of solutions.

REFERENCES
1. W. Bogaerts, A. Van Haute and M.J. Brabers, Proc. 8th Int. Congress on Metallic Corrosion, p. 31, Mainz (1981).
2. W. Bogaerts, A. Van Haute, M. Brabers and P. Vanslembrouck, Research Symposium CORROSION/82, p. 84, Houston (1982).

CHARACTERIZATIONS OF PASSIVE FILMS FORMED ON STAINLESS STEEL IN HIGH
TEMPERATURE WATER

J. B. Lumsden and P. J. Stocker
Science Center, Rockwell International, Thousand Oaks,
CA (USA) 91360

ABSTRACT
    Surface study techniques were used to investigate films on Type 304 stain-
less steel which were formed during exposure to high purity water at 288°C.
The results indicated that the film chemistry depended strongly upon the concentration of the dissolved O_2 in the water. Films formed in water having
8 ppm O_2 were stoichiometric mixed oxides; whereas, those formed in water with
10 ppb O_2 were highly defective oxyhydroxides. The latter films are not as
protective as the stoichiometric oxides.

INTRODUCTION
    Previous work (1) has shown that solution annealed Type 304 stainless steel
is not susceptible to stress corrosion cracking (SCC) in high purity water at
288°C provided that the dissolved oxygen concentration is not below 200 ppm.
However, if the oxygen concentration in this solution is below this level,
transgranular SCC can occur.
    Stress corrosion cracking in stainless steels is best correlated with the
film rupture model (2). The high purity water system offers the opportunity of
investigating the properties of the surface films on stainless steel relative
to SCC susceptibility without the possible complicating effects of salt
films. In the work reported here, Auger electron spectroscopy (AES) and X-ray
photoelectron spectroscopy (XPS) were used to investigate the films formed on
Type 304 stainless steel resulting from exposure to high purity water at
288°C. Films were formed in solutions having dissolved oxygen levels well
within both the regimes where transgranular SCC occurs and where immunity to
SCC exists.

Experimental
    Films were formed in a once through autoclave system. All components in the
system, except the preheater and autoclave, were constructed from Type 316
stainless steel. The preheater and autoclave were made of Ti-6Al-4V. The
titanium alloy was used to prevent scavenging of O_2 at high temperatures.

The 20 ml autoclave was continuously refreshed at a flow rate of 20 ml/min with high purity water having a conductivity less than 10^{-6} mho/cm. The desired level of dissolved oxygen was obtained by purging the water in a holding tank with mixtures of argon and oxygen gases. The dissolved oxygen concentrations were continuously measured at both the inlet and outlet of the autoclave.

Before mounting in the autoclave, the surfaces of the specimens were mechanically polished down to 15 μm. They were then washed with acetone and distilled water. Films were formed at 288°C at 10 ppb and 8 ppm oxygen concentrations. Exposure times of 2 hours and 24 hours were used. The samples were cooled from 288 to 30°C in approximately 15 minutes and immediately placed in an ultrahigh vacuum system.

In-depth composition profiles were obtained using a Physical Electronics Model 590 scanning Auger microprobe. Spectra were taken using a 3 keV electron beam. A 5 keV Xe ion beam was used to sputter the films. The atomic concentration of the constituents in the film was determined using the procedure given by Palmberg et al. (3). the following Auger lines were used for the quantitative analysis O (503 eV), Cr (529 eV), Fe (703 eV) and Ni (848 eV). All XPS spectra were taken with a Physical Electronics Model 548 ESCA spectrometer using Al K_α exciting radiation. Energy calibration was based on the following line positions for clean copper metal: $Cu(2p^{3/2})$ - 932.5 eV and $Cu(3p^{3/2})$ - 75.0 eV.

RESULTS AND DISCUSSION

Figures 1 and 2 show Auger spectra from the surface of Type 304 stainless steel which had been exposed to high purity water at 288°C having 8 ppm and 10 ppb dissolved oxygen, respectively. These spectra show that there is a pronounced difference in the surface composition of the films formed under these two conditions. Exposure to the solution having the high oxygen content results in a Fe-Ni oxide; whereas, the film resulting from exposure to the 10 ppb oxygen solution is an Fe-Cr-Ni oxide.

The in-depth composition profile of the film formed by exposure to the 8 ppm oxygen solution is shown in Fig. 3. This figure gives the atomic percent versus sputtering time. The sputtering rate was estimated to be approximately 2Å/sec based on results from sputtering SiO_2 films of known thicknesses under the same conditions used in obtaining the profiles. The results shown in Fig. 3 suggest a duplex film is formed in the 8 ppm oxygen solution. The outer film is approximately 2000Å thick. The quantitative analysis of the Auger spectra suggest that this film is $(Fe,Ni)_2O_3$. As the film is sputtered, the amount of oxygen in the film decreases, iron increases slightly, nickel remains the same, and chromium begins to appear after approximately 3000Å is removed.

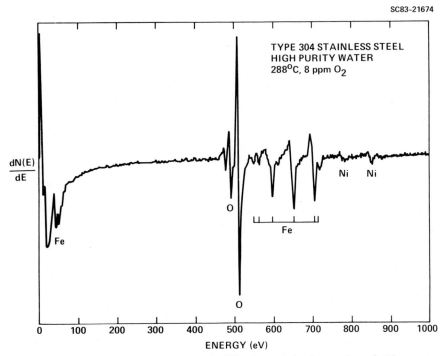

Fig. 1. AES spectrum for surface of film formed in 8 ppm O_2 solution.

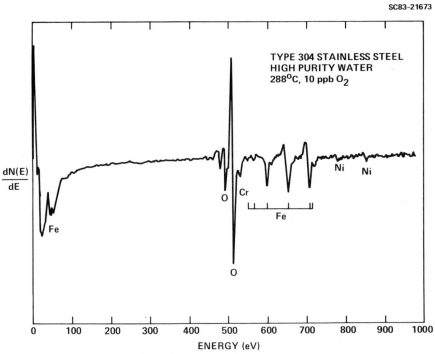

Fig. 2. AES spectrum from surface of film formed in 10 ppb O_2 solution.

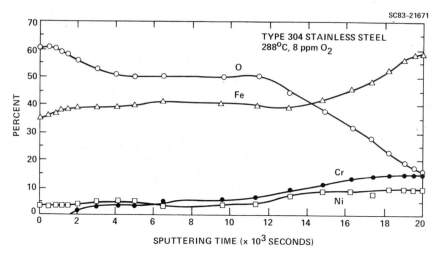

Fig. 3. In-depth composition profile of the film formed in the 8 ppm O_2 solution.

There is a broad range of approximately 2.5 μm where the film appears to have a constant composition and formula of $(Fe,Cr,Ni)_3O_4$. The broad film-alloy inter-facial region is likely a result of roughening of the alloy during film growth (4).

Figure 4 is the in-depth composition profile of the film formed on stainless steel in the 10 ppb oxygen solution. This film appears to have a high density

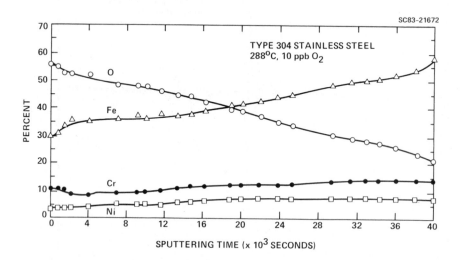

Fig. 4. In-depth composition profile of film formed in the 10 ppb O_2 solution.

of oxygen vacancies. In most situations, a highly defective film is not a good ionic diffusion barrier. The higher rate of ionic transport through the film formed in the 10 ppb oxygen solution relative to that formed in the 8 ppm oxygen solution is the likely explanation for the former being thicker.

The XPS analysis of the surfaces of the oxides provides additional details concerning their properties. The nickel spectra were the same from both oxides. A binding energy of 855.6 eV was obtained for the $Ni(2p^{3/2})$ line with an accompanying satellite at 862.1 eV. These energies are approximately those obtained for $NiFe_2O_4$ (5). The binding energy for $Cr(2p_{3/2})$ in the film formed in the 10 ppb oxygen solution was 576.9 eV. This energy is between that obtained for Cr_2O_3 and $CrOOH$ (6), which suggests a mixed oxide-hydroxide film. In accordance with the AES results, Cr was not detected by XPS in the surface layers of the film formed in the highly oxygenated water.

The iron and oxygen spectra were different for the two films. A binding energy of 711.8 eV was obtained for $Fe(2p_{3/2})$ in the film formed in the low oxygen solution. The $Fe(2p_{3/2})$ line from the film formed in the high oxygen solution had a binding energy of 711.4 eV; however, this iron spectrum also had a distinctive satellite line between the $Fe(2p_{1/2})$ and $Fe(2p_{3/2})$ peaks. This is characteristic of ferric oxide and oxy-hydroxide. The satellite peak was not observed in the iron spectrum from the film formed in the 10 ppm oxygen solution, which suggests a mixture of Fe^{+2} and Fe^{+3} ions since such a satellite is not observed in Fe_3O_4 (5).

The oxygen line from both oxides was asymmetric, which indicated that it was a convolution of more than one line. The peak positions of oxygen in films formed in the high and low oxygen solutions were 530.6 and 531.1 eV, respectively. These energies are in accordance with those obtained for metal-oxide and metal-hydroxide bonds, respectively (5).

Thus the AES and XPS analysis of the films on Type 304 stainless steel have shown that there are pronounced differences in the chemistries of films formed under conditions in which this alloy is susceptible and not susceptible to transgranular SCC. The results suggest that stoichiometric, mixed oxides form when the alloy is exposed to high purity water at 288°C containing 8 ppm oxygen; whereas, nonstoichiometric, highly defective oxy-hydroxides form when the alloy is exposed to water at the same temperature with 10 ppb dissolved oxygen. One would expect stoichiometric films to be effective ionic diffusion barriers and that metal surfaces, exposed during conditions favorable for the formation of these films, to be rapidly repassivated. However, the defective films are not good diffusion barriers and the dissolution rate of newly exposed metal surfaces would decrease relatively slowly on the film thickened. The latter conditions are considered to be favorable for SCC in the slip-dissolution model.

ACKNOWLEDGEMENT

The support of the Electric Power Research Institute (Contract No. RP1167-07) is gratefully acknowledged. The authors also wish to thank Don Cubicciotti of EPRI for his encouragement and many helpful discussions.

REFERENCES

1.   F.P. Ford and M.J. Povich, Corrosion, 35 (1979) 569-574.
2.   R.W. Staehle, in J.C. Scully (Eds.), The theory of Stress Corrosion Cracking in Alloys, North Atlantic Treaty Organization Scientific Affairs Division, Brussels, 1971, pp. 223-288.
3.   P.W. Palmberg, G.E. Riach, R.E. Weber, and N.C. McDonald, Handbook of Auger Electron Spectroscopy, Physical Electronics Industries, Edina Prairie, 1977.
4.   M. Seo, J.B. Lumsden, and R.W. Staehle, Surface Science, 50, (1975) 540-552.
5.   N.S. McIntyre and D.S. Zetaruk, Anal. Chem., 49 (1977) 1521-1532.
6.   K. Asami and K. Hashimoto, Corros. Sci., 17 (1977) 559-570.

Passivity of Metals and Semiconductors, edited by M. Froment
Elsevier Science Publishers B.V., Amsterdam — Printed in The Netherlands

STRUCTURE AND STABILITY OF THE ANODICALLY FORMED FILMS ON 304 STAINLESS STEEL
IN SULFURIC ACID

C.R. Clayton[1], Kantesh Doss[1] and J.B. Warren[2]
[1]Department of Materials Science and Engineering, State University of New York,
Stony Brook, NY 11794.
[2]Brookhaven National Laboratory, Upton, NY 11973

ABSTRACT

The structure and composition of the passive films formed on 304 stainless
steel in deaerated IN H_2SO_4 were studied by RHEED, XPS and AES. The stability
of the passive films as a function of passivation potential and passivation
time were investigated. The role of bound water in affecting the stability of
the passive films is discussed.

INTRODUCTION

Recently, Okamoto (ref. 1) has proposed a model for the structure of the
passive films formed on 304 stainless steel that lays emphasis on the role
played by bound water in affecting the stability of the passive films in acidic
sulfate and sulfate/chloride solutions at various anodic passivation potentials.
Using radiotracer techniques, Okamoto noted a decrease in the amount of bound
water in the film with increasing passivation potential especially above 0.4V
(SCE). This variation in bound water in the film was found to be related to a
compositional transition around 0.4V. According to this model, the passive
films formed on 304 stainless steel are amorphous consisting of hydrous oxides
of iron and chromium. Initially, these films are enriched in water as H_2O-M-
H_2O networks which deprotonate to HO-M-OH and finally to O-M-O. It was proposed that the passivation potential and passivation time either collectively
or separately control the deprotonation process.

In our previous RHEED studies (ref. 2) of 304 stainless steel, we have shown
that there is a high degree of crystallinity in the potentiostatically formed
films on 304 stainless steel in deaerated IN sulfuric acid. There was also a
remarkable agreement between the structure of various compounds detected in the
passive film and the nature of passive film conceived by Okamoto. But, since
these studies involved transfer of the samples into the electron microscope
through air, it was necessary to check whether such a transfer introduced an
artifact in the structural information obtained from such films owing to the

modification of the surface film by atmospheric oxygen. We, therefore carried out another set of polarization experiments in a cell enclosed in an environmental chamber that was purged with prepurified nitrogen. Subsequently, the samples were transferred in an argon saturated transfer device to the electron microscope. In order to acquire an overall picture of the nature of the passive films, we also carried out XPS and AES analysis and we have attempted to compare the information obtained in these analyses with those of RHEED results.

The major objective of the present work is to characterize the structure and composition of the passive films formed on 304 stainless steel and hence, to test the validity of the model of the structure of the passive films already presented in the literature. Another important objective of this investigation is to study the stability of the film due to aging at a given potential.

EXPERIMENTAL

Samples of 304 stainless steel were vacuum annealed at 1100°C for 3 hours, quenched in water and polished to a 1 micron Cr_2O_3 finish. To remove the surface films formed during mechanical polishing, the samples were cathodically polarised at -600 mV for 15 minutes (ca. $10^3 \mu A\ cm^{-2}$) in deaerated 1N sulfuric acid and potentiostated for 1 hour at the passivating potential of interest. Subsequently, the samples were cleaned with deaerated doubly distilled water, dried in argon and immediately transferred to the electron microscope or electron spectrometer either through the atmosphere or argon. The structural and compositional analyses were done on separate freshly prepared samples. All the RHEED patterns were obtained using a Phillips EM 300 electron microscope at 100 KeV. The structural information was obtained by calculating the interplanar spacings employing a thin film of gold as a standard to calculate the camera constant. XPS and AES analyses were performed with a V.G. Scientific ESCA3MK II system. All XPS and AES measurements were carried out in a vacuum of ca. 5×10^{-10} torr. The composition depth profile of the passive films were carried out using AES in conjunction with Ar ion etching.

RESULTS

The potentials chosen for the present passivation studies were 0.25V, 0.40V and 0.55V (vs SCE). The RHEED results are summarized in Table I which presents the structures of the various compounds detected following both aerial and Ar transfer. Typical diffraction patterns acquired from high potential and low potential films are presented in Figures 1a and 1b. It is evident from the diffraction patterns that the passive films formed on 304 stainless steel display a high degree of crystallinity. At 0.25V and 0.40V, an iron-based compound resembling the structure of Green rust II ($4Fe(OH)_2.2Fe(OH)_3.FeSO_4.XH_2O$) (ref. 3) was detected in the film. The passivation of the 304 stainless steel

TABLE I

RHEED analysis of the passive films formed on 304 stainless steel in IN sulfuric acid.

Potential	Mode of Sample Transfer	d Spacing	Phases Observed
0.25V	Aerial	3.54, 3.21, 2.85, 2.46, 2.21, 1.88, 1.7, 1.6, 1.31	Green rust II, γ-CrOOH
	Argon	5.48, 3.65, 2.75, 2.66, 2.46 2.2, 1.94, 1.71, 1.59, 1.53	Green rust II, γ-CrOOH
0.40V	Aerial	6.0, 4.24, 3.54, 3.27, 2.86, 2.57, 2.25, 2.2, 1.84, 1.75, 1.71	Green rust II, γ-CrOOH α-FeOOH
	Argon	7.48, 5.52, 3.85, 3.63, 3.43, 3.26, 2.76, 2.65, 2.44, 2.12, 2.05, 1.96, 1.74, 1.67	Green rust II, γ-CrOOH γ-FeOOH
0.55V	Aerial	3.23, 2.73, 2.35, 2.18, 1.88, 1.69, 1.43, 1.19, 1.16	γ-FeOOH, γ-CrOOH
	Argon	6.34, 3.52, 3.33, 3.13, 2.76, 2.47, 2.35, 2.19, 2.11, 2.04 1.86, 1.81, 1.71, 1.67	γ-FeOOH, γ-CrOOH

 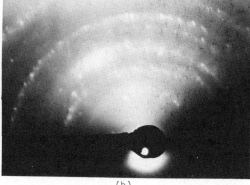

(a) (b)

Fig. 1a and 1b. RHEED patterns obtained from the passive films formed on 304 stainless steel in deaerated IN H_2SO_4 following argon transfer at (a) 0.25V (SCE), (b) 0.55V (SCE).

above 0.4V resulted in the formation of γ-FeOOH while chromium was present at all potentials as γ-CrOOH. No nickel compounds were found in the film. Figure 2 presents the XPS spectra of oxygen obtained from the 304 stainless steel passivated at 0.25V and 0.55V in IN sulfuric acid while Figures 3a and 3b show the

AES depth profiles of the stainless steel following passivation at these potentials.

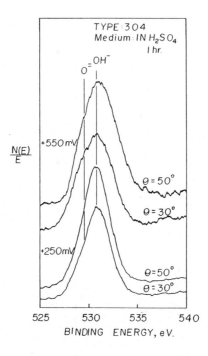

Fig. 2.   Ols XPS spectra obtained from the passive films formed on 304 stainless steel in deaerated IN H_2SO_4 at 0.25V (SCE) and 0.55V (SCE).

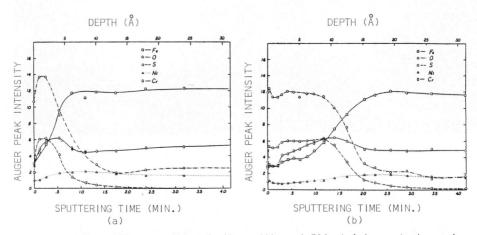

Fig. 3a and 3b.   AES composition depth profiles of 304 stainless steel passivated in deaerated IN H_2SO_4 at (a) 0.25V and (b) 0.55V (SCE).

DISCUSSION

It is apparent from the RHEED data that the structure of the passive films are strongly dependent on the passivation potential and time of passivation. The passive films formed on stainless steel are found to consist of both iron and chromium compounds in polycrystalline form as evidenced by continuous rings observed in our RHEED patterns. This is contrary to the assumption of Okamoto (ref. 1) and several other investigators (Refs. 4, 5) who contend that the passive films formed on stainless steel are amorphous. Both aerial and argon transfer resulted in the polycrystalline films. It was found that the aerial transfer of the passivated samples resulted in formation of α-FeOOH at 0.40V which was not observed in the samples transferred through argon atmosphere. It may be noted in this connection that Misawa et al (ref. 3) have shown that α-FeOOH may be formed by aerial oxidation of γ-FeOOH. Hence, it is clear that in order to obtain very reliable structural information from the RHEED data, it is important to avoid the exposure of the passivated specimens to atmospheric oxygen. Henceforth, our discussions in this section will be presented with respect to the results involving argon transfer. It is noted that the low potential films (<0.40V) contain γ-CrOOH and predominantly complex hydroxy compounds of iron with a significant amount of bound water as seen from the green rust II structure. At high potentials (>0.40V), the passive film exists in the form of oxyhydroxides (γ-FeOOH and γ-CrOOH). The XPS spectra which we have previously published (ref. 6) show the presence of Fe^{3+}, Fe^{2+} and Cr^{3+} peaks in low potential range and predominantly Fe^{3+} and Cr^{3+} peaks at high potentials showing a good agreement with the RHEED results. The relative amounts of Fe^{2+} and Fe^{3+} will be reported later in a more comprehensive paper (ref. 7). The narrow scan of O1s spectra given in Figure 2 shows the peaks corresponding to OH^- at 0.25V and O-OH at 0.55V. The AES depth profiles of the passivated stainless steel indicates an increase in film thickness with potential of passivation.

It is interesting to note that the hydroxide based green rust II phase has a significant amount of entrapped water. We also observed from AES analysis apparent enrichment of chromium at higher potentials. These results brings into question Okamoto's assumption that the bound water tends to be exclusively associated with chromium compounds in the film. Further studies in this area are necessary before this controversy can be resolved. In the passive film aging experiments reported by Okamoto, it is assumed that the chromium compounds are responsible for the high degree of deprotonation observed. To test this assumption, we carried out passivation experiments for a prolonged period (3 hours) in 1N sulfuric acid at 0.25V followed by RHEED analysis. In this work, we observed the complete transformation of γ-FeOOh from green rust II while chromium evidently remained unaltered as γ-CrOOH. Thus, the RHEED data

590

presents evidence that the deprotonation is largely associated with the highly
hydrated iron based compounds which is contrary to Okamoto's model. Despite
these contradictions, the type of bonding between metal and oxygen in the films
proposed by Okamoto for low potential and high potential films fits very well
into the structures indicated by the RHEED data. The compositional transition
observed at 0.40V was accommodated well by a structural transition wherein only
iron compounds were involved as evident from the presence of green rust II and
γ-FeOOH at 0.40V. This transition was found to be the intermediate stage between the transformation of a metastable compound (green rust II) in the film
to a more stable compound (γ-FeOOH) and hence, the film structure at 0.40V
exhibits the characteristics of both low and high potential films.

SUMMARY AND CONCLUSION

The passive films formed on 304 stainless steel were found to be polycrystalline incorporating varying amounts of bound water depending on the
potential and time of passivation. Two iron compounds were detected in the
films such as green rust II (<0.40V), γ-FeOOH (>0.40V). γ-CrOOH was observed
at each of the three potentials investigated viz., 0.25-0.55. The unstable
nature of low potential film was evident by aging experiments which showed
their transformation from the original green rust II structure to γ-FeOOH.
The passive films analyzed by RHEED following argon transfer showed very
little difference in the structure from the ones which were analyzed following
aerial transfer except for the formation of an additional phase (α-FeOOH) at
0.40V by aerial oxidation.

REFERENCES
1. G. Okamoto, Corr. Sci., 13 (1973), 473-489.
2. C.R. Clayton, K.G.K. Doss and J.B. Warren, Spring Meeting, The Electrochemical Society, Minneapolis, Minnesota, May 1981, Extended Abstracts.
3. T. Misawa, K. Hashimoto and S. Shimodiara, Corr. Sci., 14 (1974), 13.
4. E.M. Mahla and N.A. Nielsen, Trans. Electrochem. Society., 89 (1946), 167.
5. T.N. Rhodin, Corrosion, 12 (1956), 123t.
6. C.R. Clayton, K.G.K. Doss, Y.F. Wang, J.B. Warren and G.K. Hubler, Ion Implantation into Metals, Ed. V. Ashworth et al, Pergamon Press, 1982.
7. To be published in the Journal of Electrochemical Society.

ACKNOWLEDGEMENTS

This work was supported by the National Science Foundation under grant #DMR
8106499A01.

Passivity of Metals and Semiconductors, edited by M. Froment
Elsevier Science Publishers B.V., Amsterdam — Printed in The Netherlands

PASSIVITY AND BREAKDOWN OF PASSIVITY OF DIRECTED ENERGY
SOURCES MODIFIED STAINLESS STEEL

P.L. de ANNA[1], M. BASSOLI[1], G. CERISOLA[2], P.L. BONORA[2] and P. MAZZOLDI[3]

[1] Ist. Chimica Fisica Applicata dei Materiali - C.N.R. - Reparto
Corrosione c/o Ist. Chimica Ingegneria, P.le Kennedy -16129 GENOVA

[2] Istituto Chimica - Facoltà di Ingegneria dell' Università,
P.le J.F. Kennedy, Fiera del Mare Pad. D, 16129 GENOVA (Italy)

[3] GNSM-CNR Ist. di Fisica, Via Marzolo, 8 - 35100 PADOVA (Italy)

ABSTRACT

The electrochemical behaviour after surface treatment of common
industrial application ferritic stainless steel (AISI 430 and 434)
as well as ELI 1802, a high quality chromium ferritic steel was
studied. A depolarization of cathodic reaction is detected on AISI
steels as a consequence of physical treatment. The passivation
transition occurs at the same potential, but with higher current
maxima for the treated samples. Impedance data along with different
polarization potentials are presented. Surface modified samples
have similar trends in the impedance spectra, but different electrochemical parameters. Molybdenum stabilizes the passive layers on
SS (AISI 434 and ELI 1802), and its effect is enhanced by the physical surface modifications. The electrochemical behaviour in the
transpassive region is very similar to the one observed with Cr
surfaces.

INTRODUCTION

It is well known (ref. 1, 3) that the stability of the passive

oxide on Cr Stainless Steels is highly affected by many parameters,

including chemical composition, chemical homogeneity, crystal struc-

ture and surface treatment .

Amorphous and fine grain microcrystalline surfaces, in particular,

are known (ref. 6) to present high corrosion resistance. The use

of directed energy sources, such as ion implantation and pulsed

laser irradiation at various energy levels, may cause the onset of

surface modifications which are likely to allow the use of low cost

materials in aggressive enviroments as substitutes of more expensive
and/or rare ones usually employed.

In this paper we present the modified features in the electrochemical behaviour observed after surface treatment of some ferritic
stainless steels of common industrial applications, such as AISI 430
and 434, as well a of high quality Chromium ferritic steel, such as
ELI 1802.

MATERIALS AND METHODS

Three Fe-Cr alloys were used, each one containing about 17 %Cr
in weight, two of them (AISI 434 and ELI 1802) containing about 1.5 %
of Mo. The chemical composition is given in table 1. The Extra Low
Interstitial 1802 stainless steel presents a very fine and regular

TABLE 1

Chemical composition of the investigated alloy (% in weight)

Alloy	Si	Mo	Mn	Ni	Cr	S	P	C
AISI 430	0.54		0.55	0.02	16.63	0.008	0.025	0.071
AISI 434	0.80	1.42	0.35		16.43			0.075
ELI 1802		1.52	0.1		17.23			0.03

grain microstructure as compared to the other two steels, which show
inhomogeneities and carbide precipitates at the grain boundaries.

The electrodes, in the shape of cylinders, were polished to a
mirror finish (1/4 μm diamond paste). They were subjected either
to pulsed ruby laser irradiation (L.I.) (pulse duration 16 nsec.
at either 3.4 or 5.3 J·cm^{-2} energy level) or to Boron implantation
(B.I.) ($2·10^{15}$ ions·cm^{-2} dose at 50 KeV).

All the samples (untreated references and surface modified ones)
were embedded in a moulding epoxy resin and only their cross-section
(untreated or physically modified) was allowed to contact with
the electrolyte in order to constitute a rotating disk electrode.
The surface area varied, depending on the moulding operation,
between 0.1 and 0.2 cm^2.

All the figures reported refer to the <u>unit area.</u>

The electrode was rotated at the speed of 1600 rev min^{-1} so that well defined hydrodynamic conditions were established. At this rotation speed the current is considered free of mass transfer effects. The experimental medium (0.5 M H_2SO_4 solution) was thermostated at 20±0.1°C and de-aerated by high purity argon bubbling. Such an experimental setup was chosen as a result of previous tests showing that both aerated and stagnant solutions mask most features of the electrochemical behaviour of chromium steels, expecially before and during the passivation process (ref.2). The potential was measured with respect to a saturated sulphate reference electrode, in order to avoid chloride contaminations, and was referred in the figures and text to the normal hydrogen electrode (nhe).

All the samples were subjected to :

- potentiodynamic polarization from -585 mV$_{nhe}$ at a scanning rate of 720 mV h^{-1} in the positive direction (AMEL Corrograph Apparatus). Prior to the voltage scanning, the sample was held at the initial cathodic potential value until the current reached a constant value, in order to reduce all the surface oxides. The fairly good reproducibility of the I/V plots confirmed such an assumption.

- electrode impedance along with different polarization points; the experimental setup for the impedance measurements makes use of a Solartron 1172 TFA, an AMEL potentiostat and a twin channel amplifier.

Auger spectroscopic techniques were employed to check possible concentration gradients of alloying elements and/or surface pollution in the near surface layers following the physical surface treatment.

RESULTS AND DISCUSSION.

In Figs. 1 and 2 the steady state polarization curves of AISI 430 and 434 SS immersed in 0.5 M H_2SO_4 and rotated at the speed of 1600 rev min^{-1} are given. The polarization curves of ELI 1802 are given in Fig. 3. Current density values are given (mA cm^{-2}) with

594

respect to the geometric surface; an approximate measurement of
the surface profile after laser irradiation, carried out with a Scanning Electron Microscope (SEM), showed an increase of about 1.5 ÷
1.7 times the untreated metal, i.e. the metal surface area is increased by a factor at least two (ref.6). The actual current density
values are hence lower by the same factor (see Fig. 4).

The anodic polarization curves show a first activation peak on all
the samples, at the same potential value of - 200 mV$_{nhe}$, the maximum current values of which are relevant to the treated surfaces.
A second peak is observed only on the physically treated samples,
in the passive region, at about + 200 mV$_{nhe}$. Such a peak was related in the literature to the oxidation of alloying elements such as
Mn or Mo. Auger spectroscopic surface analysis did not show any
surface composition alteration between the near surface layers and
the bulk as a consequence of L.I. and B.I., except a higher Oxygen
and Nitrogen content. No explanation is therefore possible for the
presence of such a peak, if we do not attempt the hypothesis that
the electrochemical response might be more sensitive to the alloy
composition than the Auger analysis.

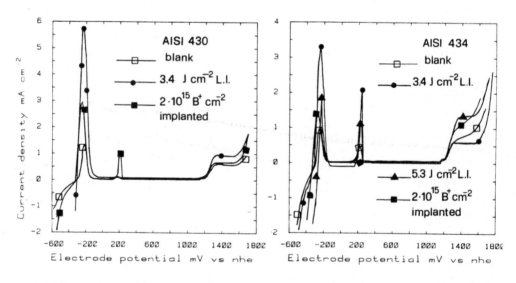

Fig. 1. AISI 430 potentiodynamic polarization curves at 720 mV h^{-1}.

Fig. 2. AISI 434 potentiodynamic polarization curves at 720 mV h^{-1}.

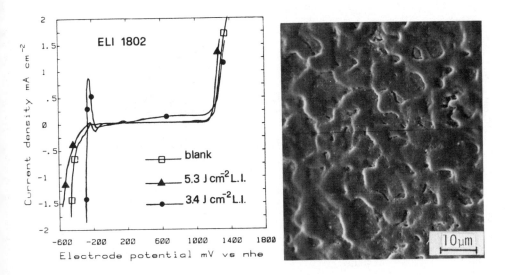

Fig. 3. ELI 1802 potentiodynamic polarization curves at 720 mV h^{-1}. Fig. 4. Morphology of AISI 430 surface after L.I. at 3.4 J cm^{-2}.

The electrode impedance (Z = R - jG) values for the untreated and treated steels along with the different polarization potentials ranges were obtained and plotted in the complex plane in which the parameter is the measuring frequency. Figures 5 a-e show, as an example , the trend of the impedance plots recorded in correspondance to the activation range (Fig. 5a, which refers to AISI 434 J cm^{-2} L.I.), to the activation-passive transition (Fig. 5b, which refers to AISI 430 untreated), to the passive range (Fig. 5c, which refers to AISI 430 3.4 J cm^{-2} L.I.) before, in the anodic sense, and after the second peak (Fig. 5d, which refers to AISI 434 2 10^{15} ions cm^{-2} B.I.), and to the transpassive range (Fig. 5e, which refers to ELI 1802 5.3 J cm^{-2} L.I.) . Table 2 summarizes some electrochemical parameters that are roughly extracted from the analysis of the experimental data. They may give , in some instances, some information about the processes occuring at the electrode.

It is possible to observe,on the activation range, that the Mo containing steels present higher R_t and lower C_{dl} values, and that the physical treatments do not noticeably affect the electrochemical parameters during the dissolution processes. Only a low frequency

596

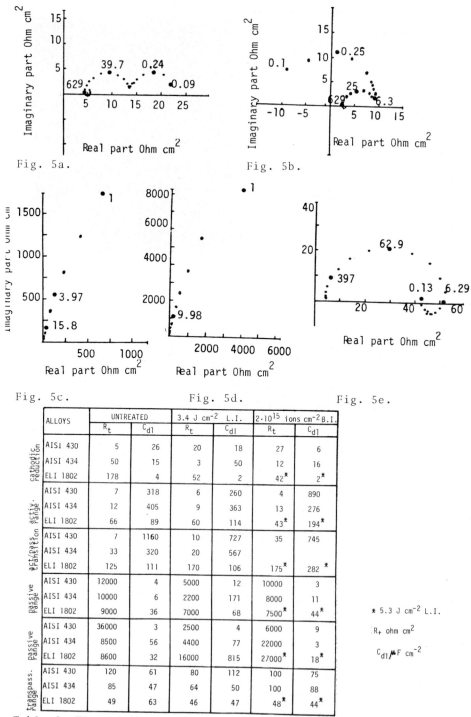

Fig. 5a.

Fig. 5b.

Fig. 5c.

Fig. 5d.

Fig. 5e.

ALLOYS	UNTREATED		3.4 J cm⁻² L.I.		2·10¹⁵ ions cm⁻²B.I.	
	R_t	C_{dl}	R_t	C_{dl}	R_t	C_{dl}
cathodic reduction						
AISI 430	5	26	20	18	27	6
AISI 434	50	15	3	50	12	16
ELI 1802	178	4	52	2	42*	2*
activ. range						
AISI 430	7	318	6	260	4	890
AISI 434	12	405	9	363	13	276
ELI 1802	66	89	60	114	43*	194*
active/pass. transition range						
AISI 430	7	1160	10	727	35	745
AISI 434	33	320	20	567		
ELI 1802	125	111	170	106	175*	282*
passive range						
AISI 430	12000	4	5000	12	10000	3
AISI 434	10000	6	2200	171	8000	11
ELI 1802	9000	36	7000	68	7500*	44*
passive range						
AISI 430	36000	3	2500	4	6000	9
AISI 434	8500	56	4400	77	22000	3
ELI 1802	8600	32	16000	815	27000*	18*
transpass. range						
AISI 430	120	61	80	112	100	75
AISI 434	85	47	64	50	100	88
ELI 1802	49	63	46	47	48*	44*

* 5.3 J cm⁻² L.I.

R_t ohm cm²

C_{dl} μF cm⁻²

Table 2. Electrochemical parameters measured, fitting the impedance data, at the corresponding polarization range.

capacitive semicircle, seen on untreated samples, disappears on irradiated ones. The higher activation peak current density has therefore to be attributed both to a surface area increase and to an enhanced reactivity of the bare metal surface following the physical treatments. The depolarization of the cathodic reactions confirms such an assumption (ref. 7).

During the activation-passivation transition, the experimental impedance spectra accounts for the theory of electrode processes underway, since the lower frequency impedance values tend towards negative values, as well as the slope of the I/E plot.

In the passive region a merely capacitive impedance plot is never observed, suggesting that the transfer of ions through the passive layer in both directions has finite values. In the first passive range ($E < 200$ mV$_{nhe}$) it is possible to extrapolate capacitive semicircles much larger for the untreated samples, so indicating more protective oxide layers than for irradiated ones. In the second passive range ($E > 200$ mV$_{nhe}$) the Mo containing steels show the best behaviour under such a stand point, and both L.I. and B.I. enhance the dielectric properties of AISI 434. The opposite trend is found on AISI 430.

At 1375 mV$_{nhe}$ the impedance data fit the spectrum of two semicircles typical of the transpassive region. The electrochemical parameters R and C are of the same magnitude of those observed in the same conditions with Chromium electrodes (ref. 5).

CONCLUSIONS

- The radiation damage introduced by directed energy sources has a beneficial effect on low purity, coarse grained stainless steel; the cathodic reactions are depolarized as a consequence of L.I. or B.I. on the same steels.
- Active-passive transition occurs at the same potential; activation currents are higher on L.I. or B.I. surfaces.
- No concentration gradients of alloying elements in the near surface layers were detected following the physical surface treatments, but the electrochemical behaviour is suggesting such

an effect.
- Mo stabilizes the passive layers on SS, and its effect is enhanced by the physical surface modifications, as it is shown by the impedance spectra in the passive range.
- In the transpassive region, all the samples exhibit the same ele‾ctrochemical behaviour, very similar to the one observed on chromium electrodes.

REFERENCES

1   M.Keddam, O.R.Mattos and H.Takenouti , Proceedings "Corrosion and Corrosion Protection Symposium",Ed. by the Electrochemical Soc. Vol. 81-8 (1981) 7-17
2   B.Alexandre, A.Caprani, J.C.Carbonier, M.Keddam and Ph.Morel, Corrosion Science, 21, (1981) 765-780.
3   I.Epelboin, M.Keddam, O.R.Mattos and H.Takenouti, Corrosion Science, 19, (1979) 1105-1112.
4   D.Schuhmann, J.Electroanalytical Chem., 17,(1968) 45-69.
5   R.D.Armstrong and M.Henderson, J.Electroanalytical Chem., 40, (1972) 121-131.
6   L.Bonora, M.Bassoli, P.L.De Anna, G.Battaglin, G.Della Mea and P.Mazzoldi, Electrochimica Acta, 25, (1980) I497-1499.
7   P.L.Bonora, M.Bassoli, G.Cerisola, P.L.De Anna, G.Della Mea and C.Tosello, Proceedings "Fundamental Aspects of Corrosion Pro-tection by Surface Modification" , Ed. by The Electrochemical Society, in press.

Passivity of Metals and Semiconductors, edited by M. Froment
Elsevier Science Publishers B.V., Amsterdam — Printed in The Netherlands

ON THE MECHANISM FOR IMPROVED PASSIVATION BY ADDITIONS OF TUNGSTEN TO AUSTENITIC
STAINLESS STEELS.

N. BUI[*], A. IRHZO[*], F. DABOSI[*], Y. LIMOUZIN-MAIRE[**]

[*] Laboratoire de Métallurgie Physique, ERA du CNRS n° 263, E.N.S.C.T.,
    31077 TOULOUSE CEDEX FRANCE
[**] Laboratoire des Organométalliques, Faculté des Sciences et Techniques de
    Saint Jérome, 13397 MARSEILLE CEDEX 4 FRANCE.

INTRODUCTION

It has been shown that alloying W increases the corrosion resistance of
austenitic 20 Cr - 25 Ni steels in acid chloride solutions (ref. 1,2). The
primary purpose of W additions to stainless steels was to improve their mechanical resistance to abrasion-corrosion in the wet process of phosphoric acid
production. While many attempts were made to clarify the role of molybdenum
additions by XPS and AES surface analysis (ref. 3-8), the mechanism for improved
passivation by alloying tungsten has not been investigated, as far as we know
in our survey of litterature. The present study is an attempt to elucidate the
role of W in the improvement of passive film stability. As W and Mo are in the
same column of the periodic table and have similar chemical properties, the question is whether alloyed W will have the same role as Mo in stainless steel.

In the present investigation, it is proposed to answer the following questions :

- Does W by itself form a protective film, without interaction with other
elements such as Cr, Ni, Fe ?

- Does dissolution of alloyed W result in the formation of WO_4^{--} ions which
will inhibit the pitting process in neutral chloride solution ? How effective is
WO_4^{2-} as pitting inhibitor ?

For these purposes, polarization measurements of pure W were carried out in
1N HCl and in 40 wt % H_3PO_4 containing 1000 ppm Cl^-. Then similar experiments
were performed to compare the effect of W either as WO_4^{--} ions dissolved in neutral chloride solution or as alloying element in the steels. ESCA measurements
were subsequently conducted on the passive films formed on the surface of

W-containing Cr-Ni steels and on Cr-Ni steels passivated in WO_4^{--} solutions. The experimental results were used to discuss the role of W on the basis of the above questions.

The five experimental austenitic 16 Cr - 14 Ni alloys containing 0 - 12 wt % W were prepared in a vacuum furnace. Their chemical compositions are given in Table I.

TABLE I

Alloy compositions (heat treatment 1150°C for 30 min and water quenched)

Alloy	Composition (wt %)					
	C	Mn	Si	Ni	Cr	W
16 - 14	0.006	0.43	0.33	14.40	16.30	0
16 - 14 - 2	0.005	0.33	0.40	14.00	16.30	2.49
16 - 14 - 4	0.006	0.30	0.41	14.10	16.50	4.81
16 - 14 - 8	0.006	0.32	0.40	14.10	16.17	8.66
16 - 14 - 12	0.006	0.30	0.43	14	16.22	12.40

EXPERIMENTAL RESULTS AND DISCUSSIONS

Electrochemical behaviour of pure W

Anodic polarization curve of pure W in 1N HCl and in 40 wt % H_3PO_4 containing 1000 ppm Cl^-, at 60°C, are shown in Fig. 1. It can be seen that corrosion potential of W is - 0.10 V, vs S.C.E., a value much higher than that of 16 Cr - 14 Ni (- 0.37 V), in 1N HCl. Pure Mo, on the other hand, also has a high corrosion potential, - 0.03 V in 1N HCl at 25°C, as measured by other authors (ref. 3). But the main feature is that W, as well as Mo, dissolves at a high rate in the passive region of the W containing 16 Cr - 14 Ni steels or 20 Cr - 25 Ni - 5 Mo steels. In other words, Mo and W alloying elements increases the pitting resistance of stainless steels in the potential region where they are highly active in pure form. This suggests that W does not protect by making a stable oxide layer but only by interaction with other oxides. Sugimoto and Sawada (ref. 3), have shown that combination of Mo and Ni, as well as those of Mo and Fe, do not result in an increase in pitting resistance. It is thought that W improves the pitting inhibition, just like Mo, when the amount of Cr is above an adequate level in the alloy. It has been shown previously that alloyed W improves the pitting resistance of 20 Cr - 25 Ni - 4 Mo steels (ref. 1). This means that W might have an additive action with Mo and might have the same mechanism as Mo in forming a passive oxide layer with chromium.

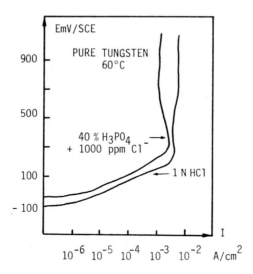

Fig. 1 - Anodic behaviour of pure tungsten

Inhibitive properties of tungstate ions

In the hypothesis that the dissolution of alloyed W results in the formation of WO_4^{2-} ions, polarization studies were carried out with 16 Cr - 14 Ni steel in 0.1 M NaCl containing Na_2WO_4 at various concentrations, in order to evaluate the inhibitive effect of WO_4^{2-} in solution (Fig. 2). It can be seen that an increase in tungstate concentration from 0 to 0.1 M results in an elevation of corrosion potential from - 360 to - 230 mV (vs SCE) and at the same time a great decrease in passive current density. The improvement of the passivity characteristics is further illustrated by the shifts of breakdown potentials to higher values : 80, 150 and 220 mV vs SCE, respectively.

For comparison, the effect of alloyed W was studied by determining the electrochemical behaviour of W-containing steels in 0.1 M NaCl without tungstate. In Fig. 3, it is shown that corrosion potential increases and passive current density decreases with increasing W content in the steels, but the effect is less pronounced than that produced by tungstate in solution. However, the effect of alloyed W in increasing the pitting potentials is greater.

These results point out that the effect of alloying W is somewhat similar to adding WO_4^{2-} in 0.1 M NaCl solutions. There is however a little difference : while tungstate in solution has the effect of shifting the passivity range (i.e. the potential difference between pitting potential and corrosion potential) in the noble direction of potential without changing it, alloying W has a greater tendency to extend this passivity range.

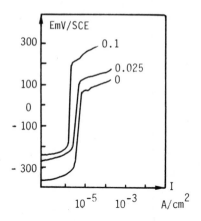

Fig. 2 - Anodic behaviour of
16 Cr - 14 Ni steels in 0.1 M NaCl
with various WO_4^{2-} additions.

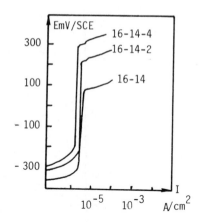

Fig. 3 - Anodic polarization curves
of W-containing steels in 0.1 M NaCl
(curve numbering refers to Table I)

The action of WO_4^{--} in solution may be explained by an adsorption process
which has already taken at the corrosion potential. According to ROSENFELD
(ref. 11), the oxide film formed through a natural process on the metal surface
changes its properties when insignificant amounts of inhibitor are adsorbed.
This modification is of structural and electrophysical nature rather than a
variation of thickness or continuity of protective films. It has been shown
that during CrO_4^{--} adsorption, chromium is reduced from + VI to the + III valency state (ref. 11) and that Mo^{VI} may be reduced to Mo^{IV} (ref. 12). Therefore
ESCA studies were undertaken to determine what W oxidation states were present
in the oxide film and whether there is a reduction of WO_4^{--} during adsorption.

ESCA study of the passive film

Prior to the examination, the passive film was developed on the surface by
maintaining the sample for 15 minutes at - 50 mV/S.C.E. a potential situated
in the middle of the passive range, as can be seen in Fig. 2.

For 16 Cr - 14 Ni steel passivated in 0.1 M NaCl + 0.05 M WO_4^{--} solution,
ESCA spectra of the surface layer (Fig. 4) shows two peaks corresponding to
tungsten, situated at 247.6 and 260.3 eV.

Spectra of pure tungsten and pure tungstate were then obtained (Fig. 5) by
putting these substances together in the sample holder of ESCA apparatus. It
can be seen that pure W is characterized by two peaks at 244.5 and 256.8 eV,
while WO_4^{--} or W (VI) have these two peaks shifted to 248.05 and 260.8 eV.

From this result, it can be stated that WO_4^{--} inhibitor is adsorbed in the
passive film under the oxidation state + VI, without electrochemical reduction.

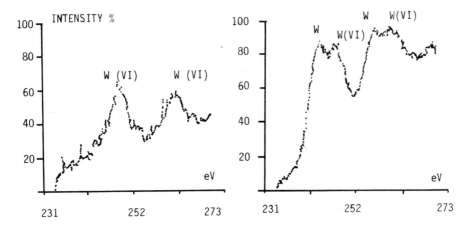

Fig. 4 - ESCA spectrum of 16 Cr - 14 Ni steel passivated in 0.1 M NaCl + 0.05 M WO_4^{2-}.

Fig. 5 - Combined ESCA spectra of pure W and $NaWO_4$.

Then the passive film developed on 16 Cr - 14 Ni - 4 W steel in 0.1 M NaCl was analysed. The spectra obtained (Fig. 6) shows that the two peaks related to tungsten in the film have the corresponding energies of 248 and 261 eV. It can be deduced that tungsten from solid solution in the stainless steel is present in the passive film under the + VI valent state.

CONCLUSIONS

The improvement of the corrosion resistance of 16 Cr - 14 Ni steel by tungsten results from the presence of W (VI) in the passive layer, either by the adsorption of dissolved tungstate in solution or by oxidation of metallic tungsten in the alloy. However the mechanism by which W enters the passive film is not well elucidated. No experimental evidence has shown for W-containing alloys whether it is a direct solid-state reaction that takes place at the surface film or whether there is a previous step of W dissolution leading to tungstate formation which later will adsorb into the passive film. Some authors favour the latter processes (ref. 13). However,

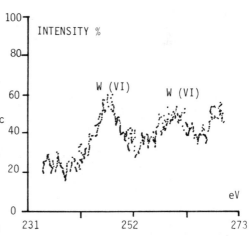

Fig. 6 - ESCA spectrum of 16 Cr - 14 Ni - 4 W passivated in 0.1 M NaCl.

this formation of soluble WO_4^{--} as intermediate step cannot account for the role
of W in the corrosion of stainless steels in acid chloride solution because
the precipitation of WO_4^{--} as sparingly soluble tungstic acid would be expected
under these conditions. In confirmation of this we observed yellow precipitates
when adding WO_4^{--} in 0.1 M HCl solution and these were subsequently identified
as sparingly soluble tungstic acid H_2WO_4. With this inactive precipitate, suppression of the pitting inhibition was noted. Furthermore, it is unlikely that
W will dissolve as WO_4^{--} in acid solution at potentials close to the corrosion
potentials but instead the following oxide formation is favoured :

$$W + 3\ H_2O \rightleftharpoons WO_3 + 6\ H^+ + 6\ e^-$$

The standard potential of this reaction is - 0.09 V (N.H.E.) or - 331 mV (S.C.E)
(ref. 14) while the corrosion potential of 16 Cr - 14 Ni - 12 W is - 330 mV
(S.C.E.) in 0.1 M HCl (ref. 10). The effectiveness of WO_3 presumably present in
the passive film can be interpreted in terms of its particular stability. WO_3
is insoluble in water and the only acid that dissolves it is hydrofluoric acid
(ref. 15). On the contrary, WO_3 is soluble in alkali hydroxide and carbonate
solutions, yielding tungstate. Thus, the selective dissolution of WO_3 into tungstate in acid solutions, as an initial step of inhibition, seems improbable. One
possible mechanism is that these elements improve the quality of the bonding at
the metal-oxide interface (ref. 11). Thus, a more efficient barrier layer is
established, impeding the ionic or electronic transfer across it. This view
can be extented to rationalize the effect of W as its physicochemical properties
are very close to those of Mo ; that is the improved pitting corrosion resistance of stainless steels in NaCl solutions by W addition to the alloy can be explained by the induced insolubility of WO_3 in the neutral region, resulting
from interaction with Cr_2O_3 and Fe_2O_3.

REFERENCES

1 - M. EL SAFTY, N. BUI, F. DABOSI, J.P. AUDOUARD, A. DESESTRET, G. VALLIER ;
     Proc. 2nd Int. Congress on Phosphorous Compounds, BOSTON, Ed. IMPHOS,
     (1980) 473-489.
2 - M. EL SAFTY ; Dr. Ing. Thesis, Institut National Polytechnique TOULOUSE,
     France (1980).
3 - K. SUGIMOTO, Y. SAWADA ; Corros. Sci., 17, (1977) 425.
4 - A.E. YANIV, J.B. LUMSDEN, R.W. STAEHLE ; J. Electrochem. Soc. 124, (1977)
     490.
5 - M. SEO, Y. MATSUMURA, N. SATO ; Trans. JIM, 20, (1979) 501.
6 - K. HASHIMOTO, K. ASAMI ; Corros. Sci. 19, (1979) 251.
7 - M. DACUNHA BELO, B. RONDOT, F. PONS, J. Le HERICY, J.P. LANGERON ;
     J. Electrochem. Soc. 124, (1977) 1317.
8 - R. BERNERON, J.C. CHARBONNIER, R. NAMDAR-IRANI, J. MANENC ; Corros. Sci.
     20, (1980) 899.
9 - I. OLEFJORD ; Mater. Sci. Eng. 42, (1980) 161.
10 - A. IRHZO ; 3rd Cycle Dr. Thesis, Institut National Polytechnique, TOULOUSE,
      France (1981).

11 - I.L. ROSENFELD ; Corrosion, 37, (1981) 371.
12 - J.N. WANKLYN ; Corros. Sci. 2I, (1981) 211.
13 - H. NAGANO, T. KUDO, Y. INAGA ; Bull. Cercle Et. Metaux, 9, (1980).
14 - M. POURBAIX ; "Atlas of Electrochemical Equilibria in Aqueous Solutions"
     Pergamon Press, New York (1966).
15 - K.C. LI, C.Y. WANG ; Tungsten, Monograph n° 130, Amer. Chem. Soc. Reinhold
     (1955).

Passivity of Metals and Semiconductors, edited by M. Froment
Elsevier Science Publishers B.V., Amsterdam — Printed in The Netherlands

PASSIVATION OF STEEL IN A BORATE BUFFER CONTAINING ORGANIC COMPOUNDS AS A MEANS
FOR IMPROVING CORROSION RESISTANCE

HIDETAKA KONNO[1] and HENRY LEIDHEISER, JR.[2]
[1]Faculty of Engineering, Hokkaido University, Sapporo 060 (Japan)
[2]Center for Surface and Coatings Research, Lehigh University, Bethlehem,
Pennsylvania 18015 (U.S.A.)

ABSTRACT
   Nineteen organic compounds have been identified which yield a corrosion-
resistant surface on iron after a two-step anodization process in borate media.
Surface analysis techniques have been employed in order to understand the
structure of the passive film.

INTRODUCTION
   It is well known that iron passivates between -0.4 and +0.9 V vs. SCE in a
neutral borate solution and at more positive potentials the passive film exhibits
breakdown. The addition of certain organic compounds to the solution, however,
allows the passivity to be maintained at more positive potentials. It has previously been shown that 8-hydroxyquinoline and some other compounds perform such
a function in a borate buffer solution (pH 7-10). A two-step anodizing process
has been discovered which results in a corrosion resistant layer on steel (ref 1).
This paper presents a list of organic compounds found to be effective and the
results of analysis of anodic surface layers formed in the presence of the most
effective three compounds, 8-hydroxyquinoline, 8-aminoquinoline, and 1,8-diamino-
naphthalene.

EXPERIMENTAL METHODS
   Anodizing was carried out on SAE 1010 steel panels abraded with No. 240 emery
paper. Electrochemical measurements and anodizing were performed using a Princeton Applied Research Co. Model 350 Corrosion Measurement Console, and potentials
quoted are with respect to a saturated calomel electrode, SCE.
   A survey of organic compounds was made by the following procedure: The steel
specimen was polarized in a 0.15M borate solution (pH 8.4) with stirring at 1.0-
1.4 V for 10 minutes after which an organic compound was added to a concentration
of 0.001 - 0.05M and anodization was continued for an additional 10 minutes. When
the anodic current was reduced markedly or a visible layer was formed on the
surface, the compound was subjected to further investigations. Specimens

TABLE 1

The list of organic compounds that passivate iron at high anode potentials and their effects on corrosion inhibition.

Compound	Conc./mM	i_a/Am^{-2}[*1]	E_{corr}/V	r_{avg}/g m^{-2}hr^{-1}[*2]
8-Aminoquinoline [AQ]	4	0.2	-0.452	0.13[*3]
Benzohydroxamic acid	10	21	-0.526	0.34
N-Benzoyl-N-phenylhydroxylamine	4	0.3	-0.434	0.48
Cupferron	10	1.3	-0.553	0.39
1,8-Diaminonaphthalene [DN]	4	0.2	-0.429	0.15[*3]
2,3-Diaminonaphthalane	2	0.3	-0.410	0.21
Diphenylguanidine	4	4.2	-0.448	0.23
4-Hydroxyquinoline	4	17	-0.415	0.26
7-Hydroxyquinoline	4	0.8	-0.440	0.25
8-Hydroxyquinoline [HQ]	4	0.3	-0.422	0.10[*3]
5-Nitro-8-hydroxyquinoline	4	5.7	-0.413	0.21
8-Nitroquinoline	4	0.7	-0.401	0.24
1-Nitroso-2-naphthol	2	0.9	-0.463	0.25
o-Phenylenediamine	20	0.8	-0.436	0.27
Phenylthiourea	4	1.9	-0.454	0.34
Quinoline	20	7.0	-0.566	0.26
Salicylaldehyde hydrazone	4	0.8	-0.474	0.23
Salicylamide	10	8.2	-0.425	0.21
Salicylanilide	2	2.8	-0.549	0.28
Blank	-	25	-0.389	0.97[*3]

*1 Anodic current density at +1.4 V vs. SCE in a borate solution (pH 8.4) containing an organic compound.

*2 Average corrosion rate after 16 hr immersion in an aerated 0.5M NaCl - 0.15M borate solution (pH 7.0).

*3 Median value

prepared in this manner are termed "__-treated". Specimens that were anodized identically in the absence of organic compounds are termed "blank".

TABLE 2

Results of gravimetric measurements after 16 hours' immersion in an aerated 0.5M NaCl-0.15M borate solution at room temperature.

Samples	Average corrosion rate/g m^{-2}hr^{-1}	
	pH 3.0	pH 7.0
HQ-treated	0.25 ± 0.07[*1]	0.09 ± 0.02[*1]
AQ-treated	0.40[*2]	0.14 ± 0.05[*1]
DN-treated	0.44[*2]	0.15 ± 0.06[*1]
Blank	1.29[*2]	0.91 ± 0.15[*1]

*1 95% confidence value.
*2 Average of three measurements.

In order to estimate the inhibitive effect of the treatment, specimens were polarized in an aerated 0.5M NaCl - 0.15M borate solution (pH 7.0). Gravimetric measurements were also made in the same media but at two different pH's, 3.0 and 7.0.

XPS and AES measurements were carried out using a Physical Electronics Model 548 electron spectrometer. Depth profiling of elements in the surface layer was carried out by sputter etching in 5.4×10^{-3} Pa argon atmosphere with a 5 kV argon ion beam that was raster-scanned in an area of about 6×6 mm^2.

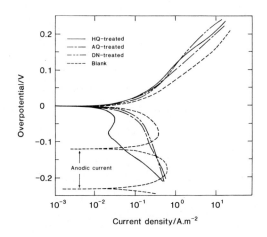

Fig. 1. Polarization curves for the steel samples measured in an aerated 0.5M NaCl - 0.15M borate solution (pH 7.0). Polarization was carried out from the corrosion potential at the rate of 0.3 mV/s in the anodic or cathodic direction using different specimens.

RESULTS AND DISCUSSION

Nineteen organic compounds as listed in Table 1 have been found which result in passivation at +1.4 V and which form an inhibitive layer on steel. These compounds do not function effectively as a corrosion inhibitor when present in an aqueous NaCl corrosion medium: The anodizing treatment is a critical feature in developing the corrosion resistance. Compounds that resulted in passivation at more positive potentials did not necessarily show good corrosion inhibition. For example, 2-mercaptobenzothiazole and thiourea when present in the borate solution exhibited typical passivation curves but the resulting surface exhibited accelerated corrosion, r_{avg} = 1.27 and 1.41 g/m^{2}hr, respectively, in an aerated NaCl solution at pH 7.0.

The most effective compounds were 8-hydroxyquinoline (HQ), 8-aminoquinoline (AQ), and 1,8-diaminonaphthalene (DN). As shown in Table 2, inhibition efficiency was 90% for the HQ-treated, 85% for the AQ-treated, and 84% for the

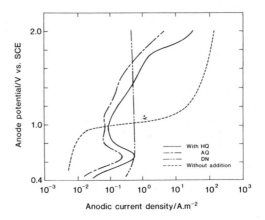

Fig. 2. Anodic polarization curves for steel, pre-passivated at +1.0 V, as determined in a 0.15M borate solution (pH 8.4) with and without 0.004M organic compounds (see Table 1 for the symbols). Scan rate was 1.0 mV/s.

DN-treated at pH 7.0, and the HQ-treated showed more than 80% inhibition at pH 3.0. Polarization measurements (Fig. 1) revealed that the anodically formed surfaces inhibited both the anodic and cathodic reactions. It was also observed that the formed layer, especially in the case of HQ, was very resistant to cathodic reduction in the chloride medium.

Anodic polarization measurements in a neutral borate solution (Fig. 2) indicated that HQ and AQ functioned as passivators of steel and shifted the passive potential region to +0.8 - +1.8 V from the normal -0.4 - +0.9 V range, while a typical passivation curve was not obtained in the presence of DN. The thickness of the HQ- or AQ-treated layer was apparently unchanged with anodizing time after several minutes, whereas the DN-treated layer becomes thicker with time and finally showed an interference color. The thick outer layer on the DN-treated steel was soluble in methanol. These results suggest that the forming mechanism was different in the case of DN as compared to HQ or AQ.

XPS measurements suggested that the outermost layer of the HQ-treated steel was covered with several layers of HQ and the active layer consisted of Fe(III)-8-hydroxyquinolate. Mössbauer spectra confirmed that the active layer consisted of 3 to 4 molecular thickness of Fe(III)-HQ complexes (ref. 2). The binding energies of N 1s electrons in the AQ- and DN-treated layers were in agreement with the values for amino-nitrogen and quinoline-nitrogen. The depth profiles of elements in the surface layer obtained by AES are shown in Fig. 3. For the HQ- and AQ-treated, iron Auger peaks appeared after a few minutes of sputtering whereas for the DN-treated they appeared after 10 min. The very low intensity of the oxygen peak at the metal-layer interface is noteworthy. The anodic oxide

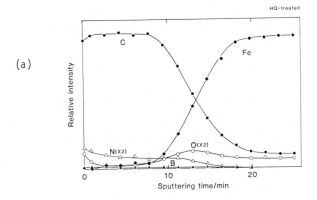

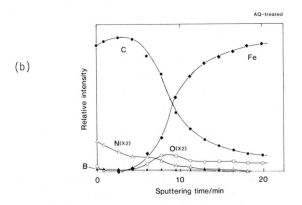

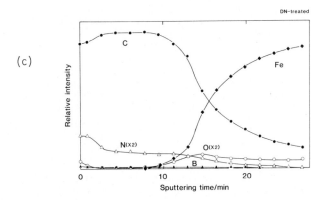

Fig. 3. The depth profile of the elements in the surface layer formed by (a) HQ-treatment, (b) AQ-treatment, (c) DN-treatment, as determined by AES combined with argon ion sputtering.

612

on the blank sample was removed within a few minutes under the same sputtering conditions, a fact which indicates that the treated samples were resistant to the ion sputtering. Unfortunately, it was not possible to calibrate the etching rate so that the thickness of the layer could not be obtained from these results.

From the above results, the following structures are suggested:

(a) $\begin{array}{c}\text{Org.}\\(\text{1 or 2 layers})\end{array}$ | $\begin{array}{c}\text{Fe-Org complexes}\\(\text{less than 5 layers})\end{array}$ | $\begin{array}{c}\text{Fe oxides}\\(\text{1 or 2 layers})\end{array}$ | Fe   Org: HQ / AQ

(b) $\begin{array}{c}\text{Org.}\\(\text{thick})\end{array}$ | $\begin{array}{c}\text{Mixture of Fe complexes and oxides}\\(\text{a few layers})\end{array}$ | Fe   Org: DN

Consequently, there may be two forming mechanisms, one is the formation of a thin iron-organic complex layer that is very stable and insoluble in water (a type of passive layer) and the second is the formation of a relatively thick organic layer by an anodic condensation reaction (ref. 3). The final structure is dependent on the properties of the organic compound used.

ACKNOWLEDGEMENT

The research reported herein has been carried out under a grant from the Office of Naval Research.

REFERENCES

1 H. Leidheiser, Jr., and H. Konno, J. Electrochem. Soc., 130 (1983) 747-753.
2 H. Leidheiser, Jr., H. Konno, and A. Vértes, Proc. Intl. Conf. Corrosion Inhibition, Dallas, Texas, 1983, NACE, Houston, Texas, in press.
3 G. Mengoli, M.T. Munari, P. Bianco, and M.M. Musiani, J. Appl. Polym. Sci., 26 (1981) 4247.

Passivity of Metals and Semiconductors, edited by M. Froment
Elsevier Science Publishers B.V., Amsterdam — Printed in The Netherlands

CRITICAL CONDITIONS IN THE PASSIVATION OF Cu-Mo AUSTENITIC STAINLESS ALLOYS IN HOT 20 % H2SO4 ACID

J.C. BAVAY[1], P. DAMIE[1], M. TRAISNEL[1] and K. VU QUANG[2]
[1]E.N.S.C. Lille, P.O. Box 108, 59652 Villeneuve d'Ascq Cedex (France)
[2]C.N.R.S., C.E.C.M., 15 Georges Urbain Street, 94400 Vitry (France)

ABSTRACT

The corrosion and passivation behaviour of several austenitic stainless alloys containing Cu and Mo, in 20 % H2SO4 acid, were assessed using electrochemical method, surface film analysis by Glow Discharge Spectrometry and weight loss testing. The beneficial effect of alloying nickel on active corrosion of alloys in deaerated solution was pointed out. In aerated solution the alloy passivation aptitude depended markedly on parameters such as temperature, electrode rotation speed and initial surface state (cathodic polarisation). The alloy behaviour was assumed to be controlled by surface film whose characteristics tightly depend on experimental conditions.

INTRODUCTION

In previous works(ref.1) corrosion resistance of several austenitic stainless steels and nickel-based alloys in deaerated 20 % H2SO4 has been investigated in the temperature range 30 - 103°C. In the present study it is shown that only some of the Cu-Mo bearing alloys(table 1) spontaneously passivate whereas the others do not, in spite of their similar electrochemical characteristics in this environment. Critical factors which impede or promote the passivation process are investigated : the temperature, the hydrodynamic condition, the dissolved oxygen and also the alloy surface treatment.

TABLE 1
Chemical composition (wt. %) of studied alloys

Alloy	Fe	Ni	Cr	Mo	Cu	Plot symbol
Z2 CNDU 17-16	58.5	16	17	5.5	3	O
Z2 NCDU 25-20	49	25	20	4.5	1.5	●
M.I.O. 20	46	29	20	2	3	Δ
INCOLOY 825	32	42	21	3	2	▲
HASTELLOY G	25.5	44	22	6.5	2	▼

614

RESULTS
Effect of temperature in deaerated 20 % H2SO4 acid

The critical current for passivation i (crit.) and then the passivation
aptitude of the five tested alloys are found to be the same at a given temperature (Fig.1a). Both alloying chromium and molybdenum greatly reduce the critical
current density for passivation of iron-base and nickel-base alloys (ref.1a).
The effect of molybdenum is about three times greater than that of chromium.
Nickel and copper have weaker effect upon the passivation aptitude (ref.2). In
agreement with the findings of these authors, the studied alloys show a similar
behaviour in spite of the difference in Cr, Mo, Ni and Cu contents.

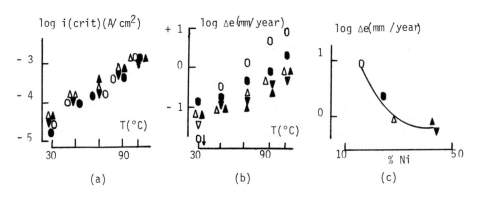

Fig.1 - Deaerated 20 % H2SO4 acid
(a) Effect of temperature on critical current for passivation
(b) Effect of temperature on corrosion rate
(c) Effect of nickel content on corrosion rate at 103°C.

The influence of temperature on the active corrosion rate Δ e (mm/ year)
- derived from weight loss testing - is shown in fig. 1b for different alloys.
Except for Z2 CNDU 17-16 steel, spontaneous passivation does not occur at 30°C
for any alloys in deaerated 20 % H2SO4 acid. However, when the temperature
increases from 30°C to 103°C (boiling point), this stainless steel is subject
to a quicker corrosion than the other alloys which possess a higher nickel
content.

The increase of nickel, molybdenum and copper contents in austenitic alloys
results in marked increases of corrosion resistance in sulfuric acid environments (ref.3). In reducing solutions, alloying copper improves markedly corrosion resistance, whereas molybdenum is less effective (ref.4).
For an equal copper content, M.I.O. alloy is more resistant to corrosion in hot
deaerated 20 % H2SO4 acid than Z2 CNDU 17-16 steel, although the molybdenum
content in the former alloy is smaller. Moreover, INCOLOY 825 and HASTELLOY G
exhibit the same corrosion rate in spite of the difference of their molybdenum

content. The significant contribution of nickel to the corrosion resistance of
alloys is pointed out in Fig. 1c.

Effects of cathodic polarization and electrode rotation speed in oxygenated 20 % H2SO4 acid

With a slight O2 gas pressure maintained above the solution and after preliminary activation by a cathodic polarization (1 mA/cm2), Z2 CNDU 17-16 electrode
passivates at 70°C when the rotation speed w is higher than 60 revolutions/mn
(Fig. 2a). However, the potential of Z2 NCDU 25-20 alloy does not shift towards
so noble value in the same conditions, even for high rotation speed. On the
other hand, passivation occurs for Z2 NCDU 25-20 and M. I.O. 20 static electrodes
(w = 0) in the absence of a cathodic polarization whereas for high rotation
speed (4000 r/mn), these alloys do not passivate, their potential remaining
in the active value (Fig. 2b). At high rotation speed, the passivation of Hastelloy G is also prevented by a cathodic treatment (Fig. 2c).

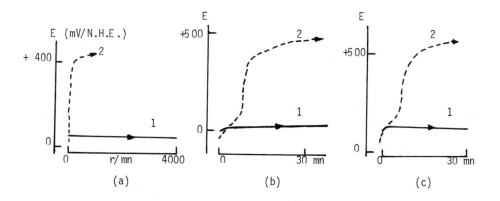

Fig.2 - Influence of rotation speed on the free corrosion potential in oxygenated 20 % H2SO4 at 70°C
(a) Preliminary activation :   1. Z2 NCDU 25-20   2. Z2 CNDU 17-16
(b) M. I.O.20 without preliminary activation :   1. 4000 r/mn   2. 0 r/mn
(c) HASTELLOY G - 4 000 r/mn - with (1) and without (2) preliminary activation
(5 mn at-25 mV/N.H.E.)

These phenomena would suggest the existence of a film due to corrosion products formed on the alloy surface during dissolution period following activation. Oxygen reduction is under diffusion control. According to the mixed
potential theory, an active-passive alloy spontaneously passivates if its critical anodic current density is less than the limiting diffusion current for
oxygen reduction. Stirring tends to increase the limiting current density and
the ease with which a given alloy passivates. It is quite likely that oxygen

diffusion is more or less hindered by the existence of a porous layer : oxygen transport takes place not only in the solution but also through the surface layer (Ref. 5-6). Therefore, when this layer exists, the effect of electrode rotation no longer occurs and oxygen reduction current is insufficient to produce passivation. However, on Z2 CNDU 17-16 alloy, the corrosion products layer is assumed to have characteristics which would allow a diffusion of dissolved O2 to the metal surface.

In addition, some unusual effects of hydrodynamic conditions may be explained by the fact that dissolution process is also under diffusion control in certain cases. In hot deaerated 20 % H2SO4, anodic current increases versus w at low constant overvoltage (Ref.1b, c). This rotation speed dependence can be associated with the marked impoverishment, at high rotation speed, in the thinned down diffusion layer, of some species which are intermediates in the redeposition process and are then necessary to the film stability.

Surface films analysis

The chemical analysis of the surface films has been carried out by using Glow Discharge Spectroscopy (G.D.S.) at IRS ID Institute (Ref. 7).

Fig. 3a shows the G.D.S. depth profile of the Z2 NCDU 25-20 alloy surface which has been exposed in deaerated 20 % H2SO4 at 70°C for 24 hours. The following informations are obtained :
- the active alloy is covered with a film
- the emission intensities due to Cr and Fe are null in the outer layer of the film and increase rapidly through the inner layers before stabilizing when the alloy matrix is reached.
- the Mo intensity exhibits a maximum in the film
- the Cu intensity shows a maximum in the external part of the film and decreases with the erosion time ; this feature is common for all alloys.
In the same conditions, G.D.S. data relative to the film formed on Hastelloy G alloy are shown in Fig. 3b. The amplitude of the emission signal due to Mo gradually increases and only stabilizes beyond the interface film/metal, in the alloy matrix. The Cr signal also increases as a function of sputtering time, but it increases more rapidly than does Mo signal. A slight enrichment with Cr is observed just before the alloy matrix is reached.
For Hastelloy G, the Cr/Mo profile varies as the inverse of that of Z2 NCDU 25-20 alloy and the peak observed on the Ni/Fe profile is greater for Hastelloy G having a higher Ni content (Fig. 3c).

In stirred oxygenated 20 % H2SO4, Z2 NCDU 25-20 alloy spontaneously passivates at room temperature but not at 70°C, contrary to the case of Hastelloy G alloy. Passive films formed at room temperature on Z2 NCDU 25-20 alloy and at 70°C on Hastelloy G alloy present the same characteristics, namely chromium and copper enrichments in the inner part (Fig. 3d). G.D.S. analysis show that oxygenation does not substantially modify the distribution of alloy components in the Z2 NCDU 25-20 film formed at 70°C ; but corrosion rate is lower in oxygenated solution, 0.19 mm/year instead of 0.48 mm/year in deaerated solution (24 hours immersion test).

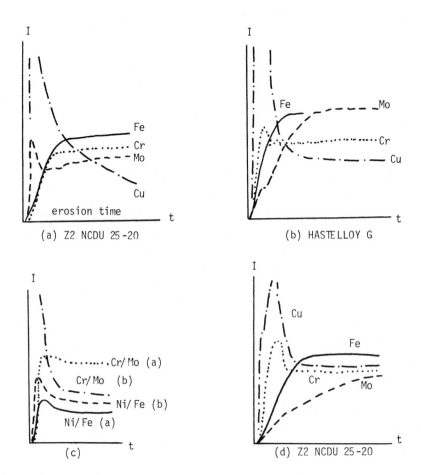

Fig.3 - G.D.S. depth profile of the surface films formed in 20 % H2SO4 (a)(b)(c) Deaerated solution at 70°C (d). Aerated solution at room temperature

The results suggest that layers formed during active dissolution have some protective effects which depend on the alloy composition. In hot deaerated 20 % H2SO4, the difference of active corrosion rate of austenitic Mo-Cu bearing alloys can be explained by a chemical nature change of the surface layers.

CONCLUSION

In 20 % H2SO4 acid, the passivation of Cu-Mo austenitic stainless alloys is very sensitive to the experimental parameters like temperature, hydrodynamic condition, dissolved oxygen and also treatment of alloy surface. The passivation process is assumed to be controlled by the characteristics of the alloy surface film. These characteristics tightly depend on the alloy composition and the tested conditions.

REFERENCES

1 J.C. Bavay, P. Damie, M. Traisnel and K. Vu Quang
   a) Journal of Applied Electrochemistry 10 (1980) 703
   b) Métaux - Corrosion - Industrie 661 (1980) 269
   c) Proceedings of the 8th International Congress on Metallic Corrosion, Mainz
   (Sept. 1981)

2 B. Baroux and P. Maîtrepierre. Revue de Métallurgie (February 1981) 145

3 A.J. Sedriks. Corrosion of stainless steels (Wiley Publication)

4 W.Z. Friend. Corrosion of nickel and nickel-base alloys (Wiley Publication)

5 A. Bonnel, F. Dabosi, C. Deslouis, M. Duprat, M. Keddam and B. Tribollet.
   J. Electrochem. Soc. 4 (1983) 753.

6 W.J. Lorenz and F. Mansfeld. Corrosion Science

7 R. Berneron, J.C. Charbonnier, J. Manec and R. Namdar-Irani. Corrosion
   Science 20 (1980) 899.

Passivity of Metals and Semiconductors, edited by M. Froment
Elsevier Science Publishers B.V., Amsterdam — Printed in The Netherlands

A STUDY OF THE FILM ON AUSTENITIC STAINLESS STEEL AT THE PASSIVE-TRANSPASSIVE
TRANSITION

G. RONDELLI[1], J. KRUGER[2], J. J. RITTER[2] and U. BERTOCCI[2]

[1]Istituto Tecnologia Materiali del CNR, Via Induno 10, 20092 Cinisello (Italy)

[2]Corrosion Group, National Bureau of Standards, Washington D.C. 20234(U.S.A.)

ABSTRACT

Anodization of 304 stainless steel in sulfuric acid solutions at the passive-transpassive transition was studied. The change of film with potentials, as well as with time at selected potentials, was followed by means of ellipsometry and a.c. impedance techniques. The composition of the films so obtained was checked with X-ray photoelectron spectroscopy. Pitting tests, carried out on specimens prepared in 3.5 M H_2SO_4 at potentials at which maximum film thickness was observed, indicated a small increase of pitting potential.

INTRODUCTION

Some authors (ref.1) have claimed that a protective film can be formed by anodization on austenitic stainless steels. According to them, this film is best formed at certain potentials in the transpassive region in 1.5-3.5 M H_2SO_4, and improves resistance to localized corrosion. The main process for the formation of the film is claimed to be the preferential dissolution of Fe. Surface analysis has shown that the average Cr/Fe ratio is three times higher than in not anodized film.

Other authors (ref.2-3) found that in 0.5 M H_2SO_4 the film formed on 18-8 austenitic stainless steels in the transpassive range is rich in Fe.

I. Olefjord and B. Elfstrom (ref.4) have analyzed by means of ESCA and Auger techniques the passive film formed on stainless steels in sulfuric acid and found that it is enriched on chromium. Mo and Ni content of the passive films is not markedly different from their concentrations in the alloy.

C. Leygraf et al. (ref.5) polarized Fe-18 Cr-3 Mo single crystals in sulfuric acid solutions in the passive and near passive range region for different durations at constant potential. They found a selective dissolution at all potentials and times and an enrichment of iron in the dissolution products; as a result of this an enrichment of chromium is found not only in the passive film but also in the alloy phase.

In order to obtain a better understanding of the processes occuring on austenitic stainless steel in sulfuric acid solution in the potential range near transpassivity we follow these processes by ellipsometric and impedance

techniques. We also carried out surface analysis of the anodized specimens
by X-ray Photoelectron Spectroscopy (X.P.S.).

Also, the effect of anodization on localized corrosion resistance was
evaluated by a potentiodynamic test in a solution containing chloride ions.

MATERIALS

304 austenitic stainless steel was studied. Before each   experiment
the specimens, obtained from a rod of 1/2" in diameter, were mounted in epoxy
resin, wet grinded with abrasive paper down to 600 mesh, then polished with
diamond paste down to grit size 1 μm.

Specimens for pitting potential evaluation and X.P.S. analysis were prepared
in the cell for ellipsometric measurements, so as to follow the film growth,
stopping when the maximum thickness was achieved.

All the potentials are referred to the Satured Calomel Electrode.

RESULTS AND DISCUSSION

Impedance

Impedance measurements were carried out both using a periodic signal and
white noise as input. In the first case measurements were performed both with
static and rotating electrodes. (R.D.E.).

Fig. 1 is a Nyquist plot of the impedance taken in static conditions,
showing a progressive decrease of the electrode resistance with increasing
potential. At 1050 mV an inductive loop begins to form. Below 950 mV no changes
of the electrode impedance with time were detected. At 975 mV however, the
impedance slowly evolves with time as shown, for instance,in Fig. 2. The effect
of electrode potential and of time at the same potential on rotating disk
electrode are shown respectively in Fig. 3 and Fig. 4.

Substantial changes in the impedance diagrams are observed when the
electrode is being rotated, indicating a contribution of transport in solution
at the lowest frequencies.For this reason, the most significant data for
discussion the effect of the electrode potential on the impedance, as well as
the effect of time at the critical potential (975-980 mV vs SCE), are those
taken with the RDE. There the absolute impedance of the electrode decreases
with increasing potential from 10 K Ω at 500 mV. to 1 K Ω at 1000mV.

Some of the data obtained with noise techniques were elaborated in order to
have R and C plots vs. frequency assuming an equivalent circuit consisting of
R and C in parallel (after subtraction of the solution resistance, which was of
the order of 3 Ω). Although the R and C are not frequency-indipendent,
indicating the inadequacy of the equivalent circuit employed, the results show
the trend both as a function of potential and as a function of time of

polarization at constant potential. The value of C ranges from 80 $\mu F/cm^2$ at 500 mV, to 280 $\mu F/cm^2$ at 1000 mV. Such numbers indicate that the capacitance in question is not across the passive film, but is related to a charge-transfer reaction, very likely at the film solution interface. Absorption probably accounts for the large values at which are found in the transpassive region, and agrees with the incipient inductive loop observed in the Nyquist plot at 1050 mV. At the critical potential of 980 mV the capacitance increases with time reaching a value of the order of 400 $\mu F/cm^2$ after 22 h. The range of frequencies employed does not allow to estimate accurately the increase of the electrode resistance with time, but it is at least a factor of 10. This result can be considered in agreement with the increase of resistance to pitting caused by anodization at this potential.However the large capacitances values found could indicate an adsorption(perhaps Cr^{3+} or Cr^{6+}) for the most relevant effect.

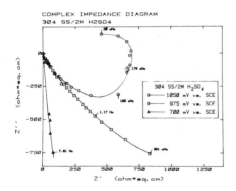

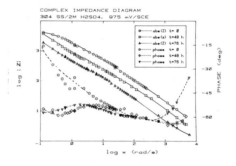

Fig. 1. Diagrams obtained for 304 ss in 2 M H_2SO_4 at different potentials. Immaginary part vs. real part of impedance.

Fig. 2. Diagrams obtained for 304 ss in 2 M H_2SO_4 at 975 mV for different intervals of time. Absolute impedance vs.frequency and phase vs. frequency.

Ellipsometry

In 2 M H_2SO_4 for potential less than 940 mV no changes in film properties could be detected by ellipsometry. We tried to increase the potential towards more positive valuesbut up to 940 mV no changes occured. At 970 mV a large change in Δ values can be obtained. The maximum film thickness ($\sim$ 18 nm) is reached after 1000 minutes, but for longer exposures at this potential dissolution of the film occurs as indicated by the increase of Δ values. An increase of $\delta\Delta$ vs time at potentials larger than 1050 mV is observed, indicating dissolution of the passive film.

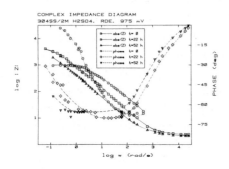
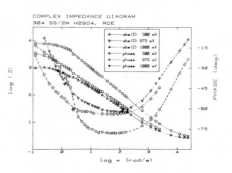

Fig. 3. Diagrams for 304 ss in 2 M
H_2SO_4 at different potentials.
Absolute impedance vs.frequency
and phase vs. frequency.

Fig. 4. Diagrams obtained for 304 ss in
2 M H_2SO_4 at 975 mV for different
intervals of time. Absolute impedance
vs. frequency and phase vs. frequency.

Some ellipsometric measurements were also taken in 3.5 M H_2SO_4. The results
were similar to those obtained in 2 M H_2SO_4 except that the potentials are
shifted towards higher values. No changes were observed below 1000 mV, and
film growth occured at 1070 mV, as shown in Fig. 5. There the time for maximum
thickness ($\sim$16 nm) is only 40 minutes. At 1125 mV, dissolution of the passive
film takes place.

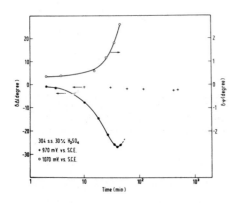
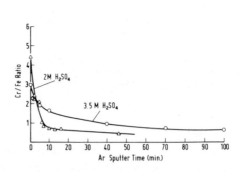

Fig. 5. $\delta\Delta$ vs. time and $\delta\psi$ vs.time
for 304 ss polarized in 3.5 M H_2SO_4 at
different potentials.

Fig. 6. X.P.S. analysis of 304 ss
specimens polarized 1000 min.at 970 mV
in 2 M H_2SO_4 and specimens polarized
40 min. at 1070 mV in 3.5 M H_2SO_4.

Localized corrosion resistance tests

The preparation of the anodized films which were tested for localized corrosion resistance, as well as for surface composition consisted in holding the specimen   at 970 mV for about 1000 min in 2M H_2SO_4 and at 1070 mV for 40 min.in 3.5 M H_2SO_4 . After anodization, the samples were dipped in 5% Na_2MoO_4, which, according to Guo-Zhu (ref.1) helps in healing defects present in the film. Some of the results of Tab.I are shown not as a single potential value for pitting but instead as a potential range. In the last case crevice corrosion as well as pitting was observed.

For the specimens anodized in 3.5 M H_2SO_4 no crevice corrosion was observed and as Tab.I shows, the pitting potentials are slightly higher than in the case of not anodized specimens; on the contrary for the specimens anodized in 2M H_2SO_4 no significant improvement of the localized corrosion resistance was observed

TABLE I

Pitting potentials in a solution containing 3500 p.p.m. of Cl^- (40°C).

304 not anodized	304 anodized in 3.5 M H_2SO_4
105 —— 160 mV	280 mV
110 mV	270 mV
80 —— 110 mV	280 mV
80 —— 150 mV	275 mV
170 mV	220 mV

X.P.S. Analysis

In fig.6 the experimental results obtained on specimens anodized in different conditions,are summarized.It can be seen that in both cases chromium enrichment takes place.As regards as the specimen anodized in 3.5 M H_2SO_4 in spite of the surface enrichment in Cr found by X.P.S,no significant improvement in pitting resistance was observed.   Possible explanations are that the treatment conditions conferring enhanced resistance are very narrow and have been missed in this work, or that the presence of inclusions in the steel caused weak spots in the anodized film. The shape of the Cr profile may also play an important role in determining corrosion resistance as already pointed out by others (ref.6). In the case of the specimen anodized in 2 M H_2SO_4, it is possible that the Cr enriched layer is too thin.

CONCLUSIONS

1) In sulfuric acid solutions a range of potentials in which a film grows on 304 stainless steel (∿18 nm in thickness) was found. This range is situated in the near-transpassive region. In the case of 2 M H_2SO_4 solution the range of potentials is   950-1000 mV, in the case of 3.5 M H_2SO_4 the range is

624

1050-1120 mV vs. S.C.E.

2) Impedance measurements carried out in 2M H_2SO_4 show that no significant changes occur in the passive region, below 900 mV vs. S.C.E. At higher potentials the real part of the impedance decreases and an inductive loop appears at low frequencies. In corrispondence with the increase in film thickness detected ellipsometrically, the electrode capacitance increases too, suggesting that such a capacitance is not across the oxide film, but rather it is connected with interfacial charge-transfers reactions, and probably absorption.

3) X.P.S. analysis shows an enrichment on the Cr surface content both for the specimen anodized in 2M H_2SO_4 and for the specimen anodized in 3.5 M H_2SO_4 . In the second case the Cr content decreases more slowly as we move away from the surface, than in the case of the specimen anodized in 2M H_2SO_4.

4) The resistance of the anodized specimen in 3.5 M H_2SO_4 to localized corrosion in solution containing chloride ions| seems to be slightly better than that of not anodized specimen. In particular crevice corrosion resistance is increased.

REFERENCES

1 - Huang Guo-Zhu et al. "The investigation of Anodized| Film on Stainless Steel - its contribution to the Corrosion Resistance of the Steel and the Mechanism of Protection". Experimental Report of Shangai Research Institute of Materials, 1981.

2 - T. Shibata and G.Okamoto, Boshuku Gijutsu, 21 (1972) 236.

3 - G. Okamoto, Corrosion Sci., 13 (1973) 471-489.

4 - I. Olefjord and B. Elfstrom, Corrosion, 38 (1982) 46-52.

5 - G.Leygraf et al., Corrosion Sci., 19 (1979) 343-357.

6 - G. Hultquist and C. Leygraf, Mat.Science and Eng., 42 (1980) 199-206.

Passivity of Metals and Semiconductors, edited by M. Froment
Elsevier Science Publishers B.V., Amsterdam — Printed in The Netherlands

ETUDE DU MECANISME DE PASSIVITE D'ACIERS 17-13 A TENEUR VARIABLE EN MOLYBDENE
DANS DES SOLUTIONS D'ACIDE PHOSPHORIQUE A 30% EN P_2O_5.

A. GUENBOUR and A. BEN BACHIR
Laboratoire de Chimie Physique Appliquée, Faculté des Sciences
AV. IBN BATOUTA. Rabat MAROC.

RESUME

L'etude du mécanisme de passivation d'aciers austénitiques 17/13 Mo a été
réalisée en milieu acide phosphorique à 30% en P_2O_5 pur ou pollué d'ions
F^- et Cl^-.
Des études potentiocinétiques ont montré que l'addition du molybdène améliore
la passivité des alliages.
Les couches de passivité sont etudiées par spectroscopie ESCA.
L'addition de Molybdène dans l'alliage affecte sensiblement la nature et la
composition des films de passivité que l'on décompose en deux couches super-
posées.

INTRODUCTION

Le rôle bénéfique des additions de Molybdène sur l'amélioration de la résistance à la corrosion des aciers austénitiques a été mis en evidence par de
nombreux auteurs.

Ce rôle se traduit d'une manière générale par une polarisation de la réaction
anodique. Toutefois il n'est pas possible de préciser à quelle étape de la
réaction anodique le molybdène intervient (Transfert d'électrons, transfert
d'ions entre le métal et la solution ou adsorption).

Malgré de nombreuses études, le mécanisme d'action du Molybdène est loin
d'être élucidé.

Cependant des études réalisées dans des milieux autres que l'acide phosphorique ont permis d'avancer certaines hypothèses. Nous retiendrons parmi ces
travaux les suivants :

Selon certains auteurs le Molybdène agit en augmentant l'épaisseur du film
de passivité(1.2), d'autres travaux (3.4) montrent que l'épaisseur du film de
passivité depend du potentiel et non de la teneur de l'acier en molybdène.

D'après Ambrose et Coll(5) le Molybdène agit en augmentant la cinétique
de repassivation, hypothèse controversée par Haskimoto et Asami, selon eux(4)
c'est le chrome qui est responsable de la repassivation par formation
d'hydroxyde de Chrome, le Molybdène peut également améliorer les propriétés
de la couche de passivité par un enrichissement de la surface en Mo VI ce

dernier stabilise les oxyhydroxydes de Cr^{III} et renforce les liaisons métal oxyde à l'interface(2.6).

D'autre part Nielson et Rhodin(7) expliquent l'effet de Molybdène par son aptitude à former des polyacides. Enfin l'hypothèse la plus invoquée pour expliquer l'effet bénéfique des additions du Molybdène dans les aciers met en jeu l'adsorption des ions MoO_4^{2-} qui se concentrent dans les piqûres par dissolution du Molybdène lors des premiers stades de l'attaque(8).

Nous nous proposons dans cet exposé de donner les résultats d'une étude des films de passivité à l'aide de la Spectroscopie ESCA et de recouper cette étude avec des essais electrochimiques en vue de contribuer à élucider le mécanisme d'action du molybdène sur la passivité des alliages austénitiques en milieu phosphorique.

Essais Potentiocinétiques réalisés sur les alliages 17-13-Mo à 60 et 80°C.

Des tracés potentiocinétiques sont réalisés sur les alliages austénitiques à teneur variable en molybdène, les résultats d'une étude systématique tenant compte des impuretés, de la température et de la concentration du milieu ont été publiés précédemment(9.10). Pour compléter, ce travail, des courbes de polarisation anodique des alliages ont été tracées dans H_3PO_4 à 40%. L'addition de molybdène est connue pour son action bénéfique vis à vis de la corrosion localisé. Ainsi les solutions étudiées sont polluées d'ions Cl^-. Les fig 1(a et b) représentent les résultats obtenus en fonction des teneurs en molybdène des alliages. L'addition croissante du molybdène abaisse la valeur de l'intensité   du

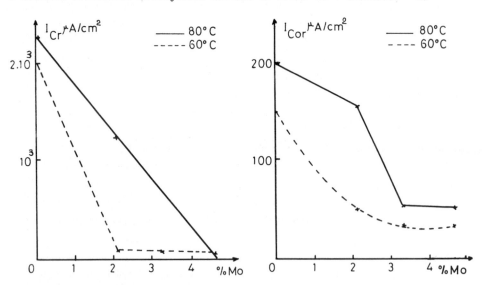

Fig.1. Acier 17/13/Mo dans H_3PO_4 +3000ppm d'ions Cl^-. Influence des additions du Molybdène sur :

(a) la densité du courant critique (b) la vitesse de corrosion du courant

courant critique et du courant de corrosion.

On note une diminution rapide de I_{Cor} et I_p pour une addition de 2,1% de
Molybdène dans l'acier de référence 17/13;des additions supplémentaires entrai-
nent encore des diminutions de ces paramètres électrochimiques mais dans des
proportions moindres au delà de 2% en Mo.

On constate en même temps un anoblissement des potentiels de corrosion avec
l'augmentation de la teneur de l'alliage en molybdène.

En présence de fluorures on obtient des résultats similaires en fonction
de la teneur en molybdène.

Ainsi le rôle bénéfique du molybdène est confirmé dans ces milieux phosphoriques.

Mais par quel mécanisme le molybdène agit sur la passivité ? Pour contribuer aux recherches sur ce mécanisme d'action du molybdène nous nous sommes
proposé d'examiner le comportement électrochimique du molybdène métallique
d'une part et d'analyser les couches de passivité d'autre part.

Comportement electrochimique du molybdène en milieu phosphorique

Des essais potentiocinétique sont réalisés sur des plaques de Molybdène
pur (99,9%) dans H_3PO_4 à 40% pollué de 1000 et 3000 ppm en Cl^- à 60°C.
La fig(2) indique que le molybdène ne se passive pas dans ces conditions.

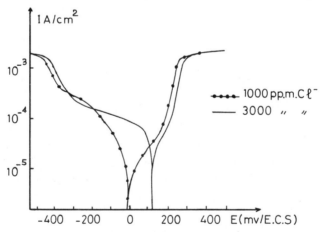

Fig.2. Courbe de polarisation du Molybdène dans H_3PO_4(40%) à 60°C.

L'analyse par spectroscopie ESCA du film epais de couleur bleu sombre qui
recouvre la surface du métal a montré qu'il s'agit de MoO_3 oxyde poreux qui
ne peut conférer une passivité au molybdène. Ainsi l'absence de passivité
propre du métal nous indique qu'il est impossible d'expliquer le mécanisme
d'action du molybdène par un transposition de ses propriétés intrinsèques,mais

plus certainement par son interaction avec les autres élements de l'alliage pour former la couche de passivité.

Analyse des films de passivité formés à l'abandon

Nous avons examiné la nature des couches superficielles formées à partir des aciers 17/13/3Mo après immersion de 16 heures dans l'acide phosphorique pollué de 1000ppm d'ion chlorure.

Résultats et discussions

Les spectres photoélectroniques révèlent la présence de chlore de phosphore, d'oxygène, de fer,de chrome,de Nickel,de Silicium et de Molybdène dans la couche de passivité.

Le tableau(I) regroupe les différentes espèces présentes dans le film de passivité.

	Concentration relative des espèces dans le film de passivité (en pourcentage atomique).			
	Cr^{3+}	$Fe^{3+}+ Fe^{2+}$	Si^{4+}	Mo^{6+}
17-13	24,2	56,9	18,9	—
17-13-3	56,2	5,5	29,7	8,6

On constate un enrichissement relatif important des couches superficielles en Si^{4+}, Cr^{3+} et Mo^{6+} par rapport à la teneur des métaux correspondant à ces ions dans la matrice.

L'enrichissement des couches superficielles de l'alliage exempt de Mo en phosphates, en oxygène (OH^-) et Fe^{3+} présume que le film de passivité est constitué essentiellement de phosphates de fer et dans une faible proportion de phosphates de chrome.

Dans le cas de l'alliage allié au molybdène on note un enrichissement de la surface en Cr^{3+}, Mo^{6+} et Si^{4+} ce qui a pour conséquence le renforcement de l'action passivante du chrome et la formation de composés molybdosiliciques. Le suivi de la variation de concentration relative des éléments des couches superficielles formées sur l'alliage en fonction du temps d'erosion peut nous renseigner sur le mécanisme de passivation;la fig(3) illustre l'évolution des concentrations des éléments à partir de la surface.

On distingue ainsi deux zones

-Zone externe de passivité d'épaisseur 7 à 8 monocouches qui se compose de deux couches distinctes :

.Couche interne riche en Fe^{3+} et Cr^{3+} et contenant des phosphates et de l'oxygène en concentration importante.

.Couche externe d'épaisseur 3 monocouches riche en Si^{4+} et Mo^{6+} oxgène et phosphate mais pauvre en Fe^{3+} et Cr^{3+}.

-Zone transitoire dont l'analyse ne représente pas les phénomènes réels car elle intègre une épaisseur de matière située entre deux zones fondamentalement différentes (couche passive et matrice métallique).

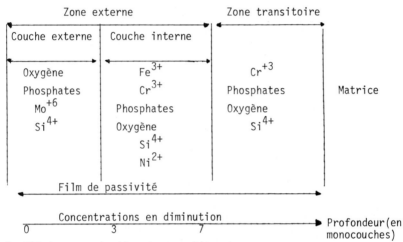

Fig.3. Profil de concentration des constituants du film de passivité.

En conclusion de ce paragraphe on peut dire que la présence exclusive du molybdène dans les couches externes du film de passivité indiquerait que des complexes molybdiques probablement sous forme MoO_4^{2-} se forment à partir de composés du molybdène en solution et non directement à partir du métal.

On peut alors penser que des ions molybdates en solution aqueuse peuvent exercer un effet analogue à celui du molybdène en solution solide.

Dans ce but nous avons réalisé des essais qui consistent à ajouter des ions molybdates sous forme Na_2MoO_4 dans des milieux phosphoriques et à caractériser leur rôle.

Ces essais nous ont permis de conclure qu'une addition de 0,05M d'ions MoO_4^{2-} dans le milieu phosphorique entraîne un anoblissement du potentiel de corrosion de l'acier exempt de Mo ainsi qu'une diminution de i_{Cor} et i_{Cr} les spectres ESCA indiquent également la présence de Mo^{VI} dans les couches de passivité.

CONCLUSION

Ces résultats et l'étude bibliographique nous permettent d'avancer le mécanisme d'action suivant:
Le molybdène se dissout à l'état d'oxydation +3 au cours de l'étape qui précède
la passivation, une fois en solution cette espèce s'oxyde sous forme d'ion
HMO_4^- (stable aux pH étudiés) qui s'adsorbe à la surface de l'alliage et réagit
avec les différentes espèces pour former des phosphates mixtes mais surtout pour
former des dérivés hydratés plus protecteurs dont le principal pourrait être
$H_4(12Mo\ O_3, Si\ O_2, nH_2O)(11)$.
Ces composés sont stables aux pH inférieurs à 3.
Remarquons enfin que ces résultats ne contredisent pas la possibilité d'augmentation de l'épaisseur de la couche de passivité.

Remerciements

Nous remercions MH. GRIMBLOT et GENGEMBRE du laboratoire de Catalyse de
l'UST de LILLE de nous avoir permis d'effectuer les essais ESCA.

BIBLIOGRAPHIE

1 J. Horvath, HH. Uhlig J. Electro. Soc. 115, 791 (1968).
2 K. Sugimoto, Y. Sawada Corr.Sci. 17. 425 (1977).
3 K. Hashimoto, K. Asami Corr. Sci 19. 251 (1979).
4 K. Hashimoto, K. Asami, K. Teramoto Corr Sci 19.3 (1979).
5 J.R. Ambrose, T. Kodamer Corr. 33. 155 (1977).
6 T.N. Rhodin, Corro. 12. 465 (1956).
7 N.A. Neilson, I.N. Rhodin Z. electrochem 62. 707 (1958).
8 K.M. Sakashita, N. Sato Corr Sci 17,473 (1977).
9 N. Bui, A. IRHZO, F. Dabosi, A. Guenbour, A. Ben Bachir
       Science des matériaux en cours de publication (1983).
10 A. Guenbour Thèse 3^{e}cycle Rabat (1983).
11 Michael. N. Hull. J. electro Anal Chem 38, 149 (1972).

Passivity of Metals and Semiconductors, edited by M. Froment
Elsevier Science Publishers B.V., Amsterdam — Printed in The Netherlands

CORROSION POTENTIAL OSCILLATIONS OF STAINLESS STEELS IN AQUEOUS CHLORIDE SOLUTIONS

A. ATRENS

Brown Boveri Research Center, CH-5405 Baden (Switzerland)

ABSTRACT

The conditions leading to corrosion potential oscillations in aqueous chloride solutions are discussed in terms of chloride ion concentration, oxygen concentration and temperature. The corrosion potentials of stainless steels in chloride solutions for conditions significantly removed from those leading to pitting and crevice corrosion are constant and to a first approximation largely independent of temperature and chloride ion concentration. With increasing temperature and/or chloride ion concentration or with decreasing corrosion resistance there comes a situation when the corrosion potential begins to oscillate, between two end values roughly corresponding to E_p and E_{pp}. These potential oscilations are often associated with pitting and crevice corrosion, although there are clear cases when no corrosion attack was visible even after prolonged exposure. The oscillations are attributed to unstable passivity. A model is developed in terms of the formation and breakup of metastable pit nuclei.

INTRODUCTION

The corrosion potential of a stainless alloy in an aqueous solution is an important quantity. Its relationship to the pitting potential, Ep, and the protection potential, Epp, determines whether or not pitting occurs. Similarly, there is a critical potential for stress corrosion cracking in chloride solutions; at potentials more negative than this critical potential no stress corrosion cracking is expected. However, to apply the above understanding requires, not only the measurement of the pitting and critical potentials, it also requires a knowledge of the behaviour of the corrosion potential. That is the subject of the present paper.

EXPERIMENTAL PROCEEDURE

The experimental proceedure has been described in detail previously (ref.1). In summary, a special apparatus was used that would allow the exposure of a deaerated specimen surface to a deaerated solution, and would allow measurement of the corrosion potential from the moment of specimen-solution contact. Alternatively, potentiodynamic determinations of E_p and E_{pp} could be carried out after the specimen had been in the solution for some time. A EG & G Princeton Applied Research Model 331-3 Corrosion Measurement System was used.

The specimens were machined from commercially available alloys and then polished as for optical microscopy. Specimen mounting and masking ensured that only one flat specimen face 1cm^2 in area was in contact with the solution. The stainless steels studied included representative alloys from all classes of stainless steels: namely martensitic, precipitation hardened, austenitic, duplex and ferritic stainless steels. These steels can be ranked according to a measure of their corrosion resistance given by % Cr + 3.3 % Mo.

RESULTS

Three different types of behaviour have been observed for stainless steels immersed in a variety of chloride solutions: piting, stable passivity and corrosion potential oscilations.

Pitting behaviour is characterized by final corrosion potential values which are slowly drifting. Rapid changes in corrosion potential may have occurred previously. For example, Fig.s 1 and 2 gives typical examples of the time dependence of the corrosion potential for a typical martensitic stainless steel containing 12 % Cr 1 % Mo. Fig. 1 gives two typical examples of the very rapid change often observed in the first few instants after specimen-solution contact. The subsequent changes in the corrosion potentials for these two cases are shown in Fig. 2; in each case there is further change. In each case, after the experiment the specimen surface contained pits and crevice corrosion attack. Similarly, the specimen in deaerated 10^{-2} N NaCl showed a corrosion potential which evinced a relatively rapid change. This specimen was also found to have undergone pitting and crevice corrosion.

Stable passivity is also characterized by steady corrosion potential values: for example, the specimen exposed to deaerated 5 x 10^{-5} N NaCl solution in Fig. 2.

A fundamentally different type of behaviour is shown in Fig. 3; corrosion potential oscillations are displayed after immersion in aerated deionate, (in this case caused most probably by the residual chloride remaining in the apparatus). Such corrosion potential oscilations have been observed for a whole range of stainless steels (ranging from a typical martensitic stainless steel containing 12% Cr 1% Mo to the highly corrosion resistant superferrite containing 28% Cr, 4% Mo 2% Ni), in chloride solutions of various concentrations (ranging from 10^{-4} N to 4 N), temperatures (between 20°C and 95°C), and degrees of aeration (from a few ppb to air saturated), (ref.1). Inevitably, these corrosion potential oscilations occured for environmental conditions of intermediate severity, between those conditions leading to stable passivity and those leading to pitting, as illustrated in Fig. 4. The severity of the environment is significantly increased by increases in temperature, chloride ion concentration and

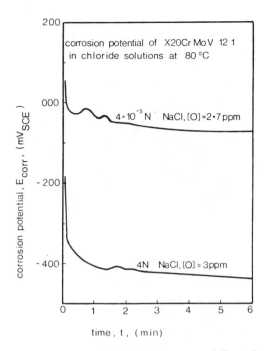

Fig. 1. The corrosion potential decreases rapidly after specimen-solution contact.

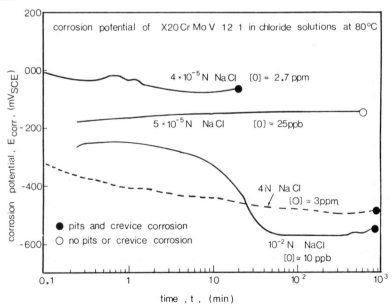

Fig. 2. The time dependence of the corrosion potential in various environments. Note that steady or slowly drifting corrosion potentials are observed for conditions leading to both pitting and also stable passivity.

634

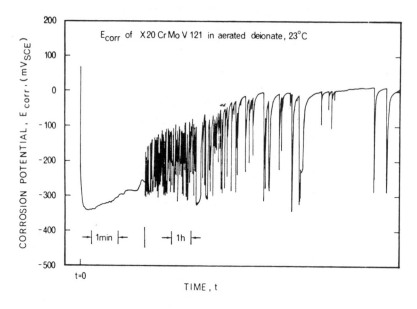

Fig. 3. Corrosion potential oscilations. The aerated deionate will have contained a small chloride contamination (perhaps of the order of 2×10^{-4} N NaCl) from NaCl remaining in the apparatus despite repeated washing.

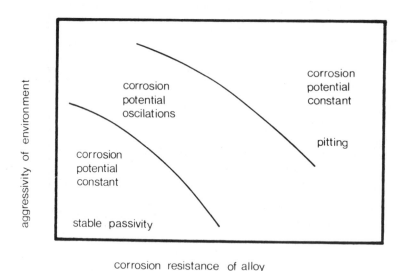

Fig. 4. The corrosion potential oscilations occured for environmental conditions of intermediate severity, between those conditions leading to stable passivity and those leading to pitting.

oxygen content.

DISCUSSION

The observations can be explained as follows. The corrosion potentials for conditions far removed from those leading to pitting, seem to be, to a first approximation, largely independent of temperature and chloride ion concentration; there being a difference of $\sim$ 200 mV between aerated and deaerated solutions. Thus, for metal-environment couples with little or no pitting tendency, the corrosion potential is nearly a constant. However, with increasing temperature and/or chloride ion concentration or with decreasing corrosion resistance, there comes a state where the corrosion potential begins to oscillate. These corrosion potential oscillations are between two end values which roughly correspond to E_p and E_{pp}, and are often associated with pitting and crevice corrosion. There are however, clear cases when no corrosion attack was visible even after prolonged immersion.

Corrosion potential oscillations have also been previously observed. For example Syrett et al. (ref.2) measured large potential fluctuations but could provide no detailed explanation at that time. Alderisio et al. (ref.3) observed transitory phenomena characterized by periodic oscilations of the corrosion potential for specimens galvanostated with a small constant applied current. Some parts of the corrosion potential curves are similar to those reported here. They attributed the oscilations to a successive activation and repassivation of pits. Hladky and Dawson (ref.4) also mention corrosion potential oscilations in passing.

It appears most likely that corrosion potential oscilations of the type observed in the present study are to be attributed to unstable passivity. That is, they occur when the metal environment couple is such as to cause the corrosion potential to lie between the pitting potential E_p and the protection potential E_{pp}.

The mechanism proposed is as follows. The interaction of the smooth specimen with the solution is such as to polarize the specimen to some positive value. However, before this potential is reached, localized breakdown of passivity occurs in the form of pit nuclei. These nuclei polarize the specimen to negative potentials and require a certain amount of current from a cathodic reaction, which current must be supplied by the rest of the specimen. When this current can no longer be delivered, the pit nuclei are no longer stable and must break up and the specimen surface repassivates. Thereupon, the interaction of the smooth, repassivated specimen surface with the solution will polarize the specimen to a positive value and the cycle will repeat leading to the observed oscilations. Examination of the potentiokinetic polarization curves, reveal that between E_p and E_{pp}, essentially the same current flows regardless of the applied potential (provided there is no pitting). This emphasises the instability of the situation; any pertubation will easily move the corrosion potential from E_p to E_{pp} or vice, versa. This mechanisms can furthermore only occur in the absence of "strong" electrochemical reactions, as these would tend to dominate the corrosion potential, (examples being the presence of such reduceable species as Fe^{3+} or Cu^{2+}).

In the proposed mechanism, pit nuclei have been used for the reason that, in a significant number of cases, no corrosion attack could be observed after prolonged oscillations. On the other hand, there were cases when pits and crevice corrosion were observed after the experiment. In these cases, the pits and/or crevice corrosion become unstable and the metal surface repassivates. This repassivated surface, acts essentially as a fresh smooth surface.

When however, the conditions are such that pits and/or crevice corrosion are stable, the final corrosion potential remains at a low negative value representative of active pitting and/or crevice corrosion. In these cases the corrosion potential shows merely a change from the initial more positive value to a final more negative value as illustrated in Fig.s 1 and 2, without any oscillations.

ACKNOWLEDGEMENTS

The author thanks R. Mueller, K. Mueller and C. Wuethrich for useful comments and R. Haering for skillful help with the experimentation.

REFERENCES

1 A. Atrens, Corrosion in press.
2 B.C. Syrett, R. Viswanathan, S.S. Wing and J.E. Wittig, Corrosion, vol. 38, p. 273 (1982).
3 A. Alderisio, B. Brevaglieri, A. Conti and G. Signorelli, Annali id Chimica, vol. 68, p. 405 (1978).
4 K. Hladky and J.L. Dawson, Corrosion Science, vol. 21, p. 317 (1981).

Passivity of Metals and Semiconductors, edited by M. Froment
Elsevier Science Publishers B.V., Amsterdam — Printed in The Netherlands

ATMOSPHERIC CORROSION OF STAINLESS STEELS

SATOSHI ITO, MASAO YABUMOTO, HIROYASU OMATA, TOMOMI MURATA
Fundamental Research Labs., Nippon Steel Corp.
1618 Ida, Nakahara-ku, Kawasaki City, Japan

ABSTRACT

A new quantitative method in laboratory was devised to investigate atmospheric corrosion of stainless steels. The potentials of stainless steels, changing with time, were measured up to the irreversible potential fall, when the surface was under thin water layer containing Cl^- ion. This method revealed that rust eventually develops on the surface at the irreversible potential fall after the periodical potential oscillations which may corresponds to micro pit formation and repassivation.

INTRODUCTION

Stainless steels are presently widely used for exterior applications such as architectures, cars, cargo containers and so on, because of their high corrosion resistance and maintenance free quality. They are, however, prone to atmospheric corrosion, and, eventually, ruining the aesthetic appeal of the material when exposed to marine or industrially polluted atmospheres.

In order to simulate the atmospheric corrosion of stainless steels in laboratories, various test methods have been employed: for instance, salt spray test, wet and dry cycle test. The conventional tests provide rather qualitative results, which do not have good correlations with field performance unfortunately.

The primary objective of this study is to develop the accelerating test method based on the process of atmospheric corrosion.

EXPERIMENTAL

Type 430 and 304 stainless steels with various surface treatments were employed for this study. A schematic model of atmospheric corrosion is illustrated in Fig. 1. The rain or condensed moisture in air form a drop of water containing Cl^- ion and/or SO_2 on stainless steel surfaces. Upon drying, dissolved ions get concentrated and, at the same time, diffusion of dissolved O_2 can be accelerated through thinner layer at the periphery of a water droplet. Then, corrosion occurs at active sites like nonmetallic inclusions and grain boundaries (ref. 1)

638

1) In order to simu-
late the process of
atmospheric corrosion an
experimental set-up was
designed as shown in
Fig. 2. A woven cotton
cloth was placed on the
surface of a given
stainless steel specimen
(30 x 30mm area) keeping
the rest of the cloth
immersed in 0.5 M NaCl solution to absorb the solution
by capillary phenomenon, and
a thin layer of the solution
on the stainless steel surface, where drying and condensation occur. The potential between the specimen
and reference electrode can
be measured through a lin-
kage of the solution. This

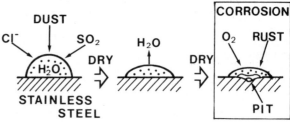

Fig. 1. A Schematic model of atmospheric corrosion
of stainless steels during service

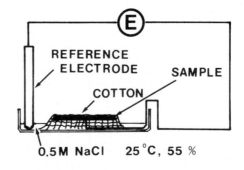

Fig. 2. The experimental apparatus

system gives an analogous condition to actual atmospheric corrosion in either
Cl⁻ or SO₂ containing environments. Then, the experimental apparatus was enclosed in a cell with a humidity controlling air thermostat which controlls
the temperature of the system at 25°C and relative humidity at 55%.

2) The pitting potential measurements were carried out in the usual manner
at the potential scanning rate 20 mV/min, in 3.5% NaCl solution at 30°C.

3) In order to examine the field performance of the atmospheric corrosion of
stainless steels, the same stainless steels as utilized in laboratory were
actually exposed to the sea shore environments for 1 week to 8 weeks. After
the exposure, the samples were lightly brashed and washed in running water,
dried, then the features of rusting on the specimen surface were visually
examined by the rating number based on a standardized charts of the rusted
samples

4) Corrosion products and passive film structure and its chemistry were
analysed by way of AES, ESCA etc. when needed.

RESULTS AND DISCUSSION

Fig. 3 demonstrates the relationship between the pitting potential measured in 3.5% NaCl solution and the rating of the same stainless steels actually exposed to the sea shore. It doesn't give so good a correlation that it means that the pitting potential is not necessarily a measure of atmospheric corrosion of stainless steels.

Then, the new type of method depicted above in Fig. 2 was devised to obtain quantitative results strongly correlated to the corrosion.

Fig. 4 shows a typical result of measured potential change with time. The potential shows oscillation periodically in between 0.1 and 0.4 V vs. Ag/AgCl, which may correspond to the local breakdown of passivity and repassivation processes. After 10 to 100 hours depending on the alloy chemistry and surface conditions, the potential fells to the level of -0.1V ~ -0.3V vs AgCl/Ag. This whole potential fall is most likely the initiation of corrosion since the rust formation can be identified

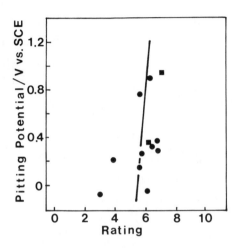

Fig. 3. Correlation between pitting potential and rating by field exposure test

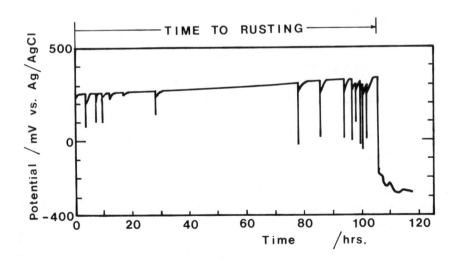

Fig. 4 A typical example of measured potential change with time, showing the potential fall to initiate rusting after several oscillations (Type 430).

approximately 10 hours after the
fall.

The following processes may
be involved during drying of a
thin water layer.

(1) Enrichment of Cl⁻ ion
occurs.

(2) The corrosion potential
at around active sites
like inclusions and grain
boundaries exceeds the
local pitting potential
due to the local change
in solution chemistry.
Then, the passive film
breakdown occurs corres-
ponding sometime to micropits
formation.

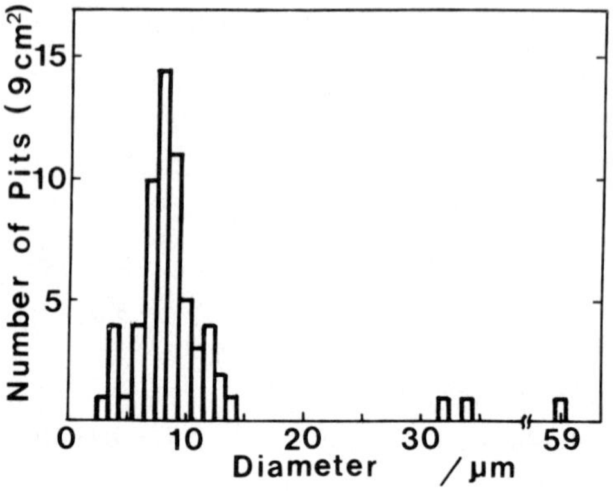

(3) The above process gives rise
to a temporary potential fall
which goes below the pitting
potential, and then repassivation occurs. This results
in the observed potential
oscillations. In fact as
shown in Fig. 5, the pit size
distribution of the already
rusted specimen in the laboratory method indicates the
existence of some critical
micro pit size (20 to 25
diameter) as has been reported
by Hisamatsu et al. (ref. 2).
Moreover, the numbers of micro
pits were of the same order of
the numbers of the potential
oscillations. Therefore it is
considered that under thin
water layer, the micro pits
formation and repassivation

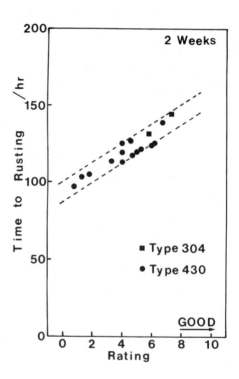

Fig. 5. Pit size distribution in the area under cotton cloth (Type 430)

Fig. 6 Correlation between time to rusting shown in Fig. 4, and rating by field
exposure test

process occures, giving potential oscillations up to the time of the irreversible potential fall which may mean the pit having over critical
size initiates propagation.

(4) Finally, the potential falls down to -0.1 ~ 0.3 V vs. Ag/AgCl. It may
be due to irreversible breakdown of the passive film by acidification
resulting from the hydration of dissolved metallic ions, and condensation (Fig. 7) or to pitting type corrosion (Fig. 8).

Thus, rust appears on the stainless steel surfaces.

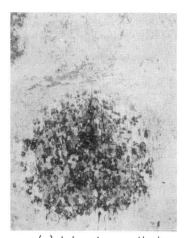 

(a) Laboratory method   (b) Exposure test

Fig. 7 A manner of corrosion in comparing the laboratory
method with field exposure test (Type 430)

 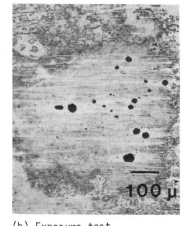

(a) Laboratory method   (b) Exposure test

Fig. 8 A manner of corrosion pitting in comparing the
laboratory method with field exposure test (Type 430)

Now, this experimental method simulating atmospheric corrosion of stainless steels can be utilized as an evaluation method of the field performance in terms of the time to the potential fall as a measure of the actual time to rusting. Fig. 6 shows the obtained relationship between the time to potential fall vs. relative visual rating of rusting on the same stainless steel actually exposed to the sea shore environments for 2 weeks. According to the correlation demonstrated in Fig. 6, this method provides a useful tool to predict the field performance of a given stainless steel rather quantitatively together with informations with respect to the breakdown and repassivation of passive films.

Additionally, the manners of rusting on the stainless steel surfaces in the laboratory method were very similar to those in the field exposure tests, when the same steels were utilized. It is evident that this method is simulating actual atmospheric corrosion of stainless steels.

CONCLUSION

1) A new quantitative method simulating actual atmospheric corrosion provides a useful tool to predict the field performance of a given stainless steels.

2) When the stainless steels are in service in atmospheric environments, the passive film may be repeatedly broken down to form micro pits. Eventually, it is subjected to irreversible break down which in some case corresponds to growing over the critical size of the pits.

REFERENCES
1. K.E. Johnson, Br. Corros. J., 15 (1980) 123
2. Y. Hisamatsu, Boshoku G-jutsu, 21 (1972) 503

Passivity of Metals and Semiconductors, edited by M. Froment
Elsevier Science Publishers B.V., Amsterdam — Printed in The Netherlands

ELECTROCHEMICAL PREVENTION OF LOCALIZED CORROSION OF STAINLESS STEELS

N. AZZERRI, F. MANCIA and A. TAMBA

Centro Sperimentale Metallurgico S.p.A. - P.O. Box 10747 - ROME-EUR (Italy)

ABSTRACT

The pitting and crevice corrosion behaviour of stainless steels is investigated. The actual meaning and the practical importance of the protection potential are examined and conclusions are drawn taking into account both the potential and non-potential dependent initiation and growth of corrosion phenomena. In this concern, the theoretical background and practical conditions for localized corrosion prevention by cath-anodic protection are discussed.

The pitting and crevice corrosion behaviour of stainless steels as a function of temperature and chloride content in industrial waters has been considered in connection with localized environmental modification occurring because of the attack propagation (ref. 1).

Experimental stability diagrams expressed in terms of electrode potential vs NaCl content in water have been constructed on the basis of the electrochemical parameters obtained from potentiodynamic measurements (ref. 2,3). The hysteresis technique (ref. 4) has been applied in deaerated NaCl aqueous solutions (from 200 up to 35000 ppm) at pH $\cong$ 7 and at pH $\leqslant$ pHd (pHd is the depassivation pH (ref. 5)).

Fig. 1 shows the stability diagram of AISI 304 stainless steel at 22°C; the diagram allows the prediction of the localized corrosion behaviour of the steel in function of the electrode potential and the chloride content and states the conditions to be fulfilled in order to prevent both initiation and propagation of localized corrosion. A detailed description of the matter, concerning four commercial stainless steels at various temperatures, is given elsewhere (ref. 1).

The validity of experimental stability diagrams has been checked as a function of potential by long term potentiostatic tests (30-40 days) in flowing industrial water on AISI 304 steel samples on which a severe crevice geometry (0-ring assembly) was set up (ref. 6).

644

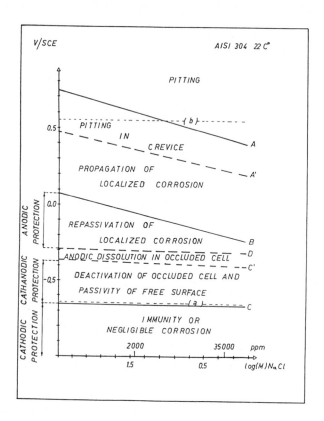

Fig. 1. Experimental stability diagram (potential vs. NaCl content in water) for prediction and control of localized corrosion of AISI 304 stainless steel at 22°C

Localized corrosion has never been observed to initiate and grow on samples polarized within the stable passivity region, i.e. at potentials under the line B, as shown in Fig. 2a. Nevertheless, in the presence of initiated localized corrosion, i.e. on samples undergoing crevice corrosion purposely allowed to start (through an anodic current of 4 mA . cm^{-2} passed for 22.5 h), the further attack is successfully stifled only by shifting the polarizing potential towards more negative values, precisely below line C' (see Fig. 2b), where the deactivation mechanism takes place (ref. 1,3). In the absence of external polarization, free-corrosion potentials of pre-corroded samples have ranged in the field of anodic dissolution (see Fig. 3), thus allowing the further propagation of localized corrosion.

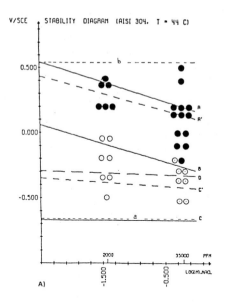

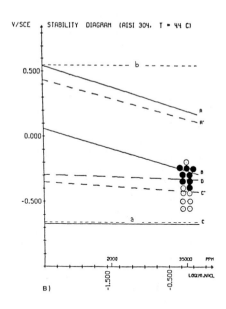

Fig. 2a. Localized corrosion(pitting and/or crevice) behaviour observed on crevice bearing (O-Ring) samples after at least 500 h of exposure in flowing Cl⁻ - containing water under polarization

● corrosion;   ○ no corrosion

Fig. 2b. Behaviour of localized corrosion propagation on precorroded samples (AISI 304) as a function of the polarizing potential

● further propagation

○ arrest of propagation

Moreover, the achievement of localized corrosion protection by external polarization has been assessed in slowly flowing natural sea water (temperature about 30°C). In Fig. 4 protection current at various potentials as a function of time are shown; mass losses due to corrosion propagation measured after 166 days are also given. It is apparent that only below line C' (V < - 450 mV vs SCE) a durable protection is attained; mass loss data are consistent with the above results.

Localized corrosion protection, therefore, takes place into two different potential regions, i.e. between lines B and D or lines C' and C. In practice, however, in the former region (field of anodic protection since free surface is kept passive and surface under crevice might remain passive or repassivate) some hazard can be encountered, since the B-D region is narrowed progressively in consequence of the chemical modifications occurring inside pits or crevices and in any case severe attack, due to non negligible leakage passive current and/or to uncontrolled potential drop for electrolyte resistance, can take place after an incubation time depending on actual crevice geometry severity.

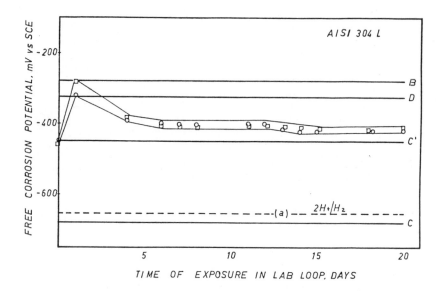

Fig. 3. Free-corrosion potential of precorroded (anodic etch at 4 mA . cm^{-2} for 22.5 h) samples (AISI 304) as a function of time of exposure in flowing NaCl (35000 ppm) solution at 44°C

Localized corrosion prevention, therefore is better achieved between C and C' lines where free (or non corroding) surface is still passive while inside already existing pits or crevices the deactivation mechanism operates. This kind of true and durable protection can be defined as "cath-anodic" protection as free surface are passive (as in anodic protection) while hindered surfaces in crevice or under deposits, as a results of the local chemical modification, are protected cathodically.

Therefore, the line C', which is slightly influenced by chloride content and is fairly independent of pH modification inside corroding pits and crevices, represents the upper potential limit for reliable localized corrosion protection.

In this circumstance, however, as time laps under cathodic polarization between C and C' lines, a progressive build-up of alkaline products could occur (as proved by experimental pH measurements in stifled crevice), changing the protection mechanism from the deactivation to the repassivation of the metal in the crevice ·

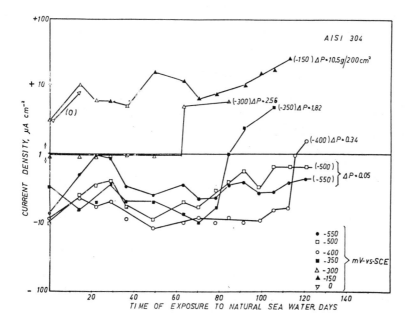

Fig. 4. Current density arising from the polarization at different potential values (given in parentheses) as a function of time of immersion in natural sea water at about 30°C. The mass loss ($\triangle$P) of samples, measured at the end of experiments, is also reported (in g/200 cm^2)

Finally, it is worth emphasizing that "cath-anodic" protection requires a very low impressed current as corroding sites only have to be protected cathodically. In any case the amount of impressed current required to drive the potential just below line C' depends on the actual redox potential of the stainless steel-environment system and the steel composition. The latter has a definite effect on the cath-anodic protection current since it affects the potential level of the line C' and,therefore, the protection current; noble elements (Ni, Cu, Mo) in the steel shift line C' towards positive values (see Fig. 5), thus claiming for lower protection current, while chromium has a reverse effect. A set of experimental stability diagrams for different stainless steel grades is available (ref. 1).

648

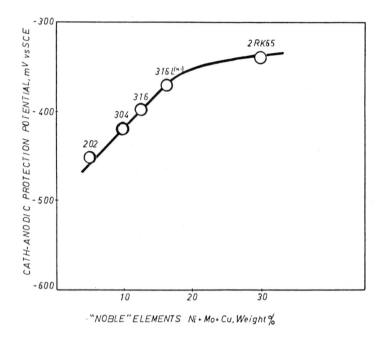

Fig. 5. Influence of alloying elements on the cath–anodic protection potential
of stainless steels in sea water.

REFERENCES

1 N. Azzerri, F. Mancia and A. Tamba, Corros. Sci.,22, (1982), 675
2 M.J. Johnson, in "Localized Corrosion. Cause of Metal Failure",ASTM STP 516
   (1972), 262
3 K.K. Starr, E.D. Verink, Jr., M. Pourbaix, Corrosion,32, (1976), 47
4 M. Pourbaix et al., Corros. Sci.,3 (1963), 239
5 J.L. Crolet, J.M. Defranous, Corros. Sci.,13, (1973), 575
6 A. Tamba, F. Mancia and N. Azzerri, Proc. Conf. Eurocorr/82, Budapest,
   Oct 18–22, 1982

Passivity of Metals and Semiconductors, edited by M. Froment
Elsevier Science Publishers B.V., Amsterdam — Printed in The Netherlands

ANODIC BEHAVIOUR OF Fe-31% Ni ALLOY IN THE MULTI STEADY STATES REGION
CORRESPONDING TO THE TRANSITION BETWEEN ACTIVE AND PASSIVE STATES, IN NORMAL
SULPHURIC ACID.

F. WENGER and J. GALLAND

Laboratoire de Corrosion et Fragilisation par l'Hydrogène

Ecole Centrale des Arts et Manufactures. 92290 Chatenay Malabry (France).

ABSTRACT

The anodic polarisation curves of a Fe-31 Ni (wt.%) alloy have been studied
with a rotating disk electrode, at rotation speed high enough to eliminate the
influence of mass transfer on kinetics. A z-shaped curve was found and showed
a transition between active and passive states. In the part of the curve
corresponding to this transition, active and passive areas coexist at the
surface of the metal. We determine in this work, the current density
distribution and the mean potential on the active and passive surfaces along
the transition region. We compare our results with those predicted by a model
which was developped to explain the z-shaped curves found at pure iron.

INTRODUCTION

In a former work (ref.1), we investigated the steady state anodic current
voltage curve of a Fe-31 Ni (wt.%) alloy rotating disk electrode, in normal
sulphuric acid. We showed that the curves have some features already observed
by Epelboin et al. (ref.2) at pure iron electrodes, in the same medium : the
curve we obtained for rotation speed high enough to eliminate the influence of
mass transfer, is a z-shaped curve (see Fig.1). Over a wide range of potentials
(several hundreds of millivolts), there are three steady current values at each
potential. The high and low values correspond respectively to the active
(A-B : dissolution) and passive (E-F) states. The intermediate values give a
reversible transition region (B-C) between active and passive states.

Epelboin et al. obtained a continuous transition region. On the contrary,
this part of our curve is discontinuous : a gap (C-D) of about 50 millivolts is
found at low current values. It corresponds to an important change of the
anodic mechanism. Along the transition region (B-C), active and passive areas
coexist at the surface of the disk electrode. The dissolution is localized on
a disk shaped area surrounded by a passive ring. Near the inner edge of this
ring the grain boundaries are attacked but the grains remain passive. The
intensity I_{ig} of this intergranular dissolution decreases steeply on the ring
surface as the distance to the inner edge increases. The diameter of the active
disk diminishes as one travels down the transition region, until the cathodic

650

boundary is reached. The total current I is the sum of I_a (intensity on the active disk), I_{ig} (intergranular dissolution) and I_p (intensity on the passive ring). Along the transition region, I_p is always negligible.

At the cathodic end of the passive region, over one hundred of millivolts (D-E), the total current increases as the potential becomes more cathodic. This current corresponds to an intergranular dissolution occuring uniformly on the whole surface of the electrode.

Some years ago, C.G. Law, Jr. and J. Newman have proposed a model (ref.3) to explain the results of Epelboin et al. This model takes into account the effect of potential distribution on the active part of the disk electrode and a local polarisation relation showing a discontinuous change from the active to the passive state. A Tafel's law is used to describe the charge transfer process, and mass transfer limitations may be included, if necessary, in the kinetic expressions. This model predicts the z-shape of the curves and the localized dissolution in the transition region.

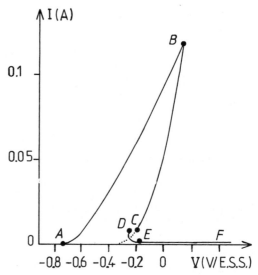

Fig.1. Anodic steady state current - voltage curve. Fe-31% Ni in H_2SO_4 1N. Diameter of the rotating disk : 5.0 mm

In order to clarify the mechanism of dissolution and passivation of the Fe-31% Ni alloy, we have compared in this work our experimental results with the behaviour predicted by the model.

EXPERIMENTAL METHODS

The metal is a Fe-31 Ni (wt%) alloy. The cristallographic structure is f.c.c.(austenitic) :

Element	C	Mn	Si	S	P	Ni
wt %	0.011	0.15	0.02	0.013	0.002	31.0

The working electrode is a rotating disk electrode. Several diameters have been studied between 0.5 to 5.0 mm in order to obtain the transition region at low current values, and to find the relation between the resistance of the electrolyte and the diameter of the disk. The rotation speed is 314 rd.s^{-1}.

This speed is high enough to eliminate the influence of mass transfer all along the transition region.

The reference electrode is a S.S.E. (Hg/Hg$_2$SO$_4$/sat. K$_2$SO$_4$). The electrolyte is a 1N H$_2$SO$_4$ solution.

A negative impedance converter (N.I.C.) was necessary to obtain the whole z-shaped curve. This device and its use have been extensively described by Epelboin et al. (ref.2).

The resistance of the electrolyte between the working electrode and the reference electrode was obtained from impedance measurements at high frequencies. The impedance measurement device includes a Solartron 1172 frequency response analyser, and was described in many other works (ref.4).

In order to analyse the current density distribution and to measure the total current flowing through the active area, the localized dissolution of the the surface was studied at different points of the transition region, where different values of the amount of electricity were imposed. For each value, the outline of the active area has been surveyed with a microscope. The depth of field was small enough for the depth of attack to be determine within 1 μm.

We have compared the theoretical Faraday's law with data obtained in the dissolution region (A-B) of the curve, at high current values and for different amounts of electricity, to verify the validity of this method.

RESULTS AND DISCUSSION

The measurements of the volume of dissolved metal on the active disk cannot give the exact value, but only the lowest limit of I_a. We could show that I_{ig} is smaller than two and higher than one milliampere, all along the transition region. The experimental error on the value of I_a is small enough not to change deeply our results.

Average current density $\overline{J}_a$:

We found the following relation between the diameter d of the active disk and the current I_a :

$$I_a = B(d + d_0)^2$$

where B = 4.4 10^{-1} A/cm^2 , and d$_0$ = 0.017 cm

This result agrees with the theoretical model which predicts an increase of the average current density $\overline{J}_a$, when I_a decreases. Here we find :

$$\overline{J}_a = \frac{4B}{\pi} + \frac{8Bd_0}{\pi d} + \frac{4Bd_0^2}{\pi d^2} \qquad (2)$$

This relation is only valid for diameters of the active disk from .5 to 5.0 mm
The limit of $\bar{J}_a$ for $I_a = 0$ cannot be deduced from the curve we obtained (see
Fig.2). According to the model, the $\bar{J}_a$ upper limit could be six times the value
of $\bar{J}_a$ at the top of the transition region (B) :

$\bar{J}_a(I_a = 0) = 3.6$ A/cm²

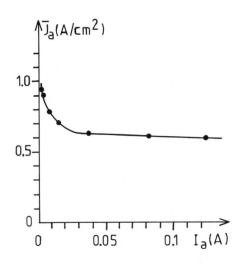

Fig.2. Evolution of the average
current density $\bar{J}_a$ with I_a.

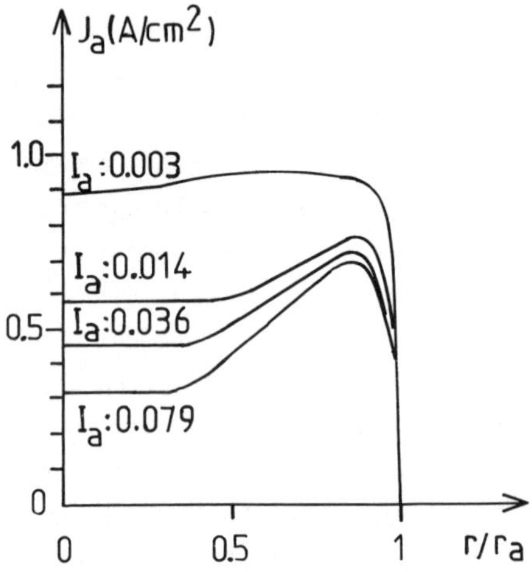

Fig.3. Variation of J_a across the
active disk at different intensities.
I_a is given in amperes.

Current density distribution :

From Faraday's law, the local current density J_a is expressed as a function
of the depth h, and the duration Δt of the attack :

$$J_a = zF \frac{\rho\ h}{M\ \Delta t}$$

with : F = 96500 c.   ;   z = 2   ; ρ = 8.0 g/cm³ (measured)
            M = 0.7 M(Fe) + 0.3 M(Ni) = 56.7
            M(Fe) and M(Ni) : atomic weights of iron and nickel respectively.
Figure 3 gives the current density distribution as a function of r/r_a at
different current values. r is the distance to the center of the electrode and
r_a = d/2.
    We find here a result predicted by the model : the current density
distribution becomes more uniform as the current decreases.

Shape of the transition region :

We fitted the experimental transition region to a parabola :

$$I_a = A(V - E_0)^2 \tag{4}$$

where $A = (0.53 \pm 0.02)$ A/V^2 and $E_0 = -(0.32 \pm 0.01)$ Volts/E.S.S.

According to the model, the transition region should be a parabola, only when the average current density remains constant along the curve (kinetics controlled by mass transfer).

The theoretical expression of I_a can be written :

$$I_a = \frac{C}{\overline{J}_a} (V - E_0)^2 \tag{5}$$

and the curve deviates from a parabola. But in fact C is not constant with V : it varies from C_0 for a primary distribution to 1.52 C_0 for a uniform distribution of current density : in our experiments, the relative increase of $\overline{J}_a$ is 1.57 along the curve and is compensated by the simultaneous increase of C. The deviation from the parabolic shape is not visible.

Potential of the active surface :

E is the average electrochemical potential of the active disk. It is calculated from the voltage V imposed between the metal and the reference electrode :

$$E = V - R_e I_a \tag{6}$$

where $R_e I_a$ is the ohmic drop through the resistance of the electrolyte between the active disk and the reference electrode.

With rotating disk electrodes of varied diameters, we found the following experimental relation between R_e and d :

$$R_e = \frac{D}{d} \text{, with } D = 26 \; 10^{-3} \Omega.m \tag{7}$$

This evolution of R_e with d agrees with the theoretical expression of

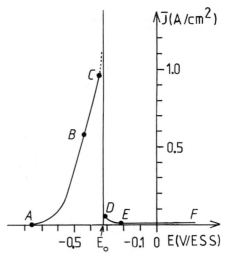

Fig.3. Anodic steady state current density-potential curve. Fe-31% Ni in H_2SO_4 1N.

654

the primary resistance of a disk (ref.3).

But when the current distribution is uniform, the theoretical expression of the primary resistance (ref.3) allows us to write :

$$R_e = \frac{0.81\ D}{d} \qquad (8)$$

We could only estimate the value of E taking into account equation (7) for I$\geq$40 mA and equation (8) for I$\leq$16 mA. Between 16 and 40 mA an intermediate value was chosen (0.9).

The whole curve giving J_a as a function of E has been represented on the figure 3. The passive region has also been represented. The potential E_0 (- 0.32 V/E.S.S.) corresponds to the anodic limit of the active state, and to the cathodic limit of the passive region.

CONCLUSION

In this work, we found that the z-shaped anodic steady state polarisation curve of a Fe-31% Ni rotating disk electrode can be described by a simple model taking into account a discontinuous transition between active and passive states and the non uniform distribution of the potential across the surface. This non uniform distribution of the potential gives rise to the coexistence of an active disk and a passive ring on the surface of the metal.

The breaking off of the passive film occurs at the cathodic limit of the passive region, on grain boundaries. This intergranular dissolution has not been observed with pure iron, and seems to be connected with the presence of nickel.

This study will allow us to carry out a further work which is intended for elucidating the influence of nickel on the mechanism of dissolution and passivation of iron-nickel alloys with different nickel contents.

REFERENCES

1 C. Dagbert, F. Wenger, J. Galland, P. Azou.
   Mem. Sci. Rev. Mét. Juillet 1979, 461.
2 I. Epelboin, C. Gabrielli, M. Keddam, J.C. Lestrade and H. Takenouti
   J. Electrochem. Soc., 119 (1972), 1632.
3 C.G. Law, Jr. and J. Newman.
   J. Electrochem. Soc., 126 (1979), 2153.
4 C. Gabrielli, M. Keddam,
   Electrochimica Acta 19 (1974), 355.

Passivity of Metals and Semiconductors, edited by M. Froment
Elsevier Science Publishers B.V., Amsterdam — Printed in The Netherlands

PIT GROWTH MEASUREMENTS ON STAINLESS STEELS

F. HUNKELER and H. BOEHNI

Institute of Materials Chemistry and Corrosion, Swiss Federal Institute of
Technology, ETH-Hönggerberg, CH-8093 Zürich, Switzerland

ABSTRACT

Pit and crevice growth of different stainless steels, investigated in chloride containing solutions of different pH values at room temperature, can be
described by the same time law. Depending on the solution composition and on the
potential the pit growth is either diffusion or ohmic controlled. Pit growth
measurements and microscopical observations indicate that the pit surface electrolyte is not necessarily saturated. Below critical limits of current density
and potential crevice corrosion is the only stable form of localized corrosion.
Above those limits pitting and crevice corrosion are possible.

INTRODUCTION

The meaning of the critical pitting and crevice potentials measured by diffe-

rent methods is still under discussion. Much attention has been paid to the un-

derstanding of the factors controlling the initiation of pits and crevices. But

under practical conditions initiation of localized corrosion can often not com-

pletely be excluded. Therefore the knowledge of the factors determining the pro-

pagation rate is an undoubted help not only for a fundamental understanding of

localized corrosion phenomena but also for practical application to design.

EXPERIMENTAL

The method for the determination of pit growth rates has been described else-

where (ref.1,2). The materials used in this study were: x8 Cr17 (type 430),

x5 CrNi 18-9 (type 304) and x2 CrNiMo 18-12-2.5 (type 316L). To get the pit

growth kinetics tests were made with 0.1, 0.2 and 0.3 mm thick foils. The expe-

riments were carried out in slightly stirred deaerated chloride containing solu-

tions of different pH values at room temperature. To reduce the induction time

and to prevent crevice corrosion the specimens for the repassivation experiments

were i) etched for 1 min in conc. HCl ii) covered with an adhesive tape with a

circular opening iii) activated for 1 min with a small drop of conc. $FeCl_3$ and

cleaned with a soft paper and then immediately fixed in the cell.

RESULTS AND DISCUSSION

Fig. 1 shows that pit and crevice growth strongly depend on the potential and that the propagation rate in both cases is time dependent. As it can be seen in Fig. 2 the pH of the solution has no significant influence on the crevice and pit growth rate. On the other hand the propagation rate at the same potential is much faster in the 1.0 m than in the 0.1 m chloride solution. Of great importance is the observation of the same time law for pit and crevice corrosion (the same slope in Fig. 2).

The pit and crevice growth curves in Fig. 1 and 2 generally follow Eq. 1

$$d = a \cdot t^b \quad [mm] \tag{1}$$

The exponent b varies between 0.53 and 0.7 indicating that b can be regarded as more or less constant. Similar values for b were found for stainless steels by other authors (ref.3-6). The factor a (=pit depth at t=1 min) depends on the solution composition and on the potential.

From Eq. 1 the pit and crevice current density i_p can be calculated (see ref.1 and 2)

$$i_p = a \cdot b \cdot c \cdot t^{b-1} = i_p^o \cdot t^{b-1} \quad [mA/cm^2] \tag{2}$$

For steels the conversion factor c in Eq. 2 is c= 1 mm/min $\doteq 52.56 \times 10^3$ mA/cm^2 (ref.7). The initial pit and crevice current density i_p^o (=current density at t=1 min) of the curves shown in Fig. 2 varies between app. 70 and 1200 mA/cm^2.

Combining Eq. 1 and 2 it follows

$$d \cdot i_p = a \cdot i_p^o \times 10^{-4} \cdot t^{2b-1} = a^2 \cdot b \cdot c \times 10^{-4} \cdot t^{2b-1} \quad [A/cm] \tag{3}$$

The products $a \cdot i_p^o$ of the curves in Fig. 2 have values between 0.15×10^{-4} and 40×10^{-4} A/cm. According to Vetter and Strehblow (ref.8) and Beck and Chan (ref.9) $d \cdot i_p$ is given either by Eq. 5 or Eq. 6.

$$(d \cdot i_p)_\Omega = \frac{\sigma \cdot \Delta\varepsilon_\Omega}{a^*} \tag{4}$$

$$(d \cdot i_p)_D = \frac{z \cdot F \cdot D \cdot c_S}{a^* \cdot t_-} \tag{5}$$

where σ is the specific conductivity, $\Delta\varepsilon_\Omega$ the ohmic overpotential, a^* the geometry factor (for hemipherical pits app. 3 (ref.8)), D the diffusion coefficient of the metal ion, c_S the saturation concentration of the metal chloride, t_- the transference number. The factors a^* and t_- largely cancel for hemispherical pits (ref.9).

If in case of an ohmic controlled growth σ and $\Delta\varepsilon_\Omega$ are constant with respect

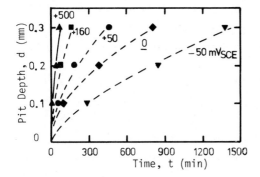

Fig. 1

Influence of the potential on pit and crevice growth of type 304 stainless steel in 0.1 m HCl

——————— pitting corrosion

— — — — crevice corrosion

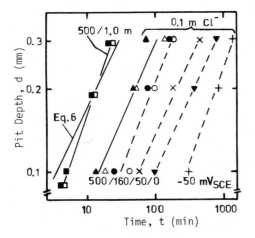

Fig. 2

Influence of the potential, chloride concentration and pH on pit and crevice growth of type 304 stainless steel

■ ▲ ● ✕ ▼ +   pH ≈ 1

□ △ ○   pH = 11

——————— pitting corrosion

— — — — crevice corrosion

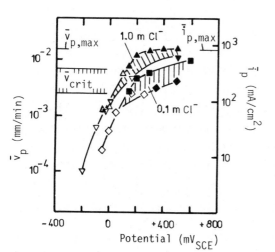

Fig. 3

Mean pit and crevice growth rate $\bar{v}_p$ versus potential of type 304 stainless steel; foil thickness: 0.2 mm; pH ≈ 1

full symbols: pitting corrosion
open symbols: crevice corrosion

▲ △ } repassivation
■ □ } experiments

▼ ▽ }
◆ ◇ } ε = const.

to time Eq. 4 leads to a growth law with b=0.5. This is approximately fulfilled
on aluminum (fef.10). Values for b higher than 0.5 are possible if σ (concentration build up in the pits) or $\Delta\varepsilon_\Omega$ (high Tafel constants) increases with time
(ref.10).

Ideally the diffusion controlled growth process (Eq. 5) would also result in
a b-value of 0.5. Several factors might cause higher b-values. One possible factor is a* which changes until a constant pit geometry is reached. An other possibility follows from the assumption in Eq. 6 that the diffusion layer thickness
is equal to the pit or crevice depth. Although this assumption is certainly not
absolutely correct it gives a rough idea of the effective diffusion layer thickness since true values are unknown.

In conclusion experimental values for b can be higher than 0.5 because of several reasons and since ohmic and diffusion controlled growth result in the same
b they do not allow one to distinguish between the different mechanisms.

From Eq. 5 the maximum product $(d \cdot i_p)_{max}$ can be estimated. Using the values
$c_S = 4.25 \times 10^{-3}$ mol/cm^3 (ref.11), $D_{Fe^{2+}} = 8.5 \times 10^{-6}$ cm^2/sec (ref.11), $\bar{z} = 2.25$
(for type 304) and $a^* \cdot t_- = 1$ gives $(d \cdot i_p)_{max} = 78.4 \times 10^{-4}$ A/cm. Assuming
b=0.5 and using Eq. 3 the constants a_{max} and $i^0_{p,max}$ for a fully diffusion controlled growth rate are: $a_{max} = 54.6 \times 10^{-3}$ mm/min$^{1/2}$ and $i^0_{p,max} = 1435$ mA/cm^2.
Inserting a_{max} in Eq. 1 the growth curve under diffusion control is obtained

$$d_{max} = 54.6 \times 10^{-3} \cdot t^{0.5} \text{ [mm]} \tag{6}$$

Comparing Eq. 6 with the experimental results (see Fig. 2) leads to the important conclusion that diffusion controlled pit growth is only realized in the
higher concentrated chloride solution. Interestingly the experimental perfora-
tion time for the 0.3 mm foil is even shorter than the calculated time indicating that either a* is smaller than 3 or that the effective diffusion layer
thickness is smaller than 0.3 mm.

For a theoretical understanding the knowledge of the minimum product
$(d \cdot i_p)_{min}$ required for pit growth is of great importance. For this purpose one
has to distinguish between crevice and pit growth. In the experiments presented
in this paper the existence and growth of pits on the open surface (that means
not at the edges of the adhesive tape or under the O-ring) was taken as criterion. In Fig. 3 the mean growth rate $\bar{v}_p$ and $\bar{i}_p$ were calculated according to
Eq. 7 and 8

$$\bar{v}_p = \frac{d}{t_p} \text{ [mm/min]; } d = \text{ foil thickness used; } t_p = \text{ perforation time measured} \tag{7}$$

$$\bar{i}_p = \bar{v}_p \cdot c \text{ [mA/cm}^2\text{]} \tag{8}$$

Fig. 3 contains two sets of experiments. In one set the potential was held
constant throughout the experiment. In the other set (repassivation experiments)
the potential was kept constant for 4 min at a high (500 mV$_{SCE}$ in the 1.0 m and
600 mV$_{SCE}$ in the 0.1 m solution resp.) and after a fast change at a lower value.
Generally the curve of the pit and crevice growth in the 0.1 m solution is
shifted to higher potentials and does not reach the plateau of the curve measured in the 1.0 m solution. The plateau for the 1.0 m solution indicates again
the diffusion controlled growth process and corresponds well with the calculated
value $\bar{i}_{p,max}$. Using $a \cdot i_p^0 = (d \cdot i_p)_{max}$ and $d = 0.2$ mm $\bar{i}_{p,max}$ is according to Eq. 9

$$\bar{i}_p = \frac{2\, i_p^0}{t^{1/2}} = \frac{2\, a\, i_p^0 \times 10^4}{d} \quad [mA/cm^2] \tag{9}$$

app. 784 mA/cm^2 or $\bar{v}_{p,max} = 14.9 \times 10^{-3}$ mm/min. Since in the 0.1 m solution the
saturation concentration can not be reached and therefore no such plateau could
be detected, the ohmic potential drop in the pits is very likely the rate controlling factor at all potentials combined with charge transfer (b>0.5).

In both solutions no pits could grow on the open surface below +150 mV$_{SCE}$
corresponding to a critical $\bar{v}_p$ of app. 2.5 to 7×10^{-3} mm/min (0.17 to
0.47 $\bar{v}_{p,max}$). This would mean that in both cases the electrolyte in the pits is
far below the saturation concentration (0.72 to 2 m FeCl$_2$). Taking the different
pit geometries into account (shallow pits in the 1.0 m and narrow pits in the
0.1 m solution) one would expect about the same lower limit. The conclusion that
the pit surface electrolyte is not necessarily saturated coincides with the microscopical observations where completely polished pit surfaces were found only
if $\bar{v}_p \approx \bar{v}_{p,max}$. At $\bar{v}_p$ values between 7×10^{-3} mm/min and $\bar{v}_{p,max}$ partly polished
surface areas were seen, whereas $\bar{v}_p$ values below app. 7×10^{-3} mm/min produced un-
polished or etched surfaces.

Fig. 3 shows that below the critical limits crevice corrosion is the only
stable form of localized corrosion. Since there is no obvious discontinuity in
the potential dependence it can be assumed that pitting and crevice corrosion
are basically the same phenomenon as it has been proposed by various authors
(see e.g. ref.12).

Considering all these facts it follows that the pits repassivate below the
critical $\bar{v}_p$ or critical $\bar{i}_p$ (corresponds to $(d \cdot i_p)_{min}$ acc. to Eq. 9) because the
ohmic potential drop $\Delta\varepsilon_\Omega$ in the pits (Eq. 4) is smaller than the difference between the control potential and the passivation potential which itself is in
turn a function of the electrolyte composition on the pit surface and of the
product $(d \cdot i_p)_D$ (Eq. 5) respectively. On the other hand crevice corrosion can

660

occur below such limits because i) $\Delta\varepsilon_\Omega$ is larger and ii) the concentration build up is higher than in the pits.

Comparative measurements showed that pit and crevice growth of the three different steels can be described by Eq. 1. At the same potential the growth rate decreases in the order: type 430 > type 304 > type 316L. This finding leads to the conclusion that not only Mo but also Ni improves remarkably the pitting and crevice resistance of stainless steels, a conclusion which could not be drawn so clearly on the basis of pitting potential measurements (ref.7,12,13) and, therefore, the pitting potentials are usually correlated with the Cr and Mo content (ref.7,13,14).

REFERENCES

1 F. Hunkeler and H. Böhni, Werkstoffe und Korr., 32 (1981) 129.
2 F. Hunkeler and H. Böhni, Corrosion, 37 (1981) 645.
3 Z. Szklarska-Smialowska, Werkstoffe und Korr., 22 (1971) 780.
4 Z. Szklarska-Smialowska et al., Corr. Sci., 12 (1972) 925.
5 J. Tousek, Corr. Sci., 18 (1978) 53.
6 T. Hakkarainen, Proc. 8th Int. Congr. Met. Corr., Mainz, 1981, p. 157.
7 A. Rahmel/W. Schwenk, Korrosion und Korrosionsschutz von Stählen, Verlag Chemie, Weinheim, 1977, p. 26,116.
8 K.J. Vetter und H.-H. Strehblow, Ber. Bunsen-Ges. physik. Chem. 74 (1970) 1024.
9 T.R. Beck and S.G. Chan, Corrosion, 37 (1981) 665.
10 F. Hunkeler und H. Böhni, submitted to Werkstoffe und Korr.
11 H.C. Kuo and D. Landolt, Electrochim. Acta, 20 (1975) 393.
12 A.J. Sedriks, Corrosion of Stainless Steels, John Wiley & Sons, New York, 1979, p. 63.
13 D. Sinigalglia et al., Werkstoffe und Korr., 31 (1980) 851.
14 G. Herbsleb, Werkstoffe und Korr., 33 (1982) 334.

ACKNOWLEDGEMENT

The authors are pleased to acknowledge the "Schweizerischer Nationalfonds zur Förderung der wissenschaftlichen Forschung" for supporting this research within its national research program "Rohstoff- und Materialprobleme".

Passivity of Metals and Semiconductors, edited by M. Froment
Elsevier Science Publishers B.V., Amsterdam — Printed in The Netherlands

ROTATING RING-DISC ELECTRODE STUDIES OF THE PASSIVATION OF LOW ALLOY STEELS

A.M. RILEY and J.M. SYKES
Department of Metallurgy & Science of Materials, University of Oxford,
Oxford OX1 3PH, UK

ABSTRACT
Current-time transients for $3\frac{1}{2}$Ni and $3\frac{1}{2}$NiCrMoV steel have been examined in carbonate and hydroxide solutions at 293K.

INTRODUCTION

When stress-corrosion cracking occurs by an anodic dissolution mechanism plastic flow at the crack tip ruptures protective oxide films and allows rapid dissolution of the essentially film-free metal, while the growth of passive film on the crack faces prevents further attack and preserves the local geometry of the crack tip (ref.1-4). Repassivation kinetics are all-important in deciding whether cracking is possible. Similarly the cracking rate will depend upon the charge which flows after each film rupture event, and hence upon the film growth process. This charge comprises two distinct components; the charge required to form new passive film and that which forms dissolved species. It has been suggested that this second component should be much larger than the first if cracking is to occur (ref.5). In practice the quantity of soluble material generated at a crack tip could be enhanced through plastic creep inhibiting oxide film growth (ref.3).

In the present study the effect of Ni, Cr, Mo and V on the repassivation behaviour of low-alloy steels is being examined to determine whether this can explain known variations in cracking behaviour (ref.6) in hydroxide or carbonate-bicarbonate solutions, or whether their influence on mechanical behaviour or grain-boundary segregation is more important.

A rotating ring-disc electrode system (ref.7) is used to establish the proportion of the total charge forming soluble species.

Preliminary results are presented here for $3\frac{1}{2}$Ni and $3\frac{1}{2}$NiCrMoV steels in 0.5M/ 1M carbonate-bicarbonate solution and 5M sodium hydroxide at 293K.

EXPERIMENTAL

Specimens were cut from fully heat-treated low alloy steel turbine rotors having a bainitic structure. Both steels contained 0.3%C by weight and 3.5%Ni. The NiCrMoV steel contains 0.7%Cr, 0.3%Mo and 0.1%V.

They were machined into 7mm diameter cylinders and mounted in epoxy resin or glass filled P.T.F.E. with a concentric ring electrode either of gold or type 304 stainless steel. The width of the insulating annulus between the disc and ring and the ring itself were made small (0.3-0.5mm) to give a fast response. The surface of the electrode was carefully ground flat and polished with 1μm diamond paste before use. A 500ml P.T.F.E. cell of conventional design with a separate compartment for a saturated calomel electrode, was employed. Potentials were measured via a Haber Luggin probe situated 15mm directly beneath the centre of the disc electrode. Solutions were made from 'Analar' grade chemicals with distilled water and were deaerated with 'white-spot' nitrogen for 45 minutes before and during experiments. The electrode was rotated at a constant 25Hz by a servo-controlled drive unit (Oxford Electrodes Ltd). A microcomputer (Research Machines 380Z) controlled the experiments, adjusting the disc potential via a digital-analogue converter connected to the disc potentiostat (Ministat, Hylton Thompson Associates). The ring potential was controlled by a second separate potentiostat. To detect iron (II) species the ring potential was held as high as was consistent with low background current (+0.8V SCE in carbonates, +0.2V in caustic). Occasionally to detect iron (III) species a lower potential of -0.5V S.C.E. was used.

The normal procedure was to reduce the pre-existing oxide film at -1.2V S.C.E. for 600 seconds, then to condition the specimen at its free corrosion potential (-1.0V S.C.E.) for 600 seconds before finally switching instantaneously to the test potential. Ring and disc currents were measured and stored by the micro-computer via a high-impedance buffer and analogue digital converter. After 600 seconds the passive film was galvanostatically reduced, the ring current (due to iron (II)) and the disc potential being monitored as a function of time.

RESULTS

Both steels exhibited broadly similar behaviour in sodium hydroxide. Cyclic voltammograms show only one distinct oxidation peak on the anodic sweep and two reduction peaks (more clearly resolved in the $3\frac{1}{2}$NiCrMoV steel). The disc current transients (Fig.1a) give rectilinear log current/log time plots (as reported for carbon steel by Macdonald (ref.8)). Their slopes vary, depending on potential but are typically of the order -1.2 to -1.3. At some potentials the slopes change abruptly during the transient. No iron (II) was detected at the ring electrode in any of these experiments.

Disc transients in the carbonate solutions were similar giving rectilinear log current/log time plots with slope of about -1.2 (Fig.1b), however soluble iron (II) species were detected at the ring, but the ring current was only a small proportion (1/100) of the total disc current (Fig.2). Soluble iron (II) species were now readily detected in cyclic voltammograms (Fig.3) at 'active' potentials and when the film was reduced.

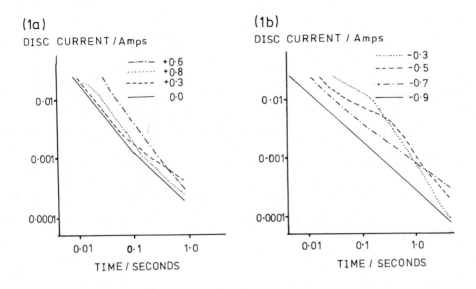

Fig.1. Log-log plots of current transients on 3½NiCrMoV steel after potential steps from -1.2V S.C.E. to the potentials indicated in (1a) 0.5M Na_2CO_3-1M $NaHCO_3$ (1b) 5M NaOH at 293K.

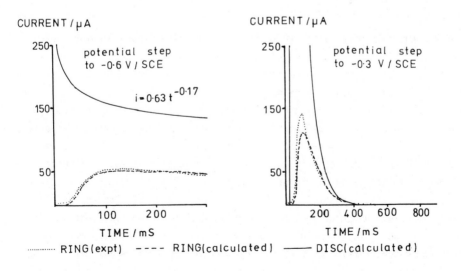

Fig.2. Ring and disc transients for potential step experiments on 3½NiCrMoV steel in carbonate-bicarbonate solution at 293K.

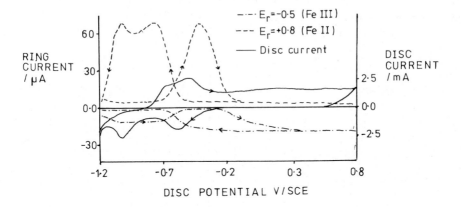

Fig.3. Cyclic voltammogram for $3\frac{1}{2}$NiCrMoV in carbonate-bicarbonate at 293K showing ring responses for iron (II) and iron (III).

During galvanostatic reduction of the passive films the ring current gradually rose until it reached a steady value almost equal to that predicted from the theoretical collection efficiency, a sudden peak then followed even larger than the constant disc current, just as a rise in disc potential indicated completion of film reduction (Fig.4).

DISCUSSION

Of the two sets of results, those in carbonate-bicarbonate are more readily interpreted. The galvanostatic reduction experiments show that small amounts of soluble iron (II) species can be determined accurately with this technique. The ratio of the charge passed in oxidising dissolved iron (II) to iron (III) at the ring to that passed in reducing Fe_2O_3 to iron (II) at the disc agreed well with the calculated collection efficiency of the electrode (0.20). The final peak on the ring response must correspond to a sudden release of iron (II) species, apparently retained in the film during earlier reduction (although Klimmeck attributes similar behaviour to pre-existing iron (II) (ref.9)). It is clear that during 'active' dissolution only a small proportion of the current produces soluble iron (II) and that an insoluble film, probably iron (II) carbonate is the main product. This conclusion is supported by the lack of any rotation speed dependance (ref.10) on the size and position of the 'active peak'. The ring response during passivation demonstrates that almost all (~ 95%) of the charge is consumed in film growth. The film growth kinetics are described by a power law for both caustic and carbonate solutions. The measured slopes (-1.2) of log i/log t plots do not correspond to logarithmic film growth (slope = -1) nor do they fit any of the available models (ref.8).

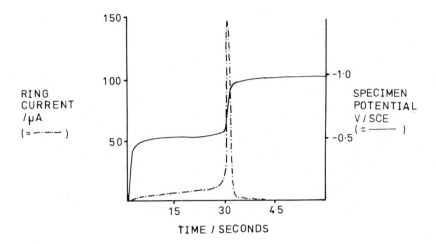

Fig.4. Specimen potential and ring current for 100μ A cm⁻² galvanostatic reduction of passive film on 3½NiCrMoV steel formed at +0.8V S.C.E. in carbonate-
bicarbonate solution at 293K.

Nucleation and growth models generate transients which rise before decaying
again, there is no evidence of this in these experiments. A model by Newman
(ref.12) in which oxide nucleation by solid-state processes and dissolution-
precipitation occur concurrently, predicts curves of the correct form and
although this is not a power law, it does show linearity in a log i/log t plot
over most of this time range. Appropriate choice of constants can give a slope
of -1.2 .

Integration of the ring response gives the total amount of soluble species
produced, but analysis of its form is more difficult. At long times (> 100ms)
the ring current and the equivalent disc current are in a simple ratio but transformation of the whole curve requires extensive calculation (ref.13) (and a
degree of intuition). The results (or alternatively trial and error solutions)
can be tested against actual ring responses by means of a numerical model
(ref.14) (Fig.2b). In some cases the flux of soluble species appears to decrease
monotonically with time but in others it may rise in the early stages.

ACKNOWLEDGEMENTS

The authors would like to thank the SERC for provision of a studentship to
A.M.R., Professor Sir Peter Hirsch for provision of laboratory facilities and
the CEGB for experimental materials. Equipment was provided by an SERC Research
Grant for which we are especially grateful.

REFERENCES

1 R.W. Staehle, Passivity and its breakdown on iron and iron base alloys, USA-Japan Seminar, 1975, 155-160.
2 R.N. Parkins, Proceedings of 8th International Congress on Metallic Corrosion, Mainz, Vol.3 (1981) 2180-2201.
3 J.C. Scully, Corr.Sci. 15 (1975) 207-224.
4 D.A. Vermilyea, J.Electrochem.Soc. (1972) 405-407.
5 J.R. Ambrose and J. Kruger, J.Electrochem.Soc. 121 (1974) 599-604.
6 R.N. Parkins, P.W. Slattery and B.S. Poulson, Corrosion, 37 (1981) 650-664.
7 W.J. Albery and M.L. Hitchman, Ring disc electrodes, Oxford University Press, Oxford (1971).
8 D.D. Macdonald and B. Roberts, Elect.Acta. 23 (1978) 557-564.
9 M Klimmeck and H.W. Wickboldt, Proceedings 8th International Congress on Metallic Corrosion, Mainz, Vol.1 (1981) 242-246.
10 A.M. Riley, D Phil Thesis Oxford University (to be submitted 1984).
11 R.D. Armstrong and J.A. Harrison, Electroanal.Chem., 36 (1972) 79-84.
12 J.R. Newman, C.E.R.L. report, RD/L/N78/75 (1975).
13 W.J. Albery, M.G. Boutelle, P.J. Colby and A.R. Hillman, J.Chem.Soc., Faraday Trans.1, 78 (1982) 2757-2763 .
14 K. Prater and A.J. Bard, J.Electrochem.Soc. 117 (1970) 207-213.

Passivity of Metals and Semiconductors, edited by M. Froment
Elsevier Science Publishers B.V., Amsterdam — Printed in The Netherlands   667

DEPASSIVATION AND REPASSIVATION OF AUSTENITIC STAINLESS STEELS CONSEQUENCES ON
STRESS CORROSION

M. HELIE[*], D. DESJARDINS, M. PUIGGALI, M.C. PETIT
Laboratoire de Mécanique Physique, E.R.A. CNRS N° 769
Université de Bordeaux I, 351 cours de la Libération,
33405 TALENCE CEDEX - France -
* since 1.12.82 : Centre d'Etudes Nucléaires de Fontenay-aux-Roses,
                  DCAEA/SCECF/SECA

ABSTRACT

    Stress corrosion cracking of stainless steels depends closely on depassi-
vation and repassivation processes. The parameters operating on these processes
(steel composition, strain rate, reagent temperature) have a great effect on
stress corrosion cracking sensivity, cracking velocity and cracks morphology.

I. INTRODUCTION

    It is well known that depassivation and repassivation play an important
part in localized corrosion processes.

    The study presented in this paper concerns the stress corrosion cracking
of austenitic stainless steels in hot chloride medium.

    The samples are made of thin steel wires, 1.5 mm in diameter and 360 mm
long. The tests are performed in concentrated magnesium chloride solutions at
various boiling temperatures (160°C, 153°C, 140°C, 130°C, 125°C, 110°C, 102°C)
to which potassium dichromate is added in some cases.

    In this paper are presented the influence of strain rate and solution tem-
perature on depassivation and repassivation processes, and the consequences on
stress corrosion cracking phenomenon.

II. EXPERIMENTAL METHODS

    The depassivation and repassivation of the tested wires are analysed in
term of current-time curves at fixed potential. The wire is placed into a "corrosion cell" with the boiling chloride solution on a tensile testing machine.
The potential is fixed and the tensile strength is applied for approximatively
five seconds, then suppressed.

    The current-time curve obtained is shown in figure 1 for an 18.10 stainless
steel in 153°C boiling magnesium chloride, at corrosion potential. First, the
anodic current rises linearly till it reaches a maximum. This corresponds to

the depassivation of the wire, due to the mechanical breakdown of the passive layer.

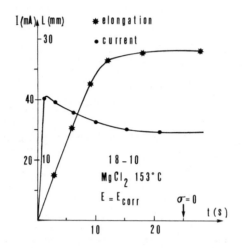

Fig.1. Mechanic depassivation test at fixed potential

The decreasing part of the curve indicates the repassivation of the steel. The fact that the current peak does not correspond to the end of the straining period shows that repassivation processes become predominant long before the complete breakdown of the passive layer. The competition between repassivation and depassivation depends on many factors. In a previous study, the influence of alloying elements has been pointed out[1] (figure 2). For 18.10 steel there is a strong current increase but it decreases quickly. In this case, there is also creation of a thick protective layer visible with the nacked eye.

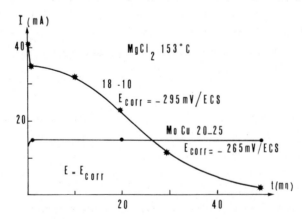

Fig.2. Current decrease after mechanical depassivation

For Mo-Cu 20.25 steel, the current increase is weaker and its decrease is then very slow.

In this paper, solution temperature and strain rate influence are analysed.

III. INFLUENCE OF STRAIN RATE ($\dot{\varepsilon}$)

The relative tests are made on the tensile testing machine, in boiling 153°C magnesium chloride on a 21.10 stainless steel at corrosion potential. The results show that competition between depassivation and repassivation depends on applied strain rate[2] (figure 3).

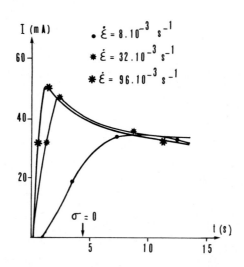

Fig.3. 21.10 steel. Influence of strain rate on current behaviour during loading

This seems to induce a strong dependance of localized corrosion processes on strain rate value. To point out this fact, corrosion tests (S.S.R.T.)[3] have been made in boiling 153°C magnesium chloride (Scheil reagent) on a 24.14 stainless steel, cathodically polarized to improve its corrosion resistance. The experimental mode is the same as in depassivation-repassivation tests but the tensile strength is maintened until wire rupture. The results are analysed in terms of maximal stress and S.E.M. micrographs of the tested wires.

Three strain rate ranges are pointed out : at low strain rates, the stress corrosion cracking does not appear and the rupture of the wire is mainly due to mechanical stress. At high strain rates, the wire shows tracks of general corrosion and the rupture is also a ductile one (figure 4).

Between these two ranges, the susceptibility of the steel to stress corrosion cracking presents a maximum, in this case, the rupture is mainly brittle, typical of cracks propagation.

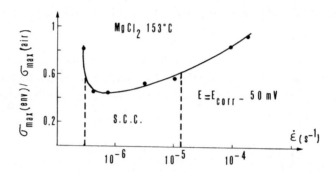

Fig.4. 24.14 steel. σ_{max}(env)/σ_{max}(air) ratio as a function of strain rate

IV. INFLUENCE OF SOLUTION TEMPERATURE

The depassivation-repassivation tests show the existence of a transitional temperature (130°C for a 18.10) [4] (figure 5).

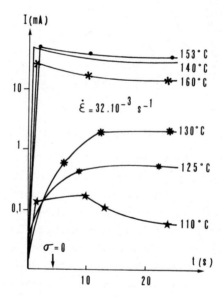

Fig.5. 18.10 steel. Current evolution during loading for various temperatures

Above this point, the current-time curve presents a marked peak separating the depassivation phase from the repassivation one. Below this transitional temperature, the anodic current increases non-linearly and ends with a plateau.

The behaviour is similar to this observed when studying strain rate influence. This transitional temperature also appears in corrosion tests performed on a constant load testing machine (creep tests). These tests have been analysed in terms of induction time, propagation time and cracking velocity (cracked area divided by propagation time) (figure 6).

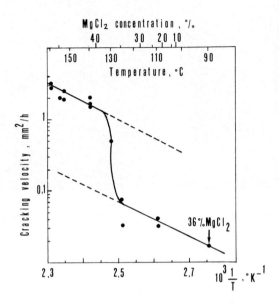

Fig.6. 18.10 steel. Temperature influence on cracking velocity

The adding of potassium dichromate to the magnesium chloride solution strongly modifies both electrochemical and mechanical properties of the passive layer. In this case, it is more difficult to obtain a frank depassivation and the repassivation rate is higher than in magnesium chloride alone.

V. CONSEQUENCES ON STRESS CORROSION CRACKING

S.E.M. micrographs show that cracks morphology depends on comparative depassivation and repassivation rates (figures 2,7 and 8). For high repassivation rates (compared to depassivation one) the cracks are thin and sharp, only the tip is subjected to anodic dissolution. In the opposite case, cracks are large, sometimes branched and blunt-edged. At temperature influence level, it has been found that below 130°C the cracking mode was mainly intergranular, and transgranular above [5] (figures 9 and 10). The cracking velocity is one hundred times higher above 130°C than below. The stress corrosion cracking phenomenon is also strongly related to strain rate. This is pointed out by the existence of a specific strain rate range allowing stress corrosion cracking to appear.

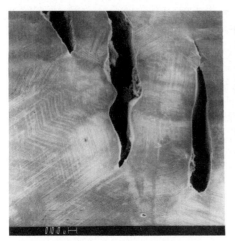

Fig.7 : 18.10 steel, MgCl$_2$ 153°C

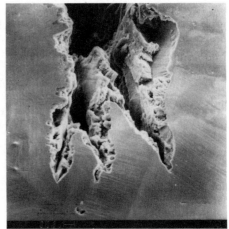

Fig.8 : Mo-Cu 20.25 steel, MgCl$_2$ 153°C

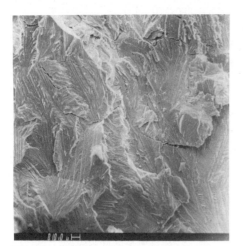

Fig.9 : 18.10 steel, MgCl$_2$ 153°C.
        Transgranular cracking

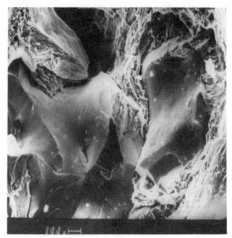

Fig.10 : 18.10 steel,
         MgCl$_2$ 125°C. Intergranular
         cracking

VI. CONCLUSION

The results obtained show that stress corrosion cracking depends on depassivation and repassivation processes. This phenomenon is then related to both solution temperature and strain rate value. The influence of these electrochemical and mechanical parameters strongly suggests that stress corrosion cracking phenomenon is controlled by anodic dissolution of mechanically depassivated areas[6].

BIBLIOGRAPHY

1 D. Desjardins, M. Puiggali, M.C. Petit, Mémoires et Etudes Scientifiques de la revue de métallurgie, mai 1981.
2 M. Helie, Thèse de spécialité, Bordeaux, 1982.
3 R.N. Parkins, The slow strain rate technique. ASTM STP 665, Ugiansky/Payer Editors, p.5, 1979.
4 M. Kowaka, T. Kudo, The sumitomo search, n°18, novembre 1977.
5 M. Takano, R.W. Staehle, Trans.J.I.M., vol.19, 1978.
6 J.C. Scully, Corrosion Science, vol.20, p. 997, 1980.

Passivity of Metals and Semiconductors, edited by M. Froment
Elsevier Science Publishers B.V., Amsterdam — Printed in The Netherlands

HYDROGEN PERMEATION THROUGH PASSIVE LAYERS ON AUSTENITIC
HIGH TEMPERATURE ALLOYS

H.P. BUCHKREMER, R. HECKER, D. STÖVER
Kernforschungsanlage Jülich, GmbH, Institut für Reaktorentwicklung,
Postfach 1913, 5170 Jülich (FRG)

ABSTRACT
Using a sample of the high temperature alloy Hastelloy X it is shown, that
chromium oxide layers of a few μm thickness reduce the permeation of hydrogen
isotopes for around three orders of magnitude. Besides the permeation rate its
pressure- and temperature dependency serves as an indicator for the quality of
an oxide barrier. For passive layers with low permeation rates the pressure
exponent increases from 0.5 to 1 and the activation energy from 60-70 kJ/Mol
to 130-170 kJ/Mol indicating completely different permeation mechanisms compared to the unscaled alloy.

INTRODUCTION

The high mobility of the hydrogen isotopes in metallic materials is well
known since a long time (ref.1) . Especially at high temperatures and pressures
a considerable amount of hydrogen permeates through metallic walls. This could
result in an increased release of radioactivity in nuclear energy-conversion
systems like the High-Temperature-Gascooled-Reactor (HTGR) or fusion devices
due to the radioactive hydrogen isotope tritium which is generated or used as
fuel respectively. In the course of the development of the HTGR for process
heat generation thin oxide-based passive layers were found as a very effective
hydrogen barrier on high temperature resistent alloys. Such layers are also
discussed for application in other conventional power systems with high impacts of hydrogen such as stirling motor, coal conversion (gasification,
liquefaction) and hydrogen storage (ref.2).

TASK DEFINITION

To operate safely heat exchanger components in a High Temperature Reactor-
system (HTR-system) it is necessary to reduce the hydrogen isotopes permeation
drastically. Special coatings and oxide layers on austenitic high temperature
alloys meet this demand. Several coatings of titanium-nitride or -carbide,
layers of pyrolitic carbon and oxide layers such as Fe, Ni, Si, Ti, Al- and
Cr-oxides were discussed or examined. In general an effective hydrogen barrier has to satisfy the conditions of drastic reduction of hydrogen permeation,

676

stability under all operating conditions, good regeneration after breakdown, simple generation and control and acceptable or neutral interaction with the experimental environment.

Pure and nearly pure chromia layers as considered here are proved to represent excellent hydrogen barriers and could reduce the permeation rate by more than three orders of magnitude, compared with the uncoated material. Another advantage is that these layers are formed "in situ", because the reacting species chromium and oxygen are present in the structural materials of the components and the gaseous environment of the reactor-circuits respectively. Therefore a damaged oxide layer is able to heal completely and their efficiency may be guaranteed for long operation times. Moreover intact oxide scales with high efficiency in reduction of permeation act as corrosion protecting layers with respect to the oxidizing and carburizing environments.

EXPERIMENTAL CONDITIONS

The passive layers are formed or stabilized respectively in our case by the so called "process gas", which is, e.g., a mixture of 42% H_2, 39% H_2O, 5 CH_4, 7% CO and 5% CO_2.

The oxide layers are generated by chemical reaction with the process gas mentioned above or by H_2/H_2O- or $Ar/H_2/H_2O$-gas mixtures. The hydrogen production rate and permeability is measured on tube-shaped samples ($\emptyset_{max}$ = 120 mm, length $\approx$ 200-300 mm, wall thickness 3-10 mm). The samples are isothermally heated and oxidized inside the tube. Other test parameters are listed in table 1.

TABLE 1

Experimental parameters in TRIPERM test facility

Test Materials	: Austenitic Wrought and Cast Alloys
Temperature Range	: $450^{\circ}C$ - $950^{\circ}C$
Oxidation Temperature	: $650^{\circ}C$ - $950^{\circ}C$
Duration of Test	:1000 h - 3000 h
Duration of Oxidation	: up to 3000 h
Atmosphere	: $Ar+H_2+H_2O$ or H_2+H_2O or "Process Gas"
Total Pressure (p)	: 1 - 30 bar (abs.)
Hydrogen Pressure (p(H_2)):	10^{-4} - 30 bar (abs.)
Steam Pressure (p(H_2O))	: 10^{-4} - 30 bar (abs.)
P(H_2)/p(H_2O)-Ratio	: 10^5 - 10^{-3}

The amounts of hydrogen either permeated or generated by the water-vapour-metal reaction are measured gaschromatographically, with a lowest detection limit of 0.1 vpm. Using the radioactive isotope tritium, it is possible to reduce this limit by means of liquid-scintillation or proportional counting systems to 10^{-8} vpm. This technique allows to examine the parallel- or counter flow of the hydrogen isotopes protium and tritium simultaneously in one apparatus even under conditions with highly variable concentrations (ref. 3).

RESULTS

Within the scope of the PNP-program (Project Nuclear Process heat) up to now the permeability and oxidation-properties of 12 high temperature materials has been examined under bare and oxidized conditions. The results described here refer to the latest examination of HASTELLOY X.

Passivation by Oxidation

After an annealing treatment at a temperature of 950° C under pure hydrogen of 30 bar for several days the test tube of HASTELLOY X was coated under an oxidizing atmosphere of a H_2/H_2O-mixture ($P_{H_2}/P_{H_2O} \sim 1$; $P_{tot} \sim 1,5$ bar). For 12 days the coating temperature was adjusted to 650° C. After this period, temperature was set to 950° C. Here a quasi-stationary permeation rate was obtained after 20 days. The flux now observed was 400 times lower compared to the rate at this temperature under uncoated conditions. This emphasizes the impeding effect of the oxide barrier for hydrogen. Figure 1 shows the run of isothermal oxidation of the experiment described.

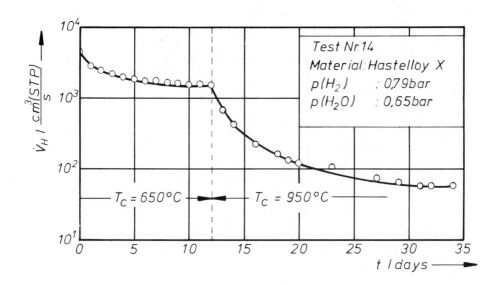

Fig. 1. Hydrogen permeation through HASTELLOY X at various conditions in function of time

678

Temperature Dependence

Whenever the temperature was reduced from 950° C to 650° C in steps of 50° C the permeation rate of protium and tritium was measured. In both cases of un-covered as well as coated samples an Arrhenius relation is valid to describe the temperature dependence of the permeation flux:

$$\text{Permeation rate } (\dot{V}) \sim \exp(-E_{act}/RT) \qquad (1)$$

E_{act} = activation energy of permeation , R = gas constant
T = Kelvin temperature

Oxide coated samples with low permeation rates clearly show higher activation energies (130-170 kJ/Mole) than uncoated ones (60-70 kJ/Mole). In Figure 2 the data of permeation measurements discussed above are plotted. In comparison with the centrifugal cast alloy IN 519 which has been reported elsewhere (ref. 4) both show very similar behaviour.

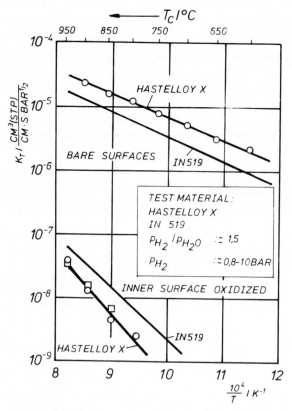

Fig. 2. Permeability of high temperature alloys as a function of reciprocal Kelvin temperature

Pressure dependence

The pressure dependence of the permeation rate especially for bare and uncoated high temperature alloys can be described by the following relation:

$$\text{Permeation rate } (\dot{V}) \sim P_{(H_2)}^{n} \tag{2}$$

$P_{(H_2)}$ = hydrogen partial pressure , n = pressure exponent

with n = 0.5 if the diffusion process in the base material is rate-controlling.
In coated test tubes however the pressure behaviour changes in a marked way.
The pressure exponent n increases in the range from 0.5 to 1.0. The experimental results for a bare and a coated testing tube of HASTELLOY X are shown
in figure 3. At less passivating oxide barriers the pressure exponent was
found between 0.6 to 0.7 (ref.4).

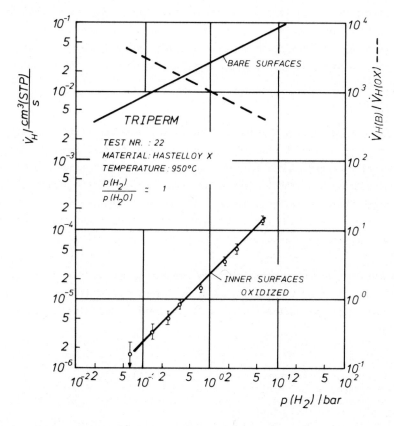

Fig. 3. Permeation rate $(\dot{V}_H)$ and impeding factor $(\dot{V}_{H(B)}/\dot{V}_{H(ox)})$ of
HASTELLOY X as a function of hydrogen partial pressure $(P_{(H_2)})$

An other specification to characterize the quality of the oxide layers is the impeding factor, which is the ratio of the permeation rates through bare and oxidized material respectively $(\dot{V}_H(B)/\dot{V}_H(ox))$. Figure 2 and 3 show that this factor is dependent on the temperature as well as the hydrogen partial pressure.

Structure of layers

The passive layers, grown on the surface of austenitic iron- or nickel-base alloys, are some μm thick and after 100-1000 h exposure to oxidizing gas reach a quasi stationary state. Metallographical check-up of this passive hydrogen barriers often pointed out a duplex layered configuration (ref. 5). A practically pure Cr_2O_3-coating is in direct contact to the bulk material, while the upper porous layer consists of spinel agglomerates of Cr, Mn and Fe. This preferably indicates in our understanding the passivating effect of pure Cr_2O_3 coatings. The description of some of the test data with the aid of a permeation model was tried and lead to the result that the permeation process of "good" passive layers would be surface controlled rather than diffusion controlled (ref. 6). A correlation between metallographically measured parameters of the oxide barrier and the hydrogen impeding factor is not yet definitely possible today.

CONCLUSIONS

The data of permeation measurements enables a decision on the "quality" of passivation layers. Important criteria are: impeding factor, temperature dependence of hydrogen permeation, pressure dependence of hydrogen permeation. Oxide coated materials with low permeation rates (high impeding factors) clearly show higher activation energies than uncoated specimens. E_{act} normally increases from ~ 60-70 kJ/Mol to ~ 130-170 kJ/Mol. Furthermore the exponent n in pressure dependence p^n changes in a characteristic pressure range from 0.5 to 1.

To our experience hydrogen permeation measurement is an extremly sensitive method for detection of the quality of oxide layers on metallic materials, not only during the production route of these coatings but also in operating facilities.

REFERENCES

1 A. Sieverts, J. Hagenacker, Phys. Chemie 68, P.115, 1909.
2. K.F.Windgassen, Annual Meeting of IOMA, Port St. Lucie, Florida(USA), 1979.
3. H.P. Buchkremer et al., 11th Symp. of Fusion Technologie 1980, Vol. 1, P. 547-552, Oxford(UK), Sept. 1980
4. D.Stöver et al., Int. Symp. on Metal-Hydrogen Systems, Miami (USA) Apr.1981.
5. O. van der Biest et al., Int. Conf. "HighTemp. Alloys", Petten(NL), 1979.
6. R. Hecker, D. Stöver, H.P. Buchkremer, KFA-Report, Jül 1771, Jülich(FRG) 1982.

Passivity of Metals and Semiconductors, edited by M. Froment
Elsevier Science Publishers B.V., Amsterdam — Printed in The Netherlands

IMPEDANCE TECHNIQUE TO STUDY PASSIVITY OF STEEL

G.N. MEHTA[1], T. TSURU[2] and S. HARUYAMA[3]

[1]Chemistry Dept., SVR College of Engineering & Technology, Surat
395 007 (India)
[2,3]Dept. of Metallurgical Engineering, Tokyo Institute of Technology, 2-12-1, Ookayama, Meguro-Ku, Tokyo-152 (Japan)

ABSTRAT
    Steel was exposed to 0.1M sodium nitrite solution. Corrosion
potentials and impedances were measured at the intervals of 3, 27
and 96 hours. The corrosion potential shifted towards the noble
direction, the interfacial capacitance remained unchanged with elapse of time. However, the extent of these changes was initially
higher, upto 27 hours and then decreased to almost half during the
next interval. While this higher extent indicated the onset and
expansion, the lesser extent reflected plugging of the defects and
the growth, of the passive film on the steel surface. The corrosion current density, calculated as supporting data from the charge
transfer resistance, decreased with ealpse of time.

INTRODUCTION
    The theory of faradaic impedance for simple electrode processes
in the presence of direct current polarization was reported by
Grahame (ref.1) and Gerischer (ref.2). The use of transfer function was made by Ichikawa and Mizuno (ref.3) in analysis of the simple electrode processes. Haruyama (ref.4) used this function to
derive the mathematical expressions for the faradaic impedances of
various mixed potential systems. Impedance measurements were reported, to evaluate the thickness and dielectric constant of passive
films (ref.5,6), to evaluate the extent of the active, passive and
transpassive regions (ref.7,8), to analyse various models proposed
for passive iron and the properties of passive film (ref.9), to
measure corrosion rate (ref.10) and to develop the a-c corrosion
monitor (ref.11,12).
    The impedance technique was employed, at different intervals of
time to study passivity of steel exposed to 0.1M $NaNO_2$ solution, in
the present work.

EXPERIMENTAL

Steel (C=0.1%) specimens of 2 x 1 cm size were soldered to copper wire leads, polished through a series of coarser to finer grades of emery papers, degreased ultrasonically in trichloroethylene, washed with redistilled water and dried in vacuum. While one side of specimen was covered with epoxy resin and allowed to set hard in vacuum, the other side of exactly 3 cm squre (sq) area was kept open in order to simplify the calculation of the charge transfer resistance (R_t) from the impedance data of two electrode system employed in the present work (ref.12). The corrosion potentials (E_{corr}) were measured using saturated calomel electrode (SCE) as the reference electrode at 3, 27 and 96 hours. At these intervals impedance measurements were also made under $\pm$ 5 mV polarization. Initially high frequency (30 K Hz) was applied and then was gradually decreased to low frequency (0.1 Hz). The sinusoidal components of voltage and current were recorded on the Memoriscope (Hitachi Model V038) at higher range of frequencies and on the chart paper of a two-channel pen recorder at lower range of frequencies. The absolute impedance (Z) was calculated from the ratio of the amplitudes of voltage and current curves. Phase shift (θ) was measured directly from voltage and current traces. Pure $NaNO_2$ was dissolved in redistilled water to prepare 0.1M solution in this work.

The equations derived earlier (ref.12-14), for calculation of the value of R_t at lower frequency arrest, the value of interfacial capacitance (C_i) from a frequency of a break-point at higher frequency arrest, the value of solution resistance (R_{sol}) at higher frequency arrest and the value of corrosion current density (i_{corr}) from R_t values, are not reproduced in this text.

RESULTS AND DISCUSSION

The results obtained in the present study are given in the Table 1. As shown in the Table 1, the value of E_{corr} shifted in the noble direction and the value of i_{corr} decreased with elapse of time which indicated passivity of steel in 0.1M $NaNO_2$ solution. A similar relation of E_{corr} and i_{corr} with time of immersion was observed and reported earlier (ref.15) for passivity of steel. Passivity of steel was further supported by the impedance data given in Table 1. A slow and gradual decrease in the C_i value indicated the coverage of the bare steel surface with passive film. Initially, there was a remarkable increase in R_t value from 110 to 200 KOhm

cm sq at 27 hours and a corresponding higher decrease in the calculated value of i_{corr}. This higher extent of 90 KOhm cm sq can be
attributed to the onset and expansion of primary oxide layer formed
on the bare steel surface. A further increase in R_t value to 240
KOhm cm sq and hence the corresponding lesser extent (40 KOhm cm
sq) , almost half of the higher extent, was observed at 96 hours of
immersion. This may be interpreted as plugging of the primary oxide
layer defects (pores). This lesser extent of increase in the value
of R_t can also be interpreted as growth (thickening) of the passive
film with elapse of time. Although, there was no direct evidence
showing the growth of the passive film, it could be generalised from a circumstantial evidence of the lesser extent of increase in
the value of R_t with elapse of time in the present study.

TABLE 1

Impedance and other electrochemical data of steel passivated in
0.1M $NaNO_2$ solution (pH=7.0, temperature $25 \pm 2^{\circ}C$).

Time of Immersion	Corrosion potential	Solution resistance	Interfacial capacitance	Charge transfer resistance	Corrosion current density
hours	E_{corr} mV vs SCE	R_{sol} Ohm cm sq	C_i μF/cm sq	R_t KOhm cm sq	i_{corr} nA/cm sq
3	− 114	25	16	110	180
27	− 17	25	11	200	100
96	+ 25	25	8	240	42

It was possible by the impedance technique to seperate R_{sol}
value from that of R_t. Comparatively lower value of R_{sol} (25 Ohm
cm sq) remained unchanged which reflected the constancy of electrical properties of the bulk solution; atleast no metal ions were
dissolved in this passivating media of 0.1M $NaNO_2$ solution. If this
were not the case, the value of R_{sol} must change. The seperation of
R_{sol} value from that of R_t was not much remarkable in this study
but in case of higher solution resistance system, it becomes quite
essential to seperate R_{sol} to arrive at accurate results.

684

CONCLUSION

The impedance technique was useful to study passivity of steel. Steel was exposed to 0.1M $NaNO_2$ solution and the passive oxide film was formed on the bare steel surface as indicated by the impedance data obtained and by shift of E_{corr} value in noble direction with elapse of time. While the higher extent of increase in R_t value and corresponding decrease in the value of i_{corr} and a slow but gradual decrease in the value of C_i was interpreted as due to onset and expansion of the primary oxide film, the lesser extent of these changes interpreted, as a circumstantial evidence, as plugging of the primary oxide film defects (pores) and as the growth of the passive film. The value of R_{sol} was essentially seperated from that of R_t but it was not much higher to add any remarkable error in calculation of corrosion rate. Moreover, R_{sol} remained unchanged suggesting the constancy of the electrical properties of the bulk solution having passivating nature.

ACKNOWLEDGEMENTS

One of the authors (GNM) gratefully acknowledges his sincere thanks to the Ministry of Education, Government of Japan and the Japanese Commission of UNESCO for the fellowship to participate in the Fifteenth International Post Graduate University Course in Chemistry and Chemical Engineering held at Tokyo Institute of Technology, Tokyo. He is highly indebted to the Principal and authorities of SVR College of Engineering and Technology, Surat (India) for the encouragement to participate in the above course. This work was carried out under the active guidance of Professor Shiro Haruyama in his laboratory during the tenure of fellowship (1979-80).

REFERENCES

1 D.C. Grahame, J. Electrochem. Soc., 99 (1952) C 370.
2 H. Gerischer, Z. Physik. Chem., 198 (1951) 236.
3 A. Ichikawa and S. Mizuno, Denki Kagaku, 28 (1960) 318.
4 S. Haruyama, Proceedings of the 5th International Congress on Metallic Corrosion, NACE, Houston, 1974, 82 pp.
5 G. Okamoto, H. Kobayashi and M. Nagayama, Z. Elektrochem., 62 (1958) 775.
6 J.L. Ord and J.H. Bartlett, J. Electrochem. Soc., 112 (1965) 160.
7 I. Epelboin and M. Keddam, J. Electrochem. Soc., 117 (1970) 1052; Electrochim. Acta., 17 (1972) 177.
8 R.D. Armstrong, J. Electroanal. Chem., 34 (1972) 387; R.D. Armstrong, M. Henderson and H.R. Thirsk, ibid., 35 (1972) 119; ibid., 39 (1972) 222; R.D. Armstrong and K. Edmonson, Electrochim, Acta, 18 (1973) 937.

9 S. Haruyama and T. Tsuru, in R.P. Frankenthal and J. Kruger (Eds.), Passivity of Metals, The Electrochemical Society, Princeton, 1978, PP. 564-584.
10 I. Epelboin, M. Keddam and H. Takenouchi, J. Appl. Electrochem., 2 (1972) 71.
11 S. Haruyama, T. Tsuru and M. Anan, Boshoku Gijutsu, 27 (1978) 449.
12 N. Krithivasan, T. Tsuru and S. Haruyama, Boschoku Gijutsu, 29 (1980) 275.
13 G.N. Mehta, T. Tsuru and S. Haruyama, Proceedings of the Symposium on Advances in Corrosion Control, SAEST, Karaikudi (India), 1982, 1.1 pp.
14 G.N. Mehta, T. Tsuru and S. Haruyama, International Conference on Corrosion Inhibition, NACE, Dallas, Texas, May 1983 (to be presented).
15 G.N. Mehta and T.P. Sastry, Indian J. Technology, 18 (1980) 513.

Passivity of Metals and Semiconductors, edited by M. Froment
Elsevier Science Publishers B.V., Amsterdam — Printed in The Netherlands

THE EFFECT OF CHROMIUM, NICKEL AND MOLYBDENUM ON PASSIVITY OF IRON ALLOYS IN ORGANIC SOLUTIONS OF SULPHURIC ACID

JACEK BANAŚ

University of Mining and Metallurgy, Institute of Foundry Engineering, 30-059 Kraków (Poland)

ABSTRACT

Experimental results showing the influence of chromium, nickel and molybdenum on the passivity of iron alloys in anhydrous solutions of sulphuric acid in dimethylformamide and in methanol are presented. Chromium and molybdenum can form stable anodic films in these solutions. As alloying components of iron alloys these metals have a beneficial effect on the anodic behaviour in anhydrous solutions of sulphuric acid. Nickel on the other hand cannot form an anodic, passive film in anhydrous solutions. Its anodic properties are analogous to the properties of iron. Therefore the presence of nickel in the iron alloy has no influence on its passivation behaviour in anhydrous solutions.

INTRODUCTION

The corrosion behaviour of iron alloys in aqueous solutions is determined mostly by its passivation ability. In such environments the passivation process can be presented by an anodic reaction with (1) :

$$Me \ + \ H_2O \ \longrightarrow \ MeO \ + \ 2H^+ \ + \ 2e \qquad\qquad (1)$$

The reaction shows that the presence of water is necessary to obtain the oxide film on a metal surface (ref.1,2). Many reports demonstrate lack of the passivity of metals in anhydrous solutions of sulphuric acid (ref.1,3-5). In anhydrous solutions, the solvent molecules or anions participate in anodic dissolution of metal. Depending upon the solvent, the metal surface undergoes a dissolution, often with selective etching, or the anodic salt-like layers formation. In some cases, the amorphous salt layers can prevent the dissolution of the metal (ref.1,6).

Many reports on the influence of alloy elements - chromium, nickel and molybdenum on the passivity of iron alloys in aqueous solutions have appeared in the literature (ref.7-9). They expla-

in the influence of these metals on the oxide film formation, as
such films can be formed in water solutions. There are no reports
on the influence of chromium, nickel and molybdenum on the anodic
behaviour of iron alloys in anhydrous environments. Anodic films
which can be formed in these solutions show a salt-like properties.

The aim of the present work is to study the influence of chromium, nickel and molybdenum on passivity of iron alloys in anhydrous solutions of sulphuric acid. The knowledge of the anodic behaviour of pure metals in these solutions is necessary for the explanation of the alloys.

EXPERIMENTAL

The investigations have been carried out with pure polycrystalline metals - iron, chromium, nickel and molybdenum, and binary
iron alloys - Fe-Cr, Fe-Ni and Fe-Mo (ref.10,12-16).Anhydrous dimethylformamide (DMF) and methanol (water content not exceeding
0.02%) were used as solvents. In order to obtain an anhydrous
H_2SO_4-DMF solution, water, determined by Karl-Fischer-method, was
stoichiometrically bound by means of oleum containing 30% SO_3 and
then an appropriate quantity of 100% H_2SO_4 was added. By diluting
with water, appropriately hydrated solutions were obtained. Anodic
polarization measurements were carried out using the interrupting
potentiostat with automatic IR compensation made by the Research
Institute of Meinsberg. These measurements were performed by means
of the potentiodynamic technique with a scan rate of 20 mV$\cdot$min^{-1}.
An aqueous saturated calomel electrode was used as reference electrode. All investigations were performed in deaerated solutions
at 20°C.

ANODIC DISSOLUTION OF IRON, CHROMIUM,NICKEL AND MOLYBDENUM IN
SOLUTIONS OF H_2SO_4-DMF AND H_2SO_4-CH$_3$OH

Iron and nickel does not form the anodic film in anhydrous solutions of H_2SO_4-DMF and H_2SO_4-CH$_3$OH (ref.4,5,10). Fig.1 presents anodic polarization curves of the metals in anhydrous solution of
1m H_2SO_4-CH$_3$OH. Similar curves can be obtained in H_2SO_4-DMF solutions. This is very difficult to obtain the minimum water concentration which is necessary to form the anodic oxide-like film. Fig.2
presents the influence of the water concentration on the anodic polarization of nickel in solutions of 1m H_2SO_4-CH$_3$OH-H$_2$O. There is
continuous change of salt-like layer to the oxide-like one, when

water content increases. Nickel shows analogous behaviour in aqueous solution of sulphuric acid with concentration of acid 4-10 $m \cdot l^{-1}$
(Fig.3). A dehydratation of solutions is observed when concentration of sulphuric acid increases. In the solutions without "free"
water (water not bounded in the first coordination shell of ions
or organic molecules) the anodic oxide film is not form usually.

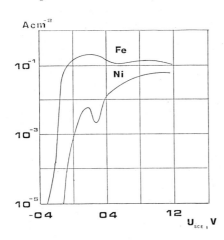

Fig.1. Anodic polarization
curves of iron and nickel in
1m H_2SO_4-CH_3OH solution.

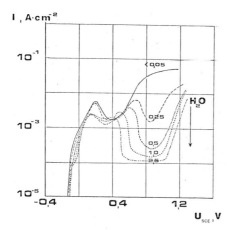

Fig.2 Influence of water concentration on the anodic polarization of nickel in solutions
of 1m H_2SO_4-CH_3OH-xH_2O (molar
concentration of water is given
on each curve).

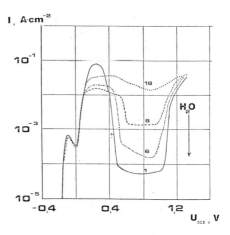

Fig.3 Influence of acid concentration on the anodic polarization of nickel in solutions
of xH_2SO_4-H_2O.(molar concentration of sulphuric acid is given
on each curve).

Contrary to iron and nickel, chromium and molybdenum can form anodic layers in anhydrous solutions of H_2SO_4-DMF and H_2SO_4-CH_3OH (Fig.4). These metals have a great ability to form a polynuclear, amorphous species with participation of not only water molecules or hydroxyl groups, but also anions too (ref.17).It is possible that in anhydrous, organic solutions of sulphuric acid this behaviour play important role in passivation of chromium and molybdenum.

The strong interaction of chromium ions with water molecules is known. The rate constant of exchange of water in the hydratation shell of the Cr^{+3} ion with the water in the bulk solution is equeal to ca. $3 \cdot 10^{-7} sec^{-1}$. For Ni^{+2} and Fe^{+3} ions these constants are $3 \cdot 10^4$ and $3 \cdot 10^3$ respectively (ref.11). The Cr^{+3} ion can bind easier water than the Ni^{+2} and Fe^{+3} ions. Therefore the chromium can form passive films in solutions with small contents of water. The increasing stability of the $Me(H_2O)_6^{+n}$ complex is as follows: $Fe^{+2} < Ni^{+2} < Fe^{+3} < Cr^{+3}$. The same order holds for ability of metals to form oxide films in solutions with a small contents of water (ref.14).

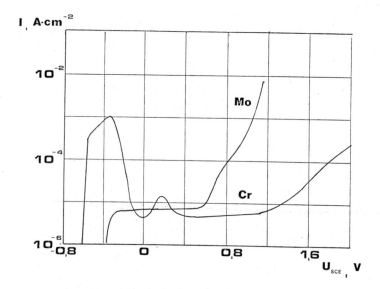

Fig.4. Anodic polarization curves of chromium and molybdenum in anhydrous solution of 1m H_2SO_4-DMF.

THE PASSIVITY OF Fe-Cr, Fe-Ni AND Fe-Mo ALLOYS IN SOLUTIONS OF H_2SO_4-DMF AND H_2SO_4-CH_3OH

In anhydrous solutions of sulphuric acid in DMF and in methanol nickel does not influence the anodic behaviour of iron alloys (Fig.5). As it has pointed out previously, nickel cannot form a anodic, stable film in anhydrous environments. On the other hand, chromium and molybdenum, elements which can form anodic layers in such solutions, have beneficial influence on the anodic properties of the iron alloys (ref.12,13). The presence of these metals in the alloy increases ability to form the anodic layer (Fig.6,7). The formation of such layer is stimulated by the diffusion of sulphuric acid molecules or sulphate ions onto the electrode surface (Fig.8,9).

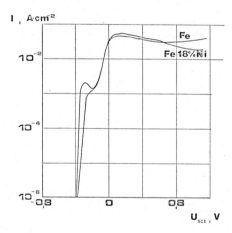

Fig.5. Anodic polarization of iron and Fe-18% Ni alloy in anhydrous solution of 1m H_2SO_4- -DMF.

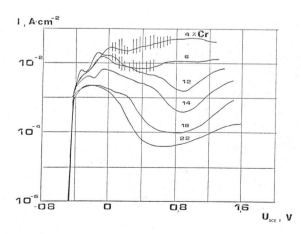

Fig.6. Influence of chromium on anodic polarization of Fe-Cr alloys in anhydrous solution of 1m H_2SO_4-DMF.

692

Probably, in anhydrous solutions the anodic layers can be formed
by direct reaction between metal surface and sulphate anions. Under this conditions only chromium and molybdenum, the metals which
can build polynuclear, amorphous salt-like compounds stimulate the
formation of anodic layers on the surface of iron alloys.

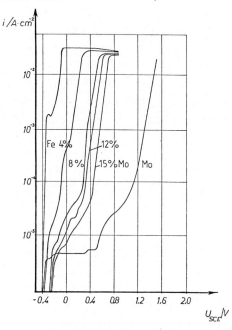

Fig.7. Influence of molybdenum on anodic polarization
of Fe-Mo alloys in anhydrous solution of 1m H_2SO_4-DMF.

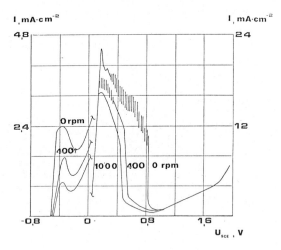

Fig.8. Effect of the rotation of a disc electrode on anodic polarization of X20Cr13 steel in anhydrous solution of 1m H_2SO_4 - DMF.

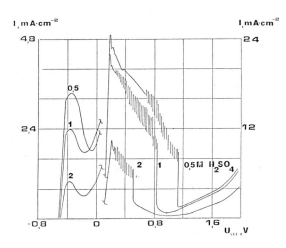

Fig.9.   Influence of acid concentration on anodic polarization
of X20Cr13 steel in anhydrous H_2SO_4-DMF solutions.

REFERENCES

1   K. Schwabe, S. Hermann, W. Oelssner, Passivity of Metals, Proc.
    of the Fourth Intern.Symp. on Passivity, New Jersey 1977, ed.
    by R. F. Frankenthal and J. Kruger.
2   J. M. Kolotyrkin, G. G. Kossyj, Zassch. Metall. 1 (1965) 252 .
3   J. Banaś, K. Schwabe,Oberfläche-Surface 20 (1979) 9, 200.
4   K. Schwabe, W. Schmidt, Corr. Sci. 10 (1970) 143.
5   M. Jänchen, K. Schwabe, Z. phys. Chem. 257 (1976) 129.
6   K. Schwabe, Z. phys. Chem. N.F. 108 (1977) 61
7   R. Olivier, Proc. of the Sixth Meeting of the Intern. Conf. for
    Electrochem. London 1955.
8   J. M. Kolotyrkin, G. M. Florianovich, Korrozia i Zasschita ot
    Korrozii, Vol. 4, Moskva 1975.
9   K. Hashimoto, N. Naka, K. Asami, T. Masumoto, Corr. Sci. 19
    (1979) 165.
10 J. Banaś, Z. phys. Chem., 262 (1981) 1105.
11 T. J. Swift, R. E. Connick, J. Chem. Phys., 17 (1962) 307.
12 J. Banaś, Electrochim. Acta 27 (1982) 1, 71.
13 J. Banaś, Corr. Sci., 22 (1982) 11, 997.
14 J. Banaś, Zesz. Nauk. AGH (Krakow) in press.
15 J. Banaś, Corr. Sci. 22 (1982) 11, 1005.
16 J. Banaś, Electrochim. Acta 27 (1882) 8, 1141.
17 C. L. Rollinson, The Chemistry of Chromium, Molybdenum and
    Tungsten, Pergamon Press, 1975.

Passivity of Metals and Semiconductors, edited by M. Froment
Elsevier Science Publishers B.V., Amsterdam — Printed in The Netherlands

ELECTROCHEMICAL MEASUREMENTS ON LOW-INDEX PLANES OF A Fe-14.5Cr-14.5Ni-2.5Mo
SINGLE CRYSTAL

S. TÄHTINEN[1], H. HÄNNINEN[1] and T. HAKKARAINEN[2]
[1]Technical Research Centre of Finland, Metals Lab., SF-02150 Espoo 15, Finland
[2]Technical Research Centre of Finland, Metallurgy Lab., SF-02150 Espoo 15,
Finland

ABSTRACT

The electrochemical behavior of low-index planes of a Fe-14.5Cr-14.5Ni-2.5Mo
single crystal in a deaerated 5N H_2SO_4 + 0.5N NaCl solution was studied.
Anisotropy was found in passivation properties whereas active dissolution
showed only slight orientation dependence. These results suggest that dissolution models alone are inadequate to interpret the crystallographic fracture
surface appearance of stress corrosion cracking in austenitic stainless steels.

INTRODUCTION

Electrochemical properties of austenitic stainless steels have been extensively studied with respect to stress corrosion cracking (SCC) susceptibility.
Cracking is reported to occur in certain potential regions: free corrosion
potential, active-passive transition and transpassive regions (ref. 1). The
dissolution and passivation of metal surface including effects of plastic
deformation on crack growth kinetics are often named as critical factors for
SCC to occur (ref. 2, 3). However, there is still discussion if the detailed
mechanism of cracking is either dissolution-controlled or hydrogen-induced
(ref. 4).

Transgranular SCC fracture surfaces are typically cleavage-like showing
clear crystallographic features. Direct electron microscopy observations (ref.
5) showed that active dissolution at the crack tip occured in <111> directions
producing corrosion slots having (110) wall planes. The macroscopic fracture
occurs then along the most favourably oriented corrosion slots with regard to
the stress axis. Thus, the separation of these corrosion slots accounts for
the crystallographic SCC morphology. However, no other studies dealing with
anisotropic dissolution mechanism are presented.

Potentiodynamic experiments (ref. 6) showed no effects of crystal
orientation of AISI 316 single crystal on open-circuit potential and only
slight differences on active anodic dissolution in boiling $MgCl_2$ solution.
However, there was a tendency for "passivation" at slow scan rates; the effect
of which was most pronounced on the (100) crystal surface. Similar results

were also reported on Fe–25Cr–20Ni single crystal (ref. 7) which showed that
this kind of "passivation" is possible only in a narrow potential range and was
dependent on potential scan rate in potentiodynamic experiments. A tendency
for "passivation" in the active dissolution range has also been reported in
boiling $MgCl_2$ solution (ref. 3, 8) and in sulfuric acid sodium cloride solution at room temperature (ref. 9).

In the present study the effect of crystallographic orientation on potentiodynamic and passivation behavior of a Fe–14.5Cr–14.5Ni–2.5Mo single crystal is
investigated in SCC producing 5N H_2SO_4 + 0.5N NaCl solution. Passive film
formation and its break-down is followed by current density measurements at
constant potentials and by open-circuit potential decay.

EXPERIMENTAL

The chemical composition of the single crystal (growth direction [013]) used
is shown in Table 1.

TABLE 1
Chemical composition.

Fe	Cr	Ni	Mo	C	Mn	Si	P	S
bal.	14.5	14.5	2.5	0.005	0.02	0.04	0.005	0.005

Specimens with low-index crystallographic faces of (100), (110), (111) and
(210) were spark-cut from the crystal. The final preparation of the specimen
surfaces was electropolishing in methanol + 15 percent HNO_3 solution at sub-
zero temperature to ensure an undeformed surface. For electrochemical measurements, specimens were mounted in resin and the free working area was defined by
laquering.

Electrochemical measurements were performed in nitrogen deaerated 5N
H_2SO_4 + 0.5N NaCl solution using a saturated calomel electrode (SCE) with a
salt bridge Luggin capillary assembly as a reference electrode. Specimens were
immersed under cathodic polarization (–800 mV) and subsequently the polarization curves were determined potentiodynamically after 15 min holding time.
Current density measurements as a function of time at constant potentials were
done after the same immersion procedure followed by a potential step. The open-
circuit potential decay was recorded after passivation treatment for 10 min at
+400 mV.

RESULTS

Potentiodynamic polarization curves determined for low-index crystal faces

are presented in Fig. 1. Corrosion potentials for various faces showed no differences. Tafel slopes were similar for the studied faces on both anodic and
cathodic regions. The cathodic overpotential was highest on the (111) and
lowest on the (210) crystal surface. The anodic dissolution rate was lowest on
the (100) and highest on the (111) crystal surface (see Fig. 2). There were no
signs of "passivation" at the active dissolution range on any of the studied
surfaces. The anodic current densities reached steady state values at -275 mV.
The (100) and (210) faces reached steady state current densities during the
experiment at -300 mV, whereas the other surfaces showed gradually increasing
currents. At -325 mV no steady state values were obtained. The (100) surface showed always the lowest dissolution rate, like in the previous observations in boiling $MgCl_2$ solution (ref. 6, 8). However, no "passivation" at
the active dissolution range was observed even at a slow scan rate of 6 mV/min.

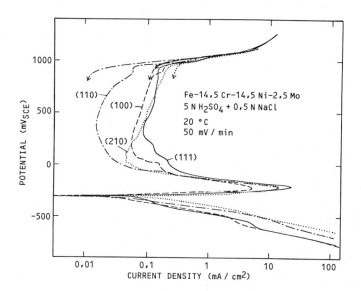

Fig. 1. Potentiodynamic polarization curves for electropolished Fe-14.5Cr-
14.5Ni-2.5Mo single crystal surfaces in deaerated 5N H_2SO_4 + 0.5N
NaCl solution.

Passive film formation was studied by current density measurements (see Fig.
3) at constant potentials of -100 mV and +400 mV corresponding to the active-
passive transition and the passive region, respectively. The observed current
density decay for the (110) surface was most rapid at both potentials and
showed a step-wise behavior at the active-passive transition potential. The
other surfaces showed almost similar current density decays with each other and
their current densities remained on markedly higher level than that for the
(110) surface.

698

The stabilities of the passive films formed during 10 min passivation at
+400 mV were followed by the open-circuit potential decay (see Fig. 4). The
open-circuit potential decay curves showed two potential arrests. The first
arrest around +150 mV was clearly detectable only for the (110) surface. The
second arrest corresponds to the active-passive transition potential at around
-100 mV. The time required for the potential to decay to the rest potential
was within 100 sec. for the (100), (111) and (210) surfaces but markedly longer
for the (110) surface.

DISCUSSION

The results indicate that the effect of anisotropy was most pronounced on
passivation behavior, whereas cathodic and active dissolution properties showed
only slight orientation dependence.

In dissolution-controlled SCC models there is a delicate balance between
film rupture and local anodic dissolution at the crack tip (ref. 2, 3). Without anisotropic dissolution the dissolution-controlled models are likely to
lead to non-crystallographic fracture surface and blunt cracks which is contrary to the typical transgranular SCC fracture surface morphology. If we sup-

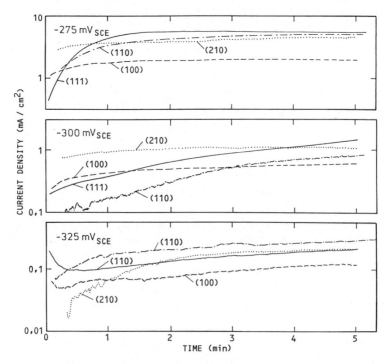

Fig. 2. Current density profiles at indicated potentials after a pretreatment
of 15 min at -800 mV$_{SCE}$ for electropolished Fe-14.5Cr-14.5Ni-2.5Mo
single crystal surfaces in deaerated 5N H$_2$SO$_4$ + 0.5N NaCl solution.

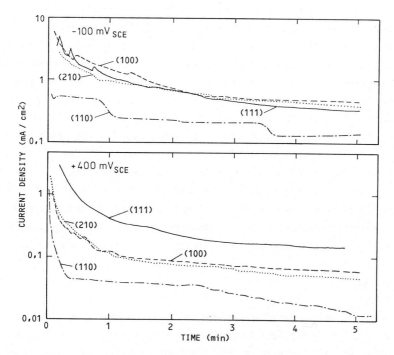

Fig. 3. Current density profiles at indicated potentials after a pretreatment of 15 min at −800 mV$_{SCE}$ for electropolished Fe–14.5Cr–14.5Ni–2.5Mo single crystal surfaces in deaerated 5N H$_2$SO$_4$ + 0.5N NaCl solution.

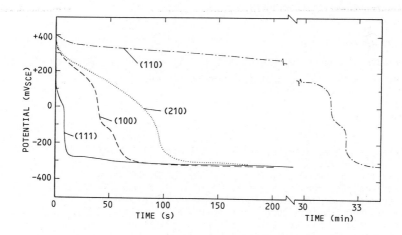

Fig. 4. Open-circuit potential decay profiles for electropolished Fe–14.5Cr–14.5Ni–2.5Mo single crystal surfaces pretreated for 10 min at +400 mV$_{SCE}$ in deaerated 5N H$_2$SO$_4$ + 0.5N NaCl solution.

pose that the planes least prone to dissolution form the crack walls, the obtained dissolution rates would favour (100) fracture surface formation. However, only slight anisotropy was observed in anodic dissolution. Therefore it
is questionable if the observed slight active dissolution anisotropy could account for dissolution-controlled SCC mechanism. The most rapid and stable passive layer formation was observed on the (110) surface. This suggests that the
combined effects of active dissolution and repassivation would favour a mixed
(100) and (110) fracture orientation. These are also the most commonly reported SCC fracture surface planes (ref. 5, 8, 10, 11). It can be anticipated,
however, that during SCC crack growth anisotropy in passivation is much less
important than anisotropy in active dissolution. This is true especially in
the vicinity of free corrosion potential which, indeed, is the most frequently
reported potential region for SCC. Thus these electrochemical measurements on
low-index crystal faces are not in favour of dissolution-controlled SCC
mechanisms.

CONCLUSIONS

The electrochemical behaviour of the Fe-14.5Cr-14.5Ni-2.5Mo single crystal
showed only slight anisotropy in active dissolution and cathodic reactions.
Anodic dissolution rates were lowest and highest on (100) and (111) crystal
surfaces, respectively. Passivation behavior showed clear anisotropy: the
(110) surface being the most readily passivating and also the most stable
passive layer formed on (110) surface. Other faces showed almost identical
dissolution and passivation behaviour.

ACKNOWLEDGEMENTS

The financial support by the Academy of Finland and the Ministry of Trade
and Industry of Finland is gratefully acknowledged.

REFERENCES

1 M. Takano and R.W. Staehle, Trans. JIM 18 (1977) 780-786.
2 J.C. Scully, Corr. Sci. 20 (1980) 997-1016.
3 A. Desestret and R. Oltra, Corr. Sci. 20 (1980) 799-820.
4 H. Hänninen, Dr. Thesis, Helsinki University of Technology, 1980.
5 G.M. Scamans and P.R. Swann, Corr. Sci. 18 (1978) 983-995.
6 M. Marek, PhD Thesis, Georgia Institute of Technology, 1970.
7 M. Ahlers and E. Riecke, Corr. Sci. 18 (1978) 21-38.
8 J.M. Silcock, Br. Corros. J. 16 (1981) 78-93.
9 T. Hakkarainen, Proc. 7th Int. Congr. on Metallic Corrosion, Rio de Janeiro,
   1978, 763-772.
10 R. Liu, N. Narita, H. Birnbaum and E.N. Pugh, Metall. Trans. 11A (1980)
   1563-1574.
11 K. Kon, S. Tsujikawa and Y. Hisamatsu, Nippon Kinzoku Gakkaishi 43 (1979)
   770-775.

Passivity of Metals and Semiconductors, edited by M. Froment
Elsevier Science Publishers B.V., Amsterdam — Printed in The Netherlands

CORROSION ATMOSPHERIQUE D'UN ACIER PATINABLE SOUDE

B.M. ROSALES[*], E.S. AYLLON et C. BONAZZOLA

CEICOR - CITEFA/CONICET - Zufriategui y Varela - (1603) Villa Martelli - Buenos Aires (ARGENTINA).

RESUME

Dans le présent travail, nous avons évalué le pouvoir passivant de la couche de rouille développée sur un cordon de soudure et des zones moins affectées par la chaleur, par rapport au même acier non-soudé.
Le cordon de la soudure réalisé avec apport développe une patine de moindres propriétés protectrices que celles formées sur les régions de l'acier soudé moins affectées par la chaleur ou les témoins non-soudés. L'analyse s'est effectuée avec les polarisations potentiocinétiques et microscopie électronique à balayage (MEB) avec microsonde électronique (EPMA).

INTRODUCTION

Les aciers faiblement alliés, désignés patinables, se caractérisent pour développer une rouille qui devient progressivement protectrice avec le temps d'exposition à l'atmosphère.

Dans des travaux précédents[1,2], nous avons proposé et constaté l'hypothèse selon laquelle le pouvoir protecteur d'une rouille est notamment dépendant de sa structure. Par contre, d'autres auteurs[3,4,5] considèrent qu'il est plutôt dépendant de sa composition chimique.

L'évaluation des propriétés protectrices des couches formées se fait avec des mesures électrochimiques, telles que le potentiel et le courant de corrosion de l'acier avec sa patine, dans des solutions aqueuses oxygénées. Ces propriétés sont contrôlées par les éléments d'alliage aussi bien que par les polluants et les conditions climatiques de l'environnement d'essai. Les éléments endogènes de l'acier qui précipitent avec les produits de corrosion du fer améliorent la compacité de la rouille en remplissant des pores et des fissures[6] . Ils arrêtent aussi l'incessant écaillage[2,7] qui conduit à la destruction complète des aciers ordinaires. La résistance des aciers patinables est la plus remarquable en atmosphères chargées en SO_2.

---

[*] Chercheur au Consejo Nacional de Investigaciones Cientificas y Tecnicas (CONICET).

CONDITIONS EXPERIMENTALES

On a utilisé de l'acier Cor Ten A (C 0,11 ; Mn 0,38 ; Si 0,23 ; S 0,02 ;
P 0,09 ; Cr 0,69 ; Ni 0,11 ; Mo 0,10 et Cu 0,38) laminé à chaud à 2 mm
d'épaisseur. La soudure avec apport a été faite avec une électrode revêtue, et
à basse teneur en hydrogène (C 0,05 ; Mn 0,65 ; Si 0,50 ; Cr 0,20 ; Ni 0,30 ;
Cu 0,40). Les échantillons de 100 x 300 mm, avec le cordon au long dans le
centre, ont été sablés, dégraissés et exposés dans notre Station d'Essais en
milieu urbain-industriel ; teneur du polluant : 0,12 mg $SO_3.dm^{-2}.jour^{-1}$. Ils
ont été orientés au nord, à 45° d'inclinaison, de façon que le cordon reste
parallèle à la direction d'écoulement de l'eau de pluie.

Au bout de cinq ans d'exposition à l'atmosphère, nous avons analysé la qualité de la patine formée sur le cordon de soudure et, par franges parallèles,
sur le reste des éprouvettes jusqu'à la zone moins affectée par la chaleur. Des
témoins non-soudés ont été exposés simultanément.

Sur des pièces de 7 x 7 mm nous avons établi des courbes de polarisation
potentiocinétiques (vitesse de polarisation : 100 $mV.min^{-1}$) dans l'eau de pluie
récoltée à la Station d'Essais où s'est faite l'exposition.

Toutes les éprouvettes ont été observées au MEB pour déterminer la morphologie des produits de corrosion formés et celle de la surface d'acier une fois
éliminée la couche de rouille. Le décapage a été fait en solution d'acide HCl
avec de l'urotropine comme inhibiteur.

La distribution des éléments d'alliage dans la rouille formée sur le cordon
a été déterminée avec EPMA.

RESULTATS

L'aspect du cordon a l'oeil nu est pareil à celui du reste de la surface.
Macroscopiquement, on peut distinguer deux régions de couleur et texture différentes. La surface de la patine complète présente une coloration marron et une
texture polie, tandis que dans des régions où la phase extérieure s'est brisée
et décollée, on trouve isolément des taches plus foncées à texture rugueuse,
correspondant à la phase intérieure.

Sur la Fig. 1 on peut voir les courbes de polarisation obtenues sur des
échantillons patinés et découpés, des différentes régions. Les courbes cathodiques ne montrent pas des effets appréciables dus à la perte de la phase extérieure ni aux différentes températures atteintes par l'acier. Les courbes
anodiques, par contre, sont très sensibles à ces deux paramètres. Sur le cordon
les courants sont plus élevés que dans les régions moins affectées par la chaleur de la soudure, en décroissant brutalement jusqu'à mettre en évidence la
passivité pour les échantillons non-soudés. On peut donc apprécier une moindre
qualité protectrice de la patine formée sur le cordon que sur le reste de la

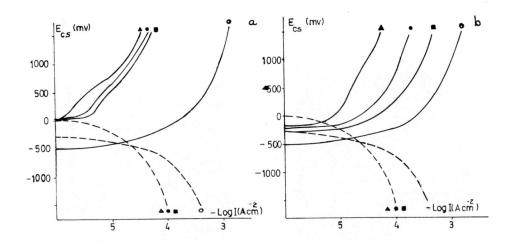

Fig. 1. Polarisation de l'acier Cor Ten A soudé, dans l'eau de pluie,
pH 5,4, dans un environnement urbain-industriel.
(a) Régions polies.   (b) Régions rugueuses.

■ – Sur le cordon de soudure.
● – A 1,5 cm du cordon.
▲ – A 5 cm du cordon et témoin non-soudé.
0 – Témoin sablé, non-patiné.
——— Non-aéré.
– – – Aéré.

surface. Par comparaison avec les courbes respectives obtenues sur des échantillons sablés, non patinés, on ne peut pas prévoir une différence intrinsèque
de réactivité de cordon. Par conséquent, les variations observées dans les
courbes anodiques, sur des échantillons patinés au bout de 5 ans d'exposition,
entre le cordon et le reste, sont dues au moindre pouvoir protecteur de la
patine sur le cordon.

Les résultats électrochimiques et de MEB (Figs 1, 2, 3 et 4) montrent un
étroit rapport entre compacité de la structure et pouvoir protecteur de la couche de rouille[1,2] . Dans tous les cas, les meilleurs comportements électrochimiques ont été trouvés pour des patines plus compactes.

Ceci peut s'interpréter en termes d'un effet majeur de barrière dans les
patines plus compactes, qui s'opposerait à la diffusion des ions Fe^{+2} solvatés,
de l'acier vers l'extérieur. Cet effet ne serait pas évident pour les courbes
cathodiques où l'espèce électrochimique qui se déplace est la petite molécule
d'oxygène. Les observations en MEB ont permis de vérifier une hypothèse préalablement proposée[8] selon laquelle les produits de corrosion atmosphérique d'un
acier seraient isomorphes de l'acier sur lequel ils s'étaient développés. La
morphologie de tels produits était donc différente s'ils s'étaient formés sur

704

Fig. 2. Sur le cordon de
soudure.

Fig. 3. A 5 cm du cordon
de soudure.

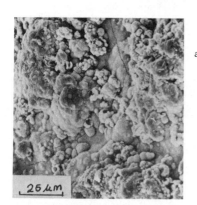

a

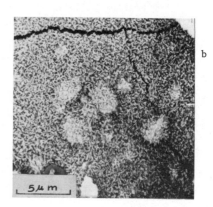

b

c

Fig. 4. a, b, c .

A 1,5 cm du cordon
de soudure.

le cordon de soudure ou dans la région moins affectée par la chaleur. Sur le cordon il y avait une nette prédominance des formes sphériques moins compactes, avec une substructure de même symétrie (Fig. 2), tandis que sur la dernière région nous avons trouvé différents types de structure compactes (Fig. 3). Sur la Fig. 4, on voit des structures sphériques avec substructure aciculaire correspondant à une région intermédiaire entre celles préalablement montrées.

Des observations sur champ des échantillons patinés nous ont permis d'interpréter le mécanisme d'initiation et propagation de la corrosion sur le cordon, par analogie à ce qui a été préalablement fait pour le même acier non-soudé dans le même milieu urbain-industriel[9], aussi bien que dans un climat marin[8]. L'attaque initiale est localisée. Pendant le développement des produits de corrosion, les régions initialement anodiques se polarisent par l'effet passivant de la rouille, en déplaçant la nucléation de nouvelles piqûres dans des régions où le dépôt est plus mince. De cette façon, l'attaque s'étend à toute la surface à des vitesses décroissantes au fur et à mesure que la patine devient plus épaisse. La patine est formée par des couches parallèles à la surface, ce qui correspondrait à des cycles d'humectation, par pluie ou rosée, et séchage. Les limites entre eux montrent, par EPMA, une précipitation sélective de Cr qui, selon ce qui avait déjà été proposé, contrôlerait le pouvoir protecteur des patines, améliorant l'agglutination des couches successives des produits de corrosion du fer[5,2]. Cet effet est moins évident sur le cordon de soudure que sur les régions moins affectées par la chaleur, spécialement dans les zones où la patine est plus poreuse, ce qui confirme l'influence décisive de la morphologie sur le pouvoir protecteur. La répartition homogène du cuivre dans la patine est associée à celle du soufre[10]. Etant uniforme sur toutes les régions étudiées, elle empêcherait l'effet nuisible de l'accumulation de ce polluant dans ces cavernes anodiques.

CONCLUSIONS

L'aspect de la patine est identique sur toute la surface des échantillons étudiés.

Les produits de corrosion développés présentent des structures isomorphes de celles de l'acier utilisé.

Le cordon de la soudure réalisé avec apport développe une patine de moindres propriétés protectrices que celle formée sur des régions moins affectées par la chaleur.

Le moindre pouvoir passivant de la patine sur le cordon de soudure s'accorde avec une structure moins compacte que celle développée sur le reste de l'échantillon. Une telle morphologie réduit l'effet bénéfique des éléments d'alliage.

REMERCIEMENTS

Le Centre de Recherches en Corrosion (CEICOR), de CITEFA, remercie spécialement pour l'appui financier offert par le Sous-Secrétariat à la Science et à la
Technique et par l'Institut Argentin de Sidérurgie, qui a rendu possible la
réalisation de ce travail.

REFERENCES

1 B.M. Rosales, E.S. Ayllon et S.L. Granese, 7th Int. Congress on Metallic
   Corrosion, oct. 1978, Rio de Janeiro, Brésil, 3, p. 1139-1151.

2 B.M. Rosales, E.S. Ayllon, R.T. Bonard, S.L. Granese et J.L. Ikehara,
   8th Int. Congress on Metallic Corrosion, set. 1981, Mainz-Allemagne, 1,
   p. 317-322.

3 T. Misawa, T. Kyuno, W. Suetaka, S. Shimodaira, Corros. Sci. 11, p. 35-48
   (1971).

4 L. de Miranda, CEBELCOR, Vol. 125, R.T. 221, oct. 1974, p. 69-77.

5 T. Misawa, K. Hashimoto et S. Shimodaira, Bosyoku Gijitsu, 23, p. 17-27
   (1974).

6 T. Moroishi, J. Satake, N. Fujino et M. Kowaka, Trans. ISIJ, 1, 390-396,
   (1971).

7 B.M. Rosales et E.S. Ayllon, Résumés 158th Meeting Electrochemical Society,
   International Symposium on Atmospheric Corrosion, 5-10 oct. 1980, Hollywood,
   Florida, USA, p. 551-552, et Atmospheric Corrosion, W.H. Ailor Ed., John
   Wiley & Sons, Inc. (1982), Chap. 35.

8 B.M. Rosales et E.S. Ayllon, 5e Congrès Int. de la Corrosion Marine et des
   Salissures, Barcelone, Espagne, mai 1980, p. 182-199.

9 B.M. Rosales et E.S. Ayllon, Résumés 158th Meeting Electrochemical Society,
   International Symposium on Atmospheric Corrosion, 5-10 oct. 1980, Hollywood,
   Florida, USA, p. 605-606, et Atmospheric Corrosion, W.H. Ailor Ed.,
   John Wiley & Sons, Inc. (1982), Chap. 28.

10 L. de Miranda, CEBELCOR, Vol. 125, R.T. 221, oct. 1974, p. 22.

Passivity of Metals and Semiconductors, edited by M. Froment
Elsevier Science Publishers B.V., Amsterdam — Printed in The Netherlands

COMPOSITION, STRUCTURE AND PROPERTIES OF PASSIVATION LAYERS ON TINPLATE

N. AZZERRÏ and L. SPLENDORINI

Centro Sperimentale Metallurgico S.p.A. - P.O. Box 10747 - ROME-EUR (Italy)

ABSTRACT

    The passive layer of cathodic dichromate (CDC) and dichromate dipped (DCD) passivated tinplate was investigated by means of electrochemical analysis. Electrochemical results were compared with X-Ray photoelectron spectroscopy (XPS), atomic absorption spectroscopy and glow discharge spectroscopy data and are discussed in relation to the solid state properties of the passive layer compounds.

Electrolytic tinplate normally undergoes a passivation treatment (ref. 1) aimed at improving some important properties of the final product such as resistance to oxidation, sulphur staining and lacquer adhesion.

The passivating treatment is generally performed (i) by simple immersion (treatment 300 or dichromate dipping (DCD))or (ii) by application of a cathodic current (treatment 311 or cathodic dichromate (CDC)).

In this work the passive layer was investigated mostly by means of electrochemical analysis based on the test of Willey and Kelsey (ref. 2) for tin oxides and the Britton's (ref. 3) and Aubrun's (ref. 4,5) methods for chromium compounds. The electrochemical results were compared with X-Ray photoelectron spectroscopy (XPS), atomic absorption spectroscopy (AAS) and glow discharge spectroscopy (GDS) data.

By XPS analysis metallic chromium was found to be present only in the CDC passivation film whilst no chromium (VI) was detected for either passivating treatment. After sputtering, anhydrous Cr_2O_3 and, in the as-received measurements, chromium (III) hydroxide species were detected: on the basis of the binding energy values, it was not possible to identify unambiguously the latter species, probably due to destructive sputtering effects, e.g. de-hydration. Tin oxides were revealed, but the oxidation state of the oxidized forms was not assigned owing to the small shift between tin (II) and tin (IV) on the basis of core-level binding energies.

In the electrochemical investigation, based on both galvanostatic (ref. 3, 5) and potentiostatic (ref. 6) measurements, in addition to the quantitative estimation of the constituents of the passive film, the chemical and structural modifications induced by the anodic treatments were evaluated. During the dissolution of chromium compounds a simultaneous oxidation of tin occurs: this oxidation is an electrochemical process which is easily controlled because it takes place according to a well-established growth rate governed by the ionic conductivity of tin oxides.

However, the anodic reactions involving chromium compounds are not well established as they occur at two different potential ranges. The amount of chromium dissolved in potentiostatic tests (ref. 6) at potentials above + 0.6 V (SCE) is potential dependent, as shown in Fig. 1; this behaviour can be due to either the influence of potential on the anodic dissolution mechanism or the presence of different electrochemically oxidizable chromium species.

The chromium dissolution reaction occurring at 0.00 V, which is evidenced in Fig. 2 by GDS analysis of residual chromium content on tinplate samples after potentiostatic polarizations, was attributed, on the basis of thermodynamic data (ref. 7), to the oxidation of metallic chromium to trivalent chromium. Probably this reaction in the potential range 0.0 ÷ 0.5 V has not been considered in the past because oxidized chromium species are removed from the tinplate surface but the amount of chromium found in the electrolyte is quite negligible owing to the low solubility of Cr^{3+} in neutral or mildly acidic solutions (ref. 7). In fact, in the above potential range the amount of chromium (Cr) is constant and fairly corresponding to its solubility limit, as already shown in Fig. 1.

Moreover GDS analysis confirms the retention of a proportion of chromium, in the state of trivalent insoluble compounds (e.g., oxi-hydroxides), on the sample surface even at high oxidation potential (>0.8 V), as already known from galvanostatic investigations (ref. 3, 5).

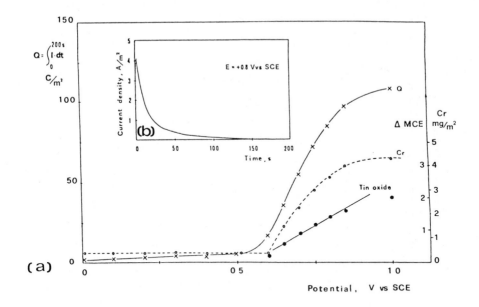

Fig. 1. (a) Electric charge (Q), dissolved chromium found in the electrolyte
(Cr) and tin oxide growth (△MCE) during potentiostatic polarization as a function of the tinplate potential; (b) Example of a current-time curve for CDC tinplate polarized potentiostatically (E = + 0.8 V (SCE)).
Potentiostatic test are carried out in a 7.5 g/l disodium phosphate solution
having pH 7.4.Tin oxides are evaluated by galvanostatic reduction in a
0.001 M HBr solution and the results are expressed in amount of electric charge
MCE (millicoulomb equivalents per square centimeter) required for oxides reduction. △MCE is the increase of tin oxides induced by the anodic potentiostatic
polarization.

The oxidation of chromium even below + 0.6 + 0.8 V, i.e. before the typical

potential plateau in the Britton's and Aubrun's galvanostatic tests, was confir-

med by further XPS investigation; therefore the electroanalytical anodic galva-

nostatic methods (ref. 3, 5) do not provide unambiguous quantitative results, al-

though they are capable of monitoring the existence of some electroactive chro-

mium compounds and can be regarded as a practical guide to the amount of surfa-

ce chromium species on the basis of empirical relationships. Conversely poten-

tiostatic tests proved to be effective in providing the potential dependence of

the electro-oxidation of different chromium species.

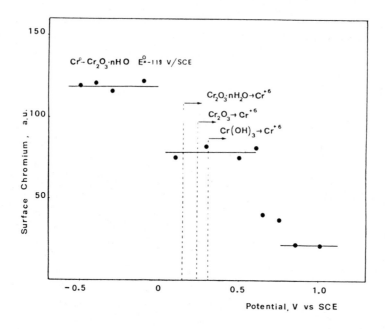

Fig. 2. Chromium remaining on the tinplate surface after potentiostatic polarization estimated by GDS analysis as a function of tinplate potential.

The importance of the chromium dissolution reaction at about 0.0 V is also
evidenced by the investigation of the semiconducting properties of the passive
layer. In fact, photopotential measurements as a function of the polarizing potential (ref. 8) show the disappearance of n-type photoeffects at about 0.0 V
and again at about + 0.6 V in the case of CDC passivated tinplate (i.e. in cor-
respondence with the occurrence of dissolution reactions). Also, with reference
to Fig. 3, the Mott-Schottky behaviour of the CDC passivated tinplate shows deviation from linearity at about − 0.1 V, which prevents the investigation of the
nobler potential region. Moreover, it is apparent that the tinplate passivation
treatment can be related to the charge carrier concentration (i.e. the number
N_D of donors): semiconducting properties appear to decrease as a function of
the chromium content in the passivation film.

The development of the preliminary investigations mentioned here on solid
state properties of passive layer components seems to be a promising area of interest; the semi-conducting behaviour displayed by some compounds in the passive layer could lead to a more straightforward correlation with final product properties such as lacquer adhesion and corrosion resistance.

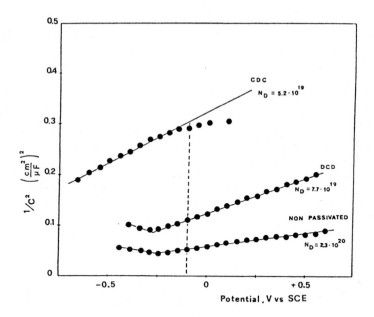

Fig. 3. The square of the reciprocal capacitance C as a function of tinplate potential (Mott–Schottky plot) in 0.5 M Na_2HPO_4 (pH 7.4; a.c. frequency, 1 kHz).

Moreover, new criteria for differentiating chromium species in the surface layer, so far based on chemical and electrochemical differential dissolution (ref. 3), valency state (ref. 4, 5) and hydration degree (ref. 9), could become available.

ACKNOWLEDGEMENTS

The present work has been carried out with the financial support of Nuova Italsider S.p.A., Genoa. The Authors wish to thank Mr. G. Savastano who performed GDS analyses on tinplate with the use of a special device purposely set up for thin samples (ref. 10).

REFERENCES

1 N. Azzerri and L. Splendorini, Imballaggio, 292, Jun (1980), 781
2 A.R. Willey and D.F. Kelsey, Anal. Chem., 30 (1958), 1804
3 S.C. Britton, Br. Corros. J., 1 (1965), 91; ibid., 10 (1975), 85
4 P.J. Aubrun and P. Roquet, J. Electrochem. Soc., 122 (1975), 861
5 P.J. Aubrun and G.A. Pennera, Rev. Metall., Nov (1976), 745
6 N. Azzerri, L. Splendorini, C. Battistoni, E. Paparazzo, Surface Technol. 15, 3 (1982), 255
7 E. Daltombe, N. Zoubov and M. Pourbaix, in M. Pourbaix (ed.) Atlas d'Equilibres Electrochimiques, Gauthier and Villars, Paris, 1963, p. 256
8 G. Brunoro and F. Zucchi, personal communication, 1981
9 S.E. Rauch Jr., R.N. Steinbicker, J. Electrochem. Soc., 120 (1973), 735
10 P. De Gregorio, G. Savastano, CSM report 4156R, Mar, 1982 (confidential)

PASSIVATION OF TIN USING CHROMATES, MOLYBDATES AND TUNGSTATES.

D.R.GABE[1] and D.BIJIMI[2]

[1]Dept. of Materials Engineering and Design, Loughborough University of
Technology, LE11 3TU, U.K.

[2]Dept. of Mechanical Engineering, Kaduna Polytechnic, Kaduna, Nigeria.

ABSTRACT

   Tin has been subjected to both anodic and cathodic polarization, using a
potentiodynamic technique, in chromate, molybdate and tungstate solutions at
various concentrations, pH and temperatures. Passive film formation has been
studied using ESCA and coulometry techniques. Chromate treatments give rise to
thinner films which contain both Cr^{VI} and Cr^{III} while molybdate and tungstate
films are thicker and contain Mo and W in the hexavalent state only; all films
also contain tin oxides.

INTRODUCTION

   Passivation treatments for tin and tin electrodeposits are well-established

commercial processes and in general use either anodic cycles for small bath

operations or cathodic cycles for continuous treatment of tin plated steel

strip in hexavalent chromium solutions (refs.1,2). Owing to its toxicity and

increasingly stringent controls on process effluents an alternative to

hexavalent chromium is desirable and the inorganic analogues of chromate –

molybdate and tungstate – are obvious choices. They have found some virtues as

inhibitors (ref.3), as passivating ions for iron (ref.4) and an application for

tin has been suggested (ref.5) but detailed studies have not yet been reported.

A preliminary study in relation to tin has been made and some of the results

will be described; a fuller account of the experimental techniques used and

results obtained will be published elsewhere (ref.6).

EXPERIMENTAL RESULTS

Polarization measurements

   The mechanism of coating formation for both anodic and cathodic cycles has

been characterized initially by means of potentiodynamic polarization curves.

Solutions containing up to 1M molybdate and tungstate have been compared with

that containing a similar concentration of chromate at otherwise standard

conditions. The effects of aeration and deaeration, pH and temperature

variation have been studied within commercially acceptable limits. Anodic

polarization in 10^{-1} M solutions at pH7 and 40^0C yields curves having
classical passive features (fig.1), the width of the passive region and its

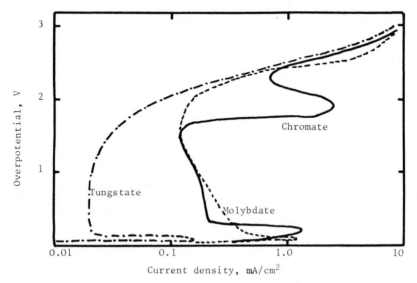

Fig.1. Anodic polarization of tin aqueous deaerated
0.1M solutions of chromate, molybdate and tungstate at
pH7 and 40^0C.

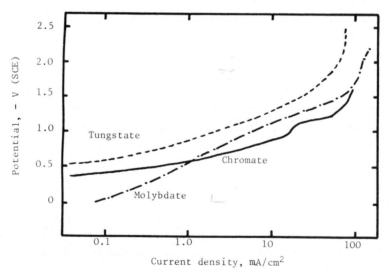

Fig.2. Cathodic polarization of tin in aqueous 1M
solutions of chromate, molybdate and tungstate at
pH9 and 40^0C.

current density level being important criteria for assessment. It should be noted that the chromate treatment yields a double passivity region, unlike the others, which has been noted previously [1]. Cathodic polarization yields a series of curves culminating in an approach to a limiting c.d. whose value is pH, temperature and concentration dependent (fig.2) and results in the co-deposition of chromium in various valency states and the incorporation of molybdenum and tungsten in the hexavalent state only.

Two criteria have been used to judge the effectiveness of polarization as a means of promoting passivity in relation to a process feasibility: ip the minimum passive current density, and ΔE the passive potential range. A low value of ip and large value of ΔE are preferable. Data is tabulated in Table 1.

TABLE 1

Comparison of passivity parameters for 0.1M Solutions ip - mA/cm^2; ΔE - V.

(a) Aerated Solutions							
pH	Temp^{0}C	CHROMATE		MOLYBDATE		TUNGSTATE	
		ip	ΔE	ip	ΔE	ip	ΔE
5	20	0.25	2.0	–	–	–	–
	40	0.20	1.8	5	2.4	–	–
	60	0.18	1.6	3.5	1.4	0.5	1.4
7	20	0.10	1.2	0.015	1.6	0.025	0.9
	40	0.16	1.4	0.15	2.0	0.025	1.8
	60	0.13	1.4	0.04	1.5	0.024	1.8
9	20	0.10	1.6	0.20	2.0	0.03	1.2
	40	0.20	1.8	0.20	2.0	0.04	1.2
	60	0.15	0.8	0.15	1.8	0.042	1.2
(b) Deaerated Solutions							
5	20	0.14	2.4	–	–	–	–
	40	0.14	1.6	7.0	2.6	–	–
	60	0.12	1.4	1.6	2.0	0.25	1.2
7	20	0.20	1.4	0.16	1.8	0.015	1.2
	40	0.20	1.4	0.14	1.8	0.02	1.0
	60	0.11	1.3	0.10	1.8	0.02	1.4
9	20	0.18	2.0	0.3	1.5	0.02	1.0
	40	0.20	1.7	0.25	2.0	0.03	1.3
	60	0.20	1.7	0.14	2.0	0.037	1.2

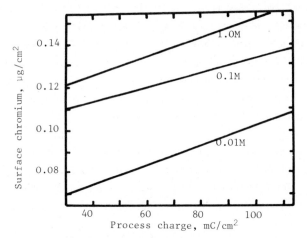

Fig.3. Effect of chromate concentration and total cathodic process charge on surface metallic chromium content. Coulometric determination after treatment at pH9 and 40^0C

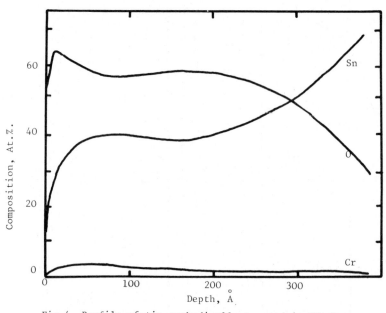

Fig.4. Profile of tin cathodically treated in 1M chromate solution at pH9, 40^0C and 300 mC/cm². E.S.C.A. measurements.

Film analysis

Compositional and constitutional studies have been carried out in a variety
of ways including coulometric analysis and E.S.C.A. or X.P.S. Coulometry was
carried out using the constant current techniques of Britton (ref.7) in which
anodic oxidation in a phosphate buffer solution at pH 7.4 oxidizes chromium in
the surface film to the soluble hexavalent form with an approximate electrochemical equivalence of $1mC \equiv 0.1\mu g$. The oxide component may be separately
estimated using cathodic coulometry to reduce surface oxides but having a less-
certain electrochemical equivalent is expressed usually in mC/cm^2 units. It
has been shown that for cathodic passivation treatments the surface chromium
content varies with pH, concentration and temperature but increases linearly
with process charge used (fig.3). E.S.C.A. has enabled in-depth profiles to
be produced showing that for such a cathodic process the chromium level is
greatest at the free surface and at a depth of 400 Å approaches zero (fig.4).
Under anodic conditions equivalent chromium levels could be achieved at much
lower charge densities. For example, using 26 mC/cm^2 at anodic polarity a
very comparable level of chromium was obtained to that attained at 300 mC/cm^2
cathodic (see fig.5).

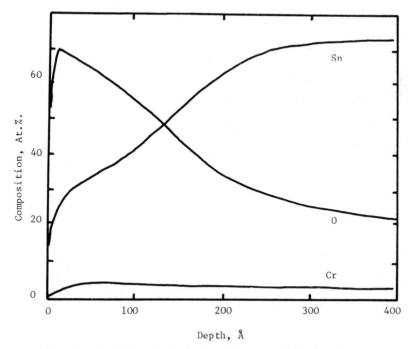

Fig. 5. Profile of tin anodically treated in 1M chromate at
pH9, 40°C and 26 mC/cm^2. E.S.C.A. measurements.

DISCUSSION

For conditions of anodic polarization it is clear that in broad terms chromates, molybdates and tungstates behave in a similar manner but that certain differences of detail can be discerned. The more concentrated solutions (e.g. > 0.1M) give most marked passivity with an optimum being observed for pH5-9 and 20°C for chromate and pH7-9 and 60°C for molybdate and tungstate.

During cathodic polarization a limiting current condition can be observed and its value is increased by increasing temperature, pH in the range 5-9 and anion concentration.

Using techniques of surface analysis the films have been shown to have a total thickness of 400-1000Å. At a depth of 50Å the trivalent form of chromium predominates (80%) and the metallic and hexavalent forms are present to about 10% each. By contrast the molybdate and tungstate solutions yield deposits which are virtually all hexavalent making the coatings essentially different in character while still revealing similar passive electrochemical behaviour.

A full account of this work will be published in due course and it will be shown that similar treatments for other metals exhibit comparable effects (ref.6).

REFERENCES

1 D.R.Gabe, Surface Tech. 5, 463 (1977).
2 N.Asserri and L.Splendorini, Imballaggio, 292, 65 (1980).
3 K.K.Chew and D.R.Gabe, Corr.Prev.Control, 26(4), 5 (1979).
4 K.Ogura and T.Majima, Electrochim. Acta, 24, 325 (1979).
5 R.E.van de Leest and G.Krijl, Thin Solid Films, 72, 237, (1980).
6 D.Bijimi and D.R.Gabe, Brit.Corr.J. 1983 (to be published).
7 S.C.Britton, "Tin versus Corrosion", Int.Tin Res.Inst., Greenford, 1975.

Passivity of Metals and Semiconductors, edited by M. Froment
Elsevier Science Publishers B.V., Amsterdam — Printed in The Netherlands

GALVANOSTATIC FORMATION AND REDUCTION OF TIN/TIN OXIDES IN Na_2SO_4
SOLUTIONS

C.V. D'ALKAINE and J. M. da SILVA
Grupo de Eletroquímica, DQ, UFSCar
Caixa Postal 676, 13560 São Carlos, SP - Brasil

ABSTRACT

The galvanostatic formation and reduction of tin/tin oxides
in oxygenated Na_2SO_4 solutions with variation of the anodic
formation and/or cathodic reduction currents were studied. Two
reduction transition times take place, showing the existence of
Sn(IV) and Sn(II) in the film. Through these results the
composition of the film is analysed.

INTRODUCTION

The electrochemical behavior of Tin has received much attention and has been comprehensively reviwed by Hampson[1], mainly in
acid and alkaline media.

In general, it was proposed that the passivity comes first
from the formation of SnO and secondly (at higher potentials),
from the growth of SnO_2 (both possibly hydratated oxides).

From the works of Gabe[2] and Hampson[3] using electron
diffraction, up to the work of Lau and Wertheim[4] using ESCA,
there is agreement on the existence of these two oxides.

The presence of SnO_2 was not postulated in the begining, but
now it is established[4,5].

In general, it is accepted that the SnO_2 is really the passivating agent[6,7]. At neutral pHs, Pourbaix diagrams[8] show very
nearly equal potentials for Sn/SnO and Sn/SnO_2 interfaces.

The studies of Sn in H_2SO_4 has been interpreted[9] also using
the concept of two kinds of passivating films based firstly on
Sn(II) and secondly on Sn(II)/Sn(IV) (in region of oxygen
evolution).

The Sn/Na_2SO_4 solution system, was studied by Sato[10]. He
proposed SnO_2 as the passivating film. The present paper shows

the results of galvanostatic technique for this case (Sn/Na_2SO_4 0,5 M oxygenated solutions). The idea of the authors is that the galvanostatic technique has not been completely explored in what the growth and reduction of passivating films are concerned, as will be attempted to be shown.

EXPERIMENTAL

The reference electrode was NCE to which all potentials are refered. The working electrode was Sn 99.85 from Merck. The electrode was polished up to emery paper 600 and before each set of measurements two different kinds of procedures were used. In both cases the electrode was aged through a series of cathodic-anodic galvanostatic transients[11] which gives an isoenergetic surface. The two procedures were different in relation to the solution. In one case it was a Na_2SO_4 solution giving rise to a non-adherent porous film formed by dissolution precipitation and which did not affect the measurements. This was showed by the fact that the second procedure in a H_2SO_4 solution, did not give rise to the non-adherent porous film formation but gave the same charge relations as those of the Na_2SO_4 procedure. The aged surfaces gave reproducible results, within the experimental error of galvanostatic measurements. A typical one is shown in Fig. 1.

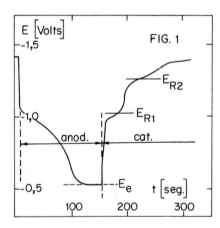 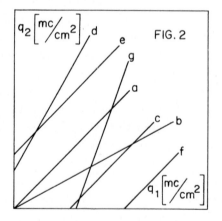

Figure 1- Galvanostatic cathodic-anodic-cathodic chronopotentiogram. Ref. electrode NCE $i_a = i_c = 60$ $\mu A/cm^2$. Na_2SO_4 0,5 M solution.

Figure 2- Theoretical different q_2 vs q_1 relations for the different possibilities of galvanostatic reduction of tin oxides.

From these results it is possible to show that during the

anodic galvanostatic growth the formation of hydratated Sn(II) and
Sn(IV) passivating films take place with two reduction transition
times corresponding to E_{R2} and E_{R1} (Fig. 1) potentials, respectively. The anodic stationary potential E_e (Fig. 1) was independent
of the current density and corresponds to the formation of pits as
it was previously established[11] and also proved in this work by
optical microscopy. The pitting potential was - 0,525 volts in
agreement with older results[11].

GENERAL ANALYSIS OF THE GALVANOSTATIC CURVES

In Fig. 2, different possibilities for the reduction of the
tin oxide are proposed, q_1 corresponds to the amount of charge
involved in the transformation Sn(IV) to Sn(II) and q_2 to that of
the passage from Sn(II) to Sn^o.

Fig. 1(a) corresponds to a film originally with Sn(IV) only.
Fig. 1(b) and (c) correspond to the same initial situation but
when during the first reduction (Sn(IV) to Sn(II)) there is a
partial reduction dissolution process. In one case the loss of
Sn(II) increases with the thickness of the film (b) and in the
other is constant (c). The film thickness increases with the
increase of q_1 and q_2. All these kinds of results give rise to
$q_1 \geqslant q_2$ values.

To make the analysis of the films initially with Sn(II), it
is convenient to define q_{or}^{2+} as the charge for the reduction of the
original Sn(II); $q_{or,d}^{2+}$ as that corresponding to the loss of the
original Sn(II) due to the reduction dissolution process and final
ly $q_{f,d}^{2+}$ which corresponds to the charge loss by the reduction -
dissolution process from that formed by the transformation Sn(IV)
to Sn(II).

It is evident that

$$q_2 = (q_1 - q_{f,d}^{2+}) + (q_{or}^{2+} - q_{or,d}^{2+}) \tag{1}$$

and with this equation it is easily to explain curves (d), (e),
(f) and (g) of Fig. 2. They differ on the value of

$$q_2' = q_{or}^{2+} - q_{f,d}^{2+} - q_{or,d}^{2+} \tag{2}$$

which could change with the film thickness.

In particular, our interest is centred in the case (g), which

corresponds to our experimentals results. It must be interpreted
in the sense that at low thickness the losses by the reduction
dissolution mechanism are higher than the original Sn(II) then,
$q_2' < 0$. At higher thickness the situation is inverted and q_2'
becomes positive.

RESULTS AND DISCUSSION

Two types of measurements of q_2 vs q_1 were made. The first
one corresponds to the situation $i_a = i_c$ (i_a the formation current,
i_c the reduction current). The second one to the situation $i_a \neq i_c$.
Typical results of the first kind are shown in Fig. 3 and 4.
Results for the second case could be seen in Fig. 5.

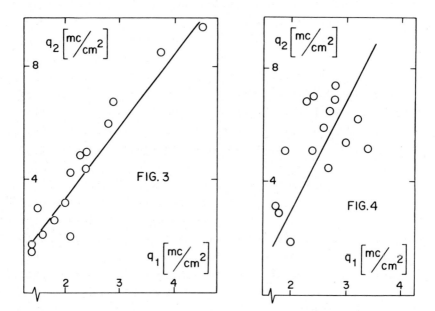

Figure 3- q_2 vs q_1 for the case of $i_a = i_c = 40$ $\mu A/cm^2$.
Figure 4- q_2 vs q_1 for the case of $i_a = i_c = 100$ $\mu A/cm^2$.

The comparison of Fig. 5 with Figs 3 and 4 shows that the
formation and the reduction currents, influence the results. The
formation current i_a through the change of the relation Sn(IV)/
Sn(II) and the reduction current i_c, through the influence on the
loss of reducible Sn(II) charge by the reduction-dissolution
process.

The fact that in all the cases the extrapolation to $q_2 = 0$

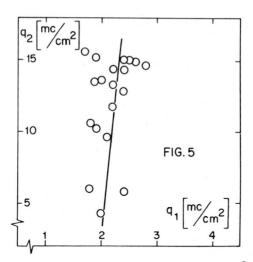

Figure 5- q_2 vs q_1 for the case of i_a=100 $\mu A/cm^2$ and i_c=47 $\mu A/cm^2$.

gives $q_1 > 0$, confirms the existence of the reduction-dissolution
mechanism which at the same time makes impossible to determine if
at lower thickness there is also original Sn(II) on the film. This
is because the actual technique does not permit to establish the
difference between q_{or}^{2+}, $q_{or,d}^{2+}$ and $q_{f,d}^{2+}$. This is also the reason
for which it is very difficult to calculate the relation Sn(IV)/
Sn(II) in the original film as a whole.

As a first aproximation (due to the fact that the O^{2-} concentration will be higher near the surface of the film) we could
consider the film composed by two layers, one external on the base
of Sn(IV) only and the other internal, a mixture of Sn(IV) and
Sn(II) oxides. Due to the fact that for the lower currents
(i_a=i_c≤60 $\mu A/cm^2$) ($\partial q_2/\partial q_1$) is of the order of 2/1, then the relation for Sn(IV)/Sn(II), for the inner film, could be considered in
this approximation near to 1/1.

Nevertheless this model must be considered a first approximation because it is highly probable that there is a continuous change of Sn(IV) and Sn(II) concentrations inside the film thickness.

In the other side, as could be seen from the comparison of
Fig. 3 and 5, the increase of the formation current produces the
increase of the Sn(II) content (under the supposition that for the
two figures losses by the reduction-dissolution mechanisms are
practically the same). This fact could also be deduced from the

724

comparison between Figs. 3 and 4, but in this case, as it was always observed, the increase is affected by an increase of the losses produced by the reduction-dissolution mechanism. Neverthless $(\partial q_2/\partial q_1)$ increases with the increase of the currents $i_a = i_c$.

Acknowledgements: The authors are gratefull to FAPESP for the financial support.

REFERENCES

01. B.N. Stirrup, N.A. Hampson. Surf. Technology 5 (1977) 429.
02. M. Pugh, D.R. Gabe, L.M. Warner. Corr. Science 7 (1967) 807.
03. N.H. Hampson, N.E. Spencer. British. Corr. J. 3 (1968) 1.
04. C.L. Lau, G.K. Wertheim. J. Vac. Sci. Technol. 15(2) (1978)622.
05. R.A. Powell, W.E. Spicer. Surf. Scien. 55 (1976) 681.
06. D.R. Gabe. Surf. Technology 5 (1977) 463.
07. B.N. Stirrup, N.H. Hampson. J. Electroanal. Chem 67 (1976) 45,57.
08. M. Pourbaix "Atlas of Electroch. Eq. in Aqueous Solutions" Pergamon Press, Oxford (1966).
09. W. Mach. A.M. Azzam, G.M. Habasch. Metallöberflache 9 (1955) 58A
10. E. Sato. Boschoku Gijtsu 17 (1968) 449.
11. C.V. D'Alkaine, S. Estevez "Anales 3º Congresso Latinoamericano de Eletroquímica y Corrosión". p. 403 INIFTA. La Plata. Argentina (1977).

COMPORTEMENT POTENTIODYNAMIQUE D'UNE ELECTRODE D'IRIDIUM ACTIVEE DANS L'ACRYLA-
MIDE ET H_2SO_4.

T.FERNANDEZ OTERO ET M.S.LARROCHA REDONDO

Facultad de Ciencias Químicas. Universidad del País Vasco. Apdo. 1072 San Sebastián. Espagne

RESUME

En milieu acide les oxydes formés aux potentiels plus anodiques que 850 mv (ECS) sur l'électrode d'iridium préalablement activée, catalysent l'oxydation de l'acrylamide (AA). Cette oxydation conduit à la passivation de l'électrode. Sur l'électrode passivée augmente considérablement la surtensión de la réaction du dégagement de l'oxygéne.

On a étudié par la méthode de la voltammétrie cyclique l'influence sur le processus de passivation de la concentration du monomère de la concentration de l'acide, de la concentration de l'acide à force ionique constante et de la temperature.

Pour expliquer le phénoméne de passivation nous proposons la formation sur la surface de l'électrode d'une mince couche de polymère.

INTRODUCTION

En milieu acide, non corrosif, en appliquant à l'électrode de Ir un traitement d'activation superficielle par balayages successifs de potentiels effectués dans certaines conditions expérimentales (1-3) on fait croitre sur l'électrode des couches d'oxyde de considérable épaisseur et de comportement rédox très reversible. Sur les voltamogrammes de l'électrode de Ir activée on détecte dans le domaine des potentiels anodiques la présence de quatre maximum de courant d'oxydation qui correspondent à la formation de quatre composés oxydés d'iridium ayant un état de valence différent. Le comportement électrochimique de la couche d'oxyde superficielle par rapport a la réaction d'oxydation de certains composés organiques simples, varie selon le domaine des potentiels anodiques (4).

Au cours de la réaction d'oxydation des composés organiques (monomère) apparaissent des radicaux et parfois des radicaux-ioniques qui peuvent déclancher un processus de polymérisation (5). La polymérisation peut se produire soit sur la surface même de l'électrode, en recouvrant celle-ci, (6), soit au sein de la solution électrolytique (7).

Dans le present étude nous essayerons de produire la polymérisation du monomère sur la surface d'une électrode de Ir préalablement activée et obtenir de ce fait la passivation de celle-ci. La couche de polymère déposée sur l'électrode serait plus au moins compacte selon les conditions expérimentales employées, à savoir: concentration du monomère dans la solution, concentration et composition de l'électrolyte et variables électriques.

PARTIE EXPERIMENTALE

Le dispositif expérimental utilise un potentiostat "Wenking ST 72", un générateur des signaux triangulaires "Wenking VSG 72" et un enrégistreur X-Y JJ Lloid Instruments PL4". Les réactifs employés sont Merk A.G.. Les solutions sont préparées avec de l'eau bidistillée.

L'électrode de travail est constituée par une lame de Ir "Johson-Matthey"dont la surface est de 0,5 cm^2. L'électrode auxiliaireest une lame de platine. L'électrode au calomel saturé est employée comme réference et tous les valeurs de potentiels sont exprimées par rapport à cette électrode.

L'électrode de Ir a été activée par des cycles succéssifs de potentiels effectués entre -230/1230 mV, ν = 5V.s^{-1}, dans H_2SO_4 1 N. L'électrode ainsi activée est transfèrée dans H_2SO_4 1 N plus acrylamide et on observe la rèponse de l'électrode aux balayages successifs de potentiels realisès dans le domaine des potentiels compris entre 100 et 1230 mV (ν = 30 mV-s^{-1}) et entre 100 et 2.600 mV (ν = 10mV. .s^{-1}).

Toutes les mesures ont été effectuées avec l'électrode ayant un degré d'activation semblable.

Entre deux mesures successives l'électrode est nettoyée en dissolvant à l'aide d'un courant alternatif la couche superficielle dans H_2SO_4 concentré et chaud.

RESULTATS ET DISCUSSION

Sur la fig. 1 est représentée une série de courbes correspondantes aux balayages de potentiels consécutifs effectués entre 100 et 1230 mV dans $H_2SO_4$1,5. 10^{-2}M + Na_2SO_4 1N et AA 0,7M. Sur cette même figure est enregistrée la courbe obtenue en milieu $H_2SO_4$1N après le traitement d'activation.

On peut remarquer les faits suivants:

- Le monomère est fortement oxydé aux potentiels supérieurs à 850 mV (voir la première courbe en présence d'AA). Les réactions de dégagement d'oxygène et de la formation de la couche d'oxyde sont partiellement inhibèes par la présence du monomère.

Dans le demi-cycle cathodique on observe la diminution d'abord et la disparition totale à la fin de l'expérience du courant de réduction de la couche d'oxyde.

- En même temps on observe, en fonction du nombre des cycles appliqués, une augmentation de la surtension pour les réactions d'oxydation anodique et de réduction cathodique de l'électrode qui se traduit par un déplacement des potentiels des pics de courant. L'effet peut être dû a l'apparition d'une résistance superficielle.

i. cathod.   i(mA.10)   i. anod.

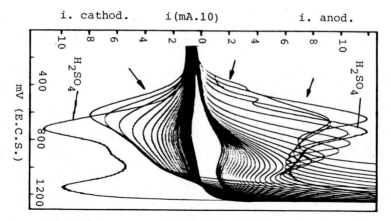

Fig. 1.- Voltamogram de réference dans $H_2SO_4$1N pour une électrode activée et voltagrames successifs dans H_2SO_4 1,5 x 10^{-2}N plus Na_2SO_4 1N et AA 0,7M (+100/ +1230 mV, ν = 30 mV.s^{-1}, T = 30ºC.)

Les produits d'oxydation du monomère restent fortement adsorbés à la surface de l'électrode. Celle-ci ne reprend pas son état superficiel initial si elle est soumise a continuation à des balayages multiples dans $H_2SO_4$1N en absence du monomère.

Sur la fig. 2 se trouvent représentées une partie des courbes de balayage correspondant au domaine des potentiels très anodiques.Les courbes on été obtenues en appliquant à l'électrode des cycles de potentiel entre 100 et 2.600 mV en faisant varier: la concentration de l'acide (courbe 2a), la concentration du monomère (courbe 2b), la concentration de l'acide à force ionique constante (courbe 2c), la température (courbe 2d).

On constate l'apparition sur les courbes d'un maximun de courant aux valeurs de potentiel très positifs,situé environ à 1340 mV.De plus, le dégagement de l'oxygène est fortement inhibé; de fait cette réaction se voit déplacée vers le potentiel d'environ 2.200 mV.

Dans le premier cycle de balayage lorsque on dépasse le potentiel de 1.600mV (minimum de courant sur les courbes) l'électrode subit une passivation totale.En effet, on peut apprécier sur les courbes de la fig. 2 la disparition totale du courant de réduction de la couche d'oxyde dans le cycle anodique enregistré en continu (courbes marquées d'un astérisque).

L'apparition du maximum de courant vers 1340 mV implique une réaction d'oxydation. Cette réaction qui s'étend dans la zone des potentiels où a lieu normalement le dégagement de l'oxygène est probablement associée a une deuxième étape d'oxydation du monomère dans lequel participent activement les oxydes de valence supérieur de l'iridium et même l'oxygène "naissant".

Le maximum à 1340 mV est donc le résultat de la superposition de deux réac-

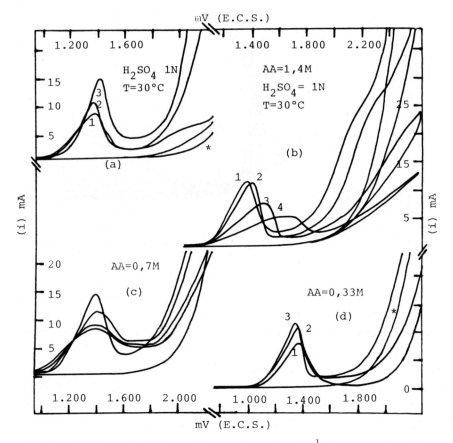

Fig. 2 .- Voltagrames (+100/2600 mV , ν = 10 mV.s^{-1}, T= 30ºC) (a) influence de la conc. d'AA (1), 2,8M ; (2) 0,35 M ; (3) 4,4 x 10^{-2}M - (b) influence de la conc. de H_2SO_4: (1) 1N, (2) 0,25 N, (3) 6,2 x 10^{-2}N, (4) 1,5 x 10^{-2}N - (c) Force ionique constante avec les memes conc. de H_2SO_4 - (d) Temperatures : (1) 30ºC, (2) 40ºC, (3) 50ºC.

tions: décharge de l'eau et oxydation du monomère, ce qui explique l'augmentation observée de l'intensité du maximum et la diminution de la surtension du dégagement de l'oxygène pour des faibles concentrations du monomère dans la solution (courbes a fig.2).

L'influence de la concentration de H_2SO_4 sur le maximum (courbes b fig.2).est probablement associé aux effets de la conductivité électrolytique. Par contre, l'influence de la variation de la concentration d'acide dans la solution à force ionique constante laisse supposer l'effet du pH sur la cinétique des réactions (courbes c fig.2).La variation avec la température est celle observée habituelle ment pour d'autres processus électrochimiques.

Il apparait d'après les résultats obtenus que dans le processus d'oxydation anodique d'AA participent activement les oxydes superficiels formés sur l'électrode par le traitement d'activation. En effet, le maximum de courant situé à 1340 mV et par la suite duquel survient la passivation totale de l'électrode n'apparait pas si les mesures sont réalisées avec une électrode polie et desactivée.

De l'ensemble de ces résultats on déduit que la passivation est due à la formation à la surface de l'électrode d'une couche couvrante de polymère. Cette couche solide commence à se former dès le début de la réaction d'oxydation du monomère, donc vers 850 mV (voir fig. 1), mais ne recouvre pas dans ces conditions de polarisation toute la surface.

Le mécanisme reactionel susceptible d'interpréter la passivation de l'électrode est difficile à établir étant donné la complexité des phenomènes qui interviennent à l'électrode: oxydation de l'électrode, oxydation du monomère par un mécanisme probablement à la fois chimique et électrochimique, formation de différents espèces radicalaires.

On peut supposer que le processus de polymérisation est déclanché par les espèces radicalaires et ioniques formées au cours d'une première ètape d'oxydation du monomère et que restent adsorbées sur l'électrode. Les molécules du monomère qui arrivent du sein de la solution à la surface, réagissent avec les espèces radicalaires adsorbées et propagent la chaîne polymérique vers la solution. Le processus de polymérisation s'achève par suite de l'interaction entre deux extrêmes de la chaîne polymérique, ou bien à cause d'une réaction chimique entre les espèces adsorbées et celles présentes dans la solution.

CONCLUSION

Les oxydes supérficiels présents sur une électrode de Ir activée aux potentiels supèrieurs a 850 mV, en milieu acide catalysent l'oxydation de l'AA.

- L'oxydation du monomère a lieu simultanement avec la décharge de l'eau. Jusqu'à une concentration de 10^{-2}M AA l'oxydation du monomère est prépondérante et inhibe celle de l'eau.

- L'oxydation du monomère dans les conditions d'une forte polarisation conduit à la passivation de l'électrode en le recouvrant d'une couche polymérique.

- Cependant, la couche de passivation n'est pas totalement isolante et la décharge de l'eau se poursuit, avec une surtension de l'ordre d'environ 1.000 mV.

BIBLIOGRAPHIE

(1) L.D.Burke et E.J.M. O'Sullivan.J.Electroanal.Chem.117,155,(1981)
(2) M.S.Cruz et T.F.Otero. An.Quim.73, 163, (1977)
(3) J.Mozota et B.E.Conway. Electrochim. Acta 28, 1 (1982)
(4) M.S.U.Zañartu, M.S.Cruz et T.F.Otero. An.Quim.79,57, (1983)
(5) G.Mengoli et G.Vidotto. J.Electroanal.Chem. 75,595, (1977)
(6) B.M.Tindiswell et D.A.Mortimer. Eur.Polym. 17, 735, (1981)
(7) G.S.Shapoval. J.Macromol. Sci.Chem. A17, 453 (1982)

Passivity of Metals and Semiconductors, edited by M. Froment
Elsevier Science Publishers B.V., Amsterdam — Printed in The Netherlands

REPASSIVATION OF TITANIUM AND TITANIUM ALLOYS DEPENDENT ON POTENTIAL AND pH

H.-J. RÄTZER-SCHEIBE

DFVLR, Institut für Werkstoff-Forschung, D 5000 Köln 90 (F.R.G.)

ABSTRACT

A theoretical model was developed to describe the repassivation of titanium and titanium alloys in aqueous solutions. To investigate repassivation, under potentiostatic conditions, anodic current transients are determined on a fresh titanium surface generated by a planing technique in electrolyte. The initial phase of current-time curves can be explained by the formation of an oxide monolayer growing two-dimensionally over the clean surface. When the surface is completely passivated by an oxide monolayer, in about 20 - 50 msec, thickening can occur due to anodic oxidation. The linear decay of the log(current) vs. log(time) plots can be explained by assuming the high-field conduction mechanism for anodic oxide growth. Variations in potential and pH yield a parallel shift of the linear log(current)-log(time) curves in the range of oxide film thickening. These effects of potential and pH can be described by the theoretical model developed.

INTRODUCTION

The experimental method to study the repassivation behaviour of newly generated surfaces has been described in detail previously (ref.1, ref.2). A rectangular area of 1.2 cm^2 of the specimen is generated by mechanically planing with a microtom directly in the electrolyte. The potentiostatic measurement of the current occurs immediately after generating an oxide free metallic surface. The current-time data are recorded and stored by a PDP 11 computer system from Digital Equipment. Then the current-time curves can be plotted on a plotter using a double-logarithmic scale.

Fig. 1 shows typical current-time data for an anodically polarized titanium electrode newly generated in a neutral electrolyte at a potential of +200 mV, in comparison with theoretical current functions described in more detail in previous publications (ref.3,ref.4). The total current I_{theor} is the algebraic sum of the partial currents I_d, I_{ox}, I_{th} and $\theta \cdot I_{th}$ respectively.

As a result of the metallic dissolution of the titanium surface a high anodic current, I_d, is observed at the beginning of the process. It is assumed that during the initial phase an oxide monolayer spreads over the surface of the electrode. The two-dimensional film-growth starts from random nuclei. This results in a current called I_{ox} which is plotted as a function of time, t, in the Fig. 1. The oxide-covered sites prevent metallic dissolution and accordingly

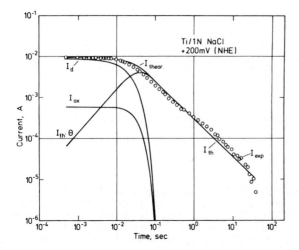

Fig. 1. Comparison of experimental data (I_{exp}) with theoretical current functions (I_d: dissolution current, I_{ox}: current for growing of an oxide monolayer, I_{th}: current for oxide film thickening, I_{theor}: total current).

the anodic dissolution current, I_d, decreases. When the oxide patches impinge and the surface is completely passivated, thickening can occur due to anodic oxidation and the current is given by I_{th}. A complete derivation of the theoretical mechanism for the thickness growth is given in the next chapter. By multiplying I_{th} and the coverage, θ, the partial thickening can be estimated when the surface is not yet completely covered by an oxide monolayer during the initial phase.

MODEL FOR OXIDE FILM THICKENING

Rate laws

Assuming ionic conduction to be the only electrode process between the charge, Q_f, being used for passive film formation per unit of area and the film thickness, D, the following relationship exists:

$$D = A \cdot Q_f \tag{1}$$

In Eq. (1) $A = \dfrac{M}{z' \cdot F \cdot \rho}$ where M = molecular weight assumed for oxide film, z' = valency of oxygen ion, F = Faraday's constant, and ρ = film density. According to Eq.(1) the thickening of an anodic passive film is proportional to the charge which must be expended for it.

If Eq.(1) is differentiated it follows:

$$\frac{dD}{dt} = A \cdot i_{th} = A \cdot (i_{anod} - i_d) \tag{2}$$

where i_{th} = current density for film thickening, i_{anod} = total anodic current density, and i_d = dissolution current density.

There are some rate laws (see column 1 in the Table) describing the increase of thickness as a function of time. The theoretical relationship is given in column 2 of the Table. The following column contains the corresponding current density-time laws derived by differentiation of charge with respect to time. Finally, the last column gives the slope of the line obtained if the current density vs. time relationships are plotted in a double-logarithmic scale.

TABLE

Parabolic, cubic, logarithmic and inverse logarithmic rate laws for film formation as a function of time. The derivative of Q_f (column 2), with respect to time, gives the anodic current density (i_{th}) (column 3) for the first three laws. The exponent of t in the i_{th} vs. t relationship (column 3) gives the slope in the log i_{th} - log t plot (column 4).

Rate law	Charge for oxide film formation (Q_f) as a function of time $Q_f = f(t)$	Current density for film thickening (i_{th}) as a function of time $i_{th} = \dot{f}(t)$	Slope of the log i_{th}- log t plot
parabolic	$Q_f = K_p \cdot t^{1/2}$	$i_{th} = \frac{1}{2} K_p \cdot t^{-1/2}$	$-\frac{1}{2}$
cubic	$Q_f = K_c \cdot t^{1/3}$	$i_{th} = \frac{1}{3} K_c \cdot t^{-2/3}$	$-\frac{2}{3}$
logarithmic	$Q_f = K_l \cdot \ln t + K_l'$	$i_{th} = K_l \cdot t^{-1}$	$- 1$
inverse logarithmic	$\frac{D_1}{D} = const. - \ln t$	$t = K_{il} \left[\frac{1}{\frac{i_{th}}{i_o} \cdot \ln \frac{i_{th}}{i_o}} - E_1 \left(\ln \frac{i_{th}}{i_o} \right) \right]$	$\approx -0.92 \left(\begin{smallmatrix} for \\ TiO_2 \end{smallmatrix} \right)$

In Fig. 2, log i_{th} - log t plots of the different rate laws are compared schematically. The hatched angled space marks the range of experimentally observed slopes between -0.76 and -0.95 for Ti and Ti-alloys. From the diagram in Fig. 2, it can be concluded that the slopes of parabolic, cubic and logarithmic rate laws, respectively, differ too widely to describe experimental results.

In the case of the inverse logarithmic rate law there is a difficulty inasmuch as the corresponding current density - time equation cannot be derived by a normal differentiation with respect to time. The equation given in column 3

734

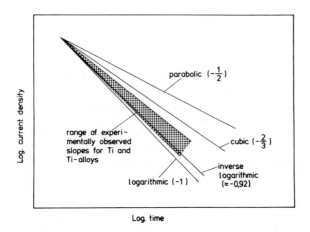

Fig. 2. Log i_{th} - log t plots of some rate laws compared with experimentally observed slopes of anodic current density transients for newly generated Ti and
Ti-alloys.

of the Table is the result of a longer derivation which will be shown later in
more detail. The slope of -0.92 follows from a calculation which used values of
the parameters for anodic formation of a TiO_2 film on titanium.

High-field conduction mechanism

Usually the oxide film thickening of valve metals obeys the high-field conduction mechanism during the first phase of anodic oxidation. Current density of
this phase, i_{th}, can be described by the following well-known equation:

$$i_{th} = n \cdot z \cdot e \cdot \nu \cdot \exp \left(- \frac{W}{RT} + \frac{zaFV}{RTD}\right) \tag{3}$$

in which n = surface density of the mobile ions at the metal surface, z·e =
charge on the ion, ν = frequency factor, W = height of the activation energy
barrier, R = gas constant, T = temperature, a = half-jump distance of the metal-
oxide interface, V = potential drop across the film, and D = film thickness.

The Eq.(3) explains the empirical relation

$$i_{th} = i_o \cdot \exp \left(B \cdot \frac{V}{D}\right) \tag{4}$$

which was demonstrated by Günterschulze and Betz (ref.5) for the anodization of
aluminium. Eq.(3) may be compared with the empirical Eq.(4) and the nature of

empirical constants in Eq.(4) may be represented as i_o = nzev·exp (-W/RT) and B = zaF/RT. B is called the inverse field-coefficient.

Eq.(3) has been solved by Cabrera and Mott (ref.6). For $D \ll D_1$ the approximate solution may be written as

$$\frac{D_1}{D} = const. - lnt \qquad (5)$$

where D_1 = zaFV/RT. D_1 being of the order of 10^{-6} to 10^{-5} cm. According to Eq.(5) the thickness D is inversely proportional to ln t ; this is, of course, the inverse logarithmic law for the oxide growth.

As mentioned above, the current density vs. time relationship cannot be simply derived by temporal differentiation in the case of the inverse logarithmic law, as it was in the other three rate laws (see Table). Beck (refs.7,8) has developed a mathematical method which yields a relationship between current density and time, as is necessary for a comparison with the measured current-time curves. By definition of a dimensionless current parameter,

$$I = \frac{i_{th}}{i_o} , \qquad (6)$$

as well as a dimensionless time parameter,

$$T = \frac{A \cdot i_o}{B \cdot V} \cdot t , \qquad (7)$$

Eq.(2) and Eq.(4) reduce to

$$dT = \frac{dD}{B \cdot V \cdot I} \qquad (8)$$

and

$$lnI = B \cdot \frac{V}{D} \qquad (9)$$

respectively.

If in Eq.(9), lnI is differentiated with respect to D, it follows that

$$\frac{dI}{I \cdot lnI} = - \frac{dD}{D} \qquad (10)$$

In Eq.(10), D and dD can be eliminated by substitution from (9) and (8), respectively. This results in:

$$dT = - \frac{dI}{I^2 (lnI)^2} \qquad (11)$$

Eq.(11) can be solved by integration and with some mathematical manipulations. Integrals and recurrence relations were used from the Handbooks of Bronstein, Semendjajew (ref.9) and of Abramowitz, Stegun (ref.10). One obtains

$$T = \frac{1}{I \cdot \ln I} - \gamma - E_1 (\ln I) + C_{int.} \tag{12}$$

in which γ is Euler's constant, 0.577.... E_1 is called the Exponential Integral and can be commonly written as

$$E_1 (x) = \int_x^\infty \frac{e^{-t}}{t} dt \tag{13}$$

Substituting of the dimensionless parameters in Eq.(12) from Eqs.(6) and (7) yields the following dimensional solution:

$$t = \frac{B \cdot V}{A \cdot i_0} \left[\frac{1}{\frac{i_{th}}{i_0} \ln \frac{i_{th}}{i_0}} - E_1 (\ln \frac{i_{th}}{i_0}) \right] \tag{14}$$

In Eq.(14) $C_{int.}$ was estimated to be approximately γ.

Titanium oxide film

In the case of an anodically formed TiO_2-film the following values from Laser et al. (ref.11) were used in the constants in Eq.(14): $A = 6.09 \cdot 10^{-5}$ cm^3/Coul, $B = 6.8 \cdot 10^{-6}$ cm/V and $i_0 = 2.53 \cdot 10^{-15}$ A/cm^2. Moreover, for a first calculation, the value of the potential drop, V, is assumed to be 1 volt. Later it will be shown that the assumption of a potential drop of this magnitude is justified in case of a TiO_2-film. Using this parameter in Eq.(14), the current density for film thickening, i_{th}, can be numerically computed as a function of time. The double-logarithmic plot of this calculation yields a line with a slope of -0.92 showing the best fit to the range of experimentally observed slopes for Ti and Ti-alloys (see Fig. 2). This result proves the assumption that the anodic thickness growth of the passive film on Ti and Ti-alloys can occur by an inverse logarithmic rate law.

In Eq.(14) the slope of log i_{th} vs. log t curves is determined only by i_0, while the parameters B, V and A merely cause a parallel shift of the lines. Therewith, in contrast to the parabolic, cubic and logarithmic rate laws whose slopes of log i_{th} - log t plots are independent of material (see Table), the corresponding slope of the inverse logarithmic rate law depends on a parameter (i_0) of the oxide forming passive film.

To derive a relationship between the potential drop across the oxide film, V, and the electrode potential, ε_{NHE}, a separate derivation must be done. According to the work of Vetter (ref.12), the following equation may be written:

$$V = \varepsilon_{NHE} - \varepsilon_F \qquad (15)$$

where ε_{NHE} = electrode potential referred to the normal hydrogen electrode and ε_F = flade potential (above this potential a passive film is formed, below this the film can be dissolved). From a thermodynamic point of view ε_F corresponds to the reversible potential of an oxide electrode. In general the following equation exists for the pH dependence of ε_F:

$$\varepsilon_F = E^O - 0.0591 \cdot pH \qquad (16)$$

where E^O = standard reversible potential.

Substituting Eq.(16) into Eq.(15) gives the potential drop dependent on potential and pH.

$$V = \varepsilon_{NHE} - E^O + 0.0591 \cdot pH \qquad (17)$$

For the anodic oxidation of the titanium electrode the following reaction can be assumed:

$$Ti + 2 H_2O = TiO_2 + 4H^+ + 4e^- \qquad (18)$$

Using data of Pourbaix (ref.13) the value of E^O for Eq.(18) is -0.899 volts. For the example of a Ti/TiO_2-electrode in 1N NaCl (pH = 5.9) at a potential of +200 mV (NHE), the potential drop is 1.45 volts as calculated from Eq.(17). Using this value of V, Eq.(14) can be numerically calculated. To compare the theoretical results with the experimental current-time data (see Fig. 1) the current density for the oxide thickening, i_{th}, has to be converted into the corresponding current, I_{th}, considering the area of electrode surface of 1.2 cm^2. No correction was made for surface roughness. Fig. 1 shows the linear log I_{th} vs. log t plot computed using Eq.(14) which describes the experimental data, I_{exp}, quite well in the range of thickness growth of the passive film.

THE EFFECT OF POTENTIAL AND pH

The effect of the applied anodic potential on Ti6Al4V in 1N NaCl is shown in Fig. 3. If the potential is raised the current increases due to higher metallic dissolution during the initial phase of the process. The experimental results and the theoretical description of this initial phase of repassivation were previously discussed (refs.3,4).

Now, the Eq.(14) derived in this work can also describe the range of linear decay of current-time data in the double-logarithmic plot for potentials more noble than -400 mV(NHE). The slight deviation of the experimental current from the theoretical straight line, at times greater than 2 sec, can be explained by film dissolution yielding an additional current as considered in Eq.(2). The

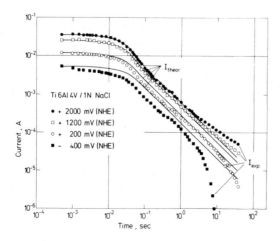

Fig. 3. Effect of electrode potential on current-time curves for fresh surfaces of Ti6Al4V in 1N NaCl.

cathodic deviation of the measured current-time values from the linear I_{theor}-t plot at current values lower than 50 µamps is due to cathodic reactions, such as oxygen reduction or hydrogen evolution, which can no longer be neglected. At the more negative potential of -400 mV(NHE) the theoretical and experimental curves differ widely during the whole period because a high cathodic current neglected in the derived theory is superimposed the anodic current. After about 10 sec, the total current becomes cathodic.

Fig. 4 shows anodic current transients for Ti6Al4V at +200 mV(NHE) in neutral 1N NaCl (pH = 5.9) and in a more acid sodium chloride solution (pH = 1.7). In the range between 50 msec and 1 sec, where the oxide film thickening starts, the effect of pH on experimental data is relatively small.

To explain the effect of pH theoretically Eq.(17) is used. Because of the factor 0.0591 in Eq.(17) variation of pH results in a relative small change of potential drop. In particular for the selected example shown in Fig. 4, the potential drop decreases at about 250 mV if the pH is varied from 5.9 to 1.7. Provided that the constants of the titanium oxide film (A, B and i_o) are independent of pH the current can be computed from Eq.(14) as a function of time in the range of film thickening. The difference of the linear part of these theoretical curves plotted in Fig. 4 is relatively small corresponding to the small variation of potential drop which follows from change of pH accordingly Eq.(17). From these considerations it can be seen that the differences in the experimental curves due to pH are in the order of magnitude as the values obtained from the theory.

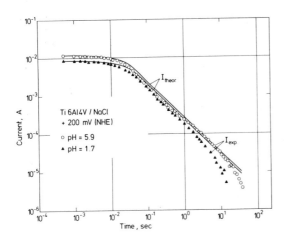

Fig. 4. Effect of pH on current-time curves for fresh surfaces of Ti6Al4V in solutions of NaCl at a potential of +200 mV(NHE).

ACKNOWLEDGEMENT

The author would like to thank U. Fuchs for his technical assistance at experiments. Appreciation is expressed to Dipl.-Math. R. Page from DFVLR-Rechen-institut who advised me at writing the computer program.

REFERENCES

1 H.-J. Rätzer-Scheibe and H. Buhl, Werkstoffe und Korrosion, 25 (1974) 889-892.
2 H.-J. Rätzer-Scheibe, Corr., 34 (1978) 437-442.
3 H.-J. Rätzer-Scheibe and H. Buhl, Werkstoffe und Korrosion, 30 (1979) 846-853.
4 H.-J. Rätzer-Scheibe, Proc. 8th ICMC, Mainz, F.R.G., 1981, Vol. I, 212 pp.
5 A. Günterschulze and H. Betz, Z. Physik, 92 (1934) 367 pp.
6 N. Cabrera and N.F. Mott, Rep. Progr. Physics, 12 (1949) 163-184.
7 T.R. Beck, J. Electrochem. Soc., 115 (1968) 890-896.
8 T.R. Beck, Reactions and Electrochemical Kinetics of Newly-Generated Metal Surfaces, Final Scientific Report, Electrochemical Technology Corporation, Seattle, 1976.
9 I.N. Bronstein and K.A. Semendjajew, Taschenbuch der Mathematik, Verlag Harri Deutsch, Frankfurt/M. und Zürich, 1964.
10 M. Abramowitz and I.A. Stegun, Handbook of Mathematical Functions, Dover Publications, New York, 1972.
11 D. Laser, M. Yaniv and S. Gottesfeld: J. Electrochem. Soc., 125 (1978) 358-365.
12 K.J. Vetter, Elektrochemische Kinetik, Springer-Verlag, Berlin-Göttingen-Heidelberg, 1961, 613 pp.
13 M. Pourbaix, Atlas of Electrochemical Equilibria in Aqueous Solutions, Pergamon Press, New York, 1976, 213 pp.

CRYSTALLINE ALUMINUM OXIDE FILMS

R. S. ALWITT and H. TAKEI

United Chemi-Con, Inc., 3000 Dundee Rd., Northbrook, IL 60062 (USA)

ABSTRACT

Thermal oxidation of Al at 525°C for short times develops an amorphous oxide with no detectable crystallites. Subsequent anodic oxidation in citrate electrolyte at 70°C produces a barrier oxide containing γ' alumina. At low voltages the film is amorphous + crystalline but at higher voltages the amorphous oxide transforms to γ' alumina so that the film is essentially a uniform crystalline layer. The growth and properties of this layer are described.

INTRODUCTION

Crystalline oxide is frequently a component of oxide films on aluminum. Thermal oxidation above 450°C for suitable times produces γ alumina crystals at the metal interface under a continuous layer of amorphous oxide (ref.1). Amorphous anodic barrier oxide may have crystalline islands of γ' alumina in the vicinity of flaws (ref.2). Anodization in borate solution after thermal oxidation at 550°C results in a barrier layer of crystalline oxide sandwiched between amorphous oxide (ref.3). If Al is reacted with boiling water to deposit a thick hydrous oxide, subsequent anodization at 90°-95°C creates a crystalline barrier layer next to the metal (ref.4) that has been identified as γ (ref.5) and as γ' (ref.6) alumina. (These oxides have a similar spinel structure, but γ' has more disorder in the cation distribution over the lattice sites (ref.7)). We report here on the properties and growth of barrier oxide films that consist essentially only of a layer of γ' alumina.

METHODS

Smooth 4N Al foil was chempolished in $HNO_3 + H_3PO_4$ and then cleaned in NaOH (ref.4). Etched 4N capacitor foil (20X surface area increase (ref.4)) was cleaned in 7% HNO_3 at 60°C for 3 min. This foil was used to get more sensitivity in weight measurements and easier handling of flakes of isolated oxide. Etched foil results are reported on a specific area basis, i.e., nominal value divided by 20.

Specimens were heated in air for 20 min. at 525°C and then anodized in 5 g/l NH_4 citrate at 70°C, 0.5 mA/cm^2. When the desired voltage (V_f) was reached, V_f was held constant for 5-7 min. (etched) or 10 min. (smooth).

Crystalline films have an electrical instability that is removed by a relaxation and reformation (ref.4). Etched specimens were relaxed for 2 min. on open circuit in the citrate bath and then reformed at 0.5 mA/cm^2 to V_f; after 1-2 minutes at V_f the cd had dropped to 25 µA/cm^2 and formation was stopped. Smooth specimens were relaxed by heating in air for 2 min. at 550°C followed by reformation. In both cases the reformation charge was 5-10% of the initial anodization charge. The electrical conductivity and dielectric properties are stabilized by these steps.

Films on smooth foils were stripped in Hg Cl$_2$ solution and examined in the TEM. Thermal films were reinforced with carbon before separation. Cross-sections were prepared with an ultramicrotone and the oxygen content was measured by nuclear microanalysis. Etched specimens were used to study several film properties vs V_f. In some cases the film was collected after dissolving the Al in the Br$_2$ -methanol. With other specimens the oxide dissolution rate was measured in 5% H$_3$PO$_4$+2%CrO$_3$ stripping solution (ref.5).

RESULTS

TEM examination of thermal film gave no evidence of crystallites, which could have been detected if 3-5 nm or larger. Low beam intensities were used because of the fragile nature of the film so this observation cannot be taken as certain verification of the absence of crystallites. The film weighed 2.5 µg/cm^2.

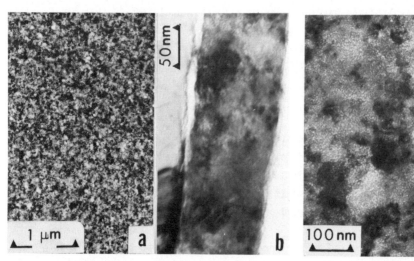

Fig. 1. TEM appearance of 140V crystalline film (a) Plan view (b) Cross-section

Fig. 2. Voids (white spots) in 140V film, 1.5 µm underfocus

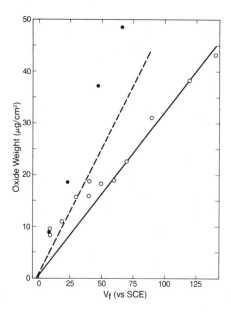

Fig. 3. Oxide weight vs. V_f. Specimens heated (o) and not heated (●).

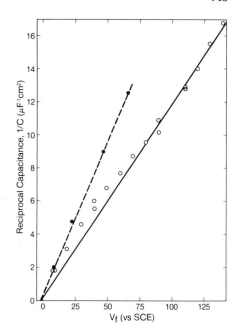

Fig. 4. Reciprocal capacitance (120Hz)
vs V_f. Same symbols as Fig. 3.

Without heat treatment the citrate anodization produces an amorphous oxide
layer. With thermal oxide present, anodization above 60V produces a layer of
γ' Al_2O_3. Fig. 1 shows a 140V film. It consists of fine crystallites and the
structure appears uniform in both lateral and normal directions. The film is
126nm thick, equivalent to 0.9 nm/V inverse field strength at 70°C. Nuclear
microanalysis gave a total oxygen coverage of 8.05x 10^{17} atoms.cm^{-2} (ref.8), so
the film density is 3.61 g/cm^3. The film contains a high density of discrete,
microscopic voids that can be revealed by the phase contrast caused by de-
focusing the TEM (ref.9). Fig. 2 shows the film at 1.5 μm underfocus. The
voids are several nm in diameter and at a density of 10^{11}-$10^{12}cm^{-2}$. The relaxation and reformation have no effect on the appearance of the void distribution,
nor any other structural features.

Figs. 3 and 4 show the film weight and capacitance (120Hz) vs V_f for etched
specimens. The solid line in Fig. 3 is drawn to intersect the abcissa at E°
for Al/Al_2O_3 and with a slope calculated from the properties of the 140V
crystalline film. The solid line in Fig. 4 is a best fit through E° and the
higher voltage points; the slope corresponds to a dielectric constant of 8.6
for the crystalline oxide. The dashed lines are drawn for typical amorphous
barrier oxide properties of 3.2 g/cm^3, 1.4nm/V, and diel const = 8.4 (ref.10).

Without heat treatment the dielectric properties are those of amorphous barrier oxide but the film weight is greater. There may be an additional porous layer at the solution side, similar to the structure after borate anodization at elevated temperature (ref.11). The heat treatment has little effect on the weight and capacitance of 10V films but at higher V_f there is a shift towards the crystalline values. The transition from amorphous to crystalline structure is complete at 60-70V.

Without heat treatment the dissolution rate in H_3PO_4-CrO_3 solution was 0.4 $\mu g/cm^2 s$, typical of amorphous oxide (Fig. 5). After rapid removal of 1 $\mu g/cm^2$, the dissolution rate of 60 and 90V films with heat treatment was 5×10^{-3} $\mu g/cm^2 s$, 100 times slower than amorphous film and typical of γ-alumina (ref.5). The 10V heated film dissolved at half the rate of the unheated 10V film, and the 20V and 40V films dissolved progressively more slowly, but still much faster than for crystalline oxide. For each of these films the rate eventually dropped to a low value like the 60V and 90V films. With the 40V film this occurred beyond 300 s (not shown).

The metal was dissolved from these specimens in Br_2- methanol and in each case additional oxide flakes were collected. The weight of this "inner" oxide is shown in Fig. 6. TEM examination in dark field and selected area electron diffraction of the 20 and 40V "inner" oxides showed them to be substantially crystalline with the γ'-structure, (Fig. 7).

Films formed through 40V are a combination of amorphous and crystalline oxide. Above 60V the films are essentially all crystalline. Between 40 and 60V there must be a transformation of amorphous to γ' oxide.

Characteristics of the amorphous oxide before the transformation are of great interest. With less intense heat conditions the formation of crystalline

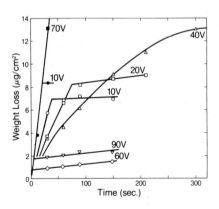

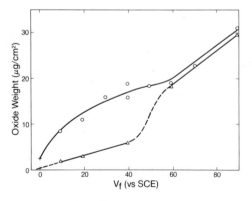

Fig. 5. Weight loss in H_3PO_4-CrO_3 at 85°C. Heated (open symbols) and not heated (filled symbols).

Fig. 6. "Inner" oxide weight ($\triangle$) and total oxide weight ($\circ$) vs. V_f: Total weight same as in Fig. 3.

oxide at low voltages can be reduced, so properties related to the amorphous film can be more clearly distinguished. After heating at 500°C/ 15 min. the thermal film appears free of crystallites by TEM. Thirty volt anodic film formed on this substrate appeared amorphous by TEM and SAED. An x-ray powder scan using point counting did detect a trace of γ' alumina, and the $2°$ peak width indicated a 5-8nm particle size (ref.12). This oxide film dissolved completely in the stripping solution, at a rate similar to that for the 40V film in Fig. 5. There was no detectable amount of inner oxide, yet with further growth such films fully transformed to γ' alumina.

DISCUSSION

The 60-140V films appear to be the first examples of anodic alumina films consisting soley of crystalline barrier oxide. This oxide has the same pro- perties as the barrier layer in composite hydrous + anodic oxide (Refs. 4, 9, 10) but its overall structure and growth process appear to be simpler and more accessible to analysis.

The inner oxide may grow either by transformation or by direct crystalline deposition, there is no specific evidence favoring one process. The decreasing dissolution rate with increasing film thickness (Fig. 5) suggests that the amorphous structure may become more ordered during growth. This occurs across the film and is not restricted to an interface. When the amorphous oxide becomes sufficiently ordered it appears to transform easily to γ' oxide, as occurred between 40 and 60V. This effect does not depend upon the presence of the inner oxide, evidenced by the results with the specimen heated 500°C/ 15 min.

Amorphous thermal oxide is thought to be identical with amorphous anodic oxide (ref.13) so it is not clear how it could cause the observed effect. The thermal film is presumably located in the interior of the anodic oxide at the boundary between oxide grown by anion transport and oxide grown by cation transport. Its only interaction with the bulk film is that the ion flux passes through this region before depositing at the interfaces, which seems insufficient to cause a structural change.

One may argue that some seed crystals are created during thermal oxidation that serve as centers for subsequent crystallization. The heat conditions used here appear to be at the threshold for γ alumina nucleation. Crystallite growth induction times of about 30 min. at 505°C and 8 min. at 530°C are reported (ref.14). According to a proposed rate equation (ref.1), heating at 525°C/ 20 min. deposits crystallites at the metal surface 7nm in diameter at a density of 3×10^9 cm^{-2}. Such a deposit would have been seen in the TEM. It may be significant that the x-ray technique, which detected crystallites

Fig. 7. Selected area electron
diffraction from inner oxide of
20V (left) and 40V (right) films.

that were not revealed by electron microscopy, requires a much larger sample
size. Perhaps the crystallites were
present at a much lower density or in a
non-uniform distribution. Even if the
presence of γ- alumina nuclei are ad-
mitted, it is not clear how they would
cause the transformation. During
thermal oxidation these crystals cause
no change in the amorphous oxide layer
and grow as an independent phase (ref.1).
Pehaps the inner oxide nucleates on
thermal oxide seed crystals.

We have observed that the amorphous transformation depends upon bath composition, e.g., the process is more sluggish in phosphate then in citrate.
It is known that anion impurities in amorphous alumina decrease the ease of
crystallization in an electron beam (ref.15). Perhaps the thermal oxide in
some way inhibits anion entry into the growing oxide.

It may be that the critical role of the thermal oxide will not be
explained by a single phenomenon but by some combination of events.

REFERENCES
1 A. F. Beck, M. A. Heine, E. J. Caule and M. J. Pryor, Corr.Sci. 7(1962)1-22.
2 K. Shimizu, S. Tajima, G. E. Thompson and G. C. Wood, Electrochim Acta, 25
   (1980) 1481-1486.
3 C. Crevecoeur and H. J. deWit, 27th I.S.E. Meeting, Zurich, 1976, Abstr. 132
4 R. S. Alwitt and C. K. Dyer, Electrochim. Acta, 23 (1978) 355-362.
5 R. S. Alwitt and W. J. Bernard, J. Electrochem, Soc., 121 (1974) 1019-1022.
6 N. F. Jackson and P.D.S. Waddell, J. Appl. Electrochem., 2 (1972) 345.
7 B. C. Lippens and J. J. Steggerda in B. G. Linsen (Ed.), Physical and
   Chemical Aspects of Adsorbents and Catalysts, Academic, New York, (1970)
   pp. 171-211.
8 J. Siejka, private communication.
9 R. S. Alwitt, C. K. Dyer and B. Noble, J. Electrochem Soc., 129 (1982)
   pp. 711-717.
10 C. K. Dyer and R. S. Alwitt, Electrochim. Acta, 23 (1978) 347-354.
11 G. A. Dorsey, J. Electrochem Soc., 116 (1969) 466-471.
12 H. Chen, private communication.
13 M. J. Dignam, J. Electrochem. Soc., 109 (1962) 183-191
14 M. J. Dignam, W. Fawcett and H. Bohni, ibid, 113 (1966) 656-662.
15 K. Shimizu, G. E. Thompson and G. C. Wood, Thin Solid Films, 77 (1981)
   pp. 313-318.

Passivity of Metals and Semiconductors, edited by M. Froment
Elsevier Science Publishers B.V., Amsterdam — Printed in The Netherlands

AUGER ELECTRON SPECTROMETRY OF POROUS LAYERS OF ANODIZED ALUMINIUM

H. TERRYN and J. VEREECKEN

Department of Metallurgy, Vrije Universiteit Brussel, Pleinlaan 2, 1050
Brussels (Belgium)

ABSTRACT

Depth composition profiles of porous anodized aluminium at different stages
of its growth have been obtained by Auger Electron Spectrometry (AES) in combination with ion etching. In order to study the growth of this layer, several
oxide layers were prepared galvanostatically at different times in sulphuric
and phosphoric acid. It has been found that the electrolyte is adsorbed and
incorporated into the whole oxide layer during its growth, and that there is an
important maximum of this amount incorporated during the transition from barrier to porous layer.

INTRODUCTION

The initial film growth and pore development sequences during the anodizing
of aluminium substrates at constant current density in phosphoric acid have
been observed by transmission electron microscopy of stripped anodic films and
ultramicrotomed sections (ref.1,2). Composition of the porous layer has been
determined using AES (ref.3-12), XPS (ref.13) and STEM (ref.14,15). AES (ref.
6) data showed that sputterprofiling has to be done carefully, to minimize electron and ion beam damage of structure and composition of the porous layer.
Electrolyte adsorption in the porous layer was found by Bador (ref.4) and Sun
(ref.9). Bador (ref.4) showed that electrolyte impurities are only located in
the outer layer adjacent to the vacuum-oxide interface. Thompson (ref.14) confirms the presence of phosphor containing and relatively pure alumina regions
within the relatively uniform barrier film formed in phosphate solution. Takahashi (ref.15) and Thompson (ref.13) show that the cell boundary band are composed of relatively pure alumina, whereas the cell material adjacant to the
bands contains incorporated phosphate species from the anodizing electrolyte.

The purpose of the present work is to determine the presence of the electrolyte incroporation during the initial film growth and pore development with use
of AES, in combination with ion etching.

EXPERIMENT

Aluminium specimens (99,5 %) were anodized in sulphuric acid galvanostatically. Afterwards specimens were rinsed with demiwater for 10 minutes.

Figure 1 shows that the galvanic potential of the aluminium anode, during anodic oxidation, changes in a characteristic way. Three different steps can be observed in the voltage-time relationship : the barrier oxide growth (step I), the pore nucleation (step II) and the regular porous film formation (step III). In order to study oxide composition during its growth, anodic oxidation was interrupted at four different times as indicated in the figures (a, b, c and d).

Subsequent composition analyses of this four layers, were performed with an Auger Electron Spectrometer (Physical Electronics model PHI - 545). Depth profiles were obtained by means of simultaneous Ar-ion etching and AES detection, using a multiplexing technique which allows sequential scanning of 6 different energy windows.

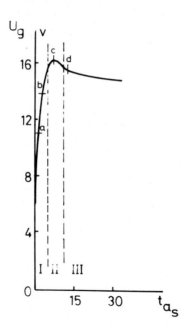

Fig. 1. Anodic oxidation voltage-time relationship
$C = 20 \% H_2SO_4$, $J = 2 A/dm^2$, $T = 20°C$

RESULTS AND DISCUSSION

AES spectra of a pure aluminium surface (a) and of a sulphuric acid anodized aluminium (b) are illustrated in figure 2. Table 1 lists the different Auger emissions.

Figure 3 represents the variation of the Auger peak amplitude of Al, S and O against the sputtering time of the four anodic layers (a, b, c and d), described above.

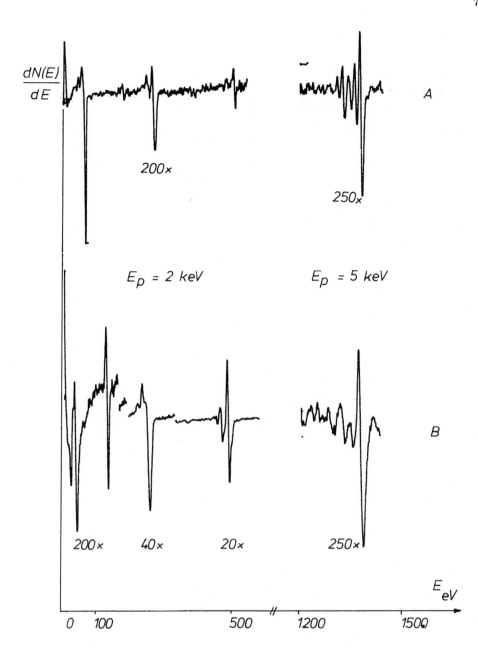

Fig. 2. AES spectrum of aluminium (a) and of sulphuric anodized aluminium film (b)
V_{mod} = 4 eV$_{p/p}$, scanrate = 4 eV/s , adsorbed current = 0,2 µA , spotsize = 3 µm

The thickness of the oxide film can be determined from the increasing Al_{LMM} and decreasing O_{KLL} signals. From these graphs it can be seen that the transition (from the oxide layer to the metal beneath) happens gradually. This is even more pronounced in layers c and d. It is already known (ref.6,9,10,11) that depth resolution is influenced by the surface roughness and the sputtering of the oxide films and the material beneath. With the use of SEM, we verified that this transition may be explained by a loss of depth resolution, rather than by a gradual chemical transition from the oxide layer to the metal.

AES Spectrum of aluminium (a)

energy	intensity	Auger transition	compound	reference
56 eV	weak		plasmon	3
67 eV	strong	LVV	Al	3
506 eV	weak	$KL_{2,3}L_{2,3}$	O	12
1348 eV	weak	KL_1L_1	Al	6
1368 eV	medium	$KL_1L_{2,3}$	Al	6
1382 eV	weak	$KL_{2,3}L_{2,3}$	Al	6
1396 eV	strong	$KL_{2,3}L_{2,3}$	Al	6

AES Spectrum of sulphuric anodized layer (b)

energy	intensity	Auger transition	compound	reference
38 eV	medium	LV_oV_o	Al_2O_3	3
56 eV	strong	LV_oV_o	Al_2O_3	3
150 eV	medium	LVV	S	-
267 eV	medium	LVV	C	-
471 eV	weak	KL_1L_1	O	12
489 eV	weak	$KL_1L_{2,3}$	O	12
505 eV	strong	$KL_{2,3}L_{2,3}$	O	12
1332 eV	medium	KL_1L_1	Al	6
1358 eV	weak	$KL_1L_{2,3}$	Al	6
1384 eV	strong	$KL_{2,3}L_{2,3}$	Al	6

TABLE 1

The small decrease of the Al signal in the beginning of the transition is due to an Auger energy shift, as a result of the Auger emission. In the case of Al_2O_3 the 56 eV peak originates from an interaction between Al and O (ref.3) and in the case of pure Al the 67 eV peak originates from the LMM Auger emission. This can be written as follows : 67 eV $L_{2,3Al}V_{Al}V_{Al}$ and 56 eV $L_{2,3Al}V_oV_o$. It has been confirmed afterwards by means of step by step sputtering analysis.

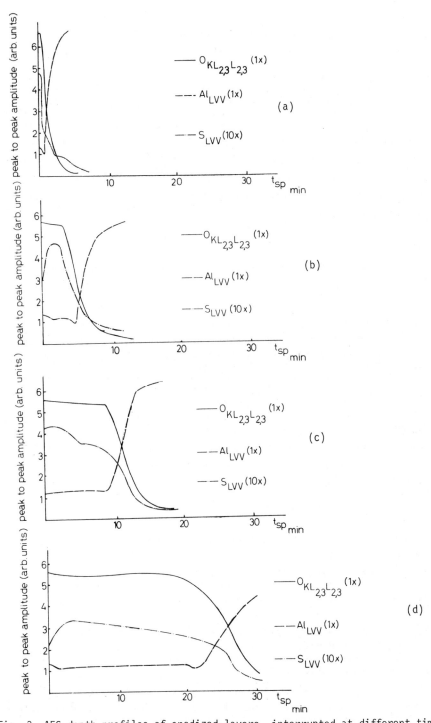

Fig. 3. AES depth profiles of anodized layers, interrupted at different times (a, b, c and d). Sputtering conditions : 3 keV Ar$^+$ 10^{-5} Torr.

752

From figure 3, it can also be noticed that sulphur is incorporated into the anodized film during the whole process of film growth (a - d).

These results are in agreement with those found by Thompson (ref.14,15) and Takahashi (ref.16). The electrolyte is found within whole oxide layer in constrast with the data presented by Bador (ref.4), who distinguishes the electrolyte only in the outer layer adjacant to the vacuum oxide interface. XPS energy shift investigation showed us that S is present as SO_4^{2-}. It can also be observed that there is a maximum of S close to the surface of the oxide layer during the transition growth from the barrier layer (b) to the porous nucleation (c). This maximum indicates the accumulation of an amount of electrolyte during the formation of the porous layer and it confirms that adsorption of anions must influence the pore nucleation.

CONCLUSIONS

In order to examine the composition of porous layers, grown on anodized aluminium, Auger Electron Spectrometry in combination with ion etching has proved to be a technique with very promising perspectives. Interpretation of the depth profiles, however, has to be done with the utmost care because of possible damage by electron beam and sputtering beam to the composition and morphology of the sample. Despite this, we were able to detect the amount of adsorbed or incorporated electrolyte during the growth of the porous layer and to confirm that this influences the pore nucleation.

ACKNOWLEDGEMENTS

The authors gratefully acknoqledge the financial support provided by AGFA-GEVAERT N.V. and I.W.O.N.L. (Institute for Encouragement of Scientific Research in Industry and Agriculture).

REFERENCES

1 G.E. Thomson and R.C. Furneaux, Nature, 272 (1978) 433-435.
2 G.E. Thomson and R.C. Furneaux, Trans. of the Inst. of Met. Finish., 56 (1978) 159-167.
3 D.T. Quinto and W.D. Robertson, Surf. Sci., 27 (1971) 645-648.
4 R. Bador, Mat. Res. Bull., 12 (1977) 197-204.
5 S. Matsuzawa and N. Baba, Electrochimica Acta, 24 (1979) 1199-1202.
6 T. Smith, Surf. Sci., 55 (1976) 601-624.
7 T.H. Allen, J. Vac. Sci. Techn., 13 (1976) 112-115.
8 G. Dufour and J.M. Mariot, Physica Scripta, 13 (1976) 370-372.
9 T. Sun and D.K. Mc Namara, Appl. Surf. Sci., 5 (1980) 406-425.
10 J.S. Solomon and D.E. Hanlin, Surf. Sci., 4 (1980) 307-323.
11 J.S. Solomon, Appl. Spec., 1 (1976) 46-49.
12 P. Légaré and C. Maire, Surf. Sci., 68 (1977) 348-356.
13 J.A. Treverton and N.C. Davies, Electrochimica Acta, 25 (1980) 1571-1576.
14 G.E. Thompson and R.C. Furneaux, J. Electrom. Soc., 125 (1978) 1480-1482.
15 G.E. Thompson and G.C. Wood, Electrochimica Acta, 26 (1981) 951-953.
16 H. Takahashi and M. Nagayama, J. Chem. Soc. Jpn., (1974), 453-458.

ETCHING AND PASSIVATION OF ALUMINUM UNDER ALTERNATING CURRENT.INFLUENCE OF
TEMPERATURE

F. BRINDEL, R. GRYNSZPAN, F. BOURELIER , J.C. BAVAY [*] and K. VU QUANG

  Centre d'Etudes de Chimie Métallurgique, C.N.R.S., 15 rue Georges Urbain,
  94400 Vitry, France

 [*] E.N.S.C. Lille, B.P. 108, 59652 Villeneuve d'Ascq

ABSTRACT

   For high purity aluminum foils etched in 1.3N HCl acid under alternating current, the surface porosity and the electrical capacitance strongly depend on
the solution temperature and on the etching time. A maximum of capacitance C_M
is observed for a given frequency value ν_M which increases with temperature. In
the temperature range 50-60°C, the etching yields a maximum of porosity and a
minimum of oxide thickness.

INTRODUCTION

   Electrolytic capacitors made of high purity aluminum foils with suitable
surface porosity are produced by controlling the pitting mechanism of chemical
or electrolytic etching.

   In a previous work (1) pitting and dissolution characteristics of 99.99 %
purity aluminum in 1.3N HCl solution has been investigated taking account of the
frequency, the amplitude of the alternating current and the etching time. The
pit density and the theoretical metal dissolution were found to decrease markedly with increasing a.c. frequency.

   The present study is devoted to the effect of the HCl solution temperature
on the characteristics of the etched foil.

EXPERIMENTAL

   99.99 % purity and 90 μm thick aluminum foils were used. The etched layer
characteristics and the capacitance of etched and anodized foils were investigated as a function of a.c. frequency ν (from 1 to 1000 Hz), a.c. amplitude I,
bath temperature T and etching time t. The etching bath was based on 1N hydrochloric acid containing small amounts of nitric and phosphoric acids. The attack was performed on one side of the electrodes consisting of 38 mm diameter
disks cut out from the above foils. These disks were cleaned in trichlorethy-
lene, in ethyl alcohol and rinsed in deionized water. They were mounted on a
Teflon holder with electrical contacts. The exposed surface (to be attacked)

was 4 cm^2. In most experiments, the maximum current density applied to electrode was 0.5 A/cm^2. The two electrodes were placed face to face inside a 2.5 l glass cell. A Teflon cylinder, provided with holes allowing the solution circulation, kept constant the distance between electrodes (5 cm) and limits the side effect on the etched sample. The sine alternating current utilized for etching was supplied by a potentiostat Tacussel PRT 20-10 X, monitored by a signal generator Tektronix 464. The anodization of etched foils is carried out in a boric acid solution at 90 C. The capacitance of etched and anodized aluminum foils was determined at 50 Hz frequency using a capacitance bridge HP 4282 A. Capacitance values were referred to unit of geometric surface area (dm^2) (different from the actual area of the etched foils).

RESULTS AND DISCUSSION

According to previous works (2-5), under alternating current, pitting and metal dissolution occur during the anodic half cycle, while reduction of protons in the cathodic half cycle induces local rise of the pH, producing the repassivation of preexisting pits. Therefore, in the next anodic half cycle, the creation of new pits results in ramification of the attack.

In a broad range of experimental conditions, our investigations show a capacitance peak C_M of etched and anodized foils corresponding to a certain frequency value ν_M (1). C_M and ν_M depend on the a.c. amplitude, on the etching time and on the solution temperature. As we increase the temperature, the capacitance peak shifts towards high frequency (fig. 2 and 3). The etching time t is 4 mn 45 sec in fig. 2 and 1 mn 11 sec in fig. 3. For an electrical charge four times smaller (i.e. : t = 1 mn 11 sec) the capacitance maximum C_M is only slightly reduced by about 15 %.

The cross sectional micrograph of aluminum foil attacked at 50 Hz shows an etch porous structure reaching 20 μm in depth, having a fine porosity (1b) for an a.c. etching at 55°C during 4 mn 45 sec (fig. 4). The scanning electron micrograph of etched foil shown in our previous work (1b) pointed out a cube etch structure for 1 Hz attacked, with etch pits developed in well defined crystal directions. This result is similar to that obtained in (6) for attack at low frequency (0.5 Hz) with a single square wave anodic half cycle. The micrographic evolution of the aluminum surface etched at 80 Hz, versus temperature, is shown fig. 5. The surface roughness (porosity) is homogeneous for the 50°C attack, while for 55°C, some blisters appear which most probably give rise to the deep pits observed at 60°C etching. Hence a temperature increase tends to destroy the fine porosity structure.

Glow Discharge Spectrometry - G.D.S. - (7) was carried out to determine (relative) concentration profile of elements (Al, O, H) in the porous layer of

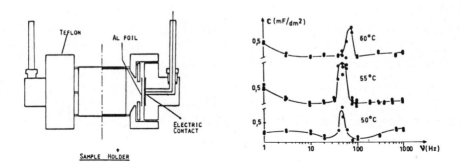

Fig. 1 (left). Sample holder utilized for etching of aluminum foil under alternating current.

Fig. 2 (right). Variation of capacitance of etched and anodized aluminum foils as a function of a.c. frequency. Etching time = 4 min 15 sec ; maximum current density = $0,5$ A/cm^2.

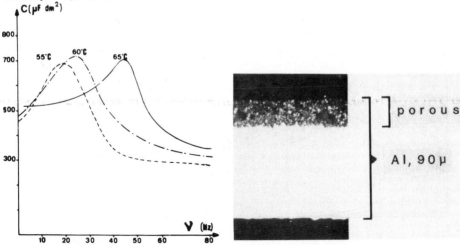

Fig. 3 (left). Variation of capacitance of etched and anodized aluminum foils as a function of a.c. frequency. Etching time = 1 min 11 sec.

Fig. 5 (right). Cross sectional micrograph of aluminum foil etched at 50 Hz (x 470) (max c.d. = $0,5$ A/cm^2, t = 4 mn 45 sec, T = 55°C).

50°C   55°C   60°C

Fig. 4. Micrographs of aluminum surface etched at 80 Hz for different temperatures (X 200).

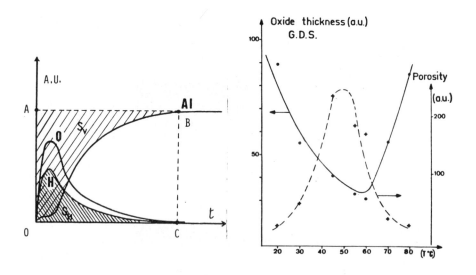

Fig. 6 (left). Concentration of elements in porous surface of aluminum etched
foil (in arbitrary unit), determined by Glow Discharge Spectrometry, versus
sputtering time.

Fig. 7 (right). Dependance of oxide thickness (derived from Glow Discharge
Spectrometry) and of porosity index, on etching temperature.

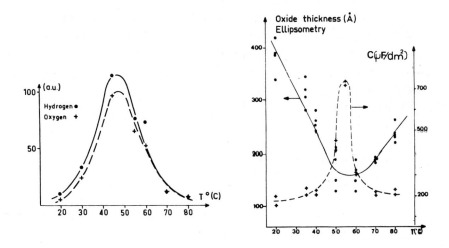

Fig. 8 (left). Amount of hydrogen and oxygen (in arbitrary unit) in the oxide, versus etching temperature.

Fig. 9 (right). Oxide thickness (derived from ellipsometry measurements) and capacitance of foils, versus etching temperature.

etched foils. A typical G.D.S. concentration profile is shown in fig. 6. A porosity indes P_0 , in arbitrary unit, could be defined by the ratio S_V/S_{ABCO} . S_V corresponds to the "void" occupied by the pores ($S_V = S_{ABCO} - S_{Al}$). Porosity index shows a maximum value for foils etched at $\simeq 50°C$ (fig. 7). G.D.S. data also exhibit a maximum of oxygen and hydrogen contents (areas under the O and H curves respectively) for temperature around 45-50°C (fig. 8). These two elements are associated to a hydrated oxide forming the etch film. We attributed this maximum of O and H contents to a maximum of total pore surface created at these temperatures.

The oxide thickness, assumed to be proportional to the ratio "oxide amount/ P_0" , is found to be minimum for the etching temperature range 55-60°C (fig.7). However the actual temperature of the etched foil surface might be slightly higher owing to the electrical current effect and to the weak solution stirring at the electrode surface. This oxide thickness minimum was confirmed by ellipsometry measurements (1) using a technique especially adapted to rough surface and developed in Research Center of Aluminum-Pechiney Voreppe. The oxide thickness ranges from 150 Å to 400 Å (fig. 9).

In conclusion, for high purity aluminum foils etched in 1.3 N HCl acid, under alternating current, the surface porosity and the electrical capacitance strongly depend on the solution temperature. In the range 50-60°C the etching yields a maximum of porosity and a minimum of oxide thickness.

REFERENCES

1 K. VU QUANG, F. BRINDEL, G. LASLAZ, R. BUTTOUDIN, a) Journées Electrochimiques Brussels, June 1981, b) Spring Meeting of Electrochem. Society, Montréal, 9-14 May 1982
2 C. ROTH, Metallob. Ang. Elek., 1 (1972), 1
3 C. BARGERON, J. Electrochem. Soc. 124 (1977), 1845
4 K. NISANCIOGLU, H. HOLTAN, Corros. Sci. 19 (1979), 537
5 G.E. THOMPSON, G.C. WOOD, Corros. Sci. 8 (1978), 721
6 C.K. DYER, R.A. ALWITT, J. Electrochem. Soc. 128, n° 2 (1981), 300
7 R. BERNERON, J.C. CHARBONNIER and al., Spectrochim. Acta 33B (1978), 665 ; Rev. Met. 78, n° 1 (1981), 73

Passivity of Metals and Semiconductors, edited by M. Froment
Elsevier Science Publishers B.V., Amsterdam — Printed in The Netherlands

ROUND TABLE DISCUSSION

STAINLESS STEELS - VALVE METALS

<u>M.B. IVES</u> : Discussion leader initiated the Discussion.

Recording the Bockris's statement in Cambridge (1970) that the theoretical things we have to understand is the nature of the passive film on iron in a Borate-Buffer solution, I am not sure we have made great progress in this direction. Looking at the posters I did not see any advance with respect to this goal. I suggest to begin with valve metals or those on which one can see the film.

<u>U. BERTOCCI</u> : Dr. Alwitt on aluminium considers the crystalline film as better than amorphous in contrast with what is assumed usually.

<u>M.B. IVES</u> : May I repete my question, what do we know more than in 1970 ?

<u>M. FROMENT</u> : We know the composition, the profiles ... where is Cr, Mo ... in the film but we don't have the relation between composition and behaviour, we have the static view, not the dynamic one.

<u>B. BARROUX</u> : Is the pitting corrosion related to the passive film ? According to M. Janik-Czachor the addition of minor elements are of interest. Doing this addition are we suppressing the weak points or improving the film quality ?

<u>H.S. ISAACS</u> : I would like to comment on the poster dealing with W addition. W seems to play an effect similar to Mo. This is a very complex problem of solution chemistry at pH > 2 . Due to the contribution of polymer anions. Maybe we pay attention at the wrong place.

<u>N. BUI</u> : My poster shows a similarity but at low pH W is passive and Mo is not. At high pH both metals are active. In neutral solution the role of W is not clear. In fact Mo and W oxides can exist in the passive film even when they dissolve into the solution.

F. DABOSI : I think we have to differentiate the semiconducting properties from the contribution of polymer anions layers and conduction / charge carriers effects. The use of impedance measurements of different types would be very helpful.

N. SATO : As to the role of Mo in the pitting and crevice corrosion of stainless steels, I would like to point out the fact that molybdenum ions dissolved from the metal, if adsorbed, change the corrosion precipitate membrane over the pit or crevice mouth from an anion-selective membrane, which will accelerate Cl^- enrichment under the membrane, to a cation-selective membrane, which will prevent Cl^- ion from entering into the pit or crevice solution.

P. MARCUS : In relation with the lecture of M. Janik-Czachor I would like to bring the attention on the role of S. We have to correlate the behaviour with the role of sulfur at an atomic scale. We have evidence of a strong bonding of S onto the surface of Ni for example. It is a very strong bonding which can weaken the Mo-Mo bond. Also S can play another role since it blocks the adsorption of OH^- ions and consequently inhibits the passive layer formation.

B. BARROUX : I would like to report an interesting result on the influence of Mo content on pitting corrosion at two different concentrations of S (20 ppm and 70 ppm). At 70 ppm a steep decrease is observed at low Mo content followed by a linear slope at higher concentrations. The same linear relationship holds down to zero Mo at 20 ppm sulfur.

P. MARCUS : That means Mo destabilizes S on the surface.

M. JANIK-CZACHOR : I have the opinion that we know more on the composition of the passive layer and specially on the role of water content.

H.H. STREHBLOW : With respect to the dynamical behaviour raised by M. Froment, I think that we have to distinguish according to the time scale. At rather long time scale we can establish some pertinent data. For instance, by combining electrochemical and surface analysis techniques we have a good chance.

K. HASHIMOTO : As concluding remarks.
           Since XPS and AES had been brought into corrosion science more than ten years ago, large progresses have been made for an understanding of the passive film mostly by means of ex-situ methods. At this symposium we have learned that in-situ methods such as EXAFS, Raman spectroscopy, modulated

reflection spectroscopy and other various methods will shed light on passivity. In this connection, one thing I would like to emphasize is that we often use the metal-environment combination suitable for our analytical equipments. This combination may not be adequate to clarify what we would like to know. Although the analysis of individual specimen is important, the purpose of corrosion scientists and engineers to use modernized equipments is to understand corrosion and related phenomena. Furthermore, sophisticated methods give us detailed and rigid results, and hence we often try to generalize our results widely for other corrosion problems. However, for the generalization we have to see carefully not only our own trees but also forest. This would be particularly important. For instance, bulk substances are quite different from thin passive films, and if we change the pH or composition of solution or alloy composition, the results obtained sometimes become significantly different.

Many questions are still remaining : What is passivity ? What is passivity breakdown ? What is pit initiation ? How can we overcome occluded cell corrosion ?

Finally, I congratulate Professor Froment and the Organizing Committee on the success of such a fruitful symposium and appreciate their warm hospitality. Six years later, we will welcome all the corrosion scientists and engineers, particularly those who are challenging directly the fundamental questions mentioned above.

SUBJECT INDEX

Acidic phosphate, 199
Acrylamide, 725
Active-passive transition, 649
Adsorption, 287
AES, 571, 579, 585, 747
Allowing effect, 405
Aluminium, 413, 741, 747, 753
Amorphous alloys, 229, 235, 311, 317, 321
Amorphous oxides, 497
Amorphous stainless steels, 305
Angle resolved XPS, 113
Anodic dissolution, 695
Anodization, 741
Artificial pit, 269
Atmospheric corrosion, 637, 701

Borate buffer, 67, 79, 89, 95
Borate solution, 199
Brass, 413
Breakdown, 229, 287, 293, 379, 393, 405, 485, 591

Capacitance, 483
Capacitor, 753
Carbonate solutions, 125
Chromate, 707, 713
Chromium, 317
Chromium oxide, 211
Chromium oxide layer, 675
Copper-zirconium alloys, 321
Crevice corrosion, 347

Depassivation, 253, 373, 667

Electrode pretreatment, 275
Electron loss spectroscopy, 457
Electron tunnelling, 23, 35
Electroreflectance, 85, 187, 211
Elements addition, 235
Ellipsommetry, 89, 95, 187, 205, 483
Etching, 521, 753
EXAFS, 131, 139, 147, 163

Ferritic alloys, 561
Ferritic stainless steels, 405
Film growth, 11, 43, 731
Fluoride, 379

Gallium arsenide, 483, 521
Glow discharge spectrometry, 613
Gold, 341

Hafnium, 503
Hafnium oxide, 35
Hastelloy, 675
Heat treatment, 317
High temperature, 393, 571, 579, 613
Hydrochloric solutions, 341
Hydrogen permeation, 675

Impedance, 51, 205, 681
Inhibition, 35

In situ cells, 131, 139, 147, 163
Intergranular corrosion, 335
Ion implantation, 35, 305, 503, 591
Iridium, 725
Iron, 51, 59, 67, 85, 89, 95, 101, 107, 131, 147, 163, 169, 199, 311, 327, 379, 477, 491
Iron based alloys, 169, 687
Iron chromium alloys, 211
Iron hydroxide, 59
Iron-nickel alloys 649
Iron oxide, 211, 491

Laser irradiation, 591
LEED, 457
Localized corrosion, 253, 531, 547, 643
Low alloys steels, 661
Low energy electron induced X-Ray Spectroscopy (LEEIXS), 181

Mass transport effect, 269, 287
Mechanical effects, 293, 373
Mild steel, 387
Models, 281, 287
Molybdate, 713
Molybdenum, 393, 561, 625
Mossbauer spectroscopy, 147

Nickel, 79, 113, 131, 275, 335, 353, 379
Nickel-cobalt alloys, 327
Nickel hydroxide layers, 73
Nickel-iron alloys, 119
Nickel-molybdenum alloys, 175
Niobium nitride, 445
Nitric solutions, 327
Noise measurements, 229, 293
Non metallic inclusions, 399, 547

Non stoichiometry, 43
Nuclear reactions, 463

Optical methods, 187
Organic compounds, 607
Organic media, 555
Organic solvents, 687

Passivation-depassivation, 341
Phosphoric solutions, 599, 625
Phosphorus, 311, 321
Photoelectrochemistry, 425, 497, 509
Pit growth, 655
Pit initiation, 217, 281, 299, 359, 387
Pitting, 275, 353, 367, 399
Pitting potential, 281
Plasma oxidation, 437
Porous layers, 747
Potassium hydroxide solution, 107
Potential oscillations, 631

Radial distribution function, 169
Raman spectroscopy, 101, 107, 187
Reflectometry, 79, 199
Repassivation, 253, 281, 367, 667, 731
RHEED, 463, 585
Ring disc electrode, 661
Ruthenium, 205

Semi-conductors, 437, 551
Semi-conductor corrosion, 425
Short range order, 169
Silicide compounds, 473
Silicon nitride, 451
Silicon oxide, 1, 11, 457, 463, 473
Silicon-silicon oxide interface, 457

Spectrophotoelectrochemistry,
477, 491
Stainless steels, 181, 347, 359,
379, 393, 399, 531, 555, 579, 585,
591, 599, 613, 619, 625, 631, 637,
643, 655, 667, 697
Statistics, 217, 299
Steels, 607, 681
STEM, 175
Stress corrosion cracking, 413, 667
Sulphur, 119

Thermal oxidation, 445, 741
Tin, 707, 713, 719
Titanium, 731
Titanium alloys, 731
Titanium oxide, 35, 503

Transfer reactions, 23, 35, 73
Transpassivity, 113, 619
Transport, 1, 463
Tungstate, 713
Tungsten, 497, 599

Uranium oxides, 125

Water, 555
Welding effect, 701
Wheathering steels, 701

XPS, 119, 125, 235, 451, 473
561, 579, 585
599, 707

Zircalloy alloys, 299